ORGANIC CHEMISTRY OF LIFE

Readings from
SCIENTIFIC AMERICAN

ORGANIC CHEMISTRY OF LIFE

With Introductions by
Melvin Calvin
University of California, Berkeley
and
William A. Pryor
Louisiana State University, Baton Rouge

NORTHWEST COMMUNITY COLLEGE

W. H. Freeman and Company
San Francisco

Library of Congress Cataloging in Publication Data

Calvin, Melvin, 1911–   comp.
    Organic chemistry of life.

    Bibliography: p.
    1. Biological chemistry—Addresses, essays,
    lectures.
I. Pryor, William A., joint comp.   II. Scientific
American.   III. Title.
QP514.2.C34   574.1'92   73–12475
ISBN 0–7167–0884–1
ISBN 0–7167–0883–3 (pbk.)

Most of the SCIENTIFIC AMERICAN articles in
Organic Chemistry of Life are available as separate
Offprints. For a complete list of more than 950
articles now available as Offprints, write to
W. H. Freeman and Company, 660 Market Street,
San Francisco, California 94104.

Printed in the United States of America

9 8 7 6 5 4 3 2 1

PREFACE

The articles in this book are designed for use as supplementary readings both for the first college course in biochemistry and for the sophomore course in organic chemistry. (Lists that indicate which articles are appropriate for use at specific stages of the biochemistry and organic chemistry courses follow the table of contents.)

The utility of this set of self-sufficient, expert, and timely articles on aspects of biochemistry and microbiology in introductory biochemistry is obvious. In addition, the sophomore course in organic chemistry is being taught to an increasing predominance of students who have majors in one of the life sciences, and it is both necessary and desirable to treat bio-organic chemistry in this course. However, because of the large number of fundamental principles of organic chemistry that command priority and the limited time available, it is impossible to do justice to biological or biochemical topics. A collection of authoritative, stimulating readings written at an introductory level can close the gap between the elementary organic textbook and a detailed biochemical treatise.

The readings that we have selected for this volume, comprising forty-seven articles first published in *Scientific American,* constitute a collection of readings that can serve as supplementary material both for the organic and for the biochemistry courses. These brief, important articles are written by experts and yet are both self-sufficient and reasonably elementary. In this collection we have chosen articles that deal with current, significant areas of biology and have a chemical emphasis.

Our personal experience is with the use of these readings in the sophomore course on organic chemistry for life-science majors. One of the most important changes in the way in which organic chemistry is taught today as opposed to the approach used prior to the 1950s is the heavy emphasis on the understanding of theory as opposed to the memorization of reactions and molecular structures. This important transformation has given students meaningful insight into organic chemistry and allows them to retain a greater comprehension of it in their professional careers. In our view, organic chemistry should be taught with this emphasis on theory for all life-science majors: it seems clear that even those biological sciences which do not operate at the molecular level today may well do so in the course of the professional careers of the present generation of students. In order to retain sufficient emphasis on modern organic chemistry in the sophomore course, it seems a virtual necessity to provide an introduction to bio-organic chemistry by the use of supplementary readings.

In our classes in sophomore organic chemistry, students are assigned individual articles from this volume as outside reading, and in lectures we often digress briefly into some of the topics covered in these articles but not in the organic text. We have found that the articles provide a convenient means for the introduction of important biochemical topics to students, most of whom are only vaguely aware of the significant role that the basic organic chemistry they are learning occupies in the modern life sciences. In the absence of such readings, students often consider many of the topics in the sophomore course to be more chemistry than they "need to know." These articles make the relevance of organic chemistry to biology very meaningful and effectively excite the students' curiosity about the subject. Our practice is to assign these articles at a rate of about one a week to be reported on in a short paper or on examinations.

This book is organized into four sections. The first, "Biological Regulators," deals with the shapes, syntheses, and mechanisms of action of the small molecules that control the biological functions in cells. The fifteen articles in this section discuss some of the currently most important classes of compounds, including steroids, cyclic AMP, hallucinogenic drugs, prostaglandins, and analgesics.

Section II, "Macromolecular Architecture," takes us into the realm of giant molecules. Biological polymers consist of proteins, carbohydrates, and nucleic acids; the thirteen articles in this section discuss a number of aspects of these biopolymers. Seven articles treat the structure of proteins, their modern, automated synthesis, their three-dimensional structure, their evolution, and their biological function. Articles 20 and 27 discuss some aspects of carbohydrate biochemistry. Articles 24–28 treat in detail the structure and synthesis of nucleic acids and their function in biology. It is noteworthy that six of the authors whose articles are included in this section have received Nobel prizes for the work described here, and all are acknowledged experts in the areas they describe.

The third section, "Cellular Architecture," although rather brief, treats an area of bio-organic chemistry that will become increasingly important in the future: the structure and functions of membranes. The biological role of membranes has been recognized only in recent years, and this is an area of intense current research.

Section IV, "Chemical Biodynamics," includes sixteen articles that discuss the dynamic changes that take place in living systems. This section starts with the mechanisms and chemical reactions involved in the production of energy in the cell. The metabolism of fats and the synthesis and metabolism of proteins and polypeptides are considered next. The genetic code is discussed in two articles and chemical paleontology in one. A series of six articles covers the role of light in biology: the chemical effects of light, the interaction of light with living matter, photosynthesis, and the mechanism of vision are treated in articles that make an interesting and coherent sequence, although they can be read independently. The final articles treat the mechanism of memory, the chemistry of aging, and the role that reactive free radicals play in degradative processes in biology including those in aging.

This volume is the successor to the earlier publication *Bio-Organic Chemistry,* with introductions by one of us and the late Margaret J. Jorgenson, a volume that included thirty-two articles published in 1968 or earlier. Our experience and that of our colleagues and students in using that volume has been used to improve the content and coverage of the present book.

For students who wish to pursue any of the topics of these articles beyond the discussion found here, additional material may be found in the bibliographies at the end of the text. In most cases, these bibliographies have been updated to include references published after the article itself originally appeared in *Scientific American.*

In order to keep this volume reasonable in size, a number of other excellent and useful articles published in *Scientific American* have been omitted. We have listed the titles of several of these articles at the end of the introductions to each of the four sections; many of them are available as Offprints from W. H. Freeman and Company.

May 1973

Melvin Calvin
William A. Pryor

CONTENTS

I   BIOLOGICAL REGULATORS

II   MACROMOLECULAR ARCHITECTURE

III CELLULAR ARCHITECTURE

IV CHEMICAL BIODYNAMICS

Note on cross-references: References to articles included in this book are noted by the title of the article and the page on which it begins; references to articles that are available as Offprints, but are not included here, are noted by the article's title and Offprint number; references to articles published by SCIENTIFIC AMERICAN, but which are not available as Offprints, are noted by the title of the article and the month and year of its publication.

SUGGESTED USE OF *ORGANIC CHEMISTRY OF LIFE* IN BIOCHEMISTRY COURSES

The following list organizes articles in this reader in the order in which they might be used in the upper-level undergraduate course in biochemistry. The topics listed are those recommended by the Biochemistry Subcommittee of the Curriculum Committee of the American Chemical Society, as published in the *Journal of Chemical Education*, 50, 203–204 (March 1973). (The number of each article as it appears in this volume is in parentheses.)

SUGGESTED USE OF *ORGANIC CHEMISTRY OF LIFE* IN ORGANIC CHEMISTRY COURSES

The following list organizes articles in this reader in the order of topics usually discussed in sophomore organic chemistry. It is meant to guide students who are taking sophomore chemistry and wish to read these articles in connection with the course. (The number of each article as it appears in this volume is in parentheses.)

INTRODUCTION

The Carbon Cycle (1)
The Chemical Elements of Life (2)

ALKANES

The Shapes of Organic Molecules (3)
Pheromones (15)
Chemical Fossils (38)

ALKENES

Pheromones (15)
Giant Molecules (16)
Molecular Isomers in Vision (44)

ALICYCLIC COMPOUNDS

Steroids (4)
Pheromones (15)
Molecular Isomers in Vision (44)

FREE RADICALS

Free Radicals in Biological Systems (47)

MOLECULAR ORBITALS

Molecular Isomers in Vision (44)

STEREOCHEMISTRY

The Shapes of Organic Molecules (3)
The Stereochemical Theory of Odor (12)
L-asparagine and Leukemia (13)

SPECTROSCOPY

Chemical Fossils (38)
The Chemical Effects of Light (39)
How Light Interacts with Living Matter (40)

CARBONYL COMPOUNDS

The Chemical Effects of Light (39)

HETEROCYCLIC COMPOUNDS

Analgesic Drugs (9)
Barbiturates (10)
The Hallucinogenic Drugs (11)
The Structure of the Hereditary Material (24)
The Genetic Code: II (36)

ACIDS

Prostaglandins (8)
The Structure of Cell Membranes (29)
The Metabolism of Fats (33)

AMINES

Alkaloids (5)
The Insulin Molecule (21)
Protein-digesting Enzymes (35)

DIAZONIUM IONS

The Genetic Code of a Virus (37)

PROTEINS

Proteins (17)
The Chemical Structure of Proteins (18)
The Automatic Synthesis of Proteins (19)
The Insulin Molecule (21)
The Structure and History of an
    Ancient Protein (22)
Protein-digesting Enzymes (35)
Memory and Protein Synthesis (45)

FATS

The Structure of Cell Membranes (29)
The Membrane of the Mitochondrion (30)
The Metabolism of Fats (33)

CARBOHYDRATES

The Three-dimensional Structure of an
    Enzyme Molecule (20)
The Recognition of DNA in Bacteria (27)
The Bacterial Cell Wall (31)
The Path of Carbon in Photosynthesis (42)
The Mechanism of Photosynthesis (43)

NUCLEIC ACIDS

Cyclic AMP (7)
The Induction of Interferon (14)
The Structure of the Hereditary Material (24)
The Nucleotide Sequence of a Nucleic Acid (25)
The Synthesis of DNA (26)
RNA-directed DNA Synthesis (28)
How Proteins Start (34)
The Genetic Code: II (36)
Ultraviolet Radiation and Nucleic Acid (41)

I

BIOLOGICAL REGULATORS

I

BIOLOGICAL REGULATORS

INTRODUCTION

In this section we consider the chemistry and biochemistry of the small chemical species—both organic and inorganic—which are vital for the functioning of the living cell. Most of the compounds discussed are small organic molecules that exercise powerful controls over other reactions of the living organism: species that have the ability to give signals to the cell to turn important chemical pathways on or off. The first article reviews those reactions which convert carbon atoms from carbon dioxide to the molecules present in living systems and back again. Next, the twenty-three elements that, in addition to carbon, are known to be essential for life are discussed; many of these elements are required in only trace amounts, but they are vital just the same. The third article introduces conformational analysis as applied to molecules as simple as cyclohexane and as complex as steroids. The next two articles discuss two important classes of bio-organic compounds: steroids and alkaloids. The sixth article introduces the idea of the releasing factor, a substance that triggers the release of other control compounds; where it comes from, how it might work, and the nature of its polypeptide structure is discussed. The next article, which discusses cyclic AMP, treats a regulator of such central importance that its discoverer was awarded a Nobel Prize in 1971; this key molecule controls many of the hormone reactions in the human body and many other processes as well. The article on prostaglandins treats a group of regulatory compounds that have been implicated in a variety of control mechanisms, often as a mediator between hormones and cyclic AMP, and appear to be the most promising new drugs since the sulfa drugs. The articles on analgesic drugs and barbiturates treat the synthesis and use of compounds, many of whose trade names will be quite familiar to students. The article on hallucinogenic drugs is popular among students and introduces many of the fascinating aspects of these drugs from both a chemical and a psychological viewpoint. The final two articles in this section, which discuss L-asparaginase and the treatment of leukemia and the use of a synthetic RNA to control virus diseases, indicate the promise and power of biochemical engineering and medicine at the molecular level.

A number of other articles dealing with the regulatory function of organic molecules have been published in *Scientific American*: S. Frank, *Carotenoids* (January 1956); S. Zuckerman, *Hormones* (March 1957); E. H. Davidson, *Hormones and Genes* (June 1965, Offprint 1013); H. E. Himcorich, *The New Psychiatric Drugs* (October 1955, Offprint 446); L. Z. Freedman, *Truth Drugs* (March 1960, Offprint 497); R. O. Roblin, Jr., *The Imitative Drugs* (April 1960); L. J. Roth and R. W. Monthei, *Radioactive Tuberculosis Drugs* (November 1956); J. Schubert, *Chelation In Medicine* (May 1966); A. H. Rose, *New Penicillins* (March 1961); F. B. Salisbury, *Plant Growth Substances* (April 1956, Offprint 110); P. K. Stumpf, *ATP* (April 1953, Offprint 41); J. D. Woodward, *Biotin* (June 1961); M. D. Kamen, *A Universal Molecule of Living Matter* (August 1958); A. J. Haagen-Smit, *Smell and Taste* (March 1952, Offprint 404); M. Jacobson and M. Beroza, *Insect Attractants* (August 1964, Offprint 189); M. A. Amerine, *Wine* (August 1964, Offprint 190); A. E. Fisher, *Chemical Stimulation of the Brain* (June 1964, Offprint 485); F. A. Fuhrman, *Tetrodotoxin* (August 1967), C. M. Williams, *Third-generation Pesticides* (July 1967, Offprint 1078); P. R. Ehrlich and P. H. Raven, *Butterflies and Plants* (June 1967, Offprint 1076); S. Clevenger, *Flower Pigments* (June 1964, Offprint 186); G. M. Edelman, *The Structure and Function of Antibodies* (August 1970, Offprint 1185); F. C. Steward, *The Control of Growth in Plant Cells* (October 1963, Offprint 167); H. O. J. Collier, *Aspirin* (November 1963, Offprint 169); H. Rasmussen and M. M. Pechet, *Calcitonin* (October 1970, Offprint 1200); R. Breslow, *The Nature of Aromatic Molecules* (August 1972); G. O. Kermode, *Food Additives* (March 1972).

1

THE CARBON CYCLE

BERT BOLIN
September 1970

*The main cycle is from carbon dioxide to living matter
and back to carbon dioxide. Some of the carbon,
however, is removed by a slow epicycle that stores
huge inventories in sedimentary rocks*

The biosphere contains a complex mixture of carbon compounds in a continuous state of creation, transformation and decomposition. This dynamic state is maintained through the ability of phytoplankton in the sea and plants on land to capture the energy of sunlight and utilize it to transform carbon dioxide (and water) into organic molecules of precise architecture and rich diversity. Chemists and molecular biologists have unraveled many of the intricate processes needed to create the microworld of the living cell. Equally fundamental and no less interesting is the effort to grasp the overall balance and flow of material in the worldwide community of plants and animals that has developed in the few billion years since life began. This is ecology in the broadest sense of the word: the complex interplay between, on the one hand, communities of plants and animals and, on the other, both kinds of community and their nonliving environment.

We now know that the biosphere has not developed in a static inorganic environment. Rather the living world has profoundly altered the primitive lifeless earth, gradually changing the composition of the atmosphere, the sea and the top layers of the solid crust, both on land and under the ocean. Thus a study of the carbon cycle in the biosphere is fundamentally a study of the overall global interactions of living organisms and their

physical and chemical environment. To bring order into this world of complex interactions biologists must combine their knowledge with the information available to students of geology, oceanography and meteorology.

The engine for the organic processes that reconstructed the primitive earth is photosynthesis. Regardless of whether it takes place on land or in the sea, it can be summarized by a single reaction: $CO_2 + 2H_2A + light \rightarrow CH_2O + H_2O + 2A + energy$. The formaldehyde molecule CH_2O symbolizes the simplest organic compound; the term "energy" indicates that the reaction stores energy in chemical form. H_2A is commonly water (H_2O), in which case $2A$ symbolizes the release of free oxygen (O_2). There are, however, bacteria that can use compounds in which A stands for sulfur, for some organic radical or for nothing at all.

Organisms that are able to use carbon dioxide as their sole source of carbon are known as autotrophs. Those that use light energy for reducing carbon dioxide are called phototrophic, and those that use the energy stored in inorganic chemical bonds (for example the bonds of nitrates and sulfates) are called chemolithotrophic. Most organisms, however, require preformed organic molecules for growth; hence they are known as heterotrophs. The nonsulfur bacteria are an unusual group that is both photosyn-

thetic and heterotrophic. Chemoheterotrophic organisms, for example animals, obtain their energy from organic compounds without need for light. An organism may be either aerobic or anaerobic regardless of its source of carbon or energy. Thus some anaerobic chemoheterotrophs can survive in the deep ocean and deep lakes in the total absence of light or free oxygen.

There is more to plant life than the creation of organic compounds by photosynthesis. Plant growth involves a series of chemical processes and transformations that require energy. This energy is obtained by reactions that use the oxygen in the surrounding water and air to unlock the energy that has been stored by photosynthesis. The process, which releases carbon dioxide, is termed respiration. It is a continuous process and is therefore dominant at night, when photosynthesis is shut down.

If one measures the carbon dioxide at various levels above the ground in a forest, one can observe pronounced changes in concentration over a 24-hour period [*see top illustration on page 7*]. The average concentration of carbon dioxide in the atmosphere is about 320 parts per million. When the sun rises, photosynthesis begins and leads to a rapid decrease in the carbon dioxide concentration as leaves (and the needles of conifers) convert carbon dioxide into organic compounds. Toward noon, as the temperature increases and the humidity decreases, the rate of respiration rises and the net consumption of carbon dioxide slowly declines. Minimum values of carbon dioxide 10 to 15 parts per million below the daily average are reached around noon at treetop level. At sunset photosynthesis ceases while respiration continues, with the result that the carbon dioxide concentration close to the

CARBON LOCKED IN COAL and oil exceeds by a factor of about 50 the amount of carbon in all living organisms. The estimated world reserves of coal alone are on the order of 7,500 billion tons. The photograph on the opposite page shows a sequence of lignite coal seams being strip-mined in Stanton, N.D., by the Western Division of the Consolidation Coal Company. The seam, about two feet thick, is of low quality and is discarded. The second seam from the top, about three feet thick, is marketable, as is the third seam, 10 feet farther down. This seam is really two seams separated by about 10 inches of gray clay. The upper is some 3½ feet thick; the lower is about two feet thick. Twenty-four feet below the bottom of this seam is still another seam (*not shown*) eight feet thick, which is also mined.

ground may exceed 400 parts per million. This high value reflects partly the release of carbon dioxide from the decomposition of organic matter in the soil and partly the tendency of air to stagnate near the ground at night, when there is no solar heating to produce convection currents.

The net productivity, or net rate of fixation, of carbon dioxide varies greatly from one type of vegetation to another. Rapidly growing tropical rain forests annually fix between one kilogram and two kilograms of carbon (in the form of carbon dioxide) per square meter of land surface, which is roughly equal to the amount of carbon dioxide in a column of air extending from the same area of the earth's surface to the top of the atmosphere. The arctic tundra and the nearly barren regions of the desert may fix as

little as 1 percent of that amount. The forests and cultivated fields of the middle latitudes assimilate between .2 and .4 kilogram per square meter. For the earth as a whole the areas of high productivity are small. A fair estimate is that the land areas of the earth fix into organic compounds 20 to 30 billion net metric tons of carbon per year. There is considerable uncertainty in this figure; published estimates range from 10 to 100 billion tons.

The amount of carbon in the form of carbon dioxide consumed annually by phytoplankton in the oceans is perhaps 40 billion tons, or roughly the same as the gross assimilation of carbon dioxide by land vegetation. Both the carbon dioxide consumed and the oxygen released are largely in the form of gas dissolved near the ocean surface. Therefore most of the carbon cycle in the sea is self-con-

tained: the released oxygen is consumed by sea animals, and their ultimate decomposition releases carbon dioxide back into solution. As we shall see, however, there is a dynamic exchange of carbon dioxide (and oxygen) between the atmosphere and the sea, brought about by the action of the wind and waves. At any given moment the amount of carbon dioxide dissolved in the surface layers of the sea is in close equilibrium with the concentration of carbon dioxide in the atmosphere as a whole.

The carbon fixed by photosynthesis on land is sooner or later returned to the atmosphere by the decomposition of dead organic matter. Leaves and litter fall to the ground and are oxidized by a series of complicated processes in the soil. We can get an approximate idea of the rate at which organic matter in the soil is

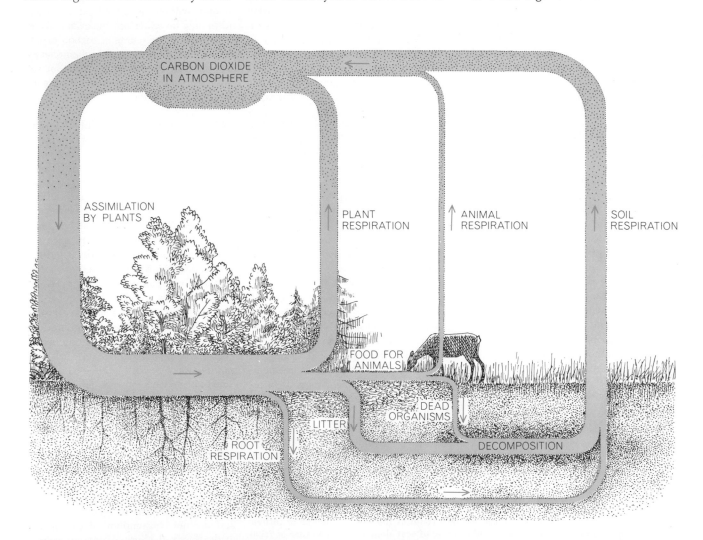

CARBON CYCLE begins with the fixation of atmospheric carbon dioxide by the process of photosynthesis, conducted by plants and certain microorganisms. In this process carbon dioxide and water react to form carbohydrates, with the simultaneous release of free oxygen, which enters the atmosphere. Some of the carbohydrate is directly consumed to supply the plant with energy; the carbon dioxide so generated is released either through the plant's leaves or through its roots. Part of the carbon fixed by plants is consumed by animals, which also respire and release carbon dioxide. Plants and animals die and are ultimately decomposed by microorganisms in the soil; the carbon in their tissues is oxidized to carbon dioxide and returns to the atmosphere. The widths of the pathways are roughly proportional to the quantities involved. A similar carbon cycle takes place within the sea. There is still no general agreement as to which of the two cycles is larger. The author's estimates of the quantities involved appear in the flow chart on page 10.

being transformed by measuring its content of the radioactive isotope carbon 14. At the time carbon is fixed by photosynthesis its ratio of carbon 14 to the non-radioactive isotope carbon 12 is the same as the ratio in the atmosphere (except for a constant fractionation factor), but thereafter the carbon 14 decays and becomes less abundant with respect to the carbon 12. Measurements of this ratio yield rates for the oxidation of organic matter in the soil ranging from decades in tropical soils to several hundred years in boreal forests.

In addition to the daily variations of carbon dioxide in the air there is a marked annual variation, at least in the Northern Hemisphere. As spring comes to northern regions the consumption of carbon dioxide by plants greatly exceeds the return from the soil. The increased withdrawal of carbon dioxide can be measured all the way up to the lower stratosphere. A marked decrease in the atmospheric content of carbon dioxide occurs during the spring. From April to September the atmosphere north of 30 degrees north latitude loses nearly 3 percent of its carbon dioxide content, which is equivalent to about four billion tons of carbon [see *bottom illustration at right*]. Since the decay processes in the soil go on simultaneously, the net withdrawal of four billion tons implies an annual gross fixation of carbon in these latitudes of at least five or six billion tons. This amounts to about a fourth of the annual terrestrial productivity referred to above (20 to 30 billion tons), which was based on a survey of carbon fixation. In this global survey the estimated contribution from the Northern Hemisphere, where plant growth shows a marked seasonal variation, constituted about 25 percent of the total tonnage. Thus two independent estimates of worldwide carbon fixation on land show a quite satisfactory agreement.

The forests of the world not only are the main carbon dioxide consumers on land; they also represent the main reservoir of biologically fixed carbon (except for fossil fuels, which have been largely removed from the carbon cycle save for the amount reintroduced by man's burning of it). The forests contain between 400 and 500 billion tons of carbon, or roughly two-thirds of the amount present as carbon dioxide in the atmosphere (700 billion tons). The figure for forests can be estimated only approximately. The average age of a tree can be assumed to be about 30 years, which implies that about 15 billion tons of carbon

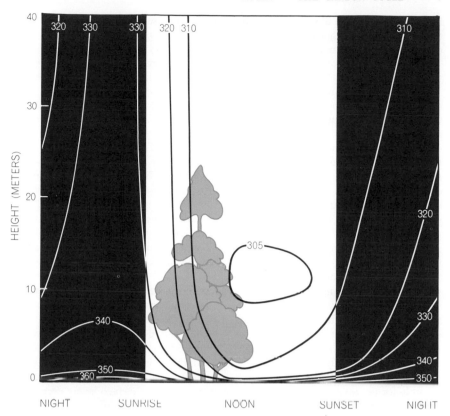

VERTICAL DISTRIBUTION OF CARBON DIOXIDE in the air around a forest varies with time of day. At night, when photosynthesis is shut off, respiration from the soil can raise the carbon dioxide at ground level to as much as 400 parts per million (ppm). By noon, owing to photosynthetic uptake, the concentration at treetop level can drop to 305 ppm.

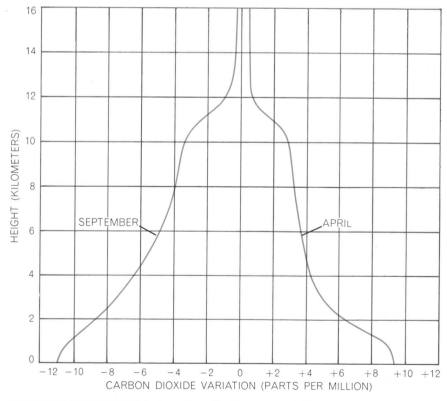

SEASONAL VARIATIONS in the carbon dioxide content of the atmosphere reach a maximum in September and April for the region north of 30 degrees north latitude. The departure from a mean value of about 320 ppm varies with altitude as shown by these two curves.

in the form of carbon dioxide is annually transformed into wood, which seems reasonable in comparison with a total annual assimilation of 20 to 30 billion tons.

The pattern of carbon circulation in the sea is quite different from the pattern on land. The productivity of the soil is mostly limited by the availability of fresh water and phosphorus, and only to a degree by the availability of other nutrients in the soil. In the oceans the overriding limitation is the availability of inorganic substances. The phytoplankton require not only plentiful supplies of phosphorus and nitrogen but also trace amounts of various metals, notably iron.

The competition for food in the sea is so keen that organisms have gradually developed the ability to absorb essential minerals even when these nutrients are available only in very low concentration. As a result high concentrations of nutrients are rarely found in surface waters, where solar radiation makes it possible for photosynthetic organisms to exist. If an ocean area is uncommonly productive, one can be sure that nutrients are supplied from deeper layers. (In limited areas they are supplied by the wastes of human activities.) The most productive waters in the world are therefore near the Antarctic continent, where the deep waters of the surrounding oceans well up and mix with the surface layers. There are similar upwellings along the coast of Chile, in the vicinity of Japan and in the Gulf Stream. In such regions fish are abundant and the maximum annual fixation of carbon approaches .3 kilogram per square meter. In the "desert" areas of the oceans, such

as the open seas of subtropical latitudes, the fixation rate may be less than a tenth of that value. In the Tropics warm surface layers are usually effective in blocking the vertical water exchange needed to carry nutrients up from below.

Phytoplankton, the primary fixers of carbon dioxide in the sea, are eaten by the zooplankton and other tiny animals. These organisms in turn provide food for the larger animals. The major part of the oceanic biomass, however, consists of microorganisms. Since the lifetime of such organisms is measured in weeks, or at most in months, their total mass can never accumulate appreciably. When microorganisms die, they quickly disintegrate as they sink to deeper layers. Soon most of what was once living tissue has become dissolved organic matter.

A small fraction of the organic particulate matter escapes oxidation and settles into the ocean depths. There it profoundly influences the abundance of chemical substances because (except in special regions) the deep layers exchange water with the surface layers very slowly. The enrichment of the deep layers goes hand in hand with a depletion of oxygen. There also appears to be an increase in carbon dioxide (in the form of carbonate and bicarbonate ions) in the ocean depths. The overall distribution of carbon dioxide, oxygen and various minor constituents in the sea reflects a balance between the marine life and its chemical milieu in the surface layers and the slow transport of substances by the general circulation of the ocean. The net effect is to prevent the ocean from becoming saturated with oxygen and to enrich the deeper strata with carbonate and bicarbonate ions.

The particular state in which we find the oceans today could well be quite different if the mechanisms for the exchange of water between the surface layers and the deep ones were either more intense or less so. The present state is determined primarily by the sinking of cold water in the polar regions, particularly the Antarctic. In these regions the water is also slightly saltier, and therefore still denser, because some of it has been frozen out in floating ice. If the climate of the earth were different, the distribution of carbon dioxide, oxygen and minerals might also be quite different. If the difference were large enough, oxygen might completely vanish from the ocean depths, leaving them to be populated only by chemibarotrophic bacteria. (This is now the case in the depths of the Black Sea.)

The time required to establish a new equilibrium in the ocean is determined by the slowest link in the chain of processes that has been described. This link is the oceanic circulation; it seems to take at least 1,000 years for the water in the deepest basins to be completely replaced. One can imagine other conditions of circulation in which the oceans would interact differently with sediments and rocks, producing a balance of substances that one can only guess at.

So far we have been concerned only with the basic biological and ecological processes that provide the mechanisms for circulating carbon through living organisms. Plants on land, with lifetimes measured in years, and phytoplankton in the sea, with lifetimes measured in weeks, are merely the innermost wheels in a biogeochemical machine that embraces the entire earth and that retains important characteristics over much longer time periods. In order to understand such interactions we shall need some rough estimates of the size of the various carbon reservoirs involved and the nature of their contents [see illustration on page 10]. In the context of the present argument the large uncertainties in such estimates are of little significance.

Only a few tenths of a percent of the immense mass of carbon at or near the surface of the earth (on the order of 20 × 10^{15} tons) is in rapid circulation in the biosphere, which includes the atmosphere, the hydrosphere, the upper portions of the earth's crust and the biomass itself. The overwhelming bulk of near-surface carbon consists of inorganic deposits (chiefly carbonates) and organic fossil deposits (chiefly oil shale, coal and petroleum) that required hundreds of

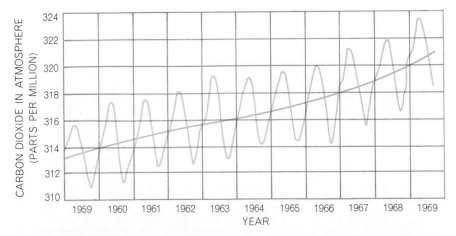

LONG-TERM VARIATIONS in the carbon dioxide content of the atmosphere have been followed at the Mauna Loa Observatory in Hawaii by the Scripps Institution of Oceanography. The sawtooth curve indicates the month-to-month change in concentration since January, 1959. The oscillations reflect seasonal variations in the rate of photosynthesis, as depicted in the bottom illustration on the opposite page. The smooth curve shows the trend.

OIL SHALE is one of the principal sedimentary forms in which carbon has been deposited over geologic time. This photograph, taken at Anvil Points, Colo., shows a section of the Green River Formation, which extends through Colorado, Utah and Wyoming. The formation is estimated to contain the equivalent of more than a trillion barrels of oil in seams containing more than 10 barrels of oil per ton of rock. Of this some 80 billion barrels is considered recoverable. The shale seams are up to 130 feet thick.

WHITE CLIFFS OF DOVER consist of almost pure calcium carbonate, representing the skeletons of phytoplankton that settled to the bottom of the sea over a period of millions of years more than 70 million years ago. The worldwide deposits of limestone, oil shale and other carbon-containing sediments are by far the largest repository of carbon: an estimated 20 quadrillion (10^{15}) tons.

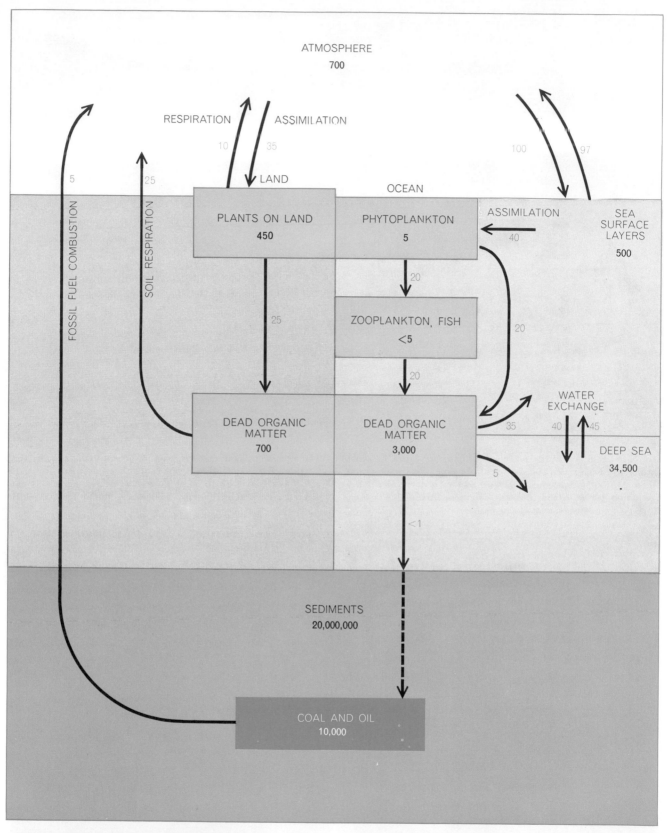

CARBON CIRCULATION IN BIOSPHERE involves two quite distinct cycles, one on land and one in the sea, that are dynamically connected at the interface between the ocean and the atmosphere. The carbon cycle in the sea is essentially self-contained in that phytoplankton assimilate the carbon dioxide dissolved in seawater and release oxygen back into solution. Zooplankton and fish consume the carbon fixed by the phytoplankton, using the dissolved oxygen for respiration. Eventually the decomposition of organic matter replaces the carbon dioxide assimilated by the phytoplank- ton. All quantities are in billions of metric tons. It will be seen that the combustion of fossil fuels at the rate of about five billion tons per year is sufficient to increase the carbon dioxide in the atmosphere by about .7 percent, equivalent to adding some two parts per million to the existing 320 ppm. Since the observed an- nual increase is only about .7 ppm, it appears that two-thirds of the carbon dioxide released from fossil fuels is quickly removed from the atmosphere, going either into the oceans or adding to the total mass of terrestrial plants. The estimated tonnages are the author's.

millions of years to reach their present magnitude. Over time intervals as brief as those of which we have been speaking—up to 1,000 years for the deep-ocean circulation—the accretion of such deposits is negligible. We may therefore consider the life processes on land and in the sea as the inner wheels that spin at comparatively high velocity in the carbon-circulating machine. They are coupled by a very low gear to more majestic processes that account for the overall circulation of carbon in its various geologic and oceanic forms.

We now know that the two great systems, the atmosphere and the ocean, are closely coupled to each other through the transfer of carbon dioxide across the surface of the oceans. The rate of exchange has recently been estimated by measuring the rate at which the radioactive isotope carbon 14 produced by the testing of nuclear weapons has disappeared from the atmosphere. The neutrons released in such tests form carbon 14 by reacting with the nitrogen 14 of the atmosphere. In this reaction a nitrogen atom ($_7N^{14}$) captures a neutron and subsequently releases a proton, yielding $_6C^{14}$. (The subscript before the letter represents the number of protons in the nucleus; the superscript after the letter indicates the sum of protons and neutrons.)

The last major atmospheric tests were conducted in 1963. Sampling at various altitudes and latitudes shows that the constituents of the atmosphere became rather well mixed over a period of a few years. The decline of carbon 14, however, was found to be rapid; it can be explained only by assuming an exchange of atmospheric carbon dioxide, enriched in carbon 14, with the reservoir of much less radioactive carbon dioxide in the sea. The measurements indicate that the characteristic time for the residence of carbon dioxide in the atmosphere before the gas is dissolved in the sea is between five and 10 years. In other words, every year something like 100 billion tons of atmospheric carbon dioxide dissolves in the sea and is replaced by a nearly equivalent amount of oceanic carbon dioxide.

Since around 1850 man has inadvertently been conducting a global geochemical experiment by burning large amounts of fossil fuel and thereby returning to the atmosphere carbon that was fixed by photosynthesis millions of years ago. Currently between five and six billion tons of fossil carbon per year are being released into the atmosphere. This would be enough to increase the amount of carbon dioxide in the air by 2.3 parts per million per year if the carbon dioxide were uniformly distributed and not removed. Within the past century the carbon dioxide content of the atmosphere has risen from some 290 parts per million to 320, with more than a fifth of the rise occurring in just the past decade [see illustration on page 8]. The total increase accounts for only slightly more than a third of the carbon dioxide (some 200 billion tons in all) released from fossil fuels. Although most of the remaining two-thirds has presumably gone into the oceans, a significant fraction may well have increased the total amount of vegetation on land. Laboratory studies show that plants grow faster when the surrounding air is enriched in carbon dioxide. Thus it is possible that man is fertilizing fields and forests by burning coal, oil and natural gas. The biomass on land may have increased by as much as 15 billion tons in the past century. There is, however, little concrete evidence for such an increase.

Man has of course been changing his environment in other ways. Over the past century large areas covered with forest have been cleared and turned to agriculture. In such areas the character of soil respiration has undoubtedly changed, producing effects that might have been detectable in the atmospheric content of carbon dioxide if it had not been for the simultaneous increase in the burning of fossil fuels. In any case the dynamic equilibrium among the major carbon dioxide reservoirs in the biomass, the atmosphere, the hydrosphere and the soil has been disturbed, and it can be said that they are in a period of transition. Since even the most rapid processes of adjustment among the reservoirs take decades, new equilibriums are far from being established. Gradually the deep oceans become involved; their turnover time of about 1,000 years and their rate of exchange with bottom sediments control the ultimate partitioning of carbon.

Meanwhile human activities continue to change explosively. The acceleration in the consumption of fossil fuels implies that the amount of carbon dioxide in the atmosphere will keep climbing from its present value of 320 parts per million to between 375 and 400 parts per million by the year 2000, in spite of anticipated large removals of carbon dioxide by land vegetation and the ocean reservoir [see illustrations on next page]. A fundamental question is: What will happen over the next 100 or 1,000 years? Clearly the exponential changes cannot continue.

If we extend the time scale with which we are viewing the carbon cycle by several orders of magnitude, to hundreds of thousands or millions of years, we can anticipate large-scale exchanges between organic carbon on land and carbonates of biological origin in the sea. We do know that there have been massive exchanges in the remote past. Any discussion of these past events and their implications for the future, however, must necessarily be qualitative and uncertain.

Although the plants on land have probably played an important role in the deposition of organic compounds in the soil, the oceans have undoubtedly acted as the main regulator. The amount of carbon dioxide in the atmosphere is essentially determined by the partial pressure of carbon dioxide dissolved in the

GIANT FERN of the genus *Pecopteris*, which fixed atmospheric carbon dioxide 300 million years ago, left the imprint of this frond in a thin layer of shale just above a coal seam in Illinois. The specimen is in the collection of the Smithsonian Institution.

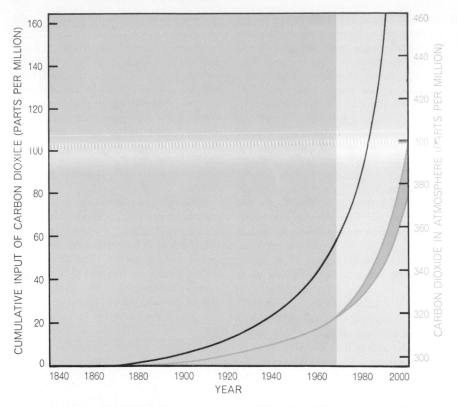

INCREASE IN ATMOSPHERIC CARBON DIOXIDE since 1860 is shown by the lower curve, with a projection to the year 2000. The upper curve shows the cumulative input of carbon dioxide. The difference between the two curves represents the amount of carbon dioxide removed by the ocean or by additions to the total biomass of vegetation on land.

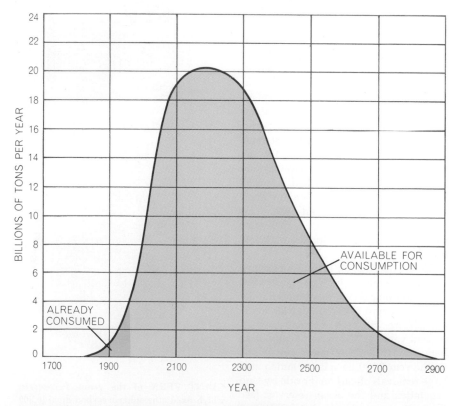

POSSIBLE CONSUMPTION PATTERN OF FOSSIL FUELS was projected by Harrison Brown in the mid-1950's. Here the fuel consumed is updated to 1960. If a third of the carbon dioxide produced by burning it all were to remain in the atmosphere, the carbon dioxide level would rise from 320 ppm today to about 1,500 ppm over the next several centuries.

sea. Over a period of, say, 100,000 years the leaching of calcium carbonates from land areas tends to increase the amount of carbon dioxide in the sea, but at the same time a converse mechanism—the precipitation and deposition of oceanic carbonates—tends to reduce the amount of carbon dioxide in solution. Thus the two mechanisms tend to cancel each other.

Over still longer periods of time—millions or tens of millions of years—the concentrations of carbonate and bicarbonate ions in the sea are probably buffered still further by reactions involving potassium, silicon and aluminum, which are slowly weathered from rocks and carried into the sea. The net effect is to stabilize the carbon dioxide content of the oceans and hence the carbon dioxide content of the atmosphere. Therefore it appears that the carbon dioxide environment, on which the biosphere fundamentally depends, may have been fairly constant right up to the time, barely a moment ago geologically speaking, when man's consumption of fossil fuels began to change the carbon dioxide content of the atmosphere.

The illustration on page 10 represents an attempt to synthetize into a single picture the circulation of carbon in nature, particularly in the biosphere. In addition to the values for inventories and transfers already mentioned, the flow chart contains other quantities for which the evidence is still meager. They have been included not only to balance the books but also to suggest where further investigation might be profitable. This may be the principal value of such an exercise. Such a flow chart also provides a semiquantitative model that enables one to begin to discuss how the global carbon system reacts to disturbances. A good model should of course include inventories and pathways for all the elements that play a significant role in biological processes.

The greatest disturbances of which we are aware are those now being introduced by man himself. Since his tampering with the biological and geochemical balances may ultimately prove injurious —even fatal—to himself, he must understand them much better than he does today. The story of the circulation of carbon in nature teaches us that we cannot control the global balances. Therefore we had better leave them close to the natural state that existed until the beginning of the Industrial Revolution. Out of a simple realization of this necessity may come a new industrial revolution.

2

THE CHEMICAL ELEMENTS OF LIFE

EARL FRIEDEN
June 1972

*Until recently it was believed that living matter
incorporated 20 of the natural elements. Now it
has been shown that a role is played by four others:
fluorine, silicon, tin and vanadium*

How many of the 90 naturally occurring elements are essential to life? After more than a century of increasingly refined investigation, the question still cannot be answered with certainty. Only a year or so ago the best answer would have been 20. Since then four more elements have been shown to be essential for the growth of young animals: fluorine, silicon, tin and vanadium. Nickel may soon be added to the list. In many cases the exact role played by these and other trace elements remains unknown or unclear. These gaps in knowledge could be critical during a period when the biosphere is being increasingly contaminated by synthetic chemicals and subjected to a potentially harmful redistribution of salts and metal ions. In addition, new and exotic chemical forms of metals (such as methyl mercury) are being discovered, and a complex series of competitive and synergistic relations among mineral salts has been encountered. We are led to the realization that we are ignorant of many basic facts about how our chemical milieu affects our biological fate.

Biologists and chemists have long been fascinated by the way evolution has selected certain elements as the building blocks of living organisms and has ignored others. The composition of the earth and its atmosphere obviously sets a limit on what elements are available. The earth itself is hardly a chip off the universe. The solar system, like the universe, seems to be 99 percent hydrogen and helium. In the earth's crust helium is essentially nonexistent (except in a few rare deposits) and hydrogen atoms constitute only about .22 percent of the total. Eight elements provide more than 98 percent of the atoms in the earth's crust: oxygen (47 percent), silicon (28 percent), aluminum (7.9 percent), iron (4.5 percent), calcium (3.5 percent), so-

dium (2.5 percent), potassium (2.5 percent) and magnesium (2.2 percent). Of these eight elements only five are among the 11 that account for more than 99.9 percent of the atoms in the human body. Not surprisingly nine of the 11 are also the nine most abundant elements in seawater [*see illustration on page 14*].

Two elements, hydrogen and oxygen, account for 88.5 percent of the atoms in the human body; hydrogen supplies 63 percent of the total and oxygen 25.5 percent. Carbon accounts for another 9.5 percent and nitrogen 1.4 percent. The remaining 20 elements now thought to be essential for mammalian life account for less than .7 percent of the body's atoms.

The Background of Selection

Three characteristics of the biosphere or of the elements themselves appear to have played a major part in directing the chemistry of living forms. First and foremost there is the ubiquity of water, the solvent base of all life on the earth. Water is a unique compound; its stability and boiling point are both unusually high for a molecule of its simple composition. Many of the other compounds essential for life derive their usefulness from their response to water: whether they are soluble or insoluble, whether or not (if they are soluble) they carry an electric charge in solution and, not least, what effect they have on the viscosity of water.

The second directing force involves the chemical properties of carbon, which evolution selected over silicon as the central building block for constructing giant molecules. Silicon is 146 times more plentiful than carbon in the earth's crust and exhibits many of the same properties. Silicon is directly below carbon in the periodic table of the elements;

like carbon, it has the capacity to gain four electrons and form four covalent bonds.

The crucial difference that led to the preference for carbon compounds over silicon compounds seems traceable to two chemical features: the unusual stability of carbon dioxide, which is readily soluble in water and always monomeric (it remains a single molecule), and the almost unique ability of carbon to form long chains and stable rings with five or six members. This versatility of the carbon atom is responsible for the millions of organic compounds found on the earth.

Silicon, in contrast, is insoluble in water and forms only relatively shorter chains with itself. It can enter into longer chains, however, by forming alternating bonds with oxygen, creating the compounds known as silicones ($-Si-O-Si-O-Si-$). Carbon-to-carbon bonds are more stable than silicon-to-silicon bonds, but not so stable as to be virtually immutable, as the silicon-oxygen polymers are. Nevertheless, silicon has recently been shown to be essential in a way as yet unknown for normal bone development and full growth in chicks.

The third force influencing the evolutionary selection of the elements essential for life is related to an atom's size and charge density. Obviously the heavy synthetic elements from neptunium (atomic number 93) to lawrencium (No. 103), along with two lighter synthetic elements, technetium (No. 43) and promethium (No. 61), were never available in nature. (The atomic number expresses the number of protons in the nucleus of an atom or the number of electrons around the nucleus.) The eight heavy elements in another group (Nos. 84 and 85 and Nos. 87 through 92) are too radioactive to be useful in living structures. Six more elements are inert gases

COMPOSITION OF UNIVERSE		COMPOSITION OF EARTH'S CRUST		COMPOSITION OF SEAWATER		COMPOSITION OF HUMAN BODY	
PERCENT OF TOTAL NUMBER OF ATOMS							
H	91	O	47	H	66	H	63
He	9.1	Si	28	O	33	O	25.5
O	.057	Al	7.9	Cl	.33	C	9.5
N	.042	Fe	4.5	Na	.28	N	1.4
C	.021	Ca	3.5	Mg	.033	Ca	.31
Si	.003	Na	2.5	S	.017	P	.22
Ne	.003	K	2.5	Ca	.006	Cl	.03
Mg	.002	Mg	2.2	K	.006	K	.06
Fe	.002	Ti	.46	C	.0014	S	.05
S	.001	H	.22	Br	.0005	Na	.03
		C	.19			Mg	.01
ALL OTHERS<.01		ALL OTHERS<.1		ALL OTHERS<.1		ALL OTHERS<.01	

Al ALUMINUM	C CARBON	Fe IRON	O OXYGEN	S SULFUR
B BORON	Cl CHLORINE	Mg MAGNESIUM	K POTASSIUM	Ti TITANIUM
Br BROMINE	He HELIUM	Ne NEON	Si SILICON	
Ca CALCIUM	H HYDROGEN	N NITROGEN	Na SODIUM	

CHEMICAL SELECTIVITY OF EVOLUTION can be demonstrated by comparing the composition of the human body with the approximate composition of seawater, the earth's crust and the universe at large. The percentages are based on the total number of atoms in each case; because of rounding the totals do not exactly equal 100. Elements in the colored boxes in the last column appear in one or more columns at the left. Thus one sees that phosphorus, the sixth most plentiful element in the body, is a rare element in inanimate nature. Carbon, the third most plentiful element, is also very scarce elsewhere.

with virtually no useful chemical reactivities: helium, neon, argon, krypton, xenon and radon. On various plausible grounds one can exclude another 24 elements, or a total of 38 natural elements, as being clearly unsatisfactory for incorporation in living organisms because of their relative unavailability (particularly the elements in the lanthanide and actinide series) or their high toxicity (for example mercury and lead). This leaves 52 of the 90 natural elements as being potentially useful.

Only three of the 24 elements known to be essential for animal life have an atomic number above 34. All three are needed only in trace amounts: molybdenum (No. 42), tin (No. 50) and iodine (No. 53). The four most abundant atoms in living organisms—hydrogen, carbon, oxygen and nitrogen—have atomic numbers of 1, 6, 7 and 8. Their preponderance seems attributable to their being the smallest and lightest elements that can achieve stable electronic configura-

tions by adding one to four electrons. The ability to add electrons by sharing them with other atoms is the first step in forming chemical bonds leading to stable molecules. The seven next most abundant elements in living organisms all have atomic numbers below 21. In the order of their abundance in mammals they are calcium (No. 20), phosphorus (No. 15), potassium (No. 19), sulfur (No. 16), sodium (No. 11), magnesium (No. 12) and chlorine (No. 17). The remaining 10 elements known to be present in either plants or animals are needed only in traces. With the exception of fluorine (No. 9) and silicon (No. 14), the remaining eight occupy positions between No. 23 and No. 34 in the periodic table [see illustration at right]. It is interesting that this interval embraces three elements for which evolution has evidently found no role: gallium, germanium and arsenic. None of the metals with properties similar to those of gallium (such as aluminum and indium) has

proved to be useful to living organisms. On the other hand, since silicon and tin, two elements with chemical activities similar to those of germanium, have just joined the list of essential elements, it seems possible that germanium too, in spite of its rarity, will turn out to have an essential role. Arsenic, of course, is a well-known poison.

Functions of Essential Elements

Some useful generalizations can be made about the role of the various elements. Six elements—carbon, nitrogen, hydrogen, oxygen, phosphorus and sulfur—make up the molecular building blocks of living matter: amino acids, sugars, fatty acids, purines, pyrimidines and nucleotides. These molecules not only have independent biochemical roles but also are the respective constituents of the following large molecules: proteins, glycogen, starch, lipids and nucleic acids. Several of the 20 amino acids contain sulfur in addition to carbon, hydrogen and oxygen. Phosphorus plays an important role in the nucleotides such as

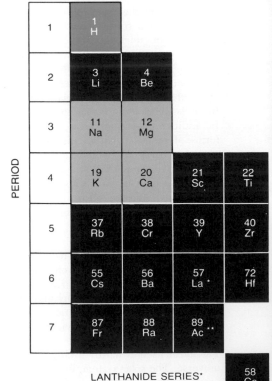

ESSENTIAL LIFE ELEMENTS, 24 by the latest count, are clustered in the upper half of the periodic table. The elements are ar-

adenosine triphosphate (ATP), which is central to the energetics of the cell. ATP includes components that are also one of the four nucleotides needed to form the double helix of deoxyribonucleic acid (DNA), which incorporates the genetic blueprint of all plants and animals. Both sulfur and phosphorus are present in many of the small accessory molecules called coenzymes. In bony animals phosphorus and calcium help to create strong supporting structures.

The electrochemical properties of living matter depend critically on elements or combinations of elements that either gain or lose electrons when they are dissolved in water, thus forming ions. The principal cations (electron-deficient, or positively charged, ions) are provided by four metals: sodium, potassium, calcium and magnesium. The principal anions (ions with a negative charge because they have surplus electrons) are provided by the chloride ion and by sulfur and phosphorus in the form of sulfate ions and phosphate ions. These seven ions maintain the electrical neutrality of body fluids and cells and also play a part in maintaining the proper liquid volume of the blood and other fluid systems. Whereas the cell membrane serves as a physical barrier to the exchange of large molecules, it allows small molecules to pass freely. The electrochemical functions of the anions and cations serve to maintain the appropriate relation of osmotic pressure and charge distribution on the two sides of the cell membrane.

One of the striking features of the ion distribution is the specificity of these different ions. Cells are rich in potassium and magnesium, and the surrounding plasma is rich in sodium and calcium. It seems likely that the distribution of ions in the plasma of higher animals reflects the oceanic origin of their evolutionary antecedents. One would like to know how primitive cells learned to exclude the sodium and calcium ions in which they were bathed and to develop an internal milieu enriched in potassium and magnesium.

The third and last group of essential elements consists of the trace elements. The fact that they are required in extremely minute quantities in no way diminishes their great importance. In this sense they are comparable to the vitamins. We now know that the great majority of the trace elements, represented by metallic ions, serve chiefly as key components of essential enzyme systems or of proteins with vital functions (such as hemoglobin and myoglobin, which respectively transports oxygen in the blood and stores oxygen in muscle). The heaviest essential element, iodine, is an essential constituent of the thyroid hormones thyroxine and triiodothyronine, although its precise role in hormonal activity is still not understood.

The Trace Elements

To demonstrate that a particular element is essential to life becomes increasingly difficult as one lowers the threshold of the amount of a substance recognizable as a "trace." It has been known for more than 100 years, for example, that iron and iodine are essential to man. In a rapidly developing period of biochemistry between 1928 and 1935 four more elements, all metals, were shown to be

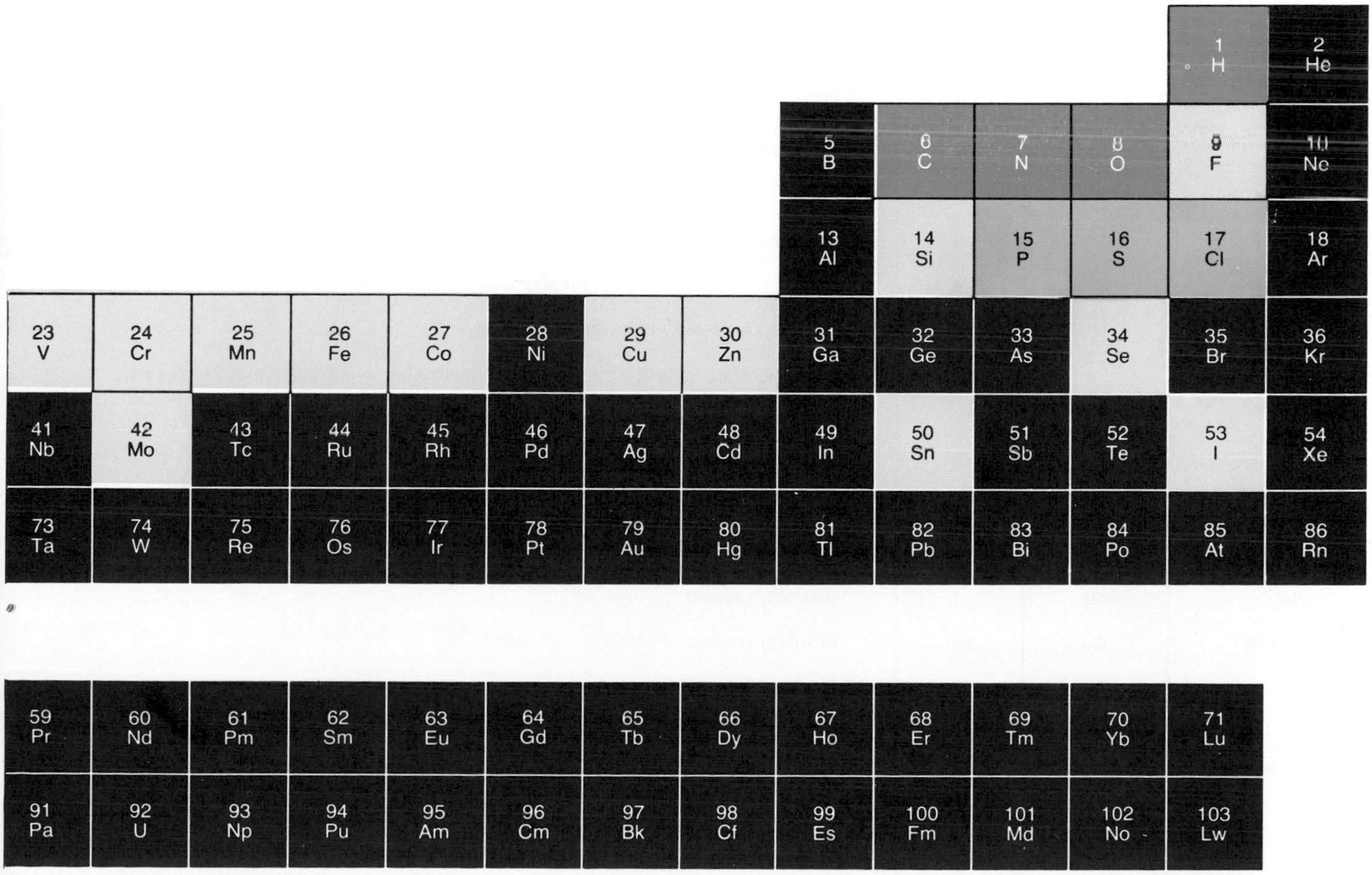

ranged according to their atomic number, which is equivalent to the number of protons in the atom's nucleus. The four most abundant elements that are found in living organisms (hydrogen, oxygen, carbon and nitrogen) are indicated by dark color. The seven next most common elements are in lighter color. The 13 elements that are shown in lightest color are needed only in traces.

ELEMENT	SYMBOL	ATOMIC NUMBER	COMMENTS
HYDROGEN	H	1	Required for water and organic compounds.
HELIUM	He	2	Inert and unused.
LITHIUM	Li	3	Probably unused.
BERYLLIUM	Be	4	Probably unused; toxic.
BORON	B	5	Essential in some plants; function unknown.
CARBON	C	6	Required for organic compounds.
NITROGEN	N	7	Required for many organic compounds.
OXYGEN	O	8	Required for water and organic compounds.
FLUORINE	F	9	Growth factor in rats; possible constituent of teeth and bone.
NEON	Ne	10	Inert and unused.
SODIUM	Na	11	Principal extracellular cation.
MAGNESIUM	Mg	12	Required for activity of many enzymes; in chlorophyll.
ALUMINUM	Al	13	Essentiality under study.
SILICON	Si	14	Possible structural unit of diatoms; recently shown to be essential in chicks.
PHOSPHORUS	P	15	Essential for biochemical synthesis and energy transfer.
SULFUR	S	16	Required for proteins and other biological compounds.
CHLORINE	Cl	17	Principal cellular and extracellular anion.
ARGON	A	18	Inert and unused.
POTASSIUM	K	19	Principal cellular cation.
CALCIUM	Ca	20	Major component of bone; required for some enzymes.
SCANDIUM	Sc	21	Probably unused.
TITANIUM	Ti	22	Probably unused.
VANADIUM	V	23	Essential in lower plants, certain marine animals and rats.
CHROMIUM	Cr	24	Essential in higher animals; related to action of insulin.
MANGANESE	Mn	25	Required for activity of several enzymes.
IRON	Fe	26	Most important transition metal ion; essential for hemoglobin and many enzymes.
COBALT	Co	27	Required for activity of several enzymes; in vitamin B_{12}.
NICKEL	Ni	28	Essentiality under study.
COPPER	Cu	29	Essential in oxidative and other enzymes and hemocyanin.
ZINC	Zn	30	Required for activity of many enzymes.
GALLIUM	Ga	31	Probably unused.
GERMANIUM	Ge	32	Probably unused.
ARSENIC	As	33	Probably unused; toxic.
SELENIUM	Se	34	Essential for liver function.
MOLYBDENUM	Mo	42	Required for activity of several enzymes.
TIN	Sn	50	Essential in rats; function unknown.
IODINE	I	53	Essential constituent of the thyroid hormones.

SOME TWO-THIRDS OF LIGHTEST ELEMENTS, or 21 out of the first 34 elements in the periodic table, are now known to be essential for animal life. These 21 plus molybdenum (No. 42), tin (No. 50) and iodine (No. 53) constitute the total list of the 24 essential elements, which are here enclosed in colored boxes. It is possible that still other light elements will turn out to be essential. The most likely candidates are aluminum, nickel and germanium. The element boron already appears to be essential for some plants.

essential: copper, manganese, zinc and cobalt. The demonstration can be credited chiefly to a group of investigators at the University of Wisconsin led by C. A. Elvehjem, E. B. Hart and W. R. Todd. At that time it seemed that these four metals might be the last of the essential trace elements. In the next 30 years, however, three more elements were shown to be essential: chromium, selenium and molybdenum. Fluorine, silicon, tin and vanadium have been added since 1970.

The essentiality of five of these last seven elements was discovered through the careful, painstaking efforts of Klaus Schwarz and his associates, initially located at the National Institutes of Health and now based at the Veterans Administration Hospital in Long Beach, Calif. For the past 15 years Schwarz's group has made a systematic study of the trace-element requirements of rats and other small animals. The animals are maintained from birth in a completely isolated sterile environment [see illustration on page 19].

The apparatus is constructed entirely of plastics to eliminate the stray contaminants contained in metal, glass and rubber. Although even plastics may contain some trace elements, they are so tightly bound in the structural lattice of the material that they cannot be leached out or be picked up by an animal even through contact. A typical isolator system houses 32 animals in individual acrylic cages. Highly efficient air filters remove all trace substances that might be present in the dust in the air. Thus the animals' only access to essential nutrients is through their diet. They receive chemically pure amino acids instead of natural proteins, and all other dietary ingredients are screened for metal contaminants.

Since the standards of purity employed in these experiments far exceed those for reagents normally regarded as analytically pure, Schwarz and his co-workers have had to develop many new analytical chemical methods. The most difficult problem turned out to be the purification of salt mixtures. Even the purest commercial reagents were contaminated with traces of metal ions. It was also found that trace elements could be passed from mothers to their offspring. To minimize this source of contamination animals are weaned as quickly as possible, usually from 18 to 20 days after birth.

With these precautions Schwarz and his colleagues have within the past several years been able to produce a new

deficiency disease in rats. The animals grow poorly, lose hair and muscle tone, develop shaggy fur and exhibit other detrimental changes [see illustration on page 20]. When standard laboratory food is given these animals, they regain their normal appearance. At first it was thought that all the symptoms were caused by the lack of one particular trace element. Eventually four different elements had to be supplied to complete the highly purified diets the animals had been receiving. The four elements proved to be fluorine, silicon, tin and vanadium. A convenient source of these

elements is yeast ash or liver preparations from a healthy animal. The animals on the deficiency diet grew less than half as fast as those on a normal or supplemented diet. Growth alone, however, may not tell the entire story. There is some evidence that even the addition of the four elements may not reverse the loss of hair and skin changes resulting from the deficiency diet.

Functions of Trace Elements

The addition of tin and vanadium to the list of essential trace metals brings

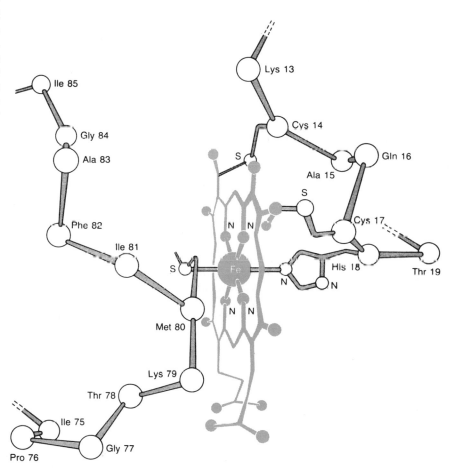

Ala	ALANINE	His	HISTIDINE	Phe	PHENYLALANINE
Cys	CYSTEINE	Ile	ISOLEUCINE	Pro	PROLINE
Gln	GLUTAMINE	Lys	LYSINE	Thr	THREONINE
Gly	GLYCINE	Met	METHIONINE		

THE METALLOENZYME CYTOCHROME c is typical of metal-protein complexes in which trace metals play a crucial role. Cytochrome c belongs to a family of enzymes that extract energy from food molecules. It consists of a protein chain of 104 amino acid units attached to a heme group (color), a rosette of atoms with an atom of iron at the center. This simplified molecular diagram shows only the heme group and several of the amino acid units closest to it. The iron atom has six coordination sites enabling it to form six bonds with neighboring atoms. Four bonds connect to nitrogen atoms in the heme group itself, and the remaining two bonds link up with amino acid units in the protein chain (histidine at site No. 18 and methionine at site No. 80). The illustration is based on the work of Richard E. Dickerson of the California Institute of Technology, in whose laboratory the complete structure of horse-heart cytochrome c was recently determined.

METAL	ENZYME	BIOLOGICAL FUNCTION
IRON	FERREDOXIN	Photosynthesis
	SUCCINATE DEHYDROGENASE	Aerobic oxidation of carbohydrates
IRON IN HEME	ALDEHYDE OXIDASE	Aldehyde oxidation
	CYTOCHROMES	Electron transfer
	CATALASE	Protection against hydrogen peroxide
	[HEMOGLOBIN]	Oxygen transport
COPPER	CERULOPLASMIN	Iron utilization
	CYTOCHROME OXIDASE	Principal terminal oxidase
	LYSINE OXIDASE	Elasticity of aortic walls
	TYROSINASE	Skin pigmentation
	PLASTOCYANIN	Photosynthesis
	[HEMOCYANIN]	Oxygen transport in invertebrates
ZINC	CARBONIC ANHYDRASE	CO_2 formation; regulation of acidity
	CARBOXYPEPTIDASE	Protein digestion
	ALCOHOL DEHYDROGENASE	Alcohol metabolism
MANGANESE	ARGINASE	Urea formation
	PYRUVATE CARBOXYLASE	Pyruvate metabolism
COBALT	RIBONUCLEOTIDE REDUCTASE	DNA biosynthesis
	GLUTAMATE MUTASE	Amino acid metabolism
MOLYBDENUM	XANTHINE OXIDASE	Purine metabolism
	NITRATE REDUCTASE	Nitrate utilization
CALCIUM	LIPASES	Lipid digestion
MAGNESIUM	HEXOKINASE	Phosphate transfer

WIDE VARIETY OF METALLOENZYMES is required for the successful functioning of living organisms. Some of the most important are given in this list. The giant oxygen-transporting molecules hemoglobin and hemocyanin are included in the list (in brackets) even though they are not strictly enzymes, that is, they do not act as biological catalysts.

to 10 the total number of trace metals needed by animals and plants. What role do these metals play? For six of the eight trace metals recognized from earlier studies (that is, for iron, zinc, copper, cobalt, manganese and molybdenum) we are reasonably sure of the answer. The six are constituents of a wide range of enzymes that participate in a variety of metabolic processes [see illustration above].

In addition to its role in hemoglobin and myoglobin, iron appears in succinate dehydrogenase, one of the enzymes needed for the utilization of energy from sugars and starches. Enzymes incorporating zinc help to control the formation of carbon dioxide and the digestion of proteins. Copper is present in more than a dozen enzymes, whose roles range from the utilization of iron to the pigmentation of the skin. Cobalt appears in enzymes involved in the synthesis of DNA and the metabolism of amino acids. Enzymes incorporating manga-

nese are involved in the formation of urea and the metabolism of pyruvate. Enzymes incorporating molybdenum participate in purine metabolism and the utilization of nitrogen.

These six metals belong to a group known as transition elements. They owe their uniqueness to their ability to form strong complexes with ligands, or molecular groups, of the type present in the side chains of proteins. Enzymes in which transition metals are tightly incorporated are called metalloenzymes, since the metal is usually embedded deep inside the structure of the protein. If the metal atom is removed, the protein usually loses its capacity to function as an enzyme. There is also a group of enzymes in which the metal ion is more loosely associated with the protein but is nonetheless essential for the enzyme's activity. Enzymes in this group are known as metal-ion-activated enzymes. In either group the role of the metal ion may be to maintain the proper confor-

mation of the protein, to bind the substrate (the molecule acted on) to the protein or to donate or accept electrons in reactions where the substrate is reduced or oxidized.

In 1968 the complete three-dimensional structure of the first metalloenzyme, cytochrome c, was published [see "The Structure and History of an Ancient Protein," By Richard E. Dickerson, beginning on page 189]. Cytochrome c, a red enzyme containing iron, is universally present in plants and animals. It is one of a series of enzymes, all called cytochromes, that extract energy from food molecules by the stepwise addition of oxygen.

The complete amino acid sequence of cytochrome c obtained from the human heart was determined some 10 years ago by a group led by Emil L. Smith of the University of California at Los Angeles and by Emanuel Margoliash of Northwestern University. The iron atom is partially complexed with an intricate organic molecule, protoporphyrin, to form a heme group similar to that in hemoglobin. Of the iron atom's six coordination sites, four are attached to the heme group through nitrogen atoms. The other two sites form bonds with the protein chain; one bond is through a nitrogen atom in the side chain of a histidine unit at site No. 18 in the protein sequence and the other bond is through a sulfur atom in the side chain of a methionine unit at site No. 80 [see illustration on preceding page].

Although the cytochrome c molecule is complicated, it is one of the simplest of the metalloenzymes. Cytochrome oxidase, probably the single most important enzyme in most cells, since it is responsible for transferring electrons to oxygen to form water, is far more complicated. Each molecule contains about 12 times as many atoms as cytochrome c, including two copper atoms and two heme groups, both of which participate in transferring the electrons.

More complicated yet is cysteamine oxygenase, which catalyzes the addition of oxygen to a molecule of cysteamine; it contains one atom each of three different metals: iron, copper and zinc. There are many other combinations of metal ions and unique molecular assemblies. An extreme example is xanthine oxidase, which contains eight iron atoms, two molybdenum atoms and two molecules incorporating riboflavin (one of the B vitamins) in a giant molecule more than 25 times the size of cytochrome c.

The metal-containing proteins of another group, the metalloproteins, closely resemble the metalloenzymes except that they lack an obvious catalytic function. Hemoglobin itself is an example. Others are hemocyanin, the copper-containing blue protein that carries oxygen in many invertebrates, metallothionein, a protein involved in the absorption and storage of zinc, and transferrin, a protein that transports iron in the bloodstream. There may be many more such compounds still unrecognized because their function has escaped detection.

The Newest Essential Elements

Much remains to be learned about the specific biochemical role of the most recently discovered essential elements. In 1957 Schwarz and Calvin M. Foltz, working at the National Institutes of Health, showed that selenium helped to prevent several serious deficiency diseases in different animals, including liver necrosis and muscular dystrophy. Rats were protected against death from liver necrosis by a diet containing one-tenth of a part per million of selenium. Comparably low doses reversed the white muscle disease observed in cattle and sheep that happen to graze in areas where selenium is scarce.

In April a group at the University of Wisconsin under J. T. Rotruck reported a direct biochemical role for selenium

Oxidative damage to red blood cells was detected in rats kept on a selenium-deficient diet. This damage was related to reduced activity of an enzyme, glutathione peroxidase, that helps to protect hemoglobin against the injurious oxidative effects of hydrogen peroxide. The enzyme uses hydrogen peroxide to catalyze the oxidation of glutathione, thus keeping hydrogen peroxide from oxidizing the reduced state of iron in hemoglobin. Oxidized glutathione can readily be converted to reduced glutathione by a variety of intracellular mechanisms. There is some reason to believe glutathione peroxidase may even contain some form of selenium acting as an integral part of the functional enzyme molecule.

The physiological importance of chromium was established in 1959 by Schwarz and Walter Mertz. They found that chromium deficiency is characterized by impaired growth and reduced life-span, corneal lesions and a defect in sugar metabolism. When the diet is deficient in chromium, glucose is removed from the bloodstream only half as fast as it is normally. In rats the deficiency is relieved by a single administration of 20 micrograms of certain trivalent chromic salts. It now appears that the chromium ion works in conjunction with insulin, and that in at least some cases

diabetes may reflect faulty chromium metabolism.

After developing the all-plastic trace-element isolator described above, Schwarz, David B. Milne and Elizabeth Vineyard discovered that tin, not previously suspected as being essential, was necessary for normal growth. Without one or two parts per million of tin in their diet, rats grow at only about two-thirds the normal rate.

The next element shown to be essential in mammals by the Schwarz group was vanadium, an element that had been detected earlier in certain marine invertebrates but whose essentiality had not been demonstrated. On a diet in which vanadium is totally excluded rats suffer a retardation of about 30 percent in growth rate. Schwarz and Milne found that normal growth is restored by adding one-tenth of a part per million of vanadium to the diet. At higher concentrations vanadium is known to have several biological effects, but its essential role in trace amounts remains to be established. A high dose of vanadium blocks the synthesis of cholesterol and reduces the amount of phospholipid and cholesterol in the blood. Vanadium also promotes the mineralization of teeth and is effective as a catalyst in the oxidation of many biological substances.

The third element most recently iden-

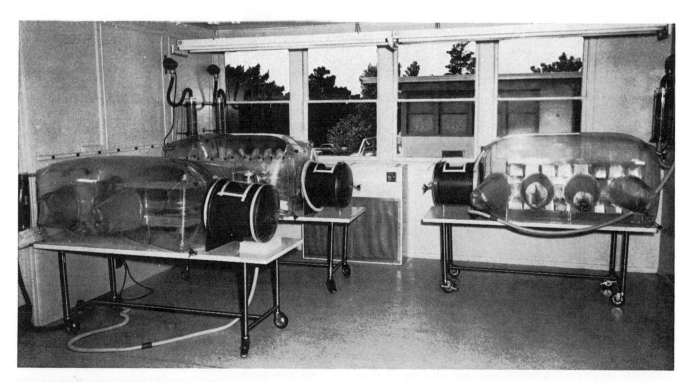

NUTRITIONAL NEEDS OF SMALL ANIMALS are studied in a trace-element isolator, a modification of the apparatus originally conceived to maintain animals in a germ-free environment. To prevent unwanted introduction of trace elements the isolator is built completely of plastics. It holds 32 animals in separate cages, individually supplied with food of precisely known composition. The system was designed by Klaus Schwarz and J. Cecil Smith of the Veterans Administration Hospital in Long Beach, Calif.

tified as being essential is fluorine. Even with tin and vanadium added to highly purified diets containing all other elements known to be essential, the animals in Schwarz's plastic cages still failed to grow at a normal rate. When up to half a part per million of potassium fluoride was added to the diet, the animals showed a 20 to 30 percent weight gain in four weeks. Although it had appeared that a trace amount of fluorine was essential for building sound teeth, Schwarz's study showed that fluorine's biochemical role was more fundamental than that. In any case fluoridated water provides more than enough fluorine to maintain a normal growth rate.

Although there were earlier clues that silicon might be an essential life element, firm proof of its essentiality, at least in young chicks, was reported only three months ago. Edith M. Carlisle of the School of Public Health at the University of California at Los Angeles finds that chicks kept on a silicon-free diet for only one or two weeks exhibit poor development of feathers and skeleton, including markedly thin leg bones. The addition of 30 parts per million of silicon to the diet increases the chicks' growth more than 35 percent and makes possible normal feathering and skeletal development. Considering that silicon is not only the second most abundant element in the earth's crust but is also similar to carbon in many of its chemical properties, it is hard to see how evolution could have totally excluded it from an essential biochemical role.

Nickel, nearly always associated with iron in natural substances, is another element receiving close attention. Also a transition element, it is particularly difficult to remove from the food used in special diets. Nickel seems to influence the growth of wing and tail feathers in chicks but more consistent data are needed to establish its essentiality. One incidental result of Schwarz's work has been the discovery of a previously unrecognized organic compound, which will undoubtedly prove to be a new vitamin.

Synergism and Antagonism

The interaction of the various essential metals can be extremely complicated. The absence of one metal in the diet can profoundly influence, either positively or negatively, the utilization of another metal that may be present. For example, it has been known for nearly 50 years that copper is essential for the proper metabolism of iron. An animal deprived of copper but not iron develops anemia because the biosynthetic machinery fails to incorporate iron in hemoglobin molecules. It has only recently been found in our laboratories at Florida State University that ceruloplasmin, the copper-containing protein of the blood, is a direct molecular link between the two metals. Ceruloplasmin promotes the release of iron from animal liver so that the iron-binding protein of the serum, transferrin, can complex with iron and transfer it to the developing red blood cells for direct utilization in the biosynthesis of hemoglobin. This represents a synergistic relation between copper and iron.

As an example of antagonism between elements one can cite the instance of copper and zinc. The ability of sheep or cattle to absorb copper is greatly reduced if too much zinc or molybdenum is present in their diet. Evidently either of the two metals can displace copper in an absorption process that probably involves competition for sites on a metal-binding protein in the intestines and liver.

The recent discoveries present many fresh challenges to biochemists. One can expect the discovery of previously unsuspected metalloenzymes containing vanadium, tin, chromium and selenium. New compounds or enzyme systems requiring fluorine and silicon may also be uncovered. The multiple and complex interdependencies of the elements suggest many hitherto unrecognized and important facts about the role and interrelations of metal ions in nutrition and in health and disease.

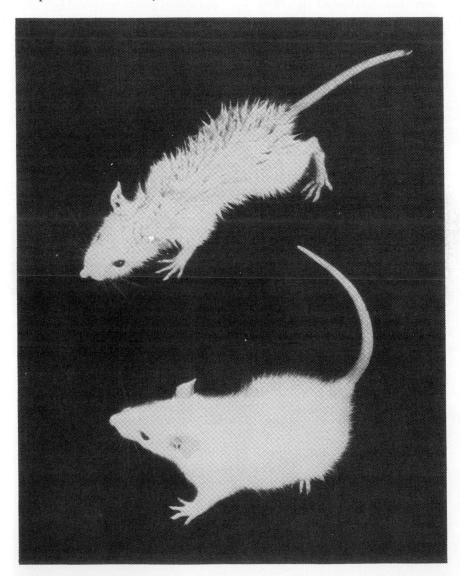

TRACE-ELEMENT DEFICIENCY developed when the rat at the top of this photograph was kept in the trace-element isolator for 20 days and fed a diet from which fluorine, tin and vanadium had been carefully excluded. The healthy animal at the bottom was fed the same diet but was kept under ordinary conditions. It was evidently able to obtain the necessary trace amounts of fluorine, tin and vanadium from dust and other contaminants.

3

THE SHAPES OF ORGANIC MOLECULES

JOSEPH B. LAMBERT
January 1970

*The new Nobel prize in chemistry was given to men
who showed that certain molecules can assume different
shapes simply by rotations around single bonds. Such
differences influence chemical reactivity*

Chemists had quite remarkable insights about the shapes of molecules many years before the electronic theory of matter was fully developed by physicists. In 1865 Friedrich August Kekulé had the inspired idea that the six carbon atoms in the benzene molecule (C_6H_6) are joined in a ring. In 1890 Hermann Sachse proposed that the ring of six carbon atoms in cyclohexane, a molecule with six more hydrogen atoms than benzene, would be free of strain if the carbon atoms were located alternately above and below the plane of the ring instead of being located in the plane itself. Such a structure has the appearance of a reclining chair. Because the benzene ring has double bonds between alternate pairs of carbon atoms, the atoms are forced to remain in a plane. Sachse's chair hypothesis, and its implications for other organic molecules, was rejected for more than 30 years until physical experiments by Walter Hückel, Jacob Boeseken and Odd Hassel of Norway provided corroboration. Sachse's model was not universally accepted, however, until after 1950, when persuasive chemical evidence in its favor was supplied by Derek H. R. Barton of the Imperial College of Science and Technology. Last month Hassel and Barton received the Nobel prize in chemistry "for their work to develop and apply the concept of conformation in chemistry."

The concept of conformation, or conformational analysis, is concerned with the different three-dimensional forms that can be assumed by certain molecules whose atoms are free to rotate around one or more bonds. Such interconvertible forms are known as conformational isomers, or conformers, to distinguish them from other isomers in which two or more molecules of the same formula have different three-dimensional structures that are intercon-

vertible only if bonds are broken and re-formed. It was Barton more than anyone else who showed that different conformational isomers can exhibit distinctly different chemical reactivities. In the past 20 years conformational analysis has been invaluable in the synthesis of pharmacological agents, notably steroids and antibiotics of the tetracycline and penicillin families. In biochemistry conformational analysis is helping to clarify the mechanisms by which enzymes promote chemical reactions in the living cell.

Some Simple Hydrocarbons

The principles of conformational analysis can be illustrated by describing the structures of several simple carbon-containing compounds. In methane (CH_4) each hydrogen atom is singly bonded to a central carbon atom and the four hydrogens form the corners of a tetrahedron. The angle formed by the C–H bonds between any two adjacent hydrogen atoms is 109.5 degrees [*see illustration on page 23*]. Since this tetrahedral shape is inflexible, methane has no conformational isomers. In ethylene (C_2H_4) the two carbon atoms are connected by a double bond, and two hydrogen atoms are bonded to each carbon in such a way that the structure is rigid and completely planar, or flat. Again no conformers are possible.

Now let us replace one hydrogen atom on each carbon atom of ethylene by a methyl group (CH_3), thus creating the hydrocarbon called 2-butene (CH_3–CH= CH–CH_3). Two different forms of 2-butene can exist: in one the methyl groups are both on the same side of the molecule (the *cis* form) and in the other the methyls are on opposite sides (the *trans* form). Rotation is not possible around the rigid double bond that connects the central carbon atoms. Inter-

conversion of the *cis* and *trans* isomers of 2-butene requires breaking the double bond, rotating one of the methyl groups 180 degrees around the residual single bond and re-forming the double bond. Since conformers must be interconvertible without breaking bonds, *cis*- and *trans*-2-butene do not qualify. Isomers that require bond breakage for interconversion are termed geometrical isomers.

Ethane (C_2H_6) is the simplest hydrocarbon that exhibits conformational isomerism. The two carbon atoms are connected by a single bond and each carbon is bonded tetrahedrally to three hydrogen atoms. In other words, the molecule consists of two methyl groups linked by a carbon-carbon single bond, around which rotation is allowed and does occur. The methyl groups thus resemble propellers rotating in a well-understood and definable fashion. A view along the carbon-carbon bond, known as the Newman projection, clearly reveals the conformational isomers [*see illustration on page 23*]. At one instant the hydrogen atoms on the front carbon eclipse those on the back. At another instant each front hydrogen falls midway between two rear hydrogens. These "eclipsed" and "staggered" arrangements in ethane are conformational isomers. Nearly all the principles of conformational analysis can be illustrated by ethane and its derivatives.

Molecules that Rotate

Until about 1935 it was thought that all rotational forms of ethane—eclipsed, staggered and intermediate—occurred with equal probability, implying that all possessed equal energy. From the work of Edward Teller, Kenneth S. Pitzer and others it became evident, however, that the staggered conformation has the least potential energy and therefore is the

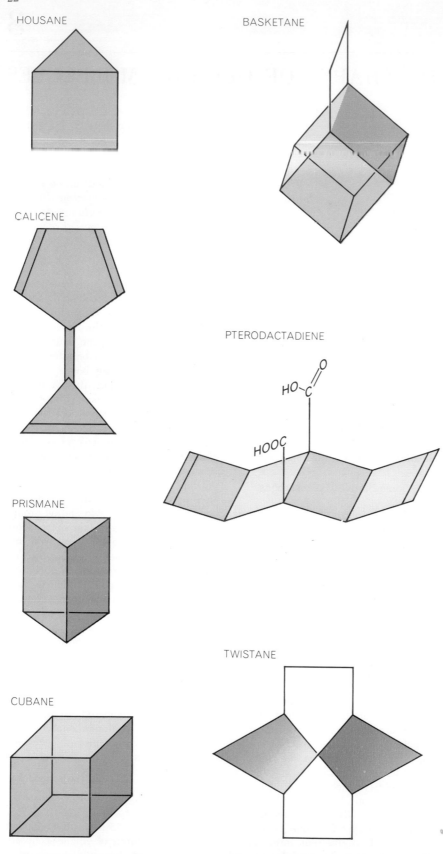

HOUSANE

BASKETANE

CALICENE

PTERODACTADIENE

PRISMANE

CUBANE

TWISTANE

DESCRIPTIVE AND HUMOROUS NAMES have been assigned to the structures of some recently synthesized organic compounds. The straight lines represent bonds between carbon atoms, which form the corners and intersections of the figures. Each carbon atom is assumed to have enough hydrogen atoms attached to it to satisfy any of its four valence bonds that are not linked to another carbon atom; thus housane would have eight hydrogens. Calicene looks like a chalice (*calix* in Latin). In pterodactadiene carboxyl groups form the "head" and the "tail" of a structure that resembles a prehistoric winged reptile, the pterodactyl.

most stable. As the hydrogen atoms move closer to the eclipsed conformation there is an increase in the repulsion between atoms and in other energy factors. The eclipsed form thus represents an unstable energy maximum [*see top illustration on page 24*]. Since a given hydrogen atom must pass three opposing hydrogen atoms in making a full 360-degree revolution, the energy barrier is said to be threefold. It follows that there are three distinct but energetically identical conformations, since any hydrogen atom can be staggered between three different pairs of opposing hydrogens. The increase in energy of the eclipsed forms is referred to as Pitzer strain.

A molecule with a similar but slightly more complex set of conformations is *n*-butane (CH_3–CH_2–CH_2–CH_3), which can be formed by adding two atoms of hydrogen to replace the double bond of either *cis*- or *trans*-2-butene. Rotation around the resulting carbon-carbon single bond (–CH_2–CH_2–) produces different conformational isomers [*see bottom illustration on page 24*]. The conformer of highest energy results when the two methyl groups eclipse each other. When the methyls are staggered 60 degrees apart, the energy drops to a certain minimum (not the lowest one), producing a stable conformer called the *gauche* isomer. When the methyl groups eclipse hydrogen atoms, the energy rises and the conformer is again unstable. The minimum of lowest energy and the most stable arrangement result when the methyl groups are 180 degrees apart, an arrangement called the *anti* isomer. Thus in a 360-degree circuit stable conformations are produced when the angles between one methyl group (imagined as being fixed at 0 degrees) and the second methyl group (imagined as being in rotation) are 60, 180 and 300 degrees. At room temperature about 80 percent of *n*-butane is in the 180-degree, or *anti*, conformation. The remaining 20 percent is divided between the two equivalent *gauche* conformers in which the methyls are 60 or 300 degrees apart.

What no one fully appreciated before the work of Barton and others is that there is a powerful relation between conformational arrangement and chemical reactivity. For example, if one of the hydrogen atoms is removed from the second carbon in *n*-butane and replaced by bromine (Br), the resulting compound is 2-bromobutane (CH_3–$CHBr$–CH_2–CH_3), which yields the same variety of conformational isomers as *n*-butane. One way to prepare *cis*- and *trans*-2-butene is to remove hydrogen bromide (HBr) from the central carbon atoms of 2-bromobutane. HBr is most easily removed when the hydrogen and bromine atoms are 180 degrees apart [see illustration on page 25]. When the methyl group rather than a hydrogen atom is *anti* to the bromine, the molecule is "sterile": HBr cannot be removed. When the methyl groups are *anti* to each other, removal of HBr yields *trans*-2-butene; when the methyls are 60 degrees apart (*gauche*),

the reaction yields *cis*-2-butene. Thus each stable conformational isomer has its own distinctive reaction path: one leads to the *trans* product, one to the *cis* product and one to no reaction at all.

Carbon Atoms in Rings

The chemistry of carbon compounds, historically referred to as organic chem-

istry, owes its complexity to the remarkable stability of the carbon-carbon bond. Carbon chains, both straight and branched, can be constructed in a seemingly infinite variety of lengths and patterns. Regardless of length or pattern, however, the conformational properties of a particular bond will be determined by the principles described above.

Let us now see what these principles

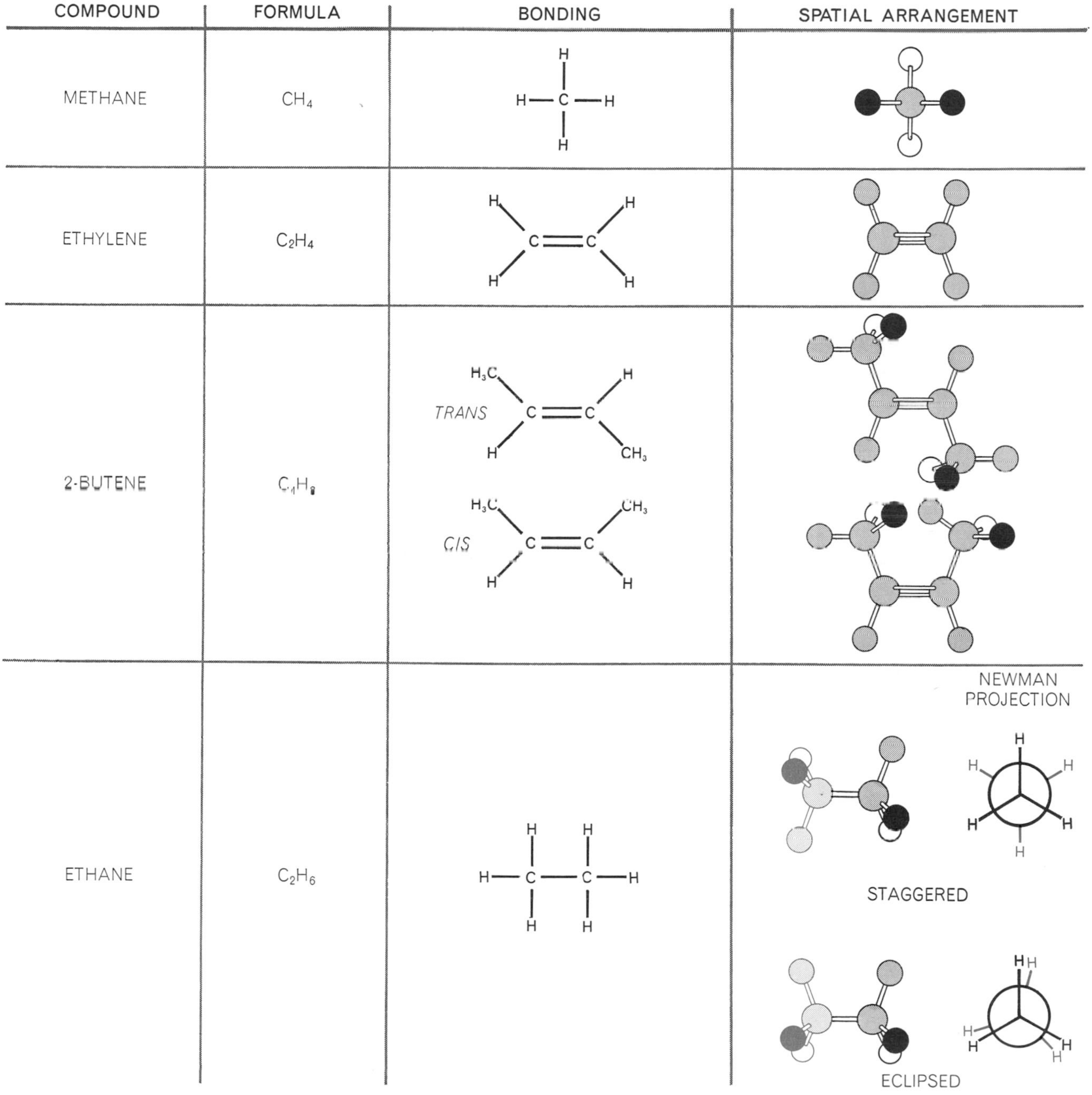

COMPOUND	FORMULA	BONDING	SPATIAL ARRANGEMENT
METHANE	CH_4		
ETHYLENE	C_2H_4		
2-BUTENE	C_4H_8	TRANS / CIS	
ETHANE	C_2H_6		NEWMAN PROJECTION / STAGGERED / ECLIPSED

RIGID AND FLEXIBLE MOLECULES are represented by four simple hydrocarbon compounds. In methane the atoms form a rigid tetrahedron; the angle between any two hydrogen atoms, with carbon at the vertex, is 109.5 degrees. In more complex hydrocarbons, particularly ring structures, this angle often cannot be conserved, producing what is called Baeyer strain. Ethylene also has a rigid structure, but all its atoms lie in a plane. In 2-butene there are two rigid arrangements: the *cis* form, in which both methyl

(CH_3) groups lie on the same side of the double bond, and the *trans* form, in which they lie on opposite sides. To interconvert these two forms the double bond must be broken and rejoined, hence they are defined as geometrical isomers. In contrast, ethane has a continuous set of conformational isomers, produced by simple bond rotation. At one extreme the hydrogen atoms are staggered 60 degrees; at the other extreme they are "eclipsed," or in line, as represented in the "Newman projections" at the lower right.

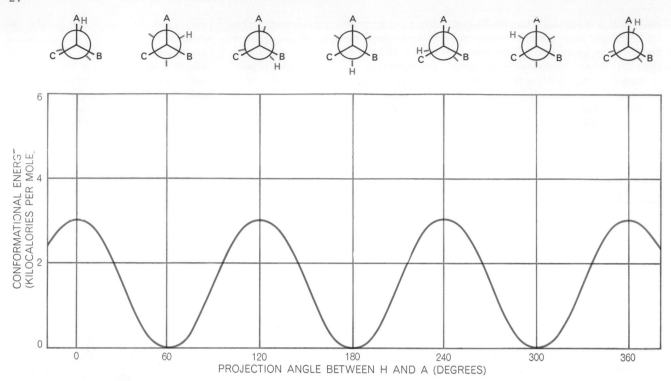

ENERGY OF ETHANE CONFORMERS traces out a simple sine curve when the front carbon atom and its three hydrogen atoms (*A, B, C*) are assumed to be stationary while the rear carbon atom and its hydrogens execute a 360-degree clockwise rotation. During this journey the molecule alternates between eclipsed states of high energy and staggered states of low energy. A given rear hydrogen (*H*) passes through three stable conformations at 60, 180 and 300 degrees that are spatially different but energetically equivalent.

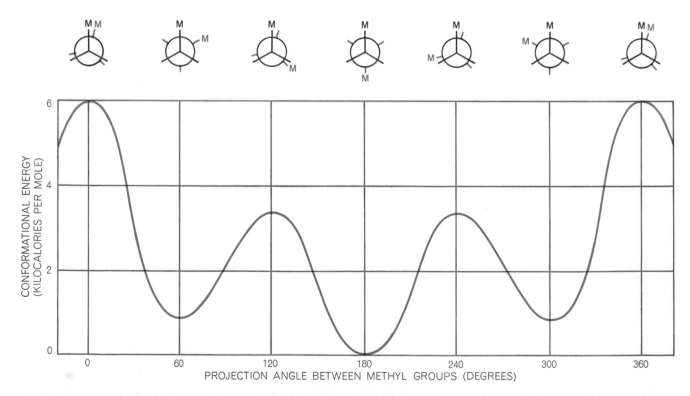

ENERGY OF N-BUTANE CONFORMERS traces out a curve similar to that for ethane. The molecule of *n,* or normal, butane (C_4H_{10}) can be thought of as a molecule of ethane in which one hydrogen atom on each carbon has been replaced by a methyl (CH_3) group. These groups are represented by the two *M*'s in the Newman projections. The stable conformation at 180 degrees is called the *anti* arrangement because the two methyls are antipodal. The slightly more energetic, and thus less stable, conformations at 60 and 300 degrees are termed *gauche.* In the *gauche* arrangements the methyl groups are closer together and consequently interfere with each other more than in the stable *anti* arrangement. Maximum steric interference occurs when the two methyl groups are fully eclipsed.

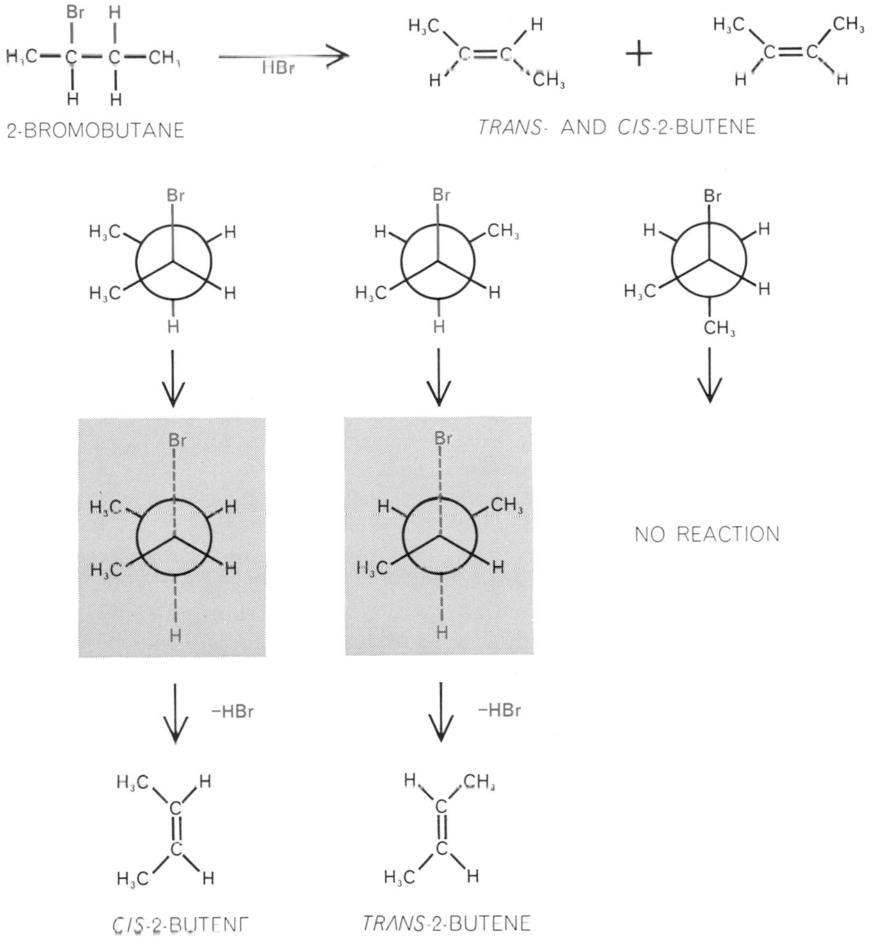

imply when a carbon chain is formed into a ring. The smallest and simplest ring is cyclopropane (C₃H₆), in which three carbon atoms form the corners of an equilateral triangle [see top diagram in illustration on next page]. The 60-degree angle between adjacent carbon atoms is far from the preferred angle of 109.5 degrees. For this reason the molecule, as Adolf von Baeyer was the first to perceive, is under considerable strain. Furthermore, when the hydrogen atoms are viewed down any carbon-carbon bond, they eclipse one another all around the ring. In spite of the strain in the cyclopropane ring, it is so rigid that no conformational isomers can exist.

The four-member analogue of this system, cyclobutane (C₄H₈), has a lesser amount of "Baeyer strain" between adjacent carbon atoms because they form 90-degree angles. Furthermore, the Pitzer, or eclipsing, strain can be relieved by deformations in the ring [see middle diagram in illustration on next page]. In the most stable arrangement opposite pairs of carbon atoms lie either above or below a plane containing the other pair. Between these stable extremes there is a continuum of unstable conformational isomers.

In the five-member system, cyclopentane (C₅H₁₀), the angle at each corner of the pentagon would be 108 degrees if the molecule were planar. Although almost no Baeyer strain would exist in the planar state, there would be strong eclipsing interactions among the subtended hydrogen atoms. If a single carbon atom, with its two hydrogens, is lifted out of the plane, only two pairs of hydrogen atoms remain eclipsed [see bottom diagram in illustration on next page]. This distortion travels around the ring like a wave, so that each carbon atom is out of the plane a fifth of the time, thereby equalizing the strain throughout the ring.

We come now to the important six-member system, cyclohexane (C₆H₁₂), which, as Sachse suspected long ago, is distinctly nonplanar. If the carbon atoms in cyclohexane were forced to lie in a plane, they would form a simple hexagon with bond angles of 120 degrees; the Baeyer strain would be considerable. There would also be strong eclipsing interactions among hydrogens on the ring. Sachse pointed out that strainless bond angles of 109.5 degrees would result if the carbon atoms were in alternate positions above and below the plane of the ring. Moreover, the hydrogen atoms would be staggered and thus would not eclipse one another [see top illustration

ACTIVE AND INERT CONFORMERS are illustrated by 2-bromobutane, a molecule of *n*-butane in which a bromine atom replaces a hydrogen atom on one of the interior carbon atoms. When 2-bromobutane reacts with a base such as potassium hydroxide, a molecule of hydrogen bromide (HBr) is removed, forming *cis*- and *trans*-2-butene. It turns out that for the reaction to take place the departing bromine and hydrogen must be opposite each other. One of three stable conformers, the one at the far right in the Newman projections, does not meet this requirement, so that it is unreactive. When the two methyls are *gauche* (diagram at left), the reaction yields *cis*-2-butene. When the methyls are *anti* (middle), the product is *trans*-2-butene. The structures in the colored panels represent transition states in which the bonds to the bromine atom and the opposite hydrogen have weakened, prior to breaking.

on page 29]. Such a chair conformation should be free of both Baeyer (angle) strain and Pitzer (eclipsing) strain.

Why did Sachse's plausible model meet with so much resistance from chemists? In the presumed planar form all the hydrogens in cyclohexane are chemically and physically indistinguishable. In Sachse's chair arrangement, however, there should be two different types of hydrogens. Half of the 12 hydrogen atoms should extend directly up and down from the average plane of the chair; the other half should lie more or less in the plane. The former hydrogens are termed axial, the latter equatorial. Each carbon atom should therefore be bonded to one axial hydrogen atom and one equatorial hydrogen atom. If another atom, such as chlorine or bromine, replaced a hydrogen atom, two isomers

should be possible: one with the replacement-atom axial, the other with it equatorial. Chemists were unable to find the two isomers.

The reason is now understood. Sachse himself described a second conformation of cyclohexane, the "boat conformation," that is free of angle strain but not of eclipsing strain. It is obtained conceptually by forcing the carbon atom that forms either the head or the foot of the chair to flip to the other side of the plane of the molecule. The chair can be re-created by performing the same operation on the carbon atom directly opposite to the first. When the original chair is inverted in this manner, all the hydrogen atoms that were axial in the first conformation become equatorial, and all the equatorial hydrogen atoms become axial. At ordinary temperatures cyclohexane chairs invert so rapidly that one cannot replace a hydrogen atom with, say, chlorine and hope to isolate two different isomers. Nonetheless, a few years ago, by working at very low temperatures, Frederick R. Jensen and his co-workers at the University of California at Berkeley succeeded in separating isomers of monochlorocyclohexane in which the chlorine atoms were either axial or equatorial.

Making Isomers Visible

A striking confirmation of the existence and interconversion of the axial and equatorial positions in cyclohexane has been provided by nuclear magnetic resonance (NMR) spectroscopy. During the past decade NMR has emerged as the organic chemist's most valuable tool for conformational analysis. Theoretically each chemically distinct hydrogen atom resonates at a different natural frequency. The frequency of the signal varies with the structural environment of the hydrogen atom; the strength of the signal varies with the total number of hydrogen atoms of a given chemical type in the molecule. As the spectrum is usually recorded, peaks correspond to resonances. Thus methane produces an NMR spectrum with only one peak because the four hydrogen atoms are indistinguishable. Ethane and ethylene also produce spectra with one peak, but the spectrum of cis-2-butene has two peaks, one large peak for the six hydrogen atoms that belong to methyl groups and a peak one-third as large for the two hydrogens attached to the carbon atoms joined by a double bond.

One would expect the NMR spectrum of cyclohexane to exhibit two peaks of equal size, one for the six axial hydrogens and one for the six equatorial hydrogens. At room temperature, however, the spectrum shows only one large peak. This observation implies that the molecular chair is inverting so rapidly that the spectrometer can record only the average signal from the two kinds of hydrogen atom. As the temperature is lowered, however, the single peak broadens and finally splits into two distinct peaks, each half the height of the original peak [see illustration on opposite page]. The two peaks correspond to the axial and equatorial hydrogen atoms in a ring that no longer appears to be inverting. When the NMR technique is used to look for the conformational isomers of cyclopentane and cyclobutane, one finds that the warping motions are so rapid that both compounds give only one sharp peak even at extremely low temperatures.

Although NMR spectroscopy clearly shows that at room temperature and above cyclohexane exists in a rapidly flipping chair conformation, neither NMR nor any other technique has demonstrated the presence of the boat conformation. Evidence for its being an intermediate during ring inversion is mainly presumptive. Because of eclipsing strain the boat conformers have more energy and therefore are less stable than the chair [see bottom illustration on page 29]. Some strain can be released if the eclipsed hydrogens are twisted away from one another. A slightly more stable form of cyclohexane known as the twist boat (or, in French terminology, the croissant) is thereby produced. The molecule twistane, for example, was synthesized specifically because it incorporates the structure of the twist boat, held in position by carbon-carbon bridges [see illustration on page 22]. This flexing motion that relieves eclipsing strain can be carried around the ring so that each carbon atom in turn serves as the bow or the stern of a boat. Thus when discussing cyclohexane conformers, it is more appropriate to speak of the flexible boat family—even of the conformational fleet —than of a single boat-shaped conformer, since so many nearly equivalent forms are interconverting.

When one or more hydrogen atoms of cyclohexane are replaced by a bulky substituent such as a methyl group, one can see more clearly how the occupation of an axial or equatorial position determines the preferred conformation of the molecule. If the methyl group is in the axial position, its hydrogen atoms are so close to nearby axial hydrogens that the hydrogens are unnaturally crowded [see illustration on page 27]. In the equatorial position the methyl group has plenty of room. As a result the axial isomer of methylcyclohexane is less stable than the equatorial isomer. At room temperature only about 5 percent of methylcyclohexane is present as the axial isomer. As one can imagine, the size of the substituent determines how much of the compound will be in the axial form or the equatorial one.

Let us now consider what happens when cyclohexane is modified so that

CYCLOPROPANE
C_3H_6

CYCLOBUTANE
C_4H_8

CYCLOPENTANE
C_5H_{10}

CYCLIC COMPOUNDS form a general class in which conformational isomerism is important. Cyclopropane, held rigid by geometrical constraints, is an exception. Its carbon-carbon bonds are under high Baeyer strain and the hydrogen atoms eclipse one another all around the ring. Eclipsed hydrogens are shown in color. In cyclobutane Baeyer strain is relieved by bending of the ring, which also serves to stagger all the hydrogen atoms. The "creases" in bent rings are shown by thin black lines; they are not bonds. The cyclopentane ring could be planar if Baeyer strain were the only consideration since the angles of a regular pentagon are 108 degrees, which is close to the preferred value of 109.5 degrees. But to relieve the eclipsing strain of hydrogen atoms one corner of the ring is pushed out of the plane so that only four hydrogens eclipse one another. If carbon atom No. 1 points upward at one instant, carbon atom No. 2 points downward the next and so on around the ring.

two hydrogen atoms on adjacent carbon atoms are replaced by two methyl groups; such a compound is 1,2-dimethylcyclohexane. Three distinct arrangements are possible: both methyls can be axial (*ax-ax*), both can be equatorial (*eq-eq*) or one can be in each position (*ax-eq*) [*see illustration on next page*]. Inasmuch as ring inversion exchanges the axial and equatorial positions, the *ax-ax* and *eq-eq* isomers must be interconvertible; similarly, the *ax-eq* isomer must interconvert with an equivalent *eq-ax* isomer.

The *eq-eq* and *ax-ax* isomers are commonly referred to as *trans* isomers because in each case one methyl group points down from the plane of the ring and the other points up. The *ax-eq* and *eq-ax* isomers are *cis* because both methyls point to the same side of the plane. The *ax-ax* form of the *trans* isomer must be almost nonexistent because two methyls in the axial position would be extremely crowded. The *cis* and *trans* forms cannot be converted into each other without breaking and re-forming a carbon-carbon bond; therefore they must be geometrical isomers.

The Multiplicity of Rings

Instead of adding two methyl groups to a cyclohexane ring one can add a complete second ring so that the two rings share a common side. The compound decalin, which is used as a paint and lacquer solvent, is one of the most important examples of a molecule with two rings "fused" in this fashion. From the point of view of each ring the other appears to be a 1,2-disubstituent [*see illustration on page 30*]. If the rings are fused in a *cis* fashion, one bond from each ring is axial and one is equatorial. Of two conceivable *trans* arrangements, *eq-eq* and *ax-ax*, only the former is structurally possible. Because axial-axial bonds point 180 degrees away from each other it is physically impossible to bridge them with only four carbon atoms (the four that complete the second ring). The rings of *cis*-decalin can invert simul-

EVIDENCE FOR CHAIR conformation in cyclohexane is supplied by nuclear magnetic resonance (NMR) spectroscopy, which supplies a separate signal for each chemically distinguishable hydrogen atom. The series of curves shows how the inversion rate of the cyclohexane ring slows down as the temperature of the molecule is lowered. Below about −60 degrees Celsius (*bottom curve*) the distinct axial and equatorial hydrogen atoms produce two very sharp peaks.

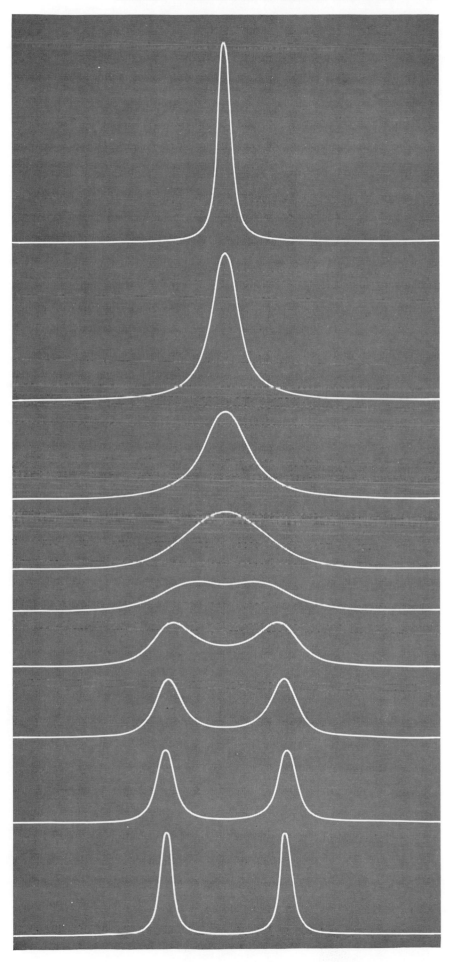

taneously to give an identical conformer in which the axial-equatorial roles of the substituent positions are reversed.

In principle it is possible to keep fusing ring on ring indefinitely. Anthracene, a well-known coal-tar derivative, consists of a sequence of three rings in a row. Because the rings in anthracene possess alternating double and single bonds, as in the benzene ring, the three rings are forced to remain in a plane. If, however, all the double bonds are hydrogenated, producing what is called a saturated molecule, the resulting three-ring product is perhydroanthracene, a molecule that has five principal conformational isomers. The most stable is the one in which all the fusion bonds are equatorial so that the structure is *trans-trans* [see illustration on page 30].

If the third ring is offset from the axis of the first two rings, in which case the points of fusion are adjacent to the carbons that join the first and second ring, the skeleton of perhydrophenanthrene is produced. Again the all-equatorial, or *trans-trans*, arrangement of bonds is the most stable. If a cyclopentane ring is added to the offset ring of the perhydrophenanthrene skeleton, one obtains perhydrocyclopentanophenanthrene, the skeleton of a family of important biological substances: the steroids. The four rings are labeled *A*, *B*, *C* and *D*, with *A* representing the six-member ring most distant from the five-member ring, which is *D*. Although many geometries of ring fusion are possible in steroids, the linkage is invariably *trans* between the *B* and *C* rings and usually *trans* between the *C* and *D* rings. The steroids are such an important class of compounds that an enormous amount of study has been devoted to them in the past 50 years. They include such natural products as cholesterol, the sex hormones, cortisone and the related adrenocortical hormones, and recently an entire family of synthetic birth-control substances.

Although rings of more than six members are common in organic chemistry, not much is known about their conformational characteristics. It is known, however, that the saturated seven-member rings resemble cyclopentane and the boat family of cyclohexane in their flexibility. The properties of larger rings will undoubtedly be closely examined during the next few years.

Some Modified Chairs

The cyclohexane chair is frequently modified when one of the carbon atoms in the ring is replaced by the atom of another element or when various substituents are added to the ring. In general one can expect the altered molecule to assume a spatial arrangement that will minimize steric interactions, that is, to adopt the "most comfortable" shape. A series of modified cyclohexane structures is illustrated on page 31.

When one of the carbon atoms is replaced by oxygen, the resulting structure, tetrahydropyran, is the basis for a large class of sugars. In the most common hexose sugar, β-D-glucose, four of the carbon atoms in the ring carry a hydroxyl (OH) group in place of one hydrogen and the fifth carbon carries a hydroxymethyl (CH_2OH) group. The chair is very little distorted, and all the substituents are equatorial. The abundance of β-D-glucose is undoubtedly due to its conformational stability. The shapes of sugar molecules were among the earliest subjects of conformational analysis; in fact, Norman Haworth first defined "conformation" as it is used today in his 1929 book on sugars.

Another "heteroatom" that creates little distortion in the cyclohexane ring is nitrogen. When a CH_2 unit is replaced by NH, the resulting structure is the piperidine ring, found in most alkaloids, a large class of polycyclic compounds that includes morphine, lysergic acid, strychnine and codeine. The piperidine chair inverts so rapidly that the H of the NH group flips continuously from equatorial to axial. At room temperature the axial and equatorial conformers are present in almost equal amounts.

When a sulfur atom rather than an oxygen or nitrogen atom is inserted in the cyclohexane ring, producing thiane, the chair is significantly distorted because the carbon-sulfur bond is appre-

METHYLCYCLOHEXANE

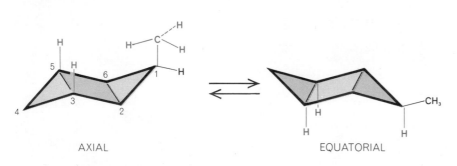

AXIAL   EQUATORIAL

1,2-DIMETHYLCYCLOHEXANE

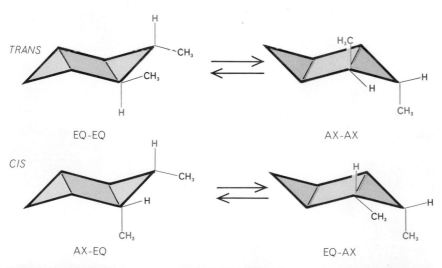

TRANS

EQ-EQ   AX-AX

CIS

AX-EQ   EQ-AX

INTERCONVERTING ISOMERS are produced when one or more methyl groups are substituted for hydrogen atoms in cyclohexane. In methylcyclohexane an axial methyl group on carbon No. 1 is crowded by axial hydrogens on carbon No. 3 and No. 5. Thus the conformer with the methyl in the equatorial position is favored. When there are two methyl groups on adjacent carbon atoms (1,2-dimethylcyclohexane), three arrangements are possible. The *trans* geometrical isomer consists of two conformational isomers, one with both methyls equatorial (*eq-eq*) and one with both axial (*ax-ax*). Only one arrangement is possible for the *cis* isomer: *ax-eq*. Ring inversion yields an identical substitution pattern (*eq-ax*).

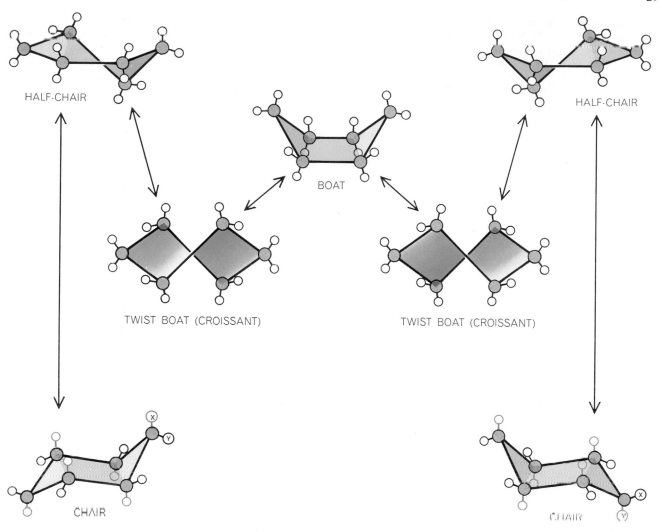

SIX-MEMBER RING OF CYCLOHEXANE, C_6H_{12}, can exist in various shapes known as conformational isomers, or conformers. The most stable conformers are the chairs; the least stable are the half-chairs (*see energy diagram below*). The hydrogen atoms in the molecule can assume two distinguishable positions: "axial," when they are more or less perpendicular to the average plane of the ring; "equatorial," when they lie close to the plane. When the chair flips from one conformation to its mirror image, axial hydrogens become equatorial and vice versa. This reversal is shown for two atoms, X and Y. In the chairs axial atoms are shown in color.

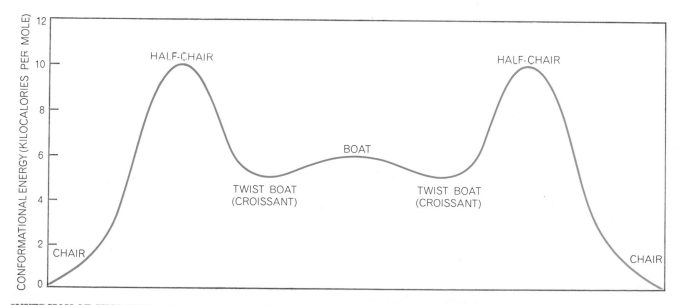

INVERSION OF CYCLOHEXANE takes place through a sequence of ring distortions involving a spectrum of energy states. At room temperature the molecule flips back and forth between the two stable chair conformations. The boat conformers are quasi-stable.

ciably longer than the carbon-carbon or carbon-nitrogen bond. Moreover, the normal C-S-C angle is close to 90 degrees, rather than 109.5 degrees. As a result the thiane ring is puckered, so that the molecule resembles a beach chair that has been raised from a reclining position to a more upright one.

The reverse distortion, a flattening, takes place when one of the carbon atoms in the cyclohexane ring is doubly bonded to an oxygen atom, forming cyclohexanone. If another oxygen atom is added to the carbon atom at the opposite side of the ring, producing 1,4-cyclohexanedione, one might expect still further flattening to occur. Instead the

molecule flips into a twist-boat structure, probably to relieve eclipsing strain that would be present if the molecule were extremely flattened. The half-chair conformation, which corresponds to the state of highest energy in the inversion of cyclohexane, can be achieved as a preferred molecular conformation by removing two hydrogen atoms from adjacent carbon atoms in cyclohexane and joining the carbons by a double bond; the resulting molecule is cyclohexene.

The Conformation of Proteins

Although steroids, alkaloids and sugars are not small molecules, the proteins

that, in the form of enzymes, catalyze the most important processes in the living cell are many times larger. Proteins are long chains of the 20-odd different amino acid molecules. The conformation of protein chains depends on complex interactions among the amino acids and on their specific sequence. In many cases the chain first develops a helical structure, and the helix as a whole folds into a conformation that probably minimizes the steric and electrostatic interactions among all the various groups. The final conformation is so important to the molecule's biological function that any alteration in shape may reduce its activity or destroy it altogether. The complete

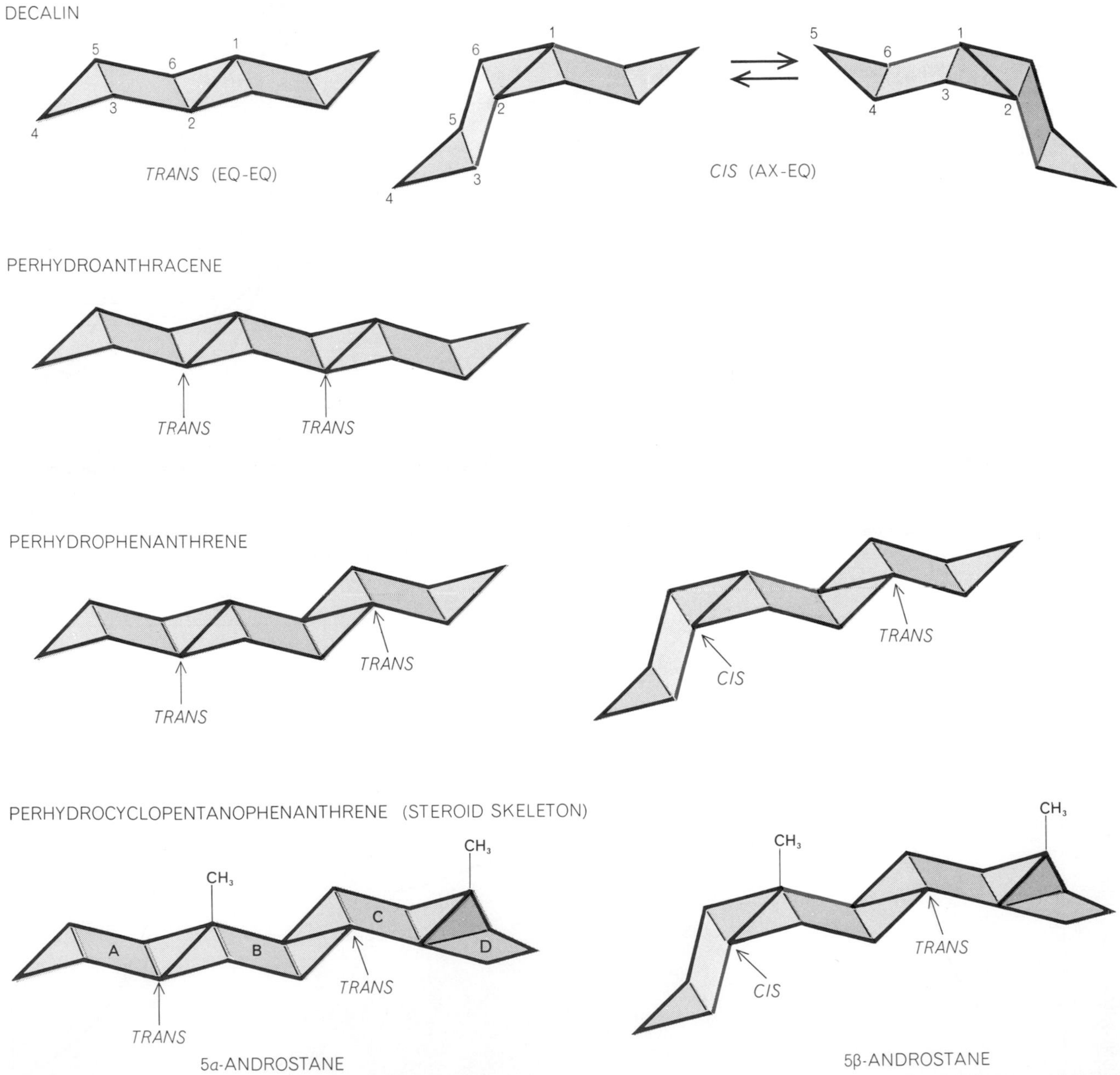

DECALIN

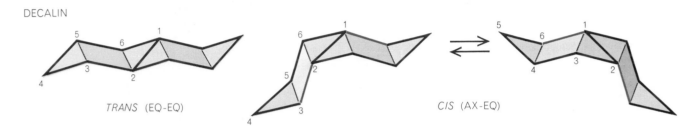

TRANS (EQ-EQ)   CIS (AX-EQ)

PERHYDROANTHRACENE

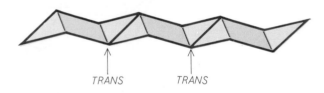

TRANS   TRANS

PERHYDROPHENANTHRENE

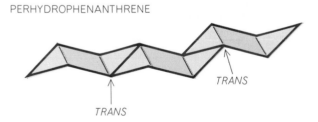   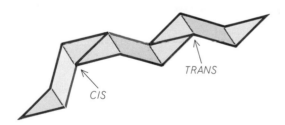

PERHYDROCYCLOPENTANOPHENANTHRENE (STEROID SKELETON)

   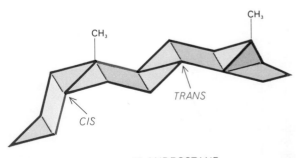

5α-ANDROSTANE   5β-ANDROSTANE

FUSION OF CYCLOHEXANE RINGS produces complex structures in which the geometry of ring fusion can be either *trans* (*eq-eq*) or *cis* (*ax-eq*). An *ax-ax trans* conformer cannot be built with six-member rings. Axial bonds are shown in color. Perhydroanthracene has four conformers, including one boat, besides the one shown. Perhydrophenanthrene also has four other conformers.

DISTORTION	COMPOUND	SHAPE
1 NONE	TETRAHYDROPYRAN	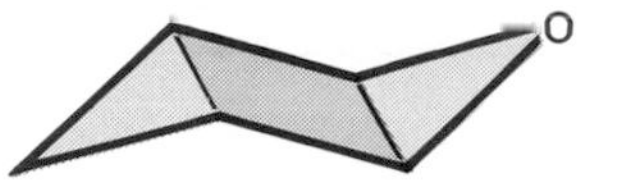
2 NONE	PIPERIDINE	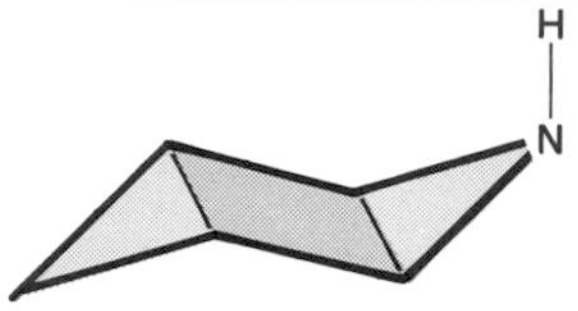
3 PUCKERING	THIANE	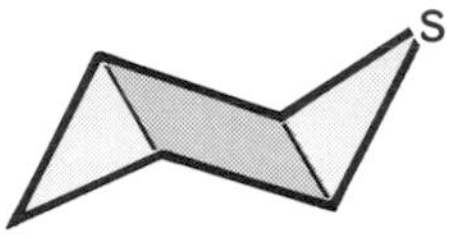
4 FLATTENING	CYCLOHEXANONE	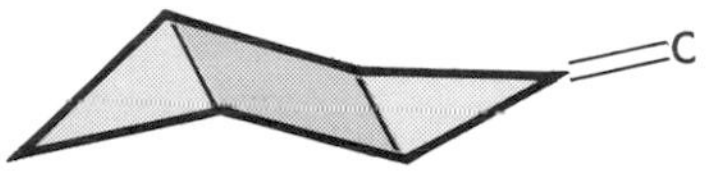
5 TWIST BOAT	1.4-CYCLOHEXANEDIONE	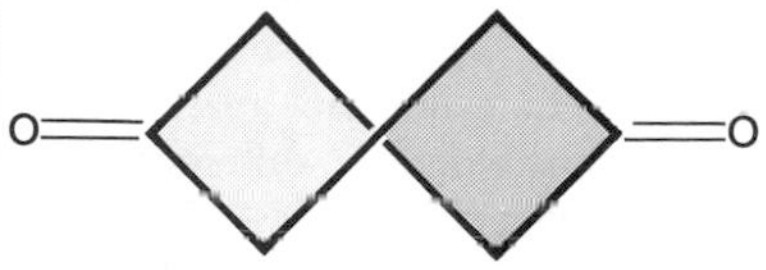
6 HALF-CHAIR	CYCLOHEXENE	

DISTORTED CHAIRS are often produced when the basic structure of cyclohexane is modified by substituent atoms. The chair may be puckered (3) or flattened (4). Addition of two external oxygen atoms produces a twist boat (5). Removal of two hydrogens creates a double bond, producing a half-chair (6). There is good evidence for all these conformations.

three-dimensional structures of about a dozen proteins have now been worked out by the X-ray-diffraction analysis of proteins in crystalline form. One of the proteins whose structure is known—the enzyme lysozyme—destroys the cell wall of bacteria by cleaving long polysaccharide chains, which consist of many sugar units tied together. For the enzyme to operate on the sugar linkages, one of its tetrahydropyran rings must first flip from its normal chair conformation to a much flattened structure [see "The Three-dimensional Structure of an Enzyme Molecule," by David C. Phillips beginning on page 170].

Flavor chemists of the Monsanto Company have recently proposed a conformational dependence between a particular protein and substances that taste sweet. When the protein combines with "sweet" molecules, its conformation changes; conceivably the change is relayed to the brain as a signal of sweetness. The intensity of the signal may be related to the amount of change.

Presumably DNA, the double-strand helical molecule that embodies the genetic message in living cells, has a structure susceptible to conformational change. One of the greatest mysteries in genetic chemistry is how the DNA helix unwinds so that each strand can act as a template for the construction of a complementary strand of DNA (or for the construction of a strand of messenger RNA, the molecule that directs the synthesis of protein molecules). Rotation around carbon-carbon, carbon-nitrogen and carbon-oxygen bonds must be involved in this elaborate and precise process. A present goal of conformational analysis is a bond by bond description of the rotational processes that accompany conformational changes in biochemical systems such as proteins and nucleic acids. Much of the current work of this type is based on X-ray-diffraction studies of crystalline solids. One of the most pressing needs today is for new methods for obtaining structural information about molecules in liquids, the state of matter in which most chemical and biochemical reactions take place.

4

STEROIDS

LOUIS F. FIESER
January 1955

Substances as diverse in their effects as vitamin D, the sex hormones and cortisone are chemically very much alike. Their study has been one of the major efforts of modern chemistry

The story of the steroids is one of the great epics of chemistry. Part of the story has come to public attention in the last few years as a result of the well-publicized struggle to synthesize the steroid hormone cortisone. But cortisone is only an episode in an enthralling history of research and discovery that has occupied more than half a century.

The steroids are a family of substances critically important to plant and animal life. They include the adrenal cortical hormones, the sex hormones, some of the vitamins, plant sterols such as ergosterol and animal sterols such as cholesterol. This array of substances, though so alike chemically that it is often difficult to tell them apart, exhibits a prodigious range of different activities.

The bile acids, for instance, aid digestion by emulsifying fats so that they can pass through the walls of the intestines. An interesting illustration of this function is that, if the normal flow of bile is obstructed by a tumor, the patient's blood does not absorb enough antihemorrhagic vitamin K_1 from his food, and an operation on such a patient is attended by the hazard that he may bleed to death unless he is given doses of the vitamin and bile salts to promote its absorption. A very different steroid function is illustrated by digitalis, a plant steroid obtained from foxglove, which stimulates the heart; still another example is a plant steroid alkaloid recently put on the market for treating high blood pressure. Certain other plant steroids rupture red blood cells and are therefore highly toxic to animals, even in extremely high dilutions. Yet many of these plant steroids, while capable of acting as powerful drugs in animals, seem to perform no direct function in the plants themselves.

Cholesterol presents a still more interesting question. This substance appears in nearly all the tissues of vertebrate animals; it is particularly abundant in the brain and spinal cord. The higher animals synthesize considerable amounts of cholesterol and have an efficient chemical mechanism for delivering it to the tissues by way of the blood. Hence it would be reasonable to suppose that cholesterol carries out some vital function. If any direct function exists, it has not yet been discovered—though cholesterol does seem to play a harmful role in hardening of the arteries. In the absence of evidence of any direct function, it has been assumed that cholesterol serves as a precursor for building the usable steroids.

Animals may obtain steroids by eating other animals, but not by eating plants, because plant sterols are not absorbed through the intestinal wall. However, an animal readily makes its own sterols. It is capable of rapidly synthesizing the complex cholesterol molecule even from simple compounds such as acetic acid.

A sterol is a white, solid, fatty sub-

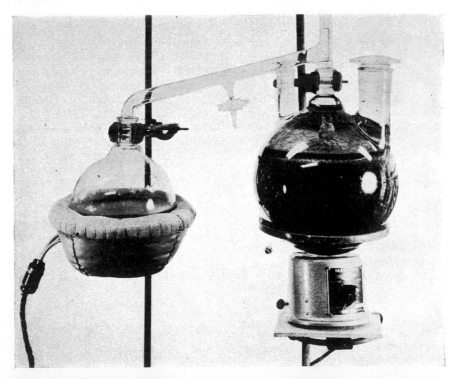

STEROIDS ARE EXTRACTED from a mixture of plant materials in the flask at the right. The flask at the left contains ether, in which steroids are soluble. When the flask at the left is heated, ether vapor rises in a distillation column above the flask at the right. As the ether condenses it flows through a tube to the bottom of the flask at the right and rises through the heavier plant material. It then fills the flask at the right and, bearing the steroids, overflows into the flask at the left. The cycle of extraction may be repeated for days.

stance—a crystalline alcohol—which can be extracted from animal or plant tissues by a process requiring several steps. First the tissues' lipids (fatty materials not soluble in water) are extracted by an oil solvent. This extract is then boiled with alkali, which splits some of the lipids into glycerol and soap The remaining "nonsaponifiable" lipids, which are principally sterols, are then captured by removal with an organic solvent. Cholesterol of about 98 per cent purity can be isolated with comparative ease from the brains or spinal cords of animals or from human gallstones, another rich source of the substance.

The subject of this article is the story of the achievement of chemists in elucidating the structure and shape of the various steroid molecules. The task presented a series of problems as baffling and complex as any encountered in the whole history of organic chemistry. Solution of the major problems came only after the concerted efforts of a large number of investigators working over a period of many years. The subject was so important and the achievements so brilliant that no fewer than six investigators in this field received Nobel prizes.

The Cholesterol Molecule

The father of sterol chemistry is Adolf Windaus of Germany, who, now 75 and retired, holds the admiration and affection of present workers in the field he pioneered. Beginning at the University of Freiburg in 1903, he devoted his career to sterol research. At the University of Göttingen he built up the world's leading school of sterol chemistry, and produced a succession of able students who, stimulated by his skill, enthusiasm and leadership, accounted for many of the discoveries and triumphs that followed his own.

Windaus started with cholesterol, partly because this substance had been isolated as early as 1770 and partly because it was the most easily obtainable of all the steroids. The formula of the cholesterol molecule is $C_{27}H_{46}O$. To understand the steroid problem it is necessary to picture the arrangement of this molecule and to review some of the fundamental concepts of organic chemistry. The cholesterol molecule is shown in the diagrams above, with the positions of its 27 carbon atoms numbered in the conventional way. (The numbering sequence is not entirely logical, because it was adopted at a time when the structure was not fully known.) Four rings of carbon atoms, plus a side chain branch-

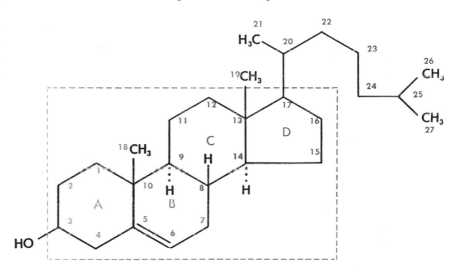

COMPLETE MOLECULAR STRUCTURE of cholesterol, a steroid abundant in the brain and spinal cord, consists of atoms of hydrogen (H), carbon (C) and oxygen (O). Seventeen of the 27 carbon atoms comprise the four rings that are characteristic of steroids.

ABBREVIATED MOLECULAR STRUCTURE of cholesterol emphasizes its salient features. The four rings are labeled with letters and the 27 carbon atoms with numbers. The solid lines indicate bonds that project upward; the dotted lines, atoms projecting to rear.

ing from the fourth, form the skeleton of the molecule. One of the major complications of steroid research is that the same skeleton is present in a host of different sterols, which differ only in the way that hydrogen or oxygen atoms or groups of them are attached to the frame; indeed, there are many variants (isomers) of cholesterol with precisely the same total number of carbon, hydrogen and oxygen atoms.

In a carbon compound the four atoms or groups of atoms attached to each carbon atom are not in one plane but are angled from it in a three-dimensional arrangement; even in the simplest case, the methane molecule, they form a tetrahedral figure. Whenever the four atoms or groups attached to the four-valent carbon atom are all different, the mole-

cule can have two mirror-image forms, known as optical isomers. A good example is the existence of two forms of lactic acid (in milk and in muscle), which are identical in all properties except that one rotates plane-polarized light to the right and the other rotates it to exactly the same degree to the left [see models on page 35].

The cholesterol skeleton has eight centers of asymmetry where such variations may occur. As a result there are 256 possible optical isomers of this molecule.

Cholesterol itself has four groups of atoms projecting on the front side of the molecule (upward from the plane of the paper as it is diagrammed here) and two hydrogen atoms projecting to the rear. The array of projecting groups on the front side forms a canopy protecting the molecule on that side. Hence most of its chemical and biological reactions occur by attack from the rear.

Dismemberment of the Molecule

At the turn of the century Windaus attacked the problem of unraveling the then unknown cholesterol molecule by breaking it down with oxidizing reactions. He found that he could open the first two rings for exploration by oxidizing respectively the hydroxyl (OH) group in the first ring and the double bond in the second ring. He also discovered two methods of splitting the side chain attached to the fourth ring: first he split off the entire eight-carbon chain with one method of oxidation, and later he found a way to remove just the three-carbon fragment at the end of the chain [see diagram below].

The latter reaction left a molecule with a five-carbon acidic side chain. And here Windaus came to a common ground with Heinrich Wieland of Freiburg, who for several years had been investigating the structure of the bile acids of cattle. The two men discovered that the five-carbon side chain Windaus obtained when he split the end group from cholesterol was identical with the side chain of bile acid. The bile acids, indeed, proved to be like sterols in most respects, differing from the sterols with which Windaus had been working only in details of arrangement of functional groups in the various rings.

From this point on, Windaus' and Wieland's lines of research helped each other at every step forward. Although the road was hard and the going slow and uncertain, the two intrepid investigators held unwaveringly to their course. They indoctrinated a number of students

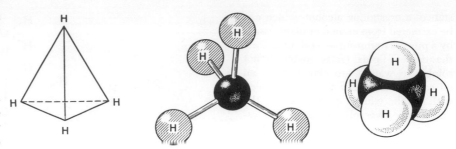

METHANE MOLECULE illustrates the fact that atoms attached to carbon project in three dimensions. The four hydrogen atoms of methane (CH₄) are arranged in a tetrahedron (*left*). The arrangement can be represented with pegs and balls (*center*) or with spheres made to conform to the known distances between the atoms of carbon and hydrogen (*right*).

in the new line of chemistry and made steady progress. By 1928 they were able to propose a tentative structural outline of the cholesterol molecule for which they jointly received the Nobel prize in chemistry.

Their formulation, as it turned out, was incorrect. The first indication that drastic revision was needed came from an unexpected source. Windaus and Wieland had supposed that the first three rings were clustered around a common carbon atom holding them together, which meant that the molecule should have a globular shape. But in 1932 the British physicist J. D. Bernal discovered by X-ray studies that the sterol molecule was long and thin. With this clue other English workers and Wieland himself were able by the end of the year to work out the correct structural formula.

Vitamins

By this time sterol chemistry had emerged from the realm of purely academic interest to a position of practical importance in medicine. It entered into medicine in connection with vitamin D,

the anti-rickets vitamin. It had been discovered that the active principle of this vitamin, found in fish-liver oils, was probably a steroid, and also that foods irradiated with sunlight could counteract a deficiency of the vitamin. The question arose: Would irradiation of an ordinary sterol convert it into an antirachitic steroid like the one in cod-liver oil?

Windaus and the Göttingen physicist R. Pohl attacked this problem with the help of ultraviolet spectroscopy, then just coming into use as a powerful tool for identifying molecular structures. They found that a substance present in minute amounts in some samples of cholesterol could be made active against rickets by irradiation, and that this activatable substance had an absorption spectrum very similar to that of ergosterol, a sterol in yeast. Windaus, and independently a research team at the National Institute for Medical Research in London, proceeded to isolate from irradiated ergosterol a new substance of high potency against rickets which was named vitamin D₂ [see formulas on page 36]. Irradiation of ergosterol effects a series of chemical changes yielding vari-

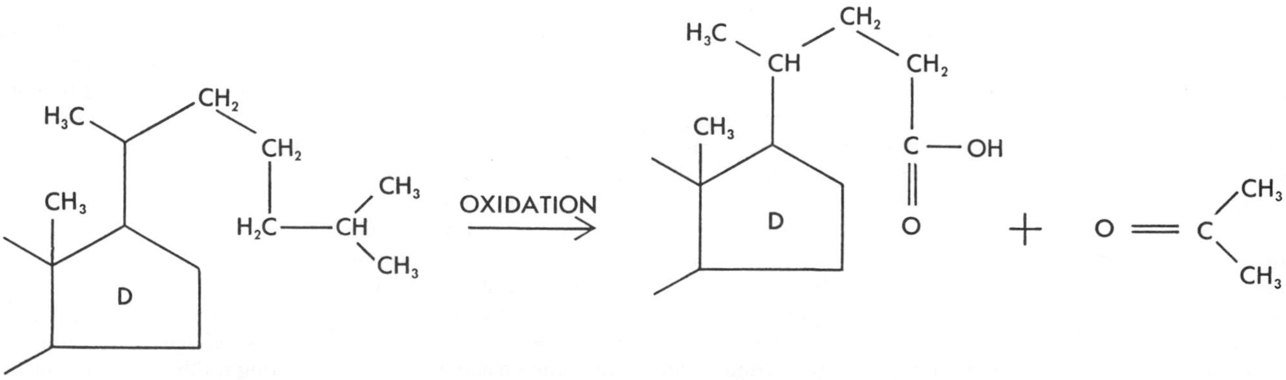

OXIDATION OF CHOLESTEROL provided much information about its structure. When cholesterol (*partly diagrammed at left*) was oxidized in one experiment, it yielded acetone (*right*) and a substance with an acid side chain of five carbon atoms (*center*).

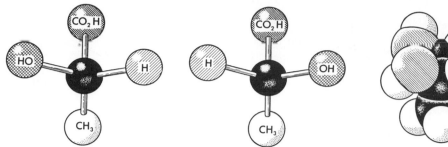

LACTIC ACID MOLECULES illustrate optical isomers. The peg-and-ball model at the left represents the molecule of lactic acid ($C_3H_6O_3$) as it is found in milk; the peg-and-ball model second from the left represents the molecule of lactic acid as it is found in muscle. The corresponding sphere models are third and fourth. The two molecules are chemically and physically identical except that one rotates plane-polarized light to the right and the other rotates it to the left. In short, the molecules are mirror images of each other.

ous isomers, of which only vitamin D_2 is active against rickets.

It seemed reasonable to suppose that D_2 was identical with the active principle in fish-liver oils, until Windaus and his associates showed that certain variants differing only slightly from D_2 were just about as potent. The most interesting variant was one prepared from cholesterol by a series of chemical operations which involved placing a double bond at the 7,8 position, as in ergosterol, and then irradiating this molecule (7-dehydrocholesterol). The product [*formula at lower right on next page*] had an antirachitic activity comparable in strength to that of D_2, as measured by bio-assay in rats. It was named vitamin D_3.

Was the natural vitamin in fish-liver oils more closely related to cholesterol than to ergosterol? This question, posed by Windaus' synthesis of D_3 and by biological investigations in the U. S., was brilliantly resolved in 1936 by Hans Brockmann of the Göttingen laboratory. By means of the sensitive method of chromatography, he isolated the vitamin of tuna-liver oil and proved that it was identical with Windaus' vitamin D_3.

Vitamins D_2, made from ergosterol, and D_3, made from cholesterol, have about the same activity against rickets in rats, but in chicks D_3 is far more potent. D_2 is still used for treating children with rickets, because it is easier to make and has satisfactory potency in human beings; however, for chickens, which are susceptible to rickets, D_3 is now manufactured in substantial quantities, both from cholesterol and from fish-oil concentrates.

The Sex Hormones

The road to chemical identification of the sex hormones was opened in 1927 when S. Aschheim and Bernhard Zondek, then in Berlin, discovered that the urine of pregnant women contained considerable amounts of a hormone which produced sexual heat when tested in mice or rats. Windaus was asked by a German chemical firm to explore this substance. Involved at the time in vitamin D research, he elected to turn over the new problem to a promising young student named Adolf Butenandt. The student plunged into the task of iso-

lating the hormone with the help of a young biologist, Erika von Ziegner, who later became his wife. In the meantime Edward A. Doisy, of the St. Louis University School of Medicine, also set out to identify the active substance. Working independently, both Butenandt and Doisy succeeded in 1929 in isolating the first known sex hormone—estrone.

With keen insight Butenandt proceeded by a roundabout route to arrive at a shrewd guess concerning the chemical structure of estrone. In pregnancy urine there was found an inactive companion of estrone, called pregnanediol. Butenandt proved that pregnanediol was related to cholic acid, a component of bile, by breaking both down to a common product. Thus he showed that pregnanediol was a steroid. By further adroit experimentation he related its companion, estrone, to bile acid and then deduced the correct structure of estrone.

It developed that estrone was not the true estrogenic hormone secreted by the ovaries but a transformed product of it excreted in the urine. The actual hormone, first made from estrone and later isolated from sow ovaries by Doisy (who extracted 12 milligrams of the substance from four tons of ovaries), is believed to be estradiol [*see formulas at top of page 37*].

The discovery and analysis of the male sex hormone testosterone followed a similar pattern. In 1931 Butenandt and Kurt Tscherning isolated from 15,000 liters of male urine 15 milligrams of a hormonal substance which they named androsterone. From this tiny pile of crystals, hardly enough to cover the tip of a small spatula, Butenandt derived a great deal of information about the nature of the molecule. He tentatively deduced its structure, and his deduction was independently proved correct by Leopold Ruzicka of Zurich, who produced androsterone by splitting off the eight-carbon side chain from a deriva-

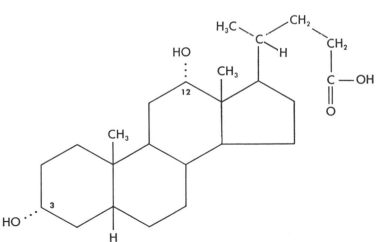

STRUCTURE OF CHOLIC ACID, found in the bile of cattle, includes the same side chain found in the oxidation product of cholesterol. The bile acids resemble steroids in other ways.

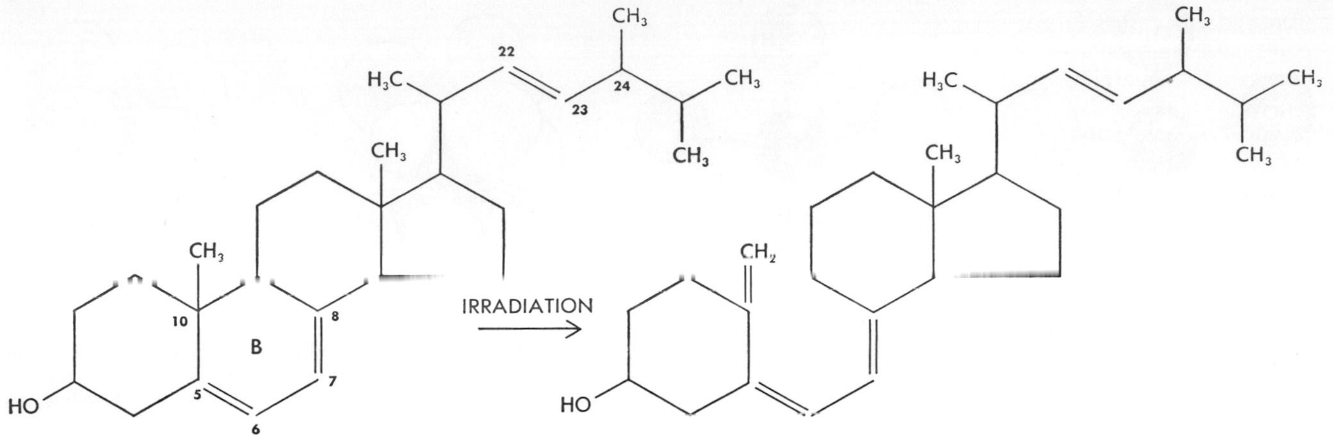

VITAMIN D₂ (*right*) was made from ergosterol (*left*), a steroid found in yeast, by exposing it to ultraviolet radiation. Ergosterol differs from cholesterol in having an extra methyl group (CH₂) at position 24 and double chemical bonds at positions 7, 8 and 22, 23.

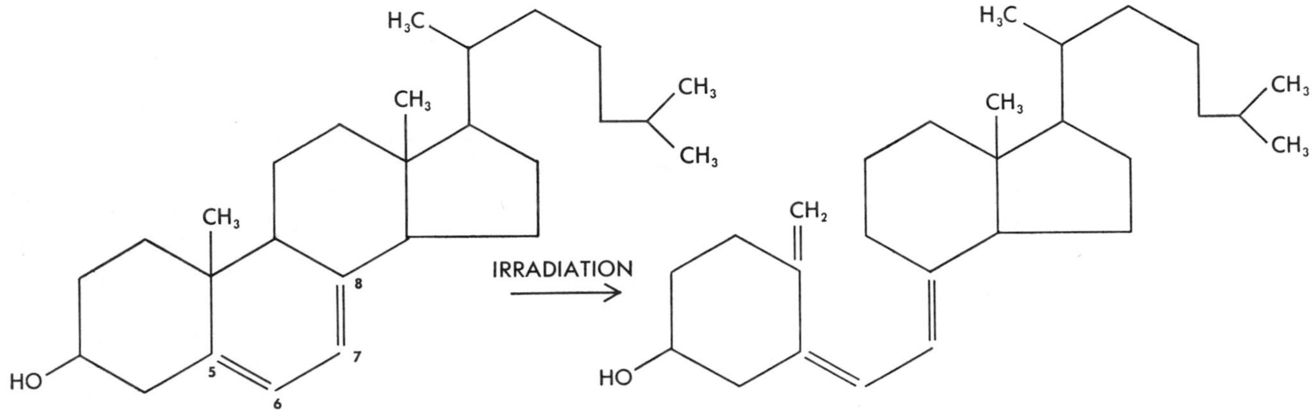

VITAMIN D₃ (*right*) was made from 7-dehydrocholesterol (*left*), also by exposing it to ultraviolet radiation. The 7-dehydro- cholesterol was prepared synthetically from cholesterol by a process adding a double chemical bond between positions 7 and 8.

tive of cholesterol through oxidation.

Androsterone was obtainable only in very small amounts, either by synthesis or by extraction from urine, but Bute- nandt, Ruzicka and others soon succeed- ed in synthesizing a related substance which could be produced in more plenti- ful yield. This substance, named de- hydroepiandrosterone, was made from cholesterol by burning off the side chain while the essential hydroxyl group at- tached to the first ring and the double bond at the 5,6 position in the second ring were protected by stable chemical combinations at those positions. With the more plentiful working material at hand, the investigators were able to synthesize a number of interesting products, some of which proved more potent than androsterone in hormonal activity. Meanwhile Ernst Laqueur in Amster- dam reported that he had isolated from steer testes a potent hormone which he called testosterone. Butenandt and Ruzicka shortly afterward synthesized testosterone from dehydroepiandroster- one. It became clear that testosterone

was the true hormone; androsterone and dehydroepiandrosterone were metabo- lites of the hormone excreted in the urine [*formulas in center on opposite page*]. A practical method is now available for converting dehydroepiandrosterone into testosterone in 81 per cent yield.

In 1934 four research groups isolated from the corpus luteum tissue in sow ovaries the pregnancy hormone—proges- terone. Butenandt obtained 20 milli- grams of the hormone from the ovaries of 50,000 sows. The structure of proges- terone, very similar to that of testoster- one [*formulas at bottom of opposite page*], was soon inferred from its chemi- cal properties and ultraviolet analysis. Butenandt promptly synthesized pro- gesterone by two methods, one of which was the oxidation of a substance called pregnenolone, a by-product of the pro- duction of testosterone from cholesterol.

Hormones from Plants

At this dramatic point in the develop- ment of the chemistry of the sex hor-

mones, there came a break in another field which was to open a more fertile route for production of the hormones. The plant steroids had been relatively neglected, though Windaus and others had published occasional reports on them. After the establishment of the structure of cholesterol in 1932, how- ever, the new surge of intensive re- search embraced active substances in plants as well as in animals. Attention was focused on the cardiac glycosides— extracts from plants which had been known for centuries as poisons. Particu- larly interesting was digitalis, the heart stimulant that had been used since 1785, when it was introduced as a treatment for dropsy. It was believed that a part of the active principle of this drug might be a steroid, and Windaus delegated the problem to one of his pupils, Rudolf Tschesche. At the Rockefeller Institute for Medical Research W. A. Jacobs and R. C. Elderfield also investigated the same question.

Within two years the two independent investigations established that steroids were indeed present in the active components of digitalis. The steroid here is attached to a sugar. Digitoxin, for example, contains a steroid of 24 carbon atoms to which is linked the rare sugar digitoxose (making the large molecule somewhat soluble in water). The steroid part has a skeleton like that of a bile acid, but one major difference is that the side chain is coiled into a ring [*see formula on next page*].

The digitalis steroids are rather special, but steroids with sugars attached are very common in the plant world. One class of these substances is known as the saponins, because water solutions of them foam like soap when shaken. The steroid part of a saponin is called a sapogenin. The structure of sapogenins, on which Windaus and his pupils and successors had worked for many years, was finally clarified by Russell E. Marker and his group at the Pennsylvania State College in an extraordinary series of studies between 1939 and 1947. The most interesting of these substances is diosgenin, which has a skeleton remarkably like that of cholesterol, with 27 carbon atoms and a double bond between the fifth and sixth carbons [*see formula at bottom of page 39*]. The discovery of this structure at once suggested that diosgenin would be useful for producing sex hormones, for which more and more uses had been found in medical therapy.

The possibility of manufacturing the hormones from plant extracts excited Marker. Production of these hormones from animal sources was still difficult and expensive, in spite of new methods of synthesis that had been discovered by various investigators. Diosgenin was abundantly available from yams (plants of the genus *Dioscorea*), and Marker in field trips to the southern U. S. and Mexico had found many other plant sapogenins (which he whimsically named for current friends and enemies: rockogenin for the late Dean Frank "Rocky" Whitmore of Pennsylvania State College, kammogenin for Oliver Kamm of Parke Davis & Company, nologenin for Carl R. Noller of Stanford University, fesogenin for the writer of this article).

Marker showed that diosgenin could easily be converted into pregnenolone, already known as a building material for progesterone. A Mexican firm named Syntex was set up to manufacture progesterone from Mexican yams. Before long Marker fell out with the management of the company and was replaced by Georg Rosenkranz, a young Hungarian-born chemist who had trained

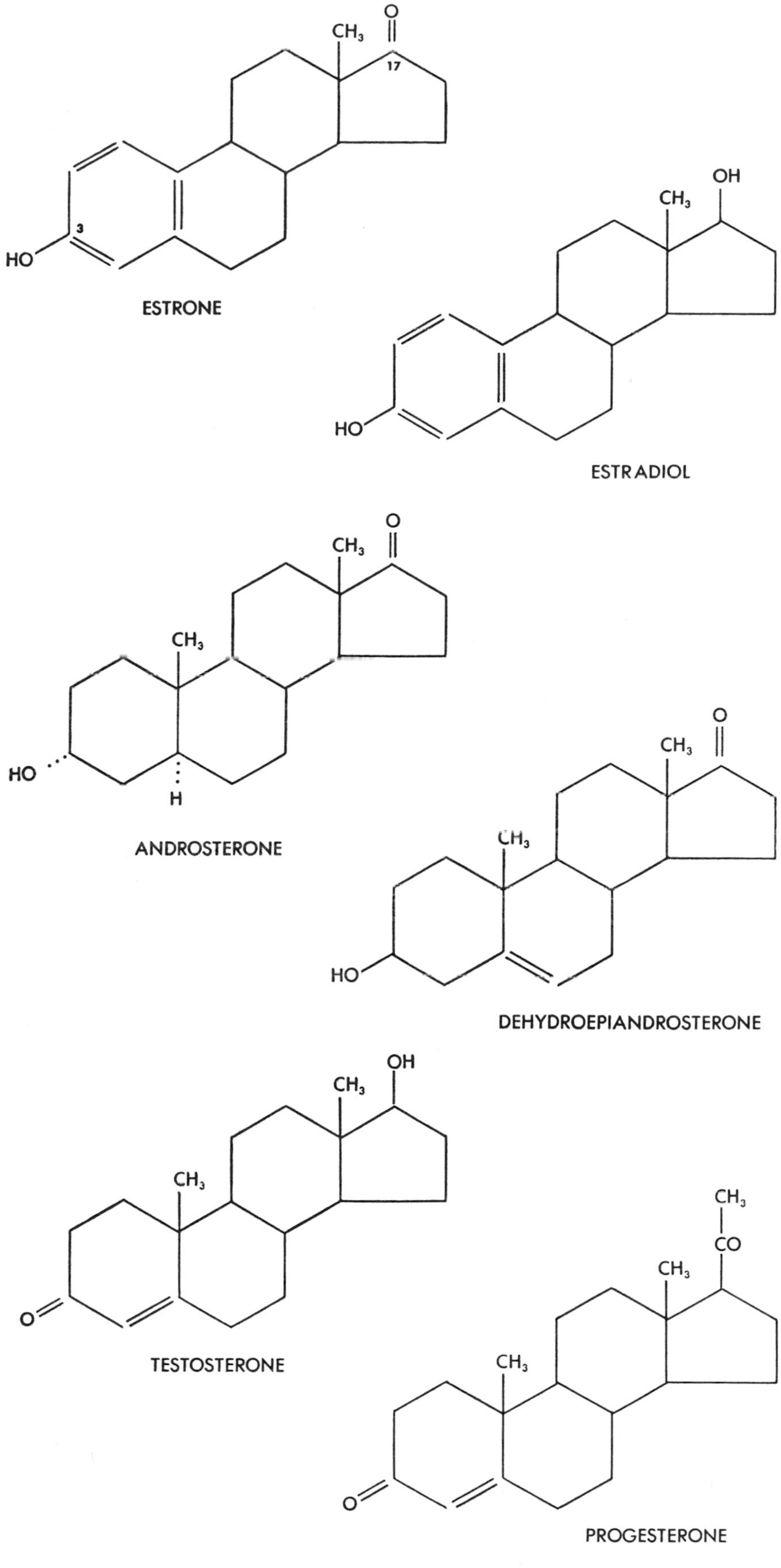

ESTRONE

ESTRADIOL

ANDROSTERONE

DEHYDROEPIANDROSTERONE

TESTOSTERONE

PROGESTERONE

SEX HORMONES share the steroid configuration. Estrone is a female hormone. Estradiol is a variant of it with higher potency. Androsterone, dehydroepiandrosterone and testosterone are male hormones. Progesterone is the female hormone found in pregnant women.

with Ruzicka in Zurich. Since Marker had not disclosed some of the essential bits of know-how to his associates, Rosenkranz had to work out the process anew. He succeeded in rediscovering the missing links, developed a system of chemical manufacture based on the training of unschooled young women to perform specific operations, and soon achieved production of progesterone on a scale surpassing previous performance. He also found a way to produce testosterone, as well as progesterone, from pregnenolone.

Rosenkranz recruited an able research staff of Mexican and foreign chemists, and he brought in as his lieutenant Carl Djerassi, a young chemist born in Bulgaria and trained at the University of Wisconsin. A first fruit of the pioneering research instituted by Rosenkranz and Djerassi was a new process for production of estrone, which involved conversion of the male hormone testosterone into the female hormone estradiol.

Cortisone

The story of the adrenal cortical hormones is so well known that I need not go into it at length here [see "Cortisone and ACTH," by George W. Gray; SCIENTIFIC AMERICAN Offprint 14]. Ever since the most active of these substances, now known as cortisone, was isolated in the laboratory of E. C. Kendall at the Mayo Clinic in 1936, the key problem in its synthesis has been to find some nonlaborious way to place an oxygen

atom at the comparatively inaccessible 11 position in the third ring [see formula on page 40]. Several workable solutions are now at hand, but for close to a decade the armies of researchers who attacked the problem found it completely baffling.

In 1942 vague rumors (now known to be unfounded) that the Luftwaffe was employing adrenal cortical hormones to improve pilots' endurance and resistance to blackout prompted the National Research Council to contract with several laboratories for research on the synthesis of Compound E (cortisone). The first objective was to introduce oxygen at position 11. Desoxycholic acid from bile seemed the best starting material for this purpose, because it carries a hydroxyl group at position 12, offering the possibility that the oxygen atom might be transposed to the neighboring 11 position. A practicable method of effecting the transposition, involving some amazing new chemistry, was devised by Kendall and his group. This made feasible production of Kendall's Compound A, which contains oxygen at position 11 but does not have Compound E's added feature of a hydroxyl group at position 17. But when, after considerable effort and expense, Merck & Company produced a quantity of Compound A by this method, the substance proved completely useless in clinical tests.

Meanwhile Lewis H. Sarett, a young Princeton-trained chemist at Merck, devised a method of introducing oxygen at position 17 and succeeded in synthesizing cortisone. After more than two

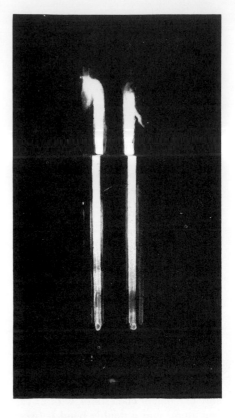

IRRADIATION of steroids is accomplished in vessel containing two ultraviolet lamps.

years of further research and chemical labor, Merck produced enough cortisone for medical trial. The result was the now classical investigation of Kendall and Philip S. Hench at the Mayo Clinic which produced dramatic relief of patients with rheumatoid arthritis.

DIGITOXIN, a plant extract used in the treatment of heart disease, consists of a sugar (left) and a steroid (right). The steroid resembles a bile acid except that the side chain is a ring, and the carbon atom at position 14 bears an OH group and is differently oriented.

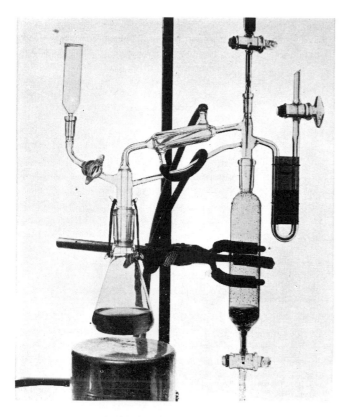

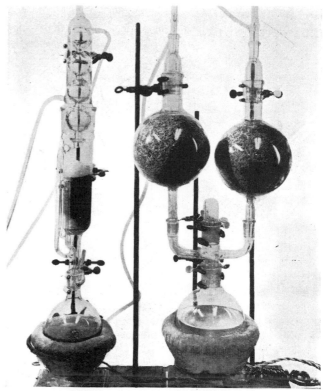

HYDROGENATION of a steroid in the vessel at left measures the number of double bonds, which take up hydrogen, in the molecule.

EXTRACTION of steroids in cranberry skins is accomplished in this apparatus. The skins are in the round vessels at upper right.

Although the yield of cortisone from animal bile was pitifully small, Merck at once undertook plant-scale production. The task was without precedent in modern chemical technology. The process, starting with bile acid and ending with pure cortisone, involved 32 separate steps. Max Tishler and his development group at Merck performed miracles of chemical technology, first to produce cortisone at all, and then to introduce a succession of improvements which reduced the price from $200 per gram to less than $20 per gram.

DIOSGENIN is also found in plants. Because the skeleton of its molecule is the same as that of cholesterol, it is uniquely useful for the synthetic production of animal hormones.

But the demand for cortisone increased and the supply of available bile began to run out. Consequently a problem already brilliantly solved had to be attacked again from some new angle. How synthesize cortisone from something other than a bile acid? Workers in the field grasped at a substance called sarmentogenin, which had been isolated in 1929 from some unknown plant seed. This molecule carries a hydroxyl group at the mystical position 11, and presumably it could be converted into cortisone without great difficulty. Expeditions to South Africa in search of sarmentogenin went out from the Swiss laboratory of Tadeus Reichstein, a leader in cortisone research, from the U. S. Public Health Service, from the National Institute of Medical Research in London, from the Merck laboratory. A plant source of sarmentogenin was found, and Reichstein recently reported its conversion to cortisone, but interest in it has diminished because of the emergence of more promising approaches.

Several groups sought a route to cortisone from the abundantly available natural sterols, particularly cholesterol, ergosterol and diosgenin. From these substances Windaus had produced certain compounds with double bonds at the 7,8 and 9,11 position. In retrospect it seems that it should have been easy to

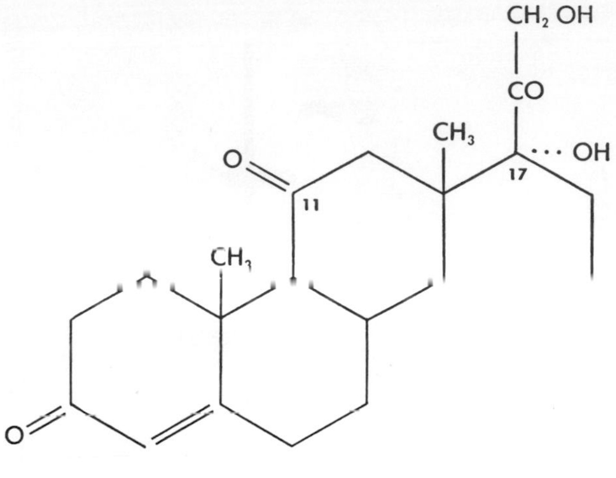

CORTISONE also has the steroid skeleton. It is characterized by the double-bonded oxygen atom attached at position 11 and by the OH group projecting to the rear at position 17.

make use of the double bond to attach an oxygen to the 11 position, but many chemists worked on the problem for two years before, in May, 1951, it suddenly broke. Within a few months success in attachment of oxygen at position 11 by various processes starting with Windaus' compounds was reported by five research teams—the writer's at Harvard University and groups at Merck, Syntex, Zurich and Glasgow.

In the same year Robert B. Woodward and his associates at Harvard made a break-through toward total synthesis of cortisone from simple materials. Woodward synthesized a steroid, the first ever manufactured outside a living organism, with a double bond at the 9,11 position. Then he carried the synthesis three steps further to form a type of molecule which,

as it happened, my associates S. Rajago-palan and Hans Heymann had already succeeded in endowing with an oxygen at the 11 position. From there it did not take long to achieve total synthesis of cortisone. A year later Sarett and his group at Merck reported another total synthesis, beautiful and efficient, which extends all the way from simple starting materials to synthetic cortisone identical in every respect with the natural hormone isolated from the adrenal gland.

But all these approaches yield at present to new processes which have enlisted microorganisms to do the critical work of placing oxygen at position 11. The Upjohn Company was the first to announce discovery of a specific fungus that converts steroids into 11-ketosteroids (oxygenated at the 11 position).

Then Syntex and E. R. Squibb & Sons found other microorganisms capable of the same performance.

The six chemists who received Nobel awards for their work in steroid research were Windaus, Wieland, Ruzicka, Butenandt, Kendall and Reichstein. Their efforts and those of the great army of their fellow workers, only a few of whom could be mentioned in this brief review, have cleared up the chemistry of practically all the known steroid hormones and vitamins.

An important unsolved problem is the function of cholesterol in the animal body. If cholesterol is the essential building material for the bile acids, steroid hormones and vitamin D_3, why does the body produce huge amounts of the precursor to make only trace amounts of the final products? Is cholesterol perhaps just an incidental by-product of the manufacture of the hormones and vitamins? There are some indications that cholesterol or a derivative of it may be involved in degenerative diseases. Cholesterol is found deposited along the walls of blood vessels in victims of hardening of the arteries and of atherosclerosis. Several recent reports suggest that a high level of cholesterol or of a cholesterol-protein complex in the blood may lead to this type of disease. Other lines of experimentation have suggested the possibility that under certain circumstances cholesterol or a companion substance may initiate cancerous growth.

It seems reasonable to hope that advances in the field of preventive medicine may emerge in the next chapter of steroid research.

5

ALKALOIDS

TREVOR ROBINSON
July 1959

This ill-defined group of plant compounds includes many that are both useful and toxic. Though most of them strongly affect human physiology, their functions in plants are still obscure

The alkaloids are a class of compounds that are synthesized by plants and are distinguished by the fact that many of them have powerful effects on the physiology of animals. Since earliest times they have served man as medicines, poisons and the stuff that dreams are made of.

The alkaloid morphine, the principal extract of the opium poppy, remains even today "the one indispensable drug." It has also had an illicit and largely clandestine history in arts and letters, politics and crime. Quinine, from cinchona bark, cures or alleviates malaria; colchicine, from the seeds and roots of the meadow saffron, banishes the pangs of gout; reserpine, from snake root, tranquilizes the anxieties of the neurotic and psychotic. The coca-leaf alkaloid cocaine, like morphine, plays Jekyll and Hyde as a useful drug and sinister narcotic. In tubocurarine, the South American arrow poison, physicians have found a powerful muscle relaxant; atropine, said to have been a favorite among medieval poisoners, is now used to dilate the pupils of the eyes and (in minute doses!) to relieve intestinal spasms; physostigmine, employed by West African tribes in trials by ordeal, has come into use as a specific for the muscular disease myasthenia gravis. Aconitine is catalogued as too toxic to use except in ineffective doses. On the other hand, caffeine and nicotine, the most familiar of all alkaloids, are imbibed and inhaled daily by a substantial fraction of the human species.

Our self-centered view of the world leads us to expect that the alkaloids must play some comparably significant role in the plants that make them. It comes as something of a surprise, therefore, to discover that many of them have no identifiable function whatever. By and large they seem to be incidental or accidental products of the metabolism of plant tissues. But this conclusion somehow fails to satisfy our anthropocentric concern. The pharmacological potency of alkaloids keeps us asking: What are they doing in plants, anyway? Investigators have found that a few alkaloids actually function in the life processes of certain plants. But this research has served principally to illuminate the subtlety of such processes.

The pharmacology of alkaloids has inspired parallel inquiry in organic chemistry. Some of the greatest figures in the field first exercised their talents on these substances. But nothing in the composition of alkaloids has been found to give them unity or identity as a group. The family name, conferred in an earlier time, literally means "alkali-like." Many alkaloids are indeed mildly alkaline and form salts with acids. Yet some perfectly respectable alkaloids, such as ricinine (found in the castor bean), have no alkaline properties at all. Alkaloids are often described as having complex structures. The unraveling of the intricate molecules of strychnine and morphine has taught us much about chemical architecture in general. Yet coniine, the alkaloid poison in the draught of hemlock that killed Socrates, has a quite simple structure. Nor is there much distinction in the characterization of alkaloids as "nitrogen-containing compounds found in plants." Proteins and the amino acids from which they are made also fit this definition.

From the chemical point of view it begins to seem that alkaloids are in a class of compounds only because we do not know enough about them to file them under any other heading. Consider the vitamin nicotinamide, the plant hormone indoleacetonitrile and the animal hormone serotonin. All these compounds occur in plants, and all contain nitrogen. We would call them alkaloids except that we have learned to classify them in more descriptive ways. As we come to know the alkaloids better, we may select other substances from this formless group and assign them to more significantly defined categories.

Though all alkaloids come from plants, not all plants produce alkaloids. Some plant families are entirely innocent of them. Every species of the poppy family, on the other hand, produces alkaloids; the opium poppy alone yields some 20 of them. The Solanaceae present a mixed picture: tobacco and deadly nightshade contain quantities of alkaloids; eggplant, almost none, the potato accumulates alkaloids in its foliage and fruits but not in its tubers. Some structurally interrelated alkaloids, such as the morphine group, occur only in plants of a single family. Nicotine, by contrast, is found not only in tobacco but in many quite unrelated plants, including the primitive horsetails. Alkaloids are often said to be uncommon in fungi, yet the ergot fungus produces alkaloids, and we might classify penicillin as an alkaloid had we not decided to call it an antibiotic. However, alkaloids do seem to be somewhat commoner among higher plants than among primitive ones.

Some 50 years ago the Swiss chemist Amé Pictet suggested that alkaloids in plants, like urea and uric acid in animals, are simply wastes—end products of the metabolism of nitrogenous compounds. But the nitrogen economy of most plants is such that they husband the element, reprocessing nitrogenous compounds of all sorts, including substances such as ammonia which are poisonous to animals. Indeed, many plants have evolved elaborate symbiotic arrangements with bacteria to secure additional nitrogen from the air. From the evolutionary standpoint the tying-up of valuable nitrogen in alkaloids seems an inefficient arrangement.

More recently investigators have come to regard alkaloids not as end products but as by-products thrown off at various points along metabolic pathways, much as substandard parts are rejected on an assembly line. That is to say, alkaloids arise when certain substances in the plant cell cross signals and make an alkaloid instead of their normal product. This idea is certainly plausible when it is applied to the alkaloids formed by the action of the commonest enzymes on the commonest metabolites. The alkaloid trigonelline, for example, is found not only in many plant seeds but also in some species of sea urchins and jellyfish. It is merely nicotinic acid with a methyl group (CH_3) added to it. Now nicotinic acid is one of the commonest components of plant cells. Compounds that can donate methyl groups are also common, as are the enzymes that catalyze such donations. A "confused" enzyme, transferring a methyl group to nicotinic acid instead of to some other substance, could thus form trigonelline by mistake [see illustration at top of page 45]. Nicotine, which has an equally wide distribution, may likewise be produced by everyday biochemical processes.

The more frequent occurrence of alkaloids in higher plants suggests another idea. More highly evolved organisms have obviously made more experiments in metabolism. Some alkaloids may represent experiments that never quite worked. Others may have originated as intermediates in once-useful processes that are no longer carried to completion. Since most alkaloids seem neither to help nor to hurt the plant, natural selection has not operated for or against them. Thus the modern plant that produces alkaloids may do so for no other reason than the persistent pattern of its genes.

Such explanations for the presence of alkaloids in plant tissues find support in what we know about the synthesis of these substances. In 1917 the noted British chemist Sir Robert Robinson showed that the structures of scores of alkaloid molecules could be built up from amino acids by postulating reactions of a few simple types: dehydration, oxidation and so on. For example, he showed that the amino acid tyrosine could easily be transformed into the alkaloid hordenine. Even the complex molecule of reserpine could be built up,

according to his scheme, from tyrosine and the amino acid tryptophan, plus a methylene group [see illustration on page 45]. More recently Robert B. Woodward of Harvard University has proposed that the same three substances, through another series of reactions, may yield the extremely complex molecule of strychnine.

During the past 40 years considerable experimental evidence has accumulated to show that these reactions are not just paper-and-pencil chemistry, but actually occur in nature. Robinson himself correctly predicted the structures of several highly complex alkaloids before these structures were worked out. Later experimenters have shown that enzyme-containing plant extracts can promote amino acid-alkaloid transformations such as the tyrosine-hordenine synthesis. Other investigators, by simply mixing together the postulated precursors of certain alkaloids, have obtained compounds of approximately the correct structure even in the absence of enzymes. Tracer experiments have furnished additional support for Robinson's scheme. If labeled amino acids are injected into alkaloid-producing plants, the plants produce labeled alkaloids. Moreover, the alkaloids contain labeled atoms at just the points that theory predicts.

We know, however, that many steps intervene between the introduction of a labeled precursor and the production of a labeled alkaloid. Moreover, a compound that yields an alkaloid when it is injected in high concentration may not be the normal precursor. Some intermediates go to form all sorts of things, and may only get to alkaloids by quite devious routes. The problem of alkaloid biosynthesis resolves itself into the task of establishing the point at which the alkaloid-producing process diverges from the other metabolic processes of the plant. In principle we might feed various labeled compounds to a plant and ascertain whether the labeled material shows up only in alkaloids or in other substances as well. But we must decide which intermediate compounds we are going to feed. It is fruitless to test a versatile intermediate like glucose, which enters into many processes. In a sense, therefore, we must know our intermediate before conducting the experiment that will identify it. One way of breaking out of this impasse may be to work backward by feeding the alkaloid itself to the plant. By building up high concentrations of alkaloid in the plant's tissues we can perhaps block the alkaloid "production line" and thus cause the im-

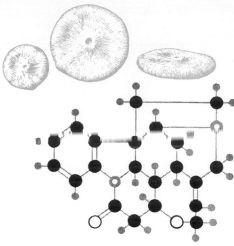

STRYCHNOS NUX-VOMICA

STRYCHNINE

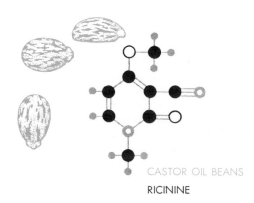

CASTOR OIL BEANS

RICININE

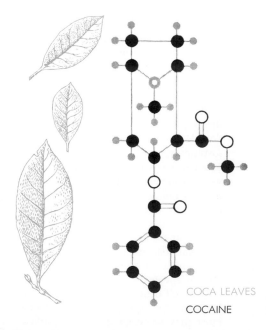

COCA LEAVES

COCAINE

● CARBON
○ OXYGEN
◎ NITROGEN
⦿ HYDROGEN

ALKALOIDS show great structural variety. Depicted in this chart are molecules of nine typical alkaloids together with

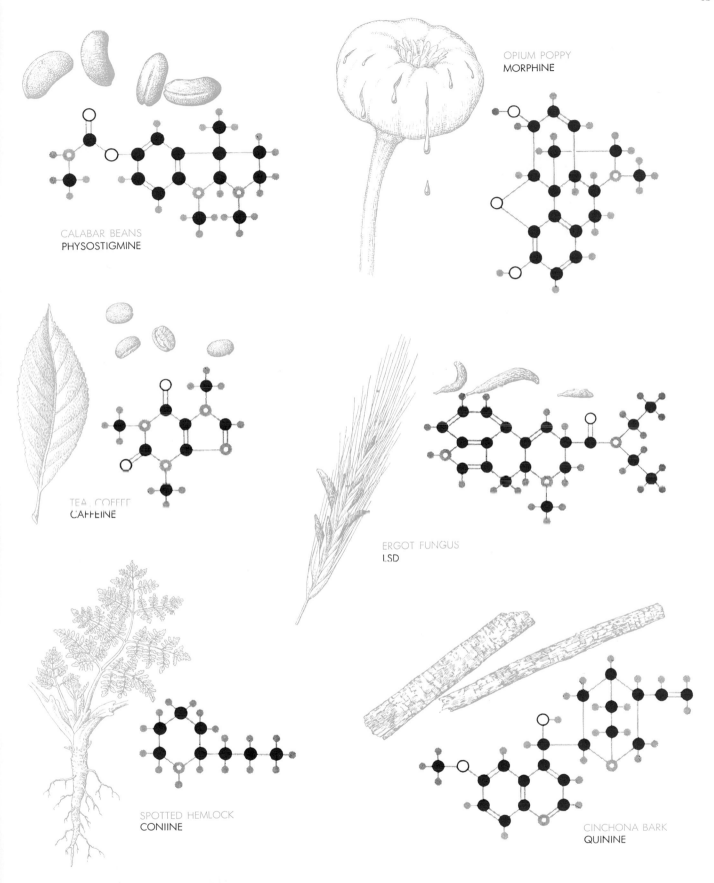

CALABAR BEANS
PHYSOSTIGMINE

OPIUM POPPY
MORPHINE

TEA, COFFEE
CAFFEINE

ERGOT FUNGUS
LSD

SPOTTED HEMLOCK
CONIINE

CINCHONA BARK
QUININE

the plants or plant substances from which they derive. At left is the key to these diagrams and those elsewhere in this article. Strychnine, a violent poison, is one of the most complex alkaloids; coniine, the poison which killed Socrates, one of the simplest. Physostigmine, a West African "ordeal poison," is now used to treat the muscular disease myasthenia gravis; LSD (lysergic acid diethylamide) produces delusions resembling those of schizophrenia. Ricinine is one of the few alkaloids that exert little effect on human beings.

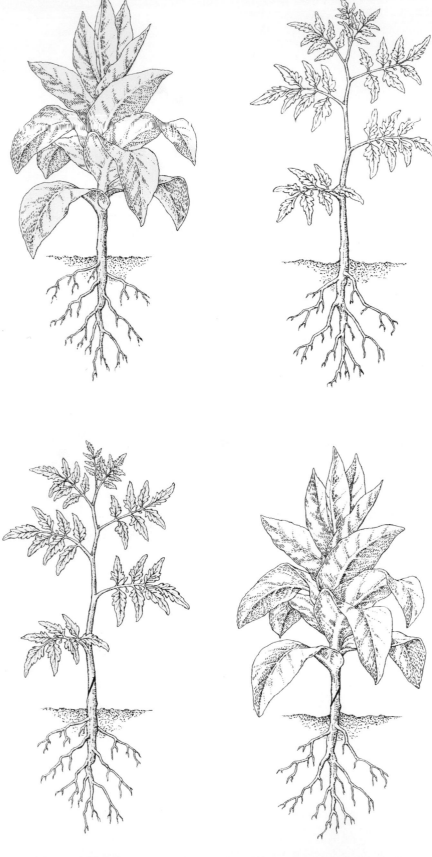

GRAFTING EXPERIMENT indicates that nicotine has no effect on plants. Tobacco plant (*top left*) produces nicotine (*color*) in its roots; the alkaloid then migrates to the leaves. Tomato plant (*top right*) produces no nicotine. A tomato top grafted to a tobacco root (*bottom left*) becomes impregnated with nicotine with no apparent ill effects; tobacco top grafted to tomato root (*bottom right*) is unaffected by the absence of alkaloid. Similar grafting experiments with other alkaloid-producing plants have with few exceptions yielded similar results.

mediate precursors of the alkaloid to accumulate to the point where they can be identified.

With this information in hand, we can go on to inquire which enzymes transform these precursors into alkaloids and whether these enzymes function only in alkaloid formation or in other metabolic processes as well. If we find, for example, that a certain enzyme catalyzes the transfer of a methyl group to an alkaloid precursor but does not function in other methylations, we will have to regard this alkaloid synthesis as a definitely programmed process, and not as a mere aberration. Our present sparse knowledge strongly suggests that at least some alkaloids are programmed. Thus ricinine contains a nitrile group (CN), which rarely occurs in living organisms. If the formation of this group is catalyzed by an enzyme that normally does some other job, we have no indication of what the other job might be. From intimate understanding of this kind we may yet help the plant physiologist to discover what there is about different plants that causes one to make reserpine, while another makes strychnine from the same starting materials.

Of course the study of any metabolic process involves not only the synthesis but also the breakdown of the substances involved. If a given alkaloid is just a waste product or by-product, it has no future and there is no breakdown to be considered. In this case it may simply accumulate in the plant's tissues. For example, quinine piles up in the bark of the cinchona tree, and nicotine in the leaves of the tobacco plant. Some ingenious grafting experiments have furnished additional evidence that many alkaloids, once synthesized, become inert and play no further role in the plant's metabolism. The tobacco plant, for example, manufactures nicotine in its roots, whence the alkaloid migrates to the leaves. However, if we graft the top of a tobacco plant to the roots of a tomato plant, which produces no nicotine, the tobacco flourishes despite the absence of the alkaloid. Conversely, a tomato top grafted to a tobacco root becomes impregnated with nicotine with no apparent ill effects.

But the alkaloids in plants are not always inactive. Hordenine, for example, is found in high concentrations in young barley plants, and gradually disappears as the plant matures. By the use of tracers Arlen W. Frank and Leo Marion of the National Research Council of Canada have found that the disappearing hordenine is converted into lignin, the "plastic" that binds the cellulose fibers in the structure of plants. To be sure, not all plants employ hordenine as an intermediate in making lignin [see "Lignin," by F. F. Nord and Walter J. Schubert; SCIENTIFIC AMERICAN, October, 1958]. But it is gratifying to find at least one case in which an alkaloid performs an identifiable function. Similarly, Edward Leete of the University of Minnesota has shown that in some plants nicotine serves as a "carrier" for methyl groups which it ultimately donates to other molecules.

Such modest findings are a far cry from the first grand-scale function assigned to alkaloids a century ago by the great German chemist Justus von Liebig. Since most alkaloids are alkaline, he proposed that plants use them to neu-tralize deleterious organic acids by forming salts with them. Many alkaloids do, in fact, occur in plants as salts of organic acids. But no one could explain why alkaloid-producing plants should elaborate poisonous acids when closely related plants manage to get along without either the acids or their metabolic antagonists. The question "Why?" still persists. Today it stimulates more modest but sometimes quite intriguing proposals.

Some experiments by the French physiologist Clément Jacquiot suggest a variant of Liebig's neutralization theory. Jacquiot has shown that the tannin produced in cultures of oak cells inhibits cell growth. The alkaloid caffeine counteracts the effects of the tannin and allows growth to proceed. Unfortunately

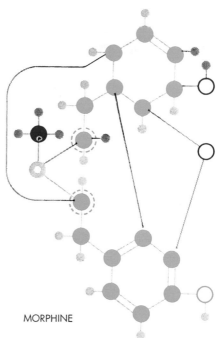

MORPHINE

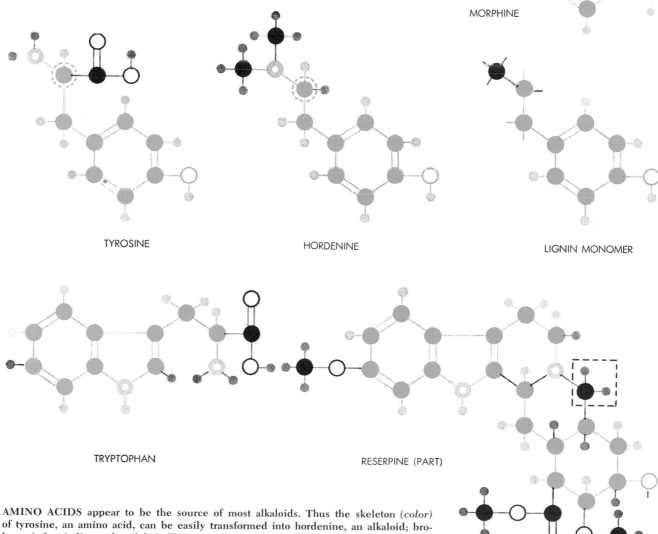

TYROSINE

HORDENINE

LIGNIN MONOMER

TRYPTOPHAN

RESERPINE (PART)

AMINO ACIDS appear to be the source of most alkaloids. Thus the skeleton (*color*) of tyrosine, an amino acid, can be easily transformed into hordenine, an alkaloid; broken circles indicate the "labeled" atoms which confirm the synthesis. Two tyrosine skeletons similarly form morphine, as shown at right; the morphine molecule shown here is distorted (*see diagram on page 43*) to emphasize its derivation. Tyrosine and tryptophan, another amino acid, could join with a methylene group (*broken square*) to form part of the molecule of reserpine, an alkaloid tranquilizer. Hordenine is one of the few alkaloids known to undergo further metabolism in plants; it is converted into one of the units that form the long-chain molecules of lignin, an essential structural material in many plants. The complete structure of the lignin monomer is not known.

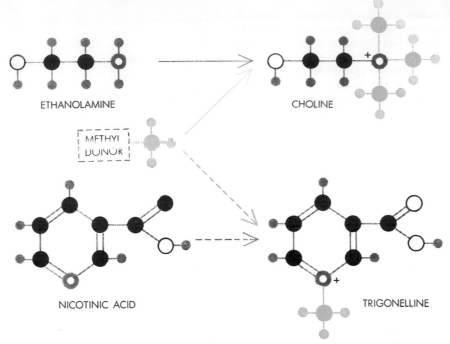

ETHANOLAMINE

CHOLINE

METHYL DONOR

NICOTINIC ACID

TRIGONELLINE

METABOLIC ACCIDENTS may account for the synthesis of some alkaloids. Changing one compound to another by adding methyl groups (color), as in the ethanolamine-choline transformation, is a common biochemical process. If a methyl group were added "by mistake" to nicotinic acid, a common plant substance, the alkaloid trigonelline would result.

this suggestion merely replaces one question with another, since the function of tannins in plants is itself unknown. The suggestion has another flaw in that oak cells produce no caffeine of their own. But here, at any rate, is one case in which caffeine is good for something other than providing a pleasant stimulant for coffee-drinkers.

Botanists and ecologists have speculated that the bitter taste of some alkaloids may discourage animals from eating a plant that contains them, and that poisonous alkaloids may kill off pathogenic organisms that attack the plant. One species of wild tomato does produce an alkaloid that protects it against Fusarium wilt, a common fungus disease of the cultivated tomato. However, the "protection" idea must be handled with caution because of its anthropocentric bias. What is unpalatable or poisonous to a man may be tasty and nourishing to a rabbit or a cutworm.

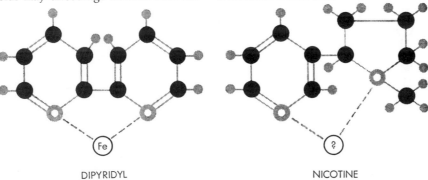

DIPYRIDYL

NICOTINE

STRUCTURE OF NICOTINE resembles that of dipyridyl, a compound that can "chelate" or bind atoms of iron. This similarity suggests that nicotine, and perhaps other alkaloids, may function as chelating agents in some plants. Whether they actually do so is not yet known.

More recently the concept of chelation, the process by which certain organic molecules "sequester" the atoms of metals, has suggested another possible alkaloid function. The structures of some alkaloids should permit them to act as chelating agents. The structure of nicotine is temptingly similar to that of dipyridyl, a common chelating agent for iron [see illustration at bottom of this page]. Such alkaloids might help a plant select one metal from the soil and reject others. Alternatively, they might facilitate the transport of the metal from the roots, where it is absorbed, to the leaves, where it is utilized. Quite a few alkaloids, including nicotine, migrate from roots to leaves, but no one has yet determined whether any of them carry metals along with them.

The structures of many alkaloids resemble those of hormones, vitamins and other metabolically active substances. This resemblance suggests that such alkaloids may function as growth regulators. In a way this hypothesis fits in with the chelation theory, since certain important growth regulators seem to owe their activity to their chelating capacity. Some alkaloids do affect growth processes: Alkaloids from the seeds of certain lupines can inhibit the germination of seeds of related species that produce no alkaloids. Presumably they help the former species to compete successfully with the latter. Similar alkaloids may function as "chemical rain gauges" which prevent germination until sufficient rain has fallen to leach them away and provide adequate moisture for plant growth. Many plants in arid and semiarid environments depend on such rain gauges for survival [see "Germination," by Dov Koller; SCIENTIFIC AMERICAN Offprint 117].

No one theory can account for the functions of so heterogeneous a group of compounds. The steroid compounds, a much smaller and structurally a far more homogeneous group, play a wide variety of physiological roles. Future research will probably reveal an even greater functional diversity among the alkaloids. Certainly we must learn a good deal more about the functions of a few alkaloids before we can safely propose generalizations about all of them.

6

THE HORMONES OF THE HYPOTHALAMUS

ROGER GUILLEMIN AND ROGER BURGUS

November 1972

The anterior pituitary gland, which controls the peripheral endocrine glands, is itself regulated by "releasing factors" originating in the brain. Two of these hormones have now been isolated and synthesized

The pituitary gland is attached by a stalk to the region in the base of the brain known as the hypothalamus. Within the past year or so, after nearly 20 years of effort in many laboratories throughout the world, two substances have been isolated from animal brain tissue that represent the first of the long sought hypothalamic hormones. Because the molecular structure of the new hormones is fairly simple the substances can readily be synthesized in large quantities. Their availability and their high activity in humans has led physiologists and clinicians to consider that the hypothalamic hormones will open a new chapter in medicine.

It has long been known that the pituitary secretes several complex hormones that travel through the bloodstream to target organs, notably the thyroid gland, the gonads and the cortex of the adrenal glands. There the pituitary hormones stimulate the secretion into the bloodstream of the thyroid hormones, of the sex hormones by the gonads and of several steroid hormones such as hydrocortisone by the adrenal cortex. The secretion of the thyroid, sex and adrenocortical hormones thus has two stages beginning with the release of pituitary hormones. Studies going back some 50 years culminated in the demonstration that the process actually has three stages: the release of the pituitary hormones requires the prior release of another class of hormones manufactured in the hypothalamus. It is two of these hypothalamic hormones that have now been isolated, chemically identified and synthesized.

One of the hypothalamic hormones acts as the factor that triggers the release of the pituitary hormone thyrotropin, sometimes called the thyroid-stimulating hormone, or TSH. Thus the hypothalamic hormone associated with TSH is called the TSH-releasing factor, or TRF. The other hormone is LRF. Here again "RF" stands for "releasing factor"; the "L" signifies that the substance releases the gonadotropic pituitary hormone LH, the luteinizing hormone. A third gonadotropic hormone, FSH (follicle-stimulating hormone), may have its own hypothalamic releasing factor, FRF, but that has not been demonstrated. It is known, however, that the hypothalamic hormone LRF stimulates the release of FSH as well as LH.

Studies are continuing aimed at characterizing several other hypothalamic hormones that are known to exist on the basis of physiological evidence but that have not yet been isolated. One of them regulates the secretion of adrenocorticotropin (ACTH), the pituitary hormone whose target is the adrenal cortex. Another hormone (possibly two hormones with opposing actions) regulates the release of prolactin, the pituitary hormone involved in pregnancy and lactation. Still another hormone (again possibly two hormones with opposing actions) regulates the release of the pituitary hormone involved in growth and structural development (growth hormone).

That the hypothalamus and the pituitary act in concert can be suspected not only from their physical proximity at the base of the brain but also from their development in the embryo. During the early embryological development of all mammals a small pouch forms in the upper part of the developing pharynx and migrates upward toward the developing brain. There it meets a similar formation, resembling the finger of a glove, that springs from the base of the primordial brain. Several months later the first pouch, now detached from the upper oral cavity, has filled into a solid mass of cells differentiated into glandular types. At this point the second pouch, still connected to the base of the brain, is rich with hundreds of thousands of nerve fibers associated with a modified type of glial cell, not too unlike the glial cells found throughout the brain. The two organs are now enclosed in a single receptacle that has formed as an open spherical cavity within the sphenoid bone, on which the brain rests.

This double organ, now ensconced in the sphenoidal bone, is the pituitary gland, or hypophysis. The part that migrated from the brain is the posterior lobe, or neurohypophysis; the part that migrated from the pharynx is the anterior lobe, or adenohypophysis. Both parts of the gland remain connected to the brain by a common stalk that goes through the covering flap of the sphenoidal cavity. For many years after the double embryological origin of the pituitary gland was recognized the role of the gland was no more clearly understood than it had been in the old days. Indeed, the name "pituitary" had been given to it in the 16th century by Vesalius, who thought that the little organ had to do with secretion of *pituita:* the nasal fluid.

We know now that the anterior lobe of the pituitary gland controls the secretion and function of all the "peripheral" endocrine glands (the thyroid, the gonads and the adrenal cortex). It also controls the mammary glands and regulates the harmonious growth of the individual. It accomplishes all this by the secretion of a series of complex protein and glycoprotein hormones. All the pituitary hormones are manufactured and secreted by the anterior lobe. Why should this master endocrine gland have migrated so far in the course of evolu-

HYPOTHALAMIC FRAGMENTS of sheep brains were the source from which the authors' laboratory extracted one milligram of TRF, the first hypothalamic hormone to be characterized and synthesized. The photograph is of about 30 frozen hypothalamic fragments; some five million such fragments, dissected from 500 tons of sheep brain tissue, were processed over a period of four years.

←

tion (a journey recapitulated in the embryo) to make contact with the brain? As we shall see, recent observations have answered the question.

The posterior lobe of the pituitary has been known for the past 50 years to secrete substances that affect the reabsorption of water from the kidney into the bloodstream. These secretions also stimulate the contraction of the uterus during childbirth and the release of milk during lactation. In the early 1950's Vincent Du Vigneaud and his co-workers at the Cornell University Medical College resolved a controversy of many years' standing by showing that the biological activities of the posterior lobe are attributable to two different molecules: vasopressin (or antidiuretic hormone) and oxytocin. The two molecules are octapeptides: structures made up of eight amino acids. Du Vigneaud's group showed that six of the eight amino acids in the two molecules are identical, which explains their closely related physicochemical properties and similar biological activity. Both hormones exhibit (in different ratios) all the major biological effects mentioned above: the reabsorption of water, the stimulation of uterine contractions and the release of milk.

As early as 1924 it was realized that the hormones secreted by the posterior lobe of the pituitary are also found in the hypothalamus: that part of the brain with which the lobe is connected by nerve fibers through the pituitary stalk. Later it was shown that the two hormones of the posterior pituitary are actually manufactured in some specialized nerve cells in the hypothalamus. They flow slowly down the pituitary stalk to the posterior pituitary through the axons, or long fibers, of the hypothalamic nerve cells [*see top illustration on page 52*]. They are stored in the posterior pituitary, which is now reduced to a storage organ rather than a manufacturing one. From it they are secreted into the bloodstream on the proper physiological stimulus.

These observations had led several

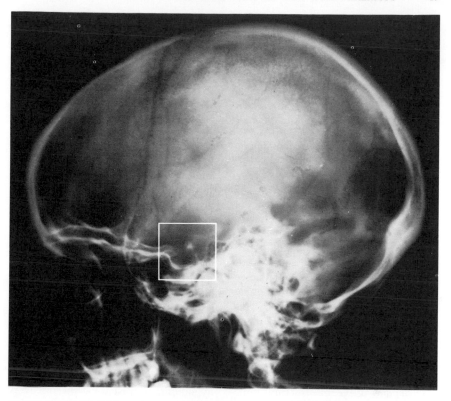

BONY RECEPTACLE in which the pituitary gland is enclosed is a cavity in the sphenoid bone, on which the base of the brain rests. White rectangle shows area diagrammed below.

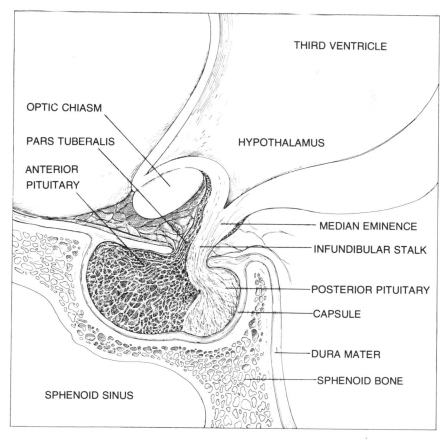

HYPOTHALAMUS AND PITUITARY are connected by a stalk that passes through the membranous lid of the receptacle in the sphenoid bone in which the pituitary rests. The double embryological origin of the two lobes of the pituitary is reflected in their differing tissues and functions and in the different ways that each is connected to the hypothalamus.

biologists, notably Ernst and Berta Scharrer, to the striking new concept of neurosecretion (the secretion of hormones by nerve cells). They suggested that specialized nerve cells might be able to manufacture and secrete true hormones, which would then be carried by the blood and would exert their effects in some target organ or tissue remote from their point of origin. The ability to manufacture hormones had traditionally been assigned to the endocrine glands: the thyroid, the gonads, the adrenals and so on. The suggestion that nerve cells could secrete hormones would endow them with a capacity far beyond their ability to liberate neurotransmitters such as epinephrine and acetylcholine at the submicroscopic regions (synapses) where they make contact with other nerve cells.

Even as these studies were in progress and these new concepts were being formulated other laboratories were reporting evidence that functions of the anterior lobe of the pituitary were somehow dependent on the structural integrity of the hypothalamic area and on a normal relation between the hypothalamus and the pituitary gland. For example, minute lesions of the hypothalamus, such as can be created by introducing small electrodes into the base of the brain in an experimental animal and producing localized electrocoagulation, were found to abolish the secretion of anterior pituitary hormones. On the other hand, the electrical stimulation of nerve cells in the same regions dramatically increased the secretion of the hormones [see illustration below].

Thus the question was presented: Precisely how does the hypothalamus regulate the secretory activity of the anterior pituitary? The results produced by electrocoagulation and electrical stimulation of the hypothalamus suggested some kind of neural mechanism. One objection to this theory was rather hard to overcome. Careful anatomical studies over many years had clearly established that there were no nerve fibers extending from the hypothalamus to the anterior pituitary. The only nerve fibers found in the pituitary stalk were those that terminate in the posterior lobe.

A way out of the dilemma was provided by an entirely different working hypothesis, suggested by the discovery in 1936 of blood vessels of a peculiar type that were shown to extend from the floor of the hypothalamus through the pituitary stalk to the anterior pituitary [see bottom illustration on page 52].

If these tiny blood vessels were cut, the secretions of the anterior pituitary would instantly decrease. If the capillary vessels regenerated across the surgical cut, the secretions resumed.

Accordingly a new hypothesis was put forward about 1945 with which the name of the late G. W. Harris of the University of Oxford will remain associated. The hypothesis proposed that hypothalamic control of the secretory activity of the anterior pituitary could be neurochemical: some substance manufactured by nerve cells in the hypothalamus could be released into the capillary vessels that run from the hypothalamus to the anterior pituitary, where it could be delivered to the endocrine cells of the gland. On reaching these endocrine cells the substance of hypothalamic origin would somehow stimulate the secretion of the various anterior pituitary hormones.

The hypothesis that pituitary function is controlled by neurohormones originating in the hypothalamus was soon well established on the basis of intensive physiological studies in several laboratories. The next problem was therefore to isolate and characterize the postulated hypothalamic hormones. It was logical to guess that the hormones might be polypeptides of small molecular weight, since it had been well established that the two known neurosecretory products of hypothalamic origin, oxytocin and vasopressin, are each composed of eight amino acids. Indeed, in 1955 it was reported that crude aqueous hypothalamic extracts designed to contain polypeptides were able specifically to stimulate the secretion of ACTH, the pituitary hormone that controls the secretion of the steroid hormones of the adrenal cortex.

It was quickly demonstrated that none of the substances known to originate in the central nervous system (such as epinephrine, acetylcholine, vasopressin and oxytocin) could account for the ACTH-releasing activity observed in the extract of hypothalamic tissue. It therefore seemed reasonable to postulate the existence and involvement in this phenomenon of a new substance designated (adreno)corticotropin-releasing factor, or CRF. Several laboratories then undertook the apparently simple task of purifying CRF from hypothalamic extracts, with the final goal of isolating it and establishing its chemical structure. Seventeen years later the task still remains to be accomplished. Technical difficulties involving the methods

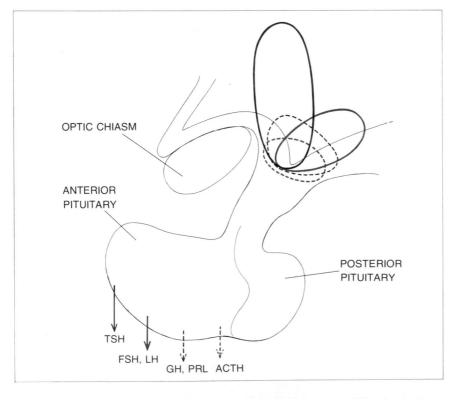

OPTIC CHIASM

ANTERIOR PITUITARY

POSTERIOR PITUITARY

TSH

FSH, LH

GH, PRL   ACTH

RELATION between the hypothalamus and anterior pituitary was established experimentally. Lesions in specific regions of the hypothalamus interfere with secretion by the anterior lobe of specific hormones; electrical stimulation of those regions stimulates secretion of the hormones. The regions associated with each hormone are mapped schematically.

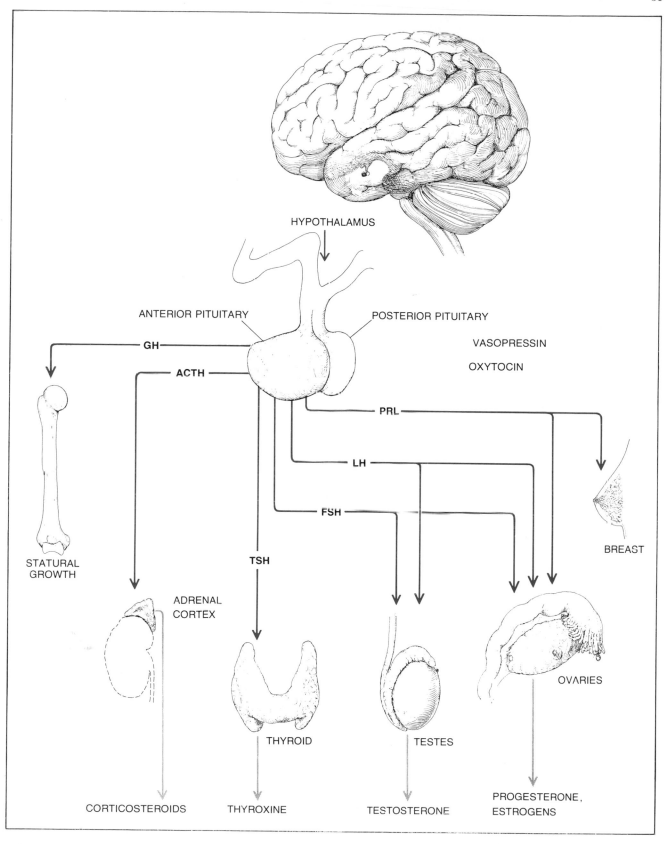

HYPOTHALAMUS

ANTERIOR PITUITARY

POSTERIOR PITUITARY

VASOPRESSIN

OXYTOCIN

GH

ACTH

PRL

LH

FSH

TSH

STATURAL
GROWTH

ADRENAL
CORTEX

THYROID

TESTES

OVARIES

BREAST

CORTICOSTEROIDS

THYROXINE

TESTOSTERONE

PROGESTERONE,
ESTROGENS

PITUITARY GLAND, connected to the hypothalamus at the base of the brain, has two lobes and two functions. The posterior lobe of the pituitary stores and passes on to the general circulation two hormones manufactured in the hypothalamus: vasopressin and oxytocin. The anterior lobe secretes a number of other hormones: growth hormone (GH), which promotes statural growth; adrenocorticotropic hormone (ACTH), which stimulates the cortex of the adrenal gland to secrete corticosteroids; thyroid-stimulating hormone (TSH), which stimulates secretions by the thyroid gland, and follicle-stimulating hormone (FSH), luteinizing hormone (LH) and prolactin (PRL), which in various combinations regulate lactation and the functioning of the gonads. Several of these anterior pituitary hormones are known to be controlled by releasing factors from the hypothalamus, two of which have now been synthesized.

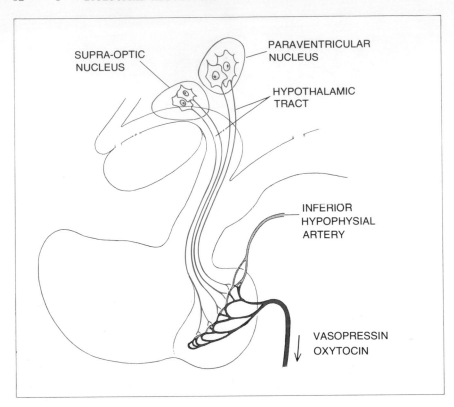

SUPRA-OPTIC
NUCLEUS

PARAVENTRICULAR
NUCLEUS

HYPOTHALAMIC
TRACT

INFERIOR
HYPOPHYSIAL
ARTERY

VASOPRESSIN
OXYTOCIN

NEURAL CONNECTIONS could not explain the relation of the hypothalamus and the anterior lobe. The only significant nerve fibers connecting hypothalamus and pituitary run from two hypothalamic centers to the posterior lobe. They transmit oxytocin and vasopressin, two hormones manufactured in the hypothalamus and stored in the posterior lobe.

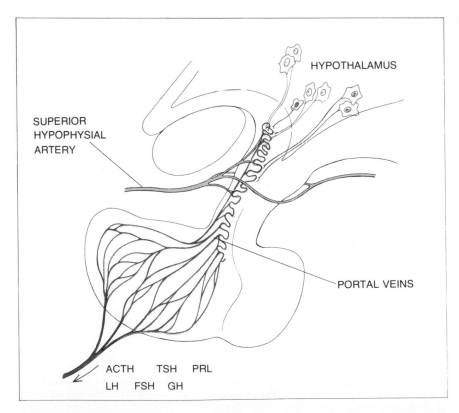

HYPOTHALAMUS

SUPERIOR
HYPOPHYSIAL
ARTERY

PORTAL VEINS

ACTH   TSH   PRL
LH   FSH   GH

VASCULAR CONNECTIONS between hypothalamus and anterior lobe were eventually discovered: a network of capillaries reaching the base of the hypothalamus supplies portal veins that enter the anterior pituitary. Small hypothalamic nerve fibers apparently deliver to the capillaries releasing factors that stimulate secretion of the anterior-lobe hormones.

of assaying for CRF, together with certain peculiar characteristics of the molecule, have defied the enthusiasm, ingenuity and hard work of several groups of investigators.

More rewarding results were obtained in a closely related effort. About 1960 it was clearly established that the same crude extracts of hypothalamic tissue were able to stimulate the secretion of not only ACTH but also the three other pituitary hormones mentioned above: thyrotropin (TSH) and the two gonadotropins (LH and FSH). TSH is the pituitary hormone that controls the function of the thyroid gland, which in turn secretes the two hormones thyroxine and triiodothyronine. LH controls the secretion of the steroid hormones responsible for the male or female sexual characteristics; it also triggers ovulation. FSH controls the development and maturation of the germ cells: the spermatozoa and the ova. In reality the way in which LH and FSH work together is considerably more complicated than this somewhat simplistic description suggests.

Results obtained between 1960 and 1962 were best explained by proposing the existence of three separate hypothalamic releasing factors: TRF (the TSH-releasing factor), LRF (the LH-releasing factor) and FRF (the FSH-releasing factor). The effort began at once to isolate and characterize TRF, LRF and FRF. Whereas it was difficult to find a good assay for CRF, a simple and highly reliable biological assay was devised for TRF. At first, however, the assays for LRF and FRF still left much to be desired.

With a good method available for assaying TRF, progress was initially rapid. Within a few months after its discovery TRF had been prepared in a form many thousands of times purer. Preparations of TRF obtained from the brains of sheep showed biological activity in doses as small as one microgram. A great deal of physiological information was obtained with those early preparations. For example, the thyroid hormones somehow inhibit their own secretion when they reach a certain level in the blood. This fact had been known for 40 years and was the first evidence of a negative feedback in endocrine regulation. Studies with TRF showed that the feedback control takes place at the level of the pituitary gland as the result of some kind of competition between the number of available molecules of thyroid hormones and of TRF. Other significant observations were made on the gonadotropin-releasing factors when purified preparations, also active at microgram levels, were injected in experimental animals, for instance to produce ovulation.

It soon became apparent, however, that the isolation and chemical characterization of TRF, LRF and FRF would not be simple. The preparations active in microgram doses were chemically heterogeneous; they showed no clear-cut indication of a major component. It was also realized that each fragment of hypothalamus obtained from the brain of a sheep or another animal contained nearly infinitesimal quantities of the releasing factors. The isolation of enough of each factor to make its chemical characterization possible would therefore require the processing of an enormous number of hypothalamic fragments. Two groups of workers in the U.S. undertook this challenge: a group headed by A. V. Schally at the Tulane University School of Medicine and our own group, first at the Baylor University College of Medicine in Houston and then at the Salk Institute in La Jolla, Calif.

Over a period of four years the Tulane group worked with extracts from perhaps two million pig brains. Our laboratory collected, dissected and processed close to five million hypothalamic fragments from the brains of sheep. Since one sheep brain has a wet weight of about 100 grams, this meant handling 500 tons of brain tissue. From this amount we removed seven tons of hypothalamic tissue (about 1.5 grams per brain). Semi-industrial methods had to be developed in order to handle, extract and purify such large quantities of material. Finally in 1968 one milligram of a preparation of TRF was obtained that appeared to be homogeneous by all available criteria.

On careful measurement the entire milligram could be accounted for by the sole presence of three amino acids: histidine, glutamic acid and proline. Moreover, the three amino acids were present in equal amounts, which suggested that we were dealing with a relatively simple polypeptide perhaps as small as a tripeptide. In the determination of peptide sequences it is customary to subject the sample to attack by proteolytic enzymes, which cleave the peptide bonds holding the polypeptide chain together in well-established ways. Pure TRF, however, was shown to be resistant to all the proteolytic enzymes used. Since we could spare only a tiny amount of our precious one-milligram sample for studies of molecular weight, we could not obtain a

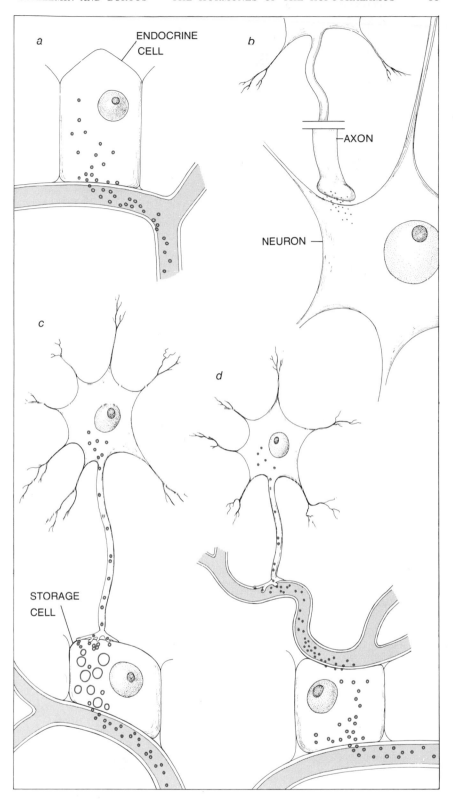

NEUROHUMORAL SECRETIONS involved in hypothalamic-pituitary interactions differ from classical hormone secretion and classical nerve-cell communication. A classical endocrine cell (such as those in the anterior pituitary or the adrenal cortex, for example) secretes its hormonal product directly into the bloodstream (a). At a classical synapse, the axon, or fiber, from one nerve cell releases locally a transmitter substance that activates the next cell (b). In neurosecretion of oxytocin or vasopressin the hormones are secreted by nerve cells and pass through their axons to storage cells in the posterior pituitary, eventually to be secreted into the bloodstream (c). Hypothalamic (releasing factor) hormones go from the neurons that secrete them into local capillaries, which carry them through portal veins to endocrine cells in the anterior lobe, whose secretions they in turn stimulate (d).

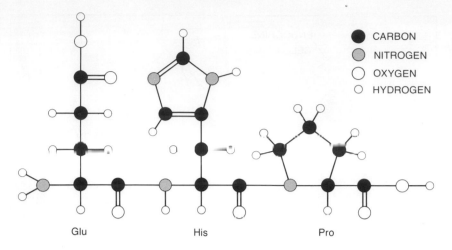

AMINO ACID CONTENT of TRF, the releasing factor for thyroid-stimulating hormone (TSH), was established: glutamic acid, histidine and proline in equal proportions. Each of six possible tripeptides was synthesized; one is diagrammed. None was active biologically.

precise value for that important measurement. On the basis of inferential evidence, however, it seemed to be reasonable to assume that the molecular weight of TRF could not be more than 1,500.

With small molecules it is often possible to use methods based on the technique of mass spectrometry to obtain in a matter of hours the complete molecular structure of the compound under investigation. Because of the minute quantities of TRF available such efforts on our part were frustrated; the mass-spectrometric methods available to us in 1969 were not sensitive enough to indicate the structure of our unknown substance. Other approaches involve the use of infrared or nuclear magnetic-resonance spectrometry, which can pro-

vide direct insight into molecular structure. Here too the techniques then available were inadequate for providing clear-cut information about polypeptide samples that weighed only a few micrograms.

Confronted with nothing but dead ends, we decided on an entirely different approach to finding the structure of TRF. That approach was first to synthesize each of six possible tripeptides composed of the three amino acids known to be present in TRF: histidine (abbreviated His), glutamic acid (Glu) and proline (Pro). The six tripeptides were then assayed for their biological activity. None showed any activity when they were injected at doses of up to a million times the level of the active natural TRF.

Was this another dead end? Not quite. Our synthetic polypeptides all had a free amino group (NH₂) at the end of the molecule designated the N terminus. We knew that in several well-characterized hormones the N-terminus end was not free; it was blocked by a small substitute group of some kind. Indeed, we had evidence from the small quantity of natural TRF that its N terminus was also blocked. To block the N terminus of our six candidate polypeptides was not difficult: we heated them in the presence of acetic anhydride, which typically couples an acetyl group (CH₃CO) to the N terminus. When these "protected" tripeptides were tested, the results were unequivocal. The biological activity of the sequence Glu-His-Pro, and that sequence alone, was qualitatively indistinguishable from the activity of natural TRF. Quantitatively, however, there was still a considerable difference between the synthetic product and natural TRF. Next it was shown that the protective effect of heating Glu-His-Pro with the acetic anhydride had been to convert the glutamic acid at the N terminus into a ring-shaped form known as pyroglutamic acid (pGlu).

We now had available gram quantities of the synthetic tripeptide pGlu-His-Pro-OH. (The OH is a hydroxyl group at the end of the molecule opposite the N terminus.) Accordingly we could bring into play all the methods that had yielded no information with the microgram quantities of natural TRF. Several of the techniques were modified, particularly with the aim of ob-

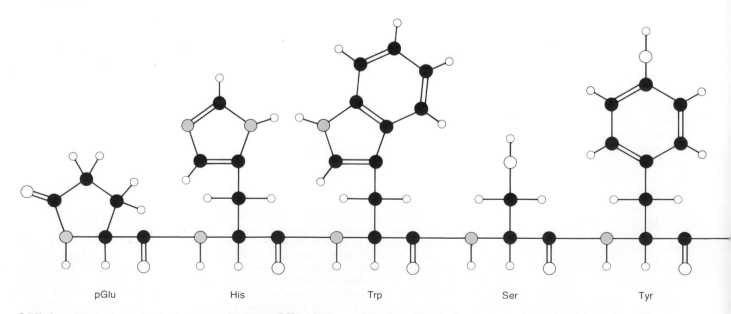

LRF, the releasing factor for the luteinizing hormone (LH), which affects the activity of the gonads, was characterized and synthesized soon after. First the hormone was isolated and its amino acid content was determined. Then their intramolecular sequence was es-

taining mass spectra of the synthetic peptide at levels of only a few micrograms.

Meanwhile, armed with knowledge about the structure of other hormones, we modified the synthetic pGlu-His-Pro-OH to pGlu-His-Pro-NH$_2$ by replacing the hydroxyl group with an amino group (NH$_2$) to produce the primary amide [*see top illustration on this page*]. This substance proved to have the same biological activity as the natural TRF. At length the complete structure of the natural TRF was obtained by high-resolution mass spectrometry. It turned out to be the structure pGlu-His-Pro-NH$_2$. The time was late 1969. Thus TRF not only was the first of the hypothalamic hormones to be fully characterized but also was immediately available by synthesis in amounts many millions of times greater than the hormone present in one sheep hypothalamus. TRF from pig brains was subsequently shown to have the same molecular structure as TRF from sheep brains.

Characterization of the hypothalamic releasing factor LRF, which controls the secretion of the gonadotropin LH, followed rapidly. Isolated from the side fractions of the programs for the isolation of TRF, LRF was shown in 1971 to be a polypeptide composed of 10 amino acids. Six of the amino acids are not found in TRF: tryptophan (Trp), serine (Ser), tyrosine (Tyr), glycine (Gly), leucine (Leu) and arginine (Arg). The full sequence of LRF is pGlu-His-Trp-Ser-Tyr-Gly-Leu-Arg-Pro-Gly-NH$_2$ [*see bottom illustration on these two pages*].

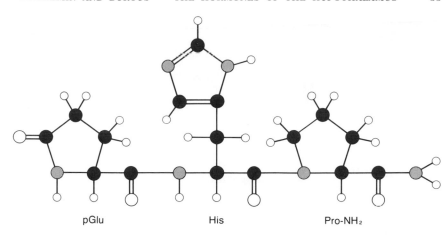

pGlu   His   Pro-NH$_2$

TRIPEPTIDES were modified in an effort to characterize the releasing factor. When the sequence glutamic acid–histidine-proline was modified by forming the glutamic acid into a ring and converting the proline end (*right*) to an amide, it was found to be TRF.

Although this structure is more complicated than the structure of TRF, it begins with the same two amino acids (pGlu-His) and has the same group at the other terminus (NH$_2$).

It turns out that LRF also stimulates the secretion of the other gonadotropin, FSH, although not as powerfully as it stimulates the secretion of LH. It has been proposed that LRF may be the sole hypothalamic controller of the secretion of the two gonadotropins: LH and FSH.

There is good physiological evidence that the hypothalamus is also involved in the control of the secretion of the other two important pituitary hormones: prolactin and growth hormone. Curiously, prolactin is as plentiful in males as in females, but its role in male physiology is still a mystery. The hypothalamic mechanism involved in the control of the secretion of prolactin or growth hormone is not fully understood. It is quite possible that the secretion of these two pituitary hormones is controlled not by releasing factors alone but perhaps jointly by releasing factors and specific hypothalamic hormones that somehow act as inhibitors of the secretion of prolactin or growth hormone. If it should turn out that inhibitory hormones rather than stimulative ones are involved in the regulation of prolactin and growth hormone, one should not be too surprised. The brain provides many examples of inhibitory and stimulative systems working in parallel.

The hypothalamic hormones TRF and LRF are both now available by synthesis in unlimited quantities. Both are

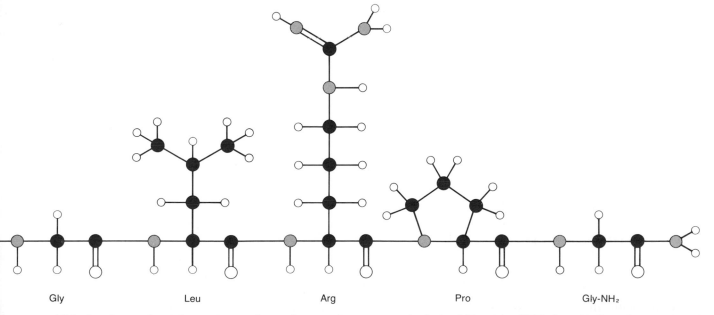

Gly   Leu   Arg   Pro   Gly-NH$_2$

tablished and reproduced by synthesis; the synthetic replicate shown here was found to have full biological activity. In addition to stimulating LH activity, LRF also stimulates the secretion of another gonadotropic hormone, FSH, although not so powerfully.

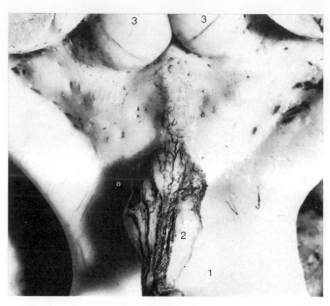

BLOOD VESSELS linking the hypothalamus and the anterior pituitary are seen in photographs made by Henri Duvernoy of the University of Besancon. The photomicrograph (*left*) shows some of the individual loops that characterize the capillary network at the base of the hypothalamus of a dog. The ascending branch of one loop is clearly seen (*1*); the loop comes close to the floor of the third ventricle (*2*) and then descends (*3*), carrying with it the releasing factors that are secreted by this region of the hypothalamus and entering the pars tuberalis of the anterior lobe (*4*). The photograph of the floor of the human hypothalamus (*right*) shows the optic chiasm (*1*), the posterior side of the pituitary stalk with its portal veins (*2*) and the mammillary bodies of the brain (*3*).

highly active in stimulating pituitary functions in humans. TRF is already a powerful tool for exploring pituitary functions in several diseases characterized by the abnormality of one or several of the pituitary secretions. There is increasing evidence that most patients with such abnormalities (primarily children) actually have normally functioning glands, since they respond promptly to the administration of synthetic hypothalamic hormones. Evidently their abnormalities are due to hypothalamic rather than pituitary deficiencies. These deficiencies can now be successfully treated by the administration of the hypothalamic polypeptide TRF.

Similarly, an increasing number of women who have no ovulatory menstrual cycle and who show no pituitary or ovarian defect begin to secrete normal amounts of the gonadotropins LH and FSH after the administration of LRF. The administration of synthetic LRF should therefore be the method of choice for the treatment of those cases of infertility where the functional defect resides in the hypothalamus-pituitary system. Indeed, ovulation can be induced in women by the administration of synthetic LRF. On the other hand, knowledge of the structure of the LRF molecule may open up an entirely novel approach to fertility control. Synthetic compounds closely related to LRF in structure may act as inhibitors of the native LRF. Two such analogues of LRF, made by modifying the histidine in the hormone, have been reported as antagonists of LRF. It is therefore possible that LRF antagonists will be used as contraceptives.

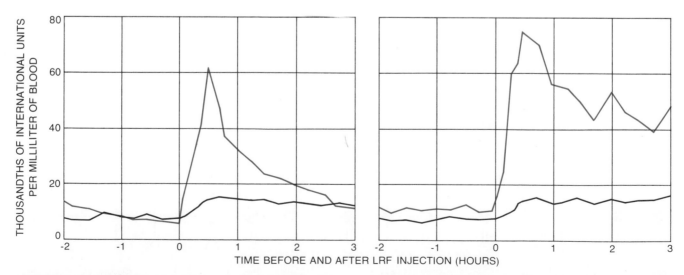

HYPOTHALAMIC HORMONES have clinical implications and applications. For example, women with no pituitary or ovarian defect respond to the administration of synthetic LRF by secreting normal amounts of the hormones LH and FSH. Curves show effect of LRF on secretion of LH (*color*) and FSH (*black*) in a normal woman on the third (*left*) and 11th (*right*) day of menstrual cycle.

7

CYCLIC AMP

IRA PASTAN
August 1972

This comparatively small molecule is a "second messenger" between a hormone and its effects within the cell. It operates in cells as diverse as bacteria and cancerous animal cells

The chemical reactions that proceed in the living cell are catalyzed by the large molecules called enzymes. If all the enzymes found within cells were working at top speed, the result would be chaos, and many mechanisms have evolved that control the speed at which these enzymes function. A small molecule that plays a key role in regulating the speed of chemical processes in organisms as distantly related as bacteria and man is cyclic-3'5'-adenosine monophosphate, more widely known as cyclic AMP. ("Cyclic" refers to the fact that the atoms in the single phosphate group of the molecule are arranged in a ring.) Among the many functions served by cyclic AMP in man and other animals is acting as a chemical messenger that regulates the enzymatic reactions within cells that store sugars and fats. Cyclic AMP has also been shown to control the activity of genes. Moreover, a precondition for one of the kinds of uncontrolled cell growth we call cancer appears to be an inadequate supply of cyclic AMP.

The first steps leading to the discovery of cyclic AMP were taken by Earl W. Sutherland, Jr., at Washington University some 25 years ago. For this work Sutherland was awarded the Nobel prize in physiology and medicine for 1971. Sutherland was trying to trace the sequence of events in a well-known physiological reaction whereby the hormone epinephrine (as adrenalin is generally known in the U.S.) causes an increase in the amount of glucose, or blood sugar, in the circulatory system. It is this reaction, usually a response to pain, anger or fear, that provides an animal with the energy either to fight or to flee. It is not a simple one-step process. Glycogen, a polymeric storage form of glucose, is held in reserve in the cells of the liver. What transforms the glycogen into glucose, which can then leave the liver and enter the bloodstream, is a series of steps involving intermediate substances. Sutherland measured the levels of these intermediates and concluded that only the initial step in the series (the transformation of glycogen into the intermediate sugar glucose-1-phosphate) was mediated by epinephrine. The transformation itself is actually catalyzed by an enzyme known as phosphorylase. Observing the activity of phosphorylase in cell-free extracts of liver tissue, Sutherland was able to enhance the enzyme's performance by first exposing to epinephrine the cells he used to make his extract.

Sutherland began to examine the properties of phosphorylase in more detail. He found that the enzyme could exist in two forms: one that degraded glycogen rapidly and one that had no effect on it. The conversion of the enzyme from the active to the inactive form was catalyzed by a second enzyme, whose only action was to remove inorganic phosphate from the phosphorylase molecule. The conversion is worth noting; it is an important example of how the activity of an enzyme can be controlled by a relatively small change in its structure.

In collaboration with the first of a number of talented co-workers, Walter D. Wosilait, Sutherland next found still another liver enzyme, phosphorylase *b* kinase, which could restore the inactive form of phosphorylase to the active state. As one might expect, the reversal was accomplished by replacement of the missing phosphate; the donor of the phosphate was a close chemical relative of cyclic AMP, adenosine triphosphate (ATP). At about the same time Edwin G. Krebs and Edmond H. Fischer of the University of Washington detected a similar activating kinase in muscle tissue. ("Kinase" is the name reserved for transformations where ATP is the phosphate-donor.)

Having established the existence of two forms of phosphorylase, Sutherland and his colleagues concluded that the speed at which glycogen was broken down in the liver was a function of the amount of the enzyme present in its active form. Sutherland and Theodore W. Rall now made preparations of ruptured liver cells. When they added epinephrine to these broken cells, they found that in spite of the damage the hormone still increased the activity of the enzyme. The experiment, although simple, was extremely significant. Never before had a hormone been observed to

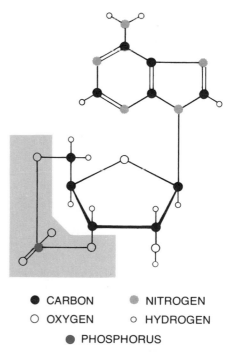

● CARBON   ● NITROGEN
○ OXYGEN   ○ HYDROGEN
● PHOSPHORUS

CYCLIC AMP is so named because the phosphate group in its molecule (*colored area*) forms a ring with the carbon atoms to which it is attached. Earl W. Sutherland, Jr., and his colleagues isolated substance in 1958.

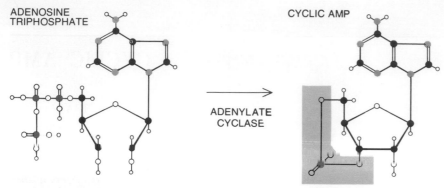

ADENOSINE TRIPHOSPHATE

CYCLIC AMP

ADENYLATE CYCLASE

FORMATION OF CYCLIC AMP takes place when an enzyme in the cell membrane, adenylate cyclase, responds to the arrival of a hormone at the membrane. The enzyme transforms molecules of adenosine triphosphate, or ATP (*left*), within the cell into cyclic AMP (*right*).

function in a preparation that contained no intact cells. Once such a reaction can be shown to occur in a cell-free preparation the investigator can go on to test various cell components one at a time to determine just which ones are affected by the hormone.

This Sutherland and his co-workers proceeded to do. They knew that the phosphorylase was present in liver-cell cytoplasm: that part of the cell outside the nucleus and inside the cell membrane. When they added epinephrine to preparations composed of cytoplasm alone, however, there was no increase in enzyme activity. The absence of response suggested that the hormone exerted its effect on some other component of the liver cell. In due course they found that this component was the cell membrane.

Exactly what was the hormone doing to the cell membrane? In an effort to find out Sutherland employed a stratagem commonly used in biochemistry. He incubated a preparation of cell membrane (which itself contains no phosphorylase) with epinephrine. He then brought the mixture to the boiling point, expecting the heat to destroy the activity of the enzyme or enzymes in the membrane that were required for epinephrine action. When he added the now denatured mixture to a cell-free preparation that contained phosphorylase but no cell membrane, the activity of the phosphorylase was increased. The epinephrine had interacted with the cell membrane during the initial incubation period, evidently causing some enzyme in the membrane to produce a heat-stable factor that enhanced the activity of phosphorylase. Unfortunately the factor—whatever it was—was present in very small amounts and was therefore difficult to identify.

Sutherland eventually collected a large enough sample of the factor to de-

termine that it belonged to the group of small molecules known as nucleotides. It did not, however, appear to be any of the known nucleotides. He wrote to Leon A. Heppel of the National Institutes of Health, who had developed many of the methods used in preparing and identifying a number of nucleotides, asking him for a quantity of an enzyme that breaks nucleotides into their component parts and thus facilitates their identification. This request set the stage for a remarkable coincidence.

It is Heppel's habit to let letters that do not require an immediate answer accumulate on his desk. He covers each few days' correspondence with a fresh sheet of wrapping paper, and every few months he clears his desk. Heppel immediately sent Sutherland the enzyme but left the letter of request on his desk. By chance the stratum just below contained a chatty letter from a friend and former colleague, David Lipkin of Washington University, describing an experiment where ATP was treated with a solution of barium hydroxide. The result was the formation of an unusual nucleotide.

Heppel remembers coming into his office one Saturday to clear his desk. He found Sutherland's and Lipkin's letters in adjacent strata and consequently reread them together. It seemed likely to him that both men had isolated the same substance, and he proceeded to put them in touch. Lipkin's chemical synthesis readily produced the nucleotide in large quantities. This made it easy to establish that the synthetic substance was structurally identical with natural cyclic AMP. It also provided an abundant supply of synthetic cyclic AMP for experimental purposes.

Taking advantage of the demonstrated ability of cyclic AMP to increase phosphorylase activity in cell-free preparations, Sutherland and his co-workers

were able to measure the amount of cyclic AMP present in a wide variety of cells. They found that it was 1,000 times less abundant than ATP, being present in cell water in a ratio of about one part per million in contrast to ATP's one part per 1,000. Although scanty in amount, cyclic AMP was present in virtually every organism they examined, from bacteria and brine-shrimp eggs to man. Among mammals it was present in almost every type of body cell. Sutherland's group went on to examine a number of tissues that are characterized by their secretion of various substances following stimulation by hormones. He discovered that the level of cyclic AMP in such cells rose soon after exposure to the hormone. Moreover, when the tissues were exposed to nothing but cyclic AMP, or to derivatives of cyclic AMP that enter cells rapidly, they produced secretions just as readily.

The cyclic AMP molecule is formed from ATP by the action of a special enzyme: adenylate cyclase. The enzyme is located in the membrane of the cell wall. Normally its activity is low and the transformation of ATP into cyclic AMP takes place at a slow rate. Let us consider what happens, however, when a hormone enters the bloodstream. The hormone acts as a "first messenger." It travels to the target cell and then binds to specific receptor sites on the outside of the cell wall. Thyroid cells have receptors that "recognize" thyrotrophin, adrenal cells have receptors that recognize adrenocorticotrophin, and so forth. The binding of the hormone to the receptor site increases the activity of the adenylate cyclase in the cell membrane; just how this occurs has not yet been established. In any event, cyclic AMP is produced as a result, utilizing the abundant supply of ATP on the inner side of the cell membrane. The cyclic AMP is then free to diffuse throughout the cell, where it acts as a "second messenger," instructing the cell to respond in a characteristic way. For example, a thyroid cell responds to this second message by secreting more thyroxine, whereas an adrenal cell responds by producing and secreting steroid hormones. In the cells of the liver the instruction results in the conversion of glycogen into glucose.

Because cyclic AMP is such a powerful regulator of cell functions the cell must be able to control its level of concentration. In most cells control is accomplished by regulating the rate of synthesis of cyclic AMP and by the actions of one or more enzymes, known as phosphodiesterases, that degrade cyclic

AMP into an inert form of adenosine monophosphate. The deactivation results from a splitting of the ester bond that joins the phosphate to the 3′ carbon of the ribose ring. The quantity of the degradative enzymes in the cell is not kept constant; apparently more can be made whenever the level of cyclic AMP in the cell is elevated for more than a few minutes. The level of cyclic AMP is also controlled by diffusion through the cell wall; this is the mechanism operating in addition to enzyme degrada-

tion in bacteria and in the cells of some animal tissues.

Krebs, whose earlier work with Fischer had established the presence of active and inactive forms of phosphorylase in muscle tissue, was able in 1968 to specify the role played by cyclic AMP in activating the enzyme. Working with Donal Walsh, he found that the cyclic AMP binds to yet another enzyme, protein kinase, which is inactive until cyclic AMP is present. The activated kinase then performs the same function for a

related enzyme, phosphorylase kinase. It is this second enzyme that at last activates the phosphorylase. The result of the final activation is the breakdown of glycogen, the storage form of glucose, in a series of steps similar to those that proceed in liver cells.

Whenever glycogen is being degraded in order to satisfy the organism's need for glucose, it would be a waste of energy to continue the synthesis of additional glycogen. This waste is avoided. A specific enzyme mediates the syn-

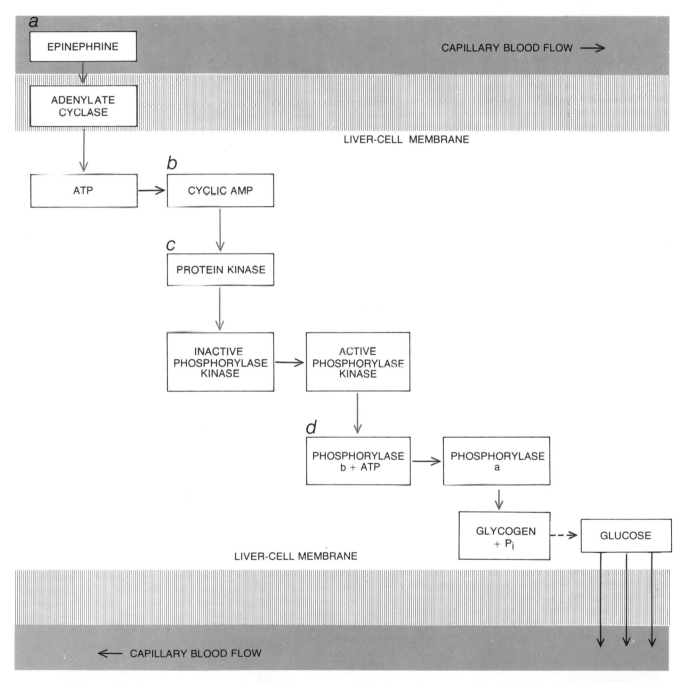

RELEASE OF BLOOD SUGAR by a glycogen storage cell in the liver is mediated by cyclic AMP. In the schematic diagram arrows in color symbolize actions and black arrows the results. "First messenger," epinephrine, arrives at the cell membrane and activates the enzyme, adenylate cyclase (*a*), causing it to convert some of the ATP present in the cytoplasm into cyclic AMP (*b*), the second messenger. The cyclic AMP then activates a protein kinase (*c*), which activates a second kinase. The second kinase (*d*) triggers a four-step sequence (*not shown*) that converts the glycogen into the assimilable sugar glucose, which then passes into the bloodstream.

thesis of glycogen. It is glycogen synthetase, and like phosphorylase and many other enzymes it has an active and an inactive form. At the time when some molecules of cyclic AMP are initiating the chain of events that leads to the breakdown of glycogen, other cyclic AMP molecules are at work converting glycogen synthetase from the active to the inactive form.

Just as cells that store glycogen have the task of supplying the fasting organism with glucose, so the task of fat cells is to satisfy the organism's need for fatty acids. Fatty acids are present in the fat cell in the storage form triglyceride. The triglyceride can ocupy as much as 90 percent of the cell's volume. Here again it is cyclic AMP that initiates the breakdown of the stored fat. In response to any of several hormonal stimuli the level of cyclic AMP in the fat cell begins to rise. As with muscle tissue, this activates a protein kinase. The kinase in turn activates a second enzyme, triglyceride lipase. On being converted to the active form the second enzyme begins to degrade the stored triglyceride into the re-

quired fatty acids. It should be noted here that protein kinases are present in the cells of many tissues other than fat and muscle; numerous other actions of cyclic AMP presumably also involve these ATP-powered enzymes.

In addition to such effects within cells cyclic AMP has been observed to stimulate the expression of genetic information. How this stimulation is accomplished in animal cells remains obscure. Almost all the detailed observations of the regulation of gene activity involve a single microorganism: the common intestinal bacterium *Escherichia coli*. There are many good reasons for molecular biologists who are engaged in genetic studies choosing to work with this simple organism. One of them is that a typical animal cell contains enough DNA to account for 10 million individual genes; *E. coli* has only enough DNA for about 10,000 genes.

Sutherland and another colleague, Richard Makman, established the presence of cyclic AMP in *E. coli* cells in 1965. By then the substance was already

known to control a variety of cell processes, and Robert Perlman and I at the National Institutes of Health guessed that it must also play an important role in the bacterium. But what role?

There were two clues. First, cultures of *E. coli* that are nourished exclusively with glucose show low levels of cyclic AMP. Second, such cultures can synthesize only very small amounts of a number of enzymes, including those needed to metabolize sugars other than glucose; this inhibition is called the "glucose effect." Putting these clues together, we reasoned that cyclic AMP was a chemical switch, so to speak; if it was present in *E. coli* in adequate quantities, it would activate the expression of those genes that are necessary for the synthesis of the missing enzymes. In order to test this speculation we added cyclic AMP to cultures of inhibited *E. coli*. The cells were then able to metabolize such sugars as maltose, lactose and arabinose in addition to a variety of other nutrients.

Now, factors that operate at the level of the gene do so by stimulating the synthesis of the messenger RNA that in

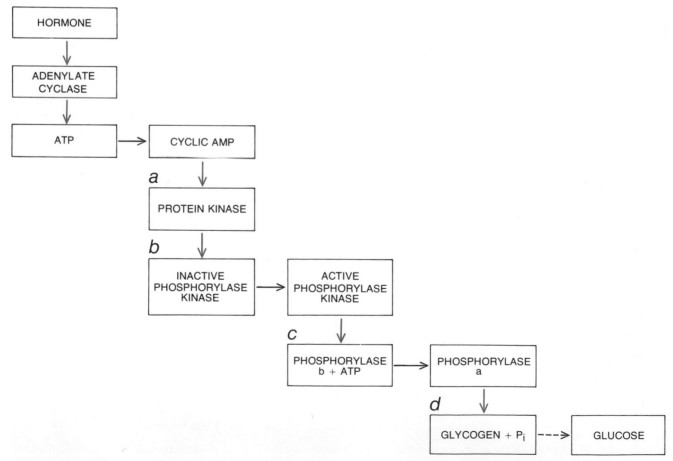

INITIAL SEQUENCE that triggers the transformation of glycogen into glucose in muscle tissue also involves cyclic AMP. First (*a*) the cyclic AMP that has been formed from ATP in the cell cytoplasm activates a protein kinase. The kinase in turn activates a second kinase (*b*) that is capable of transforming phosphorylase *b* into phosphorylase *a*. When this takes place (*c*), the transformed phosphorylase then starts the sequence (*d*) that converts glycogen into glucose. Edwin Krebs and Edmond Fischer of the University of Washington discovered the phosphorylase alteration in muscle at the same time that Sutherland found the same process in the liver.

effect reproduces the information contained in the DNA of the gene. The messenger RNA then provides the ribosomes of the cell with the information they need to construct the appropriate enzyme. In our laboratory by this time we had isolated a mutant form of *E. coli* that lacked the cell-membrane enzyme required for the synthesis of cyclic AMP. Curiously, the mutant cells were viable; apparently the presence of cyclic AMP is not absolutely crucial to survival. For our next experiment we selected cultures of the mutant strain that contained a known amount of the two messenger RNA's needed for the metabolism of two sugars: lactose and galactose. When we added cyclic AMP to the mutant cultures, we found that the quantity of the two messenger RNA's was increased but that the expression of most of the other genes in the cells was not affected. In order to learn exactly how the increase took place we now needed to study the reaction in a cell-free preparation.

At Columbia University, Geoffrey L. Zubay and his co-workers were working with a complex cell-free preparation made from *E. coli*. When they added to the preparation DNA that was greatly enriched in the genes for lactose metabolism, small amounts of one of the enzymes of lactose metabolism, beta-galactosidase, were formed. The DNA was enriched because it was derived from a bacterial virus that had acquired the lactose genes from the *E. coli* host it had once lived in and now contained it permanently. The virus is a hybrid of two other viruses, designated λ and 80, and the DNA derived from it is λ80*lac* (for lactose) DNA.

When Zubay and his co-worker Donald Chambers added cyclic AMP to the cell-free preparation, synthesis of the enzyme for lactose metabolism was greatly stimulated. Soon thereafter my colleagues B. de Crombrugghe and H. Varmus showed that the synthesis of *lac* messenger RNA was also increased. John Parks in our laboratory, following Zubay's procedure, developed a cell-free *E. coli* preparation that responded to the addition of λ*gal* (for galactose) DNA by producing one of the enzymes of galactose metabolism. Like Zubay's preparation, Parks's synthesized much more of the enzyme when cyclic AMP was added.

One might now have expected that the same result could be achieved without even using the complex cell-free preparation. It would be necessary only to mix together in the test tube appropriate quantities of DNA rich in lactose genes and of the special enzyme that

TISSUE		HORMONE	PRINCIPAL RESPONSE
FROG SKIN		MELANOCYTE-STIMULATING HORMONE	DARKENING
BONE		PARATHYROID HORMONE	CALCIUM RESORPTION
MUSCLE	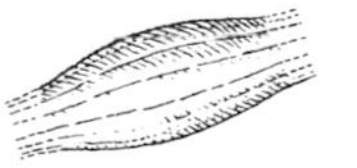	EPINEPHRINE	GLYCOGENOLYSIS
FAT	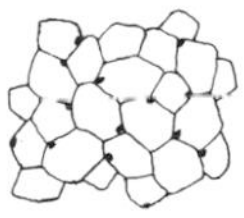	EPINEPHRINE	LIPOLYSIS
		ADRENOCORTICO-TROPHIC HORMONE	LIPOLYSIS
		GLUCAGON	LIPOLYSIS
BRAIN	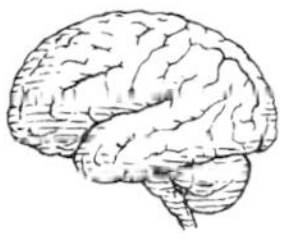	NOREPINEPHRINE	DISCHARGE OF PURKINJE CELLS
THYROID	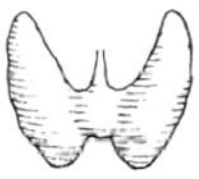	THYROID-STIMULATING HORMONE	THYROXIN SECRETION
HEART	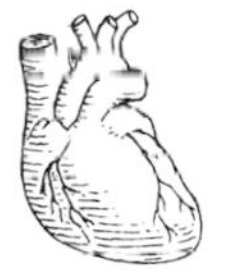	EPINEPHRINE	INCREASED CONTRACTILITY
LIVER	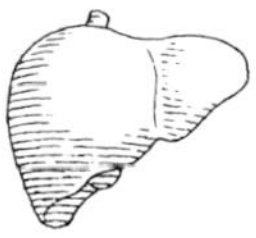	EPINEPHRINE	GLYCOGENOLYSIS
KIDNEY	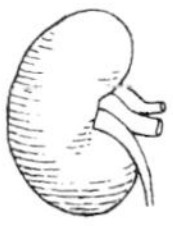	PARATHYROID HORMONE	PHOSPHATE EXCRETION
		VASOPRESSIN	WATER REABSORPTION
ADRENAL		ADRENOCORTICO-TROPHIC HORMONE	HYDROCORTISONE SECRETION
OVARY	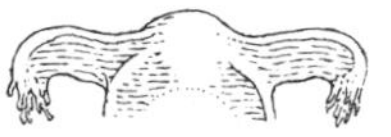	LUTEINIZING HORMONE	PROGESTERONE SECRETION

FOURTEEN EXAMPLES of hormonal activities that affect many different target tissues (*left*) have one factor in common: each causes an increase in the level of cyclic AMP in the tissue. It seems probable that all the responses are set in train by the sudden increase in level.

a

GENE FOR LACTOSE METABOLISM

PROMOTER SITE   OPERATOR SITE

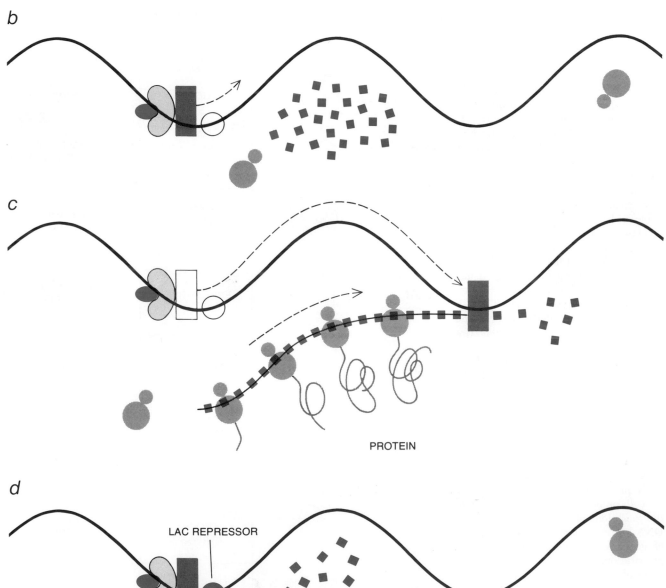

NUCLEOSIDE
TRIPHOSPHATES

CYCLIC AMP

CRP

RNA
POLYMERASE

RIBOSOMES

LAC
REPRESSOR

b

c

PROTEIN

d

LAC REPRESSOR

copies DNA into RNA, add some ATP and other nucleoside triphosphates as building blocks and cyclic AMP as mediator and the result would be the production of *lac* messenger RNA. The expectation proved to be false. Some RNA was formed in the test tube but none of it corresponded to the *lac* gene. Clearly some factor was missing. Fortunately there was a clue to its nature.

In our search for mutant strains of *E. coli* that could not make various enzymes known to be controlled by cyclic AMP we had found some mutants that produced cyclic AMP in abundance but were still unable to metabolize either lactose or several other sugars. Similar *E. coli* mutants had been isolated at the Harvard Medical School by Jonathan Beckwith and his colleagues. What was lacking in the mutants was a protein that has the ability to bind cyclic AMP. The protein, which is known either as catabolite gene activation protein (abbreviated CAP) or as cyclic AMP receptor protein (CRP), was difficult to purify but a pure form was finally prepared by my colleague Wayne B. Anderson.

De Crombrugghe added some of the pure protein to the test-tube mixture that had failed to yield *lac* messenger RNA. This time the effort was successful. Moreover, when CRP was added to a similar test-tube mixture that included λ*gal* DNA, *gal* messenger RNA was formed. These experiments, incidentally, were the first to achieve the transcription of a bacterial gene in a system containing only purified components. We could now proceed to examine in detail how the gene activity was controlled.

The initial step in the process involved the combination of cyclic AMP with CRP. The complex of the two substances was then bound to the DNA; once this took place the enzyme that copies DNA into RNA was enabled to bind to a specific site, called the "pro-

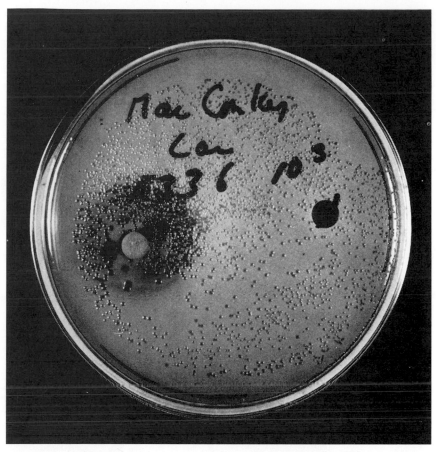

MUTANT STRAIN of the bacterium *Escherichia coli* lacks the enzyme that turns ATP into cyclic AMP. As the different reactions show, addition of cyclic AMP to one disk in the dish (*left*) enabled the colonies of bacteria nearby to metabolize certain sugars. The bacteria near the second disk (*right*), to which only 5' AMP was added, were not able to do this.

moter," at the beginning of the *lac* gene in the bacterial DNA. After that the nucleoside triphosphates initiated the *lac* transcription process. We soon found that the transcription of the genes for galactose metabolism followed the same steps.

Of the 10,000 or so *E. coli* genes, perhaps some hundreds are regulated in this way. How the others are controlled remains a mystery. The regulation of even those genes that are controlled by

cyclic AMP involves other substances as well. The *lac* transcription process provides an example. A specific protein, *lac* repressor, is bound to the DNA at a site (called the operator) at the beginning of the *lac* gene. The operator site is near the site on the DNA where the complex of cyclic AMP and CRP binds. The cell cannot now produce *lac* messenger RNA even in the presence of CRP and cyclic AMP because the repressor prevents the copying enzyme from beginning the transcription [*see illustration on opposite page*]. Only when the repressor is removed by the action of a substance closely related to lactose can the transcription take place.

This pattern of events, where cyclic AMP is able to stimulate the transcription of many genes while at the same time individual repressors can prevent certain transcriptions, gives the *E. coli* cell the flexibility it needs for the efficient utilization of the foodstuffs in its environment. *E. coli* does not store large amounts of carbohydrate and must live on what is in its immediate vicinity. If the cell is exposed, say, both to glucose (which it can already metabolize) and to

TRANSCRIPTION OF A GENE, the sequence of events shown schematically on the opposite page, occurred in a cell-free medium that was supplied (*a*) with RNA polymerase (*solid bar*), nucleoside triphosphates (*colored squares*), the protein CRP (*crescent shape*), which reversibly binds cyclic AMP (*colored ovals*), and quantities of DNA enriched with the gene for lactose metabolism (*bracketed area of helix*). Transcription begins when a combined unit of CRP and cyclic AMP activates a promoter site at the beginning of the *lac* operon (*b*); RNA polymerase now binds to the promoter site, ready to link the nucleoside triphosphates together in the correct sequence. As transcription proceeds (*c*), the RNA polymerase arranges the nucleoside triphosphates in the correct sequence. Ribosomes meanwhile begin the process of assembling the proteins that comprise the enzymes for lactose metabolism. If still another protein, *lac* repressor (*colored doughnut*), attaches itself to the operator site at the beginning of the *lac* operon (*d*), then the RNA polymerase can bind but cannot transcribe in spite of the presence of the CRP-cyclic AMP unit. This pattern, whereby cyclic AMP stimulates the transcription of many different operons, whereas the repressors are specific for a particular gene, makes for flexibility in the utilization of foodstuffs.

lactose (which it cannot), there is nothing to be gained by expending energy to produce the enzymes for lactose metabolism. As long as its supply of cyclic AMP remains at a low level the *E. coli* cell conserves its capacity for protein synthesis.

That so many different cell functions should be under the control of a single substance is remarkable. Cyclic AMP also participates in the process of visual excitation and regulates the aggregation of certain social amoebae so that they can form complex reproductive structures [see "Hormones in Social Amoebae and Mammals," by John Tyler Bonner; SCIENTIFIC AMERICAN Offprint 1145]. A good example of the substance's versa-

tility is provided by the contrast between the mechanisms of glycogen and lipid degradation on the one hand and the stimulation of gene transcription on the other. The degradation reactions involve enzymes and depend on protein phosphorylation. In *E. coli* gene transcription no phosphorylation is involved. It will be interesting to learn whether the *E. coli* mechanism is employed to control gene activity elsewhere in nature and whether cyclic AMP acts in still other ways. Most of the substance's actions in animal cells remain to be explored.

Abnormalities in the metabolism of cyclic AMP may explain the nature of certain diseases. For example, in chol-

era the bacteria responsible for the disease produce a toxin that stimulates the intestinal cells to secrete huge amounts of salt and water; it is the resulting dehydration, if it is unchecked, that is fatal [see "Cholera," by Norbert Hirschhorn and William B. Greenough III; SCIENTIFIC AMERICAN, August, 1971]. It appears that the toxin first stimulates the intestinal cells to accumulate excess cyclic AMP, whereupon the cyclic AMP instructs the cells to secrete the salty fluid.

In our own laboratory we are interested in understanding the difference between normal cells and cancer cells. Since cancer cells typically grow in an uncontrolled manner and lose their ability to carry out specialized functions, it seemed possible to us that some of their abnormal properties might be attributable to their inability to accumulate normal amounts of cyclic AMP. My colleague George Johnson and I have begun to investigate this possibility.

The cells that are commonly used for such studies are fibroblasts: cells that contribute to the formation of connective tissue. They are usually taken from the embryos of chickens, mice or other animals and allowed to grow in a nutrient fluid. After a short period of culture in a medium approximating blood serum in composition, the embryonic cells take on the appearance of the normal fibroblast cells in connective tissue and grow in bottles or dishes in a controlled manner. When the cultured cells are exposed to cancer-producing viruses or to chemical carcinogens, however, they begin to grow in an abnormal manner. They take on the appearance of tumor cells, and when they are injected into a suitable host, they usually produce tumors. The process of changing a normal cell to a cancer cell in culture is termed transformation. Among the properties of transformed cells are a change in appearance, accelerated growth, looser adherence to the container surface, alterations in the rate of production of specialized large molecules such as mucopolysaccharides and clumping on exposure to certain agglutinative plant proteins.

Now, cells that have been transformed and are then grown in the presence of cyclic AMP tend to return to normal. They grow more slowly, adhere more tenaciously to the container, synthesize certain large molecules at a faster rate and clump less when they are exposed to agglutinative plant protein. Their appearance also frequently returns toward normal. As far as we now know, the morphologic reversal occurs only in embryo cells and those derived from con-

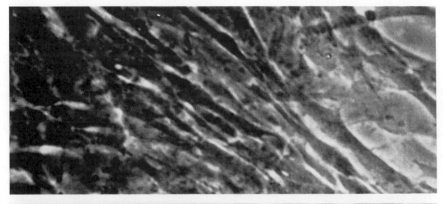

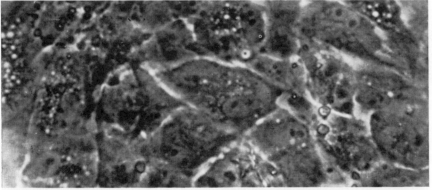

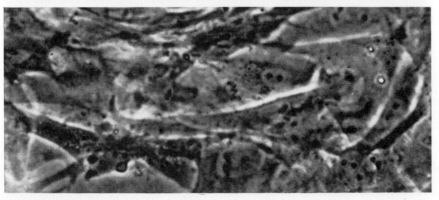

CHICK EMBRYO CELLS infected with a temperature-sensitive mutant strain of Rous sarcoma virus maintain a normal appearance (*top micrograph*) when they are cultured at a temperature of 40.5 degrees Celsius. When the temperature is reduced to 36 degrees, however, they quickly develop an abnormal appearance (*middle*). If cyclic AMP is added to the cells, their appearance remains normal (*bottom*) even after the temperature is reduced.

nective tissue; tumors from a few other kinds of cell do not show the same morphologic response.

In any event the reversal suggested to us that transformed cells might contain abnormally low levels of cyclic AMP. One of our co-workers, Jack Otten, investigated this possibility. He found that cells from chick embryos that had been transformed by exposure to Rous sarcoma virus contained much less cyclic AMP than normal chick-embryo cells. We could not be certain, however, which came first: the abnormal appearance of the transformed cells or the low level of cyclic AMP.

To settle the question we needed a way to transform cells very rapidly. We had at our disposal a mutant variety of the Rous sarcoma virus that made rapid cell transformation possible. For example, chick cells infected with the mutant virus (which had been isolated by our colleague John Bader of the National Cancer Institute) remain normal in appearance as long as the culture is kept at a temperature of 40.5 degrees Celsius. When the temperature is reduced to 36 degrees, however, the cells are rapidly transformed.

We grew cultures infected with this temperature-sensitive virus and kept them at the "normal" temperature. We incubated some of the cultures with a potent derivative of cyclic AMP and some without it. When we lowered the temperature of the cells incubated with cyclic AMP to the transformation level, they continued to look normal for some time [see illustration on opposite page]. The cultures without cyclic AMP, once the temperature was lowered, developed the characteristic transformed appearance within a few hours.

Otten measured the level of natural cyclic AMP in the readily transformed cells after their exposure to the lower temperature. He found that as soon as 20 minutes later the level had fallen greatly; this was well before the cells began to develop a transformed appearance.

This finding obviously leaves many questions unanswered. Is the phenomenon confined to tumors of connective tissue? How many of the abnormal properties of the transformed cells result from the low level of cyclic AMP? Which enzymes are responsible for lowering the level? Are there other tumor-forming viruses and chemical carcinogens that similarly lower the cell's supply of cyclic AMP? Not least, can the findings to date be exploited in a therapeutically useful manner? The search for answers continues in each of these areas.

8

PROSTAGLANDINS

JOHN E. PIKE
November 1971

These recently isolated hormone-like substances show much clinical promise. They affect a wide range of physiological processes, from the contraction of the uterus to secretion from the stomach wall

In the early 1930's several investigators studying human semen and animal seminal tissues discovered that these materials contained substances showing high physiological potency. Two gynecologists in New York, Raphael Kurzrok and Charles C. Lieb, examining the effects of fresh semen on strips of uterus from hysterectomized women patients, found that the seminal fluid caused the uterine tissue to relax or contract, depending on whether the woman was fertile (having borne children) or was sterile. At about the same time Maurice W. Goldblatt in England and Ulf S. von Euler in Sweden obtained other striking effects in experiments with human semen and with extracts from the seminal vesicular gland of sheep. They reported that the active substance could stimulate muscle tissue to contract or, on injection into an animal, could produce sharp lowering of the blood pressure. To this previously unknown, remarkably potent material von Euler gave the name prostaglandin.

Actually the term was a misnomer. The active substances had come not from the prostate gland but from the seminal vesicles, glands that also contribute to the semen. It was eventually learned that "prostaglandins" could be found in other body fluids and tissues, including the menstrual fluid of women. The prostaglandins are now known to be a family of substances showing a wide diversity of biological effects. They give promise of pharmacological uses in many medical areas. Although they resemble other known hormones in their dramatic effects, chemically they are a quite different class of compounds. Von Euler deduced from his investigations, and studies have since confirmed, that the prostaglandins are fatty acids. This was a considerable surprise, as fatty acids had not previously been suspected

of playing the kind of role the prostaglandins are now observed to perform.

For some 20 years after the original discovery of the prostaglandins little was learned about their chemistry or their specific properties. One of the reasons is that the body produces these highly potent substances in very small quantities; for example, a man normally synthesizes only about a tenth of a milligram per day of two of the most important prostaglandins. Furthermore, prostaglandins are rapidly broken down in the body by catabolic enzymes. Consequently the amount in the tissues is so small that the prostaglandins could not be obtained in quantity or studied chemically until refined techniques were developed for isolating and analyzing them.

Interest in the prostaglandins grew, however, as extracts of crude acidic lipids (fatty substances) obtained from a variety of other sources proved to owe their biological activity to their prostaglandin content. By the late 1950's biochemists began to achieve success in purifying prostaglandins and examining their molecular structure. Sune Bergström and his associate Jan Sjövall in Sweden crystallized two of these substances, called PGE_1 and PGF_1-alpha. Working with minute amounts, they were able to decipher the structure of the two compounds, including some details of their three-dimensional configuration, with the help of gas chromatography, mass spectroscopy and X-ray analysis. With Bengt I. Samuelsson and other investigators they proceeded to work out the structure of several additional prostaglandins. They were all seen to be closely akin, and the distinctive biological effects of the various compounds in the family could be related systematically to their structural

variations [*see illustration on opposite page*].

It is now well established that the prostaglandins are 20-carbon carboxylic acids (that is, incorporating the COOH group) and that they are synthesized in the body from certain polyunsaturated fatty acids by the formation of a five-member ring and the incorporation of three oxygen atoms at certain positions [*see top illustration on pages 68 and 69*]. The enzymes involved in their synthesis have been found to be widely distributed in a variety of tissues, and the synthesis of prostaglandins has been demonstrated in many tissues, from the vesicular gland of sheep to the lung tissue of guinea pigs. It has also been established that after "primary" prostaglandins have been formed from fatty-acid precursors they can be converted into other members of the family by changing the primary structure.

The linking of the prostaglandins to polyunsaturated fatty acids provided clues to some of the prostaglandins' important functions. For example, one of the common fatty-acid precursors for prostaglandin is arachidonic acid. The main source of this substance is the phospholipids, which constitute a principal component of the cell membrane. Consequently it appears that the conversion of arachidonic acid to prostaglandin may play an important role in the regulation of the membrane's functions. It seems that the cell membrane is a prime site for the formation of the prostaglandins. Since there is no indication that any cells carry a store of prostaglandin, the likelihood is that the substance is produced by the membrane as needed. The tiny amount of prostaglandin we find in most tissues may therefore represent just the amount produced in the interval between the removal of the tissue and its preparation for extrac-

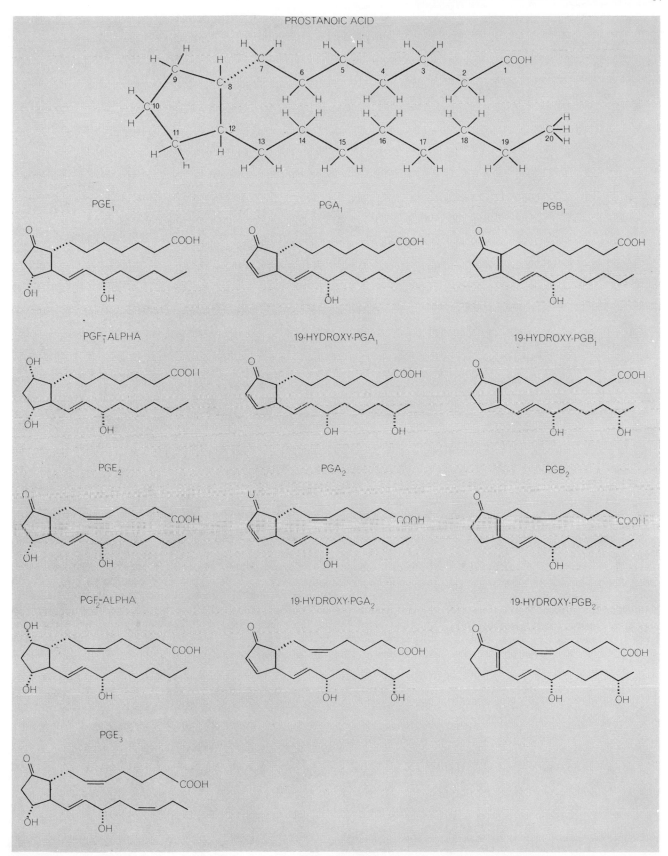

SOME IMPORTANT PROSTAGLANDINS are represented on this page by diagrams of their molecular structure. In general all the prostaglandins are variants of a basic 20-carbon carboxylic (COOH-bearing) fatty acid incorporating a five-member cyclopentane ring (*top*). Slight structural changes are responsible for quite distinct biological effects. Prostaglandins of the 1, 2 and 3 series respectively incorporate one, two and three double bonds.

The molecules designated PGE and PGF are called primary prostaglandins; the PGE structures have an oxygen atom (O) attached to the cyclopentane ring at carbon site 9, whereas the PGF structures have a hydroxyl (OH) group at the same site. Dehydration of a PGE molecule leads to either a PGA or a PGB compound. Broken bonds are those that extend below the plane of the cyclopentane ring. Hydrogen atoms are shown only in top diagram.

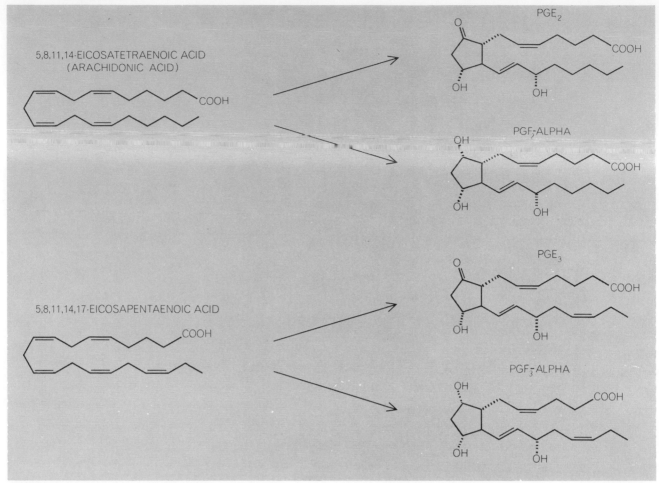

5,8,11,14-EICOSATETRAENOIC ACID
(ARACHIDONIC ACID)

PGE₂

PGF₂-ALPHA

5,8,11,14,17-EICOSAPENTAENOIC ACID

PGE₃

PGF₃-ALPHA

PROSTAGLANDINS ARE SYNTHESIZED in the body from certain polyunsaturated fatty acids by the formation of the five-member ring and the incorporation of three oxygen atoms at certain positions. For example, either PGE₂ or PGF₂-alpha can be formed directly from arachidonic acid (*top diagram at left*), whereas PGE₃ or PGF₃-alpha can be formed from a closely related fatty-acid pre-

tion of the substance. This amount typically is only about one microgram per gram of wet tissue. The seminal fluid of man or the extract from a seminal gland, on the other hand, usually contains 100 times this concentration of prostaglandins. Were it not for this rela-

tively rich source, the prostaglandins might still be awaiting discovery.

In addition to the involvement of the cell membrane, another fascinating question raised by the interrelation of fatty acids and prostaglandins has to do with diseases arising from dietary de-

ficiencies. It has long been known that certain polyunsaturated fatty acids are essential in a mammal's diet. Rats fed a fat-free diet suffer a stunting of growth, inability to reproduce and a characteristic skin lesion. The question arises: Could the administration of prostaglan-

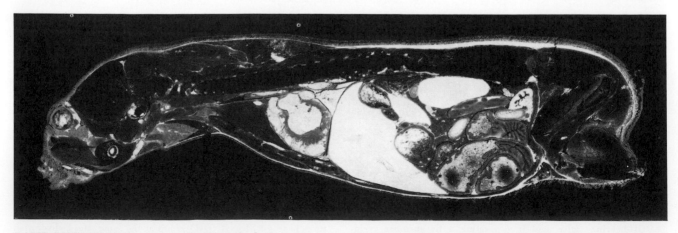

RAPID DISTRIBUTION of prostaglandin in the body is demonstrated by these two autoradiograms made by Bengt I. Samuelsson of the Royal Veterinary College and Eskil Hanson of the Astra Pharmaceutical Company, both in Stockholm. The autoradiograms show the distribution of radioactivity (*light areas*) in a female mouse (*left*) and a male mouse (*right*) 15 minutes after each had

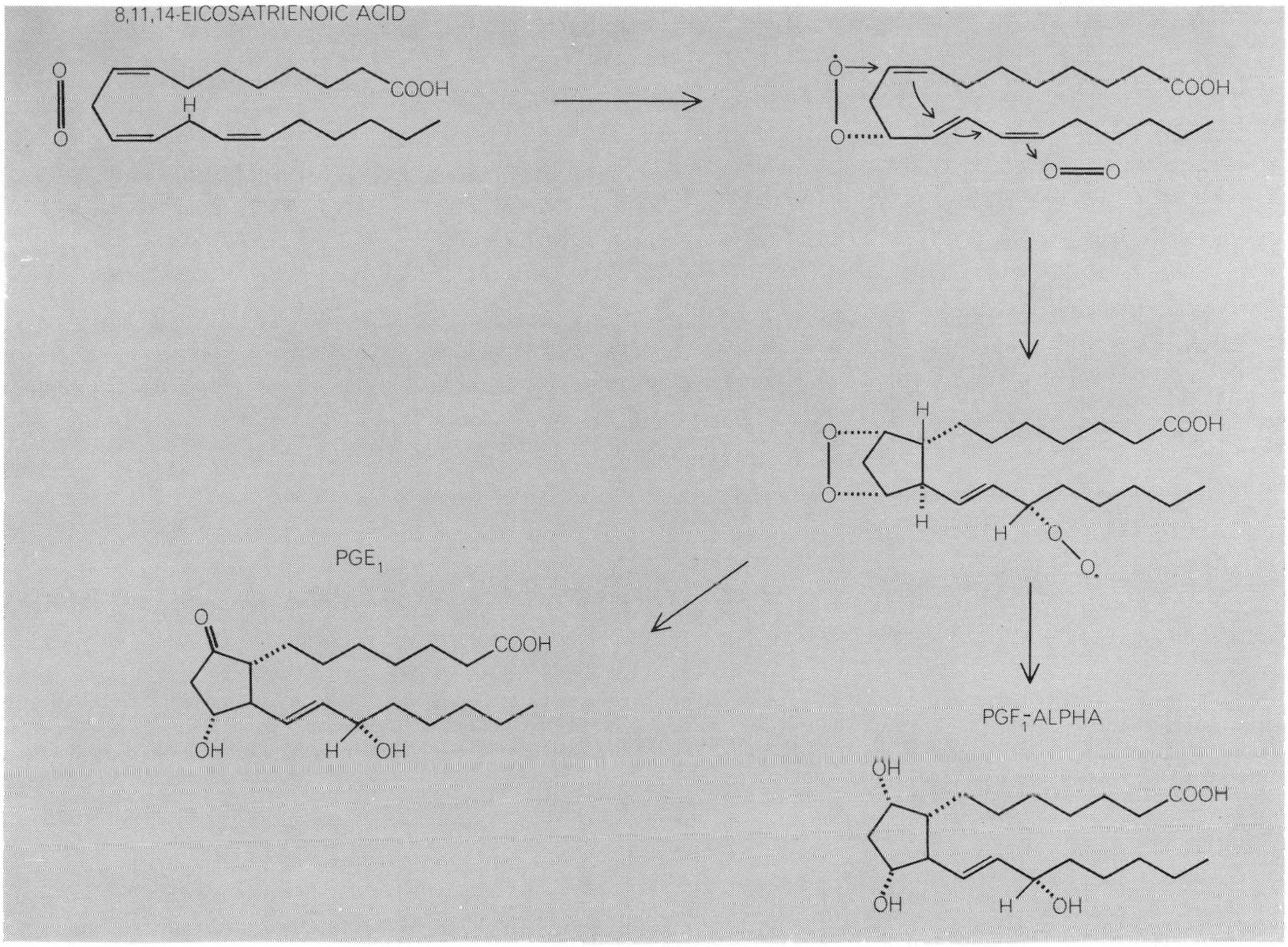

8,11,14-EICOSATRIENOIC ACID

PGE₁

PGF₁-ALPHA

cursor (*bottom diagram at left*). The probable mechanism of this type of biosynthesis is illustrated at the right. Two molecules of atmospheric oxygen provide three new oxygen atoms to a hypo-

thetical intermediate structure, which then leads to either the PGE or the PGF structure. These primary structures can thereafter be converted into other members of the prostaglandin family.

dins remedy the fatty-acid deficiency? James R. Weeks of the Upjohn Company found that treatment of deficient rats with prostaglandins had no effect on the external signs of deficiency. David A. van Dorp and his co-workers in the Netherlands have found a clear connec-

tion, however, between prostaglandins and the fatty acids that are essential in the diet. Only those fatty acids that serve as precursors for the synthesis of prostaglandins are effective in curing rats that have been subjected to a fatty-acid deficiency. This suggests that some

less obvious, perhaps metabolic, aspect of the deficiency disease in rats deprived of the essential fatty acids may be due in part to failure to synthesize necessary prostaglandins.

Once it was recognized that the prostaglandins are formed from fatty acids

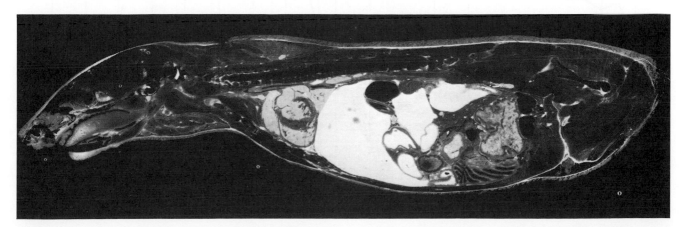

been injected intravenously with a small amount of radioactively labeled PGE₁. High concentrations of the prostaglandin are evident in the liver, the kidney and the subcutaneous connective tis-

sue of both animals and also in the uterus of the female and the thoracic duct of the male. The autoradiograms were made by placing a cross section of each animal on a radiation-sensitive plate.

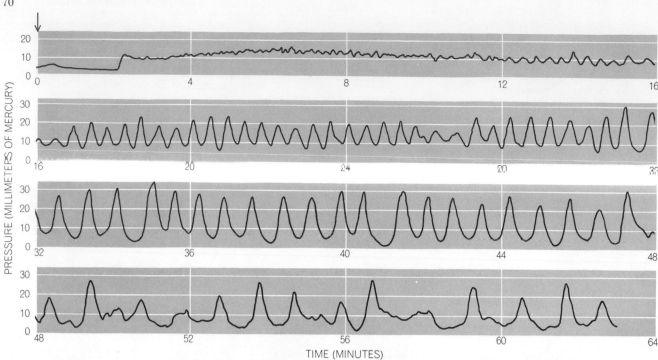

STRIKING EFFECT of prostaglandins on the female reproductive system is evidenced by the fact that uterine contractions can be stimulated by a very low dose of either PGE_2 or PGF_2-alpha. This particular trace, divided into four parts, records pressure changes in a monkey uterus following an injection of PGE_2 at the time indicated by the arrow. Prostaglandins have already been used to facilitate childbearing labor in several thousand women, and there are indications that they may be widely adopted for this purpose.

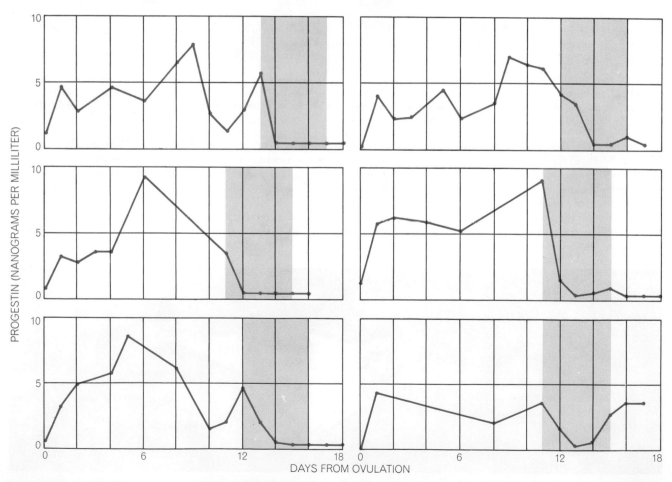

DRAMATIC REDUCTION in the secretion of progesterone by the ovaries of mated female monkeys followed the administration of PGF_2-alpha. In these graphs of the progestin levels in six such monkeys the prostaglandin was administered during the periods indicated by the colored bands. Since progesterone is needed to ensure implantation of a fertilized ovum in the wall of the uterus, there seems to be a strong possibility that prostaglandins may become important agents in helping to control population growth.

it became possible to turn to laboratory production of these scarce and expensive substances for experimental work. It was found that a fatty acid (gamma-linolenic acid) extracted from seeds of the borage plant (a European herb) could be converted into a polyunsaturated fatty acid that was a precursor for a prostaglandin biosynthesis. Enzymes for the biosynthesis were obtained from the seminal vesicular glands of sheep. This resort to artificial biosynthesis produced gram quantities of the prostaglandin and supported studies for several years. Recently chemists at the Upjohn Company and Elias J. Corey and his co-workers at Harvard University developed methods of total synthesis of prostaglandins from more common materials, so that the problem of supply of the substances for research is no longer so acute.

Incidentally, remarkably high concentrations of a prostaglandin isomer were found by Alfred J. Weinheimer and R. L. Spraggins of the University of Oklahoma in a type of coral, the sea whip or sea fan *Plexaura homomalla,* a primitive sessile animal found in coral reefs off the Florida coast. This isomer has been successfully converted to biologically important prostaglandins by Gordon Bundy, William P. Schneider and their co-workers at the Upjohn Company. Apparently prostaglandins once played more elementary roles than they do in mammals, and it is interesting to speculate on their origin and evolutionary development. Polyunsaturated fatty acids are absent in bacteria, but they do turn up in blue-green algae and some higher plants as well as in animals. The prostaglandins may represent an evolutionary step in which, thanks to the availability of molecular oxygen, these complex molecules were formed as agents for the regulatory functions of specialized cells.

The increase in the supply of prostaglandins for experiments has made possible a wide exploration of their effects and powers. These tests have demonstrated that the prostaglandins are among the most potent of all known biological materials, producing marked effects in extremely small doses, and that their existence in the body is remarkably ephemeral. For example, it is found that after a radioactively labeled prostaglandin is injected intravenously into a subject's arm, the injected material is metabolized very rapidly; in as short a time as a minute and a half 90 per cent of the radioactivity is dispersed in

NONMAMMALIAN SOURCE of prostaglandins is represented by the sea whip or sea fan *Plexaura homomalla,* a type of coral found in reefs off the Florida coast. The coral contains high concentrations of a prostaglandin precursor, which has been converted to biologically important prostaglandins by workers at the Upjohn Company. There has been considerable interest in this primitive organism as a starting material for synthesizing prostaglandins.

metabolic products, as is shown in samples of venous blood taken from the other arm. Evidently the body's cells possess a copious supply of catabolic enzymes designed to confine the action of the prostaglandins to particular areas.

The principal interest in the prostaglandins has focused on their remarkable versatility and the wide range of their effects. The effects themselves may be quite specific; for example, one prostaglandin (PGE_2) lowers blood pressure, whereas a closely related member of the family (PGF_2-alpha) raises blood pressure. In general the effects of the prostaglandins are based on certain broad powers: regulation of the activity of smooth muscles, of secretion, including some endocrine-gland secretions, or of blood flow. Through these actions they are capable of affecting many aspects of human physiology, and this accounts

for the great current interest in their pharmacological possibilities.

Particularly striking are the effects on the female reproductive system. An intravenous injection of a very low dose of either PGE_2 or PGF_2-alpha was shown by Marc Bygdeman to stimulate contraction of the uterus. This finding, together with the finding by S. M. M. Karim, now at the Makerere University School of Medicine in Uganda, that prostaglandins are present in the amniotic fluid and in the venous blood of women during the contractions of labor, suggested that the prostaglandins may play an important role in parturition. The substances have been used to facilitate childbearing labor in several thousand women. Infusion of PGE_2 at the rate of .05 microgram per kilogram per minute has been found to induce delivery within a few hours. Oral administration of PGE_2 or PGF_2-alpha has also been reported to be effective. Further studies of the most effective dosage and route of administration of the drug, as well as evaluation in comparison with other methods of inducing labor, may lay a basis for wide adoption of the use of a prostaglandin for this purpose.

The possible use of prostaglandins as agents for abortion and for inducing menstruation (in cases of menstrual failure) is also being investigated. There is evidence that prostaglandins produce these effects by some process more complex than the mere stimulation of uterine contraction. Infusion of PGF_2-alpha in female monkeys after mating has been found by Kenneth T. Kirton to produce a dramatic reduction of the secretion of progesterone by the corpus luteum of the ovary, which is needed to ensure implantation of a fertilized ovum in the wall of the uterus. One suggested explanation is that the prostaglandin may induce regression of the corpus luteum, an event that generally does not occur after an ovum is fertilized. There seems to be a strong possibility that prostaglandins or perhaps synthetic analogues may become important agents in controlling population growth.

In an entirely different area of physiology, the prostaglandins, illustrating their versatility, seem to hold promise for the prevention of peptic ulcers. Andre Robert of the Upjohn Company has shown that the prostaglandin E_1 or E_2 can inhibit gastric secretions in dogs. The dogs were first given a stimulant for secretion (histamine, pentagastrin or food) and then were infused intravenously with one of the prostaglandins. The drug drastically reduced the digestive system's secretion of acid and pepsin throughout the infusion [*see illustration on opposite page*]. It is believed to have produced this effect by changing the chemical activity within the parietal cells of the stomach. Follow-up experiments have shown further that these two prostaglandins can prevent gastric and duodenal ulcers in rats. It seems well established that ulceration of the stomach and the duodenum is generally caused by prolonged exposure of their mucous membrane to gastric juice of high acidity and peptic potency. The stomach normally produces prostaglandins of the E series, and these may serve to regulate gastric secretion under normal circumstances and thus protect the stomach wall against ulceration. If the studies on dogs and rats are confirmed in man, the administration of E prostaglandins may be a helpful treatment for ulcer patients who lack this normal protection.

A number of other biological activities of prostaglandins suggesting possible pharmacological uses are under study. Among these potential applications are:

1. Opening the airways to the lungs. It has been shown experimentally in asthmatic subjects that breathing an aerosol preparation of the prostaglandin E_1 can improve the airflow by relaxing the smooth muscle of the bronchial tubes.

2. Regulating blood pressure. In experiments on dogs it has been found that infusion into the kidney artery of a very low dose of PGE_1 or another prostaglandin called PGA_1 produces an increase in the flow of urine and the excretion of sodium ions. This finding and others indicate that prostaglandins in the body normally help to regulate blood pressure. Tests have demonstrated that the infusion of PGA_1 can lower blood pressure in patients with essential hypertension.

3. Clearing the nasal passages. Applied topically to the nose, PGE_1 has been found effective in widening the passages by constricting the blood vessels.

4. Regulating metabolism. PGE_1 has been shown to counteract the effects of

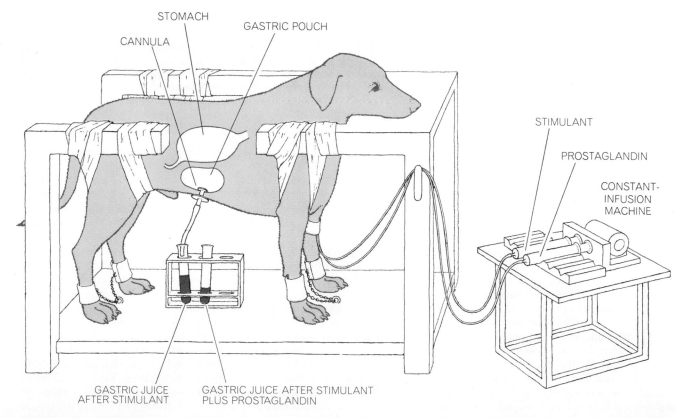

INFLUENCE OF PROSTAGLANDINS on the gastric secretions responsible for peptic ulcers was studied by Andre Robert of the Upjohn Company, who used dogs as experimental animals. The dogs were first given an intravenous infusion of a stimulant for secretion and then one containing the prostaglandin in addition. The output of gastric juice was reduced during the latter treatment.

many hormones in stimulation of metabolic processes, for example the breakdown of lipids in fatty tissues.

Many items of evidence now point strongly to the likelihood that the prostaglandins play a fundamental and critical role in the physiology of mammals. Production and release of these substances can be evoked by stimulation of nerves. Recent experiments by a group at the Royal Caroline Institute in Sweden (Per Hedqvist, Åke Wennmalm, Lennart Stjärne and Samuelsson) demonstrated a clear interrelation between the prostaglandins and the substances involved in the transmission of nerve impulses.

Testing preparations of isolated organs (the spleen of the cat and the heart of the rabbit), they found that infusion of the prostaglandin E_2 markedly inhibited the release of norepinephrine in response to nerve stimulation. In other words, the prostaglandin acted as a "brake" on the nerves' action in inciting release of the norepinephrine, a hormone that plays a key part in the transmission of nerve impulses in the sympathetic nervous system. When the preparation was infused, on the other hand, with an inhibitor of the formation of prostaglandins, the consequent removal of the brake from the tissues resulted in an abnormally large release of norepinephrine by nerve stimulation.

These and other results suggest that prostaglandins, playing a negative-feedback role, are part of the mechanism that normally controls transmission in the sympathetic nervous system. The inhibition of the production of prostaglandins in the body obviously invites further investigation. J. R. Vane and his co-workers at the Institute of Basic Medical Science of the Royal College of Surgeons recently conducted such a study and obtained results that suggest the anti-inflammatory action of aspirin and certain other agents may be explainable on the basis that they block the synthesis of prostaglandins.

Prostaglandin formation has been noted in many other apparently unrelated systems. For example, prostaglandins are formed by the lungs during anaphylaxis, by the kidney when its blood supply is restricted, by the surface of the brain when peripheral sensory nerves are stimulated, by the skin during human allergic contact eczema and during certain experimental inflammatory conditions. These diverse situations suggest that prostaglandins may play a fundamental role not only in normal physi-

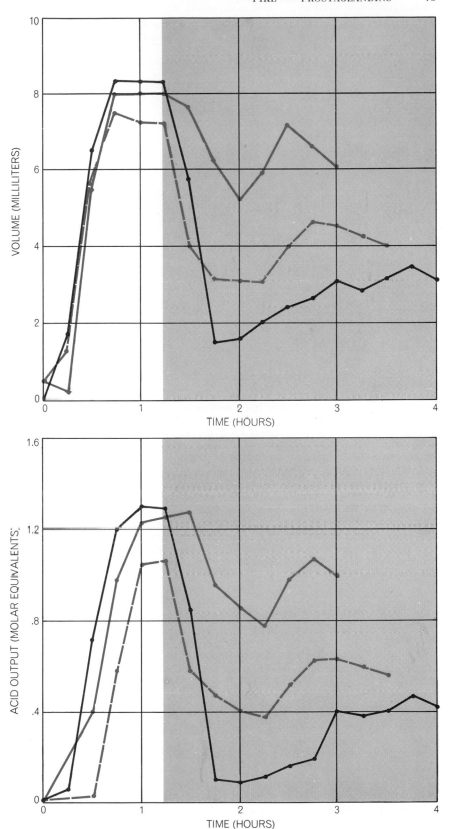

ABRUPT REDUCTIONS in gastric secretion were observed in dogs as a result of the administration of PGE$_2$. In this particular experiment the secretion of gastric juice was stimulated by the intravenous infusion of the histamine dihydrochloride at a dose rate of one milligram per hour. When PGE$_2$ was added to the infusion (*colored band*), both the volume (*top graph*) and the acidity (*bottom graph*) of the gastric secretion were reduced proportionately to the dose of PGE$_2$. In each graph solid colored curve represents a PGE$_2$ dose rate of 12.5 micrograms per minute, broken colored curve represents a dose rate of 18.75 micrograms per minute and black curve represents a dose rate of 25 micrograms per minute.

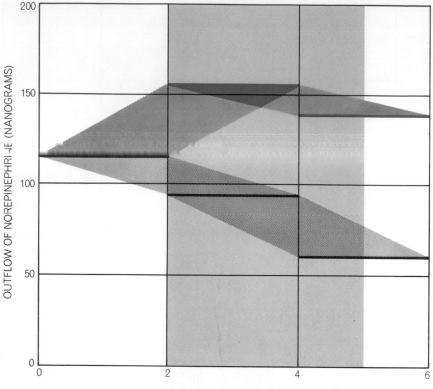

MINUTES AFTER START OF NERVE STIMULATION

INTERRELATION of the prostaglandins with the substances involved in the transmission of nerve impulses was demonstrated by a group of investigators at the Royal Caroline Institute in Sweden. In general they found that PGE_2 acted as a "brake" on the nerve's action in inciting the release of norepinephrine, a hormone that plays a key part in the transmission of nerve impulses in the sympathetic nervous system. In this experiment nerve stimulation of an isolated rabbit heart over a period of six minutes caused an outflow of norepinephrine, the amount of which was measured over three two-minute intervals and declined somewhat over this time span (*black lines*). In separate tests a known inhibitor of the formation of prostaglandin was infused during a three-minute period from the beginning of the second interval, and caused an increased outflow of norepinephrine during the second and third intervals (*colored lines*). Presumably the inhibitor removed the normal prostaglandin brake, resulting in an abnormally large release of norepinephrine.

ological functions but also in certain pathological conditions.

It appears that the prostaglandins are leading us to the meeting ground of the two main systems of communication in the body: the hormones and the nerves. They meet most directly, it seems, in the membrane of the cell. The cell membrane is indeed coming to be recognized as a most important crossroads for a great number of vital activities. Its complex biochemical apparatus controls the selective transport of all kinds of materials into and out of the cell, governs many aspects of the formation of products within the cell and is responsible for cell-to-cell communication, particularly the transmission of signals across the synapses in the nervous system, which is accomplished by chemical means. The discovery of the prostaglandins now lends new significance to the fact that the cell membrane is made up largely of phospholipids, along with proteins as the other main component. Since phospholipids supply the building materials—fatty acids—for the formation of prostaglandins, it can be supposed that the prostaglandins have a great deal to do with regulation of the functioning of the cell membrane itself. The membrane contains a common medium of communication—the chemical medium—for the chemical agents of the endocrine system and the nervous system. It will be fascinating, therefore, to explore the roles of the prostaglandins in mediating this communication, particularly with regard to their possible connection with the formation of cyclic AMP, the "second messenger," which is increasingly being recognized as a key actor in important functions such as translating the messages of specific hormones and regulating the growth and differentiation of cells.

9

ANALGESIC DRUGS

MARSHALL GATES
November 1966

For severe pain the most widely used drug is still morphine. Other useful drugs have been found, however, in the course of the search for substances with morphine's good qualities but not its bad ones

Early man, omnivorous and inquisitive, had discovered the use of opium by the time of the earliest written records. Babylonian and Egyptian writings contain many references to the value of opium preparations for the relief of pain. Hippocrates, Dioscorides, Galen and other early physicians formed a regard amounting to veneration for opium's almost miraculous power to "lull all pain and anger and bring relief to every sorrow." With the progress of medicine the respect of physicians for the drug increased. Thomas Sydenham, the 17th-century pioneer of English medicine, wrote: "Among the remedies which it has pleased Almighty God to give to man to relieve his sufferings, none is so universal and so efficacious as opium." Today, although opium is clearly no longer regarded as a universal remedy,

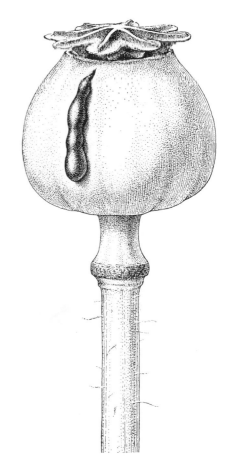

OPIUM POPPY (*Papaver somniferum*) is the source of morphine. The technique of obtaining the raw opium is to make a cut in the unripe seed capsule of the plant (*enlarged drawing at right*). The rubbery exudate that oozes from the cut is then collected and dried.

RAW OPIUM contains some 20 alkaloid substances, one of which is morphine. A typical yield of morphine from opium is 10 percent. The drug was first isolated from opium in 1805.

its active principle, morphine, is still the drug that is most commonly prescribed by physicians for the relief of severe pain.

The defects of the drug have long been well known. The ancients could scarcely have been unaware of the ad-

dicting property of opium; it appears, however, that Thomas De Quincey's *Confessions of an English Opium Eater*, published in a magazine in 1821, was the first detailed description of opiate addiction. De Quincey was a brilliant man of letters who developed an addiction to

laudanum (an opium preparation) while he was a student at Oxford. Opiate addiction is too vast a subject to be discussed in detail here. For our purposes it is enough to note that prolonged use of an opiate generally leads to a physiological and psychological dependence on the drug, and this dependence is shown by the onset of highly unpleasant and sometimes dangerous symptoms when use of the drug is stopped. In addition to the opiates' addicting property, they have a depressing effect on respiration. It is no wonder, then, that in this century chemists and clinical investigators have conducted a tireless and expensive search for an analgesic that would be as potent as morphine but free of its dangerous effects.

Opium is obtained from the opium poppy (*Papaver somniferum*) by scarifying the unripe seed capsule and collecting and drying the exudate. The dried material, pressed into bricks, is the opium of commerce. The substance is a complex mixture containing at least 20 alkaloids. Morphine, which constitutes about 10 percent of opium and is primarily responsible for its physiological effects, was first isolated in 1805 by Friedrich Sertürner, an apothecary's assistant in Paderborn, Germany. The isolation of this alkaloid was the beginning of alkaloid chemistry, which has yielded many important medicinal substances, including (to cite a recent example) the tranquilizing and blood-pressure-reducing drug reserpine, extracted from the Indian snakeroot (*Rauwolfia serpentina*).

Morphine itself became an object of increasingly intensive study. The unraveling of its structure is a fascinating story. The difficulty involved in the work is apparent from the fact that the correct basic structure of the alkaloid was not suggested until 120 years after the isolation of the drug. The workers who ascertained the structure in 1925 were John M. Gulland and Robert Robinson, then at the University of Manchester. Even at that time the subtleties of morphine's three-dimensional structure were largely unresolved.

The chemical investigations of morphine made available a number of substances closely related to it structurally. Many of them—including codeine (also occurring in opium), heroin, Dilaudid and Bentley's compound—were physiologically active [*see illustration on opposite page*]. Bentley's compound, recently prepared by the British chemist K. W. Bentley, is of particular interest. In tests on small animals it has proved to be 10,000 times more potent than

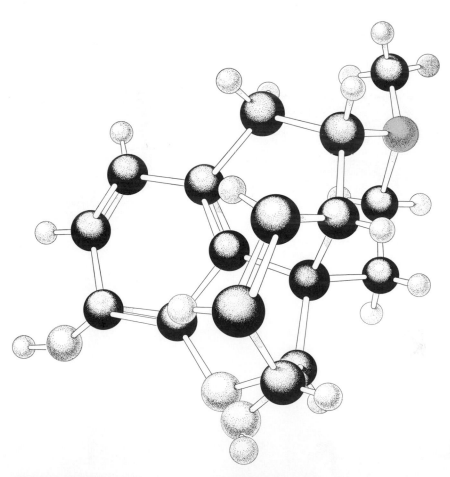

THREE-DIMENSIONAL MODEL of the morphine molecule shows the characteristic structure of the drug. The black balls in the model are carbon atoms, the dark-colored ball is a nitrogen atom, the light-colored balls are oxygens and the gray balls are hydrogens.

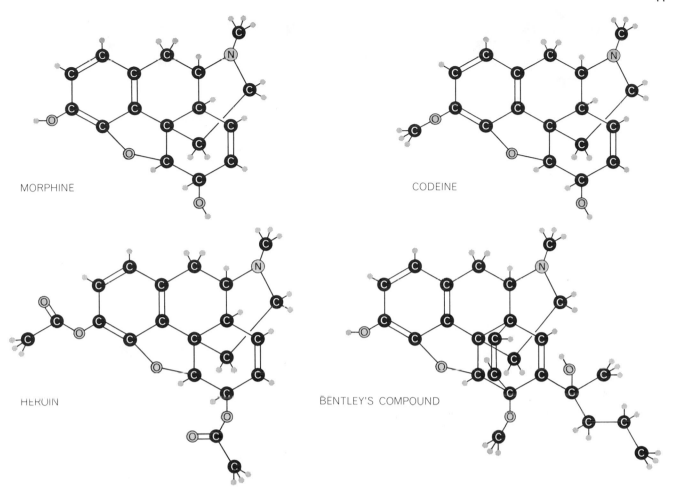

MORPHINE

CODEINE

HEROIN

BENTLEY'S COMPOUND

CHEMICAL STRUCTURES of morphine and several of its close relatives are markedly similar. Codeine, heroin and Bentley's com- pound were produced by altering the structure of morphine. In these two-dimensional diagrams the hydrogens are small gray dots.

morphine, and some of Bentley's preparations have been used to subdue elephants on African game preserves at doses of one milligram per elephant! Unfortunately the Bentley compounds have disappointed early hopes that they might, at analgesic doses, produce less respiratory depression than morphine, and they have also given evidence that they are dangerously addicting.

Of the many fragments or synthetic assemblies of fragments of the morphine molecule that have been studied, only a few have shown a significant degree of analgesic activity [see top and middle illustrations on next page]. And in these compounds, whose structure is sometimes only vaguely related to that of morphine, addiction liability seems to parallel analgesic activity. Only the substances with very mild analgesic activity are nonaddicting.

What general principles might be drawn from these analytic studies? In 1955 Nathan B. Eddy of the National Institutes of Health, examining the experimental results, attempted to iden-

tify the structural features that might account for the analgesic activity of a drug of the morphine type. The most active drugs, he noted, all seemed to be characterized by (1) the presence of a tertiary or trisubstituted amine with a small group attached to the nitrogen atom, (2) a central carbon atom, none of whose valences was occupied by a hydrogen atom, (3) a phenyl or closely analogous group attached to the central carbon atom and (4) separation of the tertiary or trisubstituted amino group from the central carbon atom by an intervening string of two carbon atoms. Events later showed that this categorical description missed the mark. Since 1955 it has been found that all four of these criteria can be violated without destroying analgesic activity.

Although a satisfactory theory of analgesic structure or action still eludes us, experimenters have developed a number of useful synthetic analgesics related to the morphine model. The oldest and perhaps the best-known is pethidine (also called meperidine, Demerol and by some 40 other names). It was

synthesized in 1939 by Otto Eisleb of Germany. Less potent than morphine, pethidine is widely used for the relief of postoperative pain and in obstetrics. It was acclaimed originally as a nonaddicting drug but was soon found to have dangerous addiction liability—a constantly recurring theme in the field of analgesic drugs. (Even heroin was originally introduced as a nonaddicting substitute for morphine.) Analgesics related to pethidine have since been prepared in great variety and in widely varying potency; pethidine itself, however, remains the most widely used of these substances.

Another family of synthetic analgesics, less closely related structurally to morphine, is the one represented by methadone, which has attracted interest recently as a relatively harmless substitute for heroin in experiments on the management of addicts. Still another, recently developed by Everett L. May of the National Institutes of Health, is the group of substances called benzomorphans. The best-known member of this group, phenazocine, is seven to 10

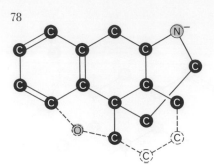

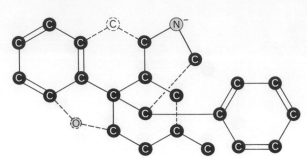

STRUCTURAL ALTERATIONS of the morphine molecule have produced several fragments that have proved to be analgesically active. Some, such as the one at left, bear a close structural relation to morphine; others, such as that at right, a less apparent relation.

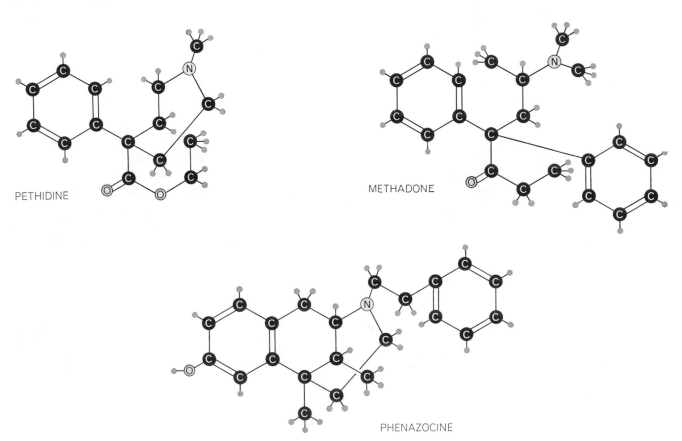

PETHIDINE

METHADONE

PHENAZOCINE

SYNTHETIC ANALGESICS have been developed by various experimenters. Pethidine, the oldest, most closely resembles the structure of morphine. Methadone is less like morphine structurally. Phenazocine, a recent development, is the best-known member of a group of substances called benzomorphans. Phenazocine is more effective than morphine against pain but is highly addicting.

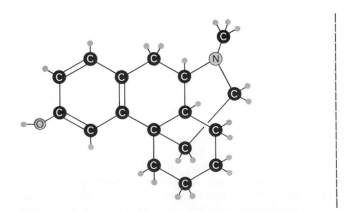

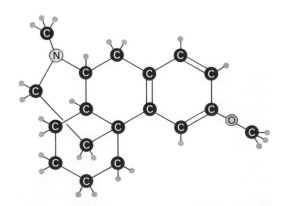

MIRROR-IMAGE MOLECULES proved to have notably different effects. The synthetic drug levorphan (*left*), with a morphine-like structure, is more potent than morphine against pain and also is strongly addicting. Dextrorphan (*right*), the right-handed form of the molecule, is neither analgesic nor addicting. The difference indicates the importance of molecular geometry in analgesic activity.

times more potent than morphine as an analgesic and possesses some other advantages, but like morphine it is dangerously addicting.

That analgesic activity depends quite specifically on molecular geometry is strikingly illustrated by the synthetic drug levorphan, a member of the group called the morphinans, which are closely akin to morphine in structure [see *bottom illustration on opposite page*]. Levorphan is several times more potent than morphine and is strongly addicting. Curiously the mirror image of this molecule (that is, the dextrorotatory, or right-handed, form) has no analgesic activity or addiction liability. (Slightly modified by the substitution of a methyl ether group for the hydroxyl group in the molecule, the dextrorotatory form has been found to be a useful cough-suppressing drug without narcotic properties and seems likely to supplant codeine for that purpose.) The geometric specificity exhibited by levorphan has been demonstrated to apply also to morphine, whose molecule is levorotatory. There is no mirror image of morphine in nature, but the Japanese chemist Kakuji Goto has synthesized a dextrorotatory counterpart of the molecule, using a method developed in my laboratory at the University of Rochester, and found that the mirror form possesses none of morphine's properties.

I have cited only a few representative examples of the synthetic analgesics; there are now a great number of them. Many have certain advantages over morphine, but none has supplanted morphine as the mainstay of physicians for the control of severe pain. The continued preference for morphine stems in part from the confidence developed from long experience with the drug; fundamentally, however, the fact remains that there is not yet available a fully satisfactory substitute that avoids morphine's principal shortcomings. Indeed, the properties of analgesic activity, addiction liability and depression of respiration seem to go together, and until recently there appeared to be no hope of separating one of these properties from another.

We now come to a most interesting development in the story. Its beginning actually goes back to a rather old discovery. Nearly 50 years ago Julius Pohl of the University of Breslau reported he had prepared a substance that suppressed the respiratory depression produced by morphine. The substance was a modification of codeine: in place of the methyl group (CH_3) attached to the nitrogen atom in codeine he substituted the unsaturated allyl group ($CH_2CH{=}CH_2$), forming a compound called N-allyl-nor-codeine. This drug, administered to an experimental animal that had received a narcotic dose of morphine, reversed morphine's respiration-depressing effect.

The significance of Pohl's finding was overlooked at the time, and it lay largely unnoticed for 27 years. In 1942 John Weijlard and A. E. Erickson at the laboratories of Merck & Co. took up this old lead and performed the same operation on the morphine molecule itself, substituting the allyl group for the methyl group. Their product, nalorphine, proved to be capable of counteracting most of the physiological effects of morphine. It not only reversed the respiratory depression but also acted as an antidote for acute morphine poisoning, and when administered to a narcotics addict it rapidly precipitated a severe withdrawal syndrome. In addition, screening tests on small animals indicated that nalorphine, in contrast to morphine, had no analgesic effect.

A surprising finding emerged, however, when nalorphine was administered to human patients. In 1954 Henry K. Beecher and Louis Lasagna conducted an experiment to examine the possibility that nalorphine, given with morphine, might reduce morphine's respiratory depression without weakening its analgesic activity. At the Massachusetts General Hospital they gave one group of patients both morphine and nalorphine and gave a control group only nalorphine. Unexpectedly nalorphine alone proved to be as effective in relieving pain in the control group as the combination of morphine and nalorphine was in the test group. This finding was quickly confirmed in experiments by Arthur S. Keats and Jane R. Telford of the Baylor University Medical School. Harris Isbell at the Addiction Research Center of the National Institute of Mental Health then proceeded to demonstrate by further tests with human subjects that nalorphine is not addicting in man.

For the first time analgesia had been divorced from addiction liability. The new drug, however, produced hallucinations when given at analgesic doses. Perhaps for this reason it did not immediately precipitate the rash of new investigations that might have been expected to follow so promising a lead. For some time only Keats and his group at Baylor actively pursued the study of nalorphine and related morphine antagonists.

In 1958 Sydney Archer and his coworkers at the Sterling-Winthrop Institute for Therapeutic Research undertook to seek nonaddicting analgesics among morphine antagonists structurally related to the benzomorphans (represented by phenazocine). Their attack was based on the supposition that the D'Amour-Smith rat-tail-flick test, one of the most widely used means of determining the analgesic effectiveness of drugs in animals, was at least as good a test for addiction liability as for analgesic activity. Therefore they reasoned that any substance active in the test should be discarded. In the test an intense beam of light is used as the pain stimulus. Normally when the light is focused on a rat's tail the animal quickly flicks its tail out of the beam; this reaction is delayed, however, if the rat has been given an analgesic drug. The test has been found to give a good measure of a drug's analgesic potency. Moreover, the results of this test on rats usually correlate well with the clinical effectiveness of a drug in man (although they failed to do so in the case of nalorphine).

From the benzomorphans Archer and his group succeeded in preparing a number of derivatives that combine analgesic activity with some degree of antagonism to morphine. Among these products is cyclazocine, an extremely interesting substance that is one of the most potent morphine antagonists discovered so far [see *illustration on next page*].

In January, 1959, our laboratory independently began a search for a nonaddicting drug among the morphine antagonists. We thought it might be possible, by modifying the allyl group in the nalorphine molecule, to transform it into a more acceptable analgesic, although perhaps somewhat less active as an antagonist of addicting drugs. As a replacement for the allyl group we chose the cyclopropylmethyl group, which is well known to affect the chemical and physical properties of a substance in much the same way as the allyl group (and which is the distinguishing feature of cyclazocine). Thomas Montzka, a graduate student working in our laboratory, prepared a number of cyclopropylmethyl derivatives of morphine-type drugs, and these substances were screened in small animals for us by Louis S. Harris, then at the Sterling-Winthrop Institute, and by Charles A.

STRUCTURE	NAME	RELATIVE ACTIVITY
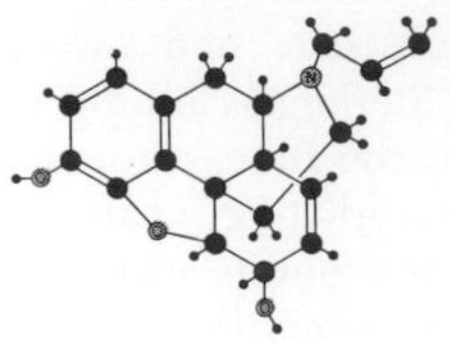	NALORPHINE	1.00
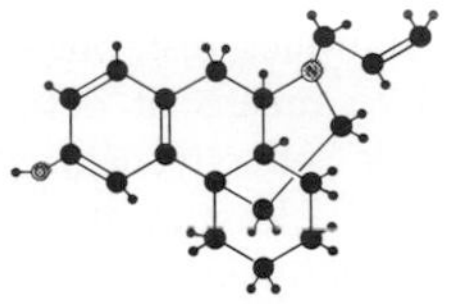	LEVALLORPHAN	2.6
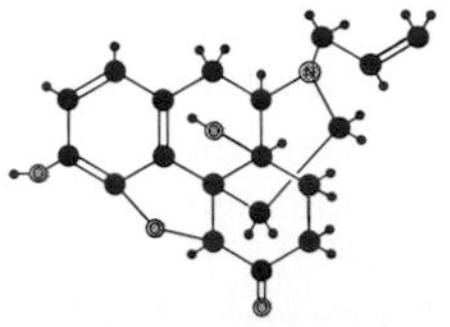	NALOXONE	18
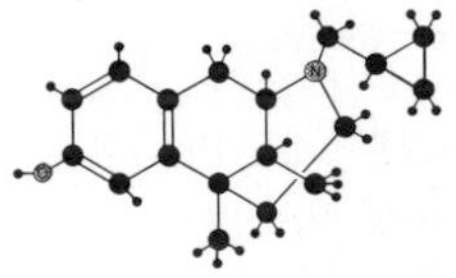	CYCLAZOCINE	28
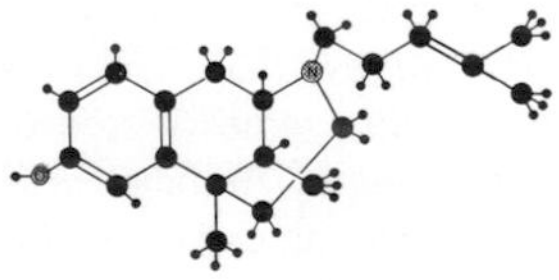	PENTAZOCINE	1/30
	N-CYCLOPROPYLMETHYL-NOR DIHYDROMORPI IINONE	10
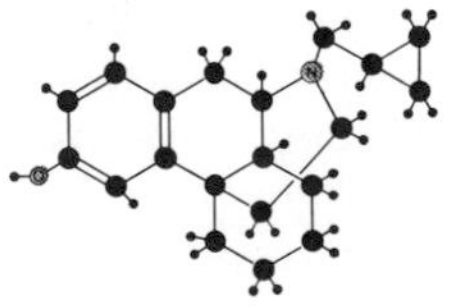	CYCLORPHAN	4
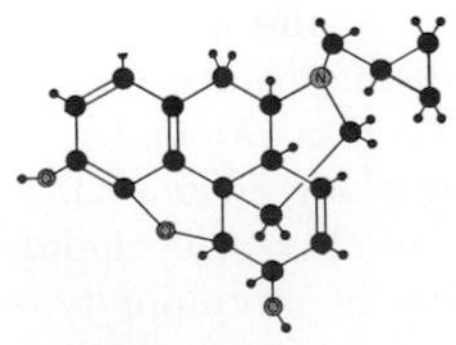	N-CYCLOPROPYLMETHYL-NOR MORPHINE	3

ANTAGONISTS TO MORPHINE suppress the addicting effects of the drug without impeding its analgesic effects. Indeed, several of them have analgesic properties in addition to their counteraction against morphine, as shown in the column headed "Relative activity."

Winter of the Merck Institute for Therapeutic Research.

The new series of substances proved to be highly active antagonists of morphine and pethidine. We selected one, which we call cyclorphan, for further study. It is a considerably more potent morphine antagonist than nalorphine and appears to have pharmacological properties quite similar to those of cyclazocine.

Both cyclazocine and cyclorphan have been found to be capable, at very low doses, of precipitating the withdrawal syndrome in monkeys that have been addicted to narcotics and also in human addicts. This capability can be taken as good evidence that the drugs will not produce addiction in man. William R. Martin, who has studied the addiction liability of cyclazocine with subjects at the Addiction Research Center of the National Institute of Mental Health, reports that it generates a mild, typical dependence that does not seem very likely to lead to serious consequences.

Cyclazocine and cyclorphan have also been examined clinically, cyclazocine extensively, cyclorphan in preliminary fashion. Both drugs have been found to be potent analgesics, perhaps 30 to 50 times more potent than morphine on a weight basis. Both, however, produce hallucinations, much less commonly than nalorphine does but still frequently enough to raise some question as to their acceptability as drugs for general use.

Perhaps the most promising of the new analgesics discovered so far is pentazocine, one of the morphine antagonists prepared from the benzomorphans by workers at the Sterling-Winthrop Institute. Although it is only weakly antagonistic to morphine, pentazocine has been shown in exhaustive clinical tests to be devoid of addiction liability. Yet it is as effective as morphine in controlling postoperative pain, labor pains, the pains of terminal cancer, cardiac pain and traumatic pains. Thus pentazocine is the first really potent analgesic that has proved in practice to be nonaddicting. It does, however, produce respiratory depression, and so it cannot be said to be the last word.

What can be said with reasonable confidence is that, in view of the recent demonstrations that analgesic potency can be separated from addiction liability, it seems not too much to hope that we shall soon have an ideal analgesic: effective, free of side effects such as respiratory depression and hallucinogenic activity—and nonaddicting.

10

BARBITURATES

ELIJAH ADAMS
January 1958

*They are among the most useful of all drugs. In small
doses they act as sedatives; in larger doses they
induce sleep; in still larger doses they are able to
produce deep anesthesia*

The barbiturates are the most versatile of all depressant drugs. They can produce the whole range of effects from mild sedation to deep anesthesia—and death. They are among the oldest of the modern drugs. Long before reserpine and chlorpromazine, the barbiturates were being used as tranquilizers. Indeed, phenobarbital has been called "the poor man's reserpine." Had phenobarbital been introduced five years ago instead of half a century ago, it might have evoked the same burst of popular enthusiasm as greeted Miltown and its fashionable contemporaries. Not that the vogue of the barbiturates is any less spectacular, in terms of total production and consumption. The U. S. people take an estimated three to four billion doses of these drugs per year, on prescription by their physicians. The barbiturates rank near the top of the whole pharmacopoeia in value to medicine. They are also a national problem.

It was nearly a century ago that a young assistant of the great chemist August Kekulé in Ghent made the first of these compounds. In 1864 this young man, Adolf von Baeyer, combined urea (an animal waste product) with malonic acid (derived from the acid of apples) and obtained a new synthetic which was named "barbituric acid." There are several stories about how it got this name. The least apocryphal version relates that von Baeyer and his fellow chemists went to celebrate the discovery in a tavern where the town's artillery garrison was also celebrating the day of Saint Barbara, the saint of artillerists. An artillery officer is said to have christened the new substance by amalgamating "Barbara" with "urea."

Chemists proceeded to produce a great variety of derivatives of barbituric acid. The medical value of the substances was not realized, however, until 1903, when two other luminaries of German organic chemistry, Emil Fischer and Joseph von Mering, discovered that one of the compounds, diethylbarbituric acid, was very effective in putting dogs to sleep. Von Mering, it is said, promptly proposed that the substance be called veronal, because the most peaceful place he knew on earth was the city of Verona.

Within a few months of their report, "A New Class of Sleep-Inducers," physicians in Europe and the U. S. took up the new drugs enthusiastically. More and more uses for them were discovered. Veronal (barbital) was soon followed by phenobarbital, sold under the trade name Luminal. In all, more than 2,500 barbiturates were synthesized in the next half-century, and of these some two dozen won an important place in medicine. By 1955 the production of barbiturates in the U. S. alone amounted to 864,000 pounds—more than enough to provide 10 million adults with a sleeping pill every night of the year.

As is true of most drugs, we still do not know how the barbiturates work or exactly how their properties are related to their chemistry. The basic structure is a ring composed of four carbon atoms and two nitrogens [*see diagrams on page 84*]. Certain side chains added to the ring increase the drug's potency; in some instances the addition of a single carbon atom transforms an inactive form of the compound into an active one. Empirical analysis of the thousands of barbiturates has given us some practical information about relations between structure and activity. But by and large the mode of action of the drugs is an unsolved problem.

We know a great deal, however, about the action itself. We can follow it best by examining the successive stages of the drugs' depressant effect on the central nervous system [*see "Anesthesia," by Henry K. Beecher; SCIENTIFIC AMERICAN, January, 1957*]. From this standpoint the barbiturates can be considered together as a group, for the differences among them are not fundamental and concern such matters as the speed and duration of the effect.

In small doses these drugs are sedatives, acting to reduce anxiety and to relieve psychogenic disorders—for example, certain types of hypertension and gastrointestinal pain. In this respect the barbiturates are yesterday's tranquilizers. They have now taken second place in popularity as sedatives to the newer tranquilizers.

At three to five times the sedative dose, the same barbiturate produces sleep. Barbiturates are still by far the most widely used drugs for this purpose, as millions of us know. Hardly any hospital patient has escaped his yellow or red capsule at evening, for it is an article of clinical faith that the patient needs chemical assistance to achieve sleep in his new environment. Another large block of users are the chronic travelers by plane and train. Finally there are the thousands of sufferers from insomnia who take the drugs habitually.

Many persons find barbiturate-induced sleep as refreshing as natural sleep. Others awake with a hangover, feeling drowsy, dizzy and suffering a headache. Tests show that, with or without symptoms, the barbiturates reduce efficiency: six to eight hours after a sleeping dose of sodium pentobarbital (Nembutal) the subjects perform below par on mental and memory tests. The various drugs act differently as sleep-

	FORMULA	NAMES	CHIEF USES	DURATION
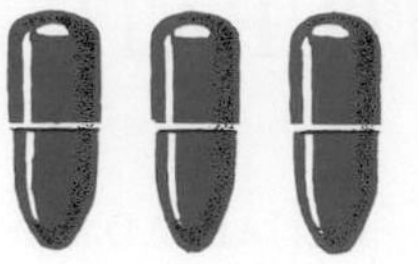	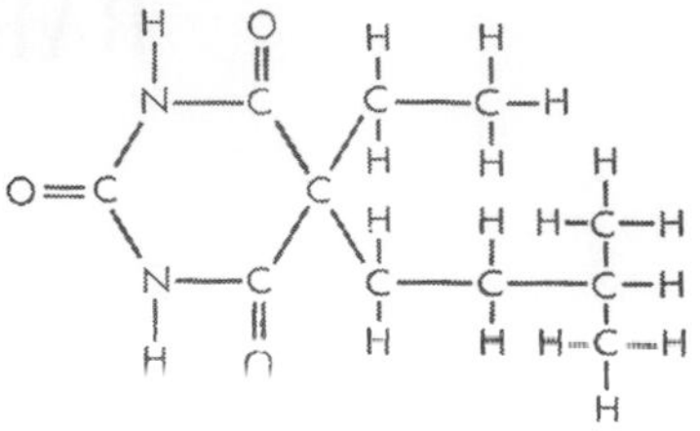	AMOBARBITAL (AMYTAL)	SEDATIVE HYPNOTIC	INTER-MEDIATE
	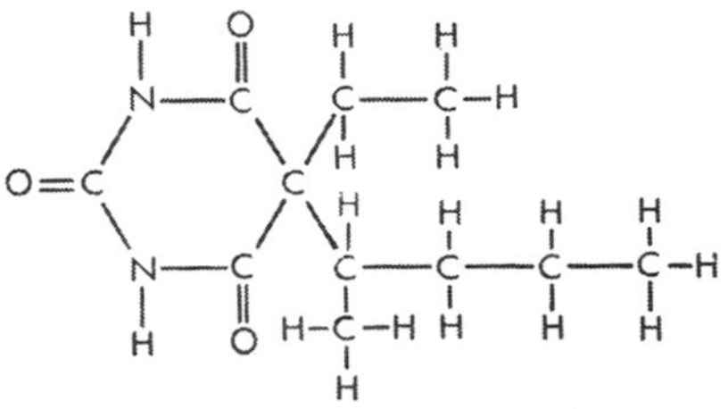	PENTOBARBITAL (NEMBUTAL)	HYPNOTIC	SHORT-ACTING
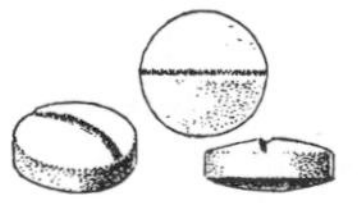	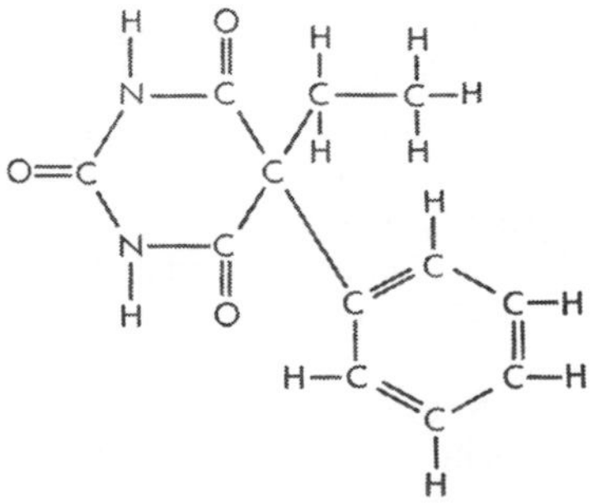	PHENOBARBITAL (LUMINAL)	HYPNOTIC SEDATIVE ANTICONVULSANT	LONG-ACTING
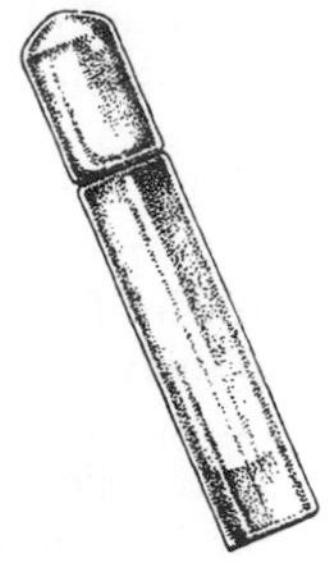	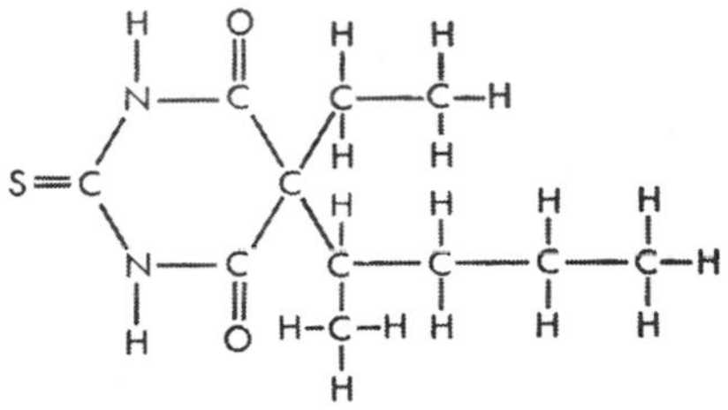	THIOPENTAL (PENTOTHAL)	ANESTHETIC	ULTRA SHORT-ACTING
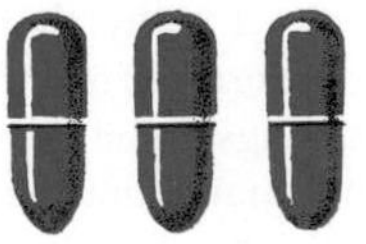	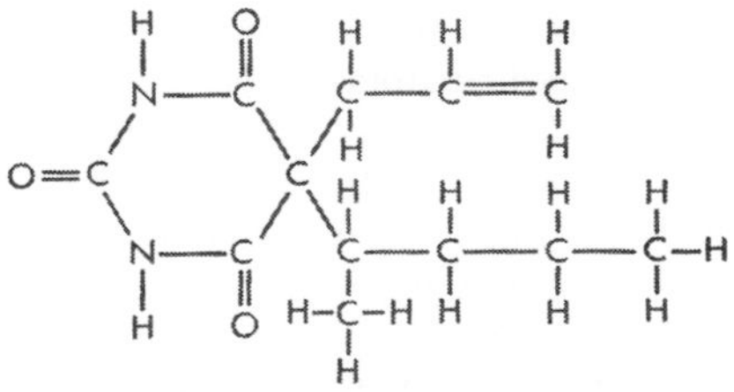	SECOBARBITAL (SECONAL)	HYPNOTIC	SHORT-ACTING

FIVE BARBITURATES are depicted at the left as they are commonly manufactured. As shown in the column of chemical structures, four of these drugs differ only in the chains of atoms attached to the carbon atom at the right side of their basic ring structure. The permutability of the basic barbiturate structure makes possible variations in speed and duration of its effect on the body.

producers. Some last for only three hours or less, others for six hours or more. The shorter-acting barbiturates (sodium pentobarbital, secobarbital) are appropriate for insomniacs who have trouble falling asleep; the longer-acting ones (barbital, phenobarbital) for people who go to sleep easily but awake after four to six hours. The latter drugs, however, are more likely to produce a hangover.

In large doses a barbiturate acts as an anesthetic. Not only does the patient become unconscious, but his spinal cord reflexes are depressed so that the muscles are relaxed and manageable for surgery. Like the gaseous anesthetics, the barbiturates depress the cerebral cortex first, then lower brain centers, next the spinal cord centers and finally the medullary centers controlling blood pressure and respiration.

The fast-acting barbiturates produce anesthesia more rapidly than ether: the patient passes from the waking state to anesthetic coma in a few moments. Sodium thiopental is the most widely used. It has important advantages over a gaseous anesthetic such as ether. Injected intravenously, it works rapidly, avoids the sense of suffocation, requires no special equipment, is free from the explosion hazard and from respiratory complications. A barbiturate anesthetic has, however, an outstanding disadvantage: the dose necessary for good muscular relaxation may seriously reduce oxygen supply to the tissues by depressing the brain center that drives respiration. Consequently for a long operation the barbiturate is often combined with a gaseous anesthetic; for a short one the dose is reduced and combined with a specific muscle-relaxing drug that has no brain-depressant action.

There are two other interesting uses of the barbiturates. One of them is in the field of psychology. As Henry Beecher observed in his article on anesthesia in SCIENTIFIC AMERICAN, an anesthetic can provide "planned access to levels of consciousness not ordinarily attainable except perhaps in dreams, in trances or in the reveries of true mystics." During World War II the barbiturates, particularly thiopental, were used for analysis and therapy for many thousands of GIs, who by this means relived and verbalized traumatic battle experiences which had been buried beyond voluntary recollection. The inhibition-relieving action of these drugs has also been employed by the police—in which application the press has given them the name of "truth serum," although they

BARBITURIC ACID (*right*), is made by combining urea (*left*) and malonic acid (*right*) with the elimination of water (*colored rectangles*). The barbiturate families arise from substitution of other substances for hydrogens at position 5 in the basic barbiturate structure.

are neither a serum nor a guaranteed truth-producer.

The other important use of the drugs is for the control of epileptic convulsions. Certain of the barbiturates—phenobarbital, mephobarbital and methabarbital —can prevent or stop these seizures by depressing brain activity. Barbiturates can control not only the generalized convulsions of genuine (idiopathic) epilepsy—a disease afflicting almost a million persons in the U. S.—but also seizures induced by stimulating drugs or by bacterial toxins such as tetanus. The barbiturates and the convulsant drugs act in opposite fashion, and, curiously, each is used as an antidote for the other. Acute barbiturate poisoning is often treated with a stimulating drug such as pentylenetetrazol (Metrazol) or picrotoxin to bring the patient out of his coma; if the dose of the stimulant turns out to be too strong, producing convulsions, this in turn is treated with a dose of a fast-acting barbiturate!

The toxic effects of barbiturates are subtle and sometimes unpredictable. For some patients even a comparatively small dose may be dangerous. The body gets rid of barbital and phenobarbital chiefly by excretion in the urine; for a person with damaged kidneys, therefore, these drugs become toxic. Pentobarbital and secobarbital, two of the most widely used barbiturates, are broken down in the liver. Given to a patient with a poorly functioning liver, they may produce a far longer sleep than desired.

Next to carbon monoxide, the barbiturates are the most popular suicide poison in the U. S. They account for one fifth of all the cases of acute drug poisoning, and most of these are suicide attempts. The barbiturates are not, as a matter of fact, a very efficient suicide agent: only about 8 per cent of those poisonees who arrive at hospitals die. But they are widely known and readily available, and they produce from 1,000

to 1,500 deaths in the U. S. each year.

Some of these deaths, though self-inflicted, are accidental. A British physician first called attention some years ago to a specific and probably common hazard. The person takes a small dose to go to sleep, and later, half asleep and confused, he swallows another, lethal dose. Some physicians now warn barbiturate-users not to keep their bottle of tablets on a night table, where they may stretch out a hand to take more while in the comatose state.

There is a comfortable margin of safety between the ordinary sleeping dose (a tenth of a gram for the average adult) and a definitely toxic dose (more than half a gram). The lethal dose is usually a gram and a half or more. Acute barbiturate poisoning has to be treated promptly. Unfortunately it is often not recognized in time, because the victim is thought to be merely in a deep sleep. The first step in treatment is to strengthen the victim's breathing, in a respirator if necessary. And a stimulant may have to be administered to restore the activity of the brain centers.

Are the barbiturates habit-forming? This much-debated question has been answered rather conclusively by recent studies. They can indeed produce addiction and chronic intoxication. The two chief criteria of addiction to a drug are a heightened tolerance to it and physical dependence on it, so that removal of the drug produces withdrawal symptoms. A morphine addict, for example, may be able to take many times the dose that would be lethal for a normal person, and he becomes acutely sick if the drug is stopped. Several years ago Havelock Fraser, Harris Isbell and their associates at the U. S. Public Health Service hospital for drug addicts in Lexington, Ky., made a thorough study of whether the barbiturates had these properties. Their investigation included experiments with human subjects who were given large doses of barbiturates over a period of months and then abruptly taken off the drug.

They found that the barbiturates acted as addicting drugs in every respect—physical and psychic. The men behaved like chronic alcoholics: they neglected their appearance and hygiene, became confused and quarrelsome, showed unpredictable mood swings and lost physical coordination and the mental discipline necessary for simple games. After abrupt withdrawal of the drug, the subjects began within a few hours to show signs of increasing apprehension and developed weakness, tremors, nausea and vomiting. In the next five days most of the subjects had convulsions like those of epilepsy and an acute psychosis

such as alcoholics suffer, with delirium and violent hallucinations.

The Lexington investigators also made similar tests on dogs. They too exhibited withdrawal symptoms. In their "canine delirium" the dogs would stare at a blank wall and move their heads, eyes and ears as if responding to imaginary animals, people or objects; even while alone in a cage a dog would growl as if being attacked.

Stories in the press have greatly exaggerated the extent of barbiturate addiction in the U. S. "Thrill pills," "goof balls," "wild geronimos," "red devils" (secobarbital), "yellow jackets" (sodium pentobarbital), "blue heaven" (amobarbital)—all these terms certainly have a currency in a limited circle of addicts,

but the number of addicts is not nearly so large as some of the stories have alleged. There are probably not more than 50,000 barbiturate addicts, compared with a million chronic alcoholics. Nevertheless, in view of the easy access to the barbiturates, the public does need to be alerted to their addictive property.

The saving fact is that it takes extraordinarily heavy use of these drugs to produce addiction. Subjects who have taken a fifth of a gram (twice the usual sleeping dose) every night for a year have shown no withdrawal symptoms after stopping the drug. In contrast, morphine, taken in the usual hospital doses for as short a time as 30 days, produces definite physical dependence. Moderate use of the barbiturates, in the doses prescribed by physicians, will not lead to addiction. Those who become addicts are probably, in the main, drug-users who turn to the barbiturates because they cannot get narcotics, alcoholics who seek relief from alcohol withdrawal and, in general, abnormal personalities who are addiction risks for any intoxication that will give psychic relief. Whether stricter Federal laws are needed to control misuse of the barbiturates has been a matter of considerable controversy.

Biologists look forward to the day when progress in medicine will make all present drugs, including the barbiturates, obsolete. Better understanding and treatment of the personal and social causes of anxiety should reduce our present reliance on chemical aids to tranquility and sleep. Meanwhile the barbiturates can teach us much about the functions of the brain and so help lead toward that more tranquil day.

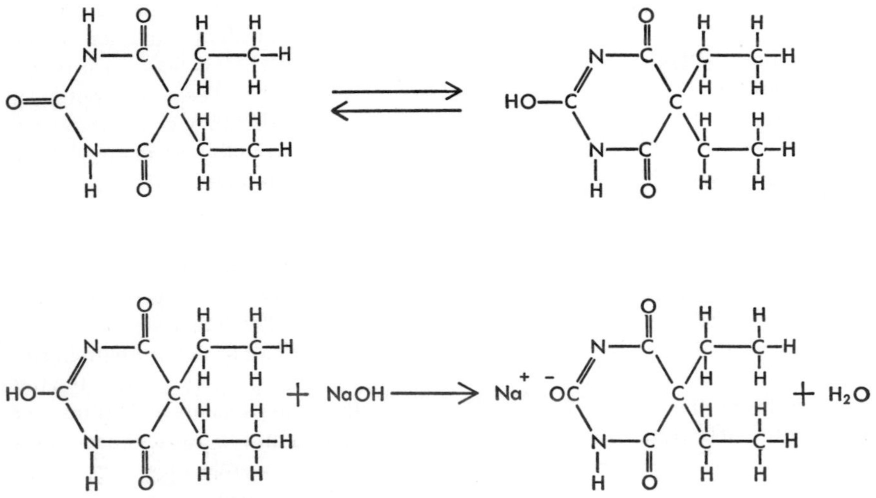

SODIUM BARBITAL (lower right) can be made from barbital (shown at top of diagram in its two forms) by addition of sodium hydroxide (NaOH) and the elimination of water.

11

THE HALLUCINOGENIC DRUGS

FRANK BARRON, MURRAY E. JARVIK AND STERLING BUNNELL, JR.

April 1964

These powerful alkaloids, tools for investigating mental illness and perhaps for treating it, have become the subject of a debate: Do their constructive potentials outweigh their admitted hazards?

Human beings have two powerful needs that are at odds with each other: to keep things the same, and to have something new happen. We like to feel secure, yet at times we like to be surprised. Too much predictability leads to monotony, but too little may lead to anxiety. To establish a balance between continuity and change is a task facing all organisms, individual and social, human and non-human.

Keeping things predictable is generally considered one of the functions of the ego. When a person perceives accurately, thinks clearly, plans wisely and acts appropriately—and represses maladaptive thoughts and emotions—we say that his ego is strong. But the strong ego is also inventive, open to many perceptions that at first may be disorganizing. Research on the personality traits of highly creative individuals has shown that they are particularly alert to the challenge of the contradictory and the unpredictable, and that they may even court the irrational in their own make-up as a source of new and unexpected insight. Indeed, through all recorded history and everywhere in the world men have gone to considerable lengths to seek unpredictability by disrupting the functioning of the ego. A change of scene, a change of heart, a change of mind: these are the popular prescriptions for getting out of a rut.

Among the common ways of changing "mind" must be reckoned the use of intoxicating substances. Alcohol has quite won the day for this purpose in the U.S. and much of the rest of the world. Consumed at a moderate rate and in sensible quantities, it can serve simultaneously as a euphoriant and tranquilizing agent before it finally dulls the faculties and puts one to sleep. In properly disposed individuals it may dissolve sexual inhibitions, relieve fear and anxiety, or stimulate meditation on the meaning of life. In spite of its costliness to individual and social health when it is used immoderately, alcohol retains its rank as first among the substances used by mankind to change mental experience. Its closest rivals in popularity are opium and its derivatives and various preparations of cannabis, such as hashish and marijuana.

This article deals with another group of such consciousness-altering substances: the "hallucinogens." The most important of these are mescaline, which comes from the peyote cactus *Lophophora williamsii;* psilocybin and psilocin, from such mushrooms as *Psilocybe mexicana* and *Stropharia cubensis;* and d-lysergic acid diethylamide (LSD), which is derived from ergot (*Claviceps purpurea*), a fungus that grows on rye and wheat. All are alkaloids more or less related to one another in chemical structure.

Various names have been applied to this class of substances. They produce distinctive changes in perception that are sometimes referred to as hallucinations, although usually the person under the influence of the drug can distinguish his visions from reality, and even when they seem quite compelling he is able to attribute them to the action of the drug. If, therefore, the term "hallucination" is reserved for perceptions that the perceiver himself firmly believes indicate the existence of a corresponding object or event, but for which other observers can find no objective basis, then the "hallucinogens" only rarely produce hallucinations. There are several other names for this class of drugs. They have been called "psychotomimetic" because in some cases the effects seem to mimic psychosis [see "Experimental Psychoses," by Six Staff Members of the Boston Psychopathic Hospital; SCIENTIFIC AMERICAN, June, 1955]. Some observers prefer to use the term "psychedelic" to suggest that unsuspected capacities of the imagination are sometimes revealed in the perceptual changes.

The hallucinogens are currently a subject of intense debate and concern in medical and psychological circles. At issue is the degree of danger they present to the psychological health of the person who uses them. This has become an important question because of a rapidly increasing interest in the drugs among laymen. The recent controversy at Harvard University, stemming at first from methodological disagreements

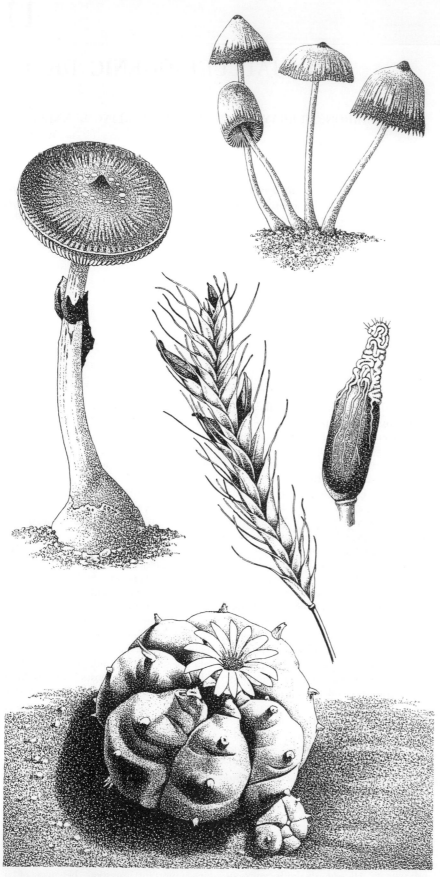

NATURAL SOURCES of the main hallucinogens are depicted. Psilocybin comes from the mushrooms *Stropharia cubensis* (*top left*) and *Psilocybe mexicana* (*top right*). LSD is synthesized from an alkaloid in ergot (*Claviceps purpurea*), a fungus that grows on cereal grains; an ergot-infested rye seed head is shown (*center*) together with a larger-scale drawing of the ergot fungus. Mescaline is from the peyote cactus *Lophophora williamsii* (*bottom*).

among investigators but subsequently involving the issue of protection of the mental health of the student body, indicated the scope of popular interest in taking the drugs and the consequent public concern over their possible misuse.

There are, on the other hand, constructive uses of the drugs. In spite of obvious differences between the "model psychoses" produced by these drugs and naturally occurring psychoses, there are enough similarities to warrant intensive investigation along these lines. The drugs also provide the only link, however tenuous, between human psychoses and aberrant behavior in animals, in which physiological mechanisms can be studied more readily than in man. Beyond this many therapists feel that there is a specialized role for the hallucinogens in the treatment of psychoneuroses. Other investigators are struck by the possibility of using the drugs to facilitate meditation and aesthetic discrimination and to stimulate the imagination. These possibilities, taken in conjunction with the known hazards, are the bases for the current professional concern and controversy.

In evaluating potential uses and misuses of the hallucinogens, one can draw on a considerable body of knowledge from such disciplines as anthropology, pharmacology, biochemistry, psychology and psychiatry.

In some primitive societies the plants from which the major hallucinogens are derived have been known for millenniums and have been utilized for divination, curing, communion with supernatural powers and meditation to improve self-understanding or social unity; they have also served such mundane purposes as allaying hunger and relieving discomfort or boredom. In the Western Hemisphere the ingestion of hallucinogenic plants in pre-Columbian times was limited to a zone extending from what is now the southwestern U.S. to the northwestern basin of the Amazon. Among the Aztecs there were professional diviners who achieved inspiration by eating either peyote, hallucinogenic mushrooms (which the Aztecs called *teo-nanacatyl*, or "god's flesh") or other hallucinogenic plants. *Teo-nanacatyl* was said to have been distributed at the coronation of Montezuma to make the ceremony seem more spectacular. In the years following the conquest of Mexico there were reports of communal mushroom rites among the Aztecs and other Indians of southern Mexico. The communal use has almost died out today, but in several

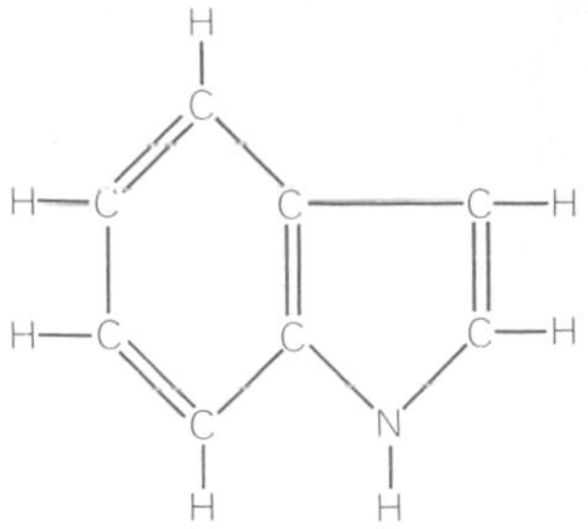

INDOLE RING

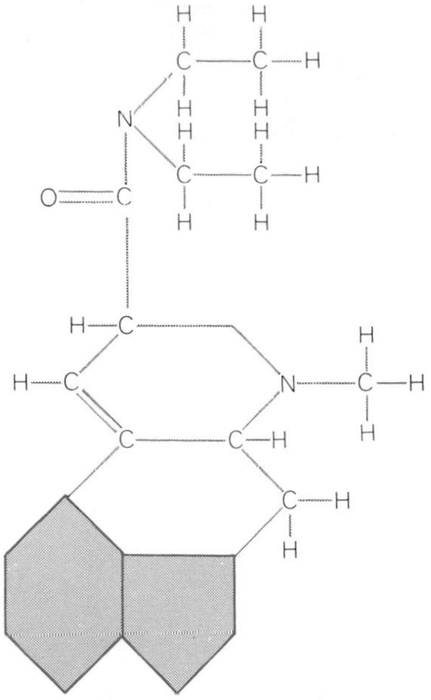

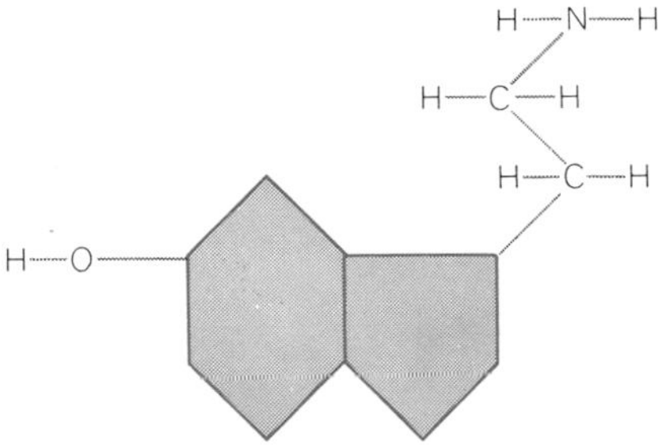

SEROTONIN

LSD

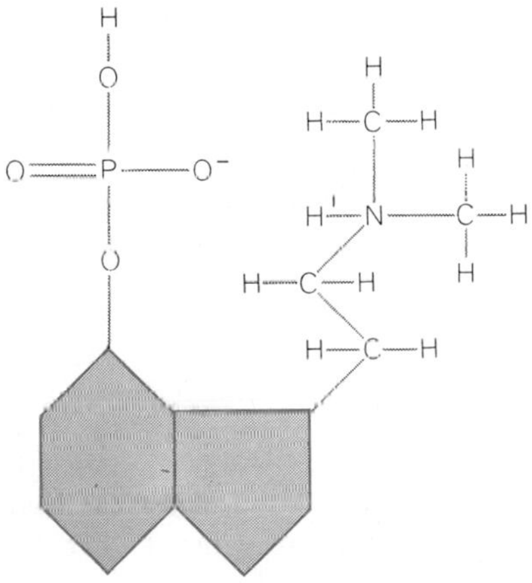

PSILOCYBIN

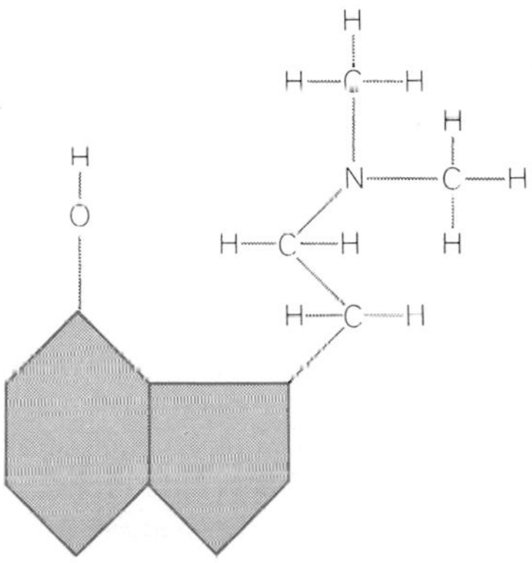

PSILOCIN

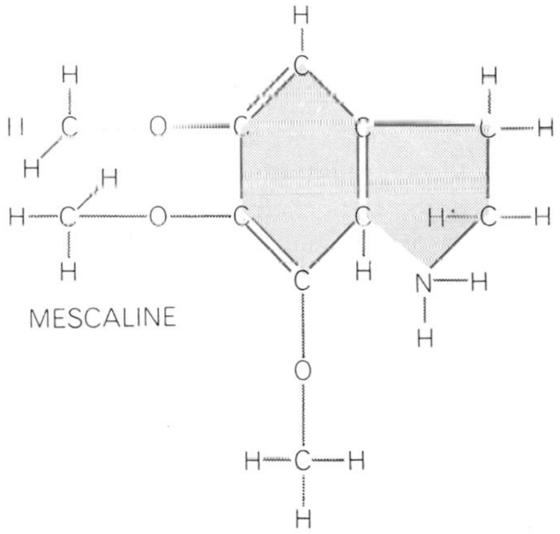

MESCALINE

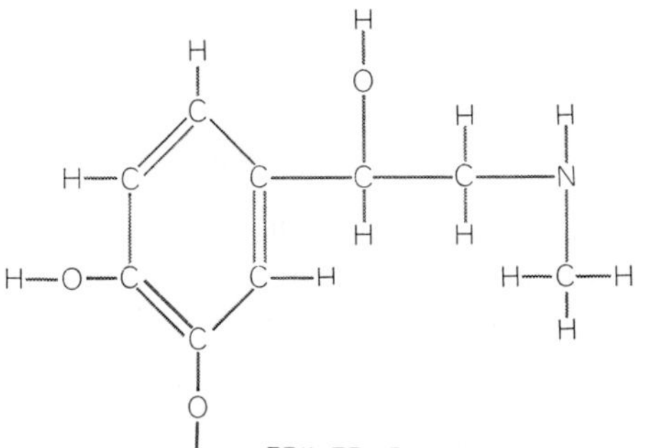

EPINEPHRINE

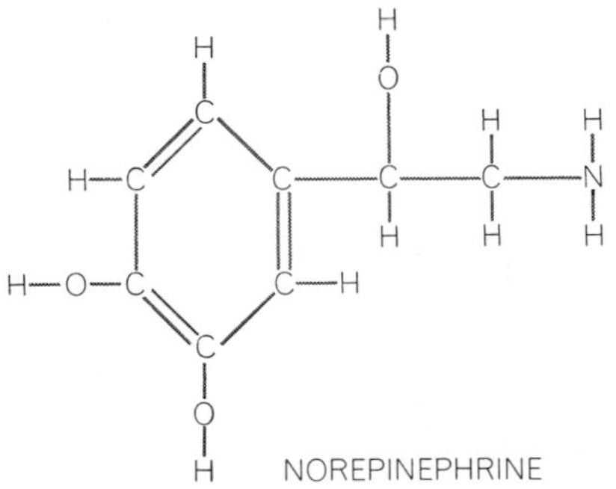

NOREPINEPHRINE

CHEMICAL RELATIONS among several of the hallucinogens and neurohumors are indicated by these structural diagrams. The indole ring (*in color at top*) is a basic structural unit; it appears, as indicated by the colored shapes, in serotonin, LSD, psilocybin and psilocin. Mescaline does not have an indole ring but, as shown by the light color, can be represented so as to suggest its relation to the ring. The close relation between mescaline and the two catechol amines epinephrine and norepinephrine is also apparent here.

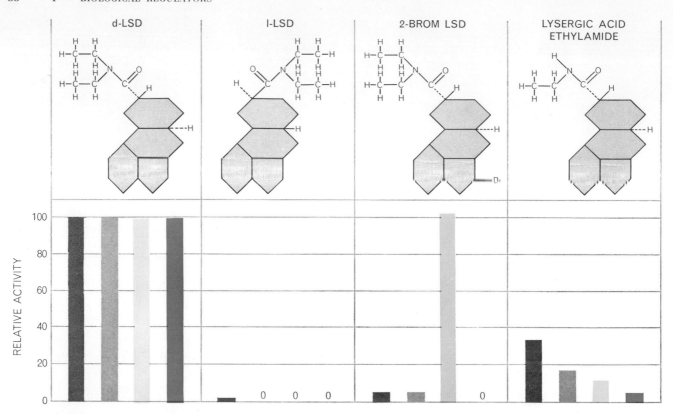

SLIGHT CHANGES in LSD molecule produce large changes in its properties. Here LSD (*left*) is used as a standard, with a "relative activity" of 100 in toxicity (*dark gray bar*), fever-producing effect (*light gray*), ability to antagonize serotonin (*light color*) and typical psychotomimetic effects (*dark color*). The stereoisomer of LSD (*second from left*) in which the positions of the side chains are reversed, shows almost no activity; the substitution of a bromine atom (*third from left*) reduces the psychotomimetic effect but not the serotonin antagonism; the removal of one of the two ethyl groups (*right*) sharply reduces activity in each of the areas.

tribes the medicine men or women (*curanderos*) still partake of *Psilocybe* and *Stropharia* in their rituals.

In the arid region between the Rio Grande and central Mexico, where the peyote cactus grows, the dried tops of the plants ("peyote buttons") were eaten by Indian shamans, or medicine men, and figured in tribal rituals. During the 19th century the Mescalero Apaches acquired the plant and developed a peyote rite. The peyotism of the Mescaleros (whence the name mescaline) spread to the Comanches and Kiowas, who transformed it into a religion with a doctrine and ethic as well as ritual. Peyotism, which spread rapidly through the Plains tribes, became fused with Christianity. Today its adherents worship God as the great spirit who controls the universe and put some of his power into peyote, and Jesus as the man who gave the plant to the Indians in a time of need. Saturday-night meetings, usually held in a traditional tepee, begin with the eating of the sacramental peyote; then the night is spent in prayer, ritual singing and introspective contemplation, and in the morning there is a communion breakfast of corn, game and fruit.

Recognizing the need for an effective organization to protect their form of worship, several peyote churches joined in 1918 to form the Native American Church, which now has about 225,000 members in tribes from Nevada to the East Coast and from the Mexican border to Saskatchewan. It preaches brotherly love, care of the family, self-reliance and abstinence from alcohol. The church has been able to defeat attempts, chiefly by the missionaries of other churches, to outlaw peyote by Federal legislation, and it has recently brought about the repeal of antipeyote legislation in several states.

The hallucinogens began to attract scholarly interest in the last decade of the 19th century, when the investigations and conceptions of such men as Francis Galton, J. M. Charcot, Sigmund Freud and William James introduced a new spirit of serious inquiry into such subjects as hallucination, mystical experience and other "paranormal" psychic phenomena. Havelock Ellis and the psychiatrist Silas Weir Mitchell wrote accounts of the subjective effects of peyote, or Anhalonium, as it was then called. Such essays in turn stimulated the interest of pharmacologists. The active principle of peyote, the alkaloid called mescaline, was isolated in 1896; in 1919 it was recognized that the molecular structure of mescaline was related to the structure of the adrenal hormone epinephrine.

This was an important turning point, because the interest in the hallucinogens as a possible key to naturally occurring psychoses is based on the chemical relations between the drugs and the neurohumors: substances that chemically transmit impulses across synapses between two neurons, or nerve cells, or between a neuron and an effector such as a muscle cell. Acetylcholine and the catechol amines epinephrine and norepinephrine have been shown to act in this manner in the peripheral nervous system of vertebrates; serotonin has the same effect in some invertebrates. It is frequently assumed that these substances also act as neurohumors in the central nervous system; at least they are present there, and injecting them into various parts of the brain seems to affect nervous activity.

The structural resemblance of mescaline and epinephrine suggested a possible link between the drug and mental illness: Might the early, excited stage of schizophrenia be produced or at least triggered by an error in metabolism that produced a mescaline-like substance? Techniques for gathering evidence on this question were not available, however, and the speculation on an "M-substance" did not lead to serious experimental work.

When LSD was discovered in 1943, its extraordinary potency again aroused interest in the possibility of finding a natural chemical activator of the schizophrenic process. The M-substance hypothesis was revived on the basis of reports that hallucinogenic effects were produced by adrenochrome and other breakdown products of epinephrine, and the hypothesis appeared to be strengthened by the isolation from human urine of some close analogues of hallucinogens. Adrenochrome has not, however, been detected in significant amounts in the human body, and it seems unlikely that the analogues could be produced in sufficient quantity to effect mental changes.

The relation between LSD and serotonin has given rise to the hypothesis that schizophrenia is caused by an imbalance in the metabolism of serotonin, with excitement and hallucinations resulting from an excess of serotonin in certain regions of the brain, and depressive and catatonic states resulting from a deficiency of serotonin. The idea arose in part from the observation that in some laboratory physiological preparations LSD acts rather like serotonin but in other preparations it is a powerful antagonist of serotonin; thus LSD might facilitate or block some neurohumoral action of serotonin in the brain.

The broad objection to the serotonin theory of schizophrenia is that it requires an oversimplified view of the disease's pattern of symptoms. Moreover, many congeners, or close analogues, of LSD, such as 2-brom lysergic acid, are equally effective or more effective antagonists of serotonin without being significantly active psychologically in man. This does not disprove the hypothesis, however. In man 2-brom LSD blocks the mental effects of a subsequent dose of LSD, and in the heart of a clam it blocks the action of both LSD and serotonin. Perhaps there are "keyholes" at the sites where neurohumors act; in the case of those for serotonin it may be that LSD fits the hole and opens the lock, whereas the psychologically inactive analogues merely occupy the keyhole, blocking the action of serotonin or LSD without mimicking their effects. Certainly the resemblance of most of the hallucinogens to serotonin is marked, and the correlations between chemical structure and pharmacological action deserve intensive investigation. The serotonin theory of schizophrenia is far from proved, but there is strong evidence for an organic factor of some kind in the disease; it may yet turn out to involve either a specific neurohumor or an imbalance among several neurohumors.

The ingestion of LSD, mescaline or psilocybin can produce a wide range of subjective and objective effects. The subjective effects apparently depend on at least three kinds of variable: the properties and potency of the drug itself; the basic personality traits and current mood of the person ingesting it, and the social and psychological context, including the meaning to the individual of his act in taking the drug and his interpretation of the motives of those who made it available. The discussion of subjective effects that follows is compiled from many different accounts of the drug experience; it should be considered an inventory of possible effects rather than a description of a typical episode.

One subjective experience that is frequently reported is a change in visual perception. When the eyes are open, the perception of light and space is affected: colors become more vivid and seem to glow; the space between objects becomes more apparent, as though space itself had become "real," and surface details appear to be more sharply defined. Many people feel a new awareness of the physical beauty of the world, particularly of visual harmonies, colors, the play of light and the exquisiteness of detail.

The visual effects are even more striking when the eyes are closed. A constantly changing display appears, its content ranging from abstract forms to dramatic scenes involving imagined people or animals, sometimes in exotic lands or ancient times. Different individuals have recalled seeing wavy lines, cobweb or chessboard designs, gratings, mosaics, carpets, floral designs, gems, windmills, mausoleums, landscapes, "arabesques spiraling into eternity," statuesque men of the past, chariots, sequences of dramatic action, the face of Buddha, the face of Christ, the Crucifixion, "the mythical dwelling places of the gods," the immensity and blackness of space. After taking peyote Silas Weir Mitchell wrote: "To give the faintest idea of the perfectly satisfying intensity and purity of these gorgeous color fruits

WATER COLORS were done, while under the influence of a relatively large dose of a hallucinogenic drug, by a person with no art training. Originals are bright yellow, purple, green and red as well as black.

90

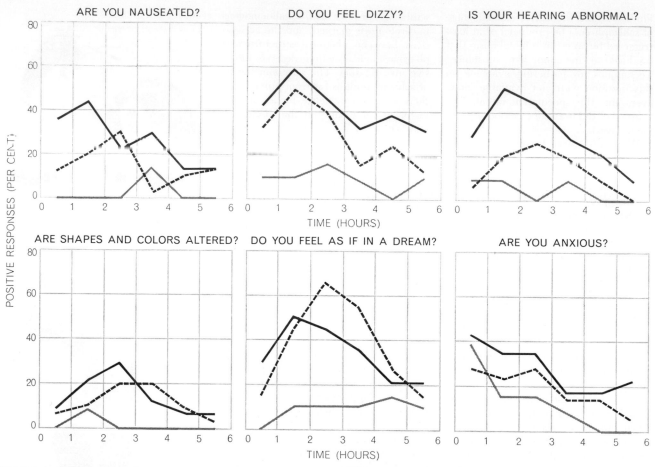

ARE YOU NAUSEATED?

DO YOU FEEL DIZZY?

IS YOUR HEARING ABNORMAL?

ARE SHAPES AND COLORS ALTERED?

DO YOU FEEL AS IF IN A DREAM?

ARE YOU ANXIOUS?

POSITIVE RESPONSES (PER CENT)

TIME (HOURS)

SUBJECTIVE REPORT on physiological and perceptual effects of LSD was obtained by means of a questionnaire containing 47 items, the results for six of which are presented. Volunteers were questioned at one-hour intervals beginning half an hour after they took the drug. The curves show the per cent of the group giving positive answers at each time. The gray curves are for those given an inactive substance, the broken black curves for between 25 and 75 micrograms and the solid black curves for between 100 and 225.

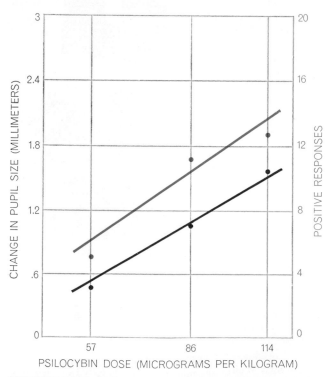

CHANGE IN PUPIL SIZE (MILLIMETERS)

POSITIVE RESPONSES

PSILOCYBIN DOSE (MICROGRAMS PER KILOGRAM)

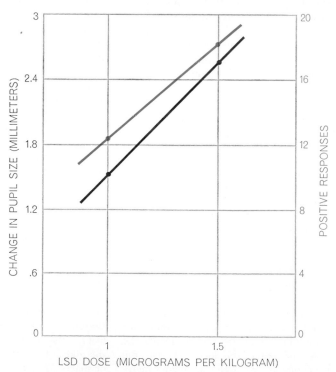

CHANGE IN PUPIL SIZE (MILLIMETERS)

POSITIVE RESPONSES

LSD DOSE (MICROGRAMS PER KILOGRAM)

OBJECTIVE AND SUBJECTIVE effects vary with dosage as shown here. The data plotted in black are for the increase in size of pupil; the number of positive responses to questions like the ones at the top of the page are shown in color. The objective and subjective measures vary in a similar manner. The data are from an experiment done by Harris Isbell of the University of Kentucky.

is quite beyond my power." A painter described the waning hours of the effects of psilocybin as follows: "As the afternoon wore on I felt very content to simply sit and stare out of the window at the snow and the trees, and at that time I recall feeling that the snow, the fire in the fireplace, the darkened and book-lined room were so perfect as to seem almost unreal."

The changes in visual perception are not always pleasant. Aldous Huxley called one of his books about mescaline *Heaven and Hell* in recognition of the contradictory sensations induced by the drug. The "hellish" experiences include an impression of blackness accompanied by feelings of gloom and isolation, a garish modification of the glowing colors observed in the "heavenly" phase, a sense of sickly greens and ugly dark reds. The subject's perception of his own body may become unpleasant: his limbs may seem to be distorted or his flesh to be decaying; in a mirror his face may appear to be a mask, his smile a meaningless grimace. Sometimes all human movements appear to be mere puppetry, or everyone seems to be dead. These experiences can be so disturbing that a residue of fear and depression persists long after the effects of the drug have worn off.

Often there are complex auditory hallucinations as well as visual ones: lengthy conversations between imaginary people, perfectly orchestrated musical compositions the subject has never heard before, voices speaking foreign languages unknown to the subject. There have also been reports of hallucinatory odors and tastes and of visceral and other bodily sensations. Frequently patterns of association normally confined to a single sense will cross over to other senses: the sound of music evokes the visual impression of jets of colored light, a "cold" human voice makes the subject shiver, pricking the skin with a pin produces the visual impression of a circle, light glinting on a Christmas tree ornament seems to shatter and to evoke the sound of sleigh bells. The time sense is altered too. The passage of time may seem to be a slow and pleasant flow or to be intolerably tedious. A "sense of timelessness" is often reported; the subject feels outside of or beyond time, or time and space seem infinite.

In some individuals one of the most basic constancies in perception is affected: the distinction between subject and object. A firm sense of personal identity depends on knowing accurately the borders of the self and on being able to distinguish what is inside from what is outside. Paranoia is the most vivid pathological instance of the breakdown of this discrimination; the paranoiac attributes to personal and impersonal forces outside himself the impulses that actually are inside him. Mystical and transcendental experiences are marked by the loss of this same basic constancy. "All is one" is the prototype of a mystical utterance. In the mystical state the distinction between subject and object disappears; the subject is seen to be one with the object. The experience is usually one of rapture or ecstasy and in religious terms is described as "holy." When the subject thus achieves complete identification with the object, the experience seems beyond words.

Some people who have taken a large dose of a hallucinogenic drug report feelings of "emptiness" or "silence," pertaining either to the interior of the self or to an "interior" of the universe—or to both as one. Such individuals have a sense of being completely undifferentiated, as though it were their personal consciousness that had been "emptied," leaving none of the usual discriminations on which the functioning of the ego depends. One man who had this experience thought later that it had been an anticipation of death, and that the regaining of the basic discriminations was like a remembrance of the very first days of life after birth.

The effect of the hallucinogens on sexual experience is not well documented. One experiment that is often quoted seemed to provide evidence that mescaline is an anaphrodisiac, an inhibitor of sexual appetite; this conclusion seemed plausible because the drugs have so often been associated with rituals emphasizing asceticism and prayer. The fact is, however, that the drugs are probably neither anaphrodisiacs nor aphrodisiacs—if indeed any drug is. There is reason to believe that if the drug-taking situation is one in which sexual relations seem appropriate, the hallucinogens simply bring to the sexual experience the same kind of change in perception that occurs in other areas of experience.

The point is that in all the hallucinogen-produced experiences it is never the drug alone that is at work. As in the case of alcohol, the effects vary widely depending on when the drug is taken, where, in the presence of whom, in what dosage and—perhaps most important of all—by whom. What happens to the individual after he takes the drug, and his changing relations to the setting and the people in it during the episode, will further influence his experience.

Since the setting is so influential in these experiments, it sometimes happens that a person who is present when someone else is taking a hallucinogenic drug, but who does not take the drug himself, behaves as though he were under the influence of a hallucinogen. In view of this effect one might expect that a person given an inactive substance he thought was a drug would respond as though he had actually received the drug. Indeed, such responses have sometimes been noted. In controlled experiments, however, subjects given an inactive substance are readily distinguishable from those who take a drug; the difference is apparent in their appearance and behavior, their answers to questionnaires and their physiological responses. Such behavioral similarities as are observed can be explained largely by a certain apprehension felt by a person who receives an inactive substance he thinks is a drug, or by anticipation on the part of someone who has taken the drug before.

In addition to the various subjective effects of the hallucinogens there are a number of observable changes in physiological function and in performance that one can measure or at least describe objectively. The basic physiological effects are those typical of a mild excitement of the sympathetic nervous system. The hallucinogens usually dilate the pupils, constrict the peripheral arterioles and raise the systolic blood pressure; they may also increase the excitability of such spinal reflexes as the knee jerk. Electroencephalograms show that the effect on electrical brain waves is usually of a fairly nonspecific "arousal" nature: the pattern is similar to that of a normally alert, attentive and problem-oriented subject, and if rhythms characteristic of drowsiness or sleep have been present, they disappear when the drug is administered. (Insomnia is common the first night after one of the drugs has been taken.) Animal experiments suggest that LSD produces these effects by stimulating the reticular formation of the midbrain, not directly but by stepping up the sensory input.

Under the influence of one of the hallucinogens there is usually some reduction in performance on standard tests of reasoning, memory, arithmetic, spelling and drawing. These findings may not indicate an inability to perform well; after taking a drug many people simply refuse to co-operate with the tester. The very fact that someone should want to

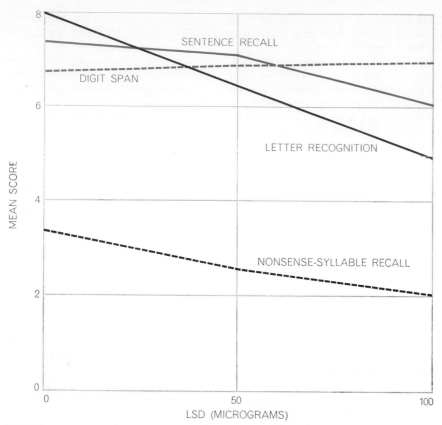

EFFECT OF LSD on memory was determined with standard tests. Curves show results of four tests for subjects given an inactive substance, 50 micrograms of the drug and 100 micrograms respectively. Effect of LSD was to decrease scores except in a test of digit-recall ability.

test them may seem absurd and may arouse either hostility or amusement. Studies by one of the authors in which tests of attention and concentration were administered to subjects who had been given different doses of LSD indicated that motivation was perhaps as important in determining scores as the subject's intellectual capacity.

The hallucinogenic drugs are not addictive—if one means by addiction that physiological dependence is established and the drug becomes necessary, usually in increasing amounts, for satisfactory physiological functioning. Some individuals become psychologically dependent on the drugs, however, and develop a "habit" in that sense; indeed, there is a tendency for those who ingest hallucinogens habitually to make the drug experience the center of all their activities. LSD, mescaline and psilocybin do produce physiological tolerance. If the same quantity of LSD is administered on three successive days, for example, it will not suffice by the third day to produce the same subjective or physiological effects; tolerance develops more slowly and less completely with mescaline and psilocybin. When an individual becomes tolerant to a given dos-

age of LSD, the ordinarily equivalent dose of psilocybin produces reduced effects. This phenomenon of cross-tolerance suggests that the two drugs have common pathways of action. Any tolerance established by daily administration of the drugs wears off rather rapidly, generally being dissipated within a few days if the drug is not taken.

The three major hallucinogens differ markedly in potency. The standard human doses—those that will cause the average adult male weighing about 150 pounds to show the full clinical effects— are 500 milligrams of mescaline, 20 milligrams of psilocybin and .1 milligram of LSD. It is assumed that in a large enough dose any of the hallucinogens would be lethal, but there are no documented cases of human deaths from the drugs alone. Death has been brought on in sensitive laboratory animals such as rabbits by LSD doses equivalent to 120 times the standard human dose. Some animals are much less susceptible; white rats have been given doses 1,000 times larger than the standard human dose without lasting harm. The maximum doses known by the authors to have been taken by human beings are 900 milligrams of mescaline, 70 milligrams

of psilocybin and two milligrams of LSD. No permanent effects were noted in these cases, but obviously no decisive studies of the upper limits of dosage have been undertaken.

There are also differences among the hallucinogens in the time of onset of effects and the duration of intoxication. When mescaline is given orally, the effects appear in two or three hours and last for 12 hours or more. LSD acts in less than an hour; some of its effects persist for eight or nine hours, and insomnia can last as long as 16 hours. Psilocybin usually acts within 20 or 30 minutes, and its full effect is felt for about five hours. All these estimates are for the standard dose administered orally; when any of the drugs is given intravenously, the first effects appear within minutes.

At the present time LSD and psilocybin are treated by the U.S. Food and Drug Administration like any other "experimental drug," which means that they can be legally distributed only to qualified investigators who will administer them in the course of an approved program of experimentation. In practice the drugs are legally available only to investigators working under a Government grant or for a state or Federal agency.

Nevertheless, there has probably been an increase during the past two or three years in the uncontrolled use of the drugs to satisfy personal curiosity or to experience novel sensations. This has led a number of responsible people in government, law, medicine and psychology to urge the imposition of stricter controls that would make the drugs more difficult to obtain even for basic research. These people emphasize the harmful possibilities of the drugs; citing the known cases of adverse reactions, they conclude that the prudent course is to curtail experimentation with hallucinogens.

Others—primarily those who have worked with the drugs—emphasize the constructive possibilities, insist that the hallucinogens have already opened up important leads in research and conclude that it would be shortsighted as well as contrary to the spirit of free scientific inquiry to restrict the activities of qualified investigators. Some go further, questioning whether citizens should be denied the opportunity of trying the drugs even without medical or psychological supervision and arguing that anyone who is mentally competent should have the right to explore the varieties of conscious experience if he can do so without harming himself or others.

The most systematic survey of the incidence of serious adverse reactions to hallucinogens covered nearly 5,000 cases, in which LSD was administered on more than 25,000 occasions. Psychotic reactions lasting more than 48 hours were observed in fewer than two-tenths of 1 per cent of the cases. The rate of attempted suicides was slightly over a tenth of 1 per cent, and these involved psychiatric patients with histories of instability. Among those who took the drug simply as subjects in experiments there were no attempted suicides and the psychotic reactions occurred in fewer than a tenth of 1 per cent of the cases.

Recent reports do indicate that the incidence of bad reactions has been increasing, perhaps because more individuals have been taking the hallucinogens in settings that emphasize sensation-seeking or even deliberate social delinquency. Since under such circumstances there is usually no one in attendance who knows how to avert dangerous developments, a person in this situation may find himself facing an extremely frightening hallucination with no one present who can help him to recognize where the hallucination ends and reality begins. Yet the question of what is a proper setting is not a simple one. One of the criticisms of the Harvard experiments was that some were conducted in private homes rather than in a laboratory or clinical setting. The experimenters defended this as an attempt to provide a feeling of naturalness and "psychological safety." Such a setting, they hypothesized, should reduce the likelihood of negative reactions such as fear and hostility and increase the positive experiences. Controlled studies of this hypothesis have not been carried out, however.

Many psychiatrists and psychologists who have administered hallucinogens in a therapeutic setting claim specific benefits in the treatment of psychoneuroses, alcoholism and social delinquency. The published studies are difficult to evaluate because almost none have employed control groups. One summary of the available statistics on the treatment of alcoholism does indicate that about 50 per cent of the patients treated with a combination of psychotherapy and LSD abstained from alcohol for at least a year, compared with 30 per cent of the patients treated by psychotherapy alone.

In another recent study the results of psychological testing before and after LSD therapy were comparable in most respects to the results obtained when conventional brief psychotherapy was employed. Single-treatment LSD therapy was significantly more effective, however, in relieving neurotic depression. If replicated, these results may provide an important basis for more directed study of the treatment of specific psychopathological conditions.

If the hallucinogens do have psychotherapeutic merit, it seems possible that they work by producing a shift in personal values. William James long ago noted that "the best cure for dipsomania is religiomania." There appear to be religious aspects of the drug experience that may bring about a change in behavior by causing a "change of heart." If this is so, one might be able to apply the hallucinogens in the service of moral regeneration while relying on more conventional techniques to give the patient insight into his habitual behavior patterns and motives.

In the light of the information now available about the uses and possible abuses of the hallucinogens, common sense surely decrees some form of social control. In considering such control it should always be emphasized that the reaction to these drugs depends not only on their chemical properties and biological activity but also on the context in which they are taken, the meaning of the act and the personality and mood of the individual who takes them. If taking the drug is defined by the group or individual, or by society, as immoral or criminal, one can expect guilt and aggression and further social delinquency to result; if the aim is to help or to be helped, the experience may be therapeutic and strengthening; if the subject fears psychosis, the drug could induce psychosis. The hallucinogens, like so many other discoveries of man, are analogous to fire, which can burn down the house or spread through the house life-sustaining warmth. Purpose, planning and constructive control make the difference. The immediate research challenge presented by the hallucinogens is a practical question: Can ways be found to minimize or eliminate the hazards, and to identify and develop further the constructive potentialities, of these powerful drugs?

NATIVE AMERICAN CHURCH members take part in a peyote ceremony in Saskatchewan, Canada. Under the influence of the drug, they gaze into the fire as they pray and meditate.

12

THE STEREOCHEMICAL THEORY OF ODOR

JOHN E. AMOORE, JAMES W. JOHNSTON, JR., AND MARTIN RUBIN
February 1964

*There is evidence that the sense of smell is based on
the geometry of molecules. Seven primary odors are
distinguished, each of them by an appropriately shaped
receptor at the olfactory nerve endings*

A rose is a rose and a skunk is a skunk, and the nose easily tells the difference. But it is not so easy to describe or explain this difference. We know surprisingly little about the sense of smell, in spite of its important influence on our daily lives and the voluminous literature of research on the subject. One is hard put to describe an odor except by comparing it to a more familiar one. We have no yardstick for measuring the strength of odors, as we measure sound in decibels and light in lumens. And we have had no satisfactory general theory to explain how the nose and brain detect, identify and recognize an odor. More than 30 different theories have been suggested by investigators in various disciplines, but none of them has passed the test of experiments designed to determine their validity.

The sense of smell obviously is a chemical sense, and its sensitivity is pro- verbial; to a chemist the ability of the nose to sort out and characterize substances is almost beyond belief. It deals with complex compounds that might take a chemist months to analyze in the laboratory; the nose identifies them instantly, even in an amount so small (as little as a ten-millionth of a gram) that the most sensitive modern laboratory instruments often cannot detect the substance, let alone analyze and label it.

Two thousand years ago the poet Lucretius suggested a simple explanation of the sense of smell. He speculated that the "palate" contained minute pores of various sizes and shapes. Every odorous substance, he said, gave off tiny "molecules" of a particular shape, and the odor was perceived when these molecules entered pores in the palate. Presumably the identification of each odor depended on which pores the molecules fitted.

It now appears that Lucretius' guess was essentially correct. Within the past few years new evidence has shown rather convincingly that the geometry of molecules is indeed the main determinant of odor, and a theory of the olfactory process has been developed in modern terms. This article will discuss the stereochemical theory and the experiments that have tested it.

The nose is always on the alert for odors. The stream of air drawn in through the nostrils is warmed and filtered as it passes the three baffle-shaped turbinate bones in the upper part of the nose; when an odor is detected, more of the air is vigorously sniffed upward to two clefts that contain the smelling organs [*see illustration on opposite page*]. These organs consist of two patches of yellowish tissue, each about one square inch in area. Embedded in the tissue are two types of nerve fiber whose endings receive and detect the odorous molecules. The chief type is represented by the fibers of the olfactory nerve; at the end of each of these fibers is an olfactory cell bearing a cluster of hairlike filaments that act as receptors. The other type of fiber is a long, slender ending of the trigeminal nerve, which is sensitive to certain kinds of molecules. On being stimulated by odorous molecules, the olfactory nerve endings send signals to the olfactory bulb and thence to the higher brain centers where the signals are integrated and interpreted in terms of the character and intensity of the odor.

From the nature of this system it is obvious at once that to be smelled at all a material must have certain basic properties. In the first place, it must be volatile. A substance such as onion soup, for example, is highly odorous because it continuously gives off vapor that can reach the nose (unless the soup is im-

PRIMARY ODOR	CHEMICAL EXAMPLE	FAMILIAR SUBSTANCE
CAMPHORACEOUS	CAMPHOR	MOTH REPELLENT
MUSKY	PENTADECANOLACTONE	ANGELICA ROOT OIL
FLORAL	PHENYLETHYL METHYL ETHYL CARBINOL	ROSES
PEPPERMINTY	MENTHONE	MINT CANDY
ETHEREAL	ETHYLENE DICHLORIDE	DRY-CLEANING FLUID
PUNGENT	FORMIC ACID	VINEGAR
PUTRID	BUTYL MERCAPTAN	BAD EGG

PRIMARY ODORS identified by the authors are listed, together with chemical and more familiar examples. Each of the primary odors is detected by a different receptor in the nose. Most odors are composed of several of these primaries combined in various proportions.

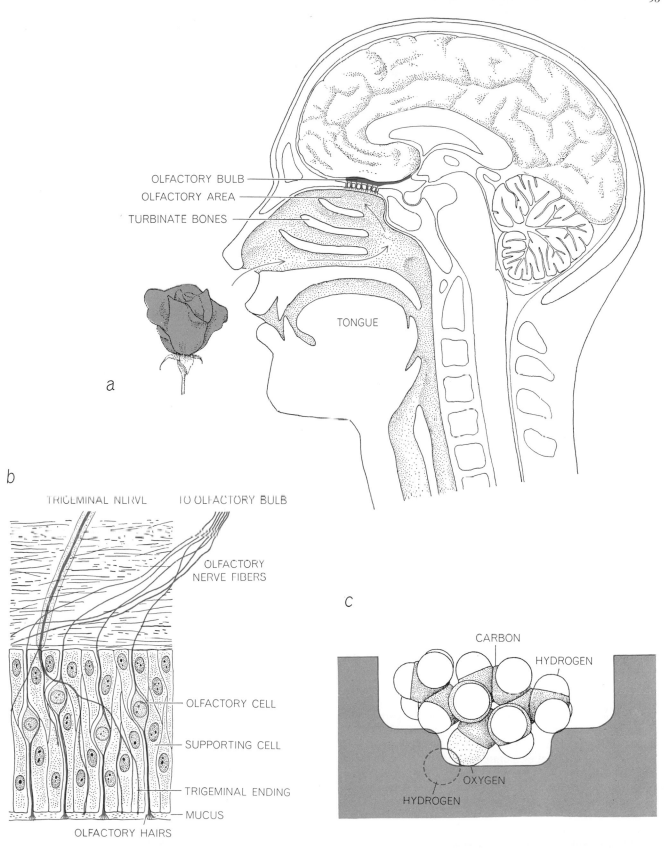

OLFACTORY BULB

OLFACTORY AREA

TURBINATE BONES

TONGUE

a

b

TRIGEMINAL NERVE   TO OLFACTORY BULB

OLFACTORY
NERVE FIBERS

c

OLFACTORY CELL

SUPPORTING CELL

TRIGEMINAL ENDING

MUCUS

OLFACTORY HAIRS

CARBON

HYDROGEN

OXYGEN

HYDROGEN

ANATOMY of the sense of smell is traced in these drawings. Air carrying odorous molecules is sniffed up past the three baffle-shaped turbinate bones to the olfactory area (*a*), patches of epithelium in which are embedded the endings of large numbers of olfactory nerves (*color*). A microscopic section of the olfactory epithelium (*b*) shows the olfactory nerve cells and their hairlike endings, trigeminal endings and supporting cells. According to the stereo-chemical theory different olfactory nerve cells are stimulated by different molecules on the basis of the size and shape or the charge of the molecule; these properties determine which of various pits and slots on the olfactory endings it will fit. A molecule of *l*-menthone is shown fitted into the "pepperminty" cavity (*c*).

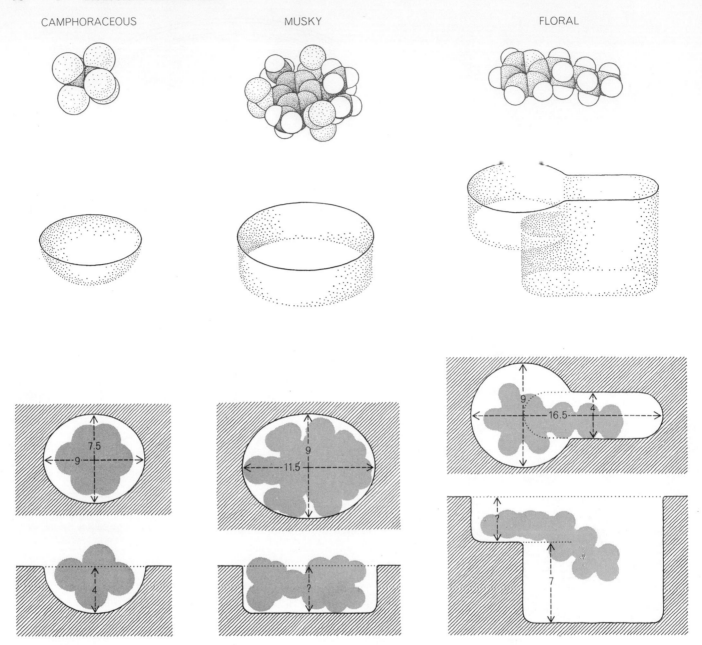

CAMPHORACEOUS   MUSKY   FLORAL

OLFACTORY RECEPTOR SITES are shown for each of the primary odors, together with molecules representative of each odor. The shapes of the first five sites are shown in perspective and (with the molecules silhouetted in them) from above and the side;

prisoned in a sealed can). On the other hand, at ordinary temperatures a substance such as iron is completely odorless because it does not evaporate molecules into the air.

The second requirement for an odorous substance is that it should be soluble in water, even if only to an almost infinitesimal extent. If it is completely insoluble, it will be barred from reaching the nerve endings by the watery film that covers their surfaces. Another common property of odorous materials is solubility in lipids (fatty substances); this enables them to penetrate the nerve endings through the lipid layer that

forms part of the surface membrane of every cell.

Beyond these elementary properties the characteristics of odorous materials have been vague and confusing. Over the years chemists empirically synthesized a wealth of odorous compounds, both for perfumers and for their own studies of odor, but instead of clarifying the properties responsible for odor these compounds seemed merely to add to the confusion. A few general principles were discovered. For instance, it was found that adding a branch to a straight chain of carbon atoms in a perfume molecule markedly increased the po-

tency of the perfume. Strong odor also seemed to be associated with chains of four to eight carbon atoms in the molecules of certain alcohols and aldehydes. The more chemists analyzed the chemical structure of odorous substances, however, the more puzzles emerged. From the standpoint of chemical composition and structure the substances showed some remarkable inconsistencies.

Curiously enough, the inconsistencies themselves began to show a pattern. As an example, two optical isomers—molecules identical in every respect except that one is the mirror image of the other —may have different odors. As another

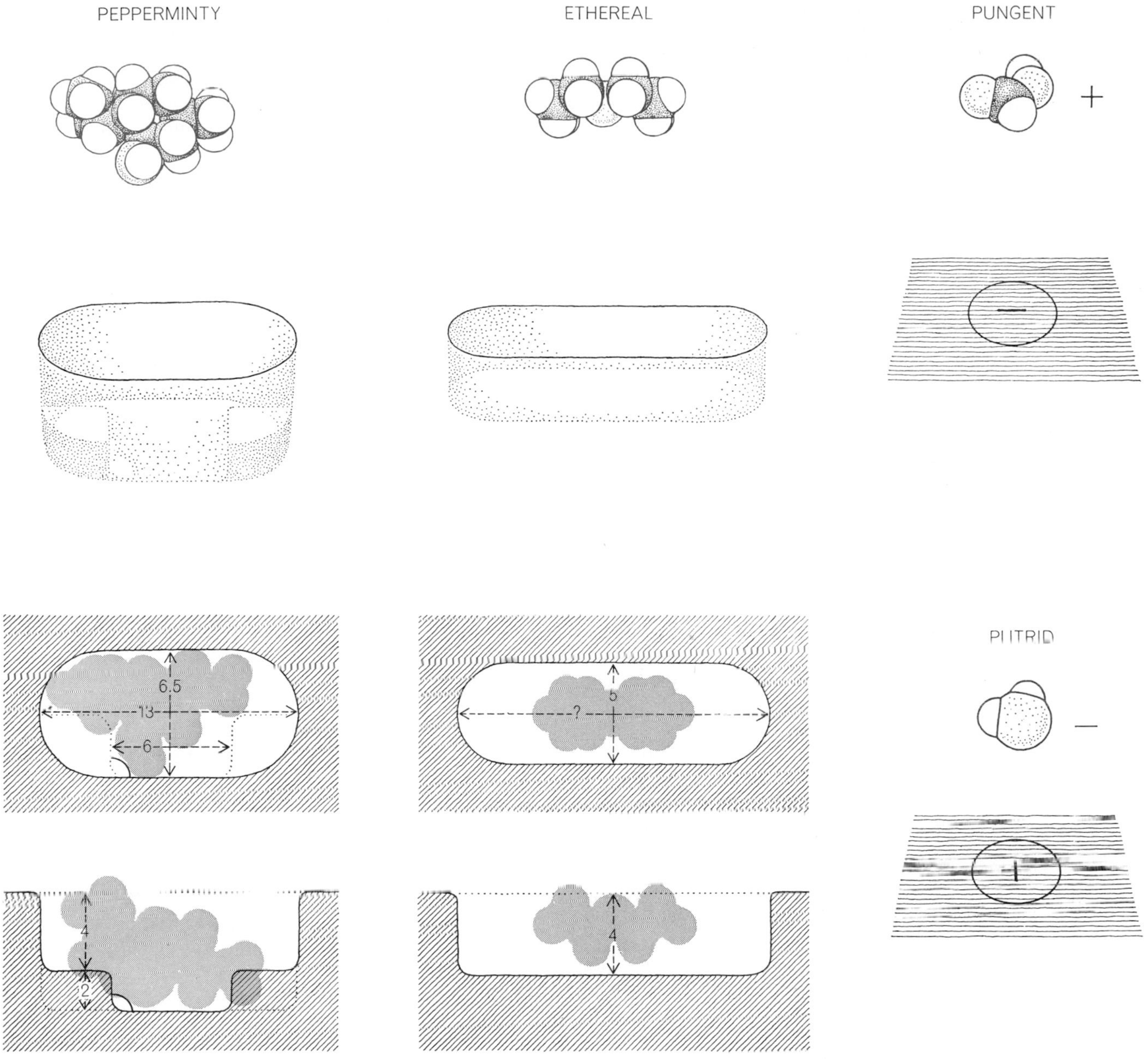

PEPPERMINTY   ETHEREAL   PUNGENT

PUTRID

known dimensions are given in angstrom units. The molecules are *l*-menthol and diethyl ether. Pungent (formic acid) and putrid (*left to right*) hexachloroethane, xylene musk, alpha-amylpyridine, (hydrogen sulfide) molecules fit because of charge, not shape.

example, in a compound whose molecules contain a small six-carbon-atom benzene ring, shifting the position of a group of atoms attached to the ring may sharply change the odor of the compound, whereas in a compound whose molecules contain a large ring of 14 to 19 members the atoms can be rearranged considerably without altering the odor much. Chemists were led by these facts to speculate on the possibility that the primary factor determining the odor of a substance might be the over-all geometric shape of the molecule rather than any details of its composition or structure.

In 1949 R. W. Moncrieff in Scotland gave form to these ideas by proposing a hypothesis strongly reminiscent of the 2,000-year-old guess of Lucretius. Moncrieff suggested that the olfactory system is composed of receptor cells of a few different types, each representing a distinct "primary" odor, and that odorous molecules produce their effects by fitting closely into "receptor sites" on these cells. His hypothesis is an application of the "lock and key" concept that has proved fruitful in explaining the interaction of enzymes with their substrates, of antibodies with antigens and of deoxyribonucleic acid with the "messenger" ribonucleic acid that presides at the synthesis of protein.

To translate Moncrieff's hypothesis into a practical approach for investigating olfaction, two specific questions had to be answered. What are the "primary odors"? And what is the shape of the receptor site for each one? To try to find answers to these questions, one of us (Amoore, then at the University of Oxford) made an extensive search of the literature of organic chemistry, looking for clues in the chemical characteristics of odorous compounds. His search resulted in the conclusion that there were seven primary odors, and in 1952 his findings were summed up in a stereochemical theory of olfaction that identified the seven odors and gave a detailed description of the size, shape and chemical affinities of the seven corresponding receptor sites.

To identify the primary odors Amoore started with the descriptions of 600 organic compounds noted in the literature as odorous. If the receptor-site hypothesis was correct, the primary odors should be recognized much more frequently than mixed odors made up of two or more primaries. And indeed, in the chemists' descriptions certain odors turned up much more commonly than others. For instance, the descriptions mentioned more than 100 compounds as having a camphor-like odor, whereas only about half a dozen were put in the category characterized by the odor of cedarwood. This suggested that in all likelihood the camphor odor was a primary one. By this test of frequency, and from other considerations, it was possible to select seven odors that stand out as probable primaries. They are: camphoraceous, musky, floral, pepperminty, ethereal (ether-like), pungent and putrid.

From these seven primaries every known odor could be made by mixing them in certain proportions. In this respect the primary odors are like the three primary colors (red, green and blue) and the four primary tastes (sweet, salt, sour and bitter).

To match the seven primary odors there must be seven different kinds of olfactory receptors in the nose. We can picture the receptor sites as ultramicroscopic slots or hollows in the nerve-fiber membrane, each of a distinctive shape and size. Presumably each will accept a molecule of the appropriate configuration, just as a socket takes a plug. Some molecules may be able to fit into two different sockets—broadside into a wide receptor or end on into a narrow one. In such cases the substance, with its molecules occupying both types of receptor, may indicate a complex odor to the brain.

The next problem was to learn the shapes of the seven receptor sites. This was begun by examining the structural formulas of the camphoraceous compounds and constructing models of their molecules. Thanks to the techniques of modern stereochemistry, which explore the structure of molecules with the aid of X-ray diffraction, infrared spectroscopy, the electron-beam probe and other means, it is possible to build a three-dimensional model of the molecule of any chemical compound once its structural formula is known. There are rules for building these models; also available are building blocks (sets of atomic units) on a scale 100 million times actual size.

As the models of the camphoraceous molecules took form, it soon became clear that they all had about the same shape: they were roughly spherical. Not only that, it turned out that when the models were translated into molecular dimensions, all the molecules also had about the same diameter: approximately seven angstrom units. (An angstrom unit is a ten-millionth of a millimeter.) This meant that the receptor site for camphoraceous molecules must be a hemispherical bowl about seven angstroms in diameter. Many of the camphoraceous molecules are rigid spheres that would inevitably fit into such a bowl; the others are slightly flexible and could easily shape themselves to the bowl.

When other models were built, shapes and sizes of the molecules representing the other primary odors were found [see *illustration on preceding two pages*]. The musky odor is accounted for by molecules with the shape of a disk about 10 angstroms in diameter. The pleasant floral odor is caused by molecules that have the shape of a disk with a flexible tail attached—a shape somewhat like a kite. The cool pepperminty odor is produced by molecules with the shape of a wedge, and with an electrically polarized group of atoms, capable of forming a hydrogen bond, near the point of the wedge. The ethereal odor is due to rod-shaped or other thin molecules. In each of these cases the receptor site in the nerve endings presumably has a shape and size corresponding to those of the molecule.

The pungent and putrid odors seem to be exceptions to the Lucretian scheme of shape-matching. The molecules responsible for these odors are of indifferent shapes and sizes; what matters in their case is the electric charge of the molecule. The pungent class of odors is produced by compounds whose molecules, because of a deficiency of electrons, have a positive charge and a strong affinity for electrons; they are called electrophilic. Putrid odors, on the other hand, are caused by molecules

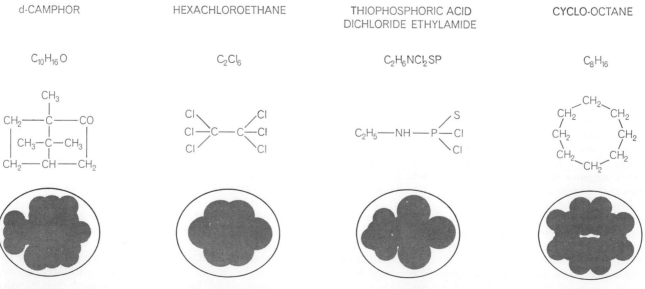

| d-CAMPHOR | HEXACHLOROETHANE | THIOPHOSPHORIC ACID DICHLORIDE ETHYLAMIDE | CYCLO-OCTANE |

$C_{10}H_{16}O$ C_2Cl_6 $C_2H_6NCl_2SP$ C_8H_{16}

UNRELATED CHEMICALS with camphor-like odors show no resemblance in empirical formulas and little in structural formulas.

Yet, because the size and shape of their molecules are similar, they all fit the bowl-shaped receptor for camphoraceous molecules.

that have an excess of electrons and are called nucleophilic, because they are strongly attracted by the nuclei of adjacent atoms.

A theory is useful only if it can be tested in some way by experiment. One of the virtues of the stereochemical theory is that it suggests some very specific and unambiguous tests. It has been subjected to six severe tests of its accuracy so far and has passed each of them decisively.

To start with, it is at once obvious that from the shape of a molecule we should be able to predict its odor. Suppose, then, that we synthesize molecules of certain shapes and see whether or not they produce the odors predicted for them.

Consider a molecule consisting of three chains attached to a single carbon atom, with the central atom's fourth bond occupied only by a hydrogen atom [see top illustration at right]. This molecule might fit into a kite-shaped site (floral odor), a wedge-shaped site (pepperminty) or, by means of one of its chains, a rod-shaped site (ethereal). The theory predicts that the molecule should therefore have a fruity odor composed of these three primaries. Now suppose we substitute the comparatively bulky methyl group (CH_3) in place of the small hydrogen atom at the fourth bond of the carbon atom. The introduction of a fourth branch will prevent the molecule from fitting so easily into a kite-shaped or wedge-shaped site, but one of the branches should still be able to occupy a rod-shaped site. As a result, the theory predicts, the ether smell should now predominate.

Another of us (Rubin) duly synthesized the two structures in his laboratory at the Georgetown University School of Medicine. The third author (Johnston), also working at the Georgetown School of Medicine, then submitted the products to a panel of trained smellers. He used an instrument called the olfactometer, which by means of valves and controlled air streams delivers carefully measured concentrations of odors, singly or mixed, to the observer. The amount of odorous vapor delivered was measured by gas chromatography. A pair of olfactometers was used, one for each of the two compounds under test, and the observer was asked to sniff alternately from each.

The results verified the predictions. The panel reported that Compound A had a fruity (actually grapelike) odor, and that Compound B, with the methyl

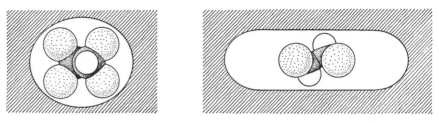

CHANGE IN SHAPE of a molecule changed its odor. The molecule at left smelled fruity because it fitted into three sites. When it was modified (*right*) by the substitution of a methyl group for a hydrogen, it smelled somewhat ethereal. Presumably the methyl branch made it fit two of the original sites less well but allowed it still to fit the ethereal slot.

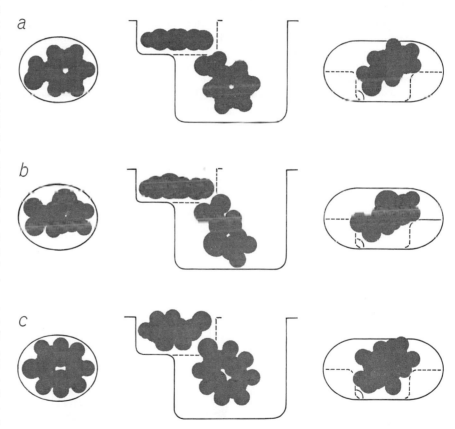

SINGLE CHEMICAL has more than one primary odor if its molecule can fit more than one site. Acetylenetetrabromide, for example, is described as smelling both camphoraceous and ethereal. It turns out that its molecule can fit either site, depending on how it lies.

COMPLEX ODORS are made up of several primaries. Three molecules with an almond odor are illustrated: benzaldehyde (*a*), alpha-nitrothiophen (*b*) and cyclo-octanone (*c*). Each of them fits (*left to right*) camphoraceous, floral (with two molecules) and pepperminty sites.

group substituted for the hydrogen atom, had a pronounced tinge of the ether-like odor. This experiment, and the theory behind it, make understandable the earlier finding that the odor of certain benzene-ring compounds changes sharply when the position of a group of atoms is shifted. The change in odor is due to the change in the over-all shape of the molecule.

A second test suggested itself. Could a complex odor found in nature be matched by putting together a combination of primary odors? Taking the odor of cedarwood oil as a test case, Amoore found that chemicals known to possess this odor had molecular shapes that would fit into the receptor sites for the camphoraceous, musky, floral and pepperminty odors. Johnston proceeded to try various combinations of these four primaries to duplicate the cedarwood odor. He tested each mixture on eight trained observers, who compared the synthetic odor with that of cedarwood oil. After 86 attempts he was able to produce a blend that closely matched the natural cedarwood odor. With the same four primaries he also succeeded in synthesizing a close match for the odor of sandalwood oil.

The next two tests had to do with the identification of pure (that is, primary) odors. If the theory was correct, a molecule that would fit only into a receptor site of a particular shape and size, and no other, should represent a primary odor in pure form. Molecules of the same shape and size should smell very much alike; those of a different primary shape should smell very different. Human subjects were tested on this point. Presented with the odors from a pair of different substances whose molecules nonetheless had the same primary shape (for example, that of the floral odor), the subjects judged the two odors to be highly similar to each other. When the pair of

substances presented had the pure molecular traits of different categories (for instance, the kite shape of the floral odor and the nucleophilic charge characteristic of putrid compounds), the subjects found the odors extremely dissimilar.

Johnston went on to make the same sort of test with honeybees. He set up an experiment designed to test their ability to discriminate between two odors, one of which was "right" (associated with sugar sirup) and the other "wrong" (associated with an electric shock). The pair of odors might be in the same primary group or in different primary groups (for example, floral and pepperminty). At pairs of scented vials on a table near the hive, the bees were first conditioned to the fact that one odor of a pair was right and the other was wrong. Then the sirup bait in the vials was replaced with distilled water and freshly deodorized scent vials were substituted for those used during the training period. The visits of the marked bees to the respective vials in search of sirup were counted. It could be assumed that they would tend to visit the odor to which they had been favorably conditioned and to avoid the one that had been associated with electric shock, provided that they could distinguish between the two.

So tested, the honeybees clearly showed that they had difficulty in detecting a difference between two scents within the same primary group (say pepperminty) but were able to distinguish easily between different primaries (pepperminty and floral). In the latter case they almost invariably chose the correct scent without delay. These experiments indicate that the olfactory system of the honeybee, like that of human beings, is based on the stereochemical principle, although the bee's smelling organ is different; it smells not with a nose but with antennae. Apparently the receptor sites on the antennae are differentiated by shape in the same way as those in the human nose.

A fifth test was made with human observers trained in odor discrimination. Suppose they were presented with a number of substances that were very different chemically but whose molecules had about the same over-all shape. Would all these dissimilar compounds smell alike? Five compounds were used for the test. They belonged to three different chemical families differing radically from one another in the internal structure of their molecules but in all five cases had the disk shape characteristic of the molecules of musky-odored substances. The observers, exposed to

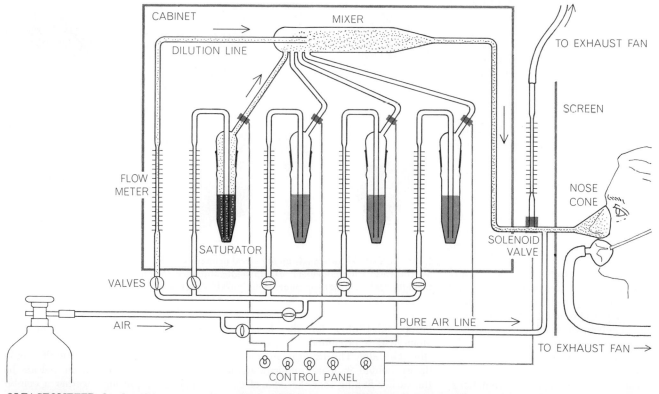

OLFACTOMETER developed by one of the authors (Johnston) mixes odors in precise proportions and delivers them to a nose cone for sampling. This schematic diagram shows the main elements. Air bubbles through a liquid in one of the saturators, picks up odorous molecules and is then diluted with pure air or mixed with air carrying other odors. The experimenter controls the solenoid valves.

the vapors of these five chemicals among many others by means of the olfactometer, did indeed pick out and identify all five as musky. By the odor test, however, they were often unable to distinguish these five quite different chemicals from one another.

Basically all this evidence in favor of the stereochemical theory was more or less indirect. One would like some sort of direct proof of the actual existence of differentiated receptor sites in the smelling organ. Recently R. C. Gesteland, then at the Massachusetts Institute of Technology, searched for such evidence. He devised a way to tap the electric impulses from single olfactory-nerve cells by means of microelectrodes. Applying his electrodes to the olfactory organ of the frog, Gesteland presented various odors to the organ and tapped the olfactory cells one by one to see if they responded with electric impulses. He found that different cells responded selectively to different odors, and his exploration indicated that the frog has about eight such different receptors. What is more, five of these receivers correspond closely to five of the odors (camphoraceous, musky, ethereal, pungent and putrid) identified as primary in the stereochemical theory! This finding, then, can be taken as a sixth and independent confirmation of the theory.

Equipped now with a tested basic theory to guide further research, we can hope for much faster progress in the science of osmics (smell) than has been possible heretofore. This may lead to unexpected benefits for mankind. For man the sense of smell may perhaps have become less essential as a life-and-death organ than it is for lower animals, but we still depend on this sense much more than we realize. One can gain some appreciation of the importance of smell to man by reflecting on how tasteless food becomes when the nose is blocked by a head cold and on how unpleasantly we are affected by a bad odor in drinking water or a closed room. Control of odor is fundamental in our large perfume, tobacco and deodorant industries. No doubt odor also affects our lives in many subtle ways of which we are not aware.

The accelerated research for which the way is now open should make it possible to analyze in fine detail the complex flavors in our food and drink, to get rid of obnoxious odors, to develop new fragrances and eventually to synthesize any odor we wish, whether to defeat pests or to delight the human nose.

CONSTANT-TEMPERATURE CABINET maintains the olfactometer parts at 77 degrees Fahrenheit. The photograph shows the interior of the cabinet, containing two units of the type diagramed on the opposite page. Several of the saturators are visible, as are two mixers (*horizontal glass vessels*), each of them connected by tubing to a nose cone at right.

13

L-ASPARAGINE AND LEUKEMIA

LLOYD J. OLD, EDWARD A. BOYSE AND H. A. CAMPBELL
August 1968

*The cells of some leukemias need an outside supply of
the amino acid L-asparagine. This means they are
vulnerable to treatment with the enzyme L-asparaginase,
which destroys the amino acid*

Cancer presents a special challenge to the investigator, not only as a hazard to life and health but also as a problem in the fundamental processes of living organisms. Intensive efforts have been made to find substances capable of destroying cancer cells selectively, and useful compounds have been discovered that will prolong the survival of cancer patients. These chemotherapeutic compounds have the deficiency that their lethal action is not confined to cancer cells: it extends in various degrees to normal cells. Their use is therefore limited by their toxicity for normal cells, particularly for cells with a high rate of multiplication, such as the cells of the blood-forming tissues and those of the intestinal lining.

The main reason for this relative lack of success is past failure to detect any absolute biochemical property that distinguishes a cancer cell from a comparable normal cell. For years cancer investigators have been looking for substances made only by cancer cells or required only by cancer cells, because such substances would offer a point of attack specifically against the cancer cell. Among the most significant developments of the past few years is the demonstration that cancer cells do indeed differ from normal cells, both in structure and in metabolism.

The evidence for differences in structure has come from studies in immunology. Components unknown in normal cells, recognizable by immunological methods as antigens, have been found in malignant cells, and this is probably characteristic of all types of cancer to which experimental animals are susceptible. Research in this area now includes ways to exploit the newfound antigenic differences, with the long-range goal of controlling cancer by immunization.

The work we shall review here is concerned more with metabolic differences than with structural ones. Evidence that the metabolism of cancer cells differs significantly from the metabolism of normal cells has been sought for some time, but only recently has such a distinction come to light. As a result of experiments undertaken in the 1950's it is now known that if malignant cells of certain types are deprived of asparagine, a common amino acid that normal cells apparently can synthesize for themselves, they die. The discovery of this metabolic defect of some malignant cells, and the recognition that it can be exploited in the treatment of cancer, is one of the landmarks of modern cancer research. We shall trace the history of this discovery and report on the prospects for its clinical application.

In 1953 John G. Kidd of the Cornell University Medical College reported experiments originally concerned with immunological responses to cancer. He wanted to see if antiserum taken from rabbits immunized with a transplanted mouse leukemia would affect the growth of the leukemia in mice. Here it should be explained that animal leukemias are maintained for study in the laboratory by transplanting the malignant cells from one animal to another; there they continue to proliferate, rendering the new host leukemic and providing more cells for subsequent transplantation. In general this is possible only because the animals used for the purpose have been inbred for many generations to the point of genetic uniformity. They no longer reject grafts of one another's tissues and consequently accept serially grafted leukemias and other cancers arising in the same inbred strain.

After transplanting a leukemia into a group of mice, Kidd gave the mice injections of the rabbit antiserum and also injections of serum taken from healthy guinea pigs. Guinea-pig serum is rich in certain proteins, known collectively as "complement," that augment antibody reactions. The guinea-pig serum was injected with the aim of increasing the effectiveness of any antibodies present in the rabbit serum.

As an experimental control Kidd also injected some leukemic mice with guinea-pig serum alone. To his surprise the leukemia of these mice regressed and in some instances disappeared permanently. He went on to establish three important facts concerning what came to be known as the "Kidd phenomenon." First, he showed that guinea-pig serum was the only serum that had this effect on leukemias. (We shall see that this statement was later qualified.) Serum from rabbits, horses and humans produced no response. Second, he established that the guinea-pig serum apparently affected only malignant cells, leaving the mouse's normal tissues unaltered and for this reason producing no toxic side effects. Third, he found that the guinea-pig serum affected some transplanted leukemias but not others. The unaffected group included all the newly transplanted leukemias, which at first made it appear that the only responsive leukemias were those that had had a long history of transplantation.

Not long after Kidd's discovery Thomas A. McCoy and his colleagues at the Samuel Roberts Noble Foundation in Ardmore, Okla., reported the results of a nutritional study of animal cancer cells grown in laboratory culture. Among other observations, McCoy and his associates noted that cultures of certain rat-tumor cells sooner or later die unless they are supplied with L-asparagine. This substance was among the first amino acids to be isolated by the pioneering biochemists of the 19th century; the

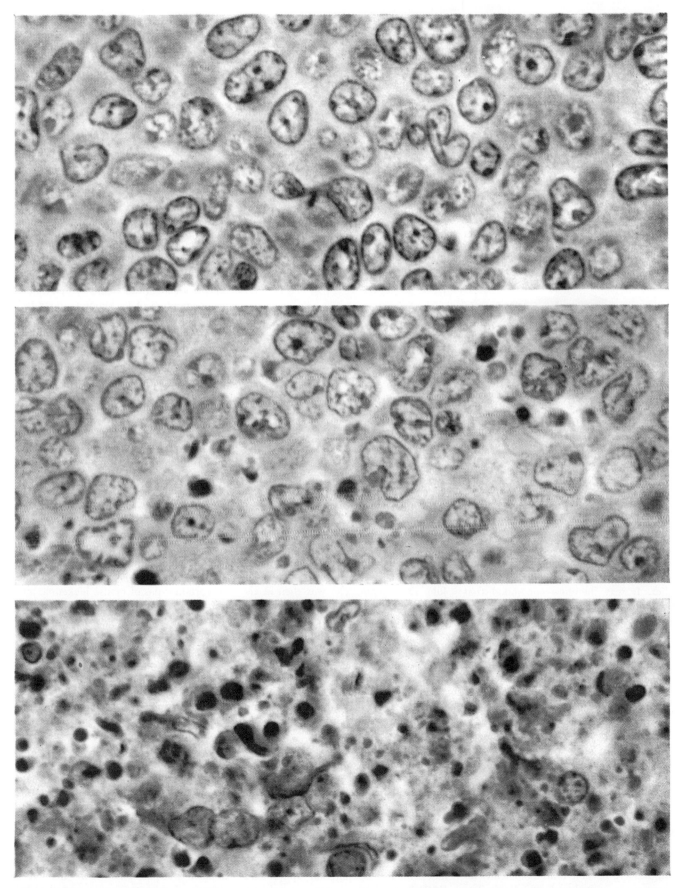

DESTRUCTION OF LEUKEMIA CELLS in mice by a single injection of the enzyme L-asparaginase is shown in this series of photomicrographs made at the Sloan-Kettering Institute for Cancer Research. In the micrograph at top is leukemic tissue taken from a mouse before it had received the injection; the round bodies are leukemia cell nuclei. In the middle micrograph is a similar sample taken from a mouse four hours after an injection; the comparatively small dark spots are leukemia cells that have died. In the micrograph at the bottom is a third sample, taken from a mouse 17 hours after an injection; most of the leukemia cells are dead.

HOST MOUSE, its shaved back swollen with a transplanted leukemia (*top*), has received a single injection of L-asparaginase. After seven days (*bottom*) the swelling has disappeared.

by a variety of physical and chemical methods, indicating that by these criteria they are one and the same. Second, he showed that this type of leukemia needed a supply of L-asparagine in order to grow in culture. In cultures lacking asparagine most of the leukemia cells were dead within a few days, although when these cultures were maintained for several weeks instead of being discarded, the few remaining cells began to grow without asparagine. When these cells were transplanted into mice once again, they no longer responded to guinea-pig serum.

Broome's incisive investigation of the Kidd phenomenon, published in *Nature* in 1961, did not receive immediate and wide recognition. Even today some reviews of cancer biochemistry make little or no mention of it. Major findings frequently go unrecognized, however, because they run contrary to established opinion—in this case the opinion that an absolute distinction between cancer cells and normal cells does not exist.

Since 1961 our group at the Sloan-Kettering Institute for Cancer Research has studied more than 100 newly transplanted mouse leukemias. This work has been directed primarily toward elucidating the antigenic structure of these leukemias in relation to their cause. It has shed further light on the Kidd phenomenon, furnishing the reason for the earlier view that only long-transplanted leukemias were affected by guinea-pig serum. First of all, it became evident that asparagine dependence is a common attribute of certain types of mouse leukemia, and that the dependence is unrelated to the transplantation history of particular leukemias. There is, however, a class of mouse leukemias that seldom exhibit dependence on asparagine. This is the class of leukemias caused by the known leukemia viruses of mice. The leukemias Kidd used for testing the effect of guinea-pig serum on new transplants came from strains of mice that carry leukemia virus and are therefore prone to the disease. Such leukemias are the most readily obtainable, but they belong, as we now realize, to an unresponsive class.

At this point the question that came to the fore was: Is responsiveness to asparaginase a peculiarity of certain rodent leukemias, or is it characteristic of cancer in many species? Before we could readily seek an answer we needed a source of asparaginase more plentiful than the guinea pig could provide. The guinea pig belongs to a superfamily of South American rodents, the Cavioidea; this superfamily includes the capybara

prefix *L* indicates that it is in the three-dimensional form designated levorotary. Although asparagine is a constituent of animal proteins, nutritional studies have shown that it is not one of the amino acids essential to an adequate animal diet. It was therefore concluded that normal animal cells can synthesize their own supply of asparagine. Indeed, the studies of several other investigators interested in the amino acid requirements of cells growing in laboratory culture had not revealed any instance in which L-asparagine needed to be added to the culture medium.

The importance of Kidd's and McCoy's findings was not widely appreciated at the time, nor was a connection perceived between them. With respect to Kidd's work the prevailing impression was that his results might in some way depend on the circumstances of prolonged transplantation. Since the leukemias that responded to guinea-pig serum in his experiments had been transplanted a large number of times, they might have become immunologically incompatible with their mouse hosts. The newly transplanted leukemias were not under suspicion of incompatibility and had not responded to the serum. Hence it was commonly supposed that the

Kidd phenomenon, although a fascinating puzzle, concerned tissue-transplantation biology rather than cancer.

Several years passed before the two apparently unrelated findings were shown to be two sides of the same coin. John D. Broome, who was then at the Cornell University Medical College, convinced himself that Kidd's results could not be attributed to immunological rejection. Searching for nonimmunological differences between guinea-pig serum and the serum of other animals, Broome found that in the 1920's A. Clementi, an investigator at the University of Rome, had discovered that the blood of guinea pigs contained an enzyme that destroys asparagine. This enzyme was absent from the blood of many other animals he tested. Broome then began experiments that left no doubt that the enzyme of guinea-pig serum, called asparaginase, is the antileukemic factor responsible for the Kidd phenomenon.

The strength of Broome's conclusions lay in his use of evidence from two independent experimental approaches. With one of the same transplanted leukemias that Kidd had studied, he showed first that the antileukemia property of guinea-pig serum could not be distinguished from asparaginase activity

(the largest of all rodents), the Patagonian hare, the paca and the agouti (all substantially larger than the guinea pig). Serums from a number of South American rodents have been tested by us and by Nelson D. Holmquist of the Louisiana State University School of Medicine. L-asparaginase activity has been found in the serum of all the members of the superfamily Cavioidea. As expected, the presence of L-asparaginase confers on these serums the ability to suppress leukemias that are known to respond to guinea-pig serum. In agouti serum the enzyme was six times as abundant as it is in guinea-pig serum. Serum from more distant relatives of the guinea pig—members of different superfamilies in the same suborder—did not contain asparaginase, showing that the occurrence of asparaginase in blood is a phylogenetic character dating back to a period before the evolutionary diversification of the Cavioidea.

Asparaginase is found in many cells of animals and plants. The Cavioidea are unusual only in that the enzyme is released from the cells and circulates in the blood serum. It seemed that the problem of supply might best be solved by extracting an effective asparaginase from microorganisms, but early attempts to accomplish this were not successful. For a time it appeared that the agouti might be the most feasible source of the enzyme in quantity. In fact, a patient was treated with agouti serum in 1965 by a group of physicians at the University of Recife in Brazil. Then Louise T. Mashburn and John C. Wriston, Jr., of the University of Delaware found that an asparaginase that was effective against

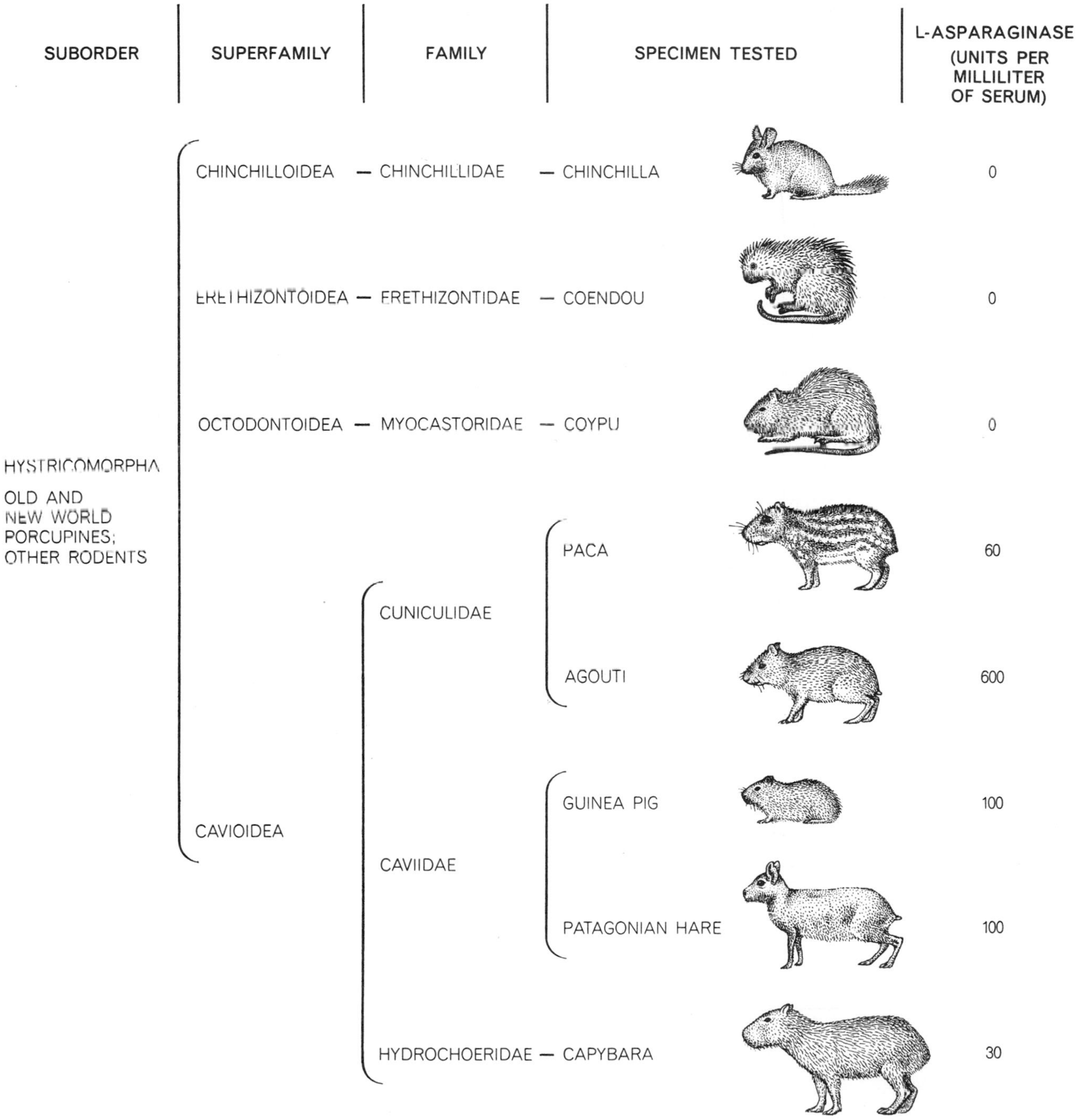

SUBORDER	SUPERFAMILY	FAMILY	SPECIMEN TESTED	L-ASPARAGINASE (UNITS PER MILLILITER OF SERUM)
	CHINCHILLOIDEA — CHINCHILLIDAE	— CHINCHILLA		0
	ERETHIZONTOIDEA — ERETHIZONTIDAE	— COENDOU		0
	OCTODONTOIDEA — MYOCASTORIDAE	— COYPU		0
HYSTRICOMORPHA OLD AND NEW WORLD PORCUPINES; OTHER RODENTS		CUNICULIDAE	PACA	60
			AGOUTI	600
	CAVIOIDEA	CAVIIDAE	GUINEA PIG	100
			PATAGONIAN HARE	100
		HYDROCHOERIDAE — CAPYBARA		30

SOUTH AMERICAN RODENTS of the suborder Hystricomorpha include five whose blood contains the enzyme L-asparaginase. Not all rodents in the group share this characteristic; the authors have not detected the enzyme in chinchilla, coendou or coypu blood.

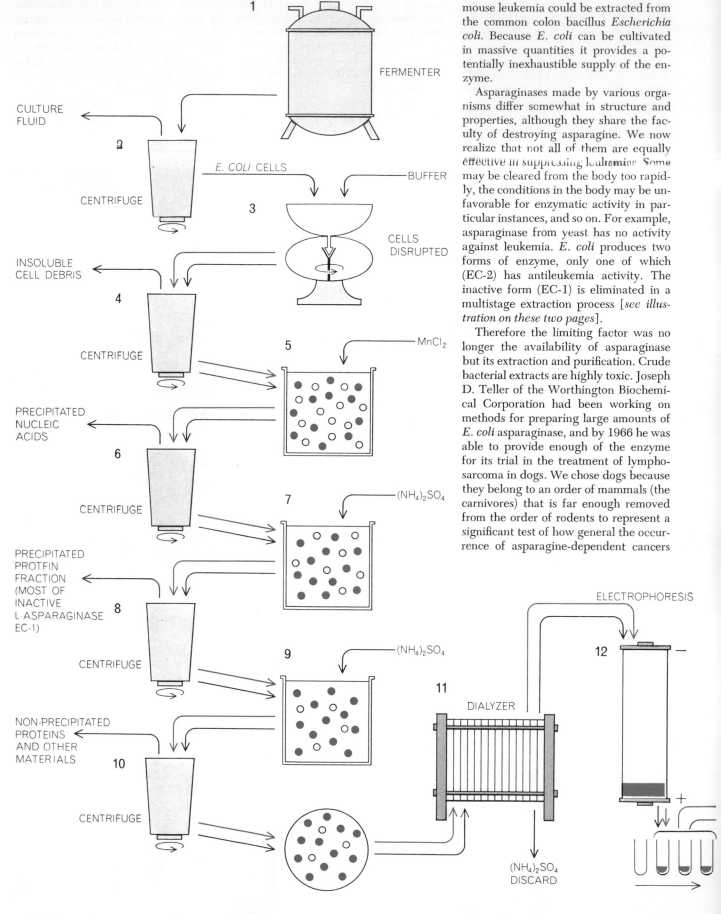

1 FERMENTER

CULTURE FLUID

2 CENTRIFUGE

E. COLI CELLS   BUFFER

3 CELLS DISRUPTED

INSOLUBLE CELL DEBRIS

4 CENTRIFUGE

5   MnCl₂

PRECIPITATED NUCLEIC ACIDS

6 CENTRIFUGE

7   (NH₄)₂SO₄

PRECIPITATED PROTEIN FRACTION (MOST OF INACTIVE L-ASPARAGINASE EC-1)

8 CENTRIFUGE

9   (NH₄)₂SO₄

NON-PRECIPITATED PROTEINS AND OTHER MATERIALS

10 CENTRIFUGE

11   DIALYZER

(NH₄)₂SO₄ DISCARD

12   ELECTROPHORESIS

mouse leukemia could be extracted from the common colon bacillus *Escherichia coli*. Because *E. coli* can be cultivated in massive quantities it provides a potentially inexhaustible supply of the enzyme.

Asparaginases made by various organisms differ somewhat in structure and properties, although they share the faculty of destroying asparagine. We now realize that not all of them are equally effective in suppressing leukemia. Some may be cleared from the body too rapidly, the conditions in the body may be unfavorable for enzymatic activity in particular instances, and so on. For example, asparaginase from yeast has no activity against leukemia. *E. coli* produces two forms of enzyme, only one of which (EC-2) has antileukemia activity. The inactive form (EC-1) is eliminated in a multistage extraction process [*see illustration on these two pages*].

Therefore the limiting factor was no longer the availability of asparaginase but its extraction and purification. Crude bacterial extracts are highly toxic. Joseph D. Teller of the Worthington Biochemical Corporation had been working on methods for preparing large amounts of *E. coli* asparaginase, and by 1966 he was able to provide enough of the enzyme for its trial in the treatment of lymphosarcoma in dogs. We chose dogs because they belong to an order of mammals (the carnivores) that is far enough removed from the order of rodents to represent a significant test of how general the occurrence of asparagine-dependent cancers

L-ASPARAGINASE CAN BE OBTAINED from bacteria rather than from animals by the process outlined here. The bacterium is

the common colon bacillus *Escherichia coli*. The bacteria are first grown in a fermenter (*1*). Then the culture fluid is removed (*2*)

is. Moreover, dogs have a high frequency of lymphosarcoma, which resembles some human cancers. In our first experiments (undertaken with Robert S. Brodey and I. J. Fidler of the University of Pennsylvania School of Veterinary Medicine) we selected three dogs in which the disease was far advanced; the dogs' lymph nodes and tonsils were greatly enlarged and the animals were scarcely able to eat or move about.

The results of treatment with asparaginase were striking. Within a week injections of the enzyme restored two of the dogs to apparently normal health, and the third dog showed considerable improvement. It is remarkable that cancerous tissue can be destroyed on such a scale, with the concomitant release of breakdown products, without evidence of toxicity.

Because asparaginase was still comparatively scarce we were not able to test the effects of massive doses or more extended periods of treatment. The lymphosarcoma reappeared in the dogs between seven and 50 days after the injections had been discontinued. Asparaginase was nonetheless available for a second course of treatment for one of the dogs, and the response of the animal was equally striking.

With evidence that asparaginase was effective against cancers of animals as distantly related as mice and dogs, hopes were raised that asparagine deprivation would prove effective in destroying some forms of cancer in man. Clinical studies undertaken last year at the Sloan-Kettering Institute and at the Wadley Institute of Molecular Biology in Dallas have shown that one kind of human leukemia, the acute lymphoblastic type, often exhibits the metabolic defect that makes cancer cells vulnerable to asparaginase. Leukemia in children is commonly of this type. To judge both from laboratory tests with cultured cells and from patients' response to treatment, the majority of such leukemias are asparagine-dependent. It is these leukemias that have been the most intensively studied so far, but there is hope that other types of human leukemia and other human cancers may also be affected by asparaginase.

The degree of benefit to be gained from asparaginase therapy, even with respect to leukemia of the acute lymphoblastic type, remains unknown. Although the enzyme causes gross manifestations of the disease to disappear for various periods of time, it is not the only substance that can accomplish this result. Certain hormones of the adrenal cortex and drugs that interfere with the synthesis of nucleic acid will also induce such remission. The distinction between asparaginase and other agents is that the enzyme has an obvious adverse effect only on malignant cells, whereas the other substances are not inherently selective in their action. Since the action of asparaginase is specific, it should follow that the patient will suffer no side effects attributable to the enzyme alone (as contrasted with those attributable to toxic contaminants). So far the supply of sufficiently pure asparaginase has been too limited to allow even a determination of how much can safely be administered in man. The dosage may well be many times the amount now being used in treatment.

Supplies of asparaginase for clinical use are likely to remain inadequate for some time. For this reason it is desirable to have tests that precede the treatment of patients and can determine in advance whether a particular cancer is asparagine-dependent. Several laboratory tests are possible. Asparagine-dependent cells of mouse leukemias and of dog lymphosarcomas may die so rapidly (within 24 hours) when they are cultured without asparagine that this alone can suffice to distinguish asparagine-requiring leukemias in these animals. Cultures of human leukemia cells, however, do not give equally clear-cut results with this simple test. Nor is the failure of cancer cells to proliferate in the absence of asparagine an adequate test in man, as it is with some transplanted leukemias in mice. The reason is that most primary cancers in man and animals do not multiply the same way in culture.

We therefore resorted to another test (first applied by L. H. Sobin and Kidd in mice) that measures the response of cultured cancer cells to asparagine deprivation. It is based on the fact that living cells continuously synthesize protein and nucleic acid. Cultures of cancer cells, some with asparagine and some without, are therefore incubated with radioactive-

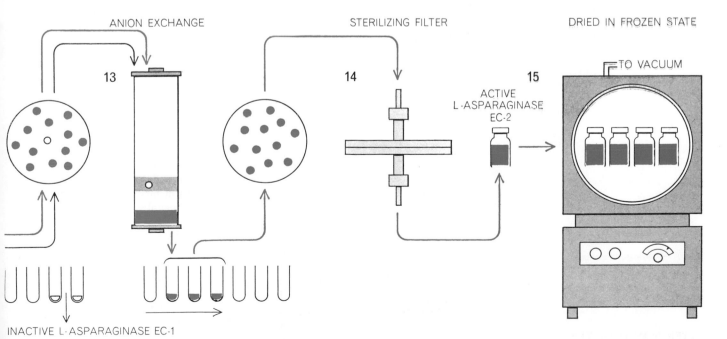

ANION EXCHANGE   STERILIZING FILTER   DRIED IN FROZEN STATE

13   14   15

TO VACUUM

ACTIVE L-ASPARAGINASE EC-2

INACTIVE L-ASPARAGINASE EC-1

and the bacterial cells are disrupted (3). Next the material other than L-asparaginase EC-2 is removed by a variety of techniques (4 through 14) including salt fractionation, electrophoresis, ion exchange and filtration. Finally enzyme is freeze-dried for storage.

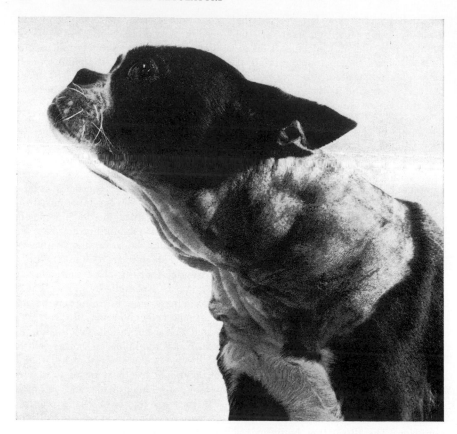

BOSTON TERRIER WITH LYMPHOSARCOMA has massively enlarged glands in the neck. (The neck is shaved.) After photograph was made dog was treated with L-asparaginase.

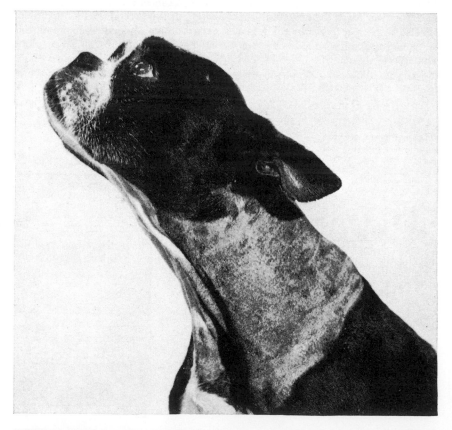

AFTER 10 DAYS OF TREATMENT the dog was restored to apparently normal health. Treatment was given by W. D. Hardy, Jr., of Sloan-Kettering Institute for Cancer Research.

ly labeled precursors of protein or nucleic acid. If the deprived culture fails to show sustained incorporation of the labeled precursors into newly synthesized protein and nucleic acid, we conclude that the cancer is asparagine-dependent. With mouse leukemias the test shows within as short a period as 30 minutes whether or not the cancer cells are subject to destruction by asparaginase. It is now being widely applied in studying the asparagine requirements of human cancer cells.

Many questions await solution. What types of human cancer, in addition to leukemia, will respond to asparaginase? Asparagine-requiring cancers of other kinds are known in rats and dogs, and it is hoped that the same may be true in man. A leading theoretical question is *why* some cells need an external supply of asparagine. Do they require more asparagine than they can synthesize? Or have they lost the capacity to synthesize the amino acid? Studies in a number of laboratories suggest that the latter is the case. It appears that asparagine-dependent cells lack an enzyme, asparagine synthetase, that in normal cells (and in cancer cells unresponsive to asparaginase) converts aspartic acid to asparagine. This suggests a parallel with microorganisms in which mutation has caused the loss of an enzyme, or the production of a defective enzyme, leading to an absolute requirement for a substance that would normally be synthesized.

The behavior of cancers may suggest that malignant cells revert to a primitive state in which they show decreased dependence on external conditions. This is the impression one gains on observing how a cancer thrives as the body wastes away. It is thus particularly gratifying to find that some cancer cells, rather than attaining metabolic independence, have lost the ability to perform a function of which normal tissues are capable.

The recognition of asparagine dependence as a property of certain cancers helps to stem the tide of pessimism generated by years of biochemical research that did not reveal absolute distinctions that would point the way to specific cancer therapies. There is clearly a possibility that other specific nutritional requirements of cancer cells may come to light, suggesting other enzymes and other approaches that may be of value in the treatment of cancer. The optimistic view is that the defect in asparagine metabolism will prove not to be the only example of its kind.

14

THE INDUCTION OF INTERFERON

MAURICE R. HILLEMAN AND ALFRED A. TYTELL

July 1971

*A synthetic RNA called poly I:C may provide
broad-spectrum protection against virus diseases.
It makes cells form interferon, the protein that defends
other cells against infection by viruses*

There are potentially three ways in which virus diseases can be controlled in man. To date the only clinically practical way has been to administer a vaccine: a preparation of killed or attenuated virus that stimulates the body to form antibodies against that virus. The specificity, or narrow spectrum, of vaccines is a limitation; it means that a different vaccine is required for each virus or strain of virus. Another possible method, and potentially one with broad-spectrum action, is chemotherapy. No safe and effective chemical agents have been discovered, however, that act against viruses in the way that sulfonamides or antibiotics act against large groups of bacteria. The third possibility is to rely on what is apparently the cell's own first line of defense against virus attack: interferon. Since 1957 it has been known that a cell infected by a virus produces a substance, interferon, that protects other cells against virus infection. Now an effective method has been found for mobilizing the body's interferon system and the way is open for the assessment of the method's safety and efficacy in clinical trials.

In 1935 Meredith Hoskins of the Rockefeller Foundation observed that infection of a monkey with an attenuated yellow-fever virus prevented death from a virulent yellow-fever virus administered at the same time. Two years later George M. Findlay and F. O. MacCallum of the Wellcome Research Institution in England found that monkeys infected with the virus of Rift Valley fever were protected against yellow-fever virus. Because it operated against a second virus given at the same time as the attenuated one and also against a different virus, this protective effect was clearly not related to known immune reactions involving the antigen-antibody response. Findlay

and MacCallum called it "virus interference," and in the next few years many examples of interference between pairs of viruses were reported. No one found a practical way to adapt the phenomenon to the prevention of viral disease, however.

Then in 1957, at the National Institute for Medical Research in England, Alick Isaacs and Jean Lindenmann made the epochal discovery that the interfering action of viruses was brought about by a substance, which they called interferon, that was produced by the virus-infected cell and that protected uninfected cells against viral infection. Interferon could be destroyed by proteolytic enzymes and so it was presumed to be a protein; it was of relatively low molecular weight, and it was more resistant to acid and alkali than most proteins. Interferon's antiviral activity was found to

SOURCE	DOSE (MICROGRAMS)	INTERFERON TITER
DOUBLE-STRAND RNA		
POLY I:C	2	640
POLY A:U	25	20
PENICILLIUM FUNICULOSUM	8	640
REOVIRUS 3	8	640
MS2 COLIPHAGE (REPLICATIVE)	8	160
MU9 MUTANT COLIPHAGE (REPLICATIVE)	2	40
RICE DWARF VIRUS	20	640
CYTOPLASMIC POLYHEDROSIS VIRUS	22	160
SINGLE-STRAND RNA		
POLY I	25	0
POLY C	20	0
MS2 COLIPHAGE	8	0
ESCHERICHIA COLI	100	0
NEWCASTLE DISEASE VIRUS	10	0
INFLUENZA VIRUS	10	0
TOBACCO MOSAIC VIRUS	40	0
YEAST RIBOSOME	1.000	0
YEAST SOLUBLE RNA	200	0
DOUBLE-STRAND DNA		
CALF THYMUS TISSUE	200	0

SCREENING OF NUCLEIC ACIDS for interferon induction yields these results. Double-strand RNA's, both natural and synthetic, were effective in microgram amounts. The titers (*right*) are the dilutions of the rabbit serum that were effective in protecting 50 percent of the cell cultures tested. Single-strand RNA's and double-strand DNA were ineffective.

be very broad: the protein was effective against essentially all viruses. On the other hand, it had a narrow spectrum in a different sense: it was effective only in the animal species in which it was produced. Mouse interferon was active only in mouse cells, chick interferon in chick cells and so on [see "Interferon," by Alick Isaacs; SCIENTIFIC AMERICAN Offprint 87].

The immediate response to the discovery of interferon was the hope that the protein might be produced in cells, purified and administered to prevent and treat viral diseases. There was even the possibility that the interferon molecule might turn out to include a simple chemical group that was responsible by itself for the antiviral activity and that could be synthesized. The first necessary step was purification, a task that was undertaken by our group at the Merck Institute for Therapeutic Research. In 1963 we were able to report a significant degree of purification of interferon produced in embryonated hens' eggs that were infected with influenza virus. The refined interferon was extremely active against viruses in test-tube experiments; it was effective as an antiviral agent, for example, in smaller quantities than the usual antibiotics are effective when they are assayed against highly susceptible bacteria. It turned out that interferon's antiviral activity stemmed from the complex molecule as a whole rather than from any simple subgroup of the molecule. Our observations were confirmed and extended by Thomas C. Merigan, Jr., of the Stanford University School of Medicine and by Karl H. Fantes of Glaxo Laboratories Ltd., in England.

It is doubtful that absolutely pure interferon has yet been isolated. Moreover, there is no practical way to produce interferon in human cells that is safe for general use in man. Even if a safe way is found, it is not likely to be practical quantitatively. That is, the relation between the required dose and the possible yield of interferon by cells is so unfavorable at present as to seem to rule out the practicality of producing human interferon, purifying it and administering it to humans. Monto Ho and Bosko Postic of the University of Pittsburgh have arrived at a similar conclusion on the basis of extrapolation from a mouse-interferon system, but we should note that not all workers share this judgment.

In view of the limitations on the use of exogenous interferon in medicine, the obvious alternative was to find a suitable inducer of interferon: a substance that could be given a patient to cause his body to manufacture its own endogenous interferon. The search for inducers in a number of laboratories turned up many different kinds of substances that stimulated interferon production in animals. These included bacteria, parasites, viruses, polysaccharides, agents such as phytohemagglutinin that promote cell division, bacterial endotoxins, synthetic plastics and other substances. For various reasons none of these agents was suitable for clinical purposes.

Our group at the Merck Institute, consisting of the authors and A. Kirk Field, George P. Lampson and Marjorie M. Nemes, concentrated on an attempt to find and then imitate the means by which viruses naturally induce interferon. We decided on two approaches. The first was to fractionate a number of virions, or virus particles, and to try to find a component or components responsible for inducing interferon. The second was to isolate the active principle from helenine, an extract of the fungus *Penicillium funiculosum* that Richard E. Shope and his colleagues at the Rockefeller Institute for Medical Research had shown would induce interferon in cell culture and induce resistance to viral infection in animals. The event that happened to put us on the right track was the finding by Werner Braun and Masayasu Nakano of Rutgers University that antibody responses in animals were enhanced by certain synthetic polynucleotides. These are analogues of the nucleic acids DNA or RNA that are made in the laboratory by combining nucleotides (the subunits of nucleic acids) in arbitrary ways instead of in the naturally occurring arrangements that encode hereditary information in living cells.

We wondered whether such polynucleotides might also induce interferon, and we tested a number of synthetic RNA's. One of them was extremely active: in small amounts (micrograms) it induced interferon in rabbits and made cell cultures and animals resistant to viral infection. The active substance was poly I:C, consisting of paired homopolymer strands of polyriboinosinic acid (poly I) and polyribocytidylic acid (poly C) [*see illustration p. 110*]. Poly A:U, a complex based on adenylic and uridylic

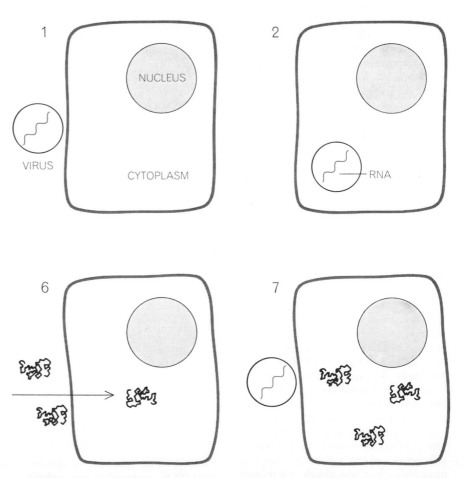

INTERFERON SYSTEM is believed to be induced and to provide protection against viral infection in two phases. The virus containing single-strand RNA attaches itself to the cell wall (*1*) and enters the cell (*2*). The outer coat of the virus is removed, leaving the single

acids, was also active in cell cultures but was much less active in animals. The individual single-strand homopolymers were not active.

Meanwhile the work on helenine was well on its way, and soon after we had established the interferon-inducing property of poly I:C we were able to report that the active principle in helenine was also a double-strand RNA. We suggested that the mold from which the helenine was derived must have been infected with a virus, which had presumably supplied the double-strand RNA. Our suspicion was confirmed when Walter J. Kleinschmidt and his associates at Eli Lilly and Company demonstrated the presence of a fungal virus in statalon, an extract of another penicillium species and, like helenine, an inducer of interferon. Later G. T. Banks and his co-workers at the Imperial College of Science and Technology in London isolated a fungal virus from *P. funiculosum* itself.

We thereupon undertook an extensive screening program, testing polynucleotides of varying origins for interferon induction. We did this by injecting the polynucleotides intravenously in rabbits, withdrawing blood and testing the ability of the blood serum to protect a cell culture against virus infection [*see illustration on page 113*]. A large number of double-strand RNA's, of natural and synthetic origin, were found to be active inducers. Single-strand RNA's were inactive, as was double-strand DNA [*see illustration on page 109*].

The question arose: Where does the double-strand RNA come from in the case of an ordinary virus infection? Whereas the genetic material is DNA in some viruses, in bacteria and in higher plants and animals, it is RNA in many viruses. In most RNA viruses it is the single-strand form of RNA that is present in the virion, not the double-strand form. We realized that such viruses therefore must not contain an interferon inducer to begin with; the interferon-inducing activity must arise only when the double-strand replicative form of RNA is produced in a virus-infected cell. It has since been demonstrated by immunofluorescence and other techniques that the double-strand form does indeed ap-

pear in the normal course of virus replication. The same thing is true of the synthetic polynucleotides: the homopolymers become active only when they are complexed to form a double strand. (Under certain conditions poly I or poly C alone may show weak interferon-inducing capacity, but such activity is trivial compared with that of poly I:C and is of no practical importance.)

The discovery that interferon is induced by double-strand RNA provided a breakthrough in the understanding of interferon induction by viruses as well as a possible new approach to the control of viral infections. The exact mechanism for interferon induction and utilization has yet to be fully elucidated, but in our working hypothesis the interferon system is conceived of as operating in two distinct phases [*see illustration below on these pages*]. In the first phase a single-strand RNA virus penetrates the cell. Its outer coat of protein is removed, releasing the RNA, which sets about producing the complementary strand of RNA that is essential to its replication. The double-strand RNA that is formed is not normal to the cell, and so in some

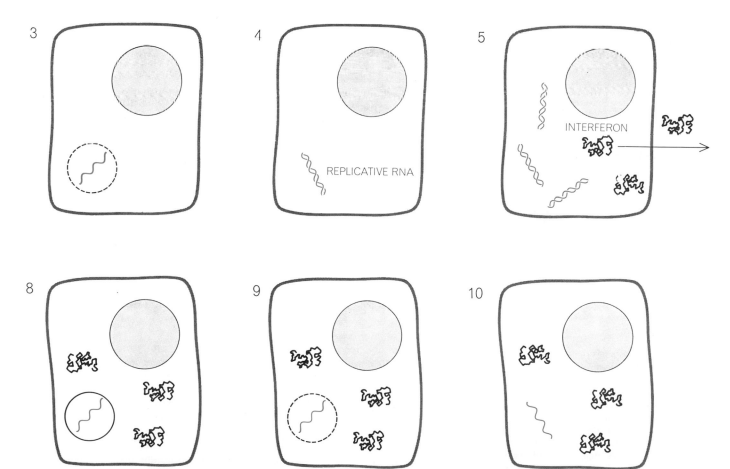

strand of RNA (3), which replicates to form a double strand (4). The presence of a foreign double-strand RNA causes the cell to synthesize interferon (5) but the RNA continues to replicate. The

interferon leaves the cell and enters a new one (6). It cannot keep a virus from entering the new cell (7, 8, 9) but somehow serves to prevent the replication of the single-strand viral RNA (10).

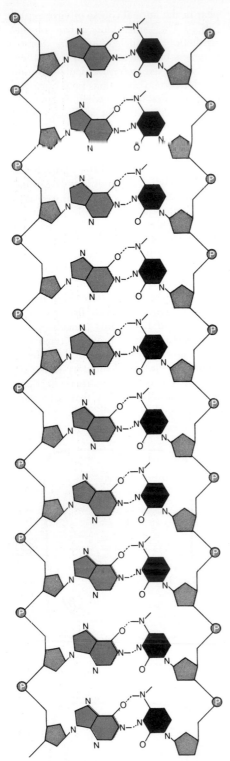

POLY I:C is a synthetic polynucleotide, a laboratory analogue of RNA formed by the complexing of two homopolymers: polyriboinosinic acid (*left*) and polyribocytidylic acid (*right*). Each strand, like the strands of a natural RNA, consists of a backbone (*gray*) of alternating phosphate (*P*) and sugar (ribose) groups to which nitrogenous bases are attached as side groups. In these synthetic homopolymers the bases are all the same: the purine hypoxanthine (*color*) and the pyrimidine cytosine (*black*). The two single strands are linked by hydrogen bonds between the bases (*broken lines*.)

undisclosed way it alerts the cell to synthesize interferon.

The interferon itself is not active against viruses. It leaves the cell in which it was produced and, in the second phase, enters uninfected cells, where it exerts antiviral activity. It may function as a "derepressor" that causes the new cell to produce a new protein that prevents the formation of viral nucleic acid when the cell is later attacked by a virus. Alternatively, it is possible that the interferon itself is altered by the new protein, perhaps enzymatically, to form an active antiviral molecule. There is only indirect evidence for the presence of this "new protein," and its precise mechanism of action in preventing viral replication is not known. The few viruses (such as the reoviruses) that contain double-strand RNA induce interferon immediately on cell penetration without the need for replication. At least one DNA virus, vaccinia, for some reason produces a double-strand RNA, which is active in inducing interferon; how most DNA viruses stimulate interferon induction is not yet known.

The usefulness of an interferon inducer cannot be predicted on the basis of a demonstration of induction alone. We went on to more meaningful tests that showed actual protection against cell destruction by a virus and prevention of disease and death in animals infected with viruses. In cell-culture trials of protective efficacy the inducer being tested is added to the culture, and about three hours later a measured challenge dose of virus is administered. Activity is demonstrated by the drug's ability to prevent viral damage to the sheet of cells in the culture. All the double-strand RNA's found to be active in inducing interferon in rabbits were also active, in microgram amounts, in preventing cell death by vesicular stomatitis virus in a wide variety of cells of avian and mammalian origin, including human cells. When the same inducers were tested in mice, all of them prevented disease and death in animals infected with the pneumonia virus of mice (PVM).

Because of the ready availability of poly I:C and its considerable promise in these early tests, further studies in our laboratories were centered on poly I:C alone. Following a common practice for appraising the potential usefulness of a drug, we established a "model system" in mice, infecting them with a standardized and ordinarily fatal dose of various viruses so that the prophylactic and therapeutic effect of poly I:C could be measured. In a typical experiment poly I:C was instilled into the nostrils of mice

before or after nasal inoculation of a standardized lethal amount of PVM virus [*see illustration on page 114*]. Nearly all untreated control mice died but all animals given poly I:C before the virus survived. So did nearly all mice that were given poly I:C two days after the virus. In other words, the poly I:C had a therapeutic as well as a prophylactic effect. Apparently poly I:C given soon enough after infection induces sufficient interferon to protect a large proportion of the cells; in effect the interferon outraces the virus, and the animal survives. Starting treatment four days after virus inoculation was too late and all the animals succumbed.

Experiments along these lines have revealed a strong protective effect of poly I:C in mice against the encephalomyocarditis virus, vaccinia virus and parainfluenza viruses. Minor protective effects were obtained in animals against the oncogenic (cancer-producing) SV-40 virus, adenovirus and Friend leukemia virus. This ability to protect against both DNA and RNA viruses and against oncogenic as well as cell-killing viruses emphasized the broad-spectrum antiviral activity of poly I:C. Workers in other laboratories have greatly extended the studies with poly I:C. They have shown poly I:C to be effective against severe herpes simplex eye infections in rabbits, Semliki Forest virus (an insect-borne virus) infections in mice, rabies in rabbits and certain respiratory virus infections such as influenza. Moreover, marked antitumor effects have been observed for poly I:C against several transplant tumors in mice, particularly the L1210 ascites tumor. Activity against experimental malaria and other parasitic diseases has also been recorded. The mechanisms of the effects against tumors and malaria are not known and may well involve properties other than interferon induction. Poly I:C, as noted above, is known to stimulate the ordinary immune mechanisms, including cell-mediated immunity, and this is believed to be important in defense against tumors.

Before any drug can be used in clinical trials in man, its safety must be assessed in animal tests, and so we carried out trials in mice, rats, dogs and monkeys with different doses and routes of injection. When the drug was given intravenously, destructive side effects were noted in the small blood vessels, liver and blood-forming organs of dogs; these effects were minor or absent following administration by the subcutaneous or respiratory routes. The toxic effects were far less in mice and rats than in dogs,

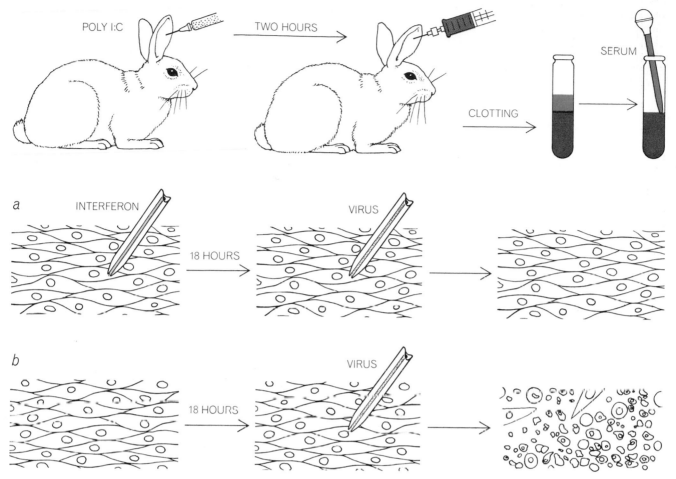

TEST FOR INDUCTION of interferon is conducted by injecting poly I:C (or any other potential inducing substance) intravenously in rabbits. After two hours the rabbits are bled and the blood is allowed to clot. Interferon induced in the rabbit is contained in the blood serum, which is tested at various dilutions by a cell-culture assay (*a*): the culture is treated with the diluted serum, incubated for 18 hours, washed and then infected with the virus of vesicular stomatitis. If the interferon level is adequate, the cells are unharmed. Meanwhile an untreated control culture is exposed to the virus in the same way (*b*). Its cells are killed by the virus.

and they were essentially absent in monkeys. The tests indicated that there would be a satisfactory margin of safety for human subjects treated with doses of poly I:C large enough to stimulate interferon.

Other tests for safety showed that poly I:C did not sensitize guinea pigs to anaphylactic reactions. The drug was not carcinogenic when injected into newborn animals, which were observed for 18 months after injection. It did not cause normal human cells to acquire cancer-like properties when they were assayed in cell culture or tested by implantation in the cheek pouch of hamsters. Finally, it did not bring about changes in the chromosomes of normal human cells grown in cell culture.

When the time came for testing in man we decided, because of the remarkable effect shown by poly I:C against certain cancer-producing viruses in animals and the striking beneficial action of the drug against certain transplanted tumors, to conduct trials in patients with

cancer. Accordingly we carried out a study in cooperation with Charles W. Young and Irwin H. Krakoff of the Memorial Hospital and Sloan-Kettering Institute for Cancer Research in New York. Twenty patients with advanced cancer of various types were given poly I:C intravenously—from two micrograms to four milligrams per kilogram of body weight. Serum samples for interferon assay were taken prior to injection and up to 72 hours thereafter. Fourteen of the 20 patients responded with the production of interferon; six failed to respond. The induction of interferon and the magnitude and duration of the interferon response did not appear to be related strongly to the size of the poly I:C dose. Interferon appeared in the blood as early as two hours after the poly I:C injection and persisted in some patients for as long as 72 hours. Peak production of interferon was usually attained between 12 and 48 hours after the administration of the drug. Unfortunately there was no significant beneficial

effect against the tumors in these patients.

Two significant problems that relate to the clinical application of interferon inducers are the relatively short duration of resistance following the administration of inducer (usually about six days in animals) and the failure to respond to a second injection of inducer within a certain period after successful induction. Fortunately this state of hyporeactivity seems to disappear at about the same time as the resistance to viral infection is lost—just when one needs to give a new dose of inducer. In the studies with human patients successful reinductions of interferon were obtained when the poly I:C injections were given six or seven days apart in two subjects and even within three days in one patient. Another patient showed lack of response to interferon induction only after the fourth of a series of daily doses of poly I:C, and this patient became fully responsive again after a 10-day period. It appears, then, that although resistance to induction of interferon does occur in man, its duration is relatively short. It may be that the administration at frequent intervals of small but therapeutically adequate doses of poly I:C will allow the continuing induction of interferon and thus maintain continuous resistance to viral infection.

Clinical reactions to poly I:C in man were remarkably few. The most consistent reaction to intravenous injection of poly I:C was fever. The temperature elevation ranged as high as seven degrees Fahrenheit, reaching a peak from six to 15 hours after injection. The fever was generally accompanied by interferon induction, although the responses did occur independently in some patients. The amount of fever was not correlated with the poly I:C dosage. Extensive laboratory observations revealed no disturbance of liver, kidney or bone-marrow functions. No effects on the blood-clotting mechanism were observed. In sum, there were no clinical symptoms that placed a limit on the doses we administered.

We were concerned about the possibility of an antibody response to the poly I:C, but tests showed that the patients did not develop antibody against either poly I:C or DNA. One related possibility remains to be investigated. It has been noted that the development of an autoimmune disease that resembles systemic lupus in man, and that appears spontaneously in a particular breed of mice, may be accelerated by multiple injections of poly I:C. The dose requirements and route of administration of poly I:C in these mice are now being studied in order to establish another safety limit for the drug in human subjects susceptible to autoimmune disease.

Most viral infections in human beings are self-limited, suggesting that natural defense mechanisms are operative in recovery from viral diseases. Antibody and cell-mediated immunity, the ordinary specific immune mechanisms, seem to come into effect rather late in the infectious process; they may be of greater importance in preventing later reinfection by the same virus or in removing virus-infected cells than in the early recovery of the host from a particular virus infection. Interferon is produced early in the process of infection and there is substantial reason to believe that it is at least one of the important factors involved in the recovery from viral disease.

The administration of exogenously produced interferon has shown very limited promise up to the present time because of limitations in source, safety and cost. The more practical approach seems to lie in the administration of interferon inducers. The fact that beneficial secondary effects, such as stimulation of the ordinary immune mechanisms, may also occur is an added reason for utilizing polynucleotides. Poly I:C shows considerable promise as a means for exploiting the interferon mechanism. The remarkably low toxicity for human beings and ready availability of this synthetic product open the door to large-scale exploratory investigations. After some final tests to rule out the danger of autoimmune disorders, poly I:C will be ready for cautious trials in human subjects for preventing infections, such as the common cold, that are caused by a variety of viruses.

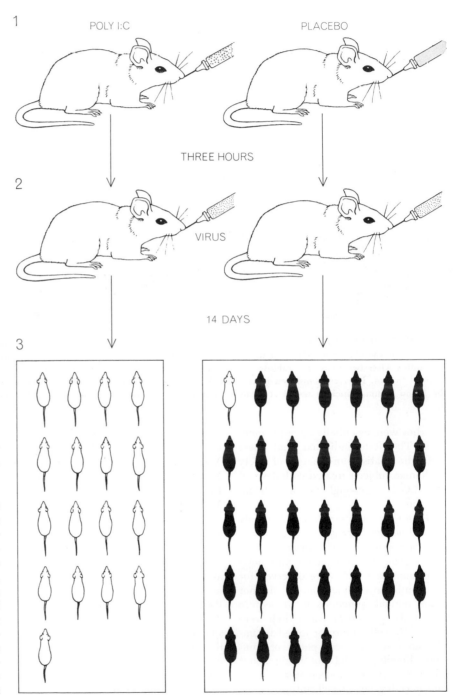

PROTECTIVE EFFECT of poly I:C was tested in laboratory animals. Poly I:C was instilled into the nostrils of some mice (*left*) and a control group (*right*) was given a placebo. After three hours both groups of mice were infected with a lethal dose of pneumonia virus of mice (PVM). The animals were then observed for 14 days. All 17 of the treated mice survived but only one of 32 mice in the control group survived. The drug was also effective when it was given two days after the virus; the survival rates were 15 of 15 treated animals and one of 25 controls. Given four days after the virus, poly I:C was ineffective.

15

PHEROMONES

EDWARD O. WILSON
May 1963

A pheromone is a substance secreted by an animal that influences the behavior of other animals of the same species. Recent studies indicate that such chemical communication is surprisingly common

It is conceivable that somewhere on other worlds civilizations exist that communicate entirely by the exchange of chemical substances that are smelled or tasted. Unlikely as this may seem, the theoretical possibility cannot be ruled out. It is not difficult to design, on paper at least, a chemical communication system that can transmit a large amount of information with rather good efficiency. The notion of such a communication system is of course strange because our outlook is shaped so strongly by our own peculiar auditory and visual conventions. This limitation of outlook is found even among students of animal behavior; they have favored species whose communication methods are similar to our own and therefore more accessible to analysis. It is becoming increasingly clear, however, that chemical systems provide the dominant means of communication in many animal species, perhaps even in most. In the past several years animal behaviorists and organic chemists, working together, have made a start at deciphering some of these systems and have discovered a number of surprising new biological phenomena.

In earlier literature on the subject, chemicals used in communication were usually referred to as "ectohormones." Since 1959 the less awkward and etymologically more accurate term "pheromones" has been widely adopted. It is used to describe substances exchanged among members of the same animal species. Unlike true hormones, which are secreted internally to regulate the organism's own physiology, or internal environment, pheromones are secreted externally and help to regulate the organism's external environment by influencing other animals. The mode of influence can take either of two general forms. If the pheromone produces a more or less immediate and reversible change

in the behavior of the recipient, it is said to have a "releaser" effect. In this case the chemical substance seems to act directly on the recipient's central nervous system. If the principal function of the pheromone is to trigger a chain of physiological events in the recipient, it has what we have recently labeled a "primer" effect. The physiological changes, in turn, equip the organism with a new behavioral repertory, the components of which are thenceforth evoked by appropriate stimuli. In termites, for example, the reproductive and soldier castes prevent other termites from developing into their own castes by secreting substances that are ingested and act through the *corpus allatum*, an endocrine gland controlling differentiation [see "The Termite and the Cell," by Martin Lüscher; SCIENTIFIC AMERICAN, May, 1953].

These indirect primer pheromones do not always act by physiological inhibition. They can have the opposite effect. Adult males of the migratory locust *Schistocerca gregaria* secrete a volatile substance from their skin surface that accelerates the growth of young locusts. When the nymphs detect this substance with their antennae, their hind legs, some of their mouth parts and the antennae themselves vibrate. The secretion, in conjunction with tactile and visual signals, plays an important role in the formation of migratory locust swarms.

A striking feature of some primer pheromones is that they cause important physiological change without an immediate accompanying behavioral response, at least none that can be said to be peculiar to the pheromone. Beginning in 1955 with the work of S. van der Lee and L. M. Boot in the Netherlands, mammalian endocrinologists have discovered several unexpected effects on the female

mouse that are produced by odors of other members of the same species. These changes are not marked by any immediate distinctive behavioral patterns. In the "Lee-Boot effect" females placed in groups of four show an increase in the percentage of pseudopregnancies. A completely normal reproductive pattern can be restored by removing the olfactory bulbs of the mice or by housing the mice separately. When more and more female mice are forced to live together, their oestrous cycles become highly irregular and in most of the mice the cycle stops completely for long periods. Recently W. K. Whitten of the Australian National University has discovered that the odor of a male mouse can initiate and synchronize the oestrous cycles of female mice. The male odor also reduces the frequency of reproductive abnormalities arising when female mice are forced to live under crowded conditions.

A still more surprising primer effect has been found by Helen Bruce of the National Institute for Medical Research in London. She observed that the odor of a strange male mouse will block the pregnancy of a newly impregnated female mouse. The odor of the original stud male, of course, leaves pregnancy undisturbed. The mouse reproductive pheromones have not yet been identified chemically, and their mode of action is only partly understood. There is evidence that the odor of the strange male suppresses the secretion of the hormone prolactin, with the result that the *corpus luteum* (a ductless ovarian gland) fails to develop and normal oestrus is restored. The pheromones are probably part of the complex set of control mechanisms that regulate the population density of animals [see "Population Density and Social Pathology," by John B. Calhoun; SCIENTIFIC AMERICAN Offprint 506].

Pheromones that produce a simple releaser effect—a single specific response mediated directly by the central nervous system—are widespread in the animal kingdom and serve a great many functions. Sex attractants constitute a large and important category. The chemical structures of six attractants are shown on page 122. Although two of the six—the mammalian scents muskone and civetone—have been known for some 40 years and are generally assumed to serve a sexual function, their exact role has never been rigorously established by experiments with living animals. In fact, mammals seem to employ musklike compounds, alone or in combination with other substances, to serve several functions: to mark home ranges, to assist in territorial defense and to identify the sexes.

The nature and role of the four insect sex attractants are much better understood. The identification of each represents a technical feat of considerable magnitude. To obtain 12 milligrams of esters of bombykol, the sex attractant of the female silkworm moth, Adolf F. J. Butenandt and his associates at the Max Planck Institute of Biochemistry in Munich had to extract material from 250,000 moths. Martin Jacobson, Morton Beroza and William Jones of the U.S. Department of Agriculture processed 500,000 female gypsy moths to get 20 milligrams of the gypsy-moth attractant gyplure. Each moth yielded only about .01 microgram (millionth of a gram) of gyplure, or less than a millionth of its body weight. Bombykol and gyplure were obtained by killing the insects and subjecting crude extracts of material to chromatography, the separation technique in which compounds move at different rates through a column packed with a suitable adsorbent substance. Another technique has been more recently developed by Robert T. Yamamoto of the U.S. Department of Agriculture, in collaboration with Jacobson and Beroza, to harvest the equally elusive sex attractant of the American cockroach. Virgin females were housed in metal cans and air was continuously drawn through the cans and passed through chilled containers to condense any vaporized materials. In this manner the equivalent of 10,000 females were "milked" over a nine-month period to yield 12.2 milligrams of what was considered to be the pure attractant.

The power of the insect attractants is almost unbelievable. If some 10,000 molecules of the most active form of bombykol are allowed to diffuse from a source one centimeter from the antennae of a male silkworm moth, a characteristic sexual response is obtained in most cases. If volatility and diffusion rate are taken into account, it can be estimated that the threshold concentration is no more than a few hundred molecules per cubic centimeter, and the actual number required to stimulate the male is probably even smaller. From this one can calculate that .01 microgram of gyplure, the minimum average content of a single female moth, would be theoretically adequate, if distributed with maximum efficiency, to excite more than a billion male moths.

In nature the female uses her powerful pheromone to advertise her presence over a large area with a minimum expenditure of energy. With the aid of published data from field experiments and newly contrived mathematical models of the diffusion process, William H. Bossert, one of my associates in the Biological Laboratories at Harvard University, and I have deduced the shape and size of the ellipsoidal space within which male moths can be attracted under natural conditions [see the bottom illustration on page 118]. When a moderate wind is blowing, the active space has a long axis of thousands of meters and a transverse axis parallel to the ground of more than 200 meters at the widest point. The 19th-century French naturalist Jean Henri Fabre, speculating on sex attraction in insects, could not bring himself to believe that the female moth could communicate over such great distances by odor alone, since "one might as well expect to tint a lake with a drop of carmine." We now know that Fabre's conclusion was wrong but that his analogy was exact: to the male moth's powerful chemoreceptors the lake is indeed tinted.

One must now ask how the male moth, smelling the faintly tinted air, knows which way to fly to find the source of the tinting. He cannot simply fly in the direction of increasing scent; it can be shown mathematically that the attractant is distributed almost uniformly after it has drifted more than a few meters from the female. Recent experiments by Ilse Schwinck of the University of Munich have revealed what is probably the alternative procedure used. When male moths are activated by the pheromone, they simply fly upwind and thus inevitably move toward the female. If by accident they pass out of the active zone, they either abandon the search or fly about at random until they pick up the scent again. Eventually, as they approach the female, there is a slight increase in the concentration of the chemical attractant and this can serve as a guide for the remaining distance.

INVISIBLE ODOR TRAILS guide fire ant workers to a source of food: a drop of sugar solution. The trails consist of a pheromone laid down by workers returning to their nest after finding a source of food. Sometimes the chemical message is reinforced by the touching of antennae if a returning worker meets a wandering fellow along the way. This is hap-

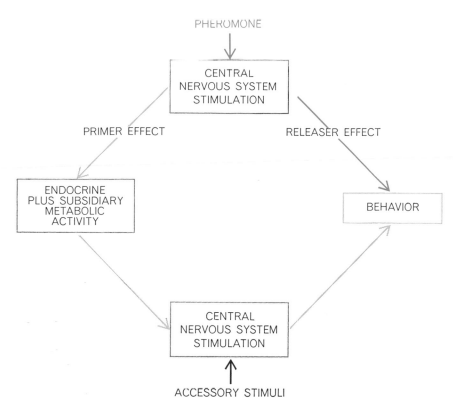

PHEROMONE

CENTRAL
NERVOUS SYSTEM
STIMULATION

PRIMER EFFECT   RELEASER EFFECT

ENDOCRINE
PLUS SUBSIDIARY
METABOLIC
ACTIVITY

BEHAVIOR

CENTRAL
NERVOUS SYSTEM
STIMULATION

ACCESSORY STIMULI

PHEROMONES INFLUENCE BEHAVIOR directly or indirectly, as shown in this schematic diagram. If a pheromone stimulates the recipient's central nervous system into producing an immediate change in behavior, it is said to have a "releaser" effect. If it alters a set of long-term physiological conditions so that the recipient's behavior can subsequently be influenced by specific accessory stimuli, the pheromone is said to have a "primer" effect.

If one is looking for the most highly developed chemical communication systems in nature, it is reasonable to study the behavior of the social insects, particularly the social wasps, bees, termites and ants, all of which communicate mostly in the dark interiors of their nests and are known to have advanced chemoreceptive powers. In recent years experimental techniques have been developed to separate and identify the pheromones of these insects, and rapid progress has been made in deciphering the hitherto intractable codes, particularly those of the ants. The most successful procedure has been to dissect out single glandular reservoirs and see what effect their contents have on the behavior of the worker caste, which is the most numerous and presumably the most in need of continuing guidance. Other pheromones, not present in distinct reservoirs, are identified in chromatographic fractions of crude extracts.

Ants of all castes are constructed with an exceptionally well-developed exocrine glandular system. Many of the most prominent of these glands, whose function has long been a mystery to entomologists, have now been identified as the source of pheromones [see illustra-

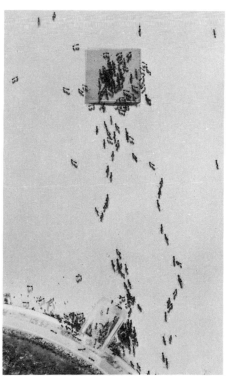

pening in the photograph at the far left. A few foraging workers have just found the sugar drop and a returning trail-layer is communicating the news to another ant. In the next two pictures the trail has been completed and workers stream from the nest in in-creasing numbers. In the fourth picture unrewarded workers return to the nest without laying trails and outward-bound traffic wanes. In the last picture most of the trails have evaporated completely and only a few stragglers remain at the site, eating the last bits of food.

ANTENNAE OF GYPSY MOTHS differ radically in structure according to their function. In the male (*left*) they are broad and finely divided to detect minute quantities of sex attractant released by the female (*right*). The antennae of the female are much less developed.

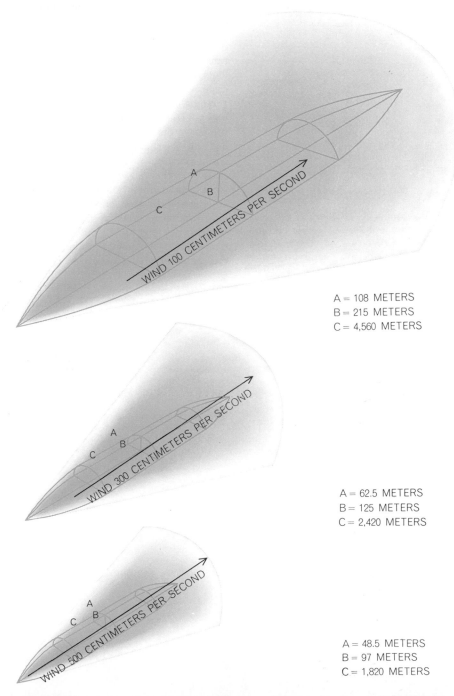

A = 108 METERS
B = 215 METERS
C = 4,560 METERS

A = 62.5 METERS
B = 125 METERS
C = 2,420 METERS

A = 48.5 METERS
B = 97 METERS
C = 1,820 METERS

ACTIVE SPACE of gyplure, the gypsy moth sex attractant, is the space within which this pheromone is sufficiently dense to attract males to a single, continuously emitting female. The actual dimensions, deduced from linear measurements and general gas-diffusion models, are given at right. Height (*A*) and width (*B*) are exaggerated in the drawing. As wind shifts from moderate to strong, increased turbulence contracts the active space.

tion on page 120]. The analysis of the gland-pheromone complex has led to the beginnings of a new and deeper understanding of how ant societies are organized.

Consider the chemical trail. According to the traditional view, trail secretions served as only a limited guide for worker ants and had to be augmented by other kinds of signals exchanged inside the nest. Now it is known that the trail substance is extraordinarily versatile. In the fire ant (*Solenopsis saevissima*), for instance, it functions both to activate and to guide foraging workers in search of food and new nest sites. It also contributes as one of the alarm signals emitted by workers in distress. The trail of the fire ant consists of a substance secreted in minute amounts by Dufour's gland; the substance leaves the ant's body by way of the extruded sting, which is touched intermittently to the ground much like a moving pen dispensing ink. The trail pheromone, which has not yet been chemically identified, acts primarily to attract the fire ant workers. Upon encountering the attractant the workers move automatically up the gradient to the source of emission. When the substance is drawn out in a line, the workers run along the direction of the line away from the nest. This simple response brings them to the food source or new nest site from which the trail is laid. In our laboratory we have extracted the pheromone from the Dufour's glands of freshly killed workers and have used it to create artificial trails. Groups of workers will follow these trails away from the nest and along arbitrary routes (including circles leading back to the nest) for considerable periods of time. When the pheromone is presented to whole colonies in massive doses, a large portion of the colony, including the queen, can be drawn out in a close simulation of the emigration process.

The trail substance is rather volatile, and a natural trail laid by one worker diffuses to below the threshold concentration within two minutes. Consequently outward-bound workers are able to follow it only for the distance they can travel in this time, which is about 40 centimeters. Although this strictly limits the distance over which the ants can communicate, it provides at least two important compensatory advantages. The more obvious advantage is that old, useless trails do not linger to confuse the hunting workers. In addition, the intensity of the trail laid by many workers provides a sensitive index of the amount of food at a given site and the rate of its depletion. As workers move to and from the food finds (consisting mostly of dead insects and sugar sources) they continuously add their own secretions to the trail produced by the original discoverers of the food. Only if an ant is rewarded by food does it lay a trail on its trip back to the nest; therefore the more food encountered at the end of the trail, the more workers that can be rewarded and the heavier the trail. The heavier the trail, the more workers that are drawn from the nest and arrive at the end of the trail. As the food is consumed, the number of workers laying trail substance drops, and the old trail fades by evaporation and diffusion, gradually constricting the outward flow of workers.

The fire ant odor trail shows other evidences of being efficiently designed. The active space within which the pheromone is dense enough to be perceived by workers remains narrow and nearly constant in shape over most of the length of the trail. It has been further deduced from diffusion models that the maximum gradient must be situated near the outer surface of the active space. Thus workers are informed of the space boundary in a highly efficient way. Together these features ensure that the following workers keep in close formation with a minimum chance of losing the trail.

The fire ant trail is one of the few animal communication systems whose information content can be measured with fair precision. Unlike many communicating animals, the ants have a distinct goal in space—the food find or nest site—the direction and distance of which must both be communicated. It is possible by a simple technique to measure how close trail-followers come to the trail end, and, by making use of a standard equation from information theory, one can translate the accuracy of their response into the "bits" of information received. A similar procedure can be applied (as first suggested by the British biologist J. B. S. Haldane) to the "waggle dance" of the honeybee, a radically different form of communication system from the ant trail [see "Dialects in the Language of the Bees," by Karl von Frisch; SCIENTIFIC AMERICAN Offprint 130]. Surprisingly, it turns out that the two systems, although of wholly different evolutionary origin, transmit about the same amount of information with reference to distance (two bits) and direction (four bits in the honeybee, and four or possibly five in the ant). Four bits of information will direct an ant or a bee into one of 16 equally probable sectors of a circle and two bits will identify one of four equally probable distances. It is conceivable that these in-

FIRE ANT WORKER lays an odor trail by exuding a pheromone along its extended sting. The sting is touched to the ground periodically, breaking the trail into a series of streaks.

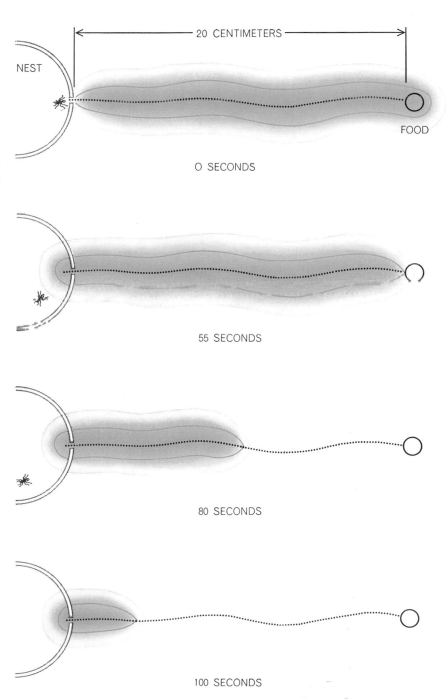

20 CENTIMETERS

NEST

FOOD

O SECONDS

55 SECONDS

80 SECONDS

100 SECONDS

ACTIVE SPACE OF ANT TRAIL, within which the pheromone is dense enough to be perceived by other workers, is narrow and nearly constant in shape with the maximum gradient situated near its outer surface. The rapidity with which the trail evaporates is indicated.

formation values represent the maximum that can be achieved with the insect brain and sensory apparatus.

Not all kinds of ants lay chemical trails. Among those that do, however, the pheromones are highly species-specific in their action. In experiments in which artificial trails extracted from one species were directed to living colonies of other species, the results have almost always been negative, even among related species. It is as if each species had its own private language. As a result there is little or no confusion when the trails of two or more species cross.

Another important class of ant pheromone is composed of alarm substances. A simple backyard experiment will show that if a worker ant is disturbed by a clean instrument, it will, for a short time, excite other workers with whom it comes in contact. Until recently most students of ant behavior thought that the alarm was spread by touch, that one worker simply jostled another in its excitement or drummed on its neighbor with its antennae in some peculiar way. Now it is known that disturbed workers discharge chemicals, stored in special glandular reservoirs, that can produce all the characteristic alarm responses solely by themselves. The chemical structure of four alarm substances is shown on page 124. Nothing could illustrate more clearly the wide differences between the human perceptual world and that of chemically communicating animals. To the human nose the alarm substances are mild or even pleasant, but to the ant they represent an urgent tocsin that can propel a colony into violent and instant action.

As in the case of the trail substances, the employment of the alarm substances appears to be ideally designed for the purpose it serves. When the contents of the mandibular glands of a worker of the harvesting ant (*Pogonomyrmex badius*) are discharged into still air, the volatile material forms a rapidly expanding sphere, which attains a radius of about six centimeters in 13 seconds. Then it contracts until the signal fades out completely some 35 seconds after the moment of discharge. The outer shell of the active space contains a low concentration of pheromone, which is actually attractive to harvester workers. This serves to draw them toward the point of disturbance. The central region of the active space, however, contains a concentration high enough to evoke the characteristic frenzy of alarm. The "alarm sphere" expands to a radius of about three centimeters in eight seconds and, as might be expected, fades out more quickly than the "attraction sphere."

The advantage to the ants of an alarm signal that is both local and short-lived becomes obvious when a *Pogonomyrmex* colony is observed under natural conditions. The ant nest is subject to almost innumerable minor disturbances. If the

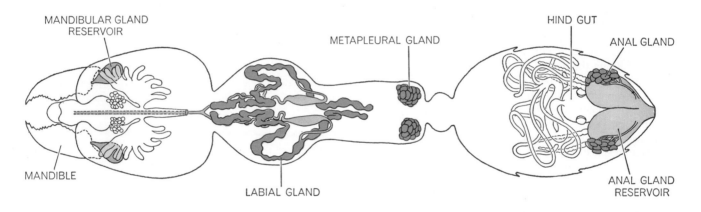

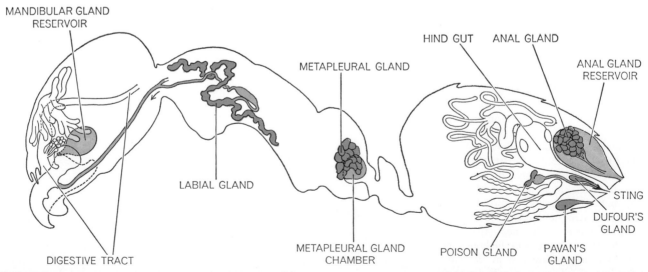

EXOCRINE GLANDULAR SYSTEM of a worker ant (*shown here in top and side cutaway views*) is specially adapted for the production of chemical communication substances. Some pheromones are stored in reservoirs and released in bursts only when needed; others are secreted continuously. Depending on the species, trail substances are produced by Dufour's gland, Pavan's gland or the poison glands; alarm substances are produced by the anal and mandibular glands. The glandular sources of other pheromones are unknown.

alarm spheres generated by individual ant workers were much wider and more durable, the colony would be kept in ceaseless and futile turmoil. As it is, local disturbances such as intrusions by foreign insects are dealt with quickly and efficiently by small groups of workers, and the excitement soon dies away.

The trail and alarm substances are only part of the ants' chemical vocabulary. There is evidence for the existence of other secretions that induce gathering and settling of workers, acts of grooming, food exchange, and other operations fundamental to the care of the queen and immature ants. Even dead ants produce a pheromone of sorts. An ant that has just died will be groomed by other workers as if it were still alive. Its complete immobility and crumpled posture by themselves cause no new response. But in a day or two chemical decomposition products accumulate and stimulate the workers to bear the corpse to the refuse pile outside the nest. Only a few decomposition products trigger this funereal response; they include certain long-chain fatty acids and their esters. When other objects, including living workers, are experimentally daubed with these substances, they are dutifully carried to the refuse pile. After being dumped on the refuse the "living dead" scramble to their feet and promptly return to the nest, only to be carried out again. The hapless creatures are thrown back on the refuse pile time and again until most of the scent of death has been worn off their bodies by the ritual.

Our observation of ant colonies over long periods has led us to believe that as few as 10 pheromones, transmitted singly or in simple combinations, might suffice for the total organization of ant society. The task of separating and characterizing these substances, as well as judging the roles of other kinds of stimuli such as sound, is a job largely for the future.

Even in animal species where other kinds of communication devices are prominently developed, deeper investigation usually reveals the existence of pheromonal communication as well. I have mentioned the auxiliary roles of primer pheromones in the lives of mice and migratory locusts. A more striking example is the communication system of the honeybee. The insect is celebrated for its employment of the "round" and "waggle" dances (augmented, perhaps, by auditory signals) to designate the location of food and new nest sites. It is not so widely known that chemical signals

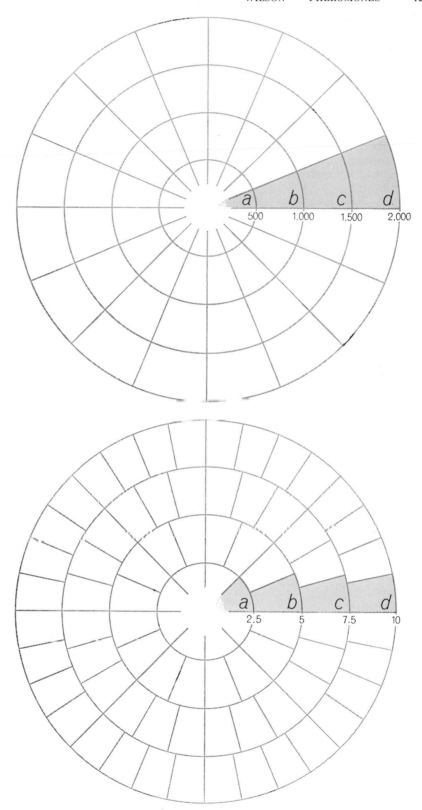

FORAGING INFORMATION conveyed by two different insect communication systems can be represented on two similar "compass" diagrams. The honeybee "waggle dance" (*top*) transmits about four bits of information with respect to direction, enabling a honeybee worker to pinpoint a target within one of 16 equally probable angular sectors. The number of "bits" in this case remains independent of distance, given in meters. The pheromone system used by trail-laying fire ants (*bottom*) is superior in that the amount of directional information increases with distance, given in centimeters. At distances *c* and *d*, the probable sector in which the target lies is smaller for ants than for bees. (For ants, directional information actually increases gradually and not by jumps.) Both insects transmit two bits of distance information, specifying one of four equally probable distance ranges.

play equally important roles in other aspects of honeybee life. The mother queen regulates the reproductive cycle of the colony by secreting from her mandibular glands a substance recently identified as 9-ketodecanoic acid. When this pheromone is ingested by the worker bees, it inhibits development of their ovaries and also their ability to manufacture the royal cells in which new queens are reared. The same pheromone serves as a sex attractant in the queen's nuptial flights.

Under certain conditions, including the discovery of new food sources, worker bees release geraniol, a pleasant-smelling alcohol, from the abdominal Nassanoff glands. As the geraniol diffuses through the air it attracts other workers and so supplements information contained in the waggle dance. When a worker stings an intruder, it discharges, in addition to the venom, tiny amounts of a secretion from clusters of unicellular glands located next to the basal plates of the sting. This secretion is responsible for the tendency, well known to beekeepers, of angry swarms of workers to sting at the same spot. One component, which acts as a simple attractant, has been identified as isoamyl acetate, a compound that has a banana-like odor. It is possible that the stinging response is evoked by at least one unidentified alarm substance secreted along with the attractant.

Knowledge of pheromones has advanced to the point where one can make some tentative generalizations about their chemistry. In the first place, there appear to be good reasons why sex attractants should be compounds that contain between 10 and 17 carbon atoms and that have molecular weights between about 180 and 300—the range actually observed in attractants so far identified. (For comparison, the weight of a single carbon atom is 12.) Only compounds of roughly this size or greater can meet the two known requirements of a sex attractant: narrow specificity, so that only members of one species will respond to it, and high potency. Compounds that contain fewer than five or so carbon atoms and that have a molecular weight of less than about 100 cannot be assembled in enough different ways to provide a distinctive molecule for all the insects that want to advertise their presence.

It also seems to be a rule, at least with insects, that attraction potency increases with molecular weight. In one series of esters tested on flies, for instance, a doubling of molecular weight resulted in as much as a thousandfold increase in efficiency. On the other hand, the molecule cannot be too large and complex or it will be prohibitively difficult for the insect to synthesize. An equally important limitation on size is

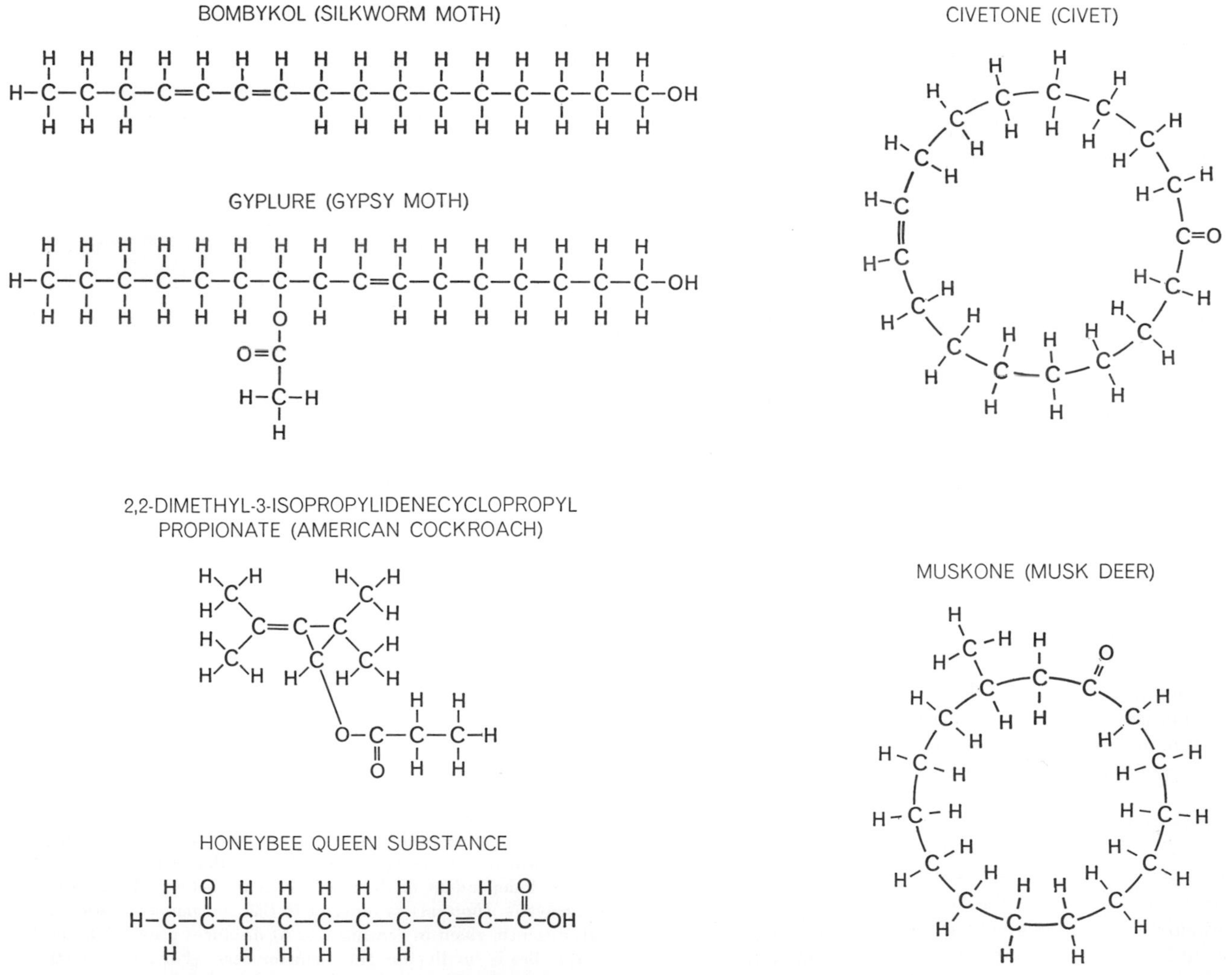

SIX SEX PHEROMONES include the identified sex attractants of four insect species as well as two mammalian musks generally believed to be sex attractants. The high molecular weight of most sex pheromones accounts for their narrow specificity and high potency.

the fact that volatility—and, as a result, diffusibility—declines with increasing molecular weight.

One can also predict from first principles that the molecular weight of alarm substances will tend to be less than those of the sex attractants. Among the ants there is little specificity; each species responds strongly to the alarm substances of other species. Furthermore, an alarm substance, which is used primarily within the confines of the nest, does not need the stimulative potency of a sex attractant, which must carry its message for long distances. For these reasons small molecules will suffice for alarm purposes. Of seven alarm substances known in the social insects, six have 10 or fewer carbon atoms and one (dendrolasin) has 15. It will be interesting to see if future discoveries bear out these early generalizations.

Do human pheromones exist? Primer pheromones might be difficult to detect, since they can affect the endocrine system without producing overt specific behavioral responses. About all that can be said at present is that striking sexual differences have been observed in the ability of humans to smell certain

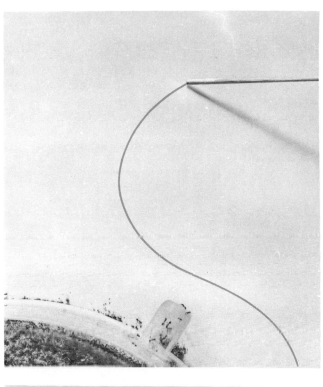

ARTIFICIAL TRAIL can be laid down by drawing a line (*colored curve in frame at top left*) with a stick that has been treated with the contents of a single Dufour's gland. In the remaining three frames, workers are attracted from the nest, follow the artificial route in close formation and mill about in confusion at its arbitrary terminus. Such a trail is not renewed by the unrewarded workers.

DENDROLASIN (*LASIUS FULIGINOSUS*)

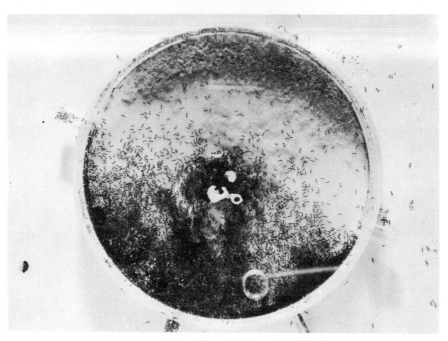

CITRAL (*ATTA SEXDENS*)

CITRONELLAL (*ACANTHOMYOPS CLAVIGER*)

2-HEPTANONE (*IRIDOMYRMEX PRUINOSUS*)

FOUR ALARM PHEROMONES, given off by the workers of the ant species indicated, have so far been identified. Disturbing stimuli trigger the release of these substances from various glandular reservoirs.

substances. The French biologist J. Le-Magnen has reported that the odor of Exaltolide, the synthetic lactone of 14-hydroxytetradecanoic acid, is perceived clearly only by sexually mature females and is perceived most sharply at about the time of ovulation. Males and young girls were found to be relatively insensitive, but a male subject became more sensitive following an injection of estrogen. Exaltolide is used commercially as a perfume fixative. LeMagnen also reported that the ability of his subjects to detect the odor of certain steroids paralleled that of their ability to smell Exaltolide. These observations hardly represent a case for the existence of human pheromones, but they do suggest that the relation of odors to human physiology can bear further examination.

It is apparent that knowledge of chemical communication is still at an early stage. Students of the subject are in the position of linguists who have learned the meaning of a few words of a nearly indecipherable language. There is almost certainly a large chemical vocabulary still to be discovered. Conceiv-

ably some pheromone "languages" will be found to have a syntax. It may be found, in other words, that pheromones can be combined in mixtures to form new meanings for the animals employing them. One would also like to know if some animals can modulate the intensity or pulse frequency of pheromone emission to create new messages. The solution of these and other interesting problems will require new techniques in analytical organic chemistry combined with ever more perceptive studies of animal behavior.

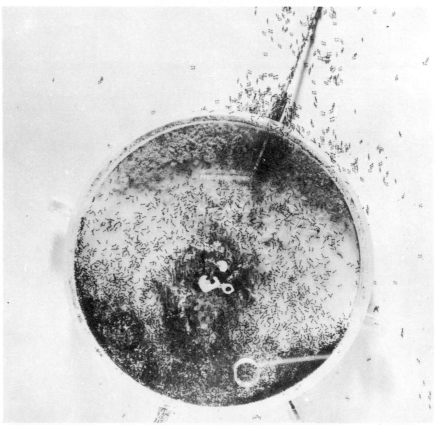

MASSIVE DOSE of trail pheromone causes the migration of a large portion of a fire ant colony from one side of a nest to another. The pheromone is administered on a stick that has been dipped in a solution extracted from the Dufour's glands of freshly killed workers.

MACROMOLECULAR ARCHITECTURE

II

MACROMOLECULAR ARCHITECTURE

INTRODUCTION

Living organisms are built of large molecules that give structure to the cell and also take part in biochemical reactions. These biopolymer molecules include proteins, nucleic acids, and polysaccharides. These materials are generally treated in a rather cursory fashion in textbooks on organic chemistry at the end of the text—too late for the life-science major; indeed, the chemistry of these materials is often given only passing attention in textbooks of biochemistry. In this section, we have included articles that cover the chemistry of proteins and nucleic acids in detail, and some discussion of the third group of polymeric materials, the polysaccharides, is given in several of the articles.

The first article of this section describes the industrial synthesis of many of the plastics in common use; students are generally curious about the chemistry of these materials whose appearance and uses they know very well. The next seven articles treat proteins, starting with an elementary description of their component amino acids and their helical structure by Doty. The article on the structure of proteins by Stein and Moore describes work that was the basis for a Nobel Prize awarded in 1972. The automatic synthesis of proteins, an important and recently developed technique that is hardly described in textbooks, is discussed by Merrifield. The next article describes the three-dimensional structure and the chemical properties of lysozyme, the first enzyme whose reactions were understood in detail in terms of the organization and architecture of its structure. The article by Thompson describes the structure and function of the insulin molecule, research for which F. Sanger was awarded the Nobel Prize. Next, Dickerson explains the detailed structure of cytochrome *c*, discusses its reactions, and gives the evolutionary development of this important Krebs-cycle electron carrier. The next article, on the utility of enzymes bound to artificial matrixes, introduces a subject that is sufficiently novel that it is not included in most textbooks, and yet it has been utilized commercially recently to convert cornstarch to invert sugar.

The series of five articles on the nucleic acids starts with an elementary introduction by F. H. C. Crick, an article describing work that produced the Watson-Crick double-helix model of DNA and revolutionized biology. Holley's article on the sequence of bases in a simple nucleic acid containing less than one hundred nucleotide units describes the first determination of the structure of such a molecule. Kornberg gives an account of work by himself and others aimed at synthesizing viral DNA. The article by Luria on the recognition of DNA in bacteria will give the student insight into the variations in the bases that can occur in the nucleic acids and the enzymatic control of DNA structure; the importance of sugars as labels by which bacteria can recognize DNA from different sources also is discussed. The articles by Crick, Holley, Kornberg, and Luria describe research for which Nobel prizes were awarded. The last article, by Temin on RNA-directed DNA synthesis, describes the recent discovery that information can flow "in reverse" from RNA to DNA, a discovery that appears to be very important in understanding the transformations in cells that lead to cancer.

The following additional articles on synthetic and biological polymer chemistry have appeared in *Scientific American;* some of these are available in Offprint form. F. O. Schmitt, *Giant Molecules in Cells and Tissues* (September 1957, Offprint 35); R. D. Preston, *Cellulose* (September 1957); B. Wunderlich, *The Solid State of Polyethylene* (November 1964); P. J. W. Debye, *How Giant Molecules Are Measured* (September 1957); G. Natta, *How Giant Molecules Are Made* (September 1957); A. V. Tobolsky, *The Mechanical Properties of Polymers* (September 1967); B. Oster, *Polyethylene* (September 1957); H. F. Mark, *The Nature of Polymeric Materials* (September 1967), G. Natta, *Precisely Constructed Polymers* (August 1961, Offprint 315); L. Bragg, *X-ray Crystallography* (July 1968, Offprint 325); J. S. Fruton, *Proteins* (June 1950, Offprint 10); L. Pauling, R. B. Corey, and R. Hayward, *The Structure of Protein Molecules* (July 1954, Offprint 31); M. B. Hoagland, *Nucleic Acids and Proteins* (December 1959, Offprint 68); J. C. Kendrew, *The Three-dimensional Structure of a Protein Molecule* (December 1961, Offprint 121); A. Champagnot, *Protein from Petroleum* (October 1965, Offprint 1020); E. Zuckerkandl, *The Evolution of Hemoglobin* (May 1965, Offprint 1012); M. O. Dayhoff, *Computer Analysis of Protein Evolution* (July 1969, Offprint 1148); C. H. Li, *The ACTH Molecule* (July 1963, Offprint 160); F. H. Crick, *Nucleic Acids* (September 1956, Offprint 54); R. L. Sinsheimer, *Single Stranded DNA* (July 1962, Offprint 128); S. Spiegelman, *Hybrid Nucleic Acids* (May 1964, Offprint 183); H. Fraenkel-Conrat, *Rebuilding a Virus* (June 1956, Offprint 9).

MONOMERS are purified in these towers at the American Cyanamid Company's Fortier Plant near New Orleans, La. In this case the monomer is acrylonitrile made from acetylene and hydrocyanic acid gas. These latter substances are here obtained from natural gas.

16

GIANT MOLECULES

HERMAN F. MARK

September 1957

*Presenting an issue on polymers: huge chains of atoms
assembled from simpler molecules. These chains, which
comprise the fabric of living matter, are now made by
man in forms unknown in nature*

Giant molecules—or high polymers, as the chemist calls them—have been feeding, clothing and housing man ever since he began to manipulate nature. Wood is a high polymer; so are meat, starch, cotton, wool and silk. They are among our oldest and most familiar materials, yet until this century they were a complete chemical mystery. As products of living things, they partake of the prejudice of nature for doing things in an elaborate way. The substances of life are made of the most complicated molecules we know. But their very complexity endows them with wonderfully versatile and powerful properties. The giant molecules therefore present a great challenge to chemists—not only to learn the secrets of their construction but also to devise new materials which nature has neglected to create. And within the past decade the chemistry of high polymers—living and nonliving—has made such rapid strides that it is today one of the most exciting fields in all science.

This development caps a century of remarkable advance in organic chemistry. Chemical understanding of living matter did not begin until 1828, when Friedrich Wöhler of Germany achieved the first test-tube synthesis of an organic substance—urea, a product of animal metabolism. His successors in the new science proceeded to work out the chemical structure and activity of a host of comparatively simple organic molecules: sugars, fats, fruit acids, soaps, alcohols, the coal and petroleum hydrocarbons, and so on. Over the past century many

thousands of scientists all over the world became absorbed in organic chemistry, developing ingenious techniques of investigation and constructing a clear theory of the behavior of the simpler organic substances, based on the behavior of the four-valent carbon atom. The theory made it possible to classify the properties of hundreds of thousands of substances, from the exhalations of gas wells to the pigments of flowers and the poisons of snakes. Organic chemistry gave birth to new synthetic products, such as dyes, perfumes, drugs, fuels, etc. Indeed, much of our present civilization—medical care, sanitation, printing, painting, photography, motor transportation, aviation—relies heavily on materials provided by "classical" organic chemistry.

All these substances were comparatively simple members of the organic family. Their bigger, more complicated relatives—proteins and the rest—got only

desultory attention from chemists, mainly because they were too difficult to deal with. The methods on which organic chemists relied for separation and analysis of organic substances—solution, melting, crystallization and the like—did not work with giant molecules. For example, cellulose, the chief component of wood, does not melt when heated; instead it hardens and decomposes. Nor can it be dissolved, except in chemicals which change it irreversibly to something different. The same is true of other organic high polymers, such as wool, silk, starch and rubber.

When chemists had the misfortune to produce large organic molecules accidentally during their experiments, they were generally crestfallen. The early literature of organic chemists is full of exasperated references to reactions which "resinified" and covered their glassware with waxy, gluey or sticky messes. These

THIS ARTICLE IS CONTINUED on page 134. On the next four pages is a chart listing the principal polymers made by man.
At the left side of the chart are the structural formulas of monomers, the simple molecules which are strung together to form polymers. In the middle of the chart is a short section of the characteristic chain of each polymer. The complete chain is made up of hundreds or even thousands of monomeric units. Some of the chains are linked to other chains, as indicated in the formulas of the phenol formaldehyde and urea formaldehyde resins. The polymers in the chart are divided into two classes: addition polymers and condensation polymers. The basis of this division is given in the illustrations on pages 134 and 135. There are only nine atoms in the chart: carbon (C), hydrogen (H), oxygen (O), nitrogen (N), sulfur (S), chlorine (Cl), fluorine (F), silicon (Si) and sodium (Na). In most cases each atom is represented by its letter symbol. In some cases groups of atoms are abbreviated, for example, CH_3 and the benzene ring. The latter structure, containing six carbon atoms to which other atoms are attached, is represented by a hexagon. A single line between two atoms represents a single chemical bond; a double line, a double bond.

ADDITION POLYMERS

MONOMER	POLYMER	PRINCIPAL USES
ETHYLENE 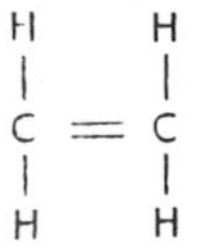	POLYETHYLENE 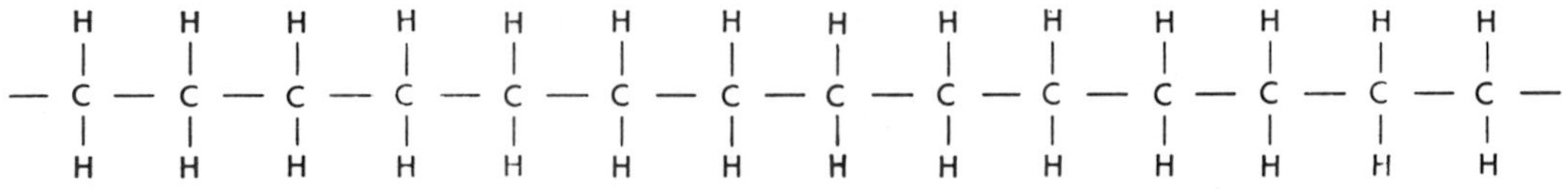	1. FILMS 2. TUBING 3. MOLDED OBJECTS 4. ELECTRICAL INSULATION
VINYL CHLORIDE 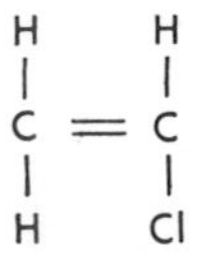	POLYVINYL CHLORIDE 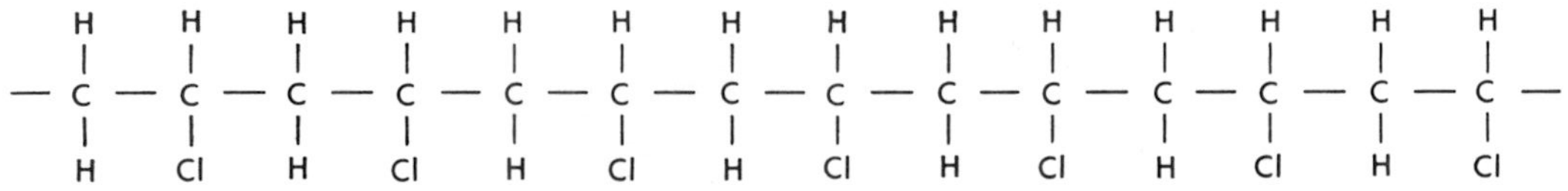	1. SHEETS 2. PHONOGRAPH RECORDS 3. COPOLYMER WITH VINYL ACETATE TO MAKE FLOOR COVERINGS, LATEX PAINTS, ETC.
ACRYLONITRILE 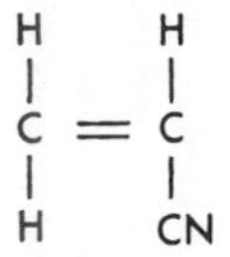	POLYACRYLONITRILE 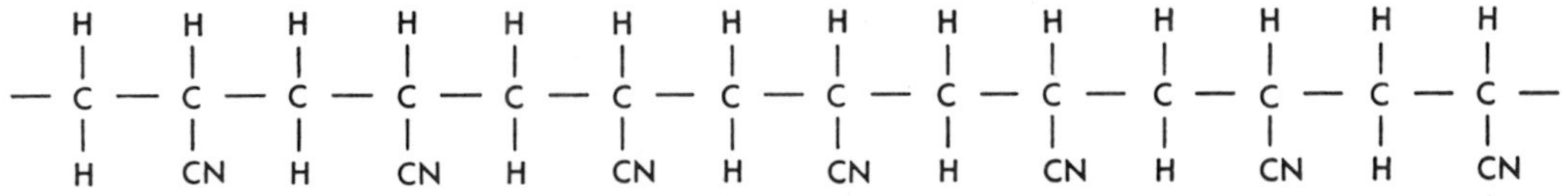	1. FIBERS; E.G., ORLON, ACRILAN
VINYL ACETATE 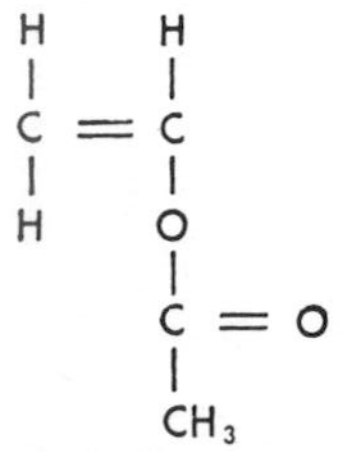	POLYVINYL ACETATE 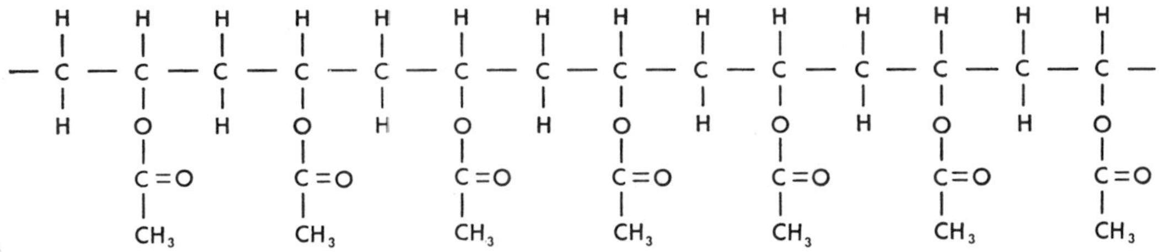	1. CHEWING GUM 2. ADHESIVES 3. TEXTILE COATINGS 4. TO MAKE POLYVINYL ALCOHOL (ON TREATMENT WITH ALKALI)
STYRENE 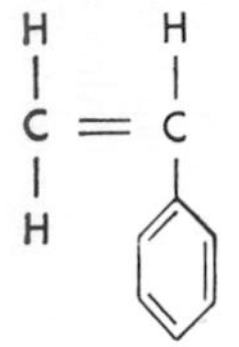	POLYSTYRENE 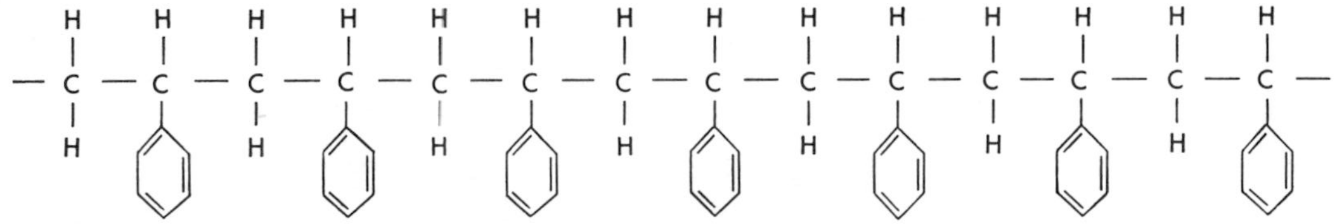	1. MOLDED OBJECTS 2. ELECTRICAL INSULATION 3. COPOLYMER WITH BUTADIENE TO MAKE BUNA-S AND GR-S RUBBER 4. TO MAKE ION-EXCHANGE RESINS (ON TREATMENT WITH SULFURIC ACID)
BUTADIENE 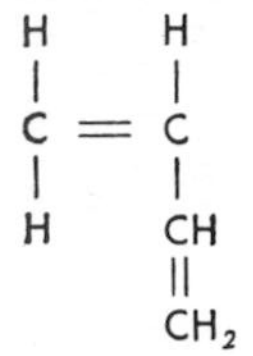	POLYBUTADIENE 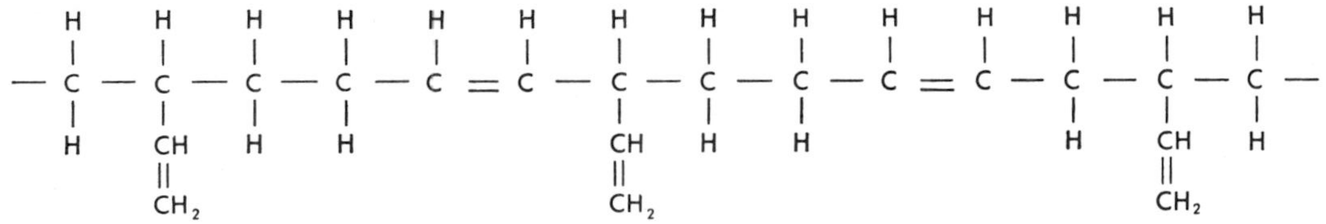	1. BUNA RUBBER

Monomer	Polymer	Uses
ISOBUTYLENE 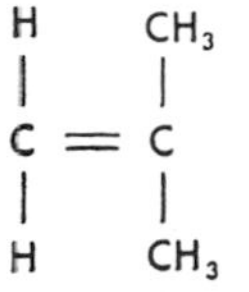	POLYISOBUTYLENE 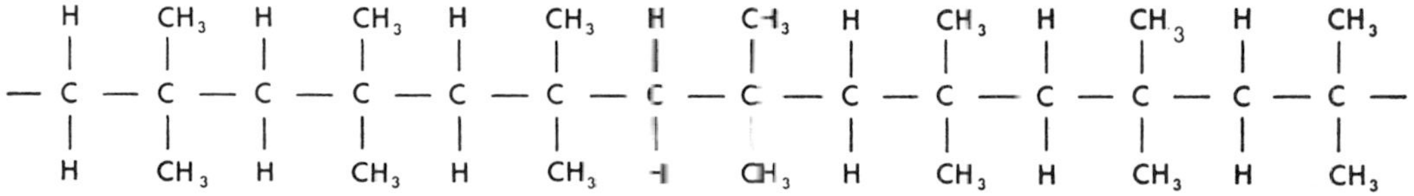	1. COLD-FLOW RUBBER 2. COPOLYMER WITH SMALL AMOUNTS OF BUTADIENE TO MAKE BUTYL RUBBER
METHYL METHACRYLATE 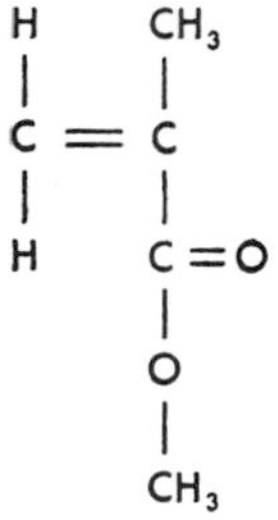	POLYMETHYL METHACRYLATE 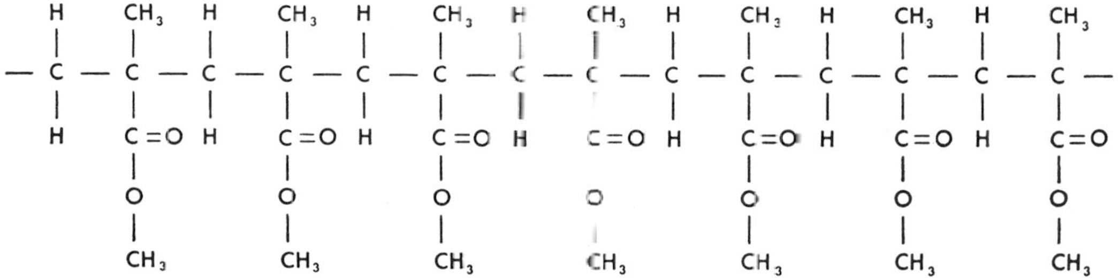	1. TRANSPARENT SHEETS, RODS, TUBING; E.G., LUCITE, PLEXIGLAS 2. PLASTICS REINFORCED WITH GLASS FIBER
VINYLIDENE CHLORIDE 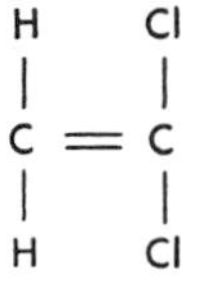	POLYVINYLIDENE CHLORIDE 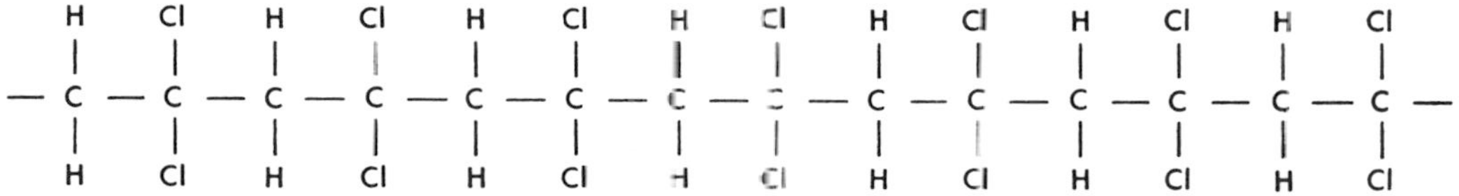	1. COPOLYMER WITH SMALL AMOUNTS OF POLYVINYL CHLORIDE TO MAKE FILMS; E.G., SARAN
CHLOROPRENE 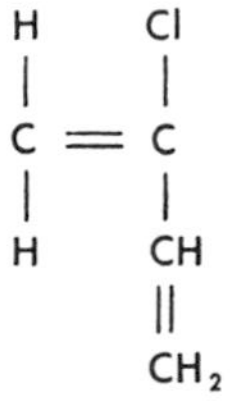	POLYCHLOROPRENE 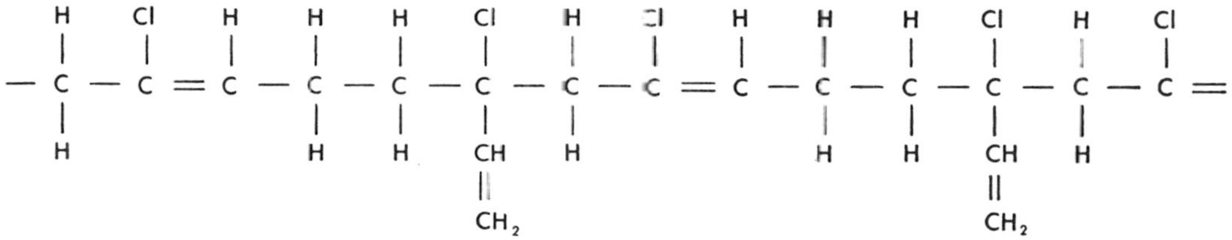	1. OIL-RESISTANT RUBBER; E.G., NEOPRENE
TETRAFLUOROETHYLENE 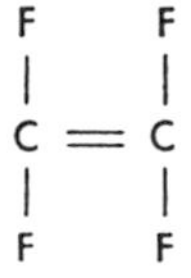	POLYTETRAFLUOROETHYLENE 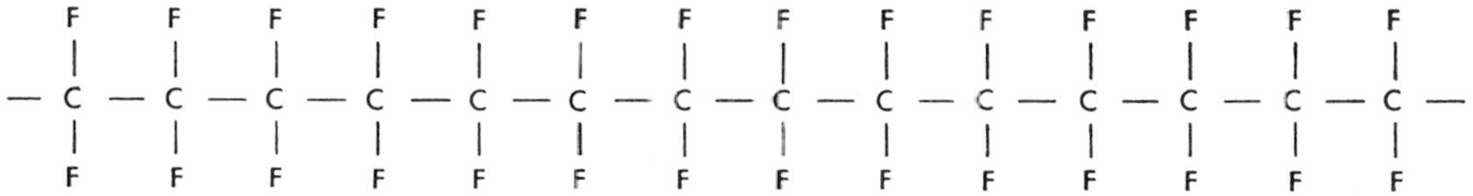	1. CHEMICALLY RESISTANT FILMS, MOLDED OBJECTS, ELECTRICAL INSULATION; E.G., TEFLON
TRIFLUOROCHLOROETHYLENE 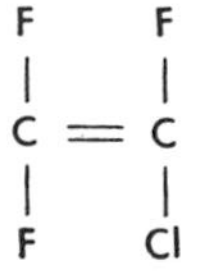	POLYTRIFLUOROCHLOROETHYLENE 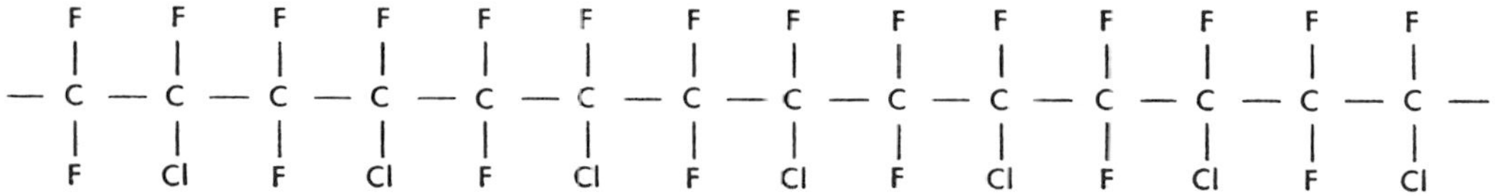	1. FILMS; E.G., KEL-F

CONDENSATION POLYMERS

MONOMERS	POLYMER	PRINCIPAL USES
ETHYLENE GLYCOL TEREPHTHALIC ACID 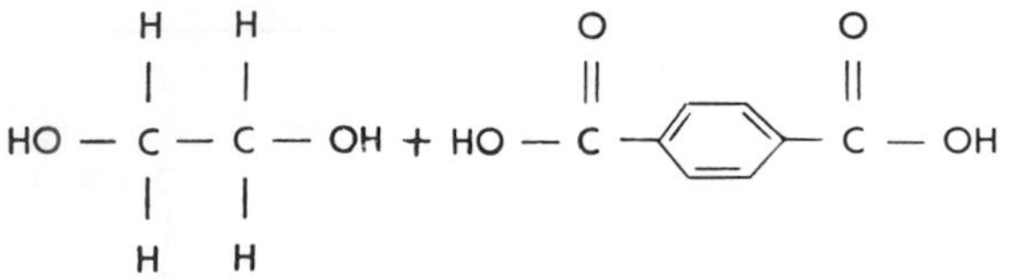	POLYETHYLENE TEREPHTHALATE 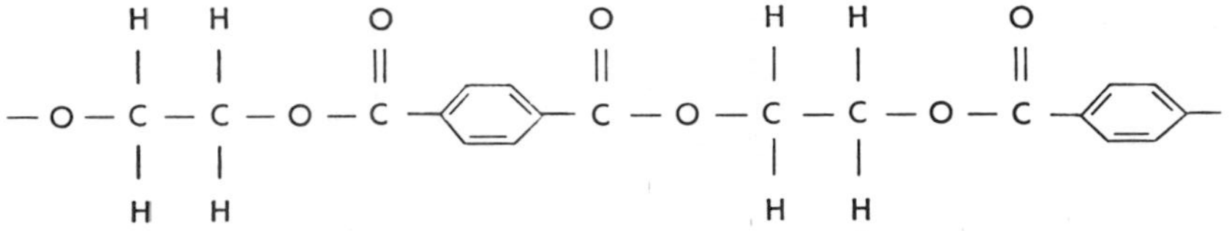	1. FILMS; E.G., MYLAR 2. FIBERS; E.G., DACRON
GLYCEROL PHTHALIC ANHYDRIDE 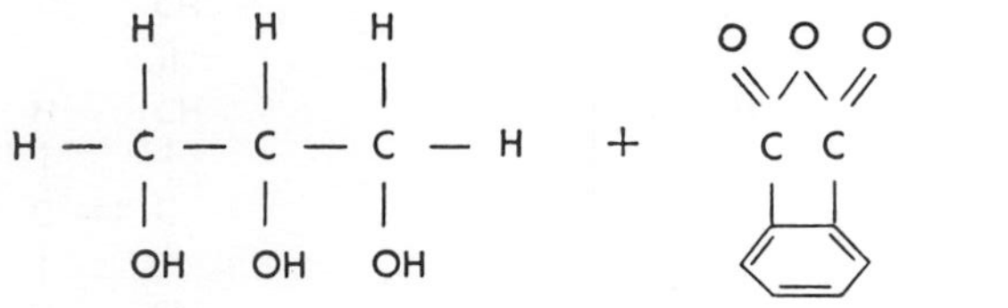	GLYPTAL RESIN 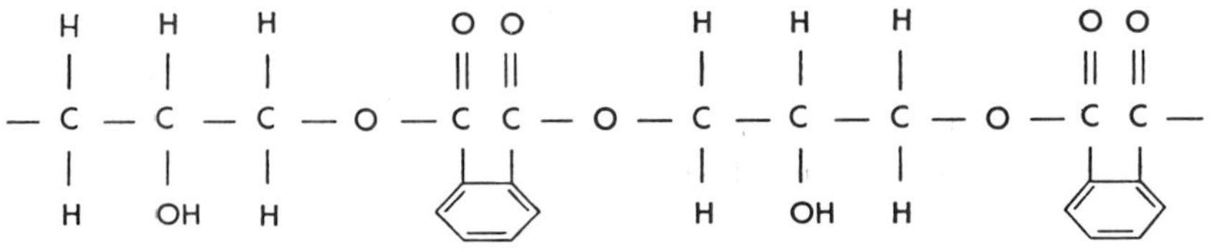	1. COATINGS 2. PLASTICIZERS FOR SHELLAC
ETHYLENE GLYCOL MALEIC ANHYDRIDE 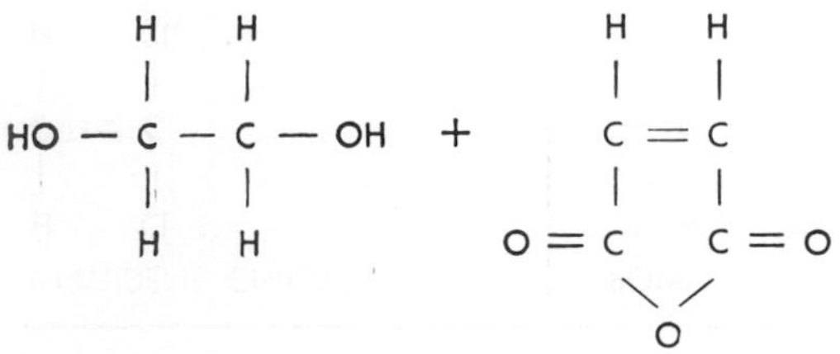	ALKYD RESIN 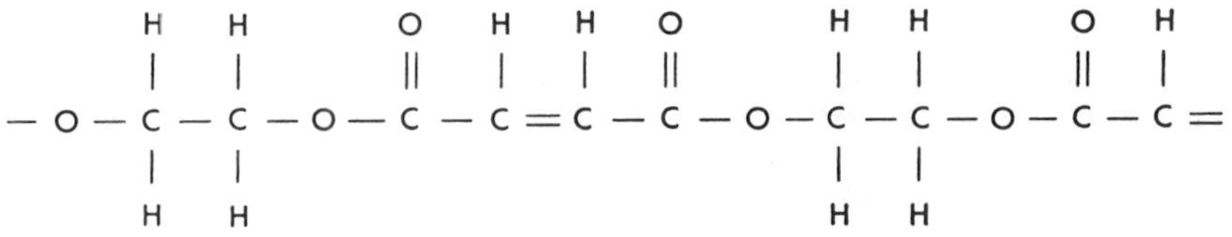	1. COATINGS
HEXAMETHYLENEDIAMINE 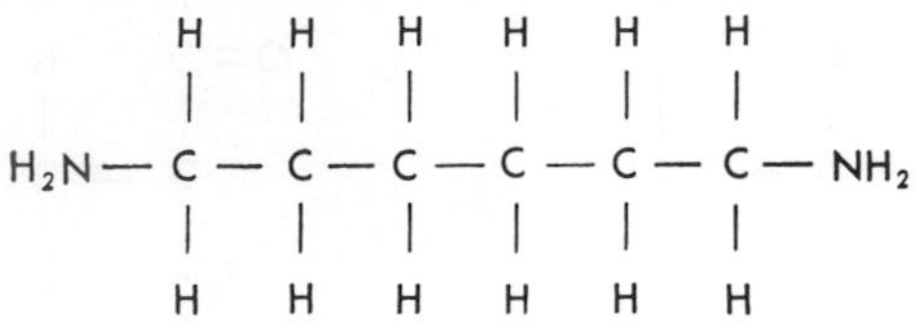 ADIPIC ACID 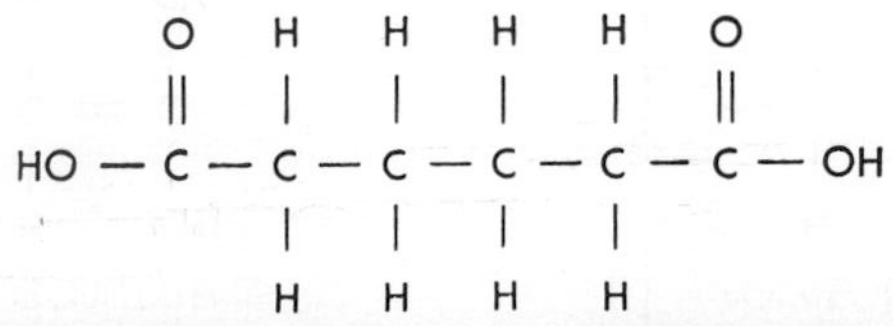	NYLON 66 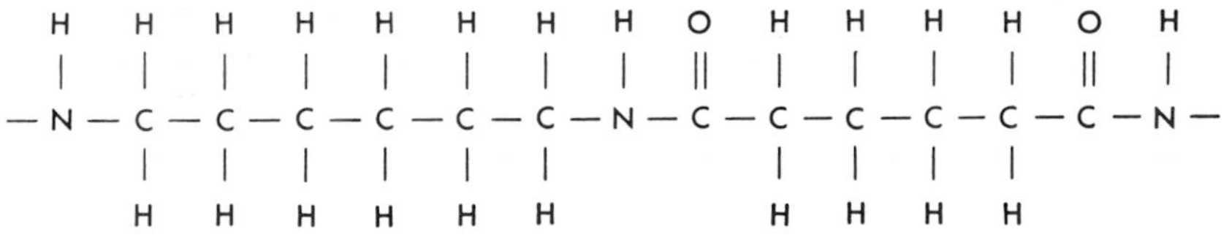	1. FIBERS 2. MOLDED OBJECTS

ETHYLENE DICHLORIDE SODIUM POLYSULFIDE

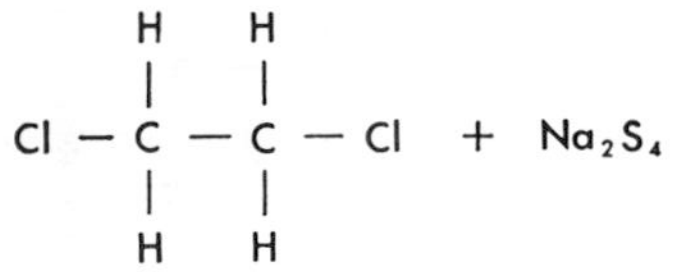

$Cl-C-C-Cl + Na_2S_4$

POLYSULFIDE RUBBER

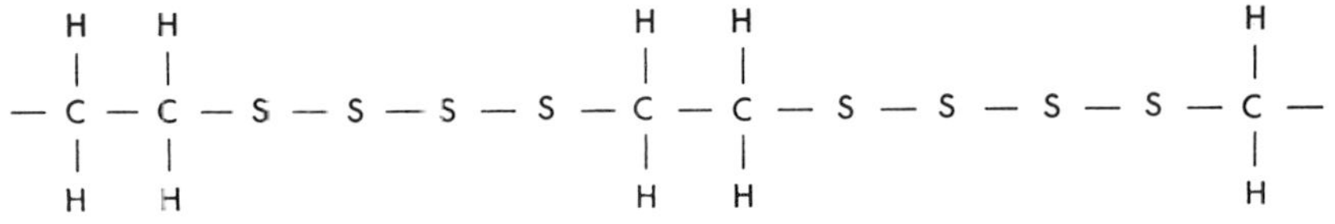

1. CHEMICALLY-RESISTANT RUBBER; E.G., THIOKOL A, GR-P

DIMETHYLSILANEDIOL

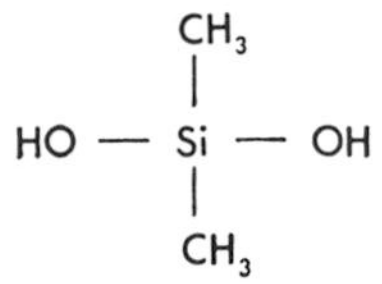

$HO-Si-OH$

SILICONE

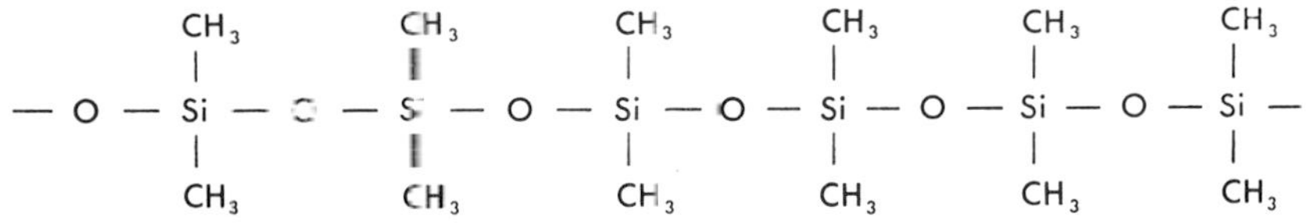

1. TEMPERATURE-RESISTANT LUBRICANTS
2. TEMPERATURE-RESISTANT RUBBER
3. WATER-REPELLENT COATINGS

PHENOL FORMALDEHYDE

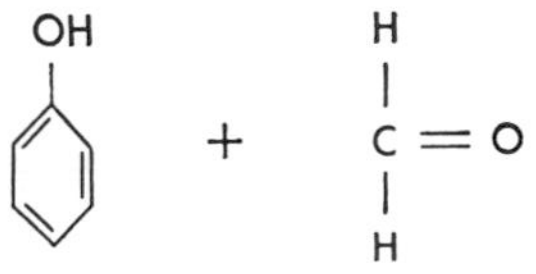

PHENOL-FORMALDEHYDE RESIN

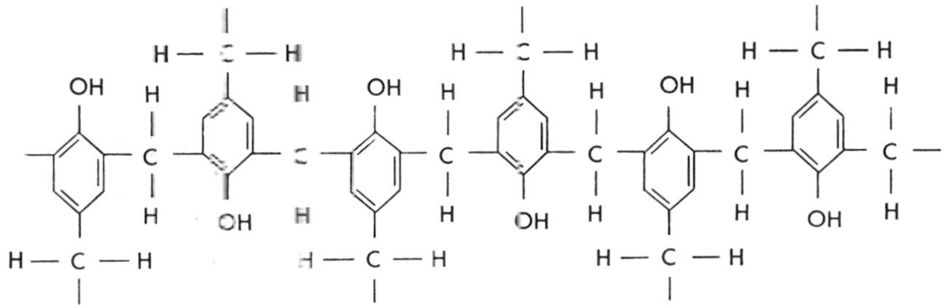

1. REINFORCED MOLDED OBJECTS; E.G., BAKELITE
2. VARNISHES
3. LACQUERS
4. ADHESIVES

UREA FORMALDEHYDE

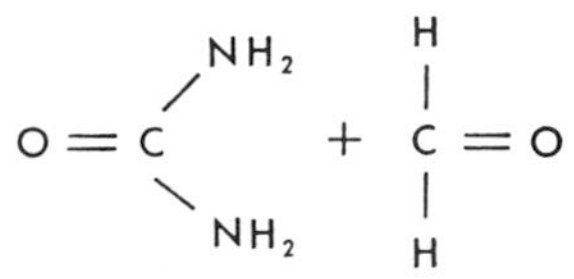

UREA-FORMALDEHYDE RESIN

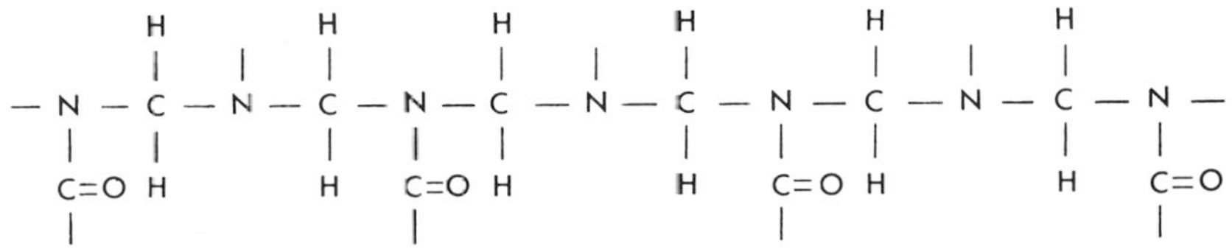

1. MOLDED OBJECTS
2. TEXTILE COATINGS
3. ADHESIVES

ETHYLENE DIISOCYANATE

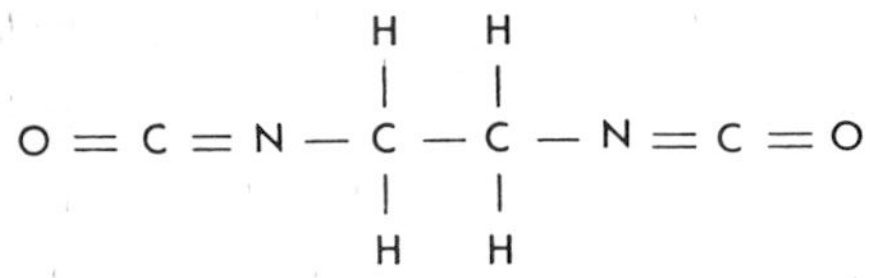

ETHYLENE GLYCOL

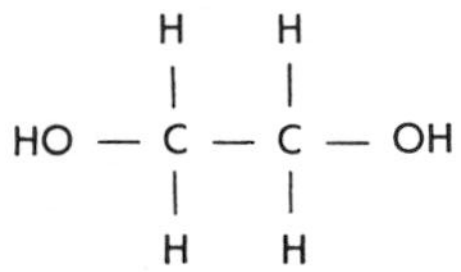

POLYURETHANE

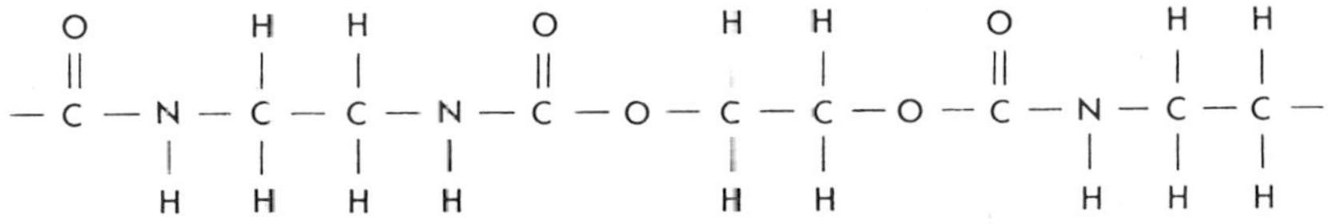

1. FIBERS
2. RUBBER

unexpected products were always a disappointment to the chemist, seeking to purify a substance in nice crystalline form, and were an unalloyed nuisance to the bottle-washer who had to clean the glassware.

In short, up to about 30 or 40 years ago the big organic molecules offered little attraction to chemists. Classical organic chemistry was full of interesting and important problems. With so many green pastures around, why should a chemist invest his career in the risky and sticky business of investigating the macromolecules?

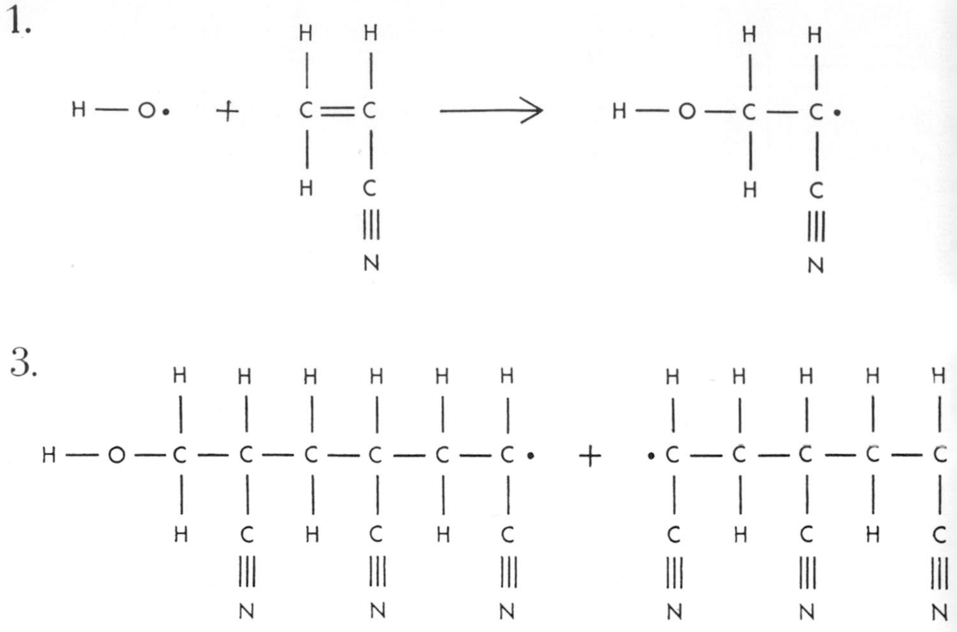

Nevertheless, in the 1920's the study of large molecules such as cellulose and rubber had already begun to look intriguing. In 1923 the writer of these lines (then 28 years old) confessed to his professor at Berlin, Wilhelm Schlenk, that he was strongly tempted to work in this new field. Schlenk, who had long been regarded as one of the most ingenious experimenters in organic chemistry, said: "If I were 20 years younger" (he was 55), "I might be very much attracted myself. Wait until you are 10 years older, and meanwhile demonstrate with 'classical' investigations that you are capable of tackling a problem of such proportions." It proved to be excellent advice.

Attempts to analyze the chemical composition of cellulose, rubber, starch and proteins had begun in the 1880's. Chemists had established that these substances, like all organic compounds, were composed mainly of carbon, hydrogen and oxygen; that cellulose was essentially a sugar compound; that starch was another carbohydrate; that natural rubber was basically a hydrocarbon; that

ADDITION POLYMERIZATION is explained by example. A free radical (H-O) combines with a monomer (acrylonitrile) in such a way that the unsatisfied valence (*dot*) of the free

proteins contained considerable amounts of nitrogen and sometimes a little sulfur or phosphorus. The investigators soon decided that the main distinguishing feature of all these substances—what made their properties so different from other organic materials—must be the size of their molecules. The insolubility of the substances and their resistance to melting argued for large molecular size, because it had been found that ordinary organic compounds such as petroleum hydrocarbons became less and less soluble and acquired higher and higher melting points as they were combined into larger and larger molecules. The mechanical strength of cotton, wool

and silk also suggested that they were made of very large, strongly coherent molecules.

It was a logical deduction that each of the big molecules was made of certain building blocks—glucose in the cases of cellulose and starch, isoprene in the case of rubber and amino acids in the case of proteins. Chemists therefore began to call these compounds "poly" something: starch, for instance, was identified as a polysaccharide, meaning that it was composed of many sugar units. This is the basis of the present general terminology for the classes of compounds with which we are concerned: a monomer is a substance which can serve as a build-

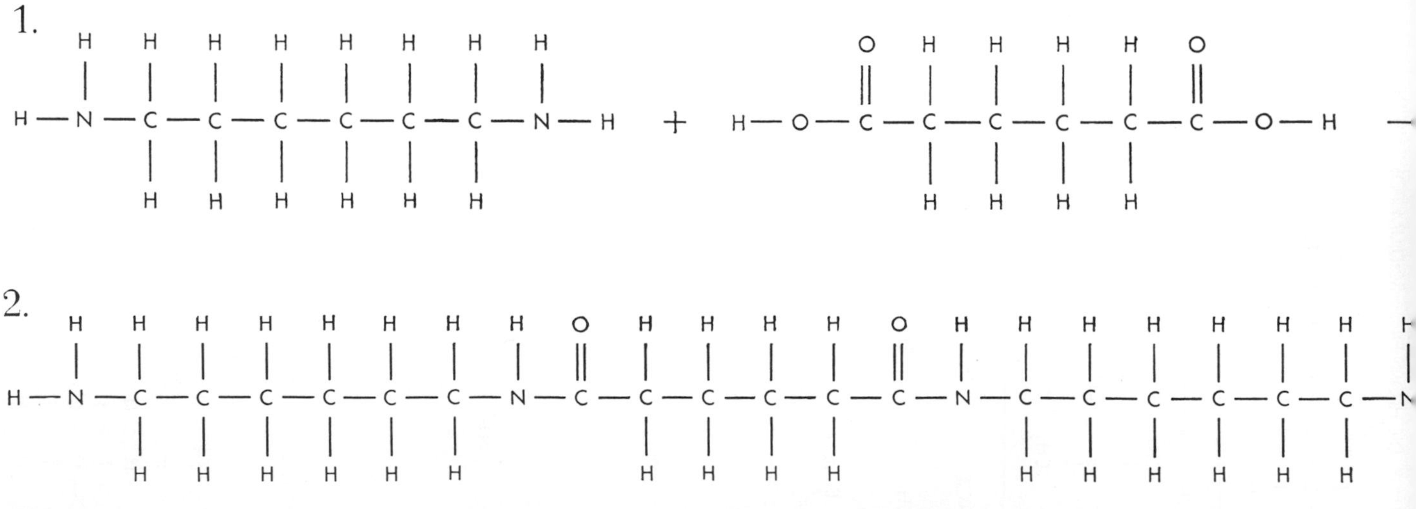

CONDENSATION POLYMERIZATION is a different process. Two molecules combine, usually with the elimination of water

(H-O-H), to form the repeating unit of the chain (1). The repeating units then combine in the same manner (2). In this example

2.

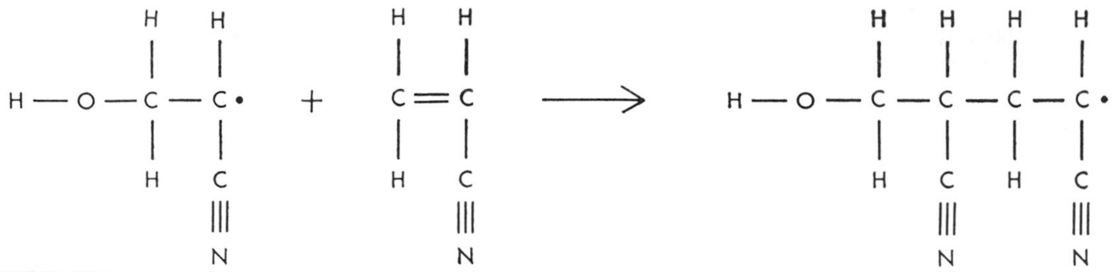

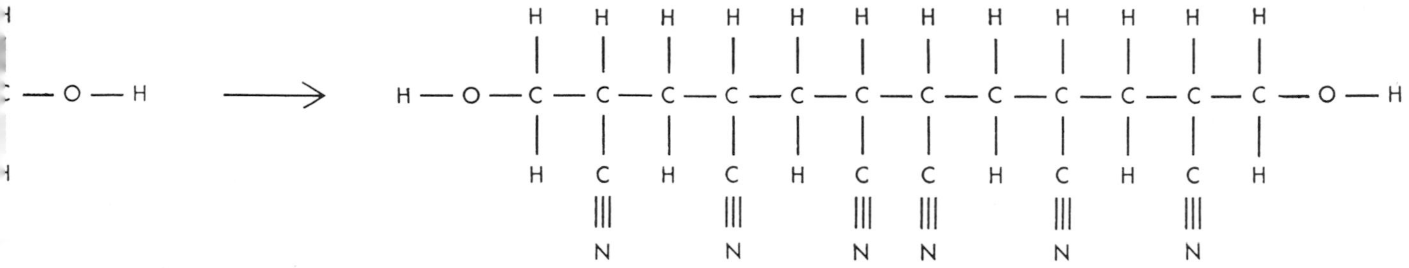

$$H-O-\overset{\overset{\displaystyle H}{|}}{\underset{\underset{\displaystyle H}{|}}{C}}-\overset{\overset{\displaystyle H}{|}}{\underset{\underset{\underset{N}{|||}}{C}}{C}}\cdot \quad + \quad \overset{\overset{\displaystyle H}{|}}{\underset{\underset{\displaystyle H}{|}}{C}}=\overset{\overset{\displaystyle H}{|}}{\underset{\underset{\underset{N}{|||}}{C}}{C}} \quad \longrightarrow \quad H-O-C-C-C-C\cdot$$

radical is transferred to the end of the monomer (1). The monomer now combines with another monomer (2). The process stops when two growing chains come together (3). In this case the free radical is obtained from the decomposition of hydrogen peroxide.

ing unit (e.g., glucose), a polymer (from the Greek, meaning "many parts") is a combination of such units; a high polymer is a very large aggregation of units, i.e., a compound of high molecular weight.

The spiritual father of high-polymer chemistry was Emil Fischer of Berlin, the great organic chemist of the late 19th and early 20th century. In his later years Fischer became fascinated by the mysterious chemistry of the organic macromolecules. Although loaded down with administrative duties for the German Academy of Sciences, Ministry of Education and National Research Council, he would retire to his private laboratory early every morning to experiment with new compounds. Sitting on a stool and watching thoughtfully as faint precipitates appeared in his reaction flasks, or white powders reluctantly crystallized out of solution, he saw visions of the chemistry which was to come 20 years later. Fischer induced some of his ablest co-workers to study rubber, starch, polypeptides, cellulose and lignin (the other major component of wood). At about the same time the great organic chemist Richard Willstätter was beginning to synthesize polysaccharides and to discover new methods for isolating lignin and enzymes. These pioneers worked largely by intuition. Willstätter was once asked, at a seminar where he reported a certain experiment, how he had happened to choose acetonitrile as the solvent, cobalt acetate as the catalyst and 75 degrees centigrade as the temperature. His answer was: "Just a thought, sir, just an idea."

Ignorance of the chemical structure of the high polymers did not, of course, debar their exploitation at the empirical level. Inventors discovered ways to convert cellulose into acetate fibers, films and coatings, into nitrated explosives and into many other useful products. Between 1870 and 1920 enterprising men (including Alfred Nobel) developed

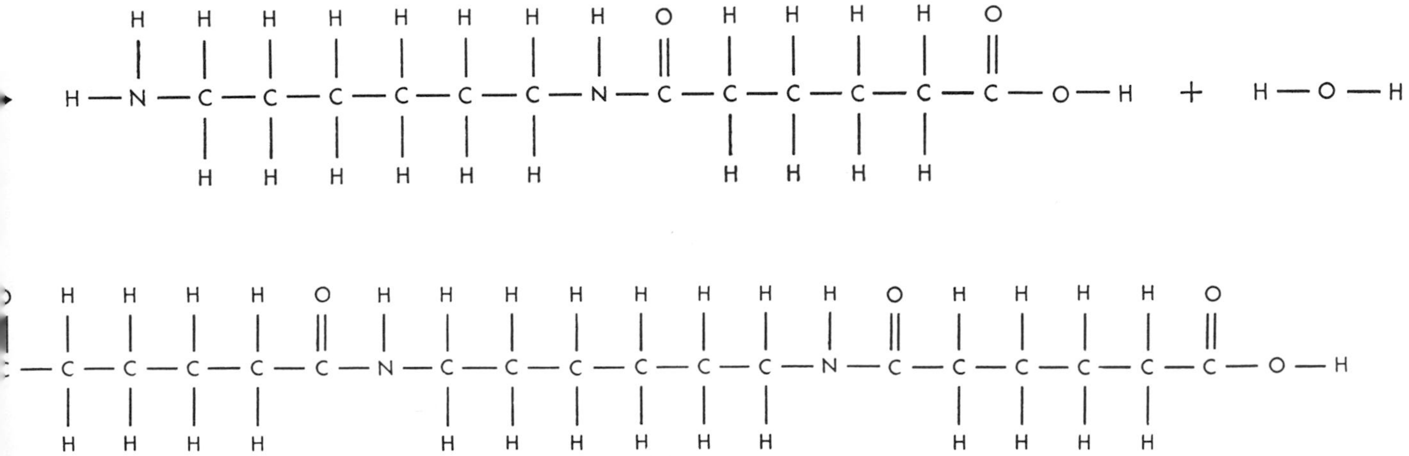

the original two molecules are those of hexamethylenediamine (left) and adipic acid (right). The polymer is nylon 66. The polymer in the example of addition polymerization is polyacrylonitrile. In both examples only a very short section of the chain is shown.

large industries based on cellulose derivatives. Meanwhile rubber also became a prominent article of commerce (and financial prosperity), thanks to Charles Goodyear's discovery, early in the 19th century, of the fact that heating it with sulfur (vulcanization) gave it useful properties. And proteins and starches likewise served as raw materials of other substantial manufacturing industries—leather, sizings, glues, adhesives, casein plastics and so on.

At the turn of the century there came an event which was to prove very significant in the development of high polymers. Leo H. Baekeland, a young Belgian chemist who had come to the U.S., took a deep interest in the sticky, resinous by-products which were such a bother to other chemists. He gave up a project on which he had been working and devoted himself to investigating the material that had fouled up his glassware. It was a gummy liquid, formed by a reaction between the common chemicals phenol and formaldehyde in solution in water. Baekeland found that by applying heat and pressure he could turn the liquid into a hard, transparent resin, which proved to be an excellent electrical insulator and to have good resistance to heat, moisture, chemicals and mechanical wear. So the synthetic plastics industry was born; other chemists went on to synthesize many other useful plastics of a similar (thermosetting) type, using formaldehyde with urea or aniline instead of phenol.

Thus by the second decade of this century factories all over the world were producing polymers in the forms of fibers, films, plastics, lacquers, coatings, adhesives and so on. Most of these were mere modifications of natural high polymers—conversions which transformed nature's substances (e.g., cellulose) into new materials of somewhat different properties. Baekeland's demonstration, on the other hand, paved the way for actual synthesis of polymers from simple materials. But all this was empirical; the basic principles governing the structure and behavior of polymers were still unknown. Chemists knew something about the "how" but not the "why."

It became increasingly important to know the why, in order to improve the products, to standarize the manufacturing processes and to reduce costs. Whenever, in such cases, the chemists of the factories tried to find out how they should best handle their systems and why they behaved as they did, they were disappointed by the lack of fundamental knowledge about polymers. Their academic colleagues in the universities had to confess that the exploration of large molecules was still in a state of infancy.

Soon after World War I a number of far-seeing leaders in science and industry recognized that a systematic exploration of polymer chemistry would pay large returns, industrially and in basic knowledge. Several of the leading organic chemists in the U.S. abandoned their successful careers in industry to gamble on full-time basic study of large molecules. Their studies were richly rewarded. One of the most fertile investigations was the memorable work of Wallace H. Carothers in the laboratories of E. I. du Pont de Nemours & Company. Supported by the vast resources of that organization and by a large group of brilliant collaborators, he developed a systematic knowledge of the chemistry of polymerization and synthesized many hundreds of polymers. This campaign produced, among other things, nylon and the synthetic rubber neoprene. While the scientific world was fascinated by the wealth and clarity of the fundamental results of the research, the du Pont Company drew great satisfaction and profit from its practical applicability. Rarely in the history of chemistry has basic research paid off so rapidly and so handsomely.

These events of the 1920's and 1930's encouraged many academic scientists and industrial researchers to turn to the large molecules. After having been a stepchild for many years, polymer chemistry became fashionable. Under the leadership of Hermann Staudinger in Germany, Thé Svedberg in Sweden and Kurt H. Meyer and Carl S. Marvel in the U.S., it moved rapidly ahead on a broad front—in experiments and in theory. Polymerization processes were developed and refined, their mechanisms explored, their products meticulously described (after having synthesized a new molecule one is naturally curious to know exactly what he has made). Very precise methods of describing the properties and behavior of polymers were developed, based on measurements of osmotic pressure, diffusion, sedimentation, light-scattering and viscosity.

Chemists could now discern a general pattern in the formation of giant molecules. Basically the high polymers were built by the linking of monomers end to end in chains, sometimes several thousand units long. But the chains then grouped themselves in two distinctly different ways. They either (1) coiled up to form a ball-shaped molecule (like a mass of intertwined spaghetti), or (2) lined up in straight, more or less rigid bundles (like wires in a cable). In general the coiled polymers had the characteristics of a rubber, while the straight-bundle type formed fibers or rigid plastics. The chemical character of the

1.

2.

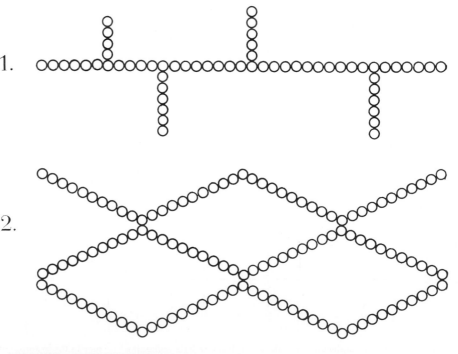

POLYMER CHAINS CAN GROW by branching (1) or can be connected by cross-linking (2). The balls in this drawing and the one on the opposite page are not atoms but monomers.

chains determined whether they would coil up at random or align themselves in bundles: if the chains were relatively rigid and contained chemically attracting groups along their length, they would attach themselves to one another side by side in bundles.

As the principles governing the properties of polymers began to shape up in the minds of the investigators, an exciting new prospect emerged. It was a matter of great intellectual satisfaction to be able to reduce the behavior of these substances to orderly laws and predict it with mathematical precision. But no less stirring was the new creative power made possible by this knowledge. The technological progress of mankind has been largely a history of putting available materials to use. It is a considerable step forward to invent the materials themselves on order. And this is the stage we have now reached in polymer chemistry. Starting from a need for some material of specified properties, we are in a position to create a new material tailored to fill that need.

As the building stones for this enterprise we now have some 40 readily available organic monomers, largely derived from coal and oil. These 40 building units can produce an almost limitless number of combinations. Already they have given us scores of important new man-made materials: all the synthetic fibers, rubbers and plastics. The production of these monomers in the U.S. now amounts to about $2 billion a year.

Polymer chemists in the U.S. and abroad are engaged in a vast effort to develop processes which will facilitate the creation of new products and reduce the cost of the present ones. They are exploring various polymerization methods, catalysts, continuous processes, and conditions which will control reactions such as very high pressures, high and low temperatures, irradiation. Knowledge of the principles underlying the chemical properties is sufficiently advanced so that the chemists can introduce into a giant molecule a monomer which will endow it with a high melting point or great resistance to solvents or high tensile strength or some other desired quality.

The products made so far can be considered only a foretaste of more spectacular ones to come. There are several frontiers inviting exploration. For example, the largest high polymers now in production have molecular weights in the neighborhood of 200,000. There is reason to believe that larger molecules would be much stronger. Consequently several industrial laboratories are looking into the possibilities for producing "super high polymers" with molecular weights in the millions. Another active frontier is the investigation of ways to raise the resistance of polymers to heat. The plastics, fibers, rubbers and coatings now made break down at temperatures of 600 degrees Fahrenheit or less. But the prospects for making high polymers which will be able to withstand substantially higher temperatures look promising: they may be based on certain highly stable organic molecules such as diphenyl oxide or diphenylmethylene, with additions of resistant elements such as fluorine, boron or silicon.

Looking much farther into the future, we can even see possibilities for synthesizing biological molecules—not to create life but to furnish aids to or substitutes for living tissues. Already we have a synthetic polymer which can serve some of the functions of blood serum. Although our chemical laboratories do not begin to approach the orderliness or perfection with which a living organism builds its high polymers, we can take hope from the fact that we have a vastly larger number of monomers at our disposal than any living system has, and therefore can make an even greater variety of products.

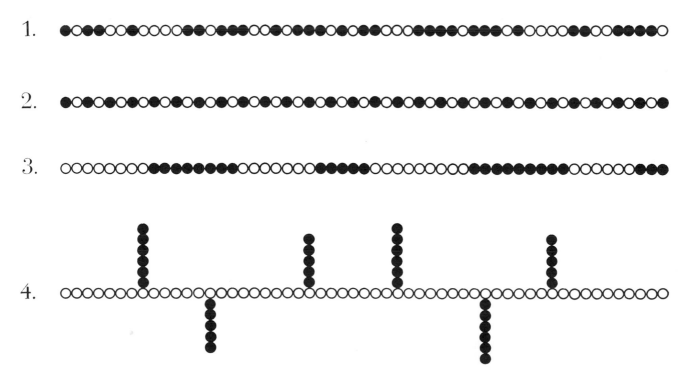

COPOLYMERS are polymer chains in which more than one kind of monomeric unit occurs. Here one kind of monomer is represented by a white ball; the other, by a black. The form of such copolymers may be random (1); alternating (2), "block" (3) or "graft" (4).

17

PROTEINS

PAUL DOTY

September 1957

The principal substance of living cells, these giant molecules have identical backbones. Each is adapted to its specific task by a unique combination of side groups, size, folding and shape

Thousands of different proteins go into the make-up of a living cell. They perform thousands of different acts in the exact sequence that causes the cell to live. How the proteins manage this exquisitely subtle and enormously involved process will defy our understanding for a long time to come. But in recent years we have begun to make a closer acquaintance with proteins themselves. We know they are giant molecules of great size, complexity and diversity. Each appears to be designed with high specificity for its particular task. We are encouraged by all that we are learning to seek the explanation of the function of proteins in a clearer picture of their structure. For much of this new understanding we are indebted to our experience with the considerably simpler giant molecules synthesized by man. High-polymer chemistry is now coming forward with answers to some of the pressing questions of biology.

Proteins, like synthetic high polymers, are chains of repeating units. The units are peptide groups, made up of the monomers called amino acids [*see diagram below*]. There are more than 20 different amino acids. Each has a distinguishing cluster of atoms as a side group [*see next two pages*], but all amino acids have a certain identical group. The linking of these groups forms the repeating peptide units in a "polypeptide" chain. Proteins are polypeptides of elaborate and very specific construction. Each kind of protein has a unique number and sequence of side groups which give it a particular size and chemical identity. Proteins seem to have a further distinction that sets them apart from other high polymers. The long chain of each protein is apparently folded in a unique configuration which it seems to maintain so long as it evidences biological activity.

We do not yet have a complete picture of the structure of any single protein. The entire sequence of amino acids has been worked out for insulin [see the article "The Insulin Molecule," by E. O. P. Thompson, beginning on page 183], the determination of several more is nearing completion. But to locate each group and each atom in the configuration set up by the folded chain is intrinsically a more difficult task; it has resisted the Herculean labors of a generation of X-ray crystallographers and their collaborators. In the early 1930s W. T. Astbury of the University of Leeds succeeded in demonstrating that two X-ray diffraction patterns, which he called alpha and beta, were consistently associated with certain fibers, and he identified a third with collagen, the protein of skin, tendons and other structural tissues of the body. The beta pattern, found in the fibroin of silk, was soon shown to arise from bundles of nearly straight polypeptide chains held tightly to one another by hydrogen bonds. Nylon and some other synthetic fibers give a similar diffraction pattern. The alpha pattern resisted decoding until 1951, when Linus Pauling and R. B. Corey of the California Institute of Technology advanced the notion, since confirmed by further X-ray diffraction studies, that it is created by the twisting of the chain into a helix. Because it is set up so naturally by the hydrogen bonds available in the backbone of a polypeptide chain [*see top diagram on page 142*], the alpha helix was deduced to be a major structural element in the configuration of most proteins. More recently, in 1954, the Indian X-ray crystallographer G. N. Ramachandran showed that the collagen pattern comes from three polypeptide helixes twisted around one another. The resolution of these master plans was theoretically and esthetically gratifying, especially since the nucleic acids, the substance of genetic chemistry, were concurrently shown to have the structure of a double helix. For all their apparent general validity, however, the master plans did not give us the complete configuration in three dimensions

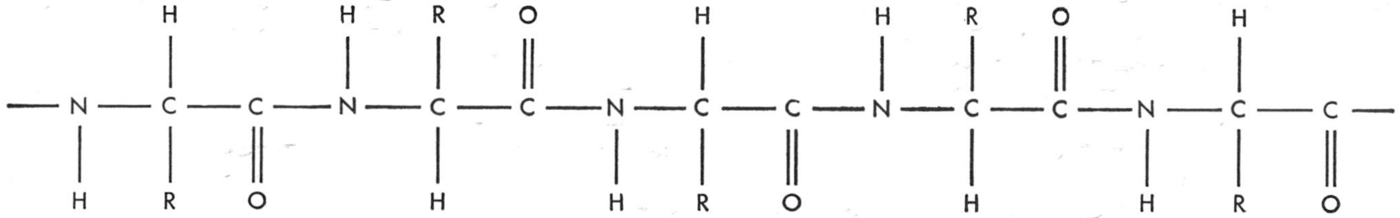

POLYPEPTIDE CHAIN is a repeating structure made up of identical peptide groups (CCONHC). The chain is formed by amino acids, each of which contributes an identical group to the backbone plus a distinguishing radical (R) as a side group.

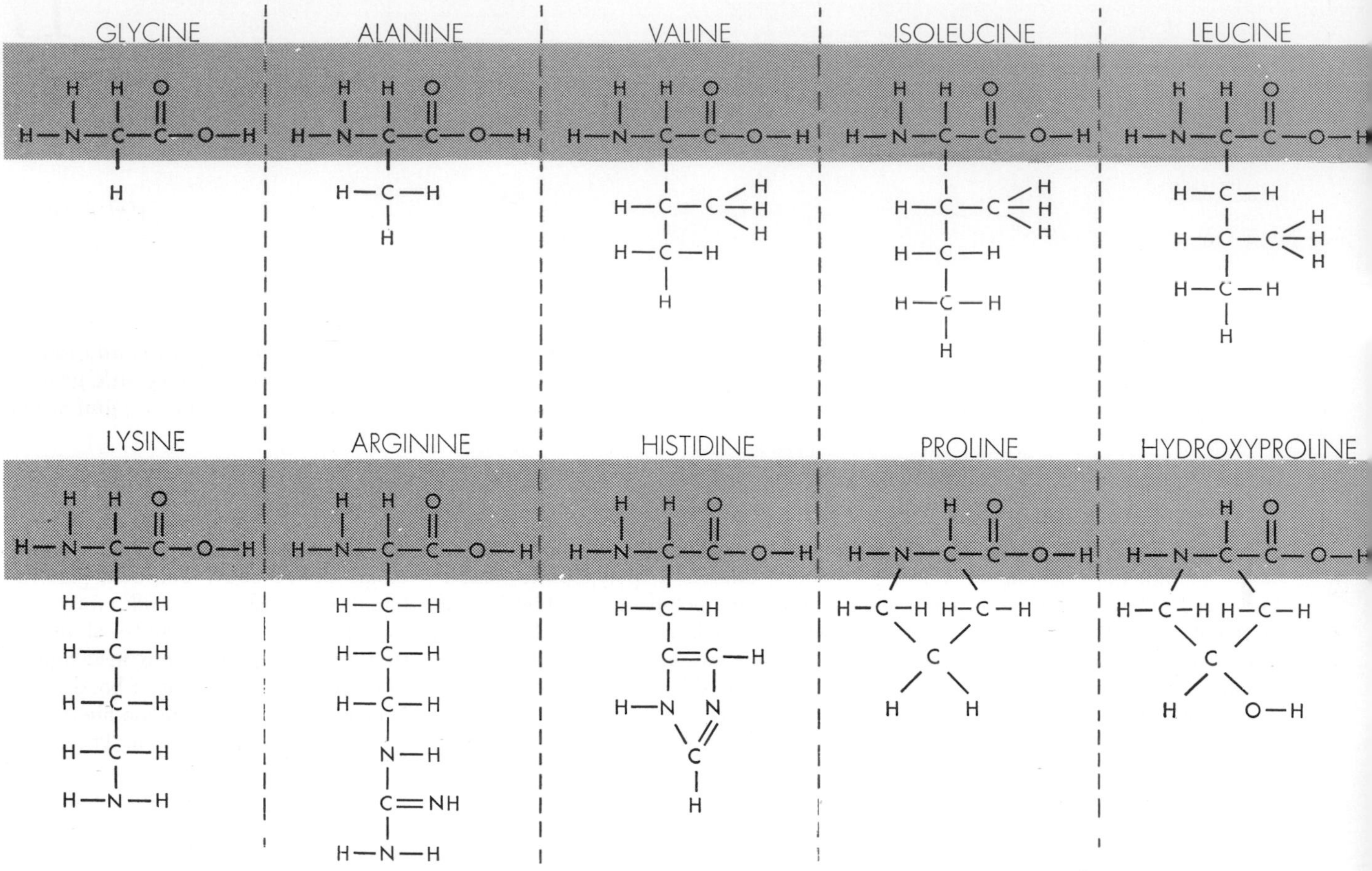

AMINO ACIDS, the 20 commonest of which are shown in this chart, have identical atomic groups (*in colored bands*) which react to form polypeptide chains. They are distinguished by their unique side groups. In forming a chain, the amino group (NH_2) of one

of any single protein.

The X-ray diffraction work left a number of other questions up in the air. Since the alpha helix had been observed only in a few fibers, there was no solid experimental evidence for its existence elsewhere. There was even a suspicion that it could occur only in fibers, where it provides an economical way to pack polypeptides together in crystalline structures. Many proteins, especially chemically active ones such as the enzymes and antibodies, are globular, not linear like those involved in fibers and structural tissues. In the watery solutions which are the natural habitat of most proteins, it could be argued, the affinity of water molecules for hydrogen bonds would disrupt the alpha helix and reduce the chain to a random coil. These doubts and suppositions have prompted investigations by our group at Harvard University in collaboration with E. R. Blout of the Children's Cancer Research Foundation in Boston. In these investigations we have em-

ployed synthetic polypeptides as laboratory models for the more complex and sensitive proteins. When Blout and coworkers had learned to polymerize them to sufficient length—100 to 1,000 amino acid units—we proceeded to observe their behavior in solution.

Almost at once we made the gratifying discovery that our synthetic polypeptides could keep their helical coils wound up in solutions. Moreover, we found that we could unwind the helix of some polypeptides by adjusting the acidity of our solutions. Finally, to complete the picture, we discovered that we could reverse the process and make the polypeptides wind up again from random coils into helixes.

The transition from the helix to the random coil occurs within a narrow range as the acidity is reduced; the hydrogen bonds, being equivalent, tend to let go all at once. It is not unlike the melting of an ice crystal, which takes place in a narrow temperature range. The reason is the same, for the ice crys-

tal is held together by hydrogen bonds. To complete the analogy, the transition from the helix to the random coil can also be induced by heat. This is a true melting process, for the helix is a one-dimensional crystal which freezes the otherwise flexible chain into a rodlet.

From these experiments we conclude that polypeptides in solution have two natural configurations and make a reversible transition from one to the other, depending upon conditions. Polypeptides in the solid state appear to prefer the alpha helix, though this is subject to the presence of solvents, especially water. When the helix breaks down here, the transition is to the beta configuration, the hydrogen bonds now linking adjacent chains. Recently Blout and Henri Lenormant have found that fibers of polylysine can be made to undergo the alpha-beta transition reversibly by mere alteration of humidity. It is tempting to speculate that a reversible alpha-beta transition may underlie the process of muscle contraction and other types of

SERINE | THREONINE | ASPARTIC ACID | GLUTAMIC ACID | TYROSINE

```
        H   H   O          H   H   O          H   H   O          H   H   O          H   H   O
        |   |   ||         |   |   ||         |   |   ||         |   |   ||         |   |   ||
   H—N—C—C—O—H      H—N—C—C—O—H      H—N—C—C—O—H      H—N—C—C—O—H      H—N—C—C—O—H
        |                  |                  |                  |                  |
      H—C—H              H—C—O—H            H—C—H              H—C—H              H—C—H
        |                  |                  |                  |
      O—H                H—C—H              C=O                H—C—H
                           |                  |                  |
                           H                O—H                C=O
                                                                 |
                                                               O—H
                                                                                  O—H
```

CYSTEINE | METHIONINE | CYSTINE | TRYPTOPHAN | PHENYLALANINE

```
        H   H   O          H   H   O          H   H   O          H   H   O          H   H   O
        |   |   ||         |   |   ||         |   |   ||         |   |   ||         |   |   ||
   H—N—C—C—O—H      H—N—C—C—O—H      H—N—C—C—O—H      H—N—C—C—O—H      H—N—C—C—O—H
        |                  |                  |                  |                  |
      H—C—H              H—C—H              H—C—H              H—C—H              H—C—H
        |                  |                  |                  |
      S—H                H—C—H                S                C=C
                           |                  |                   
                           S                  S
                           |                  |
                         C                  H—C—H
                           |                  |
                           H            H—N—C—C—O—H
                                             |   |   ||
                                             H   H   O
```

movement in living things.

Having learned to handle the poly peptides in solution we turned our attention to proteins. Two questions had to be answered first: Could we find the alpha helix in proteins in solution, and could we induce it to make the reversible transition to the random coil and back again? If the answer was yes in each case, then we could go on to a third and more interesting question: Could we show experimentally that biological activity depends upon configuration? On this question, our biologically neutral synthetic polypeptides could give no hint.

For the detection of the alpha helix in proteins the techniques which had worked so well on polypeptides were impotent. The polypeptides were either all helix or all random coil and the rodlets of the first could easily be distinguished from the globular forms of the second by use of the light-scattering technique. But we did not expect to find that any of the proteins we were going to investigate were 100 per cent helical in configura-

tion. The helix is invariably disrupted by the presence of one of two types of amino acid units. Proline lacks the hydrogen atom that forms the crucial hydrogen bond; the side groups form a distorting linkage to the chain instead. Cystine is really a double unit, and forms more or less distorting cross-links between chains. These units play an important part in the intricate coiling and folding of the polypeptide chains in globular proteins. But even in globular proteins, we thought, some lengths of the chains might prove to be helical. There was nothing, however, in the over-all shape of a globular protein to tell us whether it had more or less helix in its structure or none at all. We had to find a way to look inside the protein.

One possible way to do this was suggested by the fact that intact, biologically active proteins and denatured proteins give different readings when observed for an effect called optical rotation. In general, the molecules that exhibit this effect are asymmetrical in

atomic structure. The side groups give rise to such asymmetry in amino acids and polypeptide chains; they may be attached in either a "left-handed" or a "right-handed" manner. Optical rotation provides a way to distinguish one from the other. When a solution of amino acids is interposed in a beam of polarized light, it will rotate the plane of polarization either to the right or to the left [see diagrams at top of page 144]. Though amino acids may exist in both forms, only left-handed units, thanks to some accident in the chemical phase of evolution, are found in proteins. We used only the left-handed forms, of course, in the synthesis of our polypeptide chains.

Now what about the change in optical rotation that occurs when a protein is denatured? We knew that native protein rotates the plane of the light 30 to 60 degrees to the left, denatured protein 100 degrees or more to the left. If there was some helical structure in the protein, we surmised, this shift in rotation might be induced by the disappearance of the helical structure in the denaturation process. There was reason to believe that the helix, which has to be either left-handed or right-handed, would have optical activity. Further, although it appeared possible for the helix to be wound either way, there were grounds for assuming that nature had chosen to make all of its helices one way or the other. If it had not, the left-handed and right-handed helixes would mutually cancel out their respective optical rotations. The change in the optical rotation of proteins with denaturation would then have some other explanation entirely, and we would have to invent another way to look for helixes.

To test our surmise we measured the optical rotation of the synthetic poly-peptides. In the random coil state the polypeptides made an excellent fit with the denatured proteins, rotating the light 100 degrees to the left. The rotations in both cases clearly arose from the same cause: the asymmetry of the amino acid units. In the alpha helix configuration the polypeptides showed almost no rotation or none at all. It was evident that the presence of the alpha helix caused a

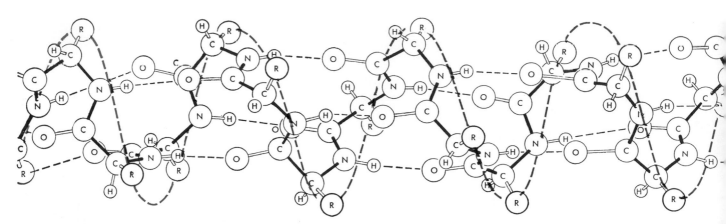

ALPHA HELIX gives a polypeptide chain a linear structure shown here in three-dimensional perspective. The atoms in the repeating unit (CCONHC) lie in a plane; the change in angle between one unit and the next occurs at the carbon to which the side group

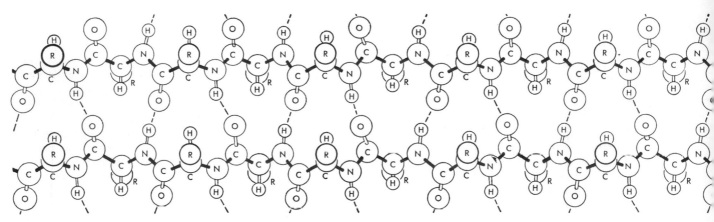

BETA CONFIGURATION ties two or more polypeptide chains to one another in crystalline structures. Here the hydrogen bonds do not contribute to the internal organization of the chain, as in the alpha helix, but link the hydrogen atoms of one chain to the oxygen

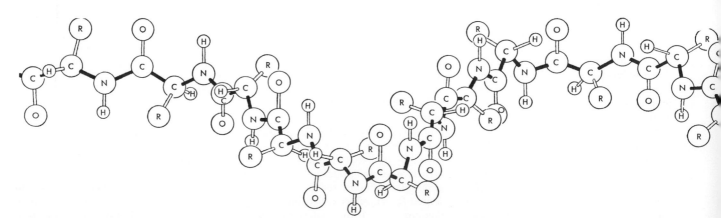

RANDOM CHAIN is the configuration assumed by the polypeptide molecule in solution, when hydrogen bonds are not formed. The flat configuration of the repeating unit remains, but the chain rotates about the carbon atoms to which the side groups are at-

counter-rotation to the right which nearly canceled out the leftward rotation of the amino acid units. The native proteins also had shown evidence of such counter-rotation to the right. The alpha configuration did not completely cancel the leftward rotation of the amino acid units, but this was consistent with the expectation that the protein structures would be helical only in part. The experiment

thus strongly indicated the presence of the alpha helix in the structure of globular proteins in solution. It also, incidentally, seemed to settle the question of nature's choice of symmetry in the alpha helix: it must be right-handed.

When so much hangs on the findings of one set of experiments, it is well to double check them by observa-

tions of another kind. We are indebted to William Moffitt, a theoretical chemist at Harvard, for conceiving of the experiment that provided the necessary confirmation. It is based upon another aspect of the optical rotation effect. For a given substance, rotation varies with the wavelength of the light; the rotations of most substances vary in the same way. Moffitt predicted that the presence of

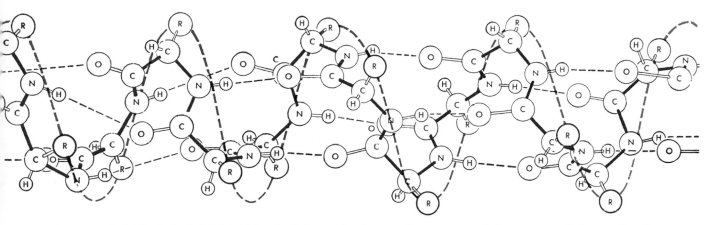

(R) is attached. The helix is held rigid by the hydrogen bond (*broken black lines*) between the hydrogen attached to the nitrogen in one group and the oxygen attached to a carbon three groups along the chain. The colored line traces the turns of the helix.

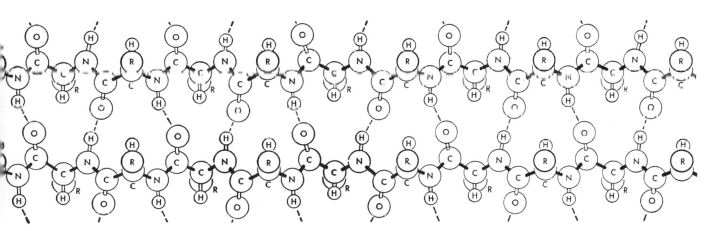

atoms in the adjoining chain. The beta configuration is found in silk and a few other fibers. It is also thought that polypeptide chains in muscle and other contractile fibers may make reversible transitions from alpha helix to beta configuration when in action.

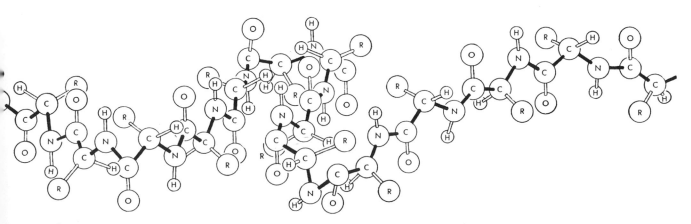

tached. The random chain may be formed from an alpha helix when hydrogen bonds are disrupted in solution. A polypeptide chain may make a reversible transition from alpha helix to random chain, depending upon the acid-base balance of the solution.

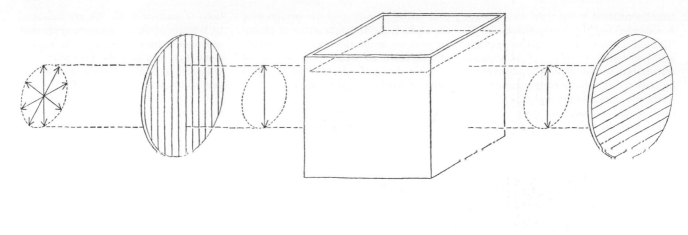

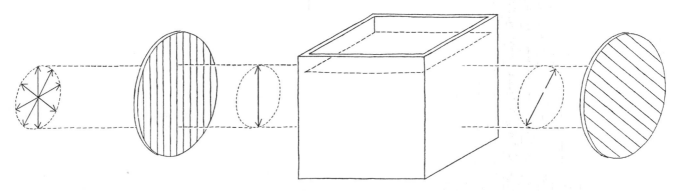

OPTICAL ROTATION is induced in a beam of polarized light by molecules having certain types of structural asymmetry. At top a beam of light is polarized in the vertical plane and transmitted unchanged through a neutral solution. At bottom asymmetrical molecules in the solution cause the beam to rotate from the vertical plane. The degree of rotation may be determined by turning the second polarizing filter (*right*) to the point at which it cuts off the beam. The alpha helix in a molecule causes such rotation.

the alpha helix in a substance would cause its rotation to vary in a different way. His prediction was sustained by observation: randomly coiled polypeptides showed a normal variation while the helical showed abnormal. Denatured and native proteins showed the same contrast. With the two sets of experiments in such good agreement, we could conclude with confidence that the alpha helix has a significant place in the structure of globular proteins. Those amino acid units that are not involved in helical configurations are weakly bonded to each other or to water molecules, probably in a unique but not regular or periodic fashion. Like synthetic high-polymers, proteins are partly crystalline and partly amorphous in structure.

The optical rotation experiments also provided a scale for estimating the helical content of protein. The measurements indicate that, in neutral solutions, the helical structure applies to 15 per cent of the amino acid units in ribonuclease, 50 per cent of the units in serum albumin and 85 per cent in tropomyosin. With the addition of denaturing agents to the solution, the helical content in each case can be reduced to zero. In

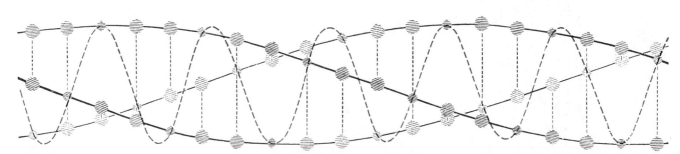

COLLAGEN MOLECULE is a triple helix. The colored broken line indicates hydrogen bonds between glycine units. The black broken lines indicate hydrogen bonds which link hydroxyproline units and give greater stability to collagens in which they are found.

some proteins the transition is abrupt, as it is in the synthetic polypeptides. On the other hand, by the use of certain solvents we have been able to increase the helical content of some proteins—in the case of ribonuclease from 15 to 70 per cent. As in the polypeptides, the transition from helix to random coil is reversible. The percentage of helical structure in proteins is thus clearly a variable. In their natural environment, it appears, the percentage at any given time represents the equilibrium between the inherent stability of the helix and the tendency of water to break it down.

In a number of enzymes we have been able to show that biological activity falls off and increases with helical content. Denaturation is now clearly identified with breakdown of configuration, certainly insofar as it involves the integrity of the alpha helix. This is not surprising. It is known that catalysts in general must have rigid geometrical configurations. The catalytic activity of an enzyme may well require that its structure meet similar specifications. If this is so, the rigidity that the alpha helix imposes on the otherwise flexible polypeptide chain must play a decisive part in establishing the biological activity of an enzyme. It seems also that adjustability of the stiffness of structure in larger or smaller regions of the polypeptide chain may modify the activity of proteins in response to their environment. Among other things, it could account for the versatility of the gamma globulins; without any apparent change in their amino acid make-up, they are able somehow to adapt themselves as antibodies to a succession of different infectious agents.

The next step toward a complete anatomy of the protein molecule is to determine which amino acid units are in the helical and which in the nonhelical regions. Beyond that we shall want to know which units are near one another as the result of folding and cross-linking, and a myriad of other details which will supply the hues and colorings appropriate to a portrait of an entity as intricate as protein. Many such details will undoubtedly be supplied by experiments that relate change in structure to change in function, like those described here.

In the course of our experiments with proteins in solution we have also looked into the triple-strand structure of collagen. That structure had not yet been resolved when we began our work, so we did not know how well it

was designed for the function it serves in structural tissues. Collagen makes up one third of the proteins in the body and 5 per cent of its total weight; it occurs as tiny fibers or fibrils with bonds that repeat at intervals of about 700 Angstroms. It had been known for a long time that these fibrils could be dissolved in mild solvents such as acetic acid and then reconstituted, by simple precipitation, into their original form with their bandings restored. This remarkable capacity naturally suggested that the behavior of collagen in solution was a subject worth exploring.

Starting from the groundwork of other investigators, Helga Boedtker and I were able to demonstrate that the collagen molecule is an extremely long and thin rodlet, the most asymmetric molecule yet isolated. A lead pencil of comparable proportions would be a yard long. When a solution of collagen is just slightly warmed, these rodlets are irreversibly broken down. The solution will gel, but the product is gelatin, as is well known to French chefs and commercial producers of gelatin. The reason the dissolution cannot be reversed was made clear when we found that the molecules in the warmed-up solution had a weight about one third that of collagen. It appeared that the big molecule of collagen had broken down into three polypeptide chains.

At about the same time Ramachandran proposed the three-strand helix as the collagen structure. Not long afterward F. H. C. Crick and Alexander Rich at the University of Cambridge and Pauline M. Cowan and her collaborators at King's College, London, worked out the structure in detail. It consists of three polypeptide chains, each incorporating three different amino acid units—proline, hydroxyproline and glycine. The key to the design is the occurrence of glycine, the smallest amino acid unit, at every third position on each chain. This makes it possible for the bulky proline or hydroxyproline groups to fit into the links of the triple strand, two of these nesting in each link with the smaller glycine unit [see diagram on page 144].

One question, however, was left open in the original model. Hydroxyproline has surplus hydrogen bonds, which, the model showed, might be employed to reinforce the molecule itself or to tie it more firmly to neighboring molecules in a fibril. Independent evidence seemed to favor the second possibility. Collagen in the skin is irreversibly broken down in a first degree burn, for example, at a

temperature of about 145 degrees Fahrenheit. This is about 60 degrees higher than the dissolution temperature of the collagen molecule in solution. The obvious inference was that hydroxyproline lends its additional bonding power to the tissue structure. Moreover, tissues with a high hydroxyproline content withstand higher temperatures than those with lower; the skin of codfish, with a low hydroxyproline content, shrivels up at about 100 degrees. Tomio Nishihara in

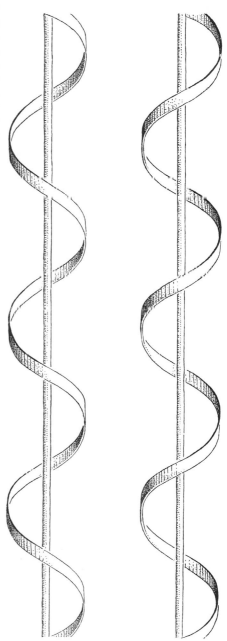

ASYMMETRY of a helix is either left-handed (left) or right-handed. Helix in proteins appears to be exclusively right-handed.

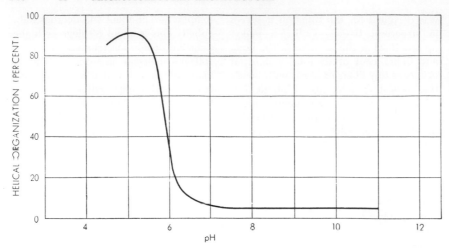

ALPHA HELIX BREAKDOWN is induced in solutions of some polypeptides when the pH (acidity or alkalinity) reaches a critical value at which hydrogen bonds are disrupted.

our laboratory has compared the breakdown temperatures of collagen molecules and tissues from various species and found that the tissue temperature is uniformly about 60 degrees higher. Thus we must conclude that the extra stability conferred by hydroxyproline goes directly to the molecule and not to the fibril.

The structure of collagen demonstrates three levels in the adaptation of polypeptide chains to fit the requirements of function. First there are the chains as found in gelatin, with their three amino acids lined up in just the right sequence. These randomly coiled and quite soluble molecules are transformed into relatively insoluble, girderlike building units when united into sets of three by hydrogen bonds. The subtly fashioned collagen molecules are still too fragile to withstand body temperatures. When arranged side by side, however, they form a crystalline structure which resists comparatively high temperatures and has fiber-like qualities with the vast range of strengths and textures required in the different types of tissues that are made of collagen.

The story of collagen, like that of other proteins, is still far from complete. But it now seems that it will rank among the first proteins whose molecular structure has been clearly discerned and related in detail to the functions it serves.

18

THE CHEMICAL STRUCTURE OF PROTEINS

WILLIAM H. STEIN AND STANFORD MOORE

February 1961

The atom-by-atom structural formulas of three proteins have now been worked out. This knowledge should help in understanding how the structures of proteins are related to their biological activity

Six years ago, after a decade of pioneering research, Frederick Sanger and his colleagues at the University of Cambridge were able to write the first structural formula of a protein—the hormone insulin. They had discovered the precise order in which the atoms are strung together in the long chains that make up the insulin molecule. With this truly epochal achievement, for which Sanger received a Nobel prize in 1958, there opened a new chapter in protein chemistry.

Some idea of the significance of this chapter for chemistry and biology can be gained from the realization that proteins comprise more than half the solid substance in the tissues of man and other mammals. Proteins are important mechanical elements and perform countless essential catalytic and protective functions. Only when the structures of large numbers of proteins have been worked out will biochemists be in a position to answer many of the fundamental questions they have long been asking. The goal is still far off, but there has been much progress in this challenging field.

Before embarking on a summary of some of the recent developments, it is well to point out that the chemical approach does not provide a complete solution to the problem of protein structure. The order of links in the chain is not the whole story. Each chain is coiled and folded in a three-dimensional pattern, no less important than the atom-by-atom sequence in determining its biological activity. Chemical methods can provide only a partial insight into this three-dimensional, or "tertiary," structure. In the past few years the spatial problem has begun to yield to X-ray analysis. The present article, however, is primarily concerned with the chemical rather than the physical line of attack.

Proteins are very large, they are complex and they are fragile. The molecule of insulin, one of the smallest proteins, contains 777 atoms. Some protein molecules are thought to be 50 times bigger, although even the size of the largest ones is not definitely settled. Fortunately most of them are constructed on the same general plan. The links, or building blocks, in the molecular chains are amino acids. In their uncombined form each of these substances consists of an amino group (NH_2) and a carboxyl group ($COOH$), both attached to the same carbon atom. Also attached to this carbon are a hydrogen atom and one of 24 different "side groups." When the amino acids link together in a protein molecule, they join end to end, the carboxyl group of one combining with the amino group of the next to form a "peptide bond" ($-CO-NH-$) and a molecule of water (HOH). Since each component has lost a molecule of water, it is called an amino acid "residue." Under the action of acids, alkalis or certain enzymes, peptide chains break apart, the $-NH-$ groups regaining a hydrogen atom and the $-CO-$ groups a hydroxyl group (OH). Thus each residue regains a molecule of water and the peptide bonds are said to have been hydrolyzed.

Among the amino acids the one called cystine is unique. Its molecule is a sort of Siamese twin containing two $-NH_2$ and two $-COOH$ groups, with identical halves of the molecule joined by a disulfide bond ($-S-S-$). One cystine molecule can therefore enter into two separate peptide chains, cross-linking them by means of the disulfide bond. Cystine can also cause a single chain to fold back on itself.

The Insulin Molecule

Much of the chemistry of proteins had been laboriously uncovered in many years of research before Sanger took up the study of insulin. His success rested largely on two advances in technique. One was the development of paper chromatography by the British chemists A. J. P. Martin and R. L. M. Synge. By means of this elegant method tiny samples of complex mixtures can be fractionated on a piece of filter paper and their components identified [see "Chromatography," by William H. Stein and Stanford Moore; Scientific American Offprint 81]. The second key to the problem was Sanger's discovery of a way to label the amino group at the end of a peptide chain. He found that a dinitrophenyl (DNP) group could be attached to free amino groups to form a yellow compound. Even when the peptide is fragmented into separate amino acids, the DNP group remains attached to the residue at the end of the chain, thus making it possible to identify this residue.

The details of Sanger's analysis have already been described in this magazine by one of his co-workers ["The Insulin Molecule," by E. O. P. Thompson, page 34]. Here we shall review their work very briefly. With the help of the DNP method they first established that the insulin molecule consists of two chains. These chains are held together by the $-S-S-$ bonds of cystine residues. By treating the hormone with a mild oxidizing agent the experimenters were able to open the disulfide bonds and thus separate the two intact chains. One proved to contain 21 amino acids; the other, 30. They then proceeded to cleave each chain into smaller pieces by treating it with acid, which hydrolyzes peptide bonds more or less at random. The fragments were separated by chromatography and by other means, labeled at their amino ends by the DNP method, broken down further, separated again, relabeled and so on. In this way the order of amino acids in a large number of small pieces was established. By shattering the chain many times and noting overlapping sequences in the various fragments the Cambridge group at last deduced the complete succession of amino acid residues in each part of the molecule.

After the completion of this Herculean task an almost equal effort was required simply to determine the pairing of the half-cystine residues. One chain was found to contain four half-residues of cystine and the other chain two. To find out which ones were paired it was necessary to break the molecule into smaller fragments containing different pairs of half-cystine residues with their disulfide bonds intact. In the process, however, the cystine halves tended to trade part-

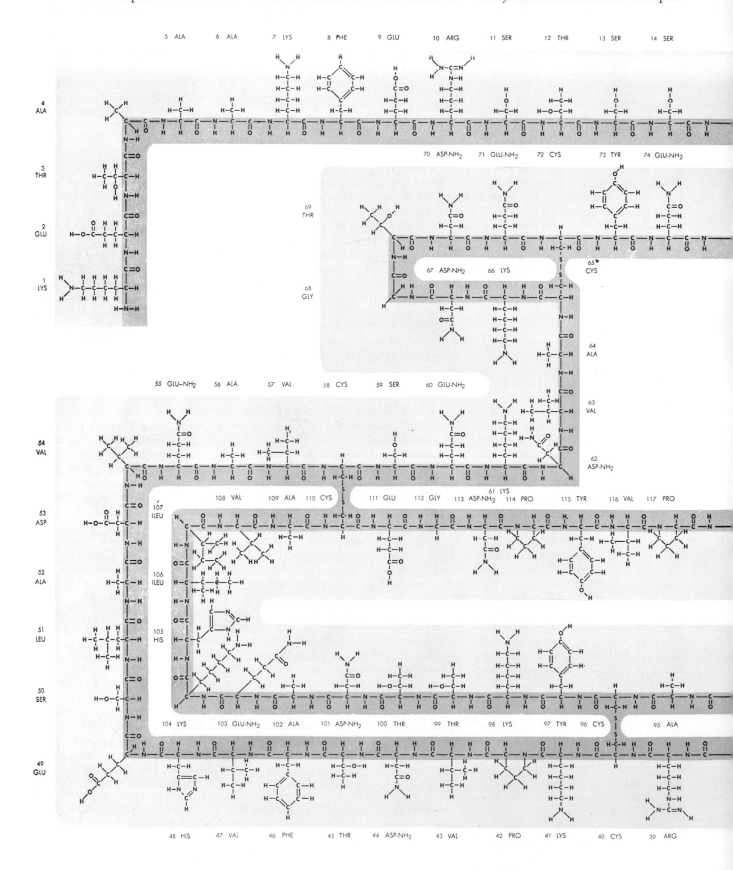

ners and produce spurious pairings. Finally a way around the difficulty was worked out, and the disulfide bonds were unequivocally pinned down.

Once Sanger had shown that the problem was solvable a number of workers began to attack the structure of larger proteins. In the six years that have passed two more molecules have been completely delineated and others are on the way. Last year the analysis of the structure of ribonuclease, an enzyme that digests ribonucleic acid, was completed in the authors' laboratory. Ribonuclease contains 124 amino acid residues. As this article was in preparation Gerhard Schramm and his associates in Germany and Heinz L. Fraenkel-Conrat, Wendell M. Stanley and their colleagues at the University of California announced that they had finished working out the structure of the 158-amino-acid-residue protein in tobacco mosaic virus.

At first sight it might seem that any protein should yield to a massive effort along the lines used so successfully with insulin. As Sanger and others realized, however, it is not so simple as that—if simple is the word for 10 years of unremitting work. Problems multiply rapidly with increasing molecular size, and an approach that was difficult and time-consuming in the case of insulin can become fruitless and interminable.

One aid to further progress has been the development of more precise methods for identifying and measuring small quantities of amino acids. Several years ago the authors undertook to apply column chromatography for this purpose. Instead of filter paper we use a five-foot

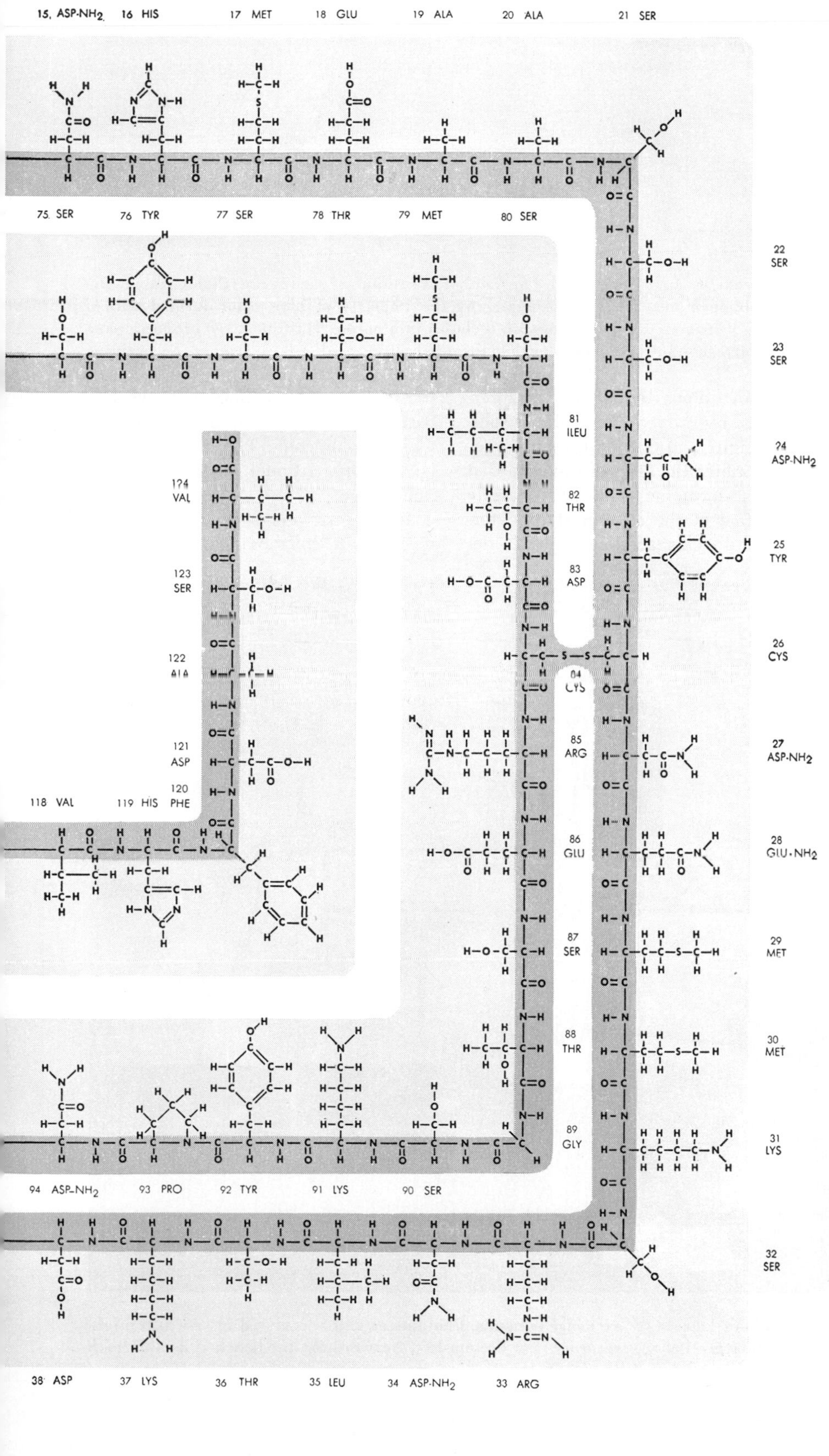

ALA	ALANINE
ARG	ARGININE
ASP	ASPARTIC ACID
ASP·NH₂	ASPARAGINE
CYS	CYSTINE
GLU	GLUTAMIC ACID
GLU·NH₂	GLUTAMINE
GLY	GLYCINE
HIS	HISTIDINE
ILEU	ISOLEUCINE
LEU	LEUCINE
LYS	LYSINE
MET	METHIONINE
PHE	PHENYLALANINE
PRO	PROLINE
SER	SERINE
THR	THREONINE
TYR	TYROSINE
VAL	VALINE

MOLECULE OF RIBONUCLEASE, an enzyme that digests the cellular substance ribonucleic acid (RNA), is diagramed in two dimensions on these two pages. In this structural formula are 1,876 atoms: 587 of carbon (C), 909 of hydrogen (H), 197 of oxygen (O), 171 of nitrogen (N) and 12 of sulfur (S). The backbone of the chain of amino acid residues is in the darker shaded area; the side chains characteristic of the various amino acids are in the lighter shaded area. The amino acid residues are numbered from 1 to 124, beginning at the amino end of the chain. Abbreviations for amino acids appearing in the diagram are indicated above.

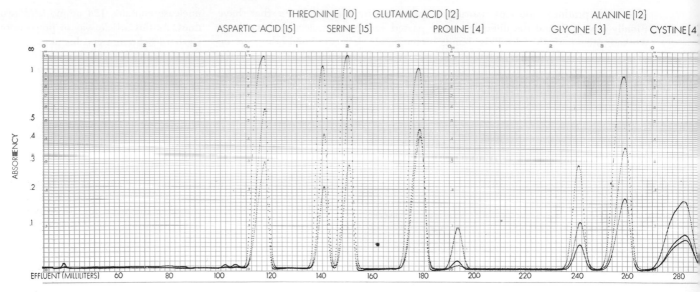

THREONINE [10]   GLUTAMIC ACID [12]   ALANINE [12]
ASPARTIC ACID [15]   SERINE [15]   PROLINE [4]   GLYCINE [3]   CYSTINE [4

AMINO ACID ANALYSIS is recorded as a series of peaks by an automatic analyzer. The trace reproduced here was obtained in the analysis of a hydrolyzed sample of ribonuclease. The names of the amino acids that were found appear above their corresponding peaks. The number of residues of each in the ribonuclease molecule, determined by the intensity of blue color formed with a special reagent, is shown in brackets. (Proline gives proportionately less color than the other amino acids.) Histidine, lysine, arginine

column of an ion-exchange resin. One or two milligrams of an amino acid mixture placed at the top of the column are washed down through the column by solutions of varying acidity. Depending on their relative affinity for the solutions and for the resin, the individual amino acids move down the column at different rates. By proper choice of salt solutions, acidity and temperature, it is possible to adjust the rates of travel so that the separate amino acids emerge from the bottom of the column at predetermined and well-spaced intervals. To detect the colorless amino acids we heat them with ninhydrin, a reagent that yields a blue color. The intensity of the color is proportional to the amount of amino acid.

In the final version of the device, developed in collaboration with D. H.

AMINO ACID ANALYZER is photographed in the authors' laboratory at the Rockefeller Institute. Vertical tubes in center are ionexchange columns. Photometer unit is enclosed in case at top left; recorder is at bottom left. Next to it on the bench is heating bath.

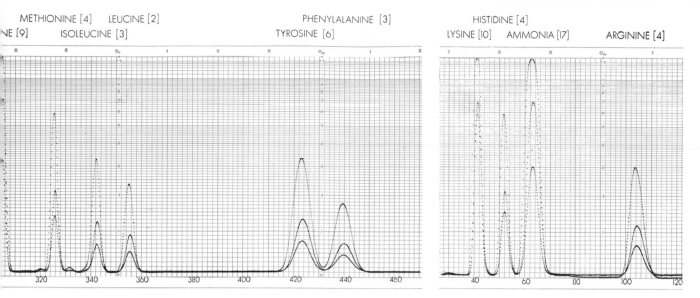

METHIONINE [4]   LEUCINE [2]   PHENYLALANINE [3]   HISTIDINE [4]
NE [9]   ISOLEUCINE [3]   TYROSINE [6]   LYSINE [10]   AMMONIA [17]   ARGININE [4]

and ammonia are determined with a separate ion-exchange column and recorded separately (*right*). The ammonia is released by the hydrolysis of residues of asparagine and glutamine in the intact ribonuclease molecule. In the process these substances are con- verted to aspartic acid and glutamic acid respectively. Horizontal scale measures milliliters of solution to have passed through the columns; vertical scale, color intensity. Different curves show absorbencies at different wave lengths and depths of solution.

Spackman, the amino acid analysis is accomplished automatically. The outflow from the column is continuously mixed with ninhydrin, sent through a heating bath and then analyzed by a photometer attached to a recorder. As it flows out of the bath and into the photometer the solution is alternately colorless (when it contains no amino acid) and blue (when an amino acid is present). A continuous plot of the intensity of the blue color shows a series of peaks, each corresponding to a particular amino acid, the area under the peak indicating the amount of that amino acid in the sample. With the automatic amino acid analyzer one operator, working part time, can carry out a complete quantitative analysis of the amino acids from a hydrolyzed protein in 24 hours. The device played an essential role in the work on ribonuclease.

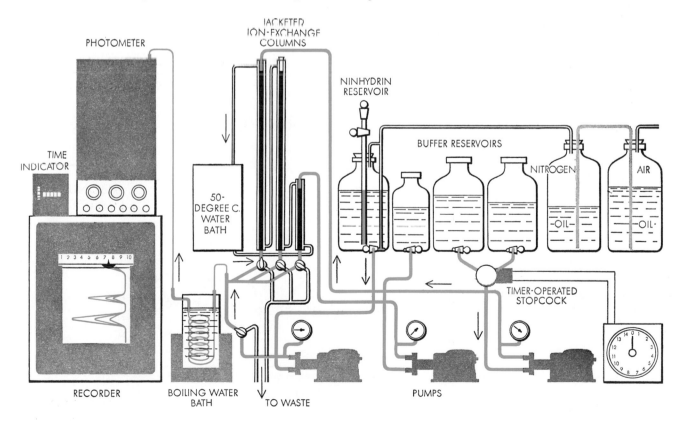

AUTOMATIC ANALYZER for amino acids is diagramed schematically. The two pumps at lower right drive salt solutions through ion-exchange columns. The third pump delivers color reagent (ninhydrin) to a stream emerging from bottom of column. The mixture passes through boiling water bath, where blue color is developed. Intensity of the color is measured by the photometer.

We chose ribonuclease for study for several reasons. The protein, first isolated from beef pancreas in 1920, was available in relatively pure form, having been crystallized by Moses Kunitz of the Rockefeller Institute in 1940. It was known to be a rather small protein, with a molecular weight of about 14,000. (The molecular weight of insulin is 5,733.) Moreover, the manner in which ribonuclease breaks down the ribonucleic acid (RNA) molecule had been worked out in several laboratories. As a result there was a chance of correlating its structure, once that was established, with its biochemical function.

As a first step in the program C. H. W. Hirs, together with the authors, further purified crystalline ribonuclease with the aid of ion-exchange resins. When a hydrolyzed sample was analyzed, it proved to contain a total of 124 amino acid residues of 17 different kinds, plus a quantity of ammonia [see illustration on pages 150 and 151]. The presence of ammonia

indicated that some of the aspartic acid and glutamic acid that showed up in the analysis came not from residues of these amino acids in the protein molecule but from the related amino acids, asparagine and glutamine. When the last two are hydrolyzed, they are converted to aspartic and glutamic acids and ammonia. Thus the intact ribonuclease molecule contains 19 different kinds of amino acid residue.

There were four residues of cystine and therefore four disulfide bonds. At about that time Christian B. Anfinsen of the National Heart Institute demonstrated, by the DNP method, that ribonuclease consists of a single peptide chain. Therefore the disulfide bonds must cause the folding together of sections of the same chain.

After opening up the disulfide bonds by oxidation, Hirs proceeded to break the 124-link chain into smaller pieces. Here Sanger's method of random hydrolysis with acid did not seem promising. It

probably would produce so many small fragments that they would be almost impossible to separate. And even if they were isolated and their structure determined, it seemed unlikely that the sequence of amino acids in the complete chain could be deduced from these small bits and pieces. For a more selective method of dissection we turned to the protein-splitting enzymes. A number of these had been purified and their mode of attack on peptide chains elucidated at the Rockefeller Institute by Max Bergmann and Joseph S. Fruton and their colleagues. The most specific is trypsin, which cleaves only bonds involving the carboxyl groups of the amino acids arginine and lysine. Others, such as chymotrypsin and pepsin, also confine their activity to certain bonds, though not so selectively as does trypsin.

Since there are 10 lysine and four arginine residues in ribonuclease, Hirs first treated the protein with trypsin. The products were separated from one an-

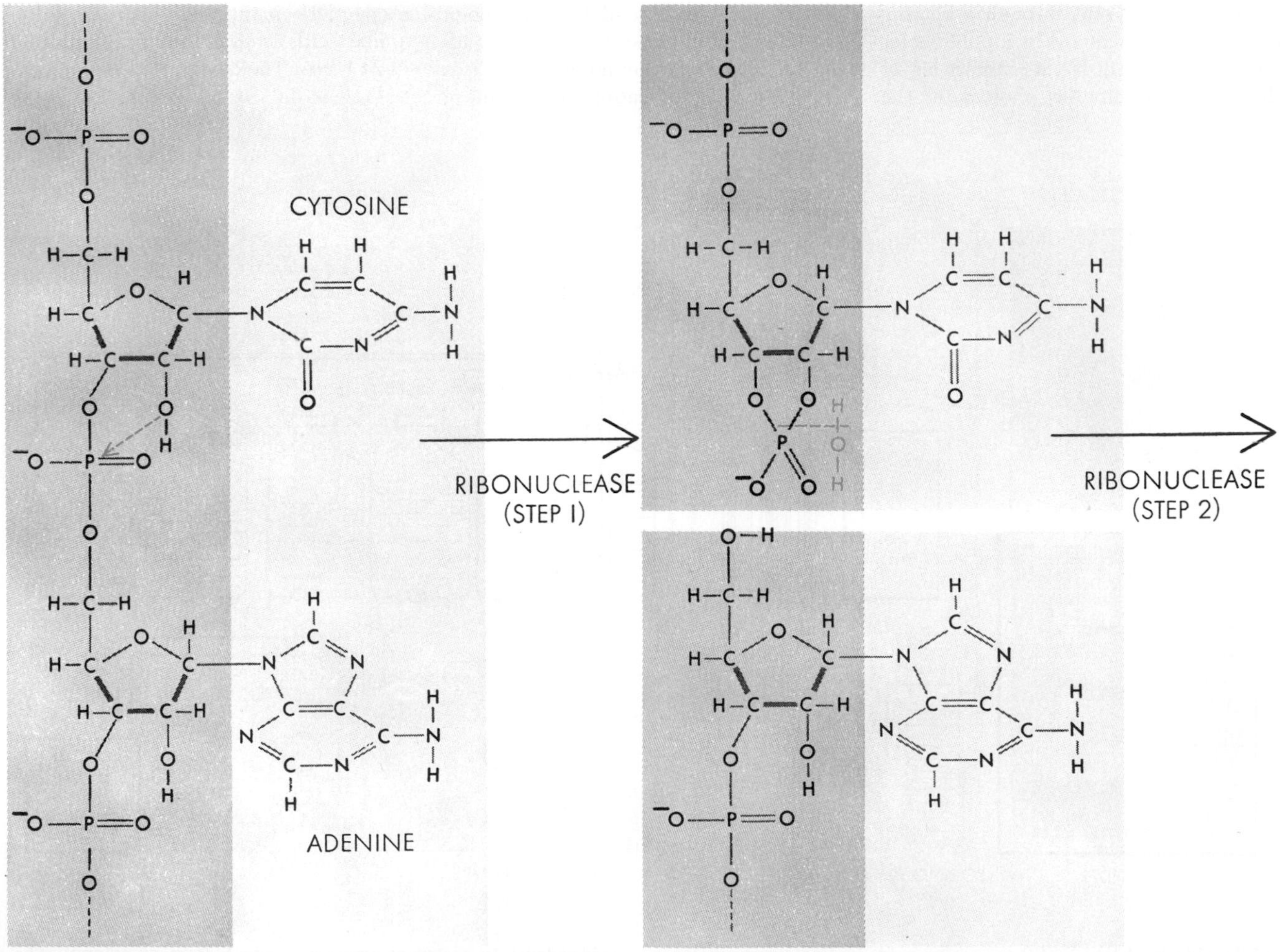

SPLITTING OF RIBONUCLEIC ACID molecule by ribonuclease takes place in two steps. Backbone of the molecule, of which a segment is shown in dark shaded area, is attacked at the phosphorus atom following cytosine (or uracil) but not adenine (or guanine). In the first step the bond between phosphorus and the oxygen atom below it opens, splitting the molecule, a different oxygen uniting

other by the use of columns of ion-exchange resins, and each was then analyzed for amino acids. All told, 13 peptide fragments were isolated, ranging from peptides with only two amino acid residues to some containing more than 20. Among them they accounted for all of the 124 amino acid residues of ribonuclease. To obtain additional fragments Hirs and J. L. Bailey also split the molecule at other points with chymotrypsin and with pepsin.

The amino acid composition of all the fragments was determined, and the residue at the amino end identified in some of them. With this knowledge it was possible, in a sort of crossword-puzzle fashion, to derive a partial structural formula that showed the order in which the various peptides produced by the different enzymatic cleavages must have been arranged in the parent molecule.

The next step was to determine the sequence of amino acid residues within the fragments. One extremely valuable

with the phosphorus (*dashed arrow*). In second step another phosphorus-oxygen bond is cleaved with addition of water (H·O·H).

tool, developed by Pehr Edman in Sweden, was a reaction that can clip off one amino acid unit at a time from the amino end of a peptide sequence. There was also a pair of enzymes that can do the same sort of job in some cases, one at the amino and the other at the carboxyl end. Using these and other means, Hirs analyzed 24 peptides completely and examined parts of many more. Each step of each manipulation was monitored on the amino acid analyzer.

After five years of work the complete sequence of the ribonuclease chain was finally established [*see illustration on pages 148 and 149*]. Spackman then undertook to determine the arrangement of the four disulfide bonds. As in the case of insulin, the problem turned out to be difficult. Two of the bonds were particularly fragile. Only after a long study did Spackman find out how to break ribonuclease into peptides that contained cystine residues with their disulfide bonds intact. Once this had been accomplished the crosslinks could be located and the complete formula written down.

Here we must emphasize that, while the complete formula is certainly correct in most respects, it must still be considered a working hypothesis. Although a great deal of quantitative data supports it, and although Anfinsen's laboratory has derived about a quarter of it independently, we still cannot be completely certain of the results. Degradative experiments, in which molecules are broken down, do not offer final proof for an organic chemical structure; the last word comes when a postulated structure is synthesized and then shown to have all the same properties as the natural product. In spite of the substantial advances of the past few years in the synthesis of complex peptides, it will probably be some time before a molecule of the size of ribonuclease is put together. Until then we must be on the lookout for surprises, because it is entirely possible that ribonuclease contains chemical linkages that are not revealed by the degradative techniques we employed. (For example, unusual linkages have already been found in peptide antibiotics by Lyman C. Craig and his associates at the Rockefeller Institute.)

In any case, our formula has one important deficiency: it is two-dimensional. As we have mentioned, the biological activity of a protein molecule usually depends not only on the sequence of its amino acids but also on how the peptide chain is coiled. Ribonuclease is no exception: it is inactivated by disruption of its three-dimensional structure. Of course, the order of amino acids in any

peptide chain must influence its spatial arrangement. Certain sequences are known to preclude certain kinds of folding. To what extent a given sequence may require a given kind of folding is not yet clear.

Although much structural information can be obtained through chemical techniques, they can go only part of the way with a molecule as big as that of a protein. For many years X-ray crystallography has been applied to the problem, recently with striking success. A description of this work must await a separate article. Here we shall merely mention that in a series of brilliant investigations British groups headed by J. C. Kendrew and Max F. Perutz have worked out, respectively, the complete spatial arrangement of the peptide chains in the oxygen-carrying proteins myoglobin and hemoglobin. The sequences of amino acids in these molecules have not yet been determined, although they doubtless will be soon. Indeed, it may prove possible to discover the order by X-ray methods alone, which would provide a valuable check on the methods of organic chemistry. When the sequences are found, it will be possible to place each amino acid residue in its proper position on the models of the coiled chains derived from X-ray studies. At that exciting moment the first true picture of a protein will have been drawn. Unfortunately the methods used by Kendrew and Perutz have not yet been successful with insulin or ribonuclease. Sooner or later, however, the difficulties will be resolved and we shall have the complete portraits of these molecules as well.

Active Sites of Enzymes

In the meantime the techniques of organic chemistry are throwing considerable light on one aspect of the relation between the structure of proteins and their function. That is the question of the "active site" of enzymes. Being proteins, enzymes are all large molecules. Often their substrates—the substances they act on—are very much smaller. For example, ribonuclease splits not only ribonucleic acid but also the comparatively small molecule cyclic cytidylic acid [*see illustration on page 155*]. In such a case only a small portion of the enzyme molecule can be in contact with the substrate when the two are joined together, as they are during the reaction catalyzed by the enzyme. This portion is known as the active site. The concept of an active site does not imply that the rest of the enzyme has no function. Sometimes part of the molecule is dispensable; sometimes it is not. The whole molecule may perhaps be something like a precision lathe, where a ton or more of machinery is required to bring a few ounces of metal in the cutting tool to bear accurately on the work. Similarly, the bulk of the enzyme molecule may be required to bring the active site into proper contact with the substrate. The problem facing the chemist is to discover which amino acid residues make up the active site and just what biological function the rest of the molecule performs.

Now that the structure of an enzyme —ribonuclease—is known, one can hope to find its active site and see how that section is related to the rest of the molecule. Workers in a number of laboratories are engaged in the effort and are making good progress. Before considering their results, it is appropriate to look at some of the earlier work on the active sites of enzymes.

A particularly illuminating series of investigations began in a most unlikely fashion with the study of nerve gases. Developed by the Germans during World War II but happily never used, these substances are phosphate esters of various organic alcohols. (An ester is the compound formed by the reaction of an acid and an alcohol.) They are extremely toxic, rapidly causing death by respiratory paralysis when they are inhaled or even absorbed through the skin. Studies of their physiological action showed that they inactivate choline esterase, an enzyme that breaks down acetylcholine in the body. This last substance plays an essential part in regulating the transmission of nerve impulses. When allowed to accumulate in excessive amounts, it deranges the nervous system, destroying its control over the breathing apparatus.

As soon as the relationship between nerve gas intoxication and choline esterase was established, several laboratories set out to discover the nature of the reaction between the two substances. It was quickly discovered that one molecule of nerve gas combined with one molecule of choline esterase to give an inactive product. The next question was: What part of the large protein molecule is involved? This could not be answered directly because the choline esterase molecule is too large to be studied in detail by the methods available. Besides, it has not been isolated in sufficient amounts. Irwin B. Wilson and David Nachmansohn of the Columbia University College of Physicians and Surgeons approached the problem from the other side; they determined how the reaction between the enzyme and the nerve gas

or other inhibitors was influenced by the exact chemical structure of the much smaller organic phosphate molecules. They could then deduce many properties of the enzyme surface. As a result of these investigations a compound was devised that could displace the nerve gas from its combination with choline esterase and so reactivate the enzyme.

Then came an important observation. A. K. Balls and E. F. Jansen of Purdue University found that an organic phosphate called DFP (for diisopropylphosphorofluoridate) also inhibits the enzyme chymotrypsin. This protein was available in large quantities and it had already been extensively studied. On examination of the reaction with DFP, Balls and Jansen found that again one molecule of phosphate combined with one enzyme molecule. Subsequently the point of attachment was identified as a specific serine residue in the enzyme molecule.

There must be something special about this particular serine residue: DFP does not react with the free amino acid serine. No more than one molecule of DFP combines with each molecule of chymotrypsin in spite of the fact that the enzyme has 25 serine residues. And any treatment that destroys the enzymatic activity of chymotrypsin, even temporarily, also destroys its ability to react with DFP. Concluding from this evidence that the reactive serine is part of the active site of the enzyme, several groups of chemists have investigated the sequence of amino acids around it. They labeled the serine with DFP containing radioactive phosphorus and then broke down the inactivated chymotrypsin with other enzymes. From the mixture of fragments a peptide containing the labeled serine was isolated. The serine residue proved to be flanked by an aspartic acid residue on one side and a glycine residue on the other.

Meanwhile several other enzymes had been found to be sensitive to DFP. In all of them the inactivation proved to involve a reaction with a single serine residue. Analysis disclosed that in two of the enzymes the serine had the same neighbors as in chymotrypsin. In two others the sequence proved to be glutamic acid followed by serine and alanine. The two arrangements are in fact much alike because glutamic acid and aspartic acid are closely related chemically, as are glycine and alanine.

Finding the same type of sequence in so many enzymes strikingly confirms a major tenet of biochemistry: a similarity in function must reflect a similarity in structure. But this cannot be the whole

story. Although the enzymes are alike in some respects, they are not identical. They do not all catalyze exactly the same reactions and so their active sites must differ in some ways. Of course the sequence aspartic acid, serine and glycine is not by itself biochemically active, nor are the larger peptide fragments that contain it. The serine in them does not react with DFP. Additional residues are required, but which ones, and how they are oriented in space with respect to the active serine, is still largely unknown.

There is persuasive evidence that the active site of these enzymes also includes a histidine residue. Yet in two of the enzyme molecules quite a number of residues on each side of the crucial trio have been identified, and histidine is not among them. If histidine does form part of the active site and is near the active serine, it must be brought there by a folding of the chain. Although still hypothetical, the idea that the amino acid units in an active site are brought into juxtaposition from different sections of a peptide chain by three-dimensional coiling is attractive. Among other things, this hypothesis explains why disrupting only the three-dimensional structure of an enzyme leads to loss of activity. The picture will become much clearer when the complete structures of the DFP-sensitive enzymes are worked out.

Ribonuclease is not sensitive to DFP, but it too can be inactivated by a specific chemical reaction involving a particular amino acid residue. The inactivating agent is either iodoacetic acid or bromoacetic acid, and it combines with a histidine residue. Two investigators in England, E. A. Barnard and W. D. Stein, have obtained evidence that the histidine residue concerned is the one at position 119, six residues from the carboxyl end of the ribonuclease chain. Neither reagent reacts with any of the other three histidines in the molecule. Unfolding the chain or otherwise destroying the activity of ribonuclease prevents the reaction with this residue. Thus the histidine residue at position 119 seems almost surely a part of the active site of ribonuclease. Other studies implicate the aspartic acid at position 121 (removing it together with the last three residues inactivates the molecule, whereas splitting off the last three alone does not) and perhaps also the lysine at position 41.

Some remarkable experiments by F. M. Richards at Yale University have provided much information about the relationship between the activity of the enzyme and its over-all structure. Using a bacterial enzyme called subtilisin, Richards succeeded in splitting the ribonuclease molecule at a single point—the bond between the alanine at position 20 and the serine at position 21. Although the peptide link had been broken, the two fragments did not separate from each other nor did the combination lose enzymatic activity. Treatment with mild acid, however, separated the altered ribonuclease into two parts, one a peptide of 20 residues, the other a large fragment of 104 residues. Neither fragment by itself exhibited activity. But when dissolved together in a neutral solution they recombined instantly, and enzymatic activity was regained. The peptide bond did not re-form under these gentle conditions. Instead the two fragments were held together by so-called

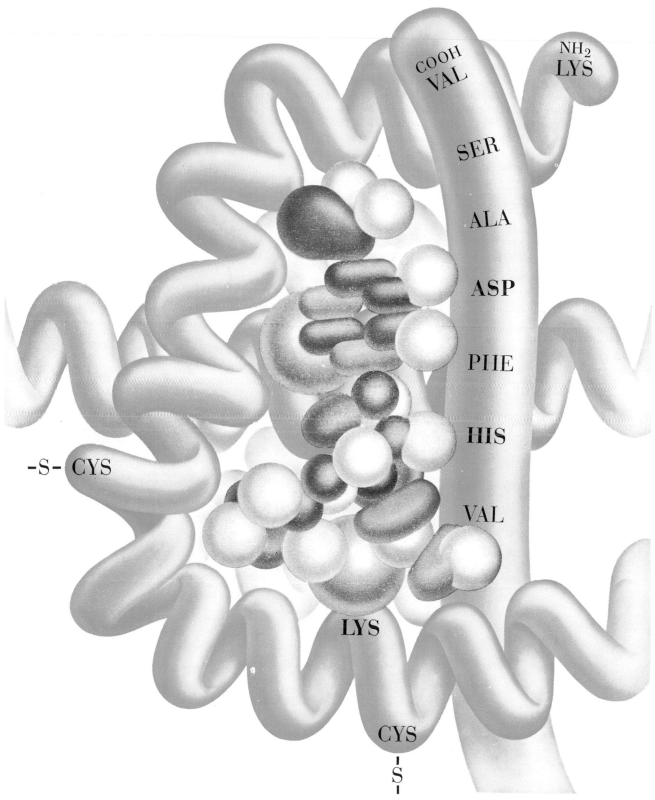

ACTIVE SITE of ribonuclease is represented in this strictly hypothetical conception. A small molecule on which the enzyme acts is shown in gray, nested in a form-fitting cavity on the enzyme surface. Peptide chain, some of which is believed to be helical and some not, is shown in color. Some of the amino acid residues thought to form part of the active site are shown in bold type.

secondary forces, probably analogous to those that unite enzymes and their substrates. The simplest explanation, though not the only one, for these results is that activity depends on the juxtaposition of one or more amino acid residues from among the first 20 with one or more in the rest of the molecule.

This brief account of the research on active sites by no means exhausts the work being done on ribonuclease, nor does it exhaust the list of enzymes under investigation. To mention another, the protein-digesting enzyme papain (found in the papaya) has been studied for years at the University of Utah School of Medicine by Emil L. Smith and his colleagues. They have gone far toward determining its structure and have, in addition, made the striking observation that more than 100 amino acid residues of the 185 in the native enzyme can be whittled away without abolishing its activity. The structures of trypsin and chymotrypsin are also being energetically studied, particularly by Hans Neurath and his associates at the University of Washington.

As structural chemistry advances, answers to a number of other fascinating biological problems begin to appear. One can study the differences among analogous protein molecules produced by different species. Sanger is the pioneer in this field, with his studies of the structure of the insulin produced by the cow, the pig and the sheep. Also of great interest are the differences in protein molecules produced by members of the same species under the influence of genetic mutation. In this area a good deal has already been learned about abnormal hemoglobins, and much more will be known when the complete structure of the hemoglobin molecule has been worked out.

Finally, the recent elucidation of the complete amino acid sequence in the protein of tobacco mosaic virus opens the way to a deeper understanding of the natural synthesis of proteins. The virus consists of protein and ribonucleic acid. According to present theory, the protein is synthesized by infected tobacco-plant cells under the direction of the viral nucleic acid. Now it may be possible to follow this process in detail.

The next few years of research on proteins should be exciting ones, both for chemists and biologists.

19

THE AUTOMATIC SYNTHESIS OF PROTEINS

R. B. MERRIFIELD

March 1968

By anchoring an amino acid to a plastic bead one can add other amino acids one by one in automatically controlled steps. This method has already been used to make the small protein insulin

The synthesis of proteins is one of the primary functions of the living cell, and the intricate series of operations by which the cell accomplishes the task has recently become known in considerable detail. Long before the biosynthesis of proteins was at all understood, at the turn of the century, the great chemist Emil Fischer believed proteins could be synthesized in the laboratory. It took a long time to accumulate the knowledge and techniques required to put together one of these enormously complex substances. Beginning in 1963, as the result of concerted work by many individuals, three groups of chemists—one in the U.S., one in Germany and one in China—succeeded in making in the laboratory a comparatively simple protein: the pancreatic hormone insulin.

The conventional process for the chemical synthesis of proteins, or of the smaller chains of amino acids called peptides, is a slow and painstaking affair. In our laboratory at Rockefeller University we set out some years ago to look for a simpler and more efficient process—one that might lend itself to automatic operation. The result was a new technique, "solid phase" peptide synthesis, in which the peptide chains are assembled not in solution but on small, solid beads of polystyrene. In 1965 the solid phase method was successfully applied to the synthesis of insulin.

The availability through controlled chemical synthesis of insulin with all the properties of the natural hormone is of great value because the road is now open for making related molecules that differ from the parent compound in precisely known ways. Such analogues should help to clarify the manner in which the natural hormone functions and may in time lead to insulin derivatives that exhibit greater or more prolonged activity for the treatment of diabetes.

Even more exciting than the synthesis of insulin is the possibility of synthesizing an enzyme—one of the proteins that catalyze metabolic processes. That goal is now in sight and will surely be attained. It will add significantly to our understanding of this important class of proteins and of the mechanisms by which the living cell carries out its essential functions.

Peptide Synthesis

To understand how the chemical synthesis of a protein can be approached it is best to look first at how peptides are synthesized. Peptides are simpler models of proteins; they consist of the same structural elements (but fewer of them) linked together in the same way. The linkage between the amino acid subunits is known as the peptide bond; a series of such bonds constitutes the primary "backbone" structure of peptides and proteins. The formation of these bonds is the principal problem in peptide synthesis and is also the first step in protein synthesis. In the case of proteins, however, there are also secondary and tertiary bonds that control the cross-linking and folding of the molecule and are responsible for its three-dimensional shape. Because peptides are shorter and lack these complicating additional bonds, they are simpler compounds to study. They have been used to develop the chemistry needed to begin the synthetic work on proteins.

Amino acids are compounds containing several reactive groups: one amino group (NH_2), one carboxyl group ($COOH$) and in many instances another reactive group located on a side chain [*see top illustration on page 168*]. In general all but one of these groups must be protected against undesired combinations during a chemical reaction if a specific, pure product of known structure is to be obtained. In order to prepare even the simplest chain of two amino acid units (a dipeptide) the basic amino group of one unit and the acidic carboxyl group of the other must be blocked. It is then possible to activate the carboxyl group of the first amino acid—that is, to increase its energy level—so that it will couple with the free amino group of the second one to form the peptide bond.

Now let us consider extending the chain to form a longer peptide. There are two general approaches. In the "fragment" method short peptide chains are built up and are combined to form the larger final molecule; in the "stepwise" method single amino acid units are added one at a time until the final molecule is completed. (In both methods the successive additions can in principle be made at either end of the molecule, although in practice there are certain limitations.) The fragment technique is the older and until recently was the more common approach. Its advantages are that more of the intermediate peptides are of small size and that there are greater differences between the properties of the reactants and of the products than there are in the stepwise procedure. On the other hand, the coupling yields are generally lower in the fragment method and there is a greater chance of unwanted side reactions.

Before any of the chain-lengthening processes can be carried out it is necessary to remove one of the blocking groups from the initial dipeptide. To "deprotect" selectively in the presence of the other protecting group (or of several such groups) and without damage to the peptide chain requires careful planning. The choice of the protecting groups and the activating, or coupling, agent for each amino acid has been a major concern of peptide chemists.

At each step of the synthesis it is usually necessary to isolate, purify and characterize the products of the reaction, and it is at this point that the greatest difficulties are often encountered. Crystallization is the classical procedure for purifying peptides, as it is for most other organic compounds. It depends on the formation of an orderly array of molecules that grows in size until it precipitates from solution. Ideally only molecules of one kind will be in the crystalline precipitate and all undesired substances will remain in solution and be washed away. Sometimes one can obtain from a reaction mixture quite pure peptides that do crystallize readily. Particularly when one is working with long peptide chains, however, the yield may often be amorphous material or crude crystalline precipitates contaminated with various by-products. One must then resort to special purification procedures that may require many days each. Consider for a moment the time and effort involved in the synthesis of a 100-unit peptide if such coupling and purification steps must be performed 99 times!

The Solid Phase Approach

To synthesize molecules of the size and complexity of proteins, it seemed clear, methods of the greatest efficiency and simplicity would have to be devel-

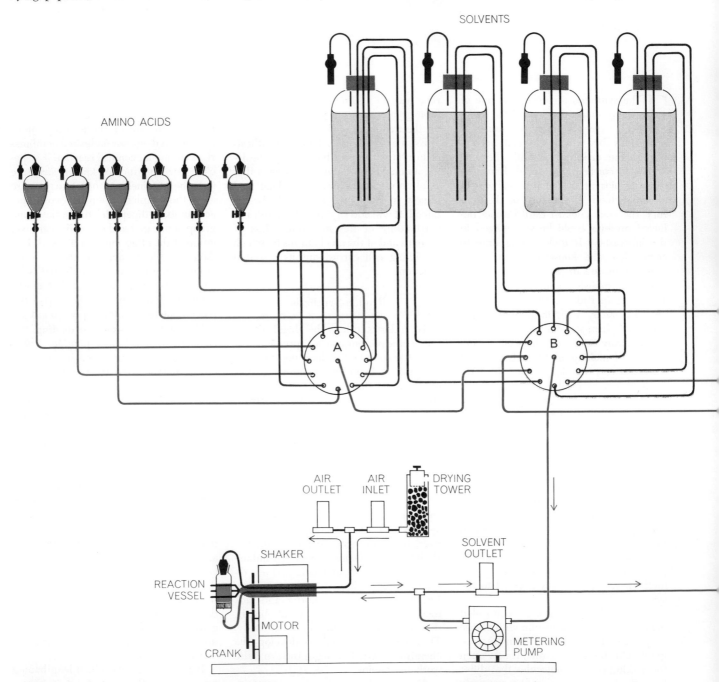

SCHEMATIC DIAGRAM of the automatic apparatus shows the "plumbing" circuits. The proper amino acid, other reagent or solvent is pumped from its reservoir through a selector valve (*A or B*) into the reaction vessel while air is displaced at the top of the vessel. A mechanical shaker rocks the vessel to mix the reactants. Solvents and by-products are removed by vacuum through the filter in the bot-

oped. In 1959, with these requirements in mind, a new approach to peptide synthesis was conceived. The new idea was to synthesize the long chains one unit at a time, but without stopping to isolate the individual intermediate peptides; to make this feasible the plan was to anchor the chain to an insoluble solid support.

The first amino acid would be chemically bonded to a solid particle and the rest of the amino acid units would be added to it stepwise in the proper order. Since the solid support would be com-

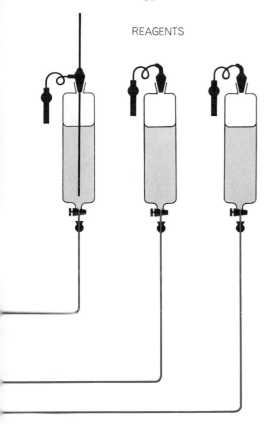

REAGENTS

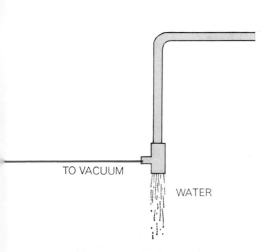

TO VACUUM

WATER

tom of the vessel while dry air is admitted at the top. One cycle of the synthesis requires the selection of 12 reagents in sequence.

pletely insoluble in the various solvents, all the intermediate peptide products would also be held in an insoluble state; they could therefore be purified simply by dissolving the unwanted by-products and reagents and washing them away. This would involve only an elementary filtration step, but it would accomplish essentially the same kind of purification as the classical recrystallization: the growing peptide would be an insoluble precipitate, whereas the undesired reagents would be in solution. Filtration is much easier and faster than crystallization. Most important, it can be done in the same way at each step, whereas crystallization is necessarily an individualized procedure that is different for each new intermediate peptide.

The general scheme for solid phase peptide synthesis is straightforward. A suitable solid support is selected and a reactive site is produced on it. The first amino acid—actually the terminal amino acid of the proposed peptide chain—is attached by its carboxyl group to the reactive site. Now the second amino acid, with all but one of its reactive groups protected, is activated and coupled to the first amino acid, leaving a protected dipeptide firmly bound to the support. The solid can be filtered and washed thoroughly to remove all the excess reagents and any by-products without the slightest danger of losing the desired peptide.

Next the protecting group on the amino end is removed and the whole process is repeated exactly as before but with a new amino acid. After the required sequence of amino acids has been assembled in this manner the peptide chain is finally removed from the support by selectively breaking the bond that has been holding the two together throughout the synthesis. Now for the first time the peptide chain is free and can be dissolved and separated from the solid support. Once it is in solution conventional purification procedures can be carried out.

Before this scheme could be developed into a workable procedure a number of rather severe requirements had to be met, having to do with the nature of the solid support, the type of bond linking the peptide to the support and the choice of protecting groups and coupling reagents. The solid support had to be completely insoluble in all the solvents that might be used in the synthetic reactions or in the washing steps; it had to be physically stable and in a convenient form to permit filtration and other manipulations; it had to have a reactive site at which the peptide chains could be at-

tached but should otherwise be chemically inert and stable; finally, in order to allow the synthesis of a sufficient quantity of peptide, it should either have a very large surface-to-volume ratio or be readily permeable to the soluble reagents.

After considerable exploration a substance that met these requirements was found. It is a polystyrene resin, a linear polymer of styrene in which the styrene chains are loosely linked together with divinylbenzene, and it is in the form of beads about 50 microns (.002 inch) in diameter. The amount of cross-linking agent was selected to give a resin of high molecular weight that would be completely insoluble but at the same time free to swell in organic solvents. This makes the beads permeable to reagents dissolved in the solvent, so that reactions can occur not only on the surface but also within the interstices of the gel-like matrix. Although we are talking about beads that are barely visible as separate particles to the unaided eye, they are actually enormous compared with the dimensions of amino acids or even of proteins: each bead can support some 10^{12} (one trillion) peptide chains!

The Anchor Bond

Polystyrene itself has no convenient reactive site for anchoring the peptide chain, but it can be readily modified in many ways to make such a site. The choice of the modification was dictated by the kind of bond needed to hold the peptide to the resin. The bond must be easy to form, it must be completely stable during the dozens of reactions involved in assembling the peptide chain and it must be readily cleaved under relatively mild conditions at the end of the synthesis. The anchor we chose was prepared by attaching chloromethyl groups ($ClCH_2$) to the six-carbon rings of the polystyrene and then reacting them with the first amino acid to form what is known as a benzyl ester [see illustration on page 160]. The benzyl group also served to protect many of the reactive side chains.

The choice of the benzyl group in turn influenced the choice of the protecting group for the amino ends of the successive amino acids. That is, it had to be possible to remove the amine protection selectively at every cycle of the synthesis without detaching the peptide from the resin or deprotecting the side chains. The protecting agent we chose was tertiary butyloxycarbonyl ("Boc"), a group that had been developed a few years earlier for use in conventional syntheses. It was sensitive to certain anhydrous acids that would not affect the ester and it could therefore be removed without disturbing the anchor.

The principal remaining problem was the peptide-forming reaction itself, which had to be rapid and must not give rise to side reactions. Most important of all, it had to go to completion; it was absolutely crucial to the success of the process that this step give essentially 100 percent yields. Suppose, for example, the second amino acid were to couple to the extent of only 90 percent, which is considered a most acceptable yield in organic chemistry. What will happen when the third amino acid is coupled? It will react with the amino end of the dipeptide to form a tripeptide, but it will also couple with the unreacted 10 percent of single amino acids to produce dipeptides that lack the second amino acid. In an ordinary synthesis the intermediate products are isolated at each stage; in the solid phase method they are simply washed free of soluble impurities, and the abnormal chain will be carried through the entire synthesis, giving at the end a mixture consisting of 90 percent of the correct peptide and 10 percent of a peptide with a missing link. In this simple case the two products could probably be separated and purified, but if incomplete reactions were to occur several times during the synthesis of a long peptide, a complex mixture would result that might not be so easy to separate [*see upper illustration on page 161*].

We therefore put great stress on finding reagents and conditions that would lead to complete coupling reactions, and we have fortunately been able to achieve that goal in practice. Many ways to activate amino acids have been developed for use in peptide synthesis. For the solid phase method the most successful procedures have been to activate with the reagent dicyclohexylcarbodiimide or to use the nitrophenyl esters of the amino acids. These activated forms are highly effective, but only if they can reach the proper sites. To ensure their rapid, unimpaired penetration into the resin, where most of the peptide chains are located, a solvent of high swelling capacity, such as methylene chloride, is necessary.

This solvent causes the beads to swell to approximately twice their original diameter, which means that the polymer chains are then distributed in eight times the initial volume. The polymer molecules occupy only about 12 percent of the total space of each swollen bead, and the remaining 88 percent is filled with solvent containing the activated amino acid molecules. The diffusion of these small molecules is relatively little inhibited by the polymer, so that the reac-

tions take place almost as fast as they would in solution.

The efficiency of coupling also depends on the concentrations of the reactants: the peptide chain and the amino acid being added to it. If one begins with equal amounts of the reactants, their concentrations will decrease to very low levels as the reaction nears completion

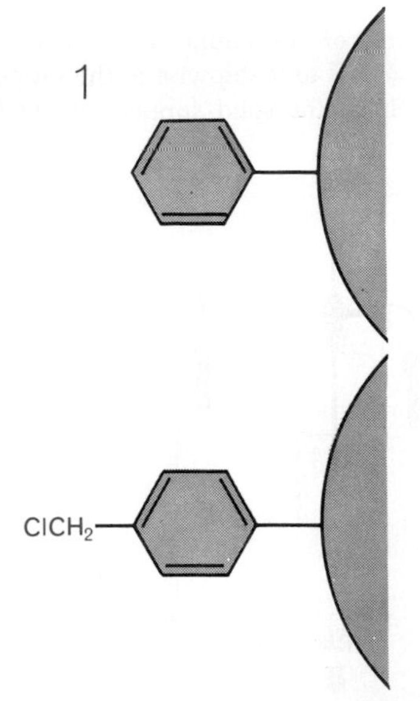

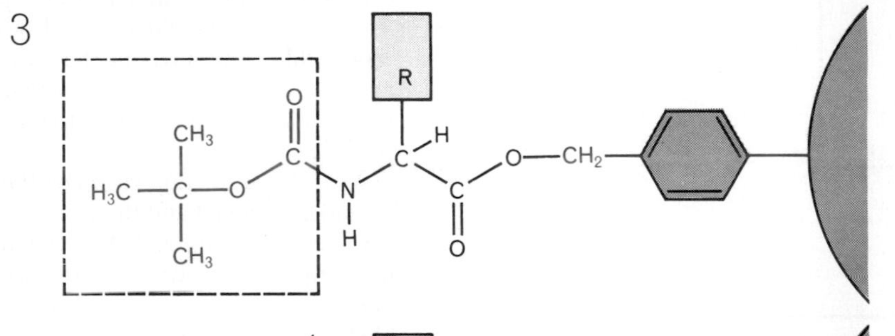

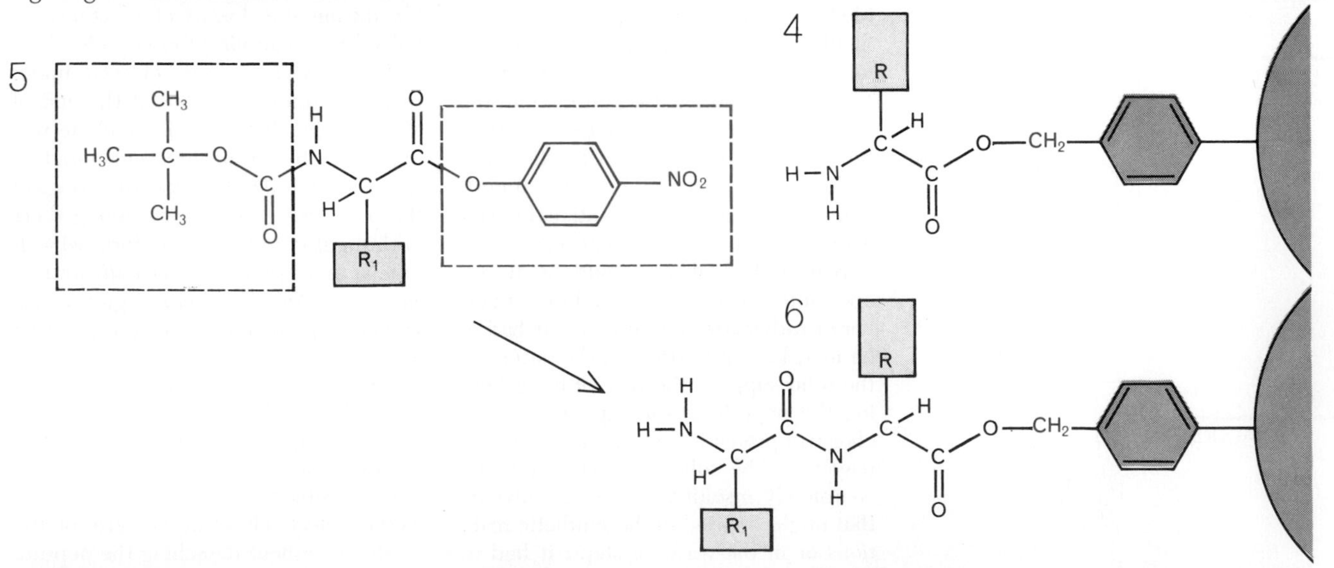

SOLID PHASE METHOD is carried out stepwise from the carboxyl end toward the amino end of the peptide. An aromatic ring of the polystyrene (*1*) is activated by attaching a chloromethyl group (*2*). The first amino acid (*black*), protected by a butyloxycarbonyl (Boc) group (*black box*), is coupled to the site (*3*) by a benzyl ester bond and is then deprotected (*4*). Subsequent amino acid units are supplied in one of two activated forms; a second unit is shown in one of these forms, the nitrophenyl ester of the amino acid (*5*). The ester (*colored box*) is eliminated as the second unit couples to the first. Then the second unit is deprotected, leaving a dipeptide (*6*).

and the reaction rate, which is proportional to the product of the concentrations, will gradually approach zero. The practical consequence is that some of the amino acids never do link up, and the reaction never quite goes to completion. This is the usual situation in conventional syntheses.

If, on the other hand, there is a rather large excess of one of the reactants, a significant rate can be maintained until essentially all the limiting reactant (the peptide chain) has entered into bond formation. To illustrate, suppose we begin with 100 parts of each reactant and

call the initial reaction rate 10,000 (100 × 100). After the reaction is 99 percent completed, one part of each reactant would remain, and the relative rate would be only 1, or 1/10,000 as fast as at the beginning. Even with a very fast initial rate it would take a long time to complete the reaction.

Suppose instead we begin with 100 parts of the peptide chain as before but with 400 parts of the activated amino acid. Now after 99 percent of the chain is used up the relative rate would still be 301 (1 × 301). In the presence of the fourfold excess of the amino acid, one

can calculate, the reaction can go to 99.99 percent of completion in the same time it would take to go to only 75 percent if equal amounts of the reactants had been used. One of the important advantages of solid phase peptide synthesis is that such an excess of the amino acid derivative can be used without complicating the subsequent purification procedure, since at the end of each reaction the excess is simply removed by filtration and washing. Thus we can force the reaction to completion and leave essentially no free, unreacted peptide chains.

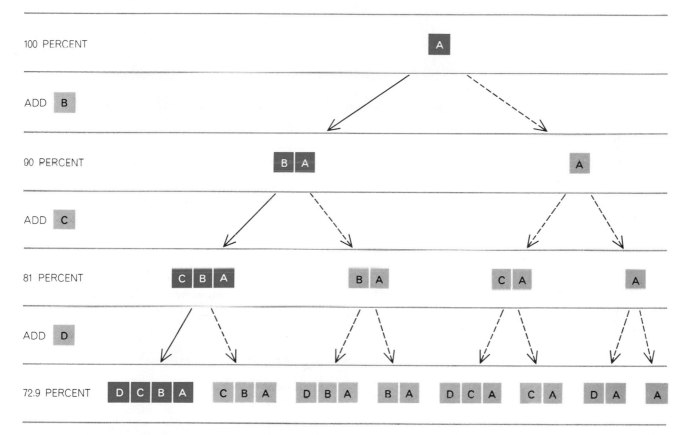

HIGH YIELD from each coupling reaction is important in the solid phase method because products of incomplete reactions persist through the filtering steps. If each amino acid were to couple with even a relatively high efficiency, say 90 percent, the yield of pure peptide (*dark color*) would be down to 72.9 percent by the time the fourth unit was added. More important, the seven different peptide fragments that lack one amino acid unit or more would have to be separated chemically. In practice, yields are close to 100 percent.

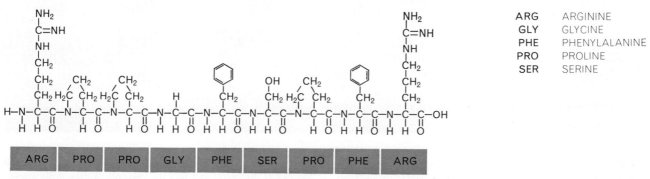

ARG	ARGININE
GLY	GLYCINE
PHE	PHENYLALANINE
PRO	PROLINE
SER	SERINE

BRADYKININ, a small peptide hormone, was one of the first peptides synthesized in the author's laboratory by the solid phase method. Its nine amino acid subunits (five different amino acids) were assembled stepwise from the carboxyl (*right*) end of the chain.

APPARATUS for the automatic synthesis of peptide chains is seen in the author's laboratory. It includes the small glass reaction vessel (*lower right*) with its attendant "plumbing" and a programming unit (*left*). The rectangular pins on the rotating drum operate switches that control the pump, valves, timers and shaker that fill and empty the vessel and mix the reagents. Amino acids are supplied from the six glass vessels (*middle right*). Solvents and other reagents are supplied from the larger containers above and at right.

The deprotection and coupling reactions just described can be repeated alternately until the desired peptide chain has been assembled on the resin beads. The final step is the cleavage of the benzyl ester bond that has been holding the chain to the resin throughout the synthesis. As indicated earlier, this bond was chosen because it is stable during the synthesis but can be selectively split at the right time without damaging the peptide chain. The resin is suspended in anhydrous trifluoroacetic acid, and dry hydrogen bromide gas is bubbled through to effect the splitting. (More recently anhydrous liquid hydrogen fluoride has been successfully employed for this step.) These reagents also remove the protecting groups on side chains. The peptide, now in a free and soluble state, is separated from its resin support by filtration and is purified. It is then ready for analysis and, where possible, for biological assay.

Those who are familiar with the mechanism of protein synthesis in the living cell will see some superficial resemblance between it and the system just described. Both depend on a particulate support (in the cell the support is the ribosome), both involve activation of the amino acid (in the cell the amino acid is activated by the energy-rich molecule adenosine triphosphate, or ATP) and both are stepwise processes. It even has been learned recently that in bacteria the synthesis starts with one end of the chain protected and that an enzyme later carries out a deprotection step just as we do in the laboratory [see the article "How Proteins Start," by Brian F. C. Clark and Kjeld A. Marcker, beginning on page 305].

The analogy should not be pushed too far, however. The natural synthesis is much more elegant and efficient than the laboratory one, and no one presumes to duplicate or even approach the complexities of the cell's scheme at this point. As a matter of fact, the chemical synthesis was not even patterned after nature; it was only in retrospect that the similarities became evident. Nevertheless, it could well be that in the future organic chemists may benefit from a more complete understanding of the way in which living systems perform their tasks.

An Automatic System

It was clear to us from the outset that an automatic mechanized process was needed for making large peptides, and that the solid phase approach would be well suited to such a process. This is so

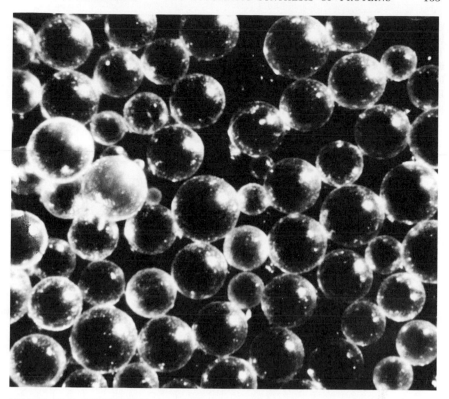

BEADS OF POLYSTYRENE, on which amino acid subunits are assembled in the "solid phase" method to make a peptide chain, are enlarged 300 diameters. They average 50 microns (.002 inch) in diameter but become about twice as large when swollen in a solvent to make them more reactive. About a trillion peptide chains can be "grown" on a single bead.

because the intermediate products in the synthesis need not be isolated but are purified by simple filtration and washing reactions that can be carried out in a single vessel; the manipulations required to transfer products from one container to another have been eliminated. Once the resin beads with an amino acid attached are placed in the vessel it is only necessary to introduce the appropriate liquid solvent or reagent, allow it to react, remove the excess reagent and by-products by filtration and then repeat the process with the next reagent.

It is easy to visualize how all these steps can be accomplished automatically by a rather simple device. We constructed a machine that consists essentially of two parts: a reaction vessel with the plumbing necessary to introduce and remove the solvents in the right order at the right times and a programmer that controls these operations. The solvents and the reagents, contained in a series of reservoirs, are selected one at a time by a specially designed rotary valve. The solvents are introduced into the bottom of the reaction vessel by a metering pump while air is displaced at the top. Valves to the vessel are then closed and a mechanical shaker mixes the reactants for a predetermined time. Next a vacuum withdraws solvent through a porous glass filter disk in the bottom of the vessel while dry air enters at the top; the beads, with the peptide attached, remain in the vessel. One cycle of the synthesis (the lengthening of the peptide chain by one amino acid) requires 12 different reagents, one of which is a protected amino acid. The next cycle calls for the same series of reagents except for a different amino acid, which is selected by a second rotary valve.

All the steps just described are controlled by a "stepping-drum" programmer. It is like an old-fashioned music box. Once the pins have been positioned on the drum to play the proper tune the machine takes over and directs the chemical synthesis. The pins activate switches that turn the pump and shaker on and off and open and close the valves at the proper times and in the proper sequence. One cycle of the synthesis requires 100 steps of the drum and takes about four hours, so that it is now possible to carry out automatically all the operations required for the assembly of a peptide chain at the rate of six amino acids a day.

After the details of the synthetic scheme had been worked out by the synthesis of small peptides the procedure was given a more demanding test: the preparation of the hormone bradykinin, a nine-amino-acid peptide with several physiological activities that served as sensitive criteria to demonstrate the identity and purity of the final product. The synthetic bradykinin was identical in all respects, both chemically and biologically, with the natural hormone. During the past four years the solid phase technique has been applied to the preparation of bradykinin analogues (nearly 100 of which have been made by J. M. Stewart at Rockefeller University) and other small peptide hormones such as angiotensin and oxytocin. It has also been used in our laboratory and others for the synthesis of the antibiotics gramicidin-S and tyrocidin, and for synthetic studies of the immunological determinants of hemoglobin and tobacco mosaic virus. It was clear that peptides containing 10 or 20 amino acids could be made by the solid phase method as well as by classical procedures. The important

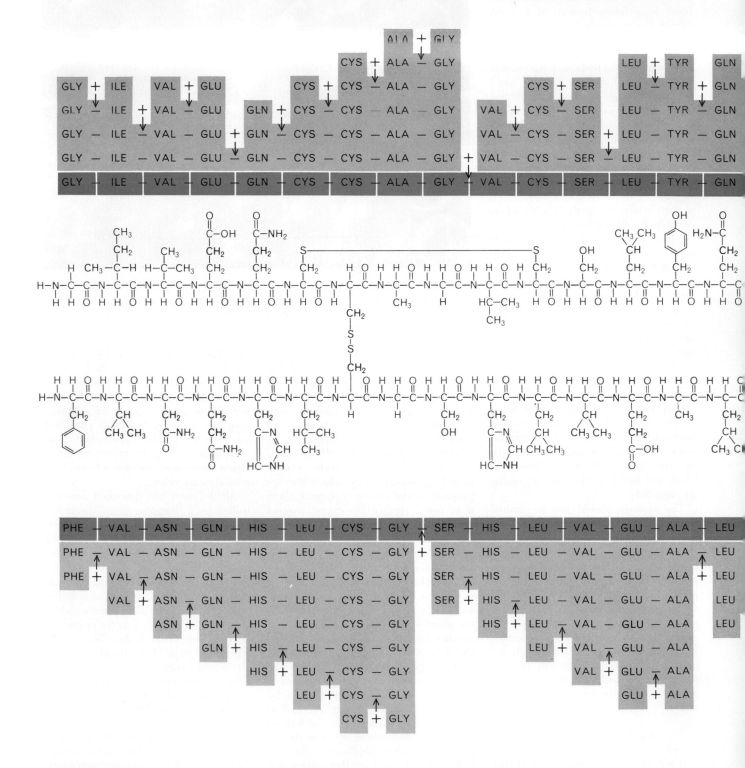

FIRST SYNTHESIS of insulin chains was accomplished by the fragment method, as illustrated here. The primary structure of the two chains is diagrammed at the center, together with the names of the amino acid units (*see key at top right*). Note the disulfide (S—S) bonds between cysteine units, two of which link the A (*top*) and B (*bottom*) chains. The general pattern of the synthesis of the

question then became whether or not molecules as large and complex as proteins could be synthesized by this method.

Synthesis of Insulin

The smallest molecule that qualifies as a true protein is insulin. It naturally became the object of intensive synthetic work by several groups of chemists when, in the late 1950's, it seemed likely that the synthesis of a protein was a feasible goal. Insulin was chosen not only for its size but also for several other important reasons. The availability of synthetic hormone would help to answer many questions about its mechanism of action. Most important, the composition and complete primary structure (the sequence of amino acid units) had become known a few years before through the work of the group led by Frederick Sanger at the University of Cambridge [see "The Insulin Molecule," by E. O. P. Thompson, beginning on page 183].

The insulin molecule is much more complex than a simple peptide such as bradykinin [see illustration on this page]. It not only has nearly six times as many amino acids but also has a greater variety of them: 17 rather than five. This introduces many new problems of side-chain protection. Particularly complicating is the presence of three disulfide-bond (S—S) cross-links between cysteine units. Insulin consists of two linear peptide chains: an A chain with 21 amino acids and a B chain with 30. They are held together by two interchain disulfide bridges, and in addition one of the chains has an intrachain disulfide loop. Furthermore, the molecule has a definite three-dimensional conformation. Although the X-ray structure has not yet

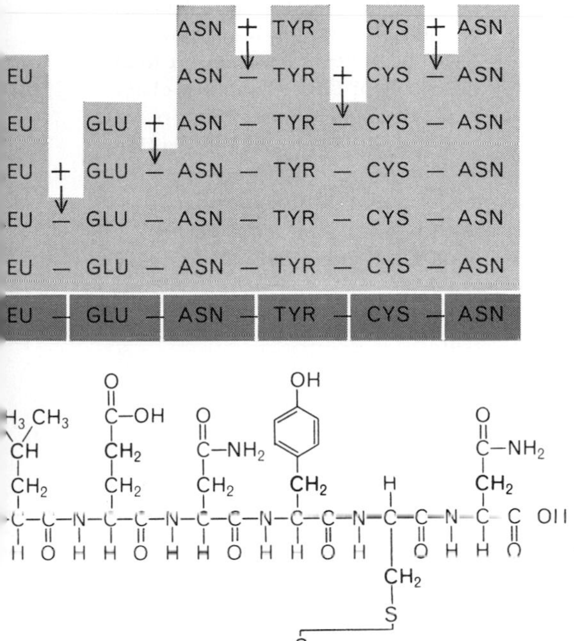

ALA	ALANINE
ARG	ARGININE
ASN	ASPARAGINE
CYS	CYSTEINE
GLN	GLUTAMINE
GLU	GLUTAMIC ACID
GLY	GLYCINE
HIS	HISTIDINE
ILE	ISOLEUCINE
LEU	LEUCINE
LYS	LYSINE
PHE	PHENYLALANINE
PRO	PROLINE
SER	SERINE
THR	THREONINE
TYR	TYROSINE
VAL	VALINE

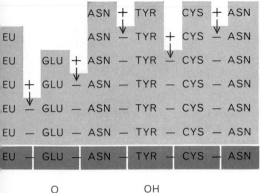

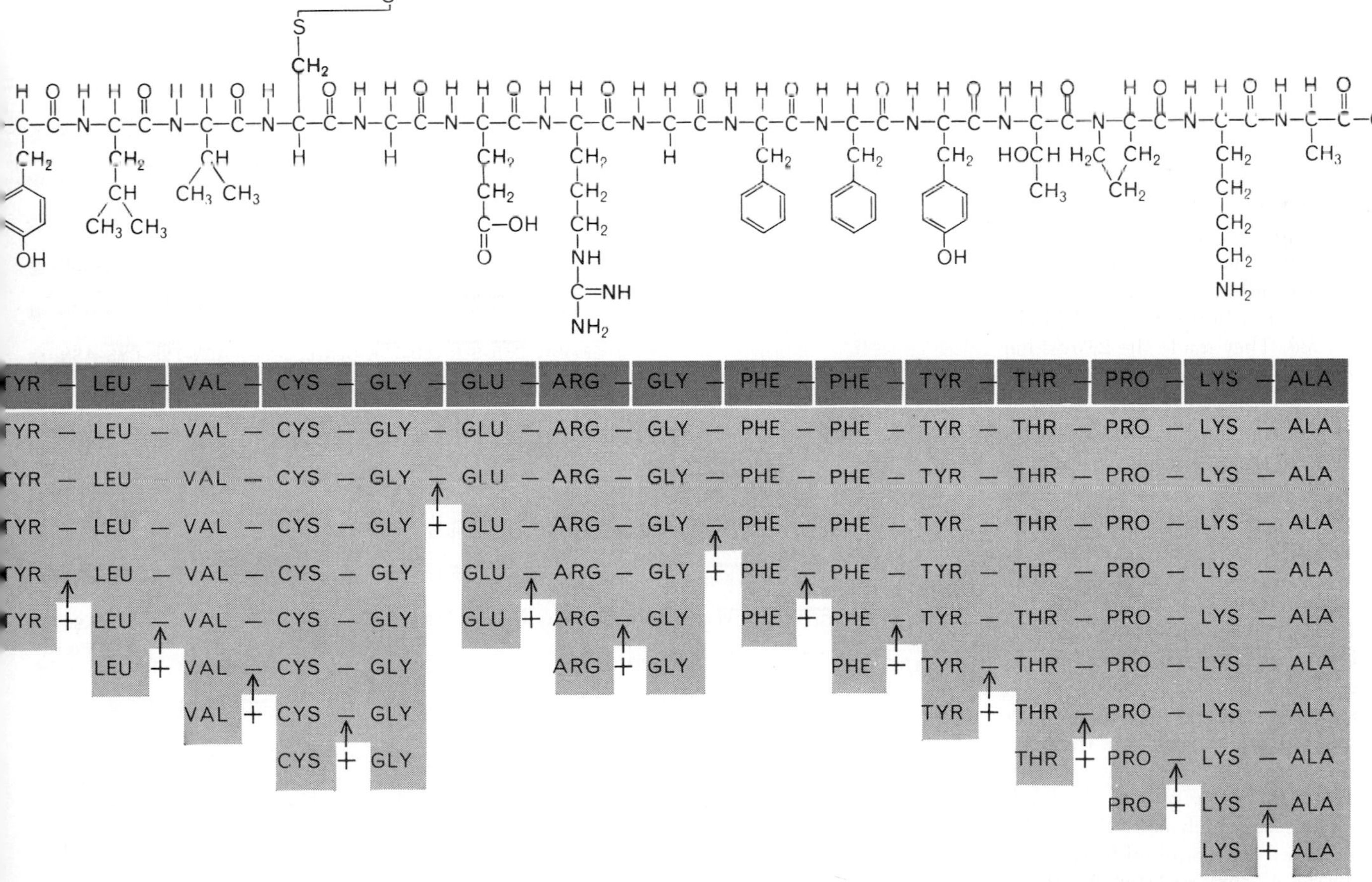

A chain by P. G. Katsoyannis' group at the University of Pittsburgh is shown above the insulin formula. The 21 amino acid units were assembled stepwise into intermediate fragments of from two to five units and the fragments were then coupled in stages to assemble the complete chain. The similar synthesis of the B chain by Helmut Zahn's group in Germany is shown below the insulin formula.

been worked out in detail, it is clear from the fact that insulin forms characteristic crystals that it is composed of molecules with a precise structure.

How can one hope to build up the long peptide chains, to form the three disulfide bonds between the correct cysteine units and then to fold the entire assembly into its proper shape? This is asking a lot, because there are many possible ways for the S—S bonds to form, and the possible variations in the conformation of the molecule are enormous. It is only possible at this time because nature comes to the chemist's aid. If we simply make the two chains with the six cysteine units all in the reduced (SH) form (in which a hydrogen atom is attached to each sulfur atom) and mix them under the proper oxidizing conditions, they will preferentially form the correct S—S bridges and fold into the characteristic insulin conformation all by themselves!

The discovery in 1960, by G. H. Dixon and A. C. Wardlaw of the University of Toronto, that this would happen provided the key peptide chemists needed to undertake the synthesis of insulin. All four laboratories that have made insulin have depended on this fact and have made the two chains separately. From the point of view of the chemist this is really peptide synthesis rather than protein synthesis, but when the two chains are combined, the final product meets all the usual criteria for a protein and justifies the conclusion that a real protein has been synthesized in the laboratory.

The first published synthesis of an individual insulin chain was made by P. G. Katsoyannis and his colleagues at the University of Pittsburgh School of Medicine in 1963. They made the 21-residue A chain of sheep insulin by the fragment method. When the synthetic A chain was linked with the natural B chain, the combination gave rise to a small but definite amount of insulin activity. Later that same year a large group of chemists at the Technische Hochschule at Aachen in Germany, under the direction of Helmut Zahn, reported the synthesis of both the A and the B chain and the successful combination of the two for the first total synthesis of insulin. The overall yields (2.9 percent for the A chain and 7 percent for the B) and the extent of the combination (.2 to 1 percent) were still low, but true insulin activity was obtained. These syntheses, which also followed the fragment approach, required 89 reaction steps for the A chain, 132 steps for the B chain and three more steps for the combination of the two. Each step, of course,

required numerous operations.

During the same period a third group was working on insulin at the Academy of Science in Shanghai and the University of Peking. Their first important contribution to the problem was the development of improved methods for the separation and recombination of natural insulin chains. Their yield was eventually increased to about 50 percent, which meant that the combination was far from a random process. Their major contribution was the preparation in 1965 of the first crystalline, all-synthetic insulin. The crystals were obtained in low yield, but they had the same form as the native molecule and, most important, the full biological activity (more than 20 units per milligram). This was a crucial element of the proof that insulin had in fact been synthesized.

Automatic Synthesis of Insulin

There remained a very real problem. Large numbers of chemists had to work for several years to produce tiny quantities of the peptides. In order to produce useful amounts of insulin and to be able to make modifications in the structure in a more efficient way, we undertook in 1965 to apply the solid phase method to the task. The results have been very encouraging. Although more than 5,000

separate operations were required to assemble the 51 amino acids into the two chains of bovine insulin, most of these were performed automatically under the control of the drum programmer, so that it was possible for one man to carry out the synthesis of both chains in only a few days.

Beginning with three grams of resin, Arnold Marglin of Rockefeller University was able to prepare approximately two grams of protected A chain. A total of eight grams of the protected B chain was made on eight grams of the resin. The reaction that detached the peptide chains from their polystyrene support also removed most of the side-chain protecting groups, leaving only the benzyl groups on the cysteine and histidine side

ASN-

CYS-ASN-

TYR-CYS-ASN-

ASN-TYR-CYS-ASN-

GLU-ASN-TYR-CYS-ASN-

LEU-GLU-ASN-TYR-CYS-ASN-

GLN-LEU-GLU-ASN-TYR-CYS-ASN-

TYR-GLN-LEU-GLU-ASN-TYR-CYS-ASN-

LEU-TYR-GLN-LEU-GLU-ASN-TYR-CYS-ASN-

SER-LEU-TYR-GLN-LEU-GLU-ASN-TYR-CYS-ASN-

CYS-SER-LEU-TYR-GLN-LEU-GLU-ASN-TYR-CYS-ASN-

VAL-CYS-SER-LEU-TYR-GLN-LEU-GLU-ASN-TYR-CYS-ASN-

SER-VAL-CYS-SER-LEU-TYR-GLN-LEU-GLU-ASN-TYR-CYS-ASN-

ALA-SER-VAL-CYS-SER-LEU-TYR-GLN-LEU-GLU-ASN-TYR-CYS-ASN-

CYS-ALA-SER-VAL-CYS-SER-LEU-TYR-GLN-LEU-GLU-ASN-TYR-CYS-ASN-

CYS-CYS-ALA-SER-VAL-CYS-SER-LEU-TYR-GLN-LEU-GLU-ASN-TYR-CYS-ASN-

GLN-CYS-CYS-ALA-SER-VAL-CYS-SER-LEU-TYR-GLN-LEU-GLU-ASN-TYR-CYS-ASN-

GLU-GLN-CYS-CYS-ALA-SER-VAL-CYS-SER-LEU-TYR-GLN-LEU-GLU-ASN-TYR-CYS-ASN-

VAL-GLU-GLN-CYS-CYS-ALA-SER-VAL-CYS-SER-LEU-TYR-GLN-LEU-GLU-ASN-TYR-CYS-ASN-

ILE-VAL-GLU-GLN-CYS-CYS-ALA-SER-VAL-CYS-SER-LEU-TYR-GLN-LEU-GLU-ASN-TYR-CYS-ASN-

GLY-ILE-VAL-GLU-GLN-CYS-CYS-ALA-SER-VAL-CYS-SER-LEU-TYR-GLN-LEU-GLU-ASN-TYR-CYS-ASN-

HBr in TFA

GLY-ILE-VAL-GLU-GLN-CYS-CYS-ALA-SER-VAL-CYS-SER-LEU-TYR-GLN-LEU-GLU-ASN-TYR-CYS-ASN

Na in NH₃

GLY-ILE-VAL-GLU-GLN-CYS-CYS-ALA-SER-VAL-CYS-SER-LEU-TYR-GLN-LEU-GLU-ASN-TYR-CYS-ASN

Na₂SO₃, Na₂S₄O₆ SO_3^- SO_3^- SO_3^- SO_3^-

GLY-ILE-VAL-GLU-GLN-CYS-CYS-ALA-SER-VAL-CYS-SER-LEU-TYR-GLN-LEU-GLU-ASN-TYR-CYS-ASN

SOLID PHASE SYNTHESIS of the A chain (*left*) and the B chain (*right*) is diagrammed. An amino acid protected by a Boc group (*vertical bar*) is coupled to a polystyrene bead (*top*), then deprotected. Activated amino acids, protected at the amino end and if necessary

chains. These could be removed by reduction with metallic sodium dissolved in liquid ammonia, a reaction that was discovered many years ago by Vincent du Vigneaud of the Cornell University Medical College and was the key to his historic syntheses of the pituitary hormones oxytocin and vasopressin. Applied to insulin, however, the sodium treatment at first broke some of the bonds between the amino acids threonine and proline in the *B* chain. Once we recognized what was happening it was possible to keep the chains from splitting by careful modification of the conditions of the reaction.

This deprotecting step left the cysteine groups in the reduced (SH) form. Although it was just this SH form that we would later want for the final oxidation step to link the two chains, the SH groups were too unstable to undergo the purification procedures that were now necessary; they were stabilized by conversion to S-sulfonates (SSO_3^-). Then the two peptide chains could be purified by three methods: filtration, which depends on molecular size; countercurrent distribution, which depends on differential solubility, and free-flow electrophoresis, which depends on electric charge. The resulting products could be shown to be homogeneous by other electrophoretic and chromatographic criteria. Amino acid analyses showed that the chains had the compositions characteristic of the *A* and *B* chains of insulin.

The final step in the synthesis of insulin was the combination of the two purified chains. First the cysteine sulfonates were converted back to the SH form. Then the chains were combined by the method developed by the Chinese, which involves the slow oxidation by air of the SH forms of both chains

ALA–●

LYS–ALA–●

PRO–LYS–ALA–●

THR–PRO–LYS–ALA–●

TYR–THR–PRO–LYS–ALA–●

PHE–TYR–THR–PRO–LYS–ALA–●

PHE–PHE–TYR–THR–PRO–LYS–ALA–●

GLY–PHE–PHE–TYR–THR–PRO–LYS–ALA–●

ARG–GLY–PHE–PHE–TYR–THR–PRO–LYS–ALA–●

GLU–ARG–GLY–PHE–PHE–TYR–THR–PRO–LYS–ALA–●

GLY–GLU–ARG–GLY–PHE–PHE–TYR–THR–PRO–LYS–ALA–●

CYS–GLY–GLU–ARG–GLY–PHE–PHE–TYR–THR–PRO–LYS–ALA–●

VAL–CYS–GLY–GLU–ARG–GLY–PHE–PHE–TYR–THR–PRO–LYS–ALA–●

LEU–VAL–CYS–GLY–GLU–ARG–GLY–PHE–PHE–TYR–THR–PRO–LYS–ALA–●

TYR–LEU–VAL–CYS–GLY–GLU–ARG–GLY–PHE–PHE–TYR–THR–PRO–LYS–ALA–●

LEU–TYR–LEU–VAL–CYS–GLY–GLU–ARG–GLY–PHE–PHE–TYR–THR–PRO–LYS–ALA–●

ALA–LEU–TYR–LEU–VAL–CYS–GLY–GLU–ARG–GLY–PHE–PHE–TYR–THR–PRO–LYS–ALA–●

GLU–ALA–LEU–TYR–LEU–VAL–CYS–GLY–GLU–ARG–GLY–PHE–PHE–TYR–THR–PRO–LYS–ALA–●

VAL–GLU–ALA–LEU–TYR–LEU–VAL–CYS–GLY–GLU–ARG–GLY–PHE–PHE–TYR–THR–PRO–LYS–ALA–●

LEU–VAL–GLU–ALA–LEU–TYR–LEU–VAL–CYS–GLY–GLU–ARG–GLY–PHE–PHE–TYR–THR–PRO–LYS–ALA–●

HIS–LEU–VAL–GLU–ALA–LEU–TYR–LEU–VAL–CYS–GLY–GLU–ARG–GLY–PHE–PHE–TYR–THR–PRO–LYS–ALA–●

SER–HIS–LEU–VAL–GLU–ALA–LEU–TYR–LEU–VAL–CYS–GLY–GLU–ARG–GLY–PHE–PHE–TYR–THR–PRO–LYS–ALA–●

GLY–SER–HIS–LEU–VAL–GLU–ALA–LEU–TYR–LEU–VAL–CYS–GLY–GLU–ARG–GLY–PHE–PHE–TYR–THR–PRO–LYS–ALA–●

CYS–GLY–SER–HIS–LEU–VAL–GLU–ALA–LEU–TYR–LEU–VAL–CYS–GLY–GLU–ARG–GLY–PHE–PHE–TYR–THR–PRO–LYS–ALA–●

LEU–CYS–GLY–SER–HIS–LEU–VAL–GLU–ALA–LEU–TYR–LEU–VAL–CYS–GLY–GLU–ARG–GLY–PHE–PHE–TYR–THR–PRO–LYS–ALA–●

HIS–LEU–CYS–GLY–SER–HIS–LEU–VAL–GLU–ALA–LEU–TYR–LEU–VAL–CYS–GLY–GLU–ARG–GLY–PHE–PHE–TYR–THR–PRO–LYS–ALA–●

GLN–HIS–LEU–CYS–GLY–SER–HIS–LEU–VAL–GLU–ALA–LEU–TYR–LEU–VAL–CYS–GLY–GLU–ARG–GLY–PHE–PHE–TYR–THR–PRO–LYS–ALA–●

ASN–GLN–HIS–LEU–CYS–GLY–SER–HIS–LEU–VAL–GLU–ALA–LEU–TYR–LEU–VAL–CYS–GLY–GLU–ARG–GLY–PHE–PHE–TYR–THR–PRO–LYS–ALA–●

VAL–ASN–GLN–HIS–LEU–CYS–GLY–SER–HIS–LEU–VAL–GLU–ALA–LEU–TYR–LEU–VAL–CYS–GLY–GLU–ARG–GLY–PHE–PHE–TYR–THR–PRO–LYS–ALA–●

PHE–VAL–ASN–GLN–HIS–LEU–CYS–GLY–SER–HIS–LEU–VAL–GLU–ALA–LEU–TYR–LEU–VAL–CYS–GLY–GLU–ARG–GLY–PHE–PHE–TYR–THR–PRO–LYS–ALA–●

HBr in NH$_3$

PHE–VAL–ASN–GLN–HIS–LEU–CYS–GLY–SER–HIS–LEU–VAL–GLU–ALA–LEU–TYR–LEU–VAL–CYS–GLY–GLU–ARG–GLY–PHE–PHE–TYR–THR–PRO–LYS–ALA

Na in NH$_3$

PHE–VAL–ASN–GLN–HIS–LEU–CYS–GLY–SER–HIS–LEU–VAL–GLU–ALA–LEU–TYR–LEU–VAL–CYS–GLY–GLU–ARG–GLY–PHE–PHE–TYR–THR–PRO–LYS–ALA

Na$_2$SO$_3$, Na$_2$S$_4$O$_6$ SO$_3^-$ SO$_3^-$

PHE–VAL–ASN–GLN–HIS–LEU–CYS–GLY–SER–HIS–LEU–VAL–GLU–ALA–LEU–TYR–LEU–VAL–CYS–GLY–GLU–ARG–GLY–PHE–PHE–TYR–THR–PRO–LYS–ALA

at the side chain (*horizontal bars*), are then coupled stepwise. When the chain is complete, it is cleaved from the bead, and most protecting groups are removed, by hydrogen bromide treatment. Sodium treatment removes groups protecting the cysteine and histidine side chains. Then the cysteines are changed from the sulfhydryl form to the stable S-sulfonate form in preparation for purification.

168

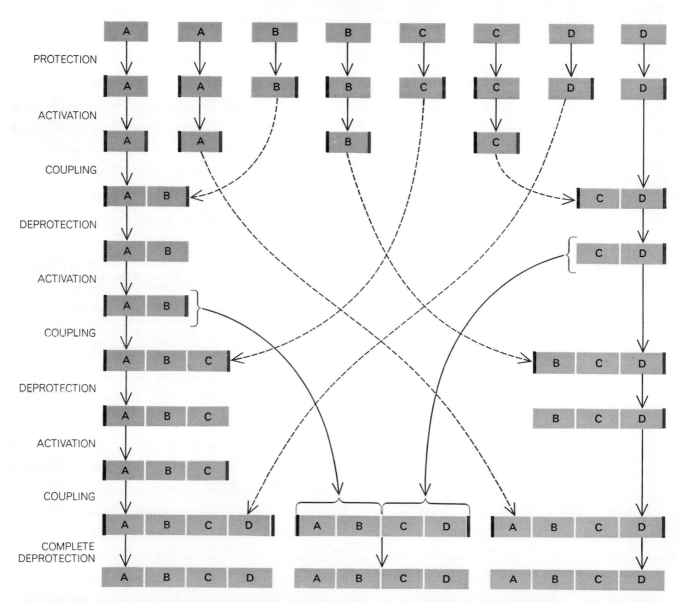

AMINO ACID (*a*) has an amino (NH_2) and a carboxyl (COOH) group separated by a carbon atom that carries a side chain (*R*). The peptide bond (*b*) forms between the carboxyl and the amino ends of two units with the elimination of a molecule of water. A series of amino acid subunits held together by such bonds (*color*) constitutes the primary ("backbone") structure of peptides and proteins (*c*).

TWO GENERAL APPROACHES to peptide synthesis are diagrammed. In the "stepwise" method amino acid units are added one at a time starting at the amino end of the peptide (*left*) or the carboxyl end (*right*) until the final peptide (of four amino acids, *A*, *B*, *C* and *D*, in this case) is assembled (*bottom left and right*). In the "fragment" method small peptides are prepared stepwise and are then combined (*long solid arrows*) to form the final peptide (*bottom center*). In each case it is necessary to protect the amino ends (*black bars*) and carboxyl ends (*gray bars*) against unwanted reactions, to activate the carboxyl ends (*dark-color bars*) for coupling, to "deprotect" selectively in preparation for the next step and finally to deprotect completely. The peptide can then be purified.

[see illustration below]. We were able to show that either of our synthetic chains could be combined with the complementary natural chain to produce biologically active, semisynthetic insulin. Then the two synthetic chains were combined to form all-synthetic insulin. The synthetic hormone was active in the standard biological assay, which is based on the amount that must be injected to lower the blood sugar enough to cause convulsions in 50 percent of a group of experimental mice. The response to the synthetic preparations was shown to be due to low blood sugar rather than to some nonspecific toxic effect because the animals recovered rapidly after they were given glucose. In addition, the synthetic material behaved like insulin in various physical and chemical tests. For example, its mobility in paper electrophoresis was the same as that of the natural hormone.

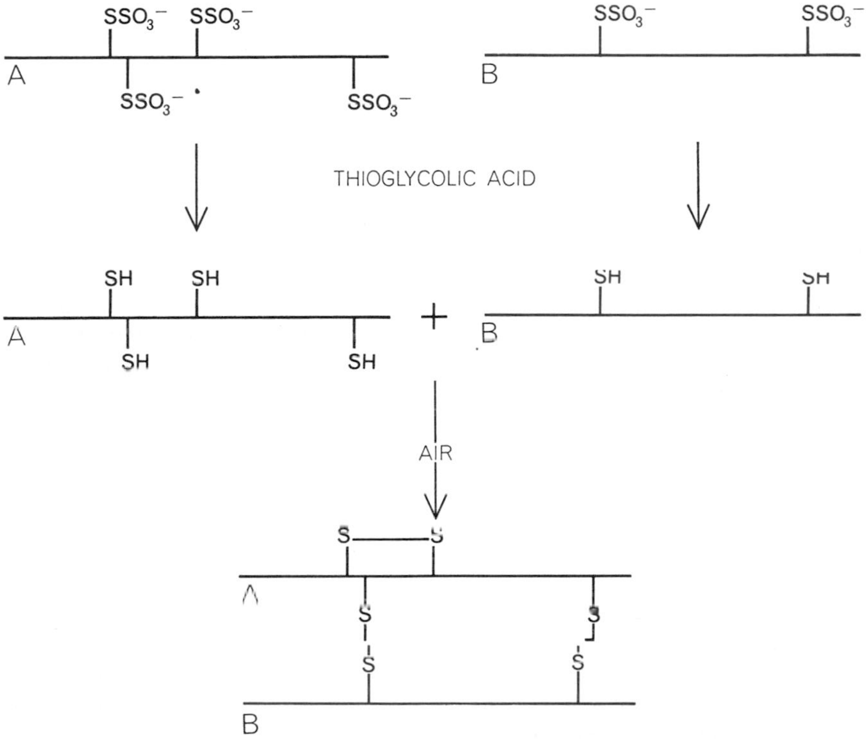

The yield and the purity of the chains themselves, which constitute the peptide synthesis portion of the work, are quite good, but our combination yields are still poor. Important progress in that regard has been made recently by Katsoyannis and his co-workers. They have modified one of the earlier Chinese methods (using the SH form of the A chain and the SSO_3^- form of the B chain) and can now obtain much improved yields in the final step of the synthesis. It is fair to conclude that the synthesis of one protein has been accomplished and that another major hurdle has been cleared by chemists in their continuing struggle to duplicate nature.

Current developments in peptide chemistry should bring within reach other small protein molecules that are of great biological interest. The structures of myoglobin, cytochrome c, ferredoxin and growth hormone are known, for example, and we can expect some of them to be synthesized. A major step toward the synthesis of living systems will come with the synthesis of virus-coat protein; that of the tobacco mosaic virus may be the first.

It is the enzymes that are probably of greatest current interest. It is important to learn how these complex protein molecules function in their control of biochemical reactions. Why are they such enormously active catalysts and why are they so specific in their action? What factors are responsible for the "active centers" of enzymes and for their specific binding sites? How does the primary structure of the protein control its three-dimensional structure and its function; how will changes in the amino acid sequence influence biological activity? Automatic solid phase syntheses should help to answer some of these questions.

COMBINATION of the A and B chains, now purified and in the S-sulfonate form, was carried out by the method developed in China. The chains were mixed and reduced to the sulfhydryl form with thioglycolic acid. On exposure to air in an alkaline solution the sulfhydryl groups oxidized slowly and the three disulfide bonds characteristic of natural insulin were formed. By-products formed by incorrect cross-linking could be removed by extraction.

20

THE THREE-DIMENSIONAL STRUCTURE OF AN ENZYME MOLECULE

DAVID C. PHILLIPS
November 1966

The arrangement of atoms in an enzyme molecule has been worked out for the first time. The enzyme is lysozyme, which breaks open cells of bacteria. The study has also shown how lysozyme performs its task

One day in 1922 Alexander Fleming was suffering from a cold. This is not unusual in London, but Fleming was a most unusual man and he took advantage of the cold in a characteristic way. He allowed a few drops of his nasal mucus to fall on a culture of bacteria he was working with and then put the plate to one side to see what would happen. Imagine his excitement when he discovered some time later that the bacteria near the mucus had dissolved away. For a while he thought his ambition of finding a universal antibiotic had been realized. In a burst of activity he quickly established that the antibacterial action of the mucus was due to the presence in it of an enzyme; he called this substance lysozyme because of its capacity to lyse, or dissolve, the bacterial cells. Lysozyme was soon discovered in many tissues and secretions of the human body, in plants and most plentifully of all in the white of egg. Unfortunately Fleming found that it is not effective against the most harmful bacteria. He had to wait seven years before a strangely similar experiment revealed the existence of a genuinely effective antibiotic: penicillin.

Nevertheless, Fleming's lysozyme has proved a more valuable discovery than he can have expected when its properties were first established. With it, for example, bacterial anatomists have been able to study many details of bacterial structure [see "Fleming's Lysozyme," by Robert F. Acker and S. E. Hartsell; SCIENTIFIC AMERICAN, June, 1960]. It has now turned out that lysozyme is the first enzyme whose three-dimensional structure has been determined and whose properties are understood in atomic detail. Among these properties is the way in which the enzyme combines with the substance on which it acts—a complex sugar in the wall of the bacterial cell.

Like all enzymes, lysozyme is a protein. Its chemical makeup has been established by Pierre Jollès and his colleagues at the University of Paris and by Robert E. Canfield of the Columbia University College of Physicians and Surgeons. They have found that each molecule of lysozyme obtained from egg white consists of a single polypeptide chain of 129 amino acid subunits of 20 different kinds. A peptide bond is formed when two amino acids are joined following the removal of a molecule of water. It is customary to call the portion of the amino acid incorporated into a polypeptide chain a residue, and each residue has its own characteristic side chain. The 129-residue lysozyme molecule is cross-linked in four places by disulfide bridges formed by the combination of sulfur-containing side chains in different parts of the molecule [see illustration on opposite page].

The properties of the molecule cannot be understood from its chemical constitution alone; they depend most critically on what parts of the molecule are brought close together in the folded three-dimensional structure. Some form of microscope is needed to examine the structure of the molecule. Fortunately one is effectively provided by the techniques of X-ray crystal-structure analysis pioneered by Sir Lawrence Bragg and his father Sir William Bragg.

ALA	ALANINE	GLY	GLYCINE	PRO	PROLINE
ARG	ARGININE	HIS	HISTIDINE	SER	SERINE
ASN	ASPARAGINE	ILEU	ISOLEUCINE	THR	THREONINE
ASP	ASPARTIC ACID	LEU	LEUCINE	TRY	TRYPTOPHAN
CYS	CYSTEINE	LYS	LYSINE	TYR	TYROSINE
GLN	GLUTAMINE	MET	METHIONINE	VAL	VALINE
GLU	GLUTAMIC ACID	PHE	PHENYLALANINE		

TWO-DIMENSIONAL MODEL of the lysozyme molecule is shown on the opposite page. Lysozyme is a protein containing 129 amino acid subunits, commonly called residues (*see key to abbreviations above*). These residues form a polypeptide chain that is cross-linked at four places by disulfide (–S–S–) bonds. The amino acid sequence of lysozyme was determined independently by Pierre Jollès and his co-workers at the University of Paris and by Robert E. Canfield of the Columbia University College of Physicians and Surgeons. The three-dimensional structure of the lysozyme molecule has now been established with the help of X-ray crystallography by the author and his colleagues at the Royal Institution in London. A painting of the molecule's three-dimensional structure appears on pages 172 and 173. The function of lysozyme is to split a particular long-chain molecule, a complex sugar, found in the outer membrane of many living cells. Molecules that are acted on by enzymes are known as substrates. The substrate of lysozyme fits into a cleft, or pocket, formed by the three-dimensional structure of the lysozyme molecule. In the two-dimensional model on the opposite page the amino acid residues that line the pocket are shown in dark green.

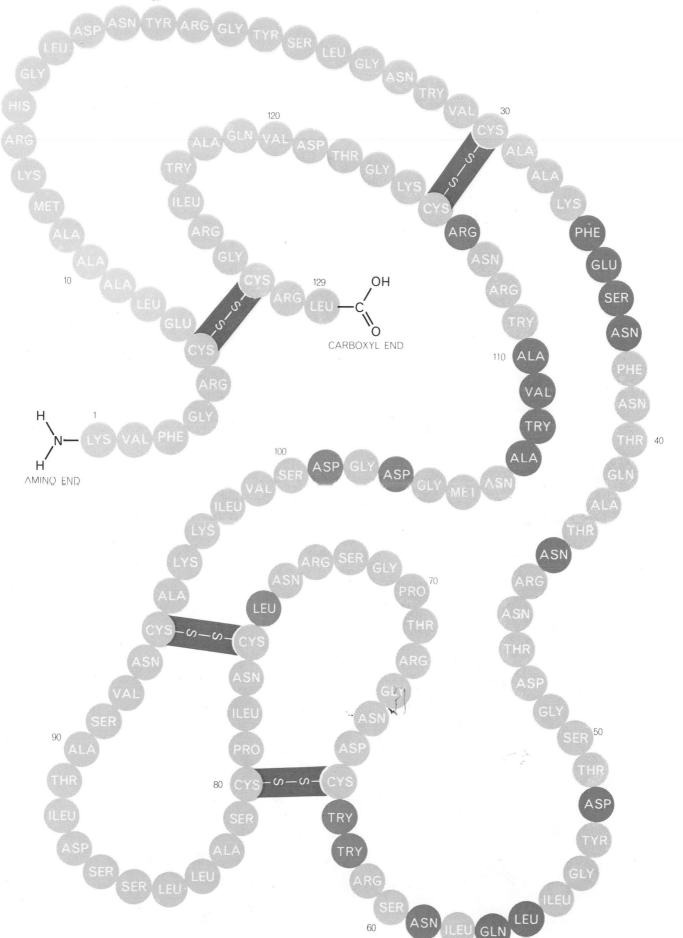

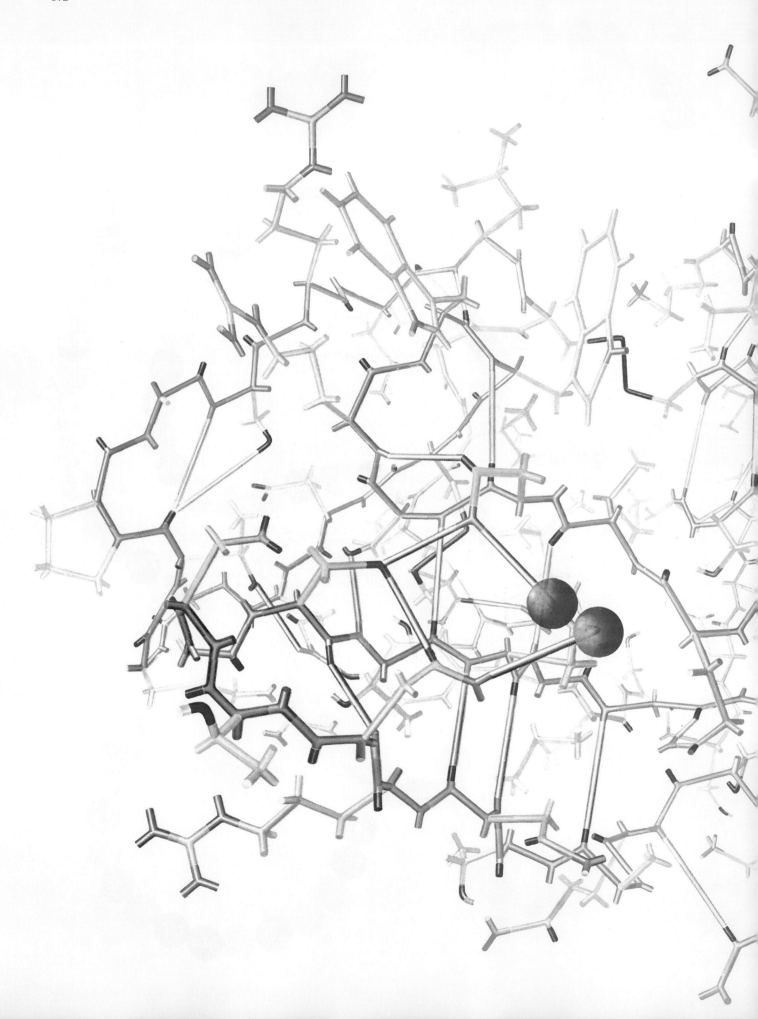

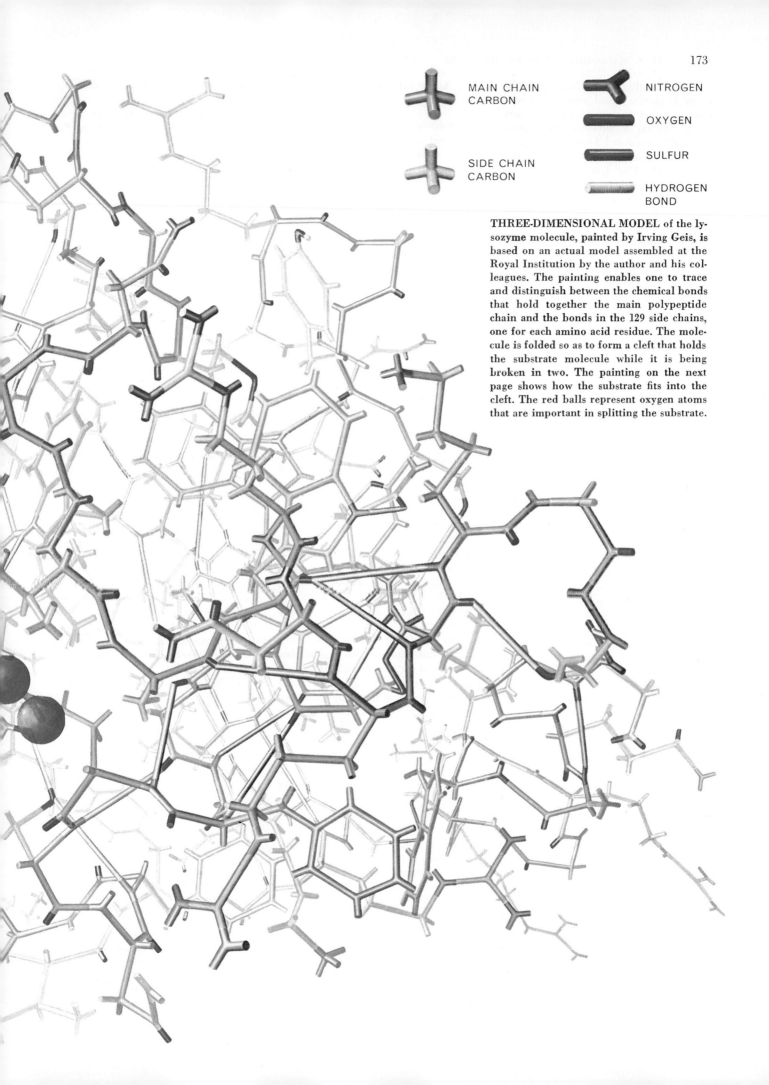

MAIN CHAIN CARBON

SIDE CHAIN CARBON

NITROGEN

OXYGEN

SULFUR

HYDROGEN BOND

THREE-DIMENSIONAL MODEL of the lysozyme molecule, painted by Irving Geis, is based on an actual model assembled at the Royal Institution by the author and his colleagues. The painting enables one to trace and distinguish between the chemical bonds that hold together the main polypeptide chain and the bonds in the 129 side chains, one for each amino acid residue. The molecule is folded so as to form a cleft that holds the substrate molecule while it is being broken in two. The painting on the next page shows how the substrate fits into the cleft. The red balls represent oxygen atoms that are important in splitting the substrate.

The difficulties of examining molecules in atomic detail arise, of course, from the fact that molecules are very small. Within a molecule each atom is usually separated from its neighbor by about 1.5 angstrom units (1.5×10^{-8} centimeter). The lysozyme molecule, which contains some 1,950 atoms, is about 40 angstroms in its largest dimension. The first problem is to find a microscope in which the atoms can be resolved from one another, or seen separately.

The resolving power of a microscope depends fundamentally on the wavelength of the radiation it employs. In general no two objects can be seen separately if they are closer together than about half this wavelength. The shortest wavelength transmitted by optical microscopes (those working in the ultraviolet end of the spectrum) is about 2,000 times longer than the distance between atoms. In order to "see" atoms one must use radiation with a much shorter wavelength: X rays, which have a wavelength closely comparable to interatomic distances. The employment of X rays, however, creates other difficulties: no satisfactory way has yet been found to make lenses or mirrors that will focus them into an image. The problem, then, is the apparently impossible one of designing an X-ray microscope without lenses or mirrors.

Consideration of the diffraction theory of microscope optics, as developed by Ernst Abbe in the latter part of the 19th century, shows that the problem can be solved. Abbe taught us that the formation of an image in the microscope can be regarded as a two-stage process. First, the object under examination scatters the light or other radiation falling on it in all directions, forming a diffraction pattern. This pattern arises because the light waves scattered from different parts of the object combine so as to produce a wave of large or small amplitude in any direction according to whether the waves are in or out of phase—in or out of step—with one another. (This effect is seen most easily in light waves scattered by a regularly repeating structure, such as a diffraction grating made of lines scribed at regular intervals on a glass plate.) In the second stage of image formation, according to Abbe, the objective lens of the microscope collects the diffracted waves and recombines them to form an image of the object. Most important, the nature of the image depends critically on how much of the diffraction pattern is used in its formation.

X-Ray Structure Analysis

In essence X-ray structure analysis makes use of a microscope in which the two stages of image formation have been separated. Since the X rays cannot be focused to form an image directly, the diffraction pattern is recorded and the image is obtained from it by calculation. Historically the method was not developed on the basis of this reasoning, but this way of regarding it (which was first suggested by Lawrence Bragg) brings out its essential features and also introduces the main difficulty of applying it. In recording the intensities of the diffracted waves, instead of focusing them to form an image, some crucial information is lost, namely the phase relations among the various diffracted waves. Without this information the image cannot be formed, and some means of recovering it has to be found. This is the well-known phase problem of X-ray crystallography. It is on the solution of the problem that the utility of the method depends.

The term "X-ray crystallography" reminds us that in practice the method was developed (and is still applied) in the study of single crystals. Crystals suitable for study may contain some 10^{15} identical molecules in a regular array; in effect the molecules in such a crystal diffract the X radiation as though they were a single giant molecule. The crystal acts as a three-dimensional diffraction grating, so that the waves scattered by them are confined to a number of discrete directions. In order to obtain a three-dimensional image of the structure the intensity of the X rays scattered in these different directions must be measured, the phase problem must be solved somehow and the measurements must be combined by a computer.

The recent successes of this method in the study of protein structures have depended a great deal on the development of electronic computers capable of performing the calculations. They are due most of all, however, to the discovery in 1953, by M. F. Perutz of the Medical Research Council Laboratory of Molecular Biology in Cambridge, that the method of "isomorphous replacement" can be used to solve the phase problem in the study of protein crystals. The method depends on the preparation and study of a series of protein crystals into which additional heavy atoms, such as atoms of uranium, have been introduced without otherwise affecting the crystal structure. The first successes of this method were in the study of sperm-whale myoglobin by John C. Kendrew of the Medical Research Council Laboratory and in Perutz' own study of horse hemoglobin. For their work the two men received the Nobel prize for chemistry in 1962 [see "The Three-dimensional Structure of a Protein Molecule," by John C. Kendrew, SCIENTIFIC AMERICAN Offprint 121, and "The Hemoglobin Molecule," by M. F. Perutz, SCIENTIFIC AMERICAN Offprint 196].

Because the X rays are scattered by the electrons within the molecules, the image calculated from the diffraction pattern reveals the distribution of electrons within the crystal. The electron density is usually calculated at a regular array of points, and the image is made visible by drawing contour lines through points of equal electron density. If these contour maps are drawn on clear plastic sheets, one can obtain a three-dimensional image by assembling the maps one above the other in a stack. The amount of detail that can be seen in such an image depends on the resolving power of the effective microscope, that is, on its "aperture," or the extent of the diffraction pattern that has been included in the formation of the image. If the waves diffracted through sufficiently high angles are included

MODEL OF SUBSTRATE shows how it fits into the cleft in the lysozyme molecule. All the carbon atoms in the substrate are shown in purple. The portion of the substrate in intimate contact with the underlying enzyme is a polysaccharide chain consisting of six ringlike structures, each a residue of an amino-sugar molecule. The substrate in the model is made up of six identical residues of the amino sugar called N-acetylglucosamine (NAG). In the actual substrate every other residue is an amino sugar known as N-acetylmuramic acid (NAM). The illustration is based on X-ray studies of the way the enzyme is bound to a trisaccharide made of three NAG units, which fills the top of the cleft; the arrangement of NAG units in the bottom of the cleft was worked out with the aid of three-dimensional models. The substrate is held to the enzyme by a complex network of hydrogen bonds. In this style of model-making each straight section of chain represents a bond between atoms. The atoms themselves lie at the intersections and elbows of the structure. Except for the four red balls representing oxygen atoms that are active in splitting the polysaccharide substrate, no attempt is made to represent the electron shells of atoms because they would merge into a solid mass.

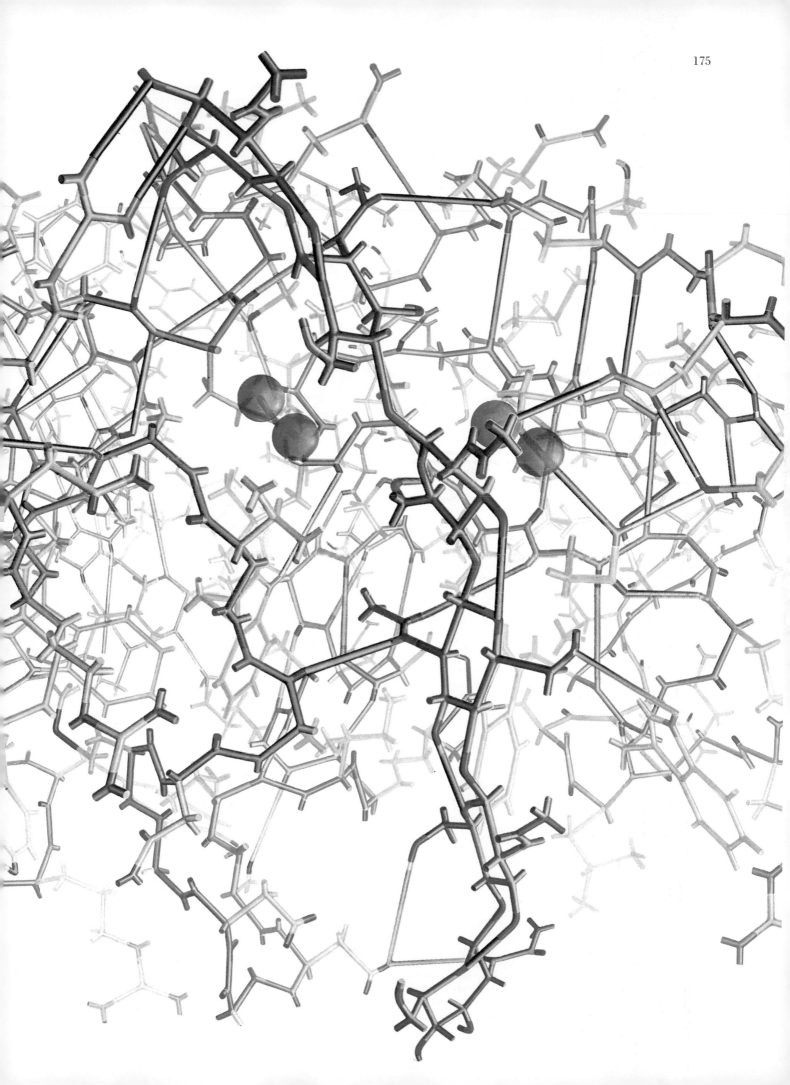

(corresponding to a large aperture), the atoms appear as individual peaks in the image map. At lower resolution groups of unresolved atoms appear with characteristic shapes by which they can be recognized.

The three-dimensional structure of lysozyme crystallized from the white of hen's egg has been determined in atomic detail with the X-ray method by our group at the Royal Institution in Lon-

don. This is the laboratory in which Humphry Davy and Michael Faraday made their fundamental discoveries during the 19th century, and in which the X-ray method of structure analysis was developed between the two world wars by the brilliant group of workers led by William Bragg, including J. D. Bernal, Kathleen Lonsdale, W. T. Astbury, J. M. Robertson and many others. Our work on lysozyme was begun in 1960

when Roberto J. Poljak, a visiting worker from Argentina, demonstrated that suitable crystals containing heavy atoms could be prepared. Since then C. C. F. Blake, A. C. T. North, V. R. Sarma, Ruth Fenn, D. F. Koenig, Louise N. Johnson and G. A. Mair have played important roles in the work.

In 1962 a low-resolution image of the structure was obtained that revealed the general shape of the molecule and

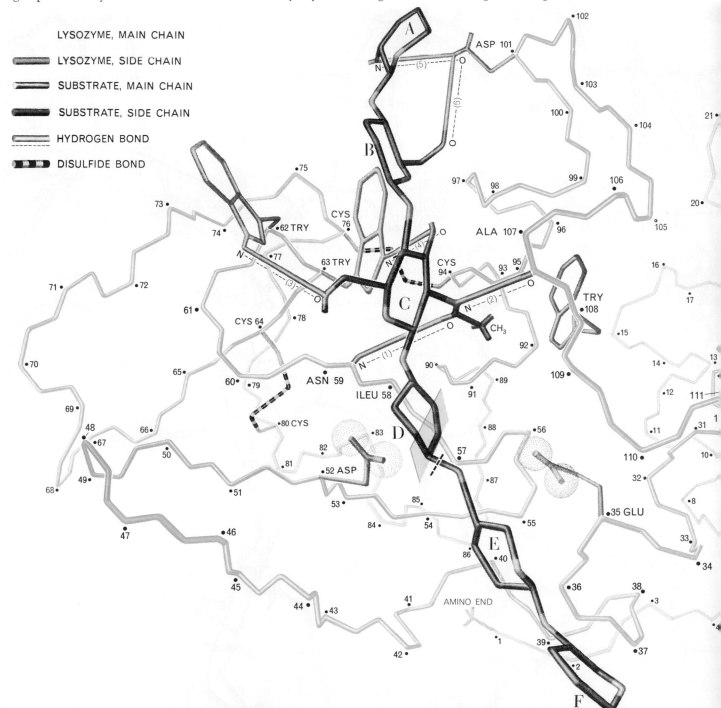

MAP OF LYSOZYME AND SUBSTRATE depicts in color the central chain of each molecule. Side chains have been omitted except for those that produce the four disulfide bonds clipping the lysozyme molecule together and those that supply the terminal connections for hydrogen bonds holding the substrate to the lysozyme. The top three rings of the substrate (A, B, C) are held to the underlying enzyme by six principal hydrogen bonds, which are identified by number to key with the description in the text. The lyso-

showed that the arrangement of the polypeptide chain is even more complex than it is in myoglobin. This low-resolution image was calculated from the amplitudes of about 400 diffraction maxima measured from native protein crystals and from crystals containing each of three different heavy atoms. In 1965, after the development of more efficient methods of measurement and computation, an image was calculated on the basis of nearly 10,000 diffraction maxima, which resolved features separated by two angstroms. Apart from showing a few well-separated chloride ions, which are present because the lysozyme is crystallized from a solution containing sodium chloride, the two-angstrom image still does not show individual atoms as separate maxima in the electron-density map. The level of resolution is high enough, however, for many of the groups of atoms to be clearly recognizable.

The Lysozyme Molecule

The main polypeptide chain appears as a continuous ribbon of electron density running through the image with regularly spaced promontories on it that are characteristic of the carbonyl groups (CO) that mark each peptide bond. In some regions the chain is folded in ways that are familiar from theoretical studies of polypeptide configurations and from the structure analyses of myoglobin and fibrous proteins such as the keratin of hair. The amino acid residues in lysozyme have now been designated by number; the residues numbered 5 through 15, 24 through 34 and 88 through 96 form three lengths of "alpha helix," the conformation that was proposed by Linus Pauling and Robert B. Corey in 1951 and that was found by Kendrew and his colleagues to be the most common arrangement of the chain in myoglobin. The helixes in lysozyme, however, appear to be somewhat distorted from the "classical" form, in which four atoms (carbon, oxygen, nitrogen and hydrogen) of each peptide group lie in a plane that is parallel to the axis of the alpha helix. In the lysozyme molecule the peptide groups in the helical sections tend to be rotated slightly in such a way that their CO groups point outward from the helix axes and their imino groups (NH) inward.

The amount of rotation varies, being slight in the helix formed by residues 5 through 15 and considerable in the one formed by residues 24 through 34. The effect of the rotation is that each NH group does not point directly at the CO group four residues back along the chain but points instead between the CO groups of the residues three and four back. When the NH group points directly at the CO group four residues back, as it does in the classical alpha helix, it forms with the CO group a hydrogen bond (the weak chemical bond in which a hydrogen atom acts as a bridge). In the lysozyme helixes the hydrogen bond is formed somewhere between two CO groups, giving rise to a structure intermediate between that of an alpha helix and that of a more symmetrical helix with a three-fold symmetry axis that was discussed by Lawrence Bragg, Kendrew and Perutz in 1950. There is a further short length of helix (residues 80 through 85) in which the hydrogen-bonding arrangement is quite close to that in the three-fold helix, and also an isolated turn (residues 119 through 122) of three-fold helix. Furthermore, the peptide at the far end of helix 5 through 15 is in the conformation of the three-fold helix, and the hydrogen bond from its NH group is made to the CO three residues back rather than four.

Partly because of these irregularities in the structure of lysozyme, the proportion of its polypeptide chain in the alpha-helix conformation is difficult to calculate in a meaningful way for comparison with the estimates obtained by other methods, but it is clearly less than half the proportion observed in myoglobin, in which helical regions make up about 75 percent of the chain. The lysozyme molecule does include, however, an example of another regular conformation predicted by Pauling and Corey. This is the "antiparallel pleated sheet," which is believed to be the basic structure of the fibrous protein silk and in which, as the name suggests, two lengths of polypeptide chain run parallel to each other in opposite directions. This structure again is stabilized by hydrogen bonds between the NH and CO groups of the main chain. Residues 41 through 45 and 50 through 54 in the lysozyme molecule form such a structure, with the connecting residues 46 through 49 folded into a hairpin bend between the two lengths of comparatively extended chain. The remainder of the polypeptide chain is folded in irregular ways that have no simple short description.

Even though the level of resolution achieved in our present image was not enough to resolve individual atoms, many of the side chains characteristic of the amino acid residues were readily identifiable from their general shape. The four disulfide bridges, for example, are marked by short rods of high electron density corresponding to the two relatively dense sulfur atoms within them. The six tryptophan residues also were easily recognized by the extended electron density produced by the large double-ring structures in their

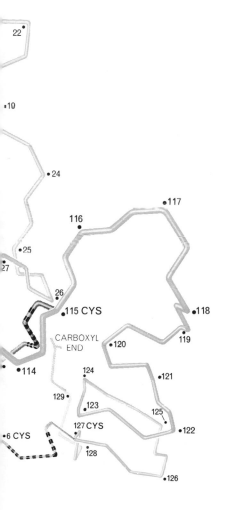

zyme molecule fulfills its function when it cleaves the substrate between the *D* and the *E* ring. Note the distortion of the *D* ring, which pushes four of its atoms into a plane.

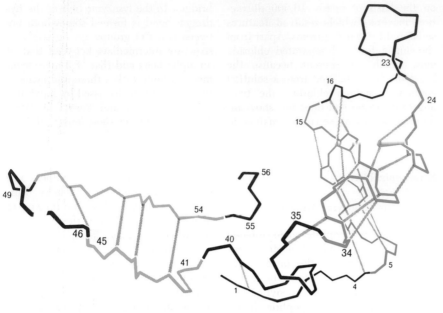

liquid. Such "polar" side chains are hydrophilic—attracted to water; they are found in aspartic acid and glutamic acid residues and in lysine, arginine and histidine residues, which have basic side groups. On the other hand, most of the markedly nonpolar and hydrophobic side chains (for example those found in leucine and isoleucine residues) are shielded from the surrounding liquid by more polar parts of the molecule. In fact, as was predicted by Sir Eric Rideal (who was at one time director of the Royal Institution) and Irving Langmuir, lysozyme, like myoglobin, is quite well described as an oil drop with a polar coat. Here it is important to note that the environment of each molecule in the crystalline state is not significantly different from its natural environment in the living cell. The crystals themselves include a large proportion (some 35 percent by weight) of mostly watery liquid of crystallization. The effect of the surrounding liquid on the protein conformation thus is likely to be much the same in the crystals as it is in solution.

It appears, then, that the observed conformation is preferred because in it the hydrophobic side chains are kept out of contact with the surrounding liquid whereas the polar side chains are generally exposed to it. In this way the system consisting of the protein and the solvent attains a minimum free energy, partly because of the large number of favorable interactions of like groups within the protein molecule and between it and the surrounding liquid, and partly because of the relatively high disorder of the water molecules that are in contact only with other polar groups of atoms.

Guided by these generalizations, many workers are now interested in the possibility of predicting the conforma-

FIRST 56 RESIDUES in lysozyme molecule contain a higher proportion of symmetrically organized regions than does all the rest of the molecule. Residues 5 through 15 and 24 through 34 (*right*) form two regions in which hydrogen bonds (*gray*) hold the residues in a helical configuration close to that of the "classical" alpha helix. Residues 41 through 45 and 50 through 54 (*left*) fold back against each other to form a "pleated sheet," also held together by hydrogen bonds. In addition the hydrogen bond between residues 1 and 40 ties the first 40 residues into a compact structure that may have been folded in this way before the molecule was fully synthesized (*see illustration at the bottom of these two pages*).

side chains. Many of the other residues also were easily identifiable, but it was nevertheless most important for the rapid and reliable interpretation of the image that the results of the chemical analysis were already available. With their help more than 95 percent of the atoms in the molecule were readily identified and located within about .25 angstrom.

Further efforts at improving the accuracy with which the atoms have been located is in progress, but an almost complete description of the lysozyme molecule now exists [*see illustration on pages 172 and 173*]. By studying it and the

results of some further experiments we can begin to suggest answers to two important questions: How does a molecule such as this one attain its observed conformation? How does it function as an enzyme, or biological catalyst?

Inspection of the lysozyme molecule immediately suggests two generalizations about its conformation that agree well with those arrived at earlier in the study of myoglobin. It is obvious that certain residues with acidic and basic side chains that ionize, or dissociate, on contact with water are all on the surface of the molecule more or less readily accessible to the surrounding

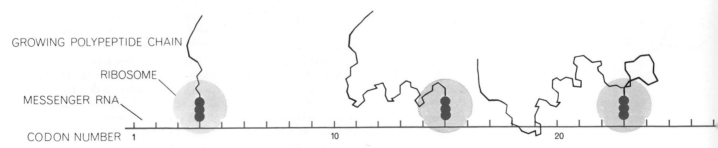

FOLDING OF PROTEIN MOLECULE may take place as the growing polypeptide chain is being synthesized by the intracellular particles called ribosomes. The genetic message specifying the amino acid sequence of each protein is coded in "messenger" ribonucleic acid (RNA). It is believed several ribosomes travel simultaneously along this long-chain molecule, reading the message as they go.

tion of a protein molecule from its chemical formula alone [see "Molecular Model-building by Computer," by Cyrus Levinthal; SCIENTIFIC AMERICAN Offprint 1043]. The task of exploring all possible conformations in the search for the one of lowest free energy seems likely, however, to remain beyond the power of any imaginable computer. On a conservative estimate it would be necessary to consider some 10^{129} different conformations for the lysozyme molecule in any general search for the one with minimum free energy. Since this number is far greater than the number of particles in the observable universe, it is clear that simplifying assumptions will have to be made if calculations of this kind are to succeed.

The Folding of Lysozyme

For some time Peter Dunnill and I have been trying to develop a model of protein-folding that promises to make practicable calculations of the minimum energy conformation and that is, at the same time, qualitatively consistent with the observed structure of myoglobin and lysozyme. This model makes use of our present knowledge of the way in which proteins are synthesized in the living cell. For example, it is well known, from experiments by Howard M. Dintzis and by Christian B. Anfinsen and Robert Canfield, that protein molecules are synthesized from the terminal amino end of their polypeptide chain. The nature of the synthetic mechanism, which involves the intracellular particles called ribosomes working in collaboration with two forms of ribonucleic acid ("messenger" RNA and "transfer" RNA), is increasingly well understood in principle, although the detailed environment of the growing protein chain remains unknown. Nevertheless,

it seems a reasonable assumption that, as the synthesis proceeds, the amino end of the chain becomes separated by an increasing distance from the point of attachment to the ribosome, and that the folding of the protein chain to its native conformation begins at this end even before the synthesis is complete. According to our present ideas, parts of the polypeptide chain, particularly those near the terminal amino end, may fold into stable conformations that can still be recognized in the finished molecule and that act as "internal templates," or centers, around which the rest of the chain is folded [see illustration at bottom of these two pages]. It may therefore be useful to look for the stable conformations of parts of the polypeptide chain and to avoid studying all the possible conformations of the whole molecule.

Inspection of the lysozyme molecule provides qualitative support for these ideas [see top illustration on opposite page]. The first 40 residues from the terminal amino end form a compact structure (residues 1 and 40 are linked by a hydrogen bond) with a hydrophobic interior and a relatively hydrophilic surface that seems likely to have been folded in this way, or in a simply related way, before the molecule was fully synthesized. It may also be important to observe that this part of the molecule includes more alpha helix than the remainder does.

These first 40 residues include a mixture of hydrophobic and hydrophilic side chains, but the next 14 residues in the sequence are all hydrophilic; it is interesting, and possibly significant, that these are the residues in the antiparallel pleated sheet, which lies out of contact with the globular submolecule formed by the earlier residues. In the light of our model of protein fold-

ing the obvious speculation is that there is no incentive to fold these hydrophilic residues in contact with the first part of the chain until the hydrophobic residues 55 (isoleucine) and 56 (leucine) have to be shielded from contact with the surrounding liquid. It seems reasonable to suppose that at this stage residues 41 through 54 fold back on themselves, forming the pleated-sheet structure and burying the hydrophobic side chains in the initial hydrophobic pocket.

Similar considerations appear to govern the folding of the rest of the molecule. In brief, residues 57 through 86 are folded in contact with the pleated-sheet structure so that at this stage of the process—if indeed it follows this course—the folded chain forms a structure with two wings lying at an angle to each other. Residues 86 through 96 form a length of alpha helix, one side of which is predominantly hydrophobic, because of an appropriate alternation of polar and nonpolar residues in that part of the sequence. This helix lies in the gap between the two wings formed by the earlier residues, with its hydrophobic side buried within the molecule. The gap between the two wings is not completely filled by the helix, however; it is transformed into a deep cleft running up one side of the molecule. As we shall see, this cleft forms the active site of the enzyme. The remaining residues are folded around the globular unit formed by the terminal amino end of the polypeptide chain.

This model of protein-folding can be tested in a number of ways, for example by studying the conformation of the first 40 residues in isolation both di-

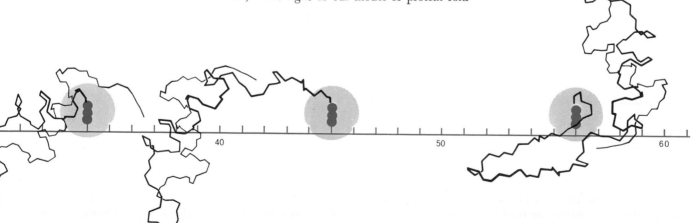

Presumably the messenger RNA for lysozyme contains 129 "codons," one for each amino acid. Amino acids are delivered to the site of synthesis by molecules of "transfer" RNA (*dark color*). The illustration shows how the lysozyme chain would lengthen as a ribosome travels along the messenger RNA molecule. Here, hypothetically, the polypeptide is shown folding directly into its final shape.

rectly (after removal of the rest of the molecule) and by computation. Ultimately, of course, the model will be regarded as satisfactory only if it helps us to predict how other protein molecules are folded from a knowledge of their chemical structure alone.

The Activity of Lysozyme

In order to understand how lysozyme brings about the dissolution of bacteria we must consider the structure of the bacterial cell wall in some detail. Through the pioneer and independent studies of Karl Meyer and E. B. Chain, followed up by M. R. J. Salton of the University of Manchester and many others, the structures of bacterial cell walls and the effect of lysozyme on them are now quite well known. The important part of the cell wall, as far as lysozyme is concerned, is made up of glucose-like amino-sugar molecules linked together into long polysaccharide chains, which are themselves cross-connected by short lengths of polypeptide chain. This part of each cell wall probably forms one enormous molecule—a "bag-shaped macromolecule," as W. Weidel and H. Pelzer have called it.

The amino-sugar molecules concerned in these polysaccharide structures are of two kinds; each contains an acetamido (–NH · CO · CH₃) side group, but one of them contains an additional major group, a lactyl side chain [see illustration below]. One of these amino sugars is known as N-acetylglucosamine (NAG) and the other as N-acetylmuramic acid (NAM). They occur alternately in the

polysaccharide chains, being connected by bridges that include an oxygen atom (glycosidic linkages) between carbon atoms 1 and 4 of consecutive sugar rings; this is the same linkage that joins glucose residues in cellulose. The polypeptide chains that cross-connect these polysaccharides are attached to the NAM residues through the lactyl side chain attached to carbon atom 3 in each NAM ring.

Lysozyme has been shown to break the linkages in which carbon 1 in NAM is linked to carbon 4 in NAG but not the other linkages. It has also been shown to break down chitin, another common natural polysaccharide that is found in lobster shell and that contains only NAG.

Ever since the work of Svante Arrhenius of Sweden in the late 19th century enzymes have been thought to work by forming intermediate compounds with their substrates: the substances whose chemical reactions they catalyze. A proper theory of the enzyme-substrate complex, which underlies all present thinking about enzyme activity, was clearly propounded by Leonor Michaelis and Maude Menton in a remarkable paper published in 1913. The idea, in its simplest form, is that an enzyme molecule provides a site on its surface to which its substrate molecule can bind in a quite precise way. Reactive groups of atoms in the enzyme then promote the required chemical reaction in the substrate. Our immediate objective, therefore, was to find the structure of a reactive complex between lysozyme and its polysaccha-

ride substrate, in the hope that we would then be able to recognize the active groups of atoms in the enzyme and understand how they function.

Our studies began with the observation by Martin Wenzel and his colleagues at the Free University of Berlin that the enzyme is prevented from functioning by the presence of NAG itself. This small molecule acts as a competitive inhibitor of the enzyme's activity and, since it is a part of the large substrate molecule normally acted on by the enzyme, it seems likely to do this by binding to the enzyme in the way that part of the substrate does. It prevents the enzyme from working by preventing the substrate from binding to the enzyme. Other simple amino-sugar molecules, including the trisaccharide made of three NAG units, behave in the same way. We therefore decided to study the binding of these sugar molecules to the lysozyme molecules in our crystals in the hope of learning something about the structure of the enzyme-substrate complex itself.

My colleague Louise Johnson soon found that crystals containing the sugar molecules bound to lysozyme can be prepared very simply by adding the sugar to the solution from which the lysozyme crystals have been grown and in which they are kept suspended. The small molecules diffuse into the protein crystals along the channels filled with water that run through the crystals. Fortunately the resulting change in the crystal structure can be studied quite simply. A useful image of the electron-density changes can be calculated from

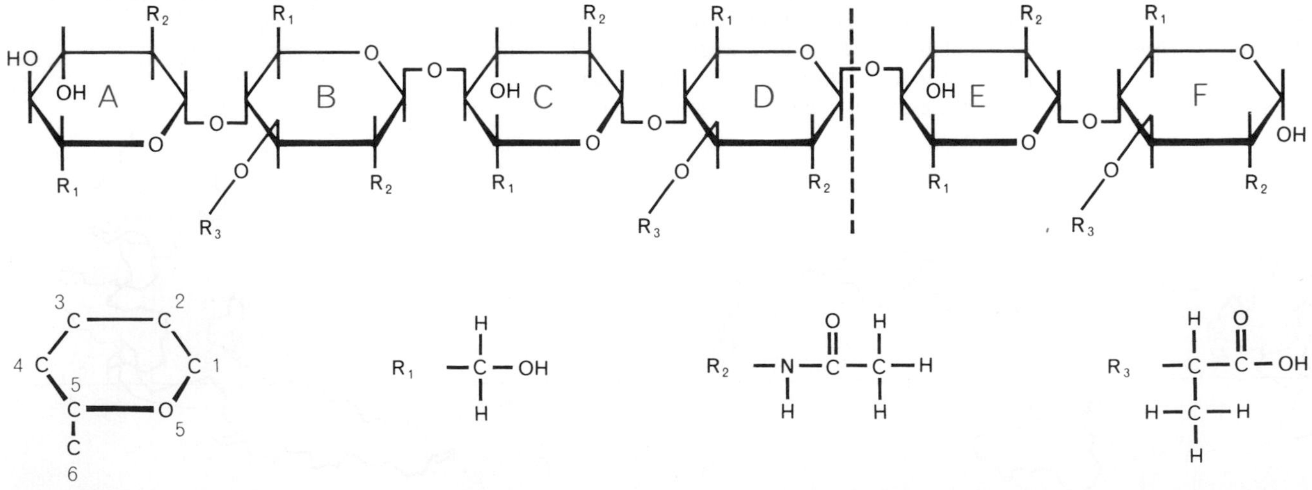

POLYSACCHARIDE MOLECULE found in the walls of certain bacterial cells is the substrate broken by the lysozyme molecule. The polysaccharide consists of alternating residues of two kinds of amino sugar: N-acetylglucosamine (NAG) and N-acetylmuramic acid (NAM). In the length of polysaccharide chain shown here A, C and E are NAG residues; B, D and F are NAM residues. The inset at left shows the numbering scheme for identifying the principal atoms in each sugar ring. Six rings of the polysaccharide fit into the cleft of the lysozyme molecule, which effects a cleavage between rings D and E (see illustration on pages 176 and 177).

measurements of the changes in amplitude of the diffracted waves, on the assumption that their phase relations have not changed from those determined for the pure protein crystals. The image shows the difference in electron density between crystals that contain the added sugar molecules and those that do not.

In this way the binding to lysozyme of eight different amino sugars was studied at low resolution (that is, through the measurement of changes in the amplitude of 400 diffracted waves). The results showed that the sugars bind to lysozyme at a number of different places in the cleft of the enzyme. The investigation was hurried on to higher resolution in an attempt to discover the exact nature of the binding. Happily these studies at two-angstrom resolution (which required the measurement of 10,000 diffracted waves) have now shown in detail how the trisaccharide made of three NAG units is bound to the enzyme.

The trisaccharide fills the top half of the cleft and is bound to the enzyme by a number of interactions, which can be followed with the help of the illustration on pages 176 and 177. In this illustration six important hydrogen bonds, to be described presently, are identified by number. The most critical of these interactions appear to involve the acetamido group of sugar residue C [third from top], whose carbon atom 1 is not linked to another sugar residue. There are hydrogen bonds from the CO group of this side chain to the main-chain NH group of amino acid residue 59 in the enzyme molecule [bond No. 1] and from its NH group to the main-chain CO group of residue 107 (alanine) in the enzyme molecule [bond No. 2]. Its terminal CH_3 group makes contact with the side chain of residue 108 (tryptophan). Hydrogen bonds [No. 3 and No. 4] are also formed between two oxygen atoms adjacent to carbon atoms 6 and 3 of sugar residue C and the side chains of residues 62 and 63 (both tryptophan) respectively. Another hydrogen bond [No. 5] is formed between the acetamido side chain of sugar residue A and residue 101 (aspartic acid) in the enzyme molecule. From residue 101 there is a hydrogen bond [No. 6] to the oxygen adjacent to carbon atom 6 of sugar residue B. These polar interactions are supplemented by a large number of nonpolar interactions that are more difficult to summarize briefly. Among the more important nonpolar interactions, however, are those between sugar residue B and the ring system of residue

62; these deserve special mention because they are affected by a small change in the conformation of the enzyme molecule that occurs when the trisaccharide is bound to it. The electron-density map showing the change in electron density when tri-NAG is bound in the protein crystal reveals clearly that parts of the enzyme molecule have moved with respect to one another. These changes in conformation are largely restricted to the part of the enzyme structure to the left of the cleft, which appears to tilt more or less as a whole in such a way as to close the cleft slightly. As a result the side chain of residue 62 moves about .75 angstrom toward the position of sugar residue B. Such changes in enzyme conformation have been discussed for some time, notably by Daniel E. Koshland, Jr., of the University of California at Berkeley, whose "induced fit" theory of the enzyme-substrate interaction is supported in some degree by this observation in lysozyme.

The Enzyme-Substrate Complex

At this stage in the investigation excitement grew high. Could we tell how the enzyme works? I believe we can. Unfortunately, however, we cannot see this dynamic process in our X-ray images. We have to work out what must happen from our static pictures. First of all it is clear that the complex formed by tri-NAG and the enzyme is not the enzyme-substrate complex involved in catalysis because it is stable. At low concentrations tri-NAG is known to behave as an inhibitor rather than as a substrate that is broken down; clearly we have been looking at the way in which it binds as an inhibitor. It is noticeable, however, that tri-NAG fills only half of the cleft. The possibility emerges that more sugar residues, filling the remainder of the cleft, are required for the formation of a reactive enzyme-substrate complex. The assumption here is that the observed binding of tri-NAG as an inhibitor involves interactions with the enzyme molecule that also play a part in the formation of the functioning enzyme-substrate complex.

Accordingly we have built a model that shows that another three sugar residues can be added to the tri-NAG in such a way that there are satisfactory interactions of the atoms in the proposed substrate and the enzyme. There is only one difficulty: carbon atom 6 and its adjacent oxygen atom in sugar residue D make uncomfortably close contacts

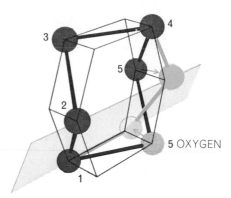

"CHAIR" CONFIGURATION (gray) is that normally assumed by the rings of amino sugar in the polysaccharide substrate. When bound against the lysozyme, however, the D ring is distorted (color) so that carbon atoms 1, 2 and 5 and oxygen atom 5 lie in a plane. The distortion evidently assists in breaking the substrate below the D ring.

with atoms in the enzyme molecule, unless this sugar residue is distorted a little out of its most stable "chair" conformation into a conformation in which carbon atoms 1, 2 and 5 and oxygen atom 5 all lie in a plane [see illustration above]. Otherwise satisfactory interactions immediately suggest themselves, and the model falls into place.

At this point it seemed reasonable to assume that the model shows the structure of the functioning complex between the enzyme and a hexasaccharide. The next problem was to decide which of the five glycosidic linkages would be broken under the influence of the enzyme. Fortunately evidence was at hand to suggest the answer. As we have seen, the cell-wall polysaccharide includes alternate sugar residues of two kinds, NAG and NAM, and the bond broken is between NAM and NAG. It was therefore important to decide which of the six sugar residues in our model could be NAM, which is the same as NAG except for the lactyl side chain appended to carbon atom 3. The answer was clear-cut. Sugar residue C cannot be NAM because there is no room for this additional group of atoms. Therefore the bond broken must be between sugar residues B and C or D and E. We already knew that the glycosidic linkage between residues B and C is stable when tri-NAG is bound. The conclusion was inescapable: the linkage that must be broken is the one between sugar residues D and E.

Now it was possible to search for the origin of the catalytic activity in the neighborhood of this linkage. Our task was made easier by the fact that John A.

Rupley of the University of Arizona had shown that the chemical bond broken under the influence of lysozyme is the one between carbon atom 1 and oxygen in the glycosidic link rather than the link between oxygen and carbon atom 4. The most reactive-looking group of atoms in the vicinity of this bond are the side chains of residue 52 (aspartic acid) and residue 35 (glutamic acid).

One of the oxygen atoms of residue 52 is about three angstroms from carbon atom 1 of sugar residue D as well as from the ring oxygen atom 5 of that residue. Residue 35, on the other hand, is about three angstroms from the oxygen in the glycosidic linkage. Furthermore, these two amino acid residues have markedly different environments. Residue 52 has a number of polar neighbors and appears to be involved in a network of hydrogen bonds linking it with residues 46 and 59 (both asparagine) and, through them, with residue 50 (serine). In this environment residue 52 seems likely to give up a terminal hydrogen atom and thus be negatively charged under most conditions, even when it is in a markedly acid solution, whereas residue 35, situated in a nonpolar environment, is likely to retain its terminal hydrogen atom.

A little reflection suggests that the concerted influence of these two amino

acid residues, together with a contribution from the distortion to sugar residue D that has already been mentioned, is enough to explain the catalytic activity of lysozyme. The events leading to the rupture of a bacterial cell wall probably take the following course [see illustration on this page].

First, a lysozyme molecule attaches itself to the bacterial cell wall by interacting with six exposed amino-sugar residues. In the process sugar residue D is somewhat distorted from its usual conformation.

Second, residue 35 transfers its terminal hydrogen atom in the form of a hydrogen ion to the glycosidic oxygen, thus bringing about cleavage of the bond between that oxygen and carbon atom 1 of sugar residue D. This creates a positively charged carbonium ion (C^+) where the oxygen has been severed from carbon atom 1.

Third, this carbonium ion is stabilized by its interaction with the negatively charged aspartic acid side chain of residue 52 until it can combine with a hydroxyl ion (OH^-) that happens to diffuse into position from the surrounding water, thereby completing the reaction. The lysozyme molecule then falls away, leaving behind a punctured bacterial cell wall.

It is not clear from this description that the distortion of sugar residue D plays any part in the reaction, but in fact it probably does so for a very interesting reason. R. H. Lemieux and G. Huber of the National Research Council of Canada showed in 1955 that when a sugar molecule such as NAG incorporates a carbonium ion at the carbon-1 position, it tends to take up the same conformation that is forced on ring D by its interaction with the enzyme molecule. This seems to be an example, therefore, of activation of the substrate by distortion, which has long been a favorite idea of enzymologists. The binding of the substrate to the enzyme itself favors the formation of the carbonium ion in ring D that seems to play an important part in the reaction.

It will be clear from this account that although lysozyme has not been seen in action, we have succeeded in building up a detailed picture of how it may work. There is already a great deal of chemical evidence in agreement with this picture, and as the result of all the work now in progress we can be sure that the activity of Fleming's lysozyme will soon be fully understood. Best of all, it is clear that methods now exist for uncovering the secrets of enzyme action.

CARBON
OXYGEN
HYDROGEN

SPLITTING OF SUBSTRATE BY LYSOZYME is believed to involve the proximity and activity of two side chains, residue 35 (glutamic acid) and residue 52 (aspartic acid). It is proposed that a hydrogen ion (H^+) becomes detached from the OH group of residue 35 and attaches itself to the oxygen atom that joins rings D and E, thus breaking the bond between the two rings. This leaves carbon atom 1 of the D ring with a positive charge, in which form it is known as a carbonium ion. It is stabilized in this condition by the negatively charged side chain of residue 52. The surrounding water supplies an OH^- ion to combine with the carbonium ion and an H^+ ion to replace the one lost by residue 35. The two parts of the substrate then fall away, leaving the enzyme free to cleave another polysaccharide chain.

21

THE INSULIN MOLECULE

E. O. P. THOMPSON
May 1955

In 1954, after a full decade of intensive work at Cambridge University, Frederick Sanger and his colleagues for the first time totally described the chemical structure of a protein

Proteins, the keystone of life, are the most complex substances known to man, and their chemistry is one of the great challenges in modern science. For more than a century chemists and biochemists have labored to try to learn their composition and solve their labyrinthine structure [see "Proteins," by Joseph S. Fruton; SCIENTIFIC AMERICAN Offprint 10]. In the history of protein chemistry the year 1954 will go down as a landmark, for last year a group of investigators finally succeeded in achieving the first complete description of the structure of a protein molecule. The protein is insulin, the pancreatic hormone which governs sugar metabolism in the body.

Having learned the architecture of the insulin molecule, biochemists can now go on to attempt to synthesize it and to investigate the secret of the chemical activity of this vital hormone, so important in the treatment of diabetes. Furthermore, the success with insulin has paved the way toward unraveling the structure of other proteins with the same techniques, and work on some of them has already begun.

The insulin achievement was due largely to the efforts of the English biochemist Frederick Sanger and a small group of workers at Cambridge University. Sanger had spent 10 years of intensive study on this single molecule. When he commenced his investigation of protein structure in 1944, he chose insulin for several reasons. Firstly, it was one of the very few proteins available in reasonably pure form. Secondly, chemists had worked out a good estimate of its atomic composition (its relative numbers of carbon, hydrogen, nitrogen, oxygen and sulfur atoms). Thirdly, it appeared that the key to insulin's activity as a hormone lay in its structure, for it contained no special components that might explain its specific behavior.

Insulin is one of the smallest proteins. Yet its formula is sufficiently formidable. The molecule of beef insulin (from cattle) is made up of 777 atoms, in the proportions 254 carbon, 377 hydrogen, 65 nitrogen, 75 oxygen and 6 sulfur. Certain general features of the organization of a protein molecule have been known for a long time, thanks to the pioneering work of the German chemist Emil Fischer and others. The atoms form building units called amino acids, which in turn are strung together in long chains to compose the molecule. Of the 24 amino acids, 17 are present in insulin. The total number of amino acid units in the molecule is 51.

Sanger's task was not only to discover the over-all chain configuration of the insulin molecule but also to learn the se-

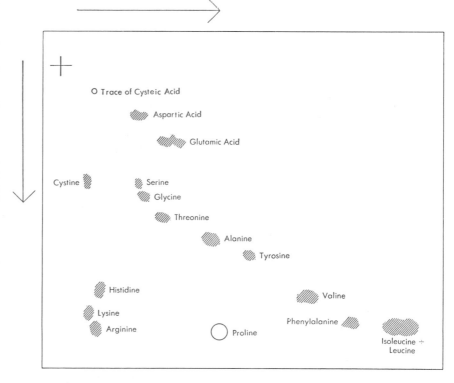

PAPER CHROMATOGRAPHY separates the 17 amino acids of insulin. In the chromatogram represented by this diagram insulin was broken down by hydrolysis and a sample of the mixture placed at the upper left on the sheet of paper. The sheet was hung from a trough filled with solvent which carried each amino acid a characteristic distance down the paper. The sheet was then turned 90 degrees and the process repeated. The amino acids, with the exception of proline, appear as purple spots when sprayed with ninhydrin.

quence of all the amino acids in the chains. The sequence is crucial: a change in the order of amino acids changes the nature of the protein. The number of possible arrangements of the amino acids of course is almost infinite. One can get some notion of the complexity of the protein puzzle by remembering that the entire English language is derived from just 26 letters (two more than the number of amino acids) combined in various numbers and sequences.

Sanger followed the time-honored method used by chemists to investigate large molecules: namely, breaking them down into fragments and then attempting to put the pieces of the puzzle together. A complete breakdown into the amino acid units themselves makes it possible to identify and measure these components. But this gives no clue to how the units are combined and arranged. To investigate the structure a protein chemist shatters the molecule less violently and then examines these larger fragments, consisting of combinations of two, three or more amino acids. The procedure is somewhat like dropping a pile of plates on the floor. The first plate may break into 10 pieces; the second plate may also give 10 pieces but with fractures at different places; the next plate may break into only eight fragments, and so on. Since the sample of protein contains billions of molecules, the experiment amounts to dropping billions of plates. The chemist then pores through this awesome debris for recognizable pieces and other pieces that overlap the breaks to show how the broken sections may be combined.

An amino acid consists of an amino group (NH_3^+), a carboxyl group (COO^-) and a side chain attached to a carbon atom. All amino acids have the amino and carboxyl groups and differ only in their side chains. In a protein molecule they are linked by combination of the carboxyl group of one unit with the amino group of the next. In the process of combination two hydrogen atoms and an oxygen atom drop out in the form of a water molecule and the link becomes CO–NH. This linkage is called the peptide bond. Because of loss of the water molecule, the units linked in the chain are called amino acid "residues." A group of linked amino acids is known as a peptide: two units form a dipeptide, three a tripeptide and so on.

When a peptide or protein is hydrolyzed—treated chemically so that the elements of water are introduced at the peptide bonds—it breaks down into amino acids. The treatment consists in heating the peptide with acids or alkalis. To break every peptide bond and reduce a protein to its amino acids it must be heated for 24 hours or more. Less prolonged or drastic treatment, known as partial hydrolysis, yields a mixture of amino acids, peptides and some unbroken protein molecules. This is the plate-breaking process by which the detailed structure of a protein is investigated.

One of the key inventions that enabled Sanger to solve the jigsaw puzzle was a method of labeling the end amino acid in a peptide. Consider a protein fragment, a peptide, which is composed of three amino acids. On hydrolysis it is found to consist of amino acids A, B and C. The question is: What was their sequence in the peptide? The first member of the three-part chain must have had a free (uncombined) amino (NH_3) group. Sanger succeeded in finding a chemical marker which could be attached to this end of the chain and would stay attached to the amino group after the peptide was hydrolyzed. The labeling material is known as DNP (for dinitrophenyl group). It gives the amino acid to which it is attached a distinctive yellow color. The analysis of the tripeptide sequence proceeds as follows. The tripeptide is treated with the labeling material and is then broken down into its three amino acids. The amino acid which occupied the end position, say B, is now identified by its yellow color. The process is repeated with a second sample of the tripeptide, but this time it is only partly hydrolyzed, so that two amino acids remain as a dipeptide derivative colored yellow. If B is partnered with, say, A in this fragment, one knows that the sequence must be BA, and the order

in the original tripeptide therefore was BAC.

Another tool that played an indispensable part in the solution of the insulin jigsaw puzzle was the partition chromatography method for separating amino acids and peptides, invented by the British chemists A. J. P. Martin and R. L. M. Synge [see "Chromatography," by William H. Stein and Stanford Moore; Scientific American Offprint 81]. Obviously Sanger's method of analysis required separation and identification of extremely small amounts of material. With paper chromatography, which isolates peptides or amino acids in spots on a piece of filter paper, it is possible to analyze a mixture of as little as a millionth of a gram of material with considerable accuracy in a matter of days. As many as 40 different peptides can be separated on a single sheet.

With the knowledge that the insulin molecule was made up of 51 amino acid units, Sanger began his attack on its structure by investigating whether the units were strung in a single long chain or formed more than one chain. Among the components of insulin were three molecules of the amino acid cystine. The cystine molecule is unusual in that it has an amino and a carboxyl group at each end [see its formula in table on page 188]. Since such a molecule could cross-link chains, its presence in insulin suggested that the protein might consist of more than one chain. Sanger succeeded in proving that there were indeed two chains, which he was able to separate intact by splitting the sulfur links in the cystine molecule. Using the DNP labeling technique, he also showed that one chain began with the amino acid glycine and the other with phenylalanine.

Sanger proceeded to break each chain

COMPLETE MOLECULE of insulin is depicted in this structural formula. Each amino acid in the molecule is represented by an abbreviation rather than its complete atomic structure. The key to these abbreviations is in the chart on page 188. The molecule consists

into fragments and study the pieces —especially overlaps which would permit him to build up a sequence. Concentrating on the beginning of the glycine chain, Sanger labeled the glycine with DNP and examined the peptide fragments produced by partial hydrolysis. In the debris of the broken glycine chains he found these sequences attached to the labeled glycine molecules: glycine-isoleucine; glycine-isoleucine-valine; glycine-isoleucine-valine–glutamic acid; glycine-isoleucine-valine–glutamic acid–glutamic acid. Thus it was evident that the first five amino acids in the glycine chain were glycine, isoleucine, valine and two glutamic acids. Similar experiments on the phenylalanine chain established the first four amino acids in that sequence: phenylalanine, valine, aspartic acid and glutamic acid.

Sanger and a colleague, Hans Tuppy, then undertook the immense task of analyzing the structure of the entire phenylalanine chain. It meant breaking down the chain by partial hydrolysis, separating and identifying the many fragments and then attempting to put the pieces of the puzzle together in proper order. The chain, made up of 30 amino acids, was by far the most complex polypeptide on which such an analysis had ever been attempted.

The bewildering mixture of products from partial breakdown of the chain—amino acids, dipeptides, tripeptides, tetrapeptides and so on—was much too complicated to be sorted out solely by chromatography. Sanger and Tuppy first employed other separation methods (electrophoresis and adsorption on charcoal and ion-exchange resins) which divided the peptide fragments into groups. Then they analyzed these simpler mixtures by paper chromatography.

They succeeded in isolating from the fractured chain 22 dipeptides, 14 tripeptides and 12 longer fragments [see chart on pages 186 and 187]. Although these were obtained only in microscopic amounts, they were identified by special techniques and the sequences of their amino acids were determined.

These were the jigsaw pieces that had to be reassembled. Just as in a jigsaw puzzle there are key pieces around which the picture grows, so in this case there were some key pieces as starting points. For instance, the chain was known to contain just one aspartic acid. Six peptides with this amino acid were found in the debris from partial breakdown of the chain [see chart]. The aspartic acid was attached to from one to four other amino acids in these pieces. Their sequences showed that in the original make-up of the chain the order must have been phenylalanine-valine–aspartic acid–glutamic acid–histidine.

Other sequences were pieced together in a similar way until five long sections of the chain were reconstructed. But this still left several gaps in the chain. Sanger and Tuppy now resorted to another method to find the missing links. They split the phenylalanine chain with enzymes instead of by acid hydrolysis. The enzyme splitting process yields longer fragments, and it leaves intact certain bonds that are sensitive to breakage by acid treatment. Thus the investigators obtained long chain fragments which bridged the gaps and revealed the missing links.

After about a year of intensive work Sanger and Tuppy were able to assemble the pieces and describe the structure of insulin's phenylalanine chain. Sanger then turned to the glycine chain and spent another year working out its

structure, with the assistance of the author of this article. The glycine chain is shorter (21 amino acids) but it provided fewer clues: there were fewer key pieces that occurred only once, and two amino acids (glutamic acid and cystine) cropped up in so many of the fragments that it was difficult to place them unequivocally in the sequence.

One detail that remained to be decided before the structure could be completed was the actual composition of two amino acids in the chain. Certain amino acids may occur in two forms: e.g., glutamic acid and glutamine. Glutamic acid has two carboxyl (COO^-) groups, whereas glutamine has an amide ($CONH_2$) group in the place of one of the carboxyls [see page 188]. The difference gives them completely different properties in the protein. Similarly there are aspartic acid and asparagine. Now acid hydrolysis changes glutamine to glutamic acid and asparagine to aspartic acid. Consequently after acid hydrolysis of a protein one cannot tell which form these amino acids had in the original chain. The question was resolved by indirect investigations, one of which involved comparing the products obtained when the same peptide was broken down by acid hydrolysis and by enzymes which do not destroy the amide groups.

By the end of 1952 the two chains were completely assembled. There remained only the problem of determining how the two chains were linked together to form the insulin molecule. But this was easier said than done. As so often happens, what looked simple in theory had complications in practice.

The bridges between the chains, as we have noted, must be cystine, because this amino acid has symmetrical bonds at both ends. The fact that insulin

of 51 amino acid units in two chains. One chain (*top*) has 21 amino acid units; it is called the glycyl chain because it begins with glycine (Gly). The other chain (*bottom*) has 30 amino acid

units; it is called the phenylalanyl chain because it begins with phenylalanine (Phe). The chains are joined by sulfur atoms (S-S). The dotted lines indicate the fragments which located the bridges.

	Phe	Val	Asp	Glu	His	Leu	CySO₃H	Gly	Ser	His	Leu	Val	Glu	Ala	Leu	Tyr
PEPTIDES FROM ACID HYDROLYZATES	Phe	Val		Glu	His		CySO₃H	Gly		His	Leu		Glu	Ala		
		Val	Asp		His	Leu					Leu	Val		Ala	Leu	
			Asp	Glu		Leu	CySO₃H		Ser	His		Val	Glu			
	Phe	Val	Asp			Leu	CySO₃H	Gly				Val	Glu	Ala		Tyr
				Glu	His	Leu			Ser	His	Leu					
		Val	Asp	Glu							Leu	Val	Glu			
					His	Leu	CySO₃H							Ala	Leu	Tyr
	Phe	Val	Asp	Glu					Ser	His	Leu	Val				Tyr
					His	Leu	CySO₃H	Gly			Leu	Val	Glu	Ala		
	Phe	Val	Asp	Glu	His				Ser	His	Leu	Val	Glu			
				Glu	His	Leu	CySO₃H			His	Leu	Val	Glu			
									Ser	His	Leu	Val	Glu	Ala		
SEQUENCES DEDUCED FROM THE ABOVE PEPTIDES	Phe	Val	Asp	Glu	His	Leu	CySO₃H	Gly								Tyr
									Ser	His	Leu	Val	Glu	Ala		
PEPTIDES FROM PEPSIN HYDROLYZATE	Phe	Val	Asp(NH₂)	Glu(NH₂)	His	Leu	CySO₃H	Gly	Ser	His	Leu					
												Val	Glu	Ala	Leu	
					His	Leu	CySO₃H	Gly	Ser	His	Leu					
PEPTIDES FROM CHYMOTRYPSIN HYDROLYZATE	Phe	Val	Asp(NH₂)	Glu(NH₂)	His	Leu	CySO₃H	Gly	Ser	His	Leu	Val	Glu	Ala	Leu	Tyr
PEPTIDES FROM TRYPSIN HYDROLYZATE																
STRUCTURE OF PHENYLALANYL CHAIN OF OXIDIZED INSULIN	Phe	Val	Asp(NH₂)	Glu(NH₂)	His	Leu	CySO₃H	Gly	Ser	His	Leu	Val	Glu	Ala	Leu	Tyr

STRUCTURE OF GLYCYL CHAIN OF OXIDIZED INSULIN

Gly · Ileu · Val · Glu · Glu(NH₂) · CySO₃H · CySO₃H · Ala · Ser · Val · CySO₃H · Ser · Leu · Tyr ·

SEQUENCE OF AMINO ACIDS in the phenylalanyl chain was deduced from fragments of the chain. The entire sequence is at the bottom above the dotted line. Each fragment is indicated by a horizontal sequence of amino acids joined by dots. The fragments are arranged so that each of their amino acids is in the vertical column above the corresponding amino acid in the entire chain.

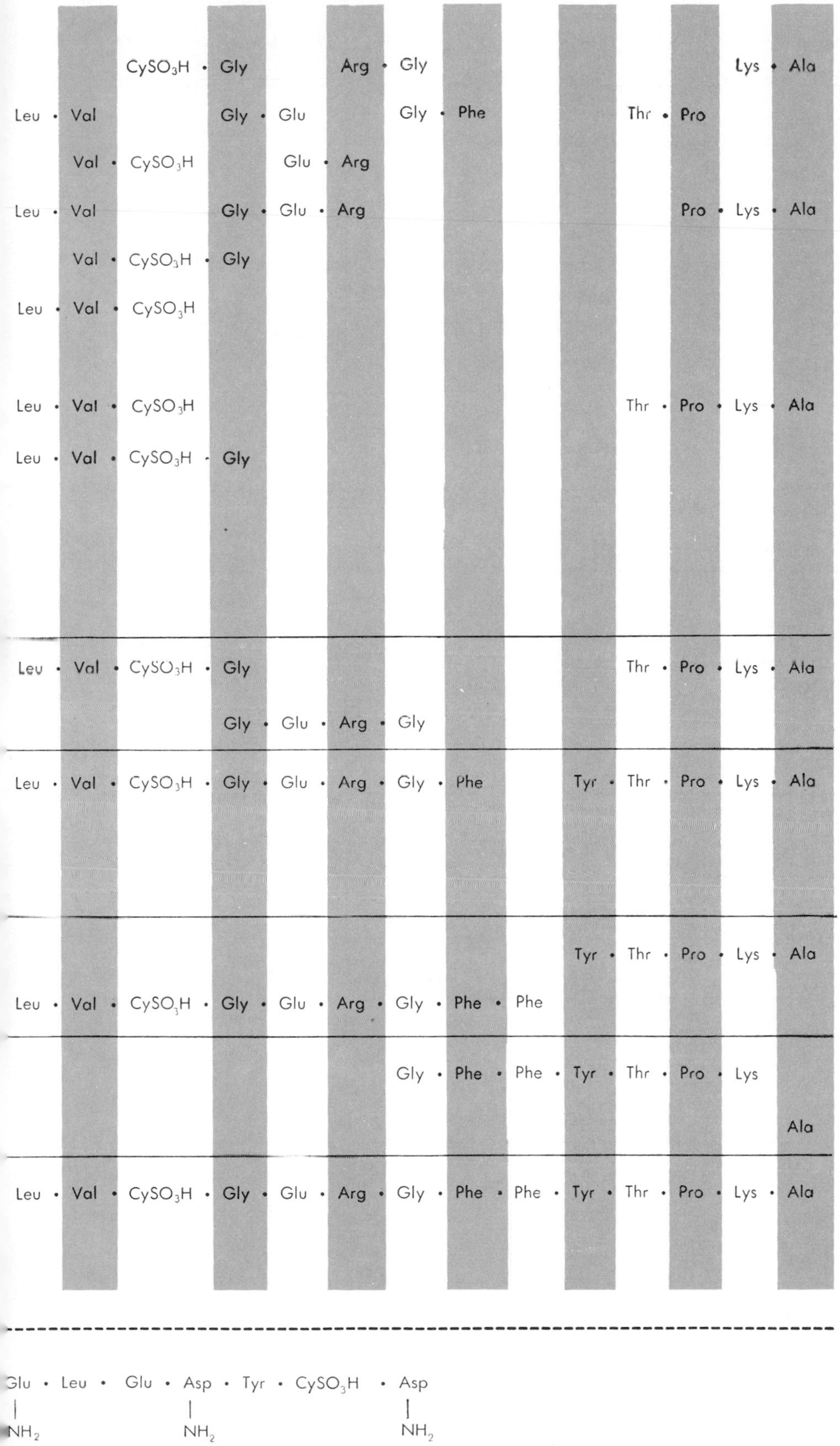

CySO₃H · Gly   Arg · Gly   Lys · Ala

Leu · Val   Gly · Glu   Gly · Phe   Thr · Pro

Val · CySO₃H   Glu · Arg

Leu · Val   Gly · Glu · Arg   Pro · Lys · Ala

Val · CySO₃H · Gly

Leu · Val · CySO₃H

Leu · Val · CySO₃H   Thr · Pro · Lys · Ala

Leu · Val · CySO₃H · Gly

Leu · Val · CySO₃H · Gly   Thr · Pro · Lys · Ala

Gly · Glu · Arg · Gly

Leu · Val · CySO₃H · Gly · Glu · Arg · Gly · Phe   Tyr · Thr · Pro · Lys · Ala

Tyr · Thr · Pro · Lys · Ala

Leu · Val · CySO₃H · Gly · Glu · Arg · Gly · Phe · Phe

Gly · Phe · Phe · Tyr · Thr · Pro · Lys

Ala

Leu · Val · CySO₃H · Gly · Glu · Arg · Gly · Phe · Phe · Tyr · Thr · Pro · Lys · Ala

- -

Glu · Leu · Glu · Asp · Tyr · CySO₃H · Asp
| | |
NH₂   NH₂   NH₂

The shorter fragments (*group at the top*) were obtained by hydrolyzing insulin with acid. The longer fragments (*groups third, fourth and fifth from the top*) were obtained with enzymes. The same method was used to deduce the sequence in the glycyl chain (*bottom*).

contains three cystine units suggested that there might be three bridges, or cross-links, between the chains. It appeared that it should be a simple matter to locate the positions of the bridges by a partial breakdown of the insulin molecule which gave cystine-containing fragments with sections of the two chains still attached to the "bridge" ends.

When Sanger began this analysis, he was puzzled to find that the cystine-containing peptides in his broken-down mixtures showed no significant pattern whatever. Cystine was joined with other amino acids in many different combinations and arrangements, as if the chains were cross-linked in every conceivable way. Sanger soon discovered the explanation: during acid hydrolysis of the insulin molecule, cystine's sulfur bonds opened and all sorts of rearrangements took place within the peptides. Sanger and his associate A. P. Ryle then made a systematic study of these reactions and succeeded in finding chemical inhibitors to prevent them.

By complex analyses which employed both acid hydrolysis and enzyme breakdown, Sanger and his co-workers L. F. Smith and Ruth Kitai eventually fitted the bridges into their proper places and obtained a complete picture of the structure of insulin [*see diagram at bottom of pages 184 and 185*]. So for the first time the biochemist is able to look at the amino-acid arrangement in a protein molecule. The achievement seems astounding to those who were working in the field 10 years ago.

To learn how insulin's structure determines its activity as a hormone is still a long, hard road. It will be difficult to synthesize the molecule, but once that has been accomplished, it will be possible to test the effect of changes in the structure on the substance's physiological behavior. Evidently slight variations do not affect it much, for Sanger has shown that the insulins from pigs, sheep and steers, all equally potent, differ slightly in structure.

The methods that proved so successful with insulin, plus some newer ones, are already being applied to study larger proteins. Among the improvements are promising new techniques for splitting off the amino acids from a peptide chain one at a time—clearly a more efficient procedure than random hydrolysis. The rate of progress undoubtedly will be speeded up as more biochemists turn their attention to the intriguing problem of relating the structure of proteins to their physiological functions.

FORMULA	NAME	ABBREVIATION	PHENYLALANYL	GLYCYL
$CH_2(NH_3{}^+) \cdot COO^-$	Glycine	Gly	3	1
$CH_3{-}CH(NH_3 \cdot COO^-$	Alanine	Ala	2	1
$CH_2OH{-}CH(NH_3{}^+) \cdot COO^-$	Serine	Ser	1	2
$CH_3 \cdot CHOH{-}CH(NH_3{}^+) \cdot COO^-$	Threonine	Thr	1	0
$\begin{array}{c}CH_3\\ \diagdown\\ CH{-}CH(NH_3{}^+) \cdot COO^-\\ \diagup\\ CH_3\end{array}$	Valine	Val	3	2
$\begin{array}{c}CH_3\\ \diagdown\\ CH \cdot CH_2{-}CH(NH_3{}^+) \cdot COO^-\\ \diagup\\ CH_3\end{array}$	Leucine	Leu	4	2
$\begin{array}{c}CH_3 \cdot CH_2\\ \diagdown\\ CH{-}CH(NH_3{}^+) \cdot COO^-\\ \diagup\\ CH_3\end{array}$	Isoleucine	Ileu	0	1
$\begin{array}{c}CH_2{-}CH_2\\ |\quad\quad |\\ CH_2\quad CH{-}COO^-\\ \diagdown\ \diagup\\ NH^+\end{array}$	Proline	Pro	1	0
$\begin{array}{c}CH{=}CH\\ \diagup\quad\quad\diagdown\\ CH\quad\quad\quad C \cdot CH_2{-}CH(NH_3{}^+) \cdot \cdot COO^-\\ \diagdown\quad\quad\diagup\\ CH{-}CH\end{array}$	Phenylalanine	Phe	3	0
$\begin{array}{c}CH{=}CH\\ \diagup\quad\quad\diagdown\\ HO{-}C\quad\quad C \cdot CH_2{-}CH(NH_3{}^+) \cdot COO^-\\ \diagdown\quad\quad\diagup\\ CH{-}CH\end{array}$	Tyrosine	Tyr	2	2
$NH_2CO \cdot CH_2{-}CH(NH_3{}^+) \cdot COO^-$	Asparagine	Asp (NH_2)	1	2
$COOH \cdot CH_2 \cdot CH_2{-}CH(NH_3{}^+) \cdot COO^-$	Glutamic Acid	Glu	2	2
$NH_2 \cdot CO \cdot CH_2 \cdot CH_2{-}CH(NH_3{}^+) \cdot COO^-$	Glutamine	Glu (NH_2)	1	2
$\begin{array}{c}NH_2{-}C{-}NH \cdot CH_2 \cdot CH_2 \cdot CH_2{-}CH(NH_3{}^+) \cdot COO^-\\ ||\\ NH\end{array}$	Arginine	Arg	1	0
$\begin{array}{c}CH{=}C \cdot CH_2{-}CH(NH_3{}^+) \cdot COO^-\\ |\quad\quad |\\ NH\quad N\\ \diagdown\ //\\ CH\end{array}$	Histidine	His	2	0
$CH_2NH_2 \cdot CH_2 \cdot CH_2 \cdot CH_2{-}CH(NH_3{}^+) \cdot COO^-$	Lysine	Lys	1	0
$\begin{array}{c}COO^-\quad\quad\quad\quad\quad\quad\quad COO^-\\ \diagdown\quad\quad\quad\quad\quad\quad\quad\diagup\\ CH{-}CH_2{-}S{-}S{-}CH_2{-}CH\\ \diagup\quad\quad\quad\quad\quad\quad\quad\diagdown\\ NH_3{}^+\quad\quad\quad\quad\quad\quad\quad NH_3{}^+\end{array}$	Cystine	CyS | CyS	2	4
			30	21

AMINO ACIDS of insulin are listed in this chart. Their chemical formulas are at the left. The dots in the formulas represent chemical bonds other than those suggested by the atoms adjacent to each other. The number of amino acid units of each kind found in the phenylalanyl chain are listed in the fourth column of the chart. The fifth column comprises a similar listing for the glycyl chain.

22

THE STRUCTURE AND HISTORY
OF AN ANCIENT PROTEIN

RICHARD E. DICKERSON
April 1972

To oxidize food molecules all organisms from yeasts to man require a variant of cytochrome c. Differences in this protein from species to species provide a 1.2-billion-year record of molecular evolution

Between 1.5 and two billion years ago a profound change took place in some of the single-celled organisms then populating our planet, a change that in time would contribute to the rise of many-celled organisms. The machinery evolved for extracting far more energy from foods than before by combining food molecules with oxygen. One of the central components of the new metabolic machinery was cytochrome *c*, a protein whose descendants can be found today in every living cell that has a nucleus. By studying the cytochrome *c* extracted from various organisms it has been possible to determine how fast the protein has evolved since plants and animals diverged into two distinct kingdoms and in fact to provide an approximate date of 1.2 billion years ago for the event. For example, the cytochrome *c* molecules in men and chimpanzees are exactly the same: in the cells of both the molecule consists of 104 amino acid units strung together in exactly the same order and folded into the same three-dimensional structure. On the other hand, the cytochrome *c* in man has diverged from the cytochrome *c* in the red bread mold *Neurospora crassa* in 44 out of 104 places, yet the three-dimensional structures of the two cytochrome *c* molecules are essentially alike. We think we can now explain how it is that so many of the 104 amino acid units in cytochrome *c* are interchangeable and also why certain units cannot be changed at all without destroying the protein's activity.

Let us try to visualize the earth before cytochrome *c* first appeared. The first living organisms on the planet were little more than scavengers, extracting energy-rich organic compounds (including their neighbors) from the water around them and releasing low-energy breakdown products. We still have the "fossils" of this life-style in the universal process of anaerobic (oxygenless) fermentation, as when a yeast extracts energy from sugar and releases ethyl alcohol, or when an athlete who exercises too rapidly converts glucose to lactic acid and gets muscle cramps. Anaerobic fermentation is part of the common biochemical heritage of all living things.

The upper limit on how much life the planet could support with only fermentation as an energy source was determined by the rate at which high-energy compounds were synthesized by nonbiological agencies: ultraviolet radiation, lightning discharges, radioactivity or heat. When some organisms developed the ability to tap sunlight for energy, photosynthesis was born and the life-carrying capacity of the earth increased enormously. This was the age of the bacteria and the blue-green algae [see "The Oldest Fossils," by Elso S. Barghoorn; SCIENTIFIC AMERICAN Offprint 895].

The more advanced forms of photosynthesis released a corrosive and poisonous gas into the atmosphere: oxygen. Some bacteria responded by retreating to oxygen-free corners of the planet, where their descendants are found today. Other bacteria and blue-green algae developed ways to neutralize gaseous oxygen by combining it with their own waste products. The next step was to harness the energy released by oxidation of these waste compounds. (If you are going to burn your garbage, you might as well keep warm by the fire.) This was the beginning of oxidation, or respiration, the second big breakthrough in increasing the supply of energy available to life on the earth.

When a yeast cell oxidizes sugars all the way to carbon dioxide and water instead of stopping short at ethyl alcohol, it gets 19 times as much energy per gram of fuel. When oxygen combines with lactic acid in the athlete's muscles and the cramps are dissipated, he receives a correspondingly greater energy return from his glucose. Any improvement in metabolism that multiplies the supply of energy available by such a large factor would be expected to have a revolutionary effect on the development of life. We now believe the specialization of cells and the appearance of multicelled plants and animals could only have come about in the presence of such a large new supply of energy.

Bacteria and blue-green algae are prokaryotes (prenuclear cells); their genetic material, DNA, is not confined within an organized nucleus, and their respiratory and photosynthetic machinery (if it is present) is similarly dispersed. Green algae and all the higher plants and animals are eukaryotes (cells with "good" nuclei); their DNA is organized within a nucleus, and their respiration is carried out in the organelles called mitochondria. In eukaryote plants photosynthesis is conducted in still other organelles called chloroplasts. Mitochondria are the powerhouse of all eukaryote cells. Their role is to break down the energy-rich molecules obtained from foods, combine them with oxygen and store the energy produced by harnessing it to synthesize molecules of adenosine triphosphate (ATP). The mitochondria of all eukaryotes are alike in their chemistry, as if once the optimum chemical mechanism had been arrived at it was never changed.

Biological oxidation involves at least a score of special enzymes that act first as acceptors and then as donors of the electrons or hydrogen atoms removed from food molecules. In the last part of the process one finds a series of cytochrome molecules (identified by various subscript letters), all of which incorporate a heme group containing iron, the same heme group found in hemoglobin. Electrons are passed down a chain of cytochrome molecules: from cytochrome b to cytochrome c_1, from cytochrome c_1 to cytochrome c, to cytochromes a and a_3 and finally to oxygen atoms, where they are combined with hydrogen ions to produce water. This is a stepwise process designed to release energy in small parcels rather than all at once. In the transfer of electrons from cytochrome b to cytochrome c_1 and again in the transfer between the cytochromes a, a_3 and oxygen, energy is channeled off to synthesize ATP, which acts as a general-purpose energy source for cell metabolism.

Most of the cytochromes are bound tightly to the mitochondrial membrane, but one of them, cytochrome c, can easily be solubilized in aqueous mediums and can be isolated in pure form. The other components can be isolated as multienzyme complexes: b and c_1 as a cytochrome reductase complex, and a and a_3 as a cytochrome oxidase. The reductase donates electrons to cytochrome c; the oxidase accepts them again. To illustrate how similar all eukaryotes really are to one another, it has been found that cytochrome c from any species of plant, animal or eukaryotic microorganism can react in the test tube with the cytochrome oxidase from any other species. Worm or primate, whale or wheat are all alike under the mitochondrial membrane.

The Evolution of Cytochrome c

Since cytochrome c is so ancient and at the same time so small and easily purified, it has received much attention from protein chemists interested in the evolutionary process. The complete amino acid sequence of cytochrome c has been determined for more than 40 species of eukaryotic life. Thirty-eight of these sequences are compared in the illustration on pages 192 and 193. We have more information on the evolution of this molecule than on the evolution of any other protein.

Emanuel Margoliash of Northwestern University and Emil Smith of the University of California at Los Angeles were among the first to notice that the amino acid sequences from various species are different and that the degree of difference corresponds quite well with

the distance that separates the two species on the evolutionary tree. Detailed computer analyses of these differences by Margoliash, by Walter Fitch of the University of Wisconsin and by others have led to the construction of elaborate family trees of living organisms entirely without recourse to the traditional anatomical data. The family trees agree remarkably well with those obtained from classical morphology; it is obvious that comparison of amino acid sequence is a powerful tool for studying the process of evolution.

Another result may at first be surprising. Cytochrome c is still evolving slowly and is doing so at a rate that is approximately constant for all species, when the rate is averaged over geological time periods. This kind of analysis of molecular evolution was first carried out on hemoglobin a decade ago by Linus Pauling and Emile Zuckerkandl at the California Institute of Technology. If we compare hemoglobin and cytochrome c, we find that cytochrome c is changing much more slowly. Why should this be? The protein chains are synthesized from instructions that are embodied in DNA, and it is in the DNA that mutations take place. Do mutations occur more often in the DNA that makes hemoglobin than in the DNA that makes cytochrome c? There is no reason to think so. The explanation therefore must

lie in the natural-selection, or screening, process that tests whether or not mutant molecules can do their job.

Before discussing the various "formulas" that have passed the test of making a successful cytochrome c, I shall describe briefly the structure of proteins. All protein molecules are built up by linking amino acids end to end. Each of the 20 different amino acids has a carboxyl group (COOH) at one end and an amino group ($-NH_2$) at the other. To link the carboxyl group of one amino acid with the amino group of another amino acid a molecule of water must be removed, producing an amide linkage ($-CO-NH-$). Because only a part (although, to be sure, the distinctive part) of an amino acid enters a protein chain, the chemist refers to it as a "residue." Thus he speaks of a glycine residue or a phenylalanine residue at such-and-such a position in a protein chain.

The carbon adjacent to the amide linkage is called the alpha carbon. It is important because each amino acid has a distinctive side chain at this position. The side chain may be nothing more than a single atom of hydrogen (as it is in the case of the amino acid glycine) or it may consist of a number of atoms, including a six-carbon "aromatic" ring (as it does in the case of phenylalanine, tryptophan and tyrosine).

The 20 amino acids can be grouped

SKELETON OF CYTOCHROME c MOLECULE is depicted in the illustration by Irving Geis on the opposite page. A variant of this protein molecule is found in the cells of every living organism that utilizes oxygen for respiration. The illustration shows in simplified form how 104 amino acid units are linked in a continuous chain that grips and surrounds a heme group, a complex rosette with an atom of iron (*Fe*) at its center. The picture is color-coded to indicate how much variation has been tolerated by evolution at each of the 104 amino acid sites in the molecule. Some species lack the 104th amino acid, and all species except vertebrates have as many as eight extra amino acids at the beginning of the chain (*see table on pages 192 and 193*). The amino acids that are most invariant throughout evolution, and presumably the most important, are shown in red and orange; the more variable sites appear in yellow-green, blue-green, blue and purple. The indispensable heme group is crimson. Each amino acid is represented only by its "alpha" carbon atom: the atom that carries a side chain unique for each of the 20 amino acids. The upper drawing at left below shows how two amino acids link up through an amide group (*colored panel*); the side chains connected to the alpha carbons (*color*) are represented by the balls labeled *R*. The lower drawing at left below shows the scheme used in the cytochrome c skeleton on the opposite page; all amide linkages ($-CO-NH-$) are omitted and the only side groups shown are those that are attached to the heme. The amino acids at the 35 invariant sites of the cytochrome c molecule (*red*) are designated in abbreviated form (*see key at right below*).

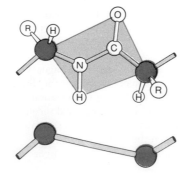

Ala	Alanine	Leu	Leucine
Asp	Aspartic acid	Lys	Lysine
Asn	Asparagine	Met	Methionine
Arg	Arginine	Phe	Phenylalanine
Cys	Cysteine	Pro	Proline
Gly	Glycine	Ser	Serine
Glu	Glutamic acid	Thr	Threonine
Gln	Glutamine	Trp	Tryptophan
His	Histidine	Tyr	Tyrosine
Ile	Isoleucine	Val	Valine

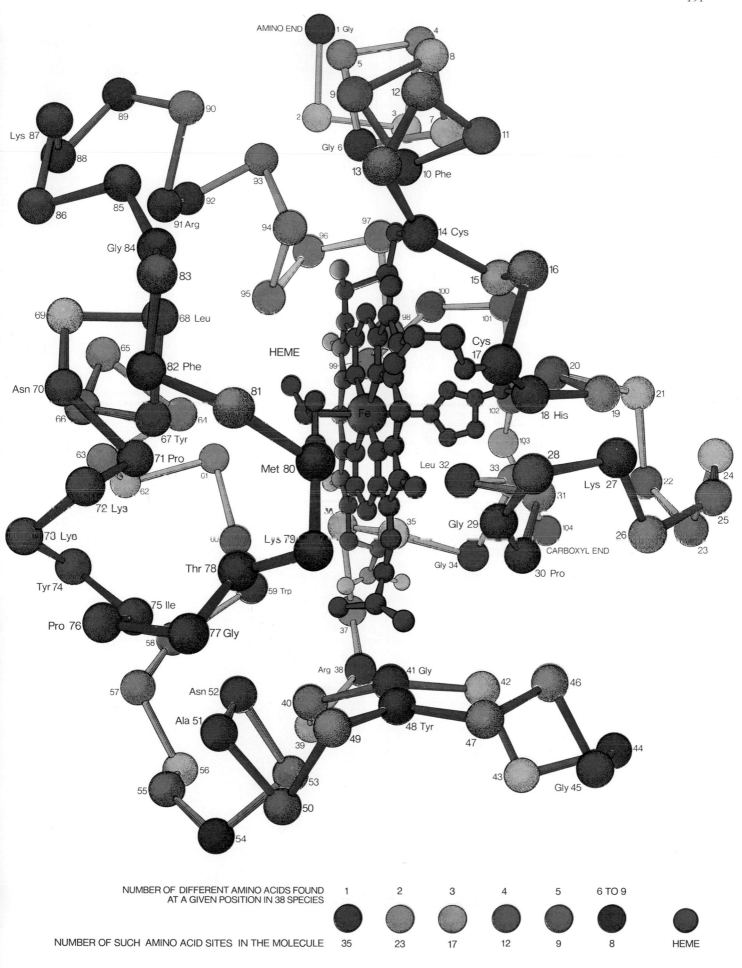

AMINO END 1 Gly

Lys 87

Gly 6

13 10 Phe

14 Cys

16

15

Cys 17

20

18 His 19 21

28

Lys 27 22

24

25

23

26

30 Pro

CARBOXYL END 104

Gly 29

33 31

Leu 32

103

102

100 101

98

97

94 96

95

93

92

91 Arg

Gly 84

83

68 Leu

69

65

82 Phe

81

Asn 70

66

67 Tyr

64

63

71 Pro

61

62

72 Lys

73 Lys

60

Met 80

Lys 79

Tyr 74

75 Ile

Thr 78

59 Trp

Pro 76

58

77 Gly

57

Ala 51

Asn 52

39

40

49

48 Tyr

47

43

Gly 45

41 Gly

42 46

44

Arg 38

37

35

36

34

53

50

54

55

56

85

86

88

89 90

99

Fe

HEME

Gly 34

NUMBER OF DIFFERENT AMINO ACIDS FOUND
AT A GIVEN POSITION IN 38 SPECIES

1   2   3   4   5   6 TO 9

NUMBER OF SUCH AMINO ACID SITES IN THE MOLECULE

35   23   17   12   9   8   HEME

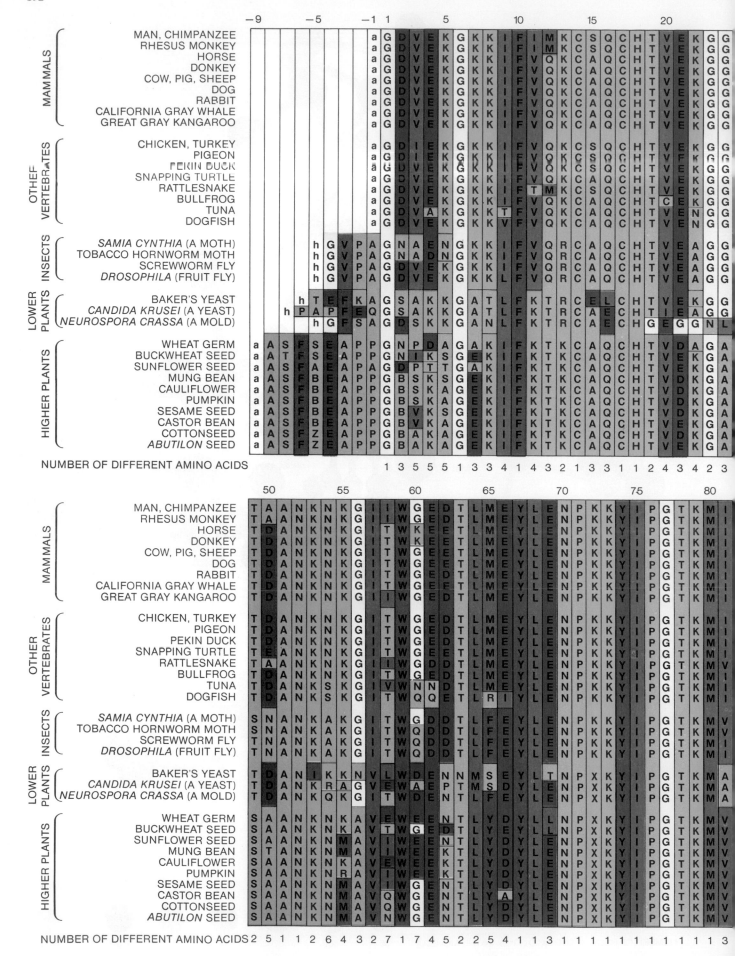

NUMBER OF DIFFERENT AMINO ACIDS

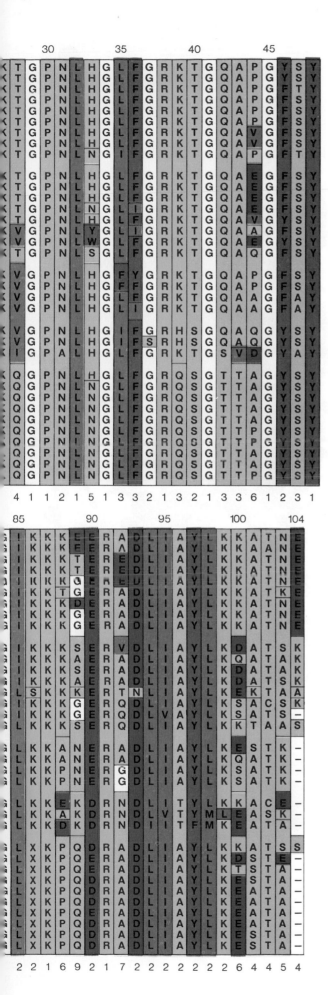

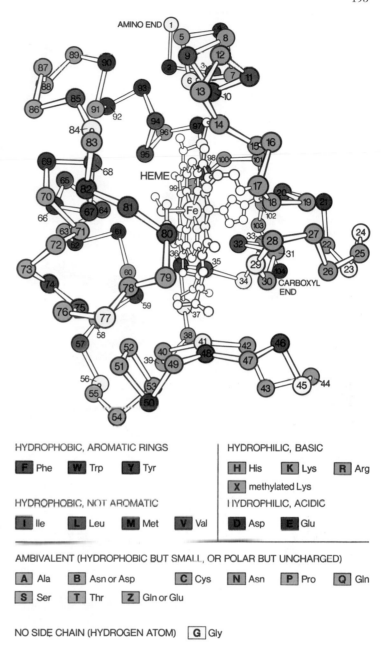

HYDROPHOBIC, AROMATIC RINGS

F Phe **W** Trp **Y** Tyr

HYDROPHILIC, BASIC

H His **K** Lys **R** Arg

X methylated Lys

HYDROPHOBIC, NOT AROMATIC

I Ile **L** Leu **M** Met **V** Val

HYDROPHILIC, ACIDIC

D Asp **E** Glu

AMBIVALENT (HYDROPHOBIC BUT SMALL, OR POLAR BUT UNCHARGED)

A Ala **B** Asn or Asp **C** Cys **N** Asn **P** Pro **Q** Gln

S Ser **T** Thr **Z** Gln or Glu

NO SIDE CHAIN (HYDROGEN ATOM) **G** Gly

COMPOSITION OF CYTOCHROME *c* IN 38 SPECIES is presented in the table at left. No other protein has been so fully analyzed for so many different organisms. The color code used here differs from the one used in the molecular skeleton on page 191. On these two pages color is employed to classify amino acids according to their chemical properties (*see key directly above*). Thus the three "oily" (hydrophobic) amino acids with aromatic benzene rings in their side chains (phenylalanine, tryptophan and tyrosine) are shown in red. Four other amino acids that are hydrophobic but nonaromatic are shown in orange. At the other extreme, amino acids that are hydrophilic, or water-loving, are shown in blue or violet. Amino acids that can be found in either aqueous or nonaqueous environments (and hence are ambivalent) are green or yellow. Polar amino acids can have asymmetric distributions of positive and negative charge. The detailed structure of the side chains of the amino acids can be found on page 195. It is easy to pick out from the table at left the amino acid sites where evolution has allowed no change or has allowed substitution only by chemically similar amino acids; these sites are identified by vertical bands of a single color. A letter *a* at the beginning of the chain indicates that a methyl group (CH₃) is attached to the amino end of the molecular chain. A letter *h* indicates that the methyl group is absent.

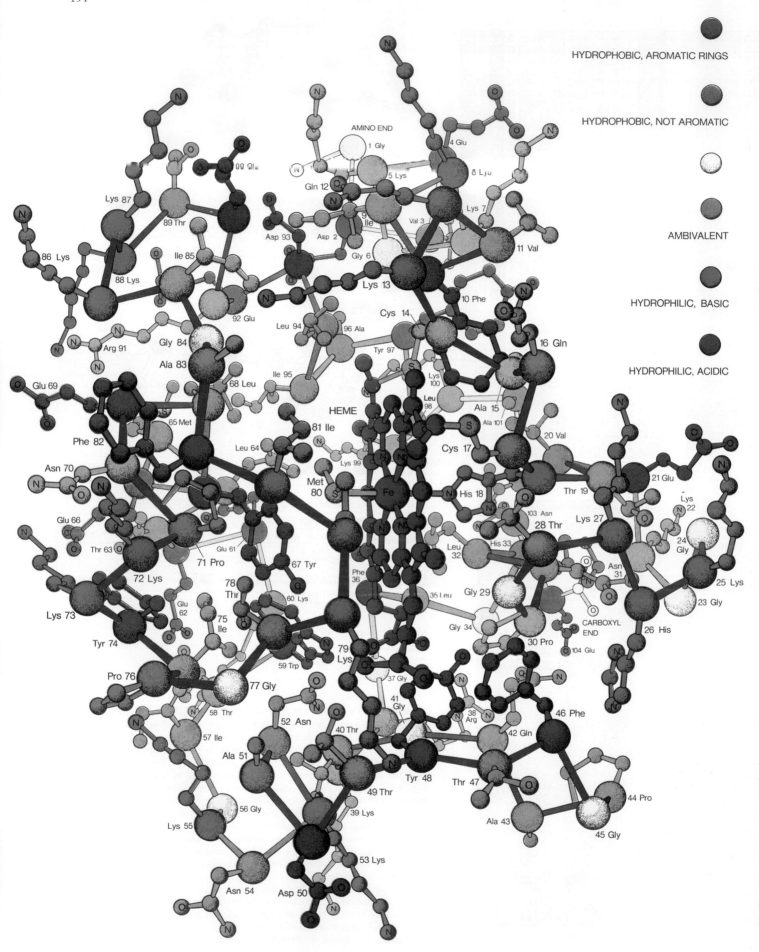

HYDROPHOBIC, AROMATIC RINGS

HYDROPHOBIC, NOT AROMATIC

AMBIVALENT

HYDROPHILIC, BASIC

HYDROPHILIC, ACIDIC

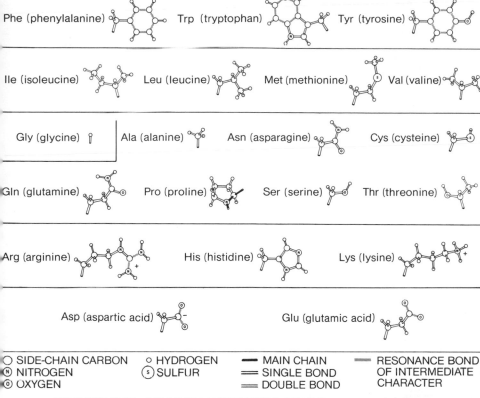

○ SIDE-CHAIN CARBON	○ HYDROGEN	━━ MAIN CHAIN	━━ RESONANCE BOND
⊗ NITROGEN	⑤ SULFUR	═══ SINGLE BOND	OF INTERMEDIATE
◎ OXYGEN		═══ DOUBLE BOND	CHARACTER

CYTOCHROME c MOLECULE WITH SIDE CHAINS appears in the illustration on the opposite page. The picture shows the structure of horse-heart cytochrome c in the oxidized state as determined through X-ray crystallography by the author and his colleagues at the California Institute of Technology. Alpha-carbon atoms are numbered and the amide groups (–CO–NH–) connecting the alpha carbons are represented only by a solid bond, as in the preceding drawings. For clarity three side chains at the "back" of the molecule have been left out: leucine 35, phenylalanine 36 and leucine 98. The color coding follows the coding in the illustration on pages 192 and 193. One can see from the three-dimensional structure that side chains in the interior of the molecule, around the heme group (crimson), tend to be hydrophobic (red and orange), whereas amino acids with hydrophilic side chains (blue and violet) are found on the outside, where they are ordinarily in contact with water. A major exception to this rule is the hydrophobic side chain of phenylalanine 82, which sits on the surface of the molecule at the left of the heme. The region between the hydrophilic chains, and above isoleucine 81, is a cavity that is apparently open to solvent molecules. Lysine 13, above this cavity, is known to interact with a large oxidase complex when cytochrome c is oxidized. The structures of the side chains of 20 amino acids appear above.

into three broad classes, depending on the character of their side chains [see illustration appearing above]. Five are hydrophilic, or water-loving, and tend to acquire either a positive or a negative charge when placed in aqueous solution; three of the five are basic in character (arginine, histidine and lysine) and the other two are acidic (aspartic acid and glutamic acid). Seven are not readily soluble in water and hence are termed hydrophobic; they include the three amino acids mentioned above that have rings in their side chains plus leucine, isoleucine, methionine and valine. The remaining eight amino acids react ambivalently to water: alanine, asparagine, cysteine, glutamine, glycine, proline, serine and threonine.

Now let us see how much the successful formulas for cytochrome c differ from species to species. The cytochrome c molecules of men and horses differ by 12 out of 104 amino acids. The cytochrome c's of the higher vertebrates—mammals, birds and reptiles—differ from the cytochrome c's of fishes by an average of 19 amino acids. The cytochrome c's of vertebrates and insects differ by an average of 27 amino acids; moreover, the cytochrome c molecules of insects and plants have a few more amino acid residues at the beginning of the chain than the equivalent molecules of vertebrates. The greatest disparity between two cytochrome c's is the one between man and the bread mold Neurospora; they differ at more than 40 percent of their amino acid positions. How can two molecules with such large differences in amino acid composition perform identical chemical functions?

We begin to see an answer when we look at where these changes are. Some parts of the amino acid sequence, as indicated on pages 192 and 193, never vary. Thirty-five of the 104 amino acid positions in cytochrome c are completely invariant in all known species, including a long sequence from residue 70 through residue 80. The 35 invariant sites are occupied by 15 different amino acids; they are shown in red in the structural drawing on page 191. Another 23 sites are occupied by only one of two different but closely similar amino acids. There are 18 different sets of interchangeable pairs at these 23 sites; they are shown in orange in the illustration. At 17 sites natural selection has evidently accepted only sets of three different amino acids; these 17 interchangeable triplets are colored yellow-green.

It was already known from sequence studies, before the X-ray structural analysis, that where such substitutions are allowed the interchangeable amino acids almost always have the same chemical character. In general all must be either hydrophilic or hydrophobic or else neutral with respect to water. Such interchanges are called conservative substitutions because they conserve the overall chemical nature of that part of the protein molecule.

In only a few places along the chain can radical changes be tolerated. Residue 89, for example, can be acidic (aspartic acid or glutamic acid), basic (lysine), polar but uncharged (serine, threonine, asparagine and glutamine), weakly hydrophobic (alanine) or devoid of a side chain (glycine). Almost the only type of side chain that appears to be forbidden at this point in the molecule is a large hydrophobic one. Such "indifferent" regions are rare, however, and cytochrome c overall is an evolutionarily conservative molecule.

We have no reason to think the gene for cytochrome c mutates more slowly than the gene for hemoglobin, or that the invariant, conservative and radical regions of the sequence reflect any difference in mutational rate within the cytochrome c gene. The mutations are presumably random, and what we see in these species comparisons are the molecules that are left after the rigid test of survivability has been applied. Invariant regions evidently are invariant because any mutational changes there are lethal and are weeded out. Conservative changes can be tolerated elsewhere as long as they preserve the essential chemical properties of the molecule at that point. Radical changes presumably indicate portions of the molecule that do not matter for the operation of the protein.

This is as far as we can go from sequence comparisons alone. The explanation of variability in terms of the essential or nonessential character of different parts of the molecule is plausible, yet science has always been plagued by plausible but incorrect hypotheses. To progress any further we need to know how the amino acid sequence is folded to make an operating molecule. In short, we need the three-dimensional structure of the protein.

The Molecule in Three Dimensions

With the active collaboration of Mar-

goliash, who was then working at the Abbott Laboratories in North Chicago, I began the X-ray-crystallographic analysis of horse-heart cytochrome c at Cal Tech in 1963, with the sponsorship of the National Science Foundation and the National Institutes of Health. As cytochrome c transfers electrons in the mitochondrion, it oscillates between an oxidized form (ferricytochrome) and a reduced form (ferrocytochrome); the iron atom in the heme group is alternately in the +3 and +2 oxidation state. We decided to begin our analysis with the oxidized form, a decision that was largely tactical since both oxidation states would ultimately be needed if we were to try to decipher the electron-

transfer process.

In X-ray crystallography one directs a beam of X rays at a purified crystal of the substance under study and records the diffraction pattern produced as the beam strikes the sample from different angles. X rays entering the sample are themselves deflected at various angles by the distribution of electron charges within the crystal. Highly sophisticated computer programs have been devised for deducing from tens of thousands of items of X ray diffraction data the three-dimensional distribution of electronic charge. From this distribution one can infer, in turn, the distribution of the amino acid side chains in the protein molecule.

We obtained our first low-resolution map of the oxidized form of horse-heart cytochrome c five years ago and the first high-resolution map three years ago. These maps have been used to construct detailed three-dimensional models of the protein. One can also feed the three-dimensional coordinates into a computer and obtain simple ball-and-stick drawings that can be viewed stereoptically, enabling one to visualize the folded chain of the protein in three dimensions [see illustrations, pages 197 and 198]. Just a year ago we calculated the first high-resolution map for the reduced form of cytochrome c. We are now improving this model and comparing the two oxidation states.

Several striking features of the amino acid sequences of cytochrome c were in the back of our minds as we worked out the first high-resolution structure. We knew that the most strongly conserved sites throughout evolution were those occupied by three distinctive types of residue: the positively charged (basic) residues of lysine; the three hydrophobic and aromatic residues of phenylalanine, tryptophan and tyrosine, and the four hydrophobic but nonaromatic residues of leucine, isoleucine, methionine and valine. These sites can now be located with the help of the illustration on page 194, whose color coding differs from the coding of the illustration on page 191. Here hydrophobic residues are shown in warm colors (red and orange) whereas neutral residues and hydrophilic residues, both basic and acidic, are shown in cool colors (green, yellow, blue and violet).

It had been known from the chemical analysis of the molecule's amino acid sequence that the basic and hydrophobic groups tend to appear in clusters along the chain. For example, basic residues are found in the regions of sites 22 through 27, sites 38 and 39, sites 53

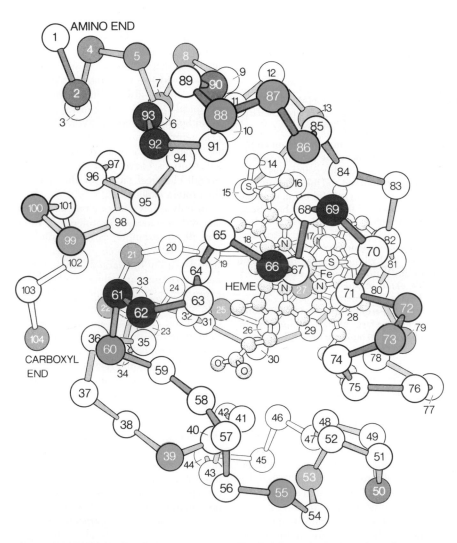

PLOT OF DISTRIBUTION OF ELECTRIC CHARGES on the back of horse-heart cytochrome c reveals that most of the 19 hydrophilic lysines (color), which carry positive charges (and hence are basic), are distributed on the two flanks of the molecule. Nine of 12 negatively charged (acidic) side chains (gray) are clustered in one zone in the upper center of the molecule. The electrically negative character of this zone has been maintained throughout evolution, although the specific locations of the acidic side groups vary. No organism, from wheat germ to man, has a cytochrome c with fewer than six acidic amino acids in this zone and no organism has more than five acidic amino acids everywhere else on the molecule. Furthermore, these extreme values are not found in the same species. It is highly likely that these charged zones participate in binding cytochrome c to other large molecules.

through 55 and sites 86 through 91. Hydrophobic residues are found in regions 9 through 11, 32 through 37, 80 through 85 and 94 through 98. The residues at sites 14 and 17 (cysteine) and site 18 (histidine) are invariant, which is understandable since they form bonds to the heme group. Less understandably, the long stretch from site 70 to site 80 is equally invariant. Before the structural evidence was available it had been suspected that methionine, at site 80, might be bonded to the iron atom on the other side of the heme from the histidine at site 18, but it was impossible to be sure from chemical evidence alone.

It was also known from chemical analysis that horse cytochrome c incorporates 12 glycines (the residues with only hydrogen as a side chain) and that these glycines were either invariant or else conserved in the great majority of species. It was known too that of the eight phenylalanines or tyrosines (with aromatic rings in their side chains) seven are either invariant in all species or replaceable only by one another. In the case of residue 36, phenylalanine or tyrosine is replaced in three species by isoleucine, whose side chain, although it is nonaromatic, is at least as large and hydrophobic as the side chains it replaces.

All these similarities and conservatisms were known before the X-ray analysis, but none could be explained in terms of structure. It was assumed that every residue had been placed where it was by natural selection and that it contained potentially important information about the working parts of the cytochrome molecule. Natural selection, however, does not act on an amino acid sequence but rather on the folded and operating molecule in its association with other biological molecules. Having a sequence without the folding instructions is like having a list of parts without a blueprint of the entire machine.

Cytochrome c and Evolution

Now that the blueprint for cytochrome c is revealed, let us look more closely at its representation on page 191. To keep the illustration simple no side chains have been included except for those that are bonded to the heme group. Moreover, along the main chain the illustration depicts only the alpha-carbon atoms from which side chains would, if they were shown, branch off. The amide groups (–CO–NH–) that connect alpha carbons are represented simply by straight lines. The picture is therefore a simplified folding diagram of

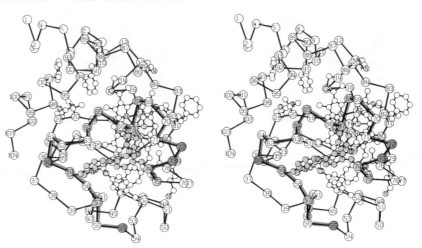

STEREOSCOPIC PAIR OF LEFT SIDE of oxidized cytochrome c molecule, drawn by computer, shows only a few key side chains for clarity. The main chain (color) from sites 55 to 75 defines a loop, the "left channel," which is filled with strongly hydrophobic side chains; their alpha carbons are in light color. Three of these side chains include aromatic rings: tryptophan 59, tyrosine 67 and tyrosine 74, also shown in light color. Alpha carbons with hydrophilic, positively charged side chains around the left channel are shown in dark color. This pair and one on following page can be viewed with standard stereoscopic viewer.

the cytochrome molecule.

We see that the flat heme group, a symmetrical rosette of carbon and nitrogen atoms with an atom of iron at its center, sits in a crevice with only one edge exposed to the outside world. If the heme participates directly in shuttling electrons in and out of the molecule, the transfer probably takes place along this edge. Cysteines 14 and 17 and histidine 18 hold the heme in place from the right as depicted, and the other heme-binding group on the left is indeed methionine 80, as had been suspected.

It was known from earlier X-ray studies of proteins that sequences of amino acids frequently fold themselves into the helical configuration known as the alpha helix; in other cases the amino acids tend to assume a rippled or corrugated configuration called a beta sheet. Cytochrome c has no beta sheets and only two stretches of alpha helix, formed by residues 1 through 11 and 89 through 101. For the most part the protein chain is wrapped tightly around the heme group, leaving little room for the alpha and beta configurations that are prominent in other proteins.

Just as one can use cytochrome c to learn about evolution, one can also use evolution to learn about cytochrome c. As I have noted, the illustration on page 191 is color-coded to indicate the amount of variability in the kind of amino acid tolerated at each site. The structure is "hot" (red and orange) in the functionally important places in the molecule when differences among species are absent or rare, and it is "cool" (green, blue

and violet) in regions that vary widely from one species to another and thus are presumably less important to a viable molecule of cytochrome c.

The heme crevice is hot, indicating that strong selection pressures tend to keep the environment of the heme group constant throughout evolution. The invariant residues 70 through 80 are also hot, and we now see that they are folded to make the left side of the molecule and the pocket in which the heme sits. The right side of the molecule is warm, consisting of sites where only one, two or three different amino acids are tolerated. The back of the molecule is its cool side; residues 58 and 60 and four more residues on the back of the alpha helix are each occupied by six or more different amino acids in various species. These are powerful clues to the important parts of the molecule, whether for electron transfer or for interaction with two large molecular complexes, the reductase and the oxidase.

How the Molecule Folds Itself

If we now turn to the illustration on page 194, which shows all the side chains of horse-heart cytochrome c, many of the evolutionary conservatisms become understandable. (As before, amide groups are still shown only as straight lines; their atomic positions are known but are not particularly relevant to this article.) In this illustration the colors are selected to classify the various sites according to the character of the amino acid tolerated (hydrophilic, hydrophobic or ambivalent); the same color coding applies in

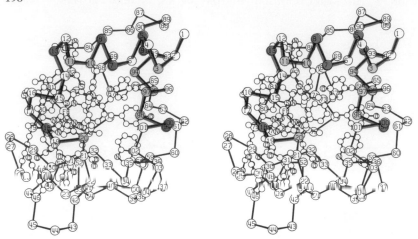

STEREOSCOPIC PAIR OF RIGHT SIDE of oxidized cytochrome *c* shows two sequences forming alpha helixes: the sequence from 1 through 11 and the sequence from 89 through 101. The two alpha helixes and the chain from 12 through 20 outline the right channel. Like the left channel it is lined with hydrophobic side groups, but it apparently contains a slot large enough to receive a hydrophobic side chain from another molecule. As in the stereoscopic drawing on the preceding page, alpha carbons with positively charged side chains around this channel are indicated in dark color; alpha carbons with strongly hydrophobic side chains are indicated in light color. The computer program for preparing the stereoscopic pictures was written by Carroll Johnson of the Oak Ridge National Laboratory.

the illustration on pages 192 and 193 showing the amino acid sequences in the cytochrome *c*'s of 38 different species.

Nonpolar, hydrophobic groups are found predominantly on the inside of the molecule, away from the external aqueous world, whereas charged groups, acidic or basic, are always on the outside. This arrangement is a good example of the "oil drop" model of a folded protein. According to this model, when an amino acid chain is synthesized inside a cell, it is helped to fold in the proper way by the natural tendency of hydrophobic, or "oily," side chains to retreat as far as possible from the aqueous environment and cluster in the center of the molecule. An even stronger statement can be made: If it is necessary for the successful operation of a protein molecule that certain portions of the polypeptide chain be folded into the interior, then natural selection will favor the retention of hydrophobic side chains at that point so that the proper folding is achieved. A charged, or hydrophilic, side chain can be pushed into the interior of a protein molecule, but a considerable price must be paid in terms of energy. Thus in most cases the presence of a charged group at a given site helps to ensure that the chain at that point will be on the outside of the folded molecule. (Charged groups inside a protein are known only in one or two cases where they play a role in the catalytic mechanism of the protein.)

We can now see the reason for the evolutionary conservatism of hydrophobic side chains, and one of the reasons for the conservatism of the hydrophilic

residue lysine: they help to make the molecule fold properly. Radical changes of side chain that prevent proper folding are lethal. No folding, no cytochrome; no cytochrome, no respiration; no respiration, no life. It is seldom that cause and effect in evolution are quite so clear-cut.

There is still more to the lysine story. The lysines are not only on the outside; they are clustered in two positively charged regions of the molecular surface, separated by another zone of negative charge. This segregation of charge has not been found in any other protein structure and, as we shall see, probably occurs because cytochrome *c* interacts with two molecular complexes (the reductase and the oxidase) rather than with small substrate molecules as an enzyme does. The charge arrangements are believed to be part of the process by which large molecules recognize each other.

Most of the 19 lysines are found on the left and right sides of the molecule, viewed from the back on page 196. The left and right sides of the molecule can be examined separately in the two stereoscopic pairs on pages 197 and 198. On the left side eight lysines surround a loop of chain from sites 55 through 75 that is tightly packed with hydrophobic groups, including the invariant tyrosine 74, tryptophan 59 and tyrosine 67 farther inside. Although we do not yet know the electron-transfer mechanism, it has been suggested that the aromatic rings of the three invariant residues could provide an inward path for the electron when cytochrome *c* is reduced. Another

eight lysines are found on the right side, on the periphery of what appears to be a true channel large enough to hold a hydrophobic side chain of another large molecule. This right-side channel is bounded by the two alpha helixes and by the continuation from the first alpha helix through residues 12 to 20. Within this channel are found two large aromatic side chains: phenylalanine 10 (which cannot change) and tyrosine 97 (which can also be phenylalanine but nothing else). In summary, on the right side is a channel lined with hydrophobic groups (including two aromatic rings) and surrounded by an outer circle of positive charges. As someone in our laboratory remarked on looking at the model, it resembles a docking ring for a spaceship.

This remark may not be entirely frivolous. It is known from chemical work that the attraction between cytochrome *c* and the cytochrome oxidase complex is largely electrostatic, involving negatively charged groups on the oxidase and positively charged basic groups on cytochrome *c*. Either the left or the right cluster of lysines must be involved in this binding. Moreover, Kazuo Okunuki of the University of Osaka has shown that if just one positive charge, lysine 13, is blocked with a bulky aromatic chemical group, the reactivity of cytochrome with its oxidase is cut in half. Chemically blocking lysine 13 means physically blocking the upper part of the heme crevice. Lysine 13 is closer to the right cluster of positive charges than to the left; thus it would appear more likely that the heme crevice and the right channel together are the portions of the molecular surface that "see" the oxidase complex.

What, then, are the roles of the positive zone on the left side and the negative patch at the rear? The positive zone, with its three aromatic rings, may be the binding site to the reductase; we know virtually nothing about the chemical nature of this binding. The negative patch may be a "trash dump," an unimportant part of the molecule's surface where there are enough negative charges to prevent an excessively positive overall charge. The fact that the six most variable amino acid sites are in this part of the molecule would support such an idea.

On the other hand, it is equally possible that this collection of negative charges has a function. Acidic amino acids are actually conserved throughout the various species, although in a subtle way that was overlooked in the earlier sequence comparisons. Selection pressures have kept this zone of the molecular surface negative, even though the individual residues that carry the negative charges differ from one species to another. Because several sections of the protein chain bend into and out of this acidic region, the conservation of negative charge is not immediately obvious if one looks only at the stretched-out sequence. This is a good illustration of the principle of molecular evolution that natural selection acts on the folded, functioning protein and not on its amino acid sequence alone.

If we look carefully at where the glycine residues are, we can appreciate why such a large number are evolutionarily invariant. The heme group is so large that 104 amino acids are barely enough to wrap around it. There are many places where a chain comes too close to the heme or to another chain for a side chain to fit in. It is just at these points that we find the glycines with their single hydrogen atom as a side chain.

The last type of conservatism, the conservatism of the aromatic side chains, is more difficult to explain. Tyrosines and phenylalanines tend to occur in nearby pairs in the folded cytochrome c molecule: residues 10 and 97 in the right channel, 46 and 48 below the heme crevice, 67 and 74 along with tryptophan 59 in the hydrophobic left channel. Only residue 36, which can be tyrosine, phenylalanine or isoleucine, seems to have merely a space-filling role on the back of the molecule; it is an "oily brick."

The three aromatic rings in the left channel may be involved in electron transfer during reduction. The two rings in the right channel could also be employed in electron transfer, or might only help to define the hydrophobic slot in the middle of that channel. Tyrosine 48 at the bottom of the molecule helps to hold the heme in place by making a hydrogen bond to one of the heme's propionic acid side chains. In cytochrome c from the tuna and the bonito, where residue 46 is tyrosine, electron-density maps have shown that this residue also holds the heme by a hydrogen bond to its other propionic acid group. These two tyrosines, along with cysteines 14 and 17, help to lock the heme in place in a way not seen in hemoglobin or myoglobin.

Phenylalanine 82 is the enigma. Never tyrosine or anything else, it extends its oily side chain out into the aqueous world on the left side of the heme crevice, where it has no visible role. A price must be paid in energy for its being

there. Why should such a large hydrophobic group be on the outside of the molecule, and why should it be absolutely unchanging through the entire course of evolution? Viewing the oxidized molecule alone, it is impossible to say, but when at the end of this article we look briefly at the recently revealed structure of reduced cytochrome, we shall see the answer fall into place at once.

The structural reasons for the evolutionary conservatism in cytochrome c throughout the history of eukaryotic life can now largely be explained. Cytochrome c is unique among the structurally analyzed proteins in that it has segregated regions of charge on its surface. The roles assigned to these regions in the foregoing discussions have been speculative and may be quite wrong. What we can be sure of is that these regions do have roles in the operation of the molecule. Chance alone, or even common ancestry, could not maintain these positive and negative regions, along with paired and exposed aromatic groups, in all species through more than a billion years of molecular evolution. The conservative sequences are shouting to us, "Look!" Now we have to be clever enough to know what to look for.

Rates of Protein Evolution

With this background we are equipped to return to a question raised earlier: What determines the rates of evolution of different proteins? We begin by making a graph where the vertical axis represents the average difference in amino acid sequence between two species of organism on two sides of an evolutionary branch point, for example the branch point between fish and reptiles or between reptiles and mammals. The horizontal axis represents the time elapsed since the divergence of the two lines as determined by the geological record. If such a graph is plotted for cytochrome c, one finds that all the branch points fall close to a straight line, indicating a constant average rate of evolutionary change [see illustration on following page].

How can this be? How can cytochrome c change at so nearly a constant rate during the long period in which the external morphology of the organism was diversifying toward the present-day cotton plant, bread mold, fruit fly, rattlesnake and chimpanzee? This is an illustration of a fundamental advantage of proteins as tools in studying evolution. Natural selection ultimately operates on populations of whole living organisms, the only criterion of success being the ability of the population to survive, reproduce and leave behind a new generation. The farther down toward the molecular level one goes in examining living organisms, the more similar they become and the less important the morphological differences are that separate a clam from a horse. One kind of chemical machinery can serve many diverse organisms. Conversely, one external change in an organism that can be acted on by natural selection is usually the effect not of a single enzyme molecule but of an entire set of metabolic pathways.

The observed uniform rate of change in cytochrome c simply means that the biochemistry of the respiratory package, the mitochondrion, is so well adjusted, and the mitochondrion is so well insulated from natural selection, that the selection pressures become smoothed out at the molecular level over time spans of millions of years. A factory can convert from making military tanks to making sports cars and keep the same machine tools and power source. Similarly, a primitive eukaryote cell line can lead to such diverse organisms as sunflowers and mammals and still retain a common metabolic chemistry, including the respiratory package that comprises cytochrome c. One of the advantages of proteins in studying the process of evolution is just this relative insulation from the immediate effects of external selection. Protein structure is farther removed from selection pressures and closer to the sources of genetic variation in DNA than gross anatomical features or inherited behavior patterns are.

The only other proteins for which enough sequence information is available to allow this kind of analysis are hemoglobin and the fibrinopeptides: the short amino acid chains left over when fibrinogen is converted to fibrin in the process of blood clotting. One hemoglobin chain consists of approximately 140 amino acids. Fibrinopeptides A and B, on the other hand, consist of only about 20 amino acids, which are cut out of fibrinogen and discarded during the clotting process. The hemoglobins and the fibrinopeptides also appear to be evolving individually at a uniform average rate, but their rates are quite different. Whereas 20 million years are required to produce a change of 1 percent in the amino acid sequence of two diverging lines of cytochrome c, the same amount of change takes a little less than six million years in hemoglobins and just over one million years in the fibrinopeptides, as indicated on page 200. The

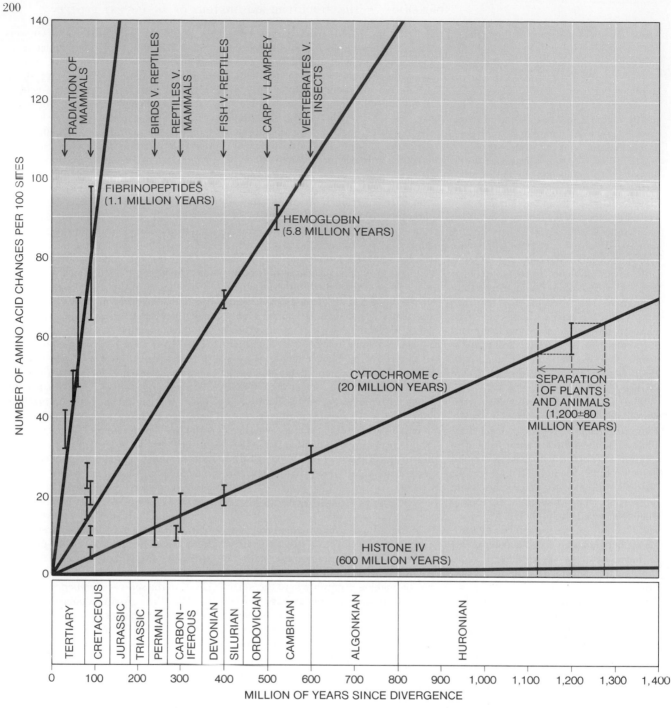

200

NUMBER OF AMINO ACID CHANGES PER 100 SITES

RADIATION OF MAMMALS

BIRDS V. REPTILES

REPTILES V. MAMMALS

FISH V. REPTILES

CARP V. LAMPREY

VERTEBRATES V. INSECTS

FIBRINOPEPTIDES (1.1 MILLION YEARS)

HEMOGLOBIN (5.8 MILLION YEARS)

CYTOCHROME c (20 MILLION YEARS)

SEPARATION OF PLANTS AND ANIMALS (1,200±80 MILLION YEARS)

HISTONE IV (600 MILLION YEARS)

TERTIARY | CRETACEOUS | JURASSIC | TRIASSIC | PERMIAN | CARBON–IFEROUS | DEVONIAN | SILURIAN | ORDOVICIAN | CAMBRIAN | ALGONKIAN | HURONIAN

0   100   200   300   400   500   600   700   800   900   1,000   1,100   1,200   1,300   1,400

MILLION OF YEARS SINCE DIVERGENCE

FIBRINOPEPTIDES

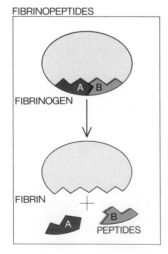

FIBRINOGEN

FIBRIN

+

A B PEPTIDES

GLOBINS

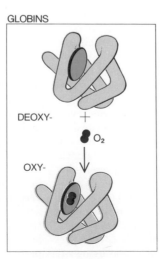

DEOXY-   +

O₂

OXY-

CYTOCHROME c

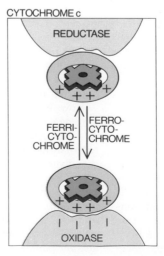

REDUCTASE

FERRI-CYTO-CHROME   FERRO-CYTO-CHROME

OXIDASE

HISTONE IV

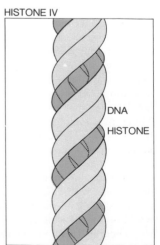

DNA

HISTONE

approximate time required for a 1 percent change in sequence to appear between diverging lines of the same protein is defined as the unit evolutionary period. That period has been roughly estimated for a number of proteins for which only two or three sequences from different species are known. Most simple enzymes evolve approximately as fast as hemoglobin and much more rapidly than cytochrome c. Does all of this mean that the genes for these proteins are mutating at different rates? Are we looking at differences in variation or in selection?

Since there is no evidence to suggest variable rates of mutation, one asks what case can be made for differences in selection pressure among the different proteins. The case appears to be quite convincing [see inset figures in illustration on page 200]. The fibrinopeptides are "spacers" that prevent fibrinogen from adopting the fibrin configuration before the clotting mechanism is triggered. As long as they can be cut out by an enzyme when the time comes for the blood to clot, they would seem to have few other requirements. Thus one would expect a fibrinogen molecule to tolerate many random changes in the fibrinopeptide spacers. If the unit evolutionary period measures not the rate of appearance of mutations but the rate of appearance of harmless mutations, then it is not surprising that a 1 percent change can occur in the sequence of fibrinopeptides in just over a million years.

A successful hemoglobin molecule has more constraints. Each hemoglobin molecule embodies four heme groups that not only bind oxygen but also cooperate in such a way that the oxygen is released more rapidly into the cell when the local acidity, created by the presence of carbon dioxide, builds up. The structural basis for this "breathing" mechanism has only recently been explained with the help of X-ray crystallography by M. F. Perutz and his co-workers at the Medical Research Council Laboratory of Molecular Biology in England. If a random mutation is five times as likely to be harmful in hemoglobin as in the fibrinopeptides, one can account for hemoglobin's having a unit evolutionary period that is five times as long.

The chances of randomly damaging cytochrome c are evidently three to four times greater than they are for hemoglobin. Why should this be, and why should the unit evolutionary period for cytochrome c be greater than the period for enzymes of comparable size? The X-ray structure has given us a clue to the answer. Cytochrome c is a small protein that interacts over a large portion of its surface with molecular complexes that are larger than itself. It is virtually a "substrate" for the reductase and oxidase complexes. A large fraction of its surface is subject to strong conservative selection pressures because of the requirement that it mate properly with other large molecules, each with its own genetic blueprint. This evidently explains why the patches of positive and negative charge are preserved so faithfully throughout the history of eukaryotic life. Hemoglobin and most enzymes, in contrast, interact principally with smaller molecules: with oxygen in the case of hemoglobin or with small substrate molecules at the active sites of enzymes. As long as these restricted regions of the molecule are preserved the rest of the molecular surface is relatively free to change. Mutations are weeded out less rigorously and sequences diverge faster.

A satisfying confirmation of these ideas comes from the amino acid sequences of histone IV, one of the basic proteins that binds to DNA in the chromosomes and that may play a role in expressing or suppressing genetic information. When molecules of histone IV from pea seedlings and calf thymus are compared, one finds that they differ in only two of their 102 amino acids. If we adopt an approximate date of 1.2 billion years ago from the cytochrome study for the divergence of plants and animals, we find that histone IV has a unit evolutionary period of 600 million years. Clearly the conservative selection pressure on histone IV must be intense. Since histone IV participates in the control processes that are at the heart of the genetic mechanism, its sensitivity to random changes is hardly surprising.

The date of 1.2 billion years ago for the divergence of plants and animals is based on the cytochrome-sequence comparisons, assuming that the observed linear rate of evolution of cytochrome c in more recent times can be extrapolated back to that remote epoch. Is this a fair extrapolation? It probably is for cytochrome c because the biochemistry of the mitochondrion evolved still earlier; the great similarity in respiratory reactions among all eukaryotes argues that there has not been much innovation in cytochrome systems since. The respiratory chain had probably "settled down" by 1.2 billion years ago. It is reassuring that the cytochrome figure of 1.2 billion years is in harmony with the relatively scarce fossil record of Precambrian life.

If one accepts the provocative suggestion that eukaryotes developed from a symbiotic association of several prokaryotes, one of which was a respiring bacterium that became the ancestor of present-day mitochondria, one is obliged to conclude that the respiratory machinery had stabilized in essentially its present form before or during this symbiosis. The same thing cannot be said for hemoglobin and its probable ancestor, myoglobin. They can provide no clue to the date when animals and plants diverged, since the globins were evolving to play several different roles during and after this period, as multicelled organisms arose. In no sense had the globins settled down 1.2 million years ago. Nevertheless, if the right proteins are selected, and if the data are not overextended, it should be possible to use the rates of protein evolution to assign times to events in the evolution of life that have left only faint traces in the geological record.

So far we have mentioned the electron-transfer mechanism of cytochrome c only in passing, virtually ignoring the structure of the reduced molecule. The mechanism is another story in itself, and one that cannot yet be written. One hopes that the clues supplied by X-ray analysis will suggest the best chemical experiments to try next, in order to learn the mechanism of the oxidation-reduction process. The reduced cytochrome structure has been obtained so recently

RATES OF EVOLUTION OF PROTEINS (opposite page) can be inferred by plotting average differences in amino acid sequences between species on two sides of an evolutionary branch point that can be dated, for example the branch point between fish and reptiles or between reptiles and mammals. The average differences (vertical axis) have been corrected to allow for the occurrence of more than one mutation at a given amino acid site. The length of the vertical data bars indicates the experimental scatter. Times since the divergence of two lines of organisms from a common branch point (horizontal axis) have been obtained from the geological record. The drawings below the graph show schematically the function (described in the text) of the molecules whose evolutionary rate of change is plotted. The rate of change is proportional to the steepness of the curve. It can be represented by a number called the unit evolutionary period, which is the time required for the amino acid sequence of a protein to change by 1 percent after two evolutionary lines have diverged. For fibrinopeptides this period is about 1.1 million years, whereas for histone IV it is 600 million years. The probable reasons for these differences are discussed in the text.

that it would be premature to base many deductions on it.

A Glimpse of Molecular Dynamics

One obvious structural feature, which undoubtedly has great physiological significance, is that in the reduced molecule the top of the heme crevice is closed. The chain from residues 80 to 83 swings to the right (as the molecule is depicted on page 194), the exposed phenylalanine 82 slips into the heme crevice to the left of the heme and nearly parallel to it, and the heme becomes less accessible to the outside world. The absolute preservation of this phenylalanine side chain throughout evolution, in an environment that is energetically unfavorable in the oxidized molecule, argues that closing of the heme crevice in the reduced molecule is important for its biological activity.

Several explanations might be offered. The aromatic ring of phenylalanine 82 may be part of the electron-transfer mechanism, or its removal from the heme crevice may be necessary to permit an electron-transferring group to enter beside the heme or just to approach the edge of the heme. At a minimum the refolding of the chain from residues 80 to 83 may be a "convulsive" motion that pushes the oxidase complex away from the protein after electron transfer is achieved by some other pathway.

This article has been speculative enough without making a choice between these or other alternatives. At this stage, as both oxidized and reduced cytochrome analyses are being extended to higher resolution, it is enough to say that we can see more refolding of the protein chain in passing between the two states than has been observed in any other protein. Phenylalanine 82 swings to an entirely new position and several other aromatic rings change orientation, including the three in the left channel. As the molecule is reduced, the right channel apparently is partly blocked by residues 20 and 21. We now have pictures of both strokes of a very ancient two-stroke molecular engine. We hope in time to be able to figure out how it operates.

ENZYMES BOUND TO ARTIFICIAL MATRIXES

KLAUS MOSBACH
March 1971

A new technique imitates the way enzymes are held in place in the living cell. Besides clarifying cell mechanisms, enzymes bound to matrixes act as biocatalysts in industry and offer a new medical tool

The complex chemistry of the living cell is engineered by thousands of different enzymes, each of which is a catalyst for a particular chemical reaction. It has gradually become clear that the cell could not function if it were simply a little bag filled with enzymes in solution. There is now much evidence that the great majority, and perhaps all, of the intracellular enzymes function either in an environment resembling a gel, or while adsorbed at interfaces, or in solid-state assemblages such as seem to exist in mitochondria and other organelles of the cell. The architectural distribution of enzymes within the cell must be extremely precise; otherwise the different enzymes, the substrates on which they act, the thousands of reaction products and the wide variety of substances that inhibit specific reactions would become chaotically mixed.

Until recently most laboratory investigations of enzymes were conducted with purified extracts in dilute aqueous solution and therefore under conditions far removed from those existing in the living cell. Probably only a few enzymes, the true extracellular enzymes, actually perform their primary biological function under such conditions. Ideally one would like to study the intracellular enzymes in their natural environment by recombining isolated enzymes and the gel-like or matrix environment with which they are normally associated inside the cell, but progress in this direction is beset with many difficulties.

An alternative and much more practical approach is to attach isolated enzymes to mechanically stable artificial matrixes such as hydrophilic (water-loving) polymers. Such systems not only can provide valuable models of how enzymes behave in their natural milieu but also can serve as efficient biological catalysts with many practical applications, including medical ones.

It is now 74 years since Eduard Buechner made the discovery, startling in its day, that a cell-free extract from yeast can ferment sugar to alcohol. Since Buechner's time it has been recognized that enzymes, which were subsequently identified as proteins, are the biocatalysts responsible for the myriad reactions that enable living cells to survive and reproduce. In 1926 James B. Sumner succeeded in crystallizing the enzyme urease, thereby demonstrating that enzymes are distinct chemical compounds and hence are amenable to detailed characterization and analysis. Subsequent progress was so rapid that by 1964 the number of enzymes listed by the International Enzyme Commission was close to 900, and by 1968 it was some 1,300. A complete three-dimensional structure has been worked out for about 20 enzymes, and one enzyme (ribonuclease) has been synthesized in the laboratory. With the recognition that enzymes in the living cell are normally attached to surfaces, the term "allotopy" (from the Greek for "other" and "position") has been introduced to describe the differences between the properties of membrane-bound enzymes and the properties of the same enzymes in solution.

In retrospect it is hard to say whether the interest in enzymes affixed to matrixes was originally stimulated by the hope that such systems would provide insight into cellular mechanisms or whether early investigators were attracted primarily by practical applications. In either case the new approach to enzyme technology has in barely half a decade initiated a remarkable volume of work. In one of the earliest references to enzyme-binding, published in 1954,

Nikolaus Grubhofer and Lotte Schleith of the Institute for Virus Research in Heidelberg describe the fixation of pepsin and other enzymes to polyaminostyrene, and they appear interested only in the technological potential of their new technique. Since then workers in many laboratories have developed a variety of artificial matrixes and enzyme-binding methods. Outstanding contributions have been made by the group under Ephraim Katchalsky at the Weizmann Institute of Science, by Jerker Porath, Rolf Axen and their co-workers at the University of Uppsala, and by a large group under Garth Kay and Malcolm Lilly at University College London.

Among the successful matrixes are cross-linked dextran gels (Sephadex), cross-linked acrylic polymers (Biogel), polyamino acids, various kinds of cellulose and even ordinary filter paper and glass. There are three principal methods for binding enzymes to matrixes: ordinary (covalent) chemical linkage, adsorption (which involves the attraction of opposite electric charges) and entrapment of the enzyme within a gel lattice whose pores are large enough to allow the molecules of substrate and product to pass freely but small enough to retain the enzyme [*see illustrations on page 206*]. A less common method is to convert the enzyme molecules themselves into insoluble matrixes by using bifunctional compounds to cross-link them into large aggregates.

Let us look first at some investigations in which enzymes bound to matrixes have been used as model systems for in vivo enzyme reactions taking place in the living cell. One property usually studied in characterizing an enzyme is the dependence of the enzymic reaction on the degree of acidity or alkalinity of

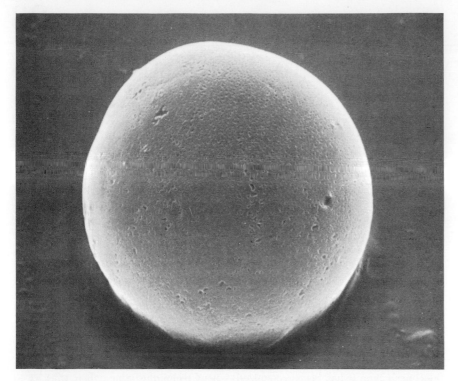

POLYMER BEAD CARRYING BOUND ENZYME looks like this under the scanning electron microscope at a magnification of 700 diameters. The bead is a copolymer of acrylamide and acrylic acid. At this low magnification the enzyme, lactate dehydrogenase, is invisible.

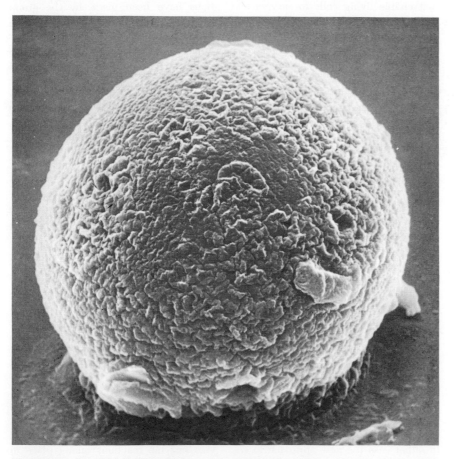

WRINKLED MATERIAL ON BEAD consists of molecules of formazan, the substance produced by the catalytic action of lactate dehydrogenase on lactate, the substrate compound. This micrograph, like the one at the top of the page, is at a magnification of 700 diameters. Both pictures were made by the author's colleague Folke Larsson of the University of Lund.

the solution containing the enzyme and its substrate. Acidity or alkalinity is measured in terms of the solution's hydrogen-ion concentration, expressed on a logarithmic scale as pH. (Values of pH below 7 indicate the high hydrogen-ion concentrations typical of acid solutions; values above 7 indicate the low concentrations typical of alkaline solutions.) This is usually done with the enzyme dissolved in an aqueous solution.

One can now compare the activity at various pH levels of an enzyme bound to a matrix with the activity of a freely dissolved enzyme. Such studies on pH-activity profiles for enzymes have been carried out in particular by Leon Goldstein of Katchalsky's group. Goldstein finds, for example, that when the enzyme trypsin is attached to a negatively charged matrix, the pH-activity profile is markedly shifted toward the alkaline side; the bound enzyme reaches 100 percent of maximum activity at apparently higher alkalinity than the unbound enzyme [see top illustration on page 207].

Evidently the negatively charged groups on the polymer matrix attract a thin "film" of positive hydrogen ions, thereby creating a microenvironment for the bound enzyme that has a higher hydrogen-ion concentration (lower pH) than the concentration in the surrounding solution where the pH is actually measured. In other words, the enzyme, bound or unbound, "prefers" to work in the same slightly alkaline environment. When the enzyme is bound to a negatively charged matrix, however, the pH of the macroenvironment of the solution is even more alkaline than the microenvironment of the "working" enzyme.

Another way to demonstrate the effect of the enzyme's microenvironment on the functioning of the enzyme is to measure the dependence of the reaction rate on the concentration of the substrate. For example, one can observe the effect of an enzyme matrix whose charge is opposite to the charge of the substrate. One can predict that the microenvironment next to the matrix will have a higher concentration of substrate than the external solution, in analogy with the hydrogen ions in the preceding example. That is indeed what happens; the enzyme bound to such a matrix works efficiently at what seems to be a low concentration of substrate. (The enzymologist would say that the apparent Michaelis constant, $K_m^{app.}$, is low.)

Such studies emphasize the importance of the microenvironment when one is considering the enzymic processes actually operating in the living cell. As an example, Israel H. Silman of the Weizmann Institute and Arthur Karlin of the Columbia University College of Physicians and Surgeons have found that the optimum pH for the activity of acetylcholine esterase in its natural state, that is, membrane-bound, is different from the optimum pH observed when the free enzyme is placed in solution. With a knowledge of the work described above they were able to attribute their unexpected findings to local pH effects within the cell membrane.

Another area in which model studies can be valuable concerns the investigation of reactions responsible for cell metabolism, for example the sequential steps by which glucose is broken down into carbon dioxide and water with the release of energy. Most of the hundreds of different enzymic reactions that take place within the cell are organized in sequences—often cyclic sequences—in which the product of one reaction serves as the substrate for the next one. Either the participating enzymes must be arranged in proper sequence inside the cell or they must be at least concentrated in certain areas. Two closely related questions immediately arise: What is the effect of the microenvironment on such systems, and how does the distance between individual enzymes in a sequence influence the efficiency of the system?

In an attempt to answer these questions my colleague Bo Mattiasson and I at the University of Lund have recently bound two enzymes, hexokinase and glucose-6-phosphate-dehydrogenase, by covalent linkages to the same polymer-bead matrix [see bottom illustration on page 207]. The product from the first enzymic reaction, glucose-6-phosphate, serves as the substrate for the second enzyme. When we compared the efficiency of this matrix-bound two-enzyme system with the efficiency of the same two enzymes unbound but present in a homogeneous solution, we found that in the initial stage the matrix-bound system was twice as effective as the solution system, and that at no stage was it less efficient than the solution system. Our interpretation is that in the matrix system, because of the close proximity of the two enzymes, the product of the first reaction, glucose-6-phosphate, is available in higher concentration for the second enzyme. The proximity of the two enzymes may not be the only contributing factor; when the matrix particles are stirred, they are surrounded by a "cage" of water molecules (the "diffusion layer") that may impede the diffusion of the product of the first reaction into the sur-

rounding medium. Thus the concentration of substrate for the second reaction may be higher in the microenvironment of the matrix than one would infer from measuring the concentration of substrate even a short distance away. A reasonable conclusion is that the rate of enzymic reactions within the living cell is determined not by the concentration of a substrate in the cell as a whole, or even in a small region of the cell, but by the concentration in the immediate vicinity of the operative enzyme and by other conditions in the microenvironment.

Let us turn our attention now from theoretical matters to some of the practi-

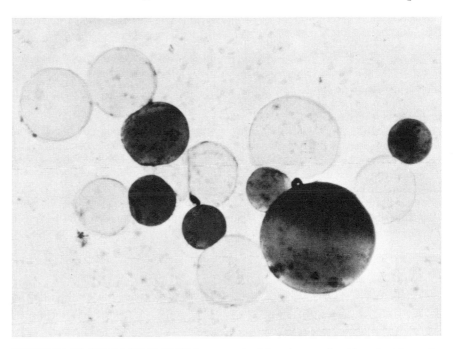

MIXTURE OF COPOLYMER BEADS appears in this micrograph with and without the attached enzyme. The dark beads are copolymers of acrylamide and acrylic acid to which the enzyme lactate dehydrogenase has been bound using a carbodiimide. The light beads lack the enzyme. The enzyme-bound beads are dark (actually blue) owing to the presence of formazan, a blue precipitate formed from the dye tetrazolium blue during the reaction.

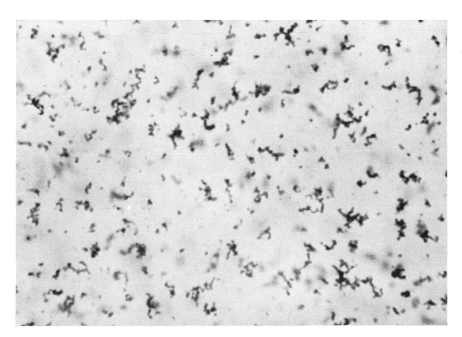

SOLUTION OF UNBOUND ENZYME, lactate dehydrogenase, when incubated with lactate, yields a product (formazan) that is evenly distributed throughout the area of the micrograph. The magnification of 70 diameters is twice that in the micrograph at the top of the page. Both micrographs were made by E. Carlemalm of the University of Lund.

cal applications that have been found for matrix-bound enzymes. In industry a growing number of processes depend on the catalytic activity provided by natural enzymes. Compared with ordinary chemical catalysts, biocatalysts offer the advantage of allowing mild reaction conditions and often the advantage of high specificity. The industrial use of biocatalysts has heretofore been restricted, however, by the high cost of the enzymes themselves and by the difficulty of separating them from the end product. By binding the enzymes to an insoluble matrix these limitations can be circumvented. A column packed with matrix-bound enzymes can be used repeatedly, and the product that emerges is uncontaminated. A further advantage is that enzymes have been reported in a number of cases to be more stable when they are bound to a matrix.

As an illustration of this approach I shall describe a project that Per-Olof Larsson and I have been working on for

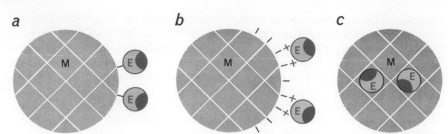

THREE METHODS OF BINDING ENZYMES to matrixes are in general use. The enzyme (*E*) can be bound to the matrix (*M*) by direct chemical linkages, known as covalent chemical bonds (*a*). The enzyme can be held in place by the attraction of unlike electric charges, the phenomenon known as adsorption (*b*), or the enzyme can be trapped within a gel lattice (*c*) whose pores are large enough to permit the substrate and the product into which it is converted to enter and leave freely. The colored regions of the enzyme represent active sites of the enzyme molecule, where the enzyme performs its catalytic work on the substrate.

some time in our laboratory in collaboration with the chemical and pharmaceutical firm of CIBA Limited. The problem had its origin some 20 years ago, well before my own involvement, when cortisol was found to be a potent drug in the treatment of rheumatoid arthritis. Cortisol belongs to the family of steroid hormones produced by the cortex of the adrenal gland. In the original process of its synthetic manufacture it was necessary to add to the starting material an oxygen or hydroxyl (OH) group at a particular molecular position designated 11-beta; this had been found essential for physiological activity. The introduction

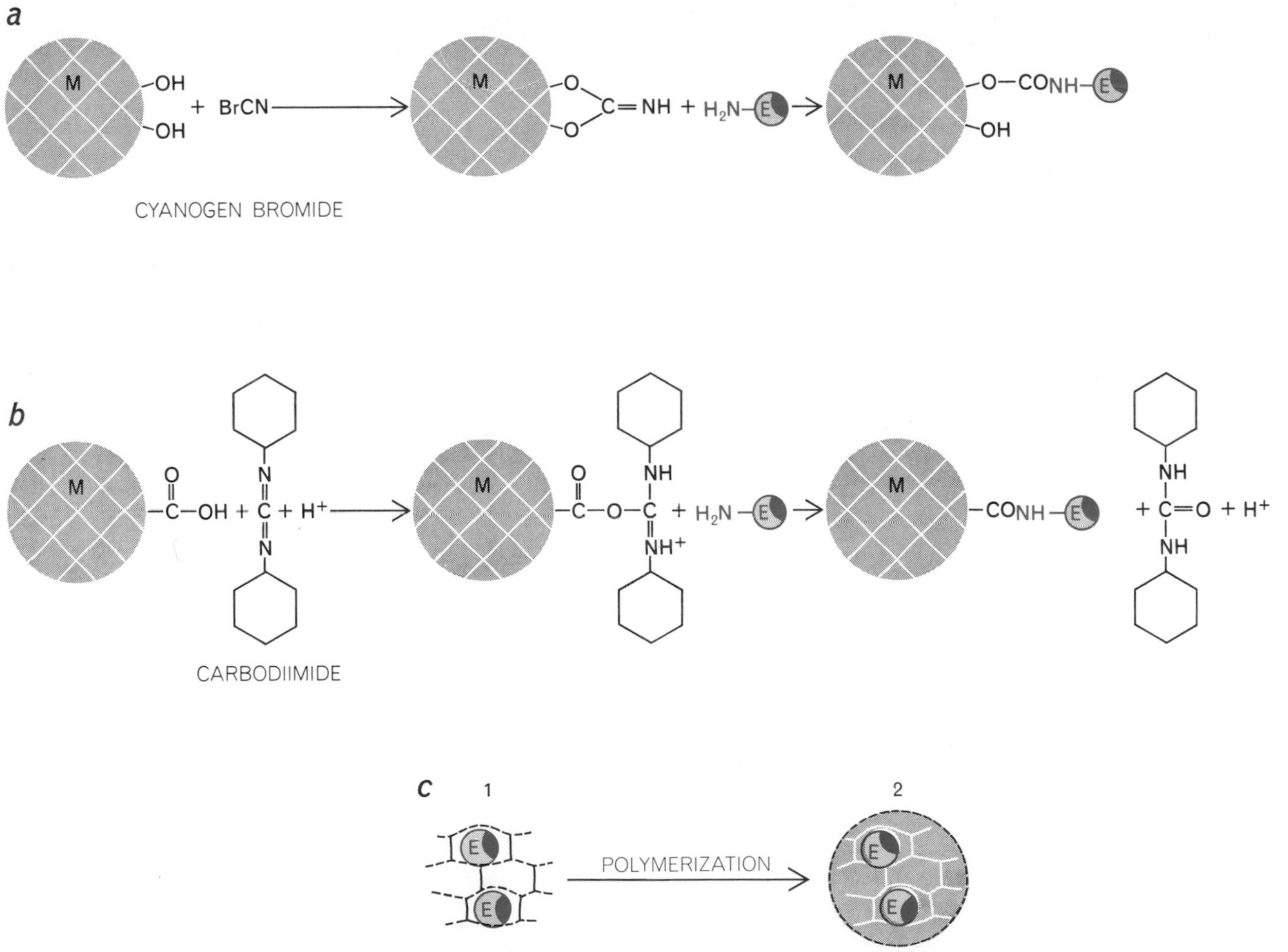

a

CYANOGEN BROMIDE

b

CARBODIIMIDE

c

POLYMERIZATION

PREPARATION METHODS are shown for binding enzymes to matrixes or trapping them in a gel. If the matrix for a covalent linkage (*a*) is a cross-linked dextran gel (Sephadex), adjacent hydroxyl (OH) groups of the gel can react with cyanogen bromide and combine with amino (H₂N–) groups of the enzyme. In *b* the enzyme has been bound to carboxyl (COOH) groups of the matrix (a cross-linked acrylic copolymer) after treatment with a carbodiimide. (The one here is dicyclohexyl carbodiimide.) In *c* the enzyme is shown in solution with monomers (*1*) and after the monomers have been polymerized to form polyacrylamide gel (*2*).

of functional groups at specific positions in complex molecules such as steroids by conventional chemical means often requires several steps, and it can give rise to undesirable side reactions and poor yields. Later it was discovered that 11-beta hydroxylation yielding cortisol from the cheap starting material Compound S could be achieved biologically in a single step with the aid of a hydroxylase enzyme present in certain fungi.

When two hydrogen atoms are removed from the 1–2 position on the A ring of cortisol, the resulting compound is prednisolone, which is superior to cortisol in relieving rheumatoid arthritis. Again it was found that the most efficient way to carry out the desired reactions, in this case the dehydrogenation of cortisol, is to use another biocatalyst: a dehydrogenase found in bacteria. In the original process the reaction step involved exposure of the substrate (cortisol) to intact microorganisms. There was then the problem of separating the end product from a large mass of bacteria. Our contribution was to isolate the desired enzyme, a steroid dehydrogenase, and trap it in a matrix consisting of a hydrophilic gel. The reaction can now be conducted in a column packed with the trapped enzyme; the flow rate is adjusted so that all the cortisol that enters is transformed to prednisolone.

As an alternative in working with enzymes that are either highly unstable or difficult to isolate we can also use intact cells trapped in a gel. This approach was chosen for handling the fungal cells that carry out the hydroxylation of Compound S to cortisol. In this way we obtained a two-step continuous transformation unit that biocatalytically converts Compound S to cortisol and cortisol to prednisolone [see illustration on opposite page].

One can imagine almost any molecule, including for instance complex antibiotics, being synthesized at least in part by passing a suitable starting material through a battery of enzyme columns, each effecting a single transformation with high efficiency. An example of what can be achieved in this way has recently been provided by Harry D. Brown and his colleagues at Columbia P&S. They arranged within one column in proper order the four different enzymes that participate in the natural breakdown of sugar: hexokinase, phosphogluco-isomerase, phosphofructokinase and aldolase. When glucose is poured into the top of the column, it is transformed in four successive steps into the expected end product: glyceraldehyde-3-phosphate.

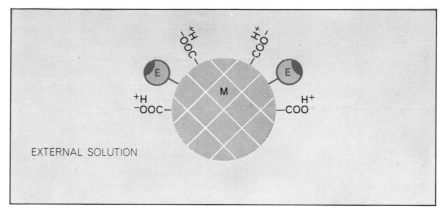

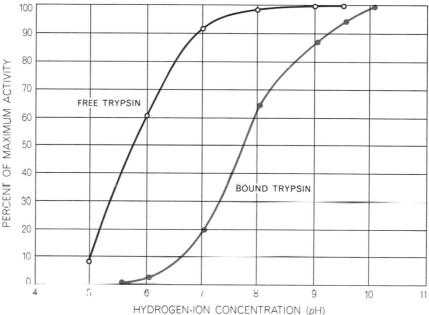

SHIFT IN ACTIVITY PROFILE results when the enzyme trypsin (E) is bound to a negatively charged matrix. As the diagram at the top shows, the matrix raises the local concentration of hydrogen ions (H$^+$) to a higher level than in the external solution where the concentration is measured. Thus (curves at bottom) the maximum activity of bound trypsin apparently takes place in more alkaline solution than when the trypsin is unbound.

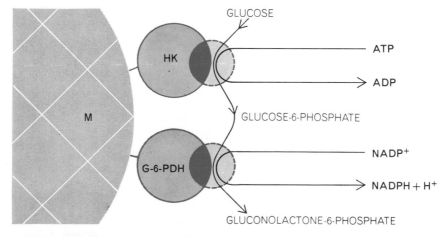

TWO ENZYMES BOUND TO ONE MATRIX prove twice as effective in the initial phase as a system in which the same enzymes are present unbound. The first enzyme, hexokinase (HK), converts glucose to glucose-6-phosphate, consuming energy and a phosphate supplied by adenosine triphosphate (ATP). The second enzyme, glucose-6-phosphate dehydrogenase (G-6-PDH), converts the product of the first reaction to gluconolactone-6-phosphate with the help of the coenzyme nicotinamide adenine dinucleotide phosphate (NADP$^+$).

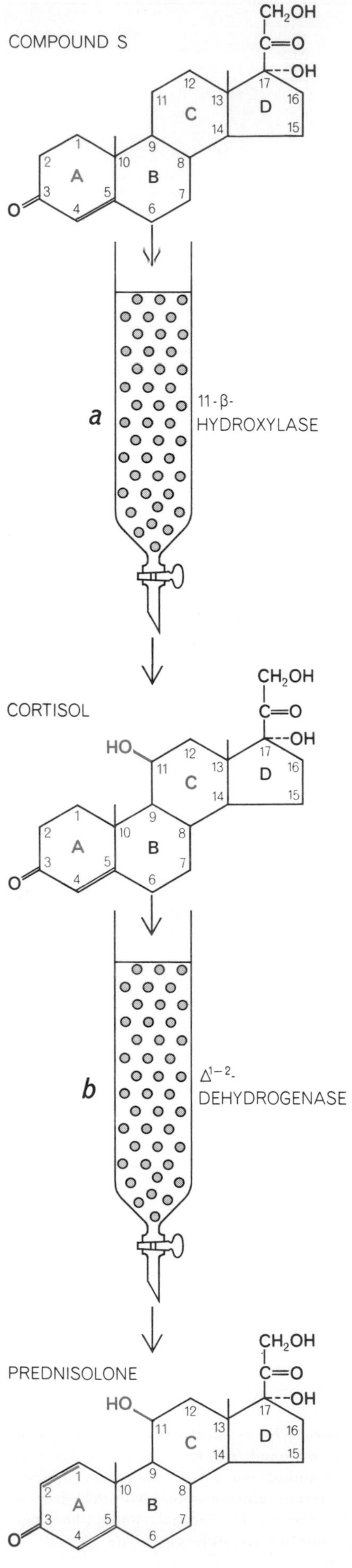

COMPOUND S

CORTISOL

PREDNISOLONE

There are now commercially available several matrix-bound enzymes, most of which break down proteins. It is fair to say that enzyme technology has received an enormous boost through the introduction of techniques for binding enzymes to carriers. The potential of the new technology will be increased still further if the stability of polymer-bound enzymes can be improved, if cheaper ways can be developed to isolate enzymes from biological materials on a large scale and if more progress can be made toward the ultimate goal of routinely synthesizing enzymes on a commercial scale using, for instance, the solid-phase synthesis technique developed by R. B. Merrifield of Rockefeller University.

Biochemical analysis is another area with great inherent potential for matrix-bound enzymes. Enzymic methods are being used increasingly as analytical tools, particularly in medicine. The use of immobilized enzyme systems has the same advantages here as in the applications mentioned above. In addition, interference from proteins present in the solution to be analyzed, a state of affairs that often creates problems in conventional enzyme analysis, can be avoided when the enzyme is securely held within a gel matrix.

Two recent developments will suggest what is being done in this area. The first involves a sensitive test for hydrogen peroxide (H_2O_2) in solution that has been developed by Howard H. Weetall and Norman Weliky of the Jet Propulsion Laboratory. In the presence of the enzyme peroxidase and hydrogen peroxide, a colorless dye is oxidized and immediately turns blue. To provide a convenient test system the enzyme is bound to the cellulose in strips of paper. The test is performed by spotting the strip with the colorless dye and with a small amount of solution suspected of contain-

ing hydrogen peroxide. Depending on the amount of hydrogen peroxide present, more or less of the dye is oxidized. The intensity of color provides a rapid and semiquantitative determination of hydrogen peroxide down to concentrations as low as .03 microgram per milliliter.

The second example is the development of enzyme electrodes by Stuart J. Updike and John W. Hicks of the University of Wisconsin. The electrode is simply covered with a thin polymer film in which the enzyme is trapped. The electrode thus represents a miniaturized chemical transducer that functions by combining an electrochemical procedure with the activity of an immobilized enzyme. George G. Guilbault and Joseph H. Montalvo of Louisiana State University have applied such an enzyme electrode to the measurement of urea in body fluids. In their device the enzyme urease is embedded in a polyacrylamide membrane coated in a layer about .1 millimeter thick on an electrode sensitive to ammonium ions (NH_4^+). In the presence of urease, urea and water react to form ammonium ions and bicarbonate ions (HCO_3^-). The concentration of ammonium ions that builds up at the surface of the electrode yields a direct measure of the urea present in the sample [see illustration at left on next page]. This has a direct analogy in the conventional determination of hydrogen ions with a glass electrode. Enzyme electrodes of this type have operated continuously at room temperature for three weeks without loss of activity.

The clinical potentials of enzyme technology using matrix-bound enzymes are also substantial. More than 120 anomalies and diseases are known that can be regarded as inborn errors of metabolism. Of these a large number represent enzyme-deficiency diseases, in which certain enzymes normally found in the body are either lacking or inactive. One of the best-known enzyme-deficiency diseases is phenylketonuria, which leads to mental retardation. The disease is believed to be caused by the lack of an enzyme that introduces a hydroxyl group into the amino acid phenylalanine, thereby converting it to tyrosine. At present children with the disease are given a costly diet free of phenylalanine.

A more direct approach would be to supply the patient with the missing enzyme. Unfortunately the foreign protein would immediately produce an adverse immunological reaction in the patient receiving it. If the enzyme could be

STEROID TRANSFORMATIONS can be carried out in two simple steps by enzymes bound to matrixes. With conventional chemical procedures the reactions would require many steps and the yields would be low. The starting material, known as Compound S, is a simple steroid obtainable from natural sources. In Step *a* the enzyme 11-beta-hydroxylase bound to a gel matrix inserts a hydroxyl (OH) group at the No. 11 position in the steroid's *C* ring. The product, cortisol, is a close relative of cortisone. In Step *b* the enzyme $\triangle^{1-2}$-dehydrogenase, also in a gel matrix, removes two hydrogen atoms from cortisol and creates a double bond (*color*) in the *A* ring. The product, prednisolone, is more potent than cortisone or cortisol in treatment of rheumatoid arthritis.

trapped in a gel, however, the reaction could be prevented. The enzymes could be enclosed in tiny semipermeable polymer beads that would allow the substrate to diffuse in and the product to diffuse out. The beads could be introduced directly into the bloodstream of the patient, where they might remain active for a considerable time. Alternatively they might be packed in a shunt chamber connected to the circulatory system. The problems to be solved, however, are not simple. For instance, enzymic processes such as the conversion of phenylalanine to tyrosine often involve the participation of coenzymes, which should be bound together with the enzyme. Recently we have been successful in binding such a coenzyme, nicotinamide adenine dinucleotide, to a polymeric matrix in such a way that its coenzymic activity is retained. I feel certain that this type of approach will be the most promising avenue for treating enzyme-deficiency diseases until such time as genetic engineering—the direct modification of the organism's genetic inheritance—can offer a solution.

Matrix-bound enzymes have been utilized for the construction of a new type of artificial kidney that was tested for the first time last year on a patient with kidney failure. As the agent for removing toxic substances dissolved in the blood, Thomas M. S. Chang of McGill University employed microcapsules consisting of tiny pellets of activated charcoal coated with a thin film of collodion. The microcapsules were packed in a chamber connected to the patient's bloodstream. The system represents a valuable alternative to the bulky and costly dialysis mechanism normally employed as an artificial kidney. The capacity of charcoal to remove toxic substances is, however, rather unspecific. The next step will be to replace the charcoal pellets with encapsulated enzymes selected for their ability to remove specific toxic substances and to leave desirable components of the bloodstream unaffected. As an example, if the enzyme urease were encapsulated, it would convert urea into ammonium ions and bicarbonate ions [see illustration at right below]. The rather toxic ammonium ions could then be removed by an ammonia absorbent or by an additional enzyme (such as glutamic dehydrogenase) that incorporates ammonia into organic nitrogen compounds.

Related techniques are being investigated with the general goal of preparing "artificial cells" or important parts of cells. Last year, for instance, Grazia L. Sessa and Charles Weissmann of the New York University School of Medicine incorporated the enzyme lysozyme in small phospholipid spheres, thereby producing a liposome: an artificial organelle with some of the properties of the organelle called the lysosome.

The general concept of binding biological material to matrixes has also had a major impact on the development of a purification technique known as affinity chromatography. Let me illustrate this technique with the following example. It is known that for most enzymes there exist specific inhibitors that function by blocking the active site of the enzyme. As a rule the complex of enzyme plus inhibitor is readily formed and just as readily dissociated.

In the course of a certain investigation Hans von Fritz and his co-workers at the University of Munich were faced with the problem of separating two enzyme inhibitors of about the same molecular weight present in the pancreas. One was nonspecific in that it inhibited both alpha-chymotrypsin and trypsin,

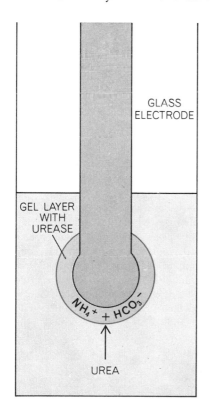

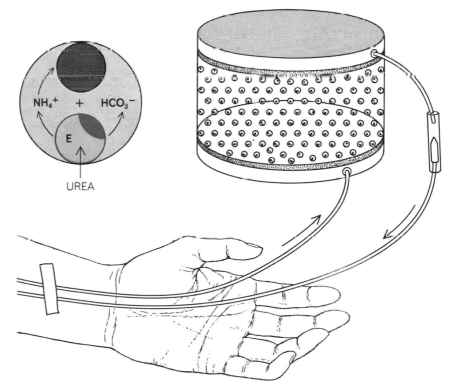

MEASUREMENT OF UREA in body fluids can be accomplished by coating the end of a glass electrode with a thin gel layer, about .1 millimeter thick, to which the enzyme urease is kept bound. The enzyme catalyzes the reaction of urea and water into ammonium ions (NH_4^+) and bicarbonate ions (HCO_3^-). The potential of the electrode is altered by the buildup of ammonium ions, providing a direct measure of the amount of urea present in the sample.

REMOVAL OF UREA from body fluids could be achieved by a new kind of artificial kidney consisting of a vessel filled with microcapsules containing the enzyme urease. The enzyme would convert urea and water into ammonium and bicarbonate ions. The microcapsules would also have to contain either an ammonia absorbent or an additional enzyme for the removal of ammonium ions. Charcoal has been used as an absorbent in clinical trials.

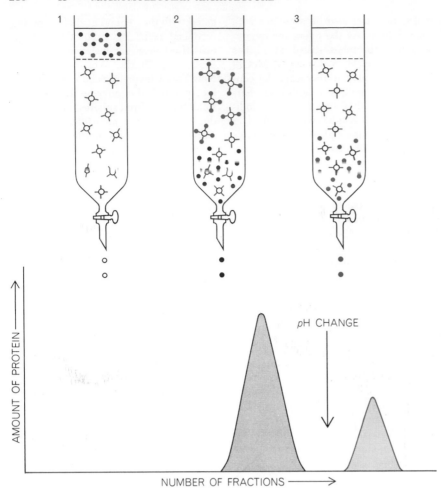

AFFINITY CHROMATOGRAPHY uses enzymes or other biological materials bound to polymer beads to separate one or more substances from a complex mixture. In this example the polymer beads carry an inhibitor that forms a complex with the enzyme alpha-chymotrypsin. In Step *1* a solution containing a small amount of the enzyme (*color*) together with several other enzymes (*black*) is poured into the column. In Step *2* alpha-chymotrypsin is retained on the beads while the other enzymes pass through freely. The liquid drawn from the column at this step will contain all the enzymes except alpha-chymotrypsin (*curves at bottom*). In Step *3* a slightly acid solution releases the alpha-chymotrypsin.

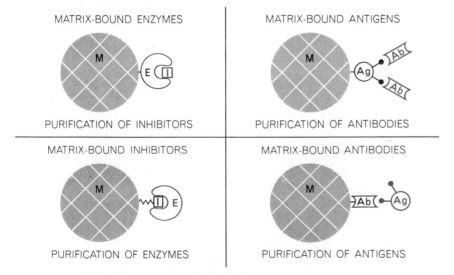

VERSATILITY OF AFFINITY CHROMATOGRAPHY is depicted schematically. In principle one can extract any substance from a complex mixture by attaching the right "grappling hook" to a suitable matrix. In these diagrams an enzyme, an antigen, an enzyme inhibitor and an antibody are used to capture substances for which they have a natural affinity.

two enzymes that cleave proteins. The other inhibitor was specific for trypsin. To separate the two the Munich workers employed a column packed with alpha-chymotrypsin bound to a matrix. When a solution containing the two inhibitors was passed through the column, the nonspecific inhibitor was retained as the inhibitor specific for trypsin passed through. Later the nonspecific inhibitor oould be freed from the column with an agent that dissociated the enzyme inhibitor complex.

One can readily see that with a reverse procedure it should be possible to pack a column with a matrix-bound inhibitor capable of removing a specific enzyme from a complex mixture of enzymes. This was first accomplished in 1953 by Leonard S. Lerman of the University of Chicago. Since then affinity chromatography has been extensively developed in several laboratories, particularly by Christian B. Anfinsen and his co-workers Pedro Cuatrecasas and Meir Wilchek at the National Institute of Arthritis and Metabolic Diseases. In one example of interest they used a covalently bound inhibitor to form a complex with alpha-chymotrypsin and thus remove it from a mixture. They discovered, however, that when the inhibitor was directly attached to the matrix (a Sepharose gel), it was not effective. In subsequent experiments they found that the inhibitor would function if the distance between the matrix and the inhibitor was extended by the insertion of a short carbon chain about seven angstroms long. Apparently this made the active site on the enzyme, the molecular diameter of which is about 40 angstroms, more accessible to the inhibitor [*see top illustration at left*].

Affinity chromatography has also shown wide usefulness in the field of immunology, where it is frequently important to select one particular type of antibody molecule out of a highly complex mixture of these key molecules of the immune system [see "The Structure and Function of Antibodies," by Gerald M. Edelman; SCIENTIFIC AMERICAN Offprint 1185]. This is accomplished quite simply by using matrixes carrying the antigen that corresponds specifically to the desired antibody. Conversely, matrix-bound antibodies can be employed to isolate their specific antigens [*see bottom illustration at left*]. Affinity chromatography thus adds to the arsenal of different purification methods available to the biologist a new and sophisticated procedure allowing the tailor-made design of materials for the separation of biological molecules.

24

THE STRUCTURE OF THE HEREDITARY MATERIAL

F. H. C. CRICK
October 1954

An account of the investigations which have led to the formulation of an understandable structure for DNA. The chemical reactions of this material within the nucleus govern the process of reproduction

Viewed under a microscope, the process of mitosis, by which one cell divides and becomes two, is one of the most fascinating spectacles in the whole of biology. No one who watches the event unfold in speeded-up motion pictures can fail to be excited and awed. As a demonstration of the powers of dynamic organization possessed by living matter, the act of division is impressive enough, but even more stirring is the appearance of two identical sets of chromosomes where only one existed before. Here lies biology's greatest challenge: How are these fundamental bodies duplicated? Unhappily the copying process is beyond the resolving power of microscopes, but much is being learned about it in other ways.

One approach is the study of the nature and behavior of whole living cells; another is the investigation of substances extracted from them. This article will discuss only the second approach, but both are indispensable if we are ever to solve the problem; indeed some of the most exciting results are being obtained by what might loosely be described as a combination of the two methods.

Chromosomes consist mainly of three kinds of chemical: protein, desoxyribonucleic acid (DNA) and ribonucleic acid (RNA). (Since RNA is only a minor component, we shall not consider it in detail here.) The nucleic acids and the proteins have several features in common. They are all giant molecules, and each type has the general structure of a main backbone with side groups attached. The proteins have about 20 different kinds of side groups; the nucleic acids usually only four (and of a different type). The smallness of these numbers itself is striking, for there is no obvious chemical reason why many more types of side groups should not occur. Another interesting feature is that no protein or nucleic acid occurs in more than one optical form; there is never an optical isomer, or mirror-image molecule. This shows that the shape of the molecules must be important.

These generalizations (with minor exceptions) hold over the entire range of living organisms, from viruses and bacteria to plants and animals. The impression is inescapable that we are dealing with a very basic aspect of living matter, and one having far more simplicity than we would have dared to hope. It encourages us to look for simple explanations for the formation of these giant molecules.

The most important role of proteins is that of the enzymes—the machine tools of the living cell. An enzyme is specific, often highly specific, for the reaction which it catalyzes. Moreover, chemical and X-ray studies suggest that the structure of each enzyme is itself rigidly determined. The side groups of a given enzyme are probably arranged in a fixed order along the polypeptide backbone. If we could discover how a cell produces the appropriate enzymes, in particular how it assembles the side groups of each enzyme in the correct order, we should have gone a long way toward explaining the simpler forms of life in terms of physics and chemistry.

We believe that this order is controlled by the chromosomes. In recent years suspicion has been growing that the key to the specificity of the chromosomes lies not in their protein but in their DNA. DNA is found in all chromosomes —and only in the chromosomes (with minor exceptions) The amount of DNA per chromosome set is in many cases a fixed quantity for a given species. The sperm, having half the chromosomes of the normal cell, has about half the amount of DNA, and tetraploid cells in the liver, having twice the normal chromosome complement, seem to have twice the amount of DNA. This constancy of the amount of DNA is what one might expect if it is truly the material that determines the hereditary pattern.

Then there is suggestive evidence in two cases that DNA alone, free of protein, may be able to carry genetic information. The first of these is the discovery that the "transforming principles" of bacteria, which can produce an inherited change when added to the cell, appear to consist only of DNA. The second is the fact that during the infection of a bacterium by a bacteriophage the DNA of the phage penetrates into the bacterial cell while most of the protein, perhaps all of it, is left outside.

The Chemical Formula

DNA can be extracted from cells by mild chemical methods, and much experimental work has been carried out to discover its chemical nature. This work has been conspicuously successful. It is now known that DNA consists of a very long chain made up of alternate sugar and phosphate groups [*see diagram on below*]. The sugar is always the same sugar, known as desoxyribose. And it is always joined onto the phosphate in the same way, so that the long chain is perfectly regular, repeating the same phosphate-sugar sequence over and over again.

But while the phosphate-sugar chain is perfectly regular, the molecule as a whole is not, because each sugar has a "base" attached to it and the base is not always the same. Four different types of base are commonly found: two of them are purines, called adenine and guanine, and two are pyrimidines, known as thymine and cytosine. So far as is known the order in which they follow one another along the chain is irregular, and probably varies from one piece of DNA to another. In fact, we suspect that the order of the bases is what confers specificity on a given DNA. Because the sequence of the bases is not known, one can only say that the *general* formula for DNA is established. Nevertheless this formula should be reckoned one of the major achievements of biochemistry, and it is the foundation for all the ideas described in the rest of this article.

At one time it was thought that the four bases occurred in equal amounts, but in recent years this idea has been shown to be incorrect. E. Chargaff and his colleagues at Columbia University, A. E. Mirsky and his group at the Rockefeller Institute for Medical Research and G. R. Wyatt of Canada have accurately measured the amounts of the bases in many instances and have shown that the relative amounts appear to be fixed for any given species, irrespective of the individual or the organ from which the DNA was taken. The proportions usually differ for DNA from different species, but species related to one another may not differ very much.

Although we know from the chemical formula of DNA that it is a chain, this does not in itself tell us the shape of the molecule, for the chain, having many single bonds around which it may rotate, might coil up in all sorts of shapes. However, we know from physical-chemical measurements and electron-microscope pictures that the molecule usually is long, thin and fairly straight, rather like a stiff bit of cord. It is only about 20 Angstroms thick (one Angstrom = one 100-millionth of a centimeter). This is very small indeed, in fact not much more than a dozen atoms thick.

The length of the DNA seems to depend somewhat on the method of preparation. A good sample may reach a length of 30,000 Angstroms, so that the structure is more than 1,000 times as long as it is thick. The length inside the cell may be much greater than this, because there is always the chance that the extraction process may break it up somewhat.

Pictures of the Molecule

None of these methods tells us anything about the detailed arrangement in space of the atoms inside the molecule. For this it is necessary to use X-ray diffraction. The average distance between bonded atoms in an organic molecule is about 1½ Angstroms; between unbonded atoms, three to four Angstroms. X-rays have a small enough wavelength (1½ Angstroms) to resolve the atoms, but unfortunately an X-ray diffraction photograph is not a picture in the ordinary sense of the word. We cannot focus X-rays as we can ordinary light; hence a picture can be obtained only by roundabout methods. Moreover, it can show clearly only the periodic, or regularly repeated, parts of the structure.

With patience and skill several English workers have obtained good diffraction pictures of DNA extracted from cells and drawn into long fibers. The first studies, even before details emerged, produced two surprises. First, they revealed that the DNA structure could take two forms. In relatively low humidity, when the water content of the fibers was about 40 per cent, the DNA molecules gave a crystalline pattern, showing that they were aligned regularly in all three dimensions. When the humidity was raised and the fibers took up more water, they increased in length by about 30 per cent and the pattern tended to become "paracrystalline," which means that the molecules were packed side by side in a less regular manner, as if the long molecules could slide over one another somewhat. The second surprising result was that DNA from different species appeared to give identical X-ray patterns, despite the fact that the amounts of the four bases present varied. This was particularly odd because of the existence of the crystalline form just mentioned. How could the structure appear so regular when the bases varied? It seemed that the broad arrangement of the molecule must be independent of the exact sequence of the bases, and it was therefore thought that the bases play no part in holding the structure together. As we shall see, this turned out to be wrong.

The early X-ray pictures showed a third intriguing fact: namely, that the repeats in the crystallographic pattern came at much longer intervals than the chemical repeat units in the molecule. The distance from one phosphate to the next cannot be more than about seven Angstroms, yet the crystallographic repeat came at intervals of 28 Angstroms in the crystalline form and 34 Angstroms

FRAGMENT OF CHAIN of desoxyribonucleic acid shows the three basic units that make up the molecule. Repeated over and over in a long chain, they make it 1,000 times as long

in the paracrystalline form; that is, the chemical unit repeated several times before the structure repeated crystallographically.

J. D. Watson and I, working in the Medical Research Council Unit in the Cavendish Laboratory at Cambridge, were convinced that we could get somewhere near the DNA structure by building scale models based on the X-ray patterns obtained by M. H. F. Wilkins, Rosalind Franklin and their co-workers at Kings' College, London. A great deal is known about the exact distances between bonded atoms in molecules, about the angles between the bonds and about the size of atoms—the so-called van der Waals' distance between adjacent non-bonded atoms. This information is easy to embody in scale models. The problem is rather like a three-dimensional jig saw puzzle with curious pieces joined together by rotatable joints (single bonds between atoms).

The Helix

To get anywhere at all we had to make some assumptions. The most important one had to do with the fact that the crystallographic repeat did not coincide with the repetition of chemical units in the chain but came at much longer intervals. A possible explanation was that all the links in the chain were the same but the X-rays were seeing every tenth link, say, from the same angle and the others from different angles. What sort of chain might produce this pattern? The answer was easy: the chain might be coiled in a helix. (A helix is often loosely called a spiral; the distinction is that a helix winds not around a cone but around a cylinder, as a winding staircase usually does.) The distance between crystallographic repeats would then correspond to the distance in the chain between one turn of the helix and the next.

We had some difficulty at first because we ignored the bases and tried to work only with the phosphate-sugar backbone. Eventually we realized that we had to take the bases into account, and this led us quickly to a structure which we now believe to be correct in its broad outlines.

This particular model contains a pair of DNA chains wound around a common axis. The two chains are linked together by their bases. A base on one chain is joined by very weak bonds to a base at the same level on the other chain, and all the bases are paired off in this way right along the structure. In the diagram on page 214, the two ribbons represent the phosphate-sugar chains, and the pairs of bases holding them together are symbolized as horizontal rods. Paradoxically, in order to make the structure as symmetrical as possible we had to have the two chains run in opposite directions; that is, the sequence of the atoms goes one way in one chain and the opposite way in the other. Thus the figure looks exactly the same whichever end is turned up.

Now we found that we could not arrange the bases any way we pleased; the four bases would fit into the structure only in certain pairs. In any pair there must always be one big one (purine) and one little one (pyrimidine). A pair of pyrimidines is too short to bridge the gap between the two chains, and a pair of purines is too big to fit into the space.

At this point we made an additional assumption. The bases can theoretically exist in a number of forms depending upon where the hydrogen atoms are attached. We assumed that for each base one form was much more probable than all the others. The hydrogen atoms can be thought of as little knobs attached to the bases, and the way the bases fit together depends crucially upon where these knobs are. With this assumption the only possible pairs that will fit in are: adenine with thymine and guanine with cytosine.

The way these pairs are formed is shown in the diagrams on page 216. The dotted lines show the hydrogen bonds, which hold the two bases of a pair together. They are very weak bonds; their energy is not many times greater than the energy of thermal vibration at room temperature. (Hydrogen bonds are the main forces holding different water molecules together, and it is because of them that water is a liquid at room temperatures and not a gas.)

Adenine must always be paired with

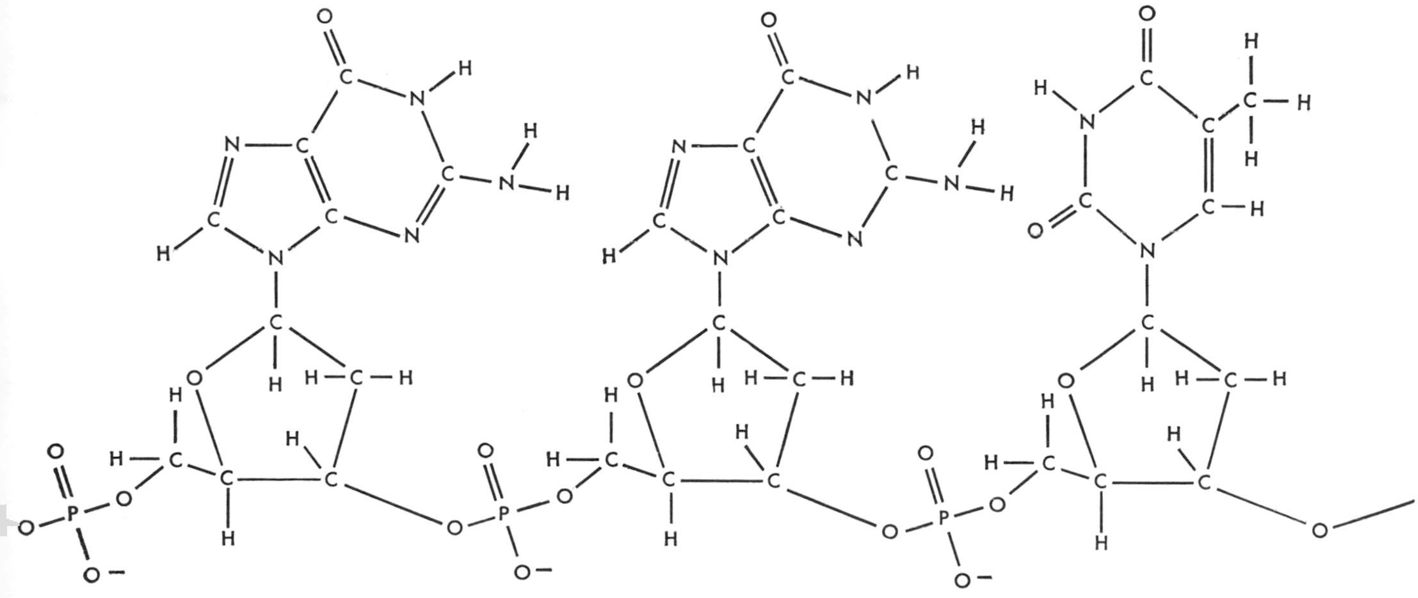

as it is thick. The backbone is made up of pentose sugar molecules (marked by the middle colored square), linked by phosphate groups (bottom square). The bases (top square), adenine, cytosine, guanine and thymine protrude off each sugar in irregular order.

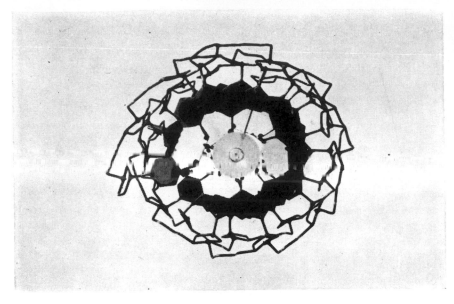

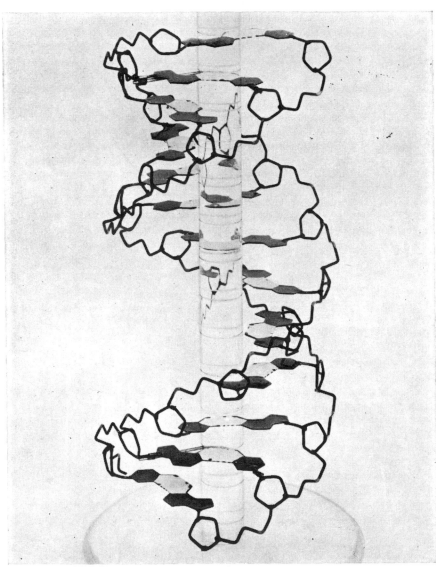

STRUCTURAL MODEL shows a pair of DNA chains wound as a helix about the fiber axis. The pentose sugars can be plainly seen. From every one on each chain protrudes a base, linked to an opposing one at the same level by a hydrogen bond. These base-to-base links act as horizontal supports, holding the chains together. Upper photograph is a top view.

thymine, and guanine with cytosine; it is impossible to fit the bases together in any other combination in our model. (This pairing is likely to be so fundamental for biology that I cannot help wondering whether some day an enthusiastic scientist will christen his newborn twins Adenine and Thymine!) The model places no restriction, however, on the sequence of pairs along the structure. Any specified pair can follow any other. This is because a pair of bases is flat, and since in this model they are stacked roughly like a pile of coins, it does not matter which pair goes above which.

It is important to realize that the specific pairing of the bases is the direct result of the assumption that both phosphate-sugar chains are helical. This regularity implies that the distance from a sugar group on one chain to that on the other at the same level is always the same, no matter where one is along the chain. It follows that the bases linked to the sugars always have the same amount of space in which to fit. It is the regularity of the phosphate-sugar chains, therefore, that is at the root of the specific pairing.

The Picture Clears

At the moment of writing, detailed interpretation of the X-ray photographs by Wilkins' group at Kings' College has not been completed, and until this has been done no structure can be considered proved. Nevertheless there are certain features of the model which are so strongly supported by the experimental evidence that it is very likely they will be embodied in the final correct structure. For instance, measurements of the density and water content of the DNA fibers, taken with evidence showing that the fibers can be extended in length, strongly suggest that there are two chains in the structural unit of DNA. Again, recent X-ray pictures have shown clearly a most striking general pattern which we can now recognize as the characteristic signature of a helical structure. In particular there are a large number of places where the diffracted intensity is zero or very small, and these occur exactly where one expects from a helix of this sort. Another feature one would expect is that the X-ray intensities should approach cylindrical symmetry, and it is now known that they do this. Recently Wilkins and his co-workers have given a brilliant analysis of the details of the X-ray pattern of the crystalline form, and have shown that they

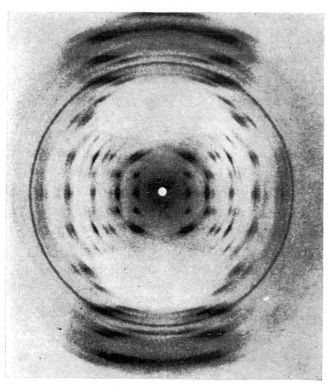

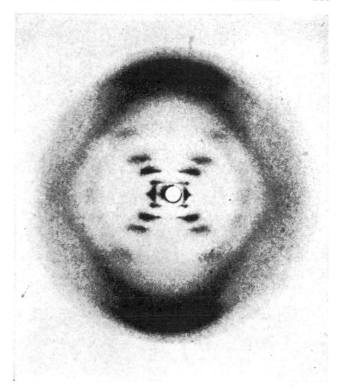

STRUCTURE A is the crystalline form of DNA found at relatively low humidity. This X-ray photograph is by H. R. Wilson.

STRUCTURE B is the paracrystalline form of DNA. The molecules are less regularly arranged. Picture is by R. E. Franklin.

are consistent with a structure of this type, though in the crystalline form the bases are tilted away from the fiber axis instead of perpendicular, as in our model. Our construction was based on the paracrystalline form.

Many of the physical and chemical properties of DNA can now be understood in terms of this model. For example, the comparative stiffness of the structure explains rather naturally why DNA keeps a long, fiber-like shape in solution. The hydrogen bonds of the bases account for the behavior of DNA in response to changes in *p*H. Most striking of all is the fact that in every kind of DNA so far examined—and over 40 have been analyzed—the amount of adenine is about equal to the amount of thymine and the guanine equal to the cytosine, while the cross-ratios (between, say, adenine and guanine) can vary considerably from species to species. This remarkable fact, first pointed out by Chargaff, is exactly what one would expect according to our model, which requires that every adenine be paired with a thymine and every guanine with a cytosine.

It may legitimately be asked whether the artificially prepared fibers of extracted DNA, on which our model is based, are really representative of intact DNA in the cell. There is every indication that they are. It is difficult to see

how the very characteristic features of the model could be produced as artefacts by the extraction process. Moreover, Wilkins has shown that intact biological material, such as sperm heads and bacteriophage, gives X-ray patterns very similar to those of the extracted fibers.

The present position, therefore, is that in all likelihood this statement about DNA can safely be made: its structure consists of two helical chains wound around a common axis and held together by hydrogen bonds between specific pairs of bases.

The Mold

Now the exciting thing about a model of this type is that it immediately suggests how the DNA might produce an exact copy of itself. The model consists of two parts, each of which is the complement of the other. Thus either chain may act as a sort of mold on which a complementary chain can be synthesized. The two chains of a DNA, let us say, unwind and separate. Each begins to build a new complement onto itself. When the process is completed, there are two pairs of chains where we had only one. Moreover, because of the specific pairing of the bases the sequence of the pairs of bases will have been duplicated exactly; in other words, the mold has not only assembled the build-

ing blocks but has put them together in just the right order.

Let us imagine that we have a single helical chain of DNA, and that floating around it inside the cell is a supply of precursors of the four sorts of building blocks needed to make a new chain. Unfortunately we do not know the makeup of these precursor units; they may be, but probably are not, nucleotides, consisting of one phosphate, one sugar and one base. In any case, from time to time a loose unit will attach itself by its base to one of the bases of the single DNA chain. Another loose unit may attach itself to an adjoining base on the chain. Now if one or both of the two newly attached units is not the correct mate for the one it has joined on the chain, the two newcomers will be unable to link together, because they are not the right distance apart. One or both will soon drift away, to be replaced by other units. When, however, two adjacent newcomers are the correct partners for their opposite numbers on the chain, they will be in just the right position to be linked together and begin to form a new chain. Thus only the unit with the proper base will gain a permanent hold at any given position, and eventually the right partners will fill in the vacancies all along the forming chain. While this is going on, the other single chain of the original pair also will

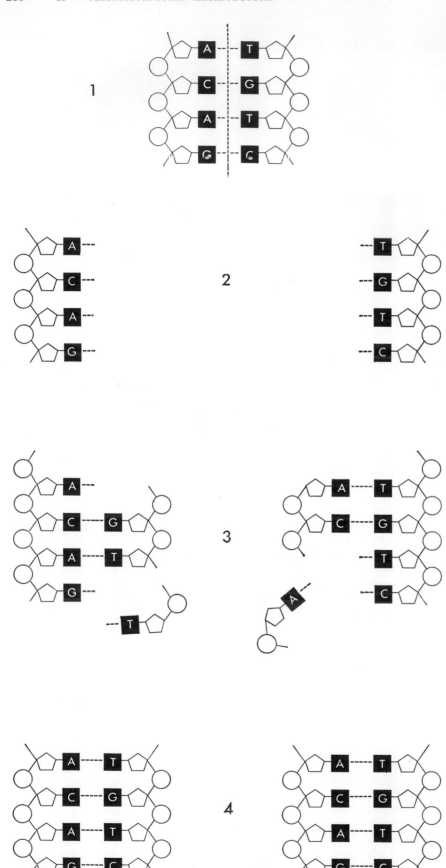

be forming a new chain complementary to itself.

At the moment this idea must be regarded simply as a working hypothesis. Not only is there little direct evidence for it, but there are a number of obvious difficulties. For example, certain organisms contain small amounts of a fifth base, 5-methyl cytosine. So far as the model is concerned, 5-methyl cytosine fits just as well as cytosine and it may turn out that it does not matter to the organism which is used, but this has yet to be shown.

A more fundamental difficulty is to explain how the two chains of DNA are unwound in the first place. There would have to be a lot of untwisting, for the total length of all the DNA in a single chromosome is something like four centimeters (400 million Angstroms). This means that there must be more than 10 million turns in all, though the DNA may not be all in one piece.

The duplicating process can be made to appear more plausible by assuming that the synthesis of the two new chains begins as soon as the two original chains start to unwind, so that only a short stretch of the chain is ever really single. In fact, we may postulate that it is the growth of the two new chains that unwinds the original pair. This is likely in terms of energy because, for every hydrogen bond that has to be broken, two new ones will be forming. Moreover, plausibility is added to the idea by the fact that the paired chain forms a rather stiff structure, so that the growing chain would tend to unwind the old pair.

The difficulty of untwisting the two chains is a topological one, and is due to the fact that they are intertwined. There would be no difficulty in "unwinding" a single helical chain, because there are so many single bonds in the chain about which rotation is possible. If in the twin structure one chain should break, the other one could easily spin around. This might relieve accumulated strain, and then the two ends of the broken chain, still being in close proximity, might be joined together again. There is even some evidence suggesting that in the process of extraction the chains of DNA may be broken in quite a number of places and that the structure nevertheless holds together by means of the hydrogen bonding, because there is never a break in both chains at the same level. Nevertheless, in spite of these tentative suggestions, the difficulty of untwisting remains a formidable one.

There remains the fundamental puzzle as to how DNA exerts its hereditary

REPLICATION mechanism by which DNA might duplicate itself is shown in diagram. A helix of two DNA chains unwinds and separates (1). Two complementary chains of DNA (2) within the cell begin to attach DNA precursor units floating loosely (3). When the proper bases are joined, two new helixes will build up (4). Letters represent the bases.

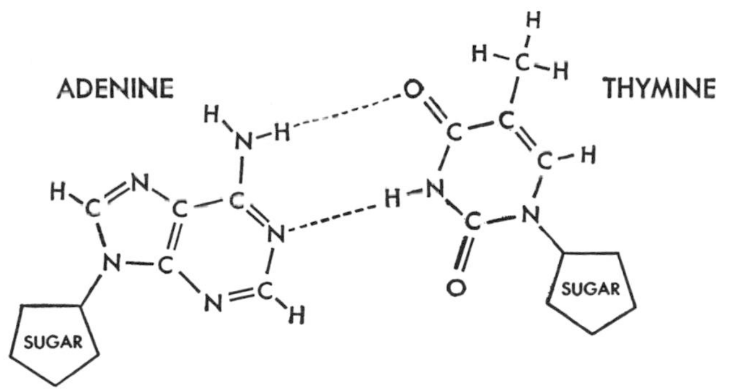

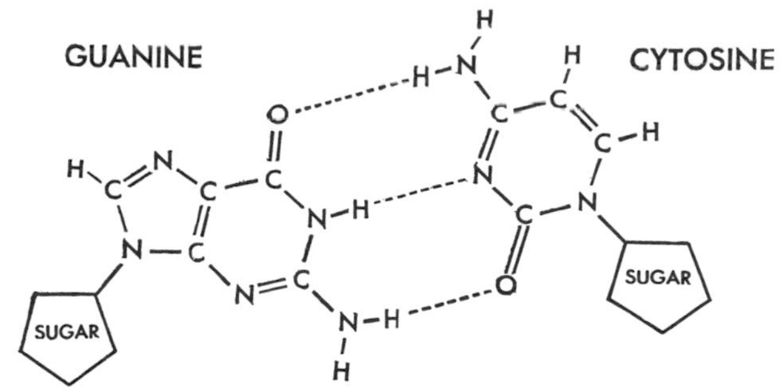

ONE LINKAGE of base to base across the pair of DNA chains is between adenine and thymine. For the structure proposed, the link of a large base with a small one is required to fit chains together.

ANOTHER LINKAGE is comprised of guanine with cytosine. Assuming the existence of hydrogen bonds between the bases, these two pairings, and only these, will explain the actual configuration.

influence. A genetic material must carry out two jobs: duplicate itself and control the development of the rest of the cell in a specific way. We have seen how it might do the first of these, but the structure gives no obvious clue concerning how it may carry out the second. We suspect that the sequence of the bases acts as a kind of genetic code. Such an arrangement can carry an enormous amount of information. If we imagine that the pairs of bases correspond to the dots and dashes of the Morse code, there is enough DNA in a single cell of the human body to encode about 1,000 large textbooks. What we want to know, however, is just how this is done in terms of atoms and molecules. In particular, what precisely is it a code for? As we have seen, the three key components of living matter—protein, RNA and DNA—are probably all based on the same general plan. Their backbones are regular, and the variety comes from the sequence of the side groups. It is therefore very natural to suggest that the sequence of the bases of the DNA is in some way a code for the sequence of the

amino acids in the polypeptide chains of the proteins which the cell must produce. The physicist George Gamow has recently suggested in a rather abstract way how this information might be transmitted, but there are some difficulties with the actual scheme he has proposed, and so far he has not shown how the idea can be translated into precise molecular configurations.

What then, one may reasonably ask, are the virtues of the proposed model, if any? The prime virtue is that the configuration suggested is not vague but can be described in terms acceptable to a chemist. The pairing of the bases can be described rather exactly. The precise positions of the atoms of the backbone is less certain, but they can be fixed within limits, and detailed studies of the X-ray data, now in progress at Kings' College, may narrow those limits considerably. Then the structure brings together two striking pieces of evidence which at first sight seem to be unrelated—the analytical data, showing the one-to-one ratios for adenine-thymine and guanine-cytosine, and the helical

nature of the X-ray pattern. These can now be seen to be two facets of the same thing. Finally, is it not perhaps a remarkable coincidence, to say the least, to find in this key material a structure of exactly the type one would need to carry out a specific replication process; namely, one showing both variety and complementarity?

The model is also attractive in its simplicity. While it is obvious that whole chromosomes have a fairly complicated structure, it is not unreasonable to hope that the molecular basis underlying them may be rather simple. If this is so, it may not prove too difficult to devise experiments to unravel it. It would, of course, help enormously if biochemists could discover the immediate precursors of DNA. If we knew the monomers from which nature makes DNA, RNA and protein, we might be able to carry out very spectacular experiments in the test tube. Be that as it may, we now have for the first time a well-defined model for DNA and for a possible replication process, and this in itself should make it easier to devise crucial experiments.

25

THE NUCLEOTIDE SEQUENCE OF A NUCLEIC ACID

ROBERT W. HOLLEY
February 1966

For the first time the specific order of subunits in one of the giant molecules that participate in the synthesis of protein has been determined. The task took seven years

Two major classes of chainlike molecules underlie the functioning of living organisms: the nucleic acids and the proteins. The former include deoxyribonucleic acid (DNA), which embodies the hereditary message of each organism, and ribonucleic acid (RNA), which helps to translate that message into the thousands of different proteins that activate the living cell. In the past dozen years biochemists have established the complete sequence of amino acid subunits in a number of different proteins. Much less is known about the nucleic acids.

Part of the reason for the slow progress with nucleic acids was the unavailability of pure material for analysis. Another factor was the large size of most nucleic acid molecules, which often contain thousands or even millions of nucleotide subunits. Several years ago, however, a family of small molecules was discovered among the ribonucleic

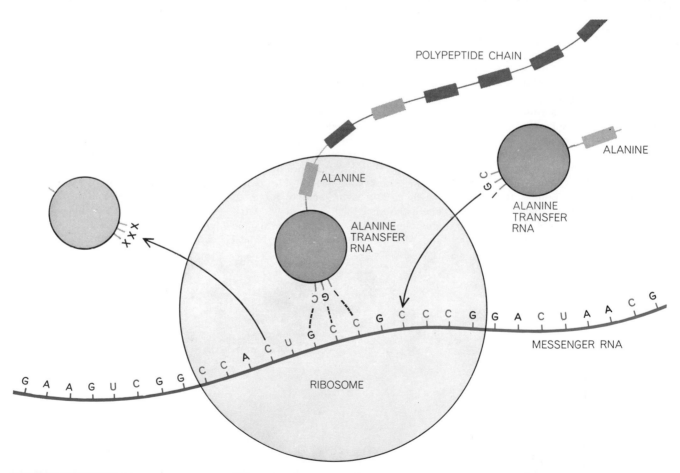

ROLE OF TRANSFER RNA is to deliver a specific amino acid to the site where "messenger" RNA and a ribosome (which also contains RNA) collaborate in the synthesis of a protein. As it is being synthesized a protein chain is usually described as a polypeptide. Each amino acid in the polypeptide chain is specified by a triplet code, or codon, in the molecular chain of messenger RNA. The diagram shows how an "anticodon" (presumably I—G—C) in alanine transfer RNA may form a temporary bond with the codon for alanine (G—C—C) in the messenger RNA. While so bonded the transfer RNA also holds the polypeptide chain. Each transfer RNA is succeeded by another one, carrying its own amino acid, until the complete message in the messenger RNA has been "read."

HYPOTHETICAL MODELS of alanine transfer ribonucleic acid (RNA) show three of the many ways in which the molecule's linear chain might be folded. The various letters represent nucleotide subunits; their chemical structure is given at the top of the next two pages. In these models it is assumed that certain nucleotides, such as C—G and A—U, will pair off and tend to form short double-strand regions. Such "base-pairing" is a characteristic feature of nucleic acids. The arrangement at the lower left shows how two of the large "leaves" of the "clover leaf" model may be folded together. The triplet I—G—C is the presumed anticodon shown in the illustration on the opposite page. The region containing the sequence G—T—Ψ—C—G may be common to all transfer RNA's.

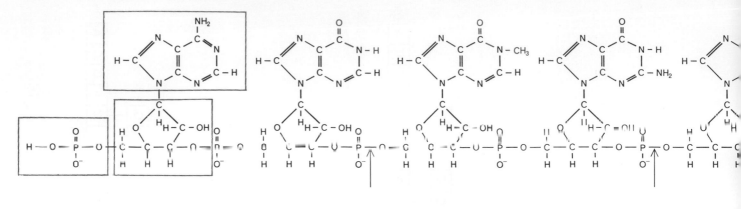

ADENYLIC ACID   INOSINIC ACID   1-METHYLINOSINIC ACID   GUANYLIC ACID   1-METHYLGUANYL

 pA   I   Iᵐ   G   Gᵐ

NUCLEOTIDE SUBUNITS found in alanine transfer RNA include the four commonly present in RNA (A, G, C, U), plus seven others that are variations of the standard structures. Ten of these 11 different nucleotide subunits are assembled above as if they were linked together in a single RNA chain. The chain begins at the left with a phosphate group (*outlined by a small rectangle*) and is followed by a ribose sugar group (*large rectangle*); the two groups alternate to form the backbone of the chain. The chain ends at the right with

acids. My associates and I at the U.S. Plant, Soil and Nutrition Laboratory and Cornell University set ourselves the task of establishing the nucleotide sequence of one of these smaller RNA molecules—a molecule containing fewer than 100 nucleotide subunits. This work culminated recently in the first determination of the complete nucleotide sequence of a nucleic acid.

The object of our study belongs to a family of 20-odd molecules known as transfer RNA's. Each is capable of recognizing one of the 20 different amino acids and of transferring it to the site where it can be incorporated into a growing polypeptide chain. When such a chain assumes its final configuration, sometimes joining with other chains, it is called a protein.

At each step in the process of protein

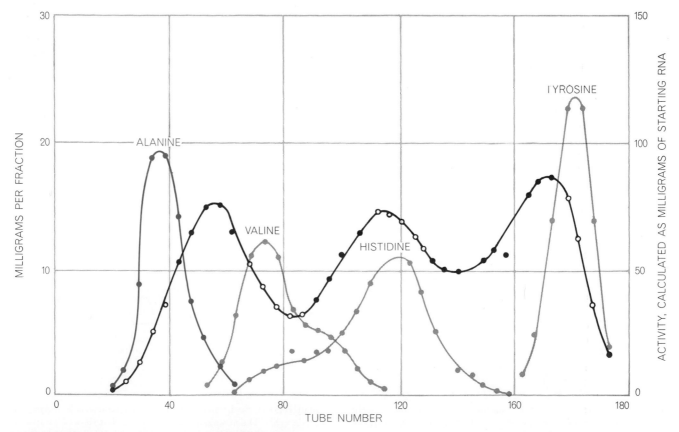

COUNTERCURRENT DISTRIBUTION PATTERN shows two steps in the separation of alanine transfer RNA, as carried out in the author's laboratory. After the first step the RNA content in various collection tubes, measured by ultraviolet absorption, follows the black curve. Biological activity, indicated by the amount of a given amino acid incorporated into polypeptide chains, follows the colored curves. Pure transfer RNA's of four types can be obtained by reprocessing the tubes designated by open circles.

▲RIBOTHYMIDYLIC ACID DIHYDROURIDYLIC ACID
T
Uʰ

MIXTURE OF URIDYLIC AND
DIHYDROURIDYLIC ACIDS

...ID N²-DIMETHYLGUANYLIC ACID   CYTIDYLIC ACID   URIDYLIC ACID   PSEUDOURIDYLIC ACID
Gᵐ
C
U
U*
Ψ OH

a hydroxyl (OH) group. Each nucleotide subunit consists of a phosphate group, a ribose sugar group and a base. The base portion in the nucleotide at the far left, adenylic acid, is outlined by a large rectangle. In the succeeding bases the atomic variations are shown in color. The base structures without color are those commonly found in RNA. Black arrows show where RNA chains can be cleaved by the enzyme takadiastase ribonuclease T1. Colored arrows show where RNA chains can be cleaved by pancreatic ribonuclease.

synthesis a crucial role is played by the structure of the various RNA's. "Messenger" RNA transcribes the genetic message for each protein from its original storage site in DNA. Another kind of RNA—ribosomal RNA—forms part of the structure of the ribosome, which acts as a jig for holding the messenger RNA while the message is transcribed into a polypeptide chain [see illustration on page 218]. In view of the various roles played by RNA in protein synthesis, the structure of RNA molecules is of considerable interest and significance.

The particular nucleic acid we chose for study is known as alanine transfer RNA—the RNA that transports the amino acid alanine. It was isolated from commercial baker's yeast by methods I shall describe later. Preliminary analyses indicated that the alanine transfer RNA molecule consisted of a single chain of approximately 80 nucleotide subunits. Each nucleotide, in turn, consists of a ribose sugar, a phosphate group and a distinctive appendage termed a nitrogen base. The ribose sugars and phosphate groups link together to form the backbone of the molecule, from which the various bases protrude [see illustration at top of these two pages].

The problem of structural analysis is fundamentally one of identifying each base and determining its place in the sequence. In practice each base is usually isolated in combination with a unit of ribose sugar and a unit of phosphate, which together form a nucleotide. Formally the problem is analogous to determining the sequence of letters in a sentence.

It would be convenient if there were a way to snip off the nucleotides one by one, starting at a known end of the chain and identifying each nucleotide as it appeared. Unfortunately procedures of this kind have such a small yield at each step that their use is limited. The alternative is to break the chain at particular chemical sites with the help of enzymes. This gives rise to small fragments whose nucleotide composition is amenable to analysis. If the chain can be broken up in various ways with different enzymes, one can determine how the fragments overlap and ultimately piece together the entire sequence.

One can visualize how this might work by imagining that the preceding sentence has been written out several times, in a continuous line, on different strips of paper. Imagine that each strip has been cut in a different way. In one case, for example, the first three words "If the chain" and the next three words "can be broken" might appear on separate strips of paper. In another case one might find that "chain" and "can" were together on a single strip. One would immediately conclude that the group of three words ending with "chain" and the group beginning with "can" form a continuous sequence of six words. The concept is simple; putting it into execution takes a little time.

For cleaving the RNA chain we used two principal enzymes: pancreatic ribonuclease and an enzyme called takadiastase ribonuclease T1, which was discovered by the Japanese workers K. Sato-Asano and F. Egami. The first enzyme cleaves the RNA chain immediately to the right of pyrimidine nucleotides, as the molecular structure is conventionally written. Pyrimidine nucleotides are those nucleotides whose bases contain the six-member pyrimidine ring, consisting of four atoms of carbon and two atoms of nitrogen. The two pyrimidines commonly found in RNA are cytosine and uracil. Pancreatic ribonuclease therefore produces fragments that terminate in pyrimidine nucleotides such as cytidylic acid (C) or uridylic acid (U).

The second enzyme, ribonuclease T1, was employed separately to cleave the RNA chain specifically to the right of nucleotides containing a structure of the purine type, such as guanylic acid (G). This provided a set of short fragments distinctively different from those produced by the pancreatic enzyme.

The individual short fragments were isolated by passing them through a thin glass column packed with diethylaminoethyl cellulose—an adaptation of a chromatographic method devised by R. V. Tomlinson and G. M. Tener of the University of British Columbia. In general the short fragments migrate through the column more rapidly than the long fragments, but there are exceptions [see illustration on next page]. The conditions most favorable for this separation were developed in our laboratories by Mark Marquisee and Jean Apgar.

The nucleotides in each fragment were released by hydrolyzing the fragment with an alkali. The individual nucleotides could then be identified by paper chromatography, paper electrophoresis and spectrophotometric analysis. This procedure was sufficient to establish the sequence of each of the dinucleotides, because the right-hand member of the pair was determined by the particular enzyme that had been used to produce the fragment. To establish the sequence of nucleotides in larger fragments, however, required special techniques.

Methods particularly helpful in the separation and identification of the fragments had been previously described by Vernon M. Ingram of the Massachusetts Institute of Technology, M. Las-kowski, Sr., of the Marquette University School of Medicine, K. K. Reddi of Rockefeller University, G. W. Rushizky and Herbert A. Sober of the National Institutes of Health, the Swiss worker M. Staehelin and Tener.

For certain of the largest fragments, methods described in the scientific literature were inadequate and we had to develop new stratagems. One of these involved the use of an enzyme (a phosphodiesterase) obtained from snake venom. This enzyme removes nucleotides one by one from a fragment, leaving a mixture of smaller fragments of all possible intermediate lengths. The mixture can then be separated into fractions of homogeneous length by passing it through a column of diethylaminoethyl cellulose [*see illustration on opposite page*]. A simple method is available for determining the terminal nucleotide at the right end of each fraction of homogeneous length. With this knowledge, and knowing the length of each fragment, one can establish the sequence of nucleotides in the original large fragment.

A summary of all the nucleotide sequences found in the fragments of transfer RNA produced by pancreatic ribonuclease is shown in Table 1 on page 224. Determination of the structure of the fragments was primarily the work of James T. Madison and Ada Zamir, who were postdoctoral fellows in my laboratory. George A. Everett of the Plant, Soil and Nutrition Laboratory helped us in the identification of the nucleotides.

Much effort was spent in determining the structure of the largest fragments and in identifying unusual nucleotides not heretofore observed in RNA molecules. Two of the most difficult to identify were 1-methylinosinic acid and 5,6-dihydrouridylic acid. (In the illustrations these are symbolized respectively by I^m and U^h.)

Because a free 5′-phosphate group (p) is found at one end of the RNA molecule (the left end as the structure is conventionally written) and a free 3′-hydroxyl group (OH) is found at the other end, it is easy to pick out from Table 1 and Table 2 the two sequences that form the left and right ends of the alanine transfer RNA molecule. The left end has the structure pG–G–G–C– and the right end the structure U–C–C–A–C–COH. (It is known, however, that the active molecule ends in C–C–AOH.)

The presence of unusual nucleotides

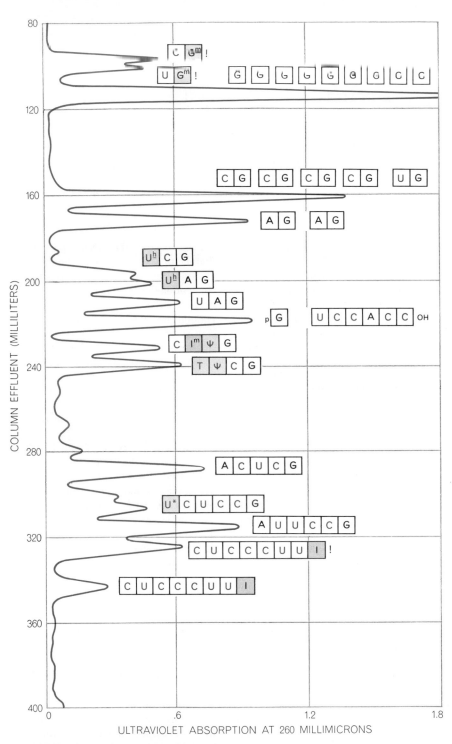

SEPARATION OF RNA FRAGMENTS is accomplished by chromatography carried out in a long glass column packed with diethylaminoethyl cellulose. The curve shows the separation achieved when the column input is a digest of alanine transfer RNA produced by takadiastase ribonuclease T1, an enzyme that cleaves the RNA into 29 fragments. The exclamation point indicates fragments whose terminal phosphate has a cyclical configuration. Such fragments travel faster than similar fragments that end in a noncyclical phosphate.

and unique short sequences made it clear that certain of the fragments found in Table 1 overlapped fragments found in Table 2. For example, there is only one inosinic acid nucleotide (I) in the molecule, and this appears in the sequence I–G–C– in Table 1 and in the sequence C–U–C–C–C–U–U–I– in Table 2. These two sequences must therefore overlap to produce the overall sequence C–U–C–C–C–U–U–I–G–C–. The information in Table 1 and Table 2 was combined in this way to draw up Table 3, which accounts for all 77 nucleotides in 16 sequences [see illustration on page 225].

With the knowledge that two of the 16 sequences were at the two ends, the structural problem became one of determining the positions of the intermediate 14 sequences. This was accomplished by isolating still larger fragments of the RNA.

In a crucial experiment John Robert Penswick, a graduate student at Cornell, found that a very brief treatment of the RNA with ribonuclease T1 at 0 degrees centigrade in the presence of magnesium ions splits the molecule at one position. The two halves of the molecule could be separated by chromatography. Analyses of the halves established that the sequences listed in the first column of Table 3 are in the left half of the molecule and that those in the second column are in the right half.

Using a somewhat more vigorous but still limited treatment of the RNA with ribonuclease T1, we then obtained and analyzed a number of additional large fragments. This work was done in collaboration with Jean Apgar and Everett. To determine the structure of a large fragment, the fragment was degraded completely with ribonuclease T1, which yielded two or more of the fragments previously identified in Table 2. These known sequences could be put together, with the help of various clues, to obtain the complete sequence of the large fragment. The process is similar to putting together a jigsaw puzzle [see illustrations on pages 226 and 227].

As an example of the approach that was used, the logical argument is given in detail for Fragment A. When Fragment A was completely degraded by ribonuclease T1, we obtained seven small fragments: three G–'s, C–G–, U–G–,U–Gᵐ– and pG–. (Gᵐ is used in the illustrations to represent 1-methyl-guanylic acid, another of the unusual nucleotides in alanine transfer RNA.) The presence of pG– shows that Frag-

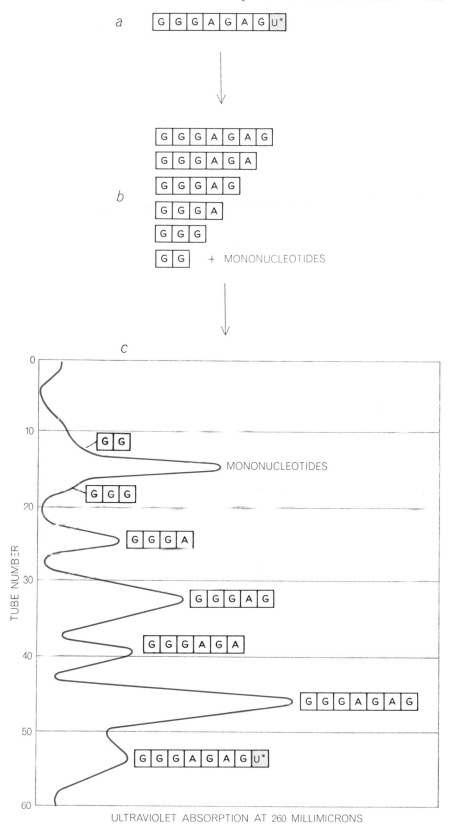

NEW DEGRADATION METHOD was developed in the author's laboratory to determine the sequence of nucleotides in fragments five to eight subunits in length. The example above begins with a fragment of eight subunits from which the terminal phosphate has been removed (a). When the fragment is treated with phosphodiesterase found in snake venom, the result is a mixture containing fragments from one to eight subunits in length (b). These are separated by chromatography (c). When the material from each peak is hydrolyzed, the last nucleoside (a nucleotide minus its phosphate) at the right end of the fragment is released and can be identified. Thus each nucleotide in the original fragment can be determined.

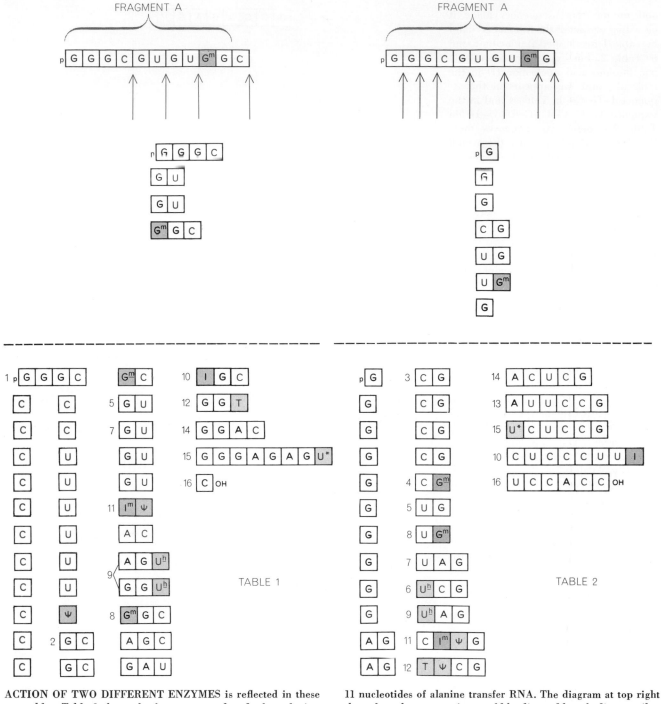

ACTION OF TWO DIFFERENT ENZYMES is reflected in these two tables. Table 1 shows the fragments produced when alanine transfer RNA is completely digested by pancreatic ribonuclease, which cleaves the molecule to the right of nucleotides containing bases with pyrimidine structures (C, U, U^h, ψ and T). The diagram at top left shows how pancreatic ribonuclease would cleave the first 11 nucleotides of alanine transfer RNA. The diagram at top right shows how the same region would be digested by takadiastase ribonuclease T1. Table 2 contains the fragments produced by this enzyme; they all end in nucleotides whose bases contain purine structures (G, G^m, $G\underline{m}$ and I). The numbers indicate which ones appear in the consolidated list in Table 3 on the opposite page.

ment A is from the left end of the molecule. Since it is already known from Table 3 that the left terminal sequence is pG—G—G—C—, the positions of two of the three G's and C—G— are known; the terminal five nucleotides must be pG—G—G—C—G—.

The positions of the remaining G—, U—G— and U—G^m— are established by the following information. Table 3 shows that the U—G^m— is present in the sequence U—G^m—G—C—. Since there is only one C in Fragment A, and its position is already known, Fragment A must terminate before the C of the U—G^m—G—C— sequence. Therefore the U—G— must be to the left of the U—G^m—, and the structure of Fragment A can be represented as pG—G—G—C—G—...U—G—...U—G^m—, with one G— remaining to be placed. If the G— is placed to the left or the right of the U—G— in this structure, it would create a G—G—U— sequence. If such a sequence existed in the molecule, it would have appeared as a fragment when the molecule was treated with pancreatic ribonuclease; Table 1 shows that it did not do so. Therefore the remaining G— must be to the right of the G^m—, and the sequence of Fragment A is pG—G—G—C—G—U—G—U—G^m—G—.

Using the same procedure, the entire structure of alanine transfer RNA was worked out. The complete nucleotide sequence of alanine transfer RNA is shown at the top of the next two pages.

The work on the structure of this molecule took us seven years from start to finish. Most of the time was consumed in developing procedures for the isolation of a single species of transfer RNA from the 20 or so different transfer RNA's present in the living cell. We finally selected a fractionation technique known as countercurrent distribution, developed in the 1940's by Lyman C. Craig of the Rockefeller Institute.

This method exploits the fact that similar molecules of different structure will exhibit slightly different solubilities if they are allowed to partition, or distribute themselves, between two nonmiscible liquids. The countercurrent technique can be mechanized so that the mixture of molecules is partitioned hundreds or thousands of times, while the nonmiscible solvents flow past each other in a countercurrent pattern. The solvent system we adopted was composed of formamide, isopropyl alcohol and a phosphate buffer, a modification of a system first described by Robert C. Warner and Paya Vaimberg of New York University. To make the method applicable for fractionating transfer RNA's required four years of work in collaboration with Jean Apgar, B. P. Doctor and Susan H. Merrill of the Plant, Soil and Nutrition Laboratory. Repeated countercurrent extractions of the transfer RNA mixture gave three of the RNA's in a reasonably homogeneous state: the RNA's that transfer the amino acids alanine, tyrosine and valine [see bottom illustration on page 220].

The starting material for the countercurrent distributions was crude transfer RNA extracted from yeast cells using phenol as a solvent. In the course of the structural work we used about 200 grams (slightly less than half a pound) of mixed transfer RNA's isolated from 300 pounds of yeast. The total amount of purified alanine transfer RNA we had to work with over a three-year period was one gram. This represented a practical compromise between the difficulty of scaling up the fractionation procedures and scaling down the techniques for structural analysis.

Once we knew the complete sequence, we could turn to general questions about the structure of transfer RNA's. Each transfer RNA presumably embodies a sequence of three subunits (an "anticodon") that forms a temporary bond with a complementary sequence of three subunits (the "codon") in messenger RNA. Each codon triplet identifies a specific amino acid [see the article "The Genetic Code: II," by Marshall W. Nirenberg, beginning on page 324].

An important question, therefore, is which of the triplets in alanine transfer RNA might serve as the anticodon for the alanine codon in messenger RNA. There is reason to believe the anticodon is the sequence I—G—C, which is found in the middle of the RNA molecule. The codon corresponding to I—G—C could be the triplet G—C—C or perhaps G—C—U, both of which act as code words for alanine in messenger RNA. As shown in the illustration on page 218, the I—G—C in the alanine transfer RNA is upside down when it makes contact with the corresponding codon in messenger RNA. Therefore when alanine transfer RNA is delivering its amino acid cargo and is temporarily held by hydrogen bonds to messenger RNA, the I would pair with C (or U) in the messenger, G would pair with C, and C would pair with G.

We do not know the three-dimensional structure of the RNA. Presumably there is a specific form that interacts with the messenger RNA and ribosomes. The illustration on page 219 shows three hypothetical structures for alanine transfer RNA that take account of the propensity of certain bases to pair with other bases. Thus adenine pairs with uracil and cytosine with guanine. In the three hypothetical structures the I—G—C sequence is at an exposed position and could pair with messenger RNA.

The small diagram on page 219 indicates a possible three-dimensional folding of the RNA. Studies with atomic models suggest that single-strand regions of the structure are highly flexible. Thus in the "three-leaf-clover" configuration it is possible to fold one side leaf on top of the other, or any of the leaves back over the stem of the molecule.

One would also like to know whether or not the unusual nucleotides are concentrated in some particular region of the molecule. A glance at the sequence shows that they are scattered throughout the structure; in the three-leaf-clo-

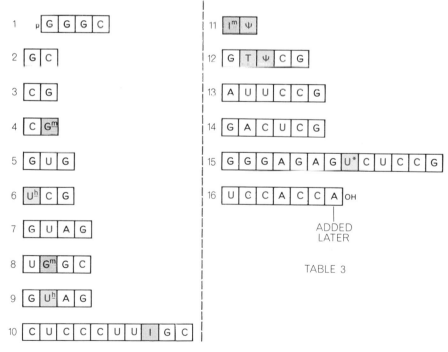

CONSOLIDATED LIST OF SEQUENCES accounts for all 77 nucleotides in alanine transfer RNA. The consolidated list is formed by selecting the largest fragments in Table 1 and Table 2 (opposite page) and by piecing together fragments that obviously overlap. Thus Fragment 15 has been formed by joining two smaller fragments, keyed by the number 15, in Table 1 and Table 2 on the opposite page. Since the entire molecule contains only one U*, the two fragments must overlap at that point. The origin of the other fragments in Table 3 can be traced in similar fashion. A separate experiment in which the molecule was cut into two parts helped to establish that the 10 fragments listed in the first column are in the left half of the molecule and that the six fragments in the second column are in the right half.

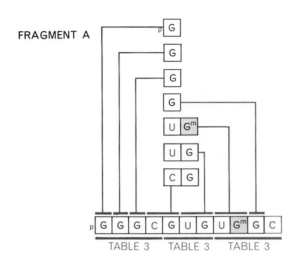

COMPLETE MOLECULE of alanine transfer RNA contains 77 nucleotides in the order shown. The final sequence required a care-

ful piecing together of many bits of information (*see illustration at bottom of these two pages*). The task was facilitated by degrada-

ver model, however, the unusual nucleotides are seen to be concentrated around the loops and bends.

Another question concerns the presence in the transfer RNA's of binding sites, that is, sites that may interact specifically with ribosomes and with

the enzymes involved in protein synthesis. We now know from the work of Zamir and Marquisee that a particular sequence containing pseudouridylic acid (Ψ), the sequence G–T–Ψ–C–G, is found not only in the alanine transfer RNA but also in the transfer RNA's for

tyrosine and valine. Other studies suggest that it may be present in all the transfer RNA's. One would expect such common sites to serve a common function; binding the transfer RNA's to the ribosome might be one of them.

Work that is being done in many

FRAGMENT A

FRAGMENT B

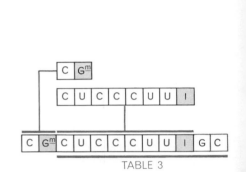

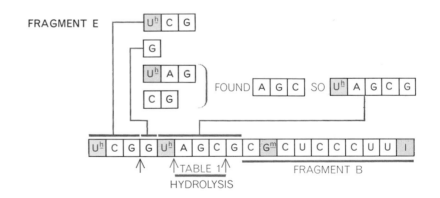

REMAINDER OF LEFT HALF OF MOLECULE

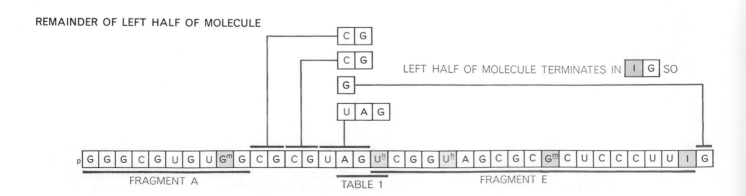

FRAGMENT C — FRAGMENT D

| C | Iᵐ | ψ | G | G | G | A | G | A | G | U* | C | U | C | C | G | G | T | ψ | C | G | A | U | U | C | C | G | G | A | C | U | C | G | U | C | C | A | C | C | A |ₒₕ

FRAGMENT F — FRAGMENT G

tion experiments that cleaved this molecule into several large fragments (*A, B, C, D, E, F, G*), and by the crucial discovery that the molecule could be divided almost precisely into two halves. The division point is marked by the "gutter" between these two pages.

laboratories around the world indicates that alanine transfer RNA is only the first of many nucleic acids for which the nucleotide sequences will be known. In the near future it should be possible to identify those structural features that are common to various transfer RNA's, and this should help greatly in defining the interactions of transfer RNA's with messenger RNA, ribosomes and enzymes involved in protein synthesis. Further in the future will be the description of the nucleotide sequences of the nucleic acids—both DNA and RNA— that embody the genetic messages of the viruses that infect bacteria, plants and animals. Much further in the future lies the decoding of the genetic messages of higher organisms, including man. The work described in this article is a step toward that distant goal.

FRAGMENT C

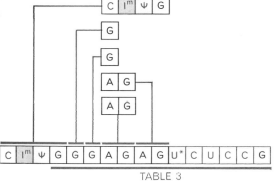

TABLE 3

FRAGMENT D

FRAGMENT F

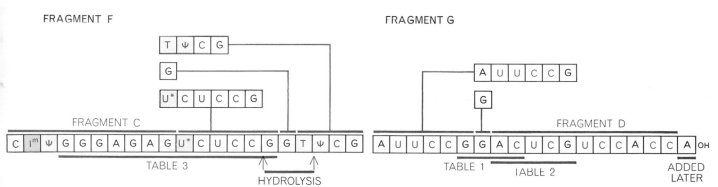

TABLE 3
HYDROLYSIS

FRAGMENT G

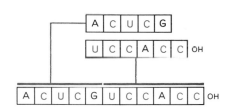

TABLE 1   TABLE 2   ADDED LATER

ASSEMBLY OF FRAGMENTS resembled the solving of a jigsaw puzzle. The arguments that established the sequence of nucleotides in Fragment *A* are described in the text. Fragment *B* contains two subfragments. The larger is evidently Fragment 10 in Table 3, which ends in G—C—. This means that the C—Gᵐ— fragment must go to the left. Fragment *E* contains Fragment *B* plus four smaller fragments. It can be shown that *E* ends with I—, therefore the four small pieces are again to the left. A pancreatic digest yielded A—G—C—, thus serving to connect Uʰ—A—G— and C—G—. A partial digestion with ribonuclease T1 removed Uʰ—C—G—, showing it to be at the far left. The remaining G— must follow immediately or a pancreatic digest would have yielded a G—G—C— sequence, which it did not. Analyses of Fragments *A* and *E* accounted for everything in the left half of the molecule except for four small pieces. The left half of the molecule was shown to terminate in I—G—, thus the remaining three pieces are between *A* and *E*. Table 1 shows that one Uʰ is preceded by A—G—, therefore U—

A—G— must be next to *E*. The two remaining C—G—'s must then fall to the left of U—A—G—. Fragment *C* contains five pieces. Table 3 (Fragment 15) shows that the two A—G—'s are next to U* and that the two G—'s are to the left of them. It is also clear that C—Iᵐ—Ψ—G— cannot follow U*, therefore it must be to the left. Fragment *D* contains two pieces; the OH group on one of them shows it to be to the right. Fragment *F* contains Fragment *C* plus three extra pieces. These must all lie to the right since hydrolysis with pancreatic ribonuclease gave G—G—T— and not G—T—, thus establishing that the single G— falls as shown. Fragment *G* gave *D* plus two pieces, which must both lie to the left (because of the terminal Cₒₕ). Table 1 shows a G—G—A—C— sequence, which must overlap the A—C— in A—C—U—C—G— and the G— at the right end of the A—U—U—C—C—G—. Fragments *F* and *G* can join in only one way to form the right half of the molecule. The molecule is completed by the addition of a final Aₒₕ, which is missing as the alanine transfer RNA is separated from baker's yeast.

26

THE SYNTHESIS OF DNA

ARTHUR KORNBERG
October 1968

*Test-tube synthesis of the double helix that controls
heredity climaxes a half-century of effort by biochemists
to re-create biologically active giant molecules outside
the living cell*

My colleagues and I first undertook to synthesize nucleic acids outside the living cell, with the help of cellular enzymes, in 1954. A year earlier James Watson and Francis Crick had proposed their double-helix model of DNA, the nucleic acid that conveys genetic information from generation to generation in all organisms except certain viruses. We attained our goal within a year, but not until some months ago—14 years later—were we able to report a completely synthetic DNA, made with natural DNA as a template, that has the full biological activity of the native material.

Our starting point was an unusual single-strand form of DNA found in the bacterial virus designated φX174. The single strand is in the form of a closed loop. When φX174 infects cells of the bacterium *Escherichia coli*, the single-strand loop of DNA serves as the template that directs enzymes in the synthesis of a second loop of DNA. The two loops form a ring-shaped double helix similar to the DNA helixes found in bacterial cells and higher organisms. In our laboratory at the Stanford University School of Medicine we succeeded in reconstructing the synthesis of the single-strand DNA copies of viral DNA and finally in making a completely synthetic double helix. The way now seems open for the synthesis of DNA from other sources: viruses associated with human disease, bacteria, multicellular organisms and ultimately the DNA of vertebrates such as mammals.

An Earlier Beginning

The story of the cell-free synthesis of DNA does not start with the revelation of the structure of DNA in 1953. It begins around 1900 with the biochemical understanding of how the fermentation of fruit juices yields alcohol. Some 40 years earlier Louis Pasteur had convinced his contemporaries that the living yeast cell played an essential role in the fermentation process. Then Eduard Buchner observed in 1897 that a cell-free juice obtained from yeast was just as effective as intact cells for converting sugar to alcohol. This observation opened the era of modern biochemistry.

During the first half of this century biochemists resolved the overall conversion of sucrose to alcohol into a sequence of 14 reactions, each catalyzed by a specific enzyme. When this fermentation proceeds in the absence of air, each molecule of sucrose consumed gives rise to four molecules of adenosine triphosphate (ATP), the universal currency of energy exchange in living cells. The energy represented by the fourfold output of ATP per molecule of sucrose is sufficient to maintain the growth and multiplication of yeast cells. When the fermentation takes place in air, the oxidation of sucrose goes to completion, yielding carbon dioxide and water along with 18 times as much energy as the anaerobic process does. This understanding of how the combustion of sugar provides energy for cell metabolism was succeeded by similar explanations of how enzymes catalyze the oxidation of fatty acids, amino acids and the subunits of nucleic acids for the energy needs of the cell.

By 1950 the enzymatic dismantling of large molecules was well understood. Little thought or effort had yet been invested, however, in exploring how the cell makes large molecules out of small ones. In fact, many biochemists doubted that biosynthetic pathways could be suc-

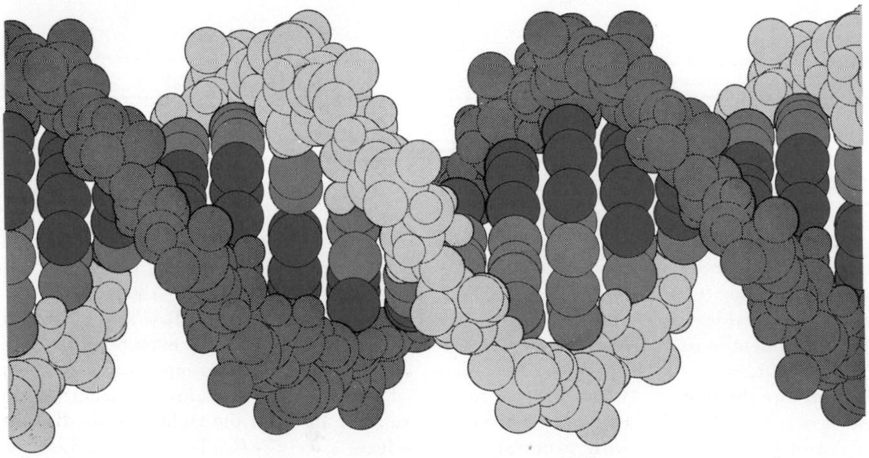

DOUBLE HELIX, the celebrated model of deoxyribonucleic acid (DNA) proposed in 1953 by James D. Watson and F. H. C. Crick, consists of two strands held together by crossties (*color*) that spell out a genetic message, unique for each organism. The Watson-Crick model explained for the first time how each crosstie consists of two subunits, called bases, that form obligatory pairs (*see illustrations on page 230*). Thus each strand of the double helix and its associated sequence of bases is complementary to the other strand and its bases. Consequently each strand can serve as a template for the reconstruction of the other strand.

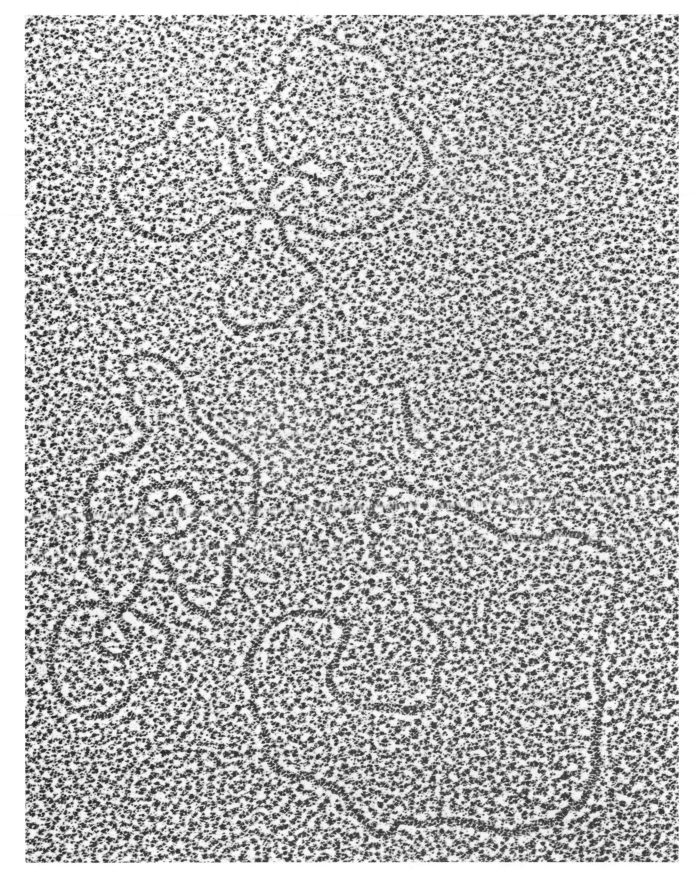

THREE CLOSED LOOPS OF DNA, each a complete double helix, are shown in this electron micrograph made in the author's laboratory at the Stanford University School of Medicine. One strand of each loop is the natural single-strand DNA of the bacterial virus φX174, which served as a template for the test-tube synthesis, carried out by enzymes, of a synthetic complementary strand. The hybrid molecules are biologically active. The enlargement is about 200,000 diameters. Each loop contains some 5,500 pairs of bases. If enlarged to the scale of the model on the opposite page, each loop of DNA would form a circle roughly 150 feet in circumference.

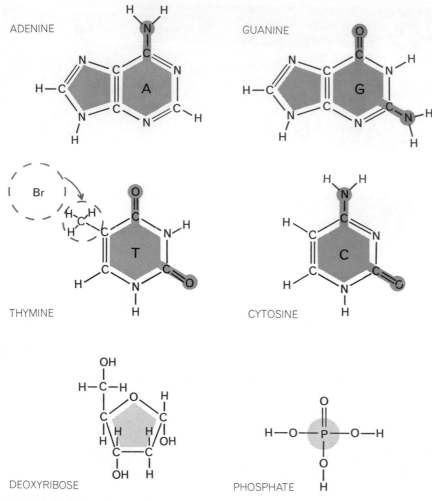

ADENINE

GUANINE

THYMINE

CYTOSINE

DEOXYRIBOSE

PHOSPHATE

DNA CONSTITUENTS are bases of four kinds, deoxyribose (a sugar) and a simple phosphate. The bases are adenine (*A*) and thymine (*T*), which form one obligatory pair, and guanine (*G*) and cytosine (*C*), which form another. Deoxyribose and phosphate form the backbone of each strand of the DNA molecule. The bases provide the code letters of the genetic message. For purposes of tagging synthetic DNA, thymine can be replaced by 5'-bromouracil, which contains a bromine atom where thymine contains a lighter CH$_3$ group.

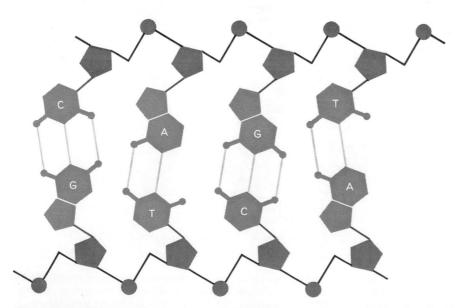

DNA STRUCTURE resembles a ladder in which the side pieces consist of alternating units of deoxyribose and phosphate. The rungs are formed by the bases paired in a special way, A with T and G with C, and held together respectively by two and three hydrogen bonds.

cessfully reconstructed in cell-free systems. Since then nearly two decades of intensive study have been devoted to the cell-free biosynthesis of large molecules. Two things above all have been made clear.

The first is that large molecules can be assembled in cell-free systems with the aid of purified enzymes and coenzymes. The second is that the routes of biosynthesis are different from those of degradation. Some biochemists had speculated that the routes of breakdown were really two-way streets whose flow might somehow be reversed. Now we know that the molecular traffic in cells flows on distinctive and divided highways. All cells have the enzymatic machinery to manufacture most of the subunits of large molecules from simple nutrients such as glucose, ammonia and carbon dioxide. Cells also have the capacity to salvage preformed subunits when they are available. On the basis of what has been learned the prospects are that in this century biochemists will assemble in the test tube complex viruses and major components of the cell. Perhaps the next century will bring the synthesis of a complete cell.

The Nucleotides

My co-workers and I were at Washington University in St. Louis when we made our first attempts to synthesize a nucleic acid in the test tube. By that time the constituents of nucleic acid were well known [*see illustrations at left*]. If one regards DNA as a chain made up of repeating links, the basic link is a structure known as a nucleotide [*see illustration on opposite page*]. It consists of a phosphate group attached to the five-carbon sugar deoxyribose, which is linked in turn to one of four different nitrogen-containing bases. The four bases are adenine (A), thymine (T), guanine (G) and cytosine (C). In the double helix of DNA the phosphate and deoxyribose units alternate to form the two sides of a twisted ladder. The rungs joining the sides consist of two bases: A is invariably linked to T and G is invariably linked to C. This particular pairing arrangement was the key insight of the Watson-Crick model. It means that if the two strands of the helix are separated, uncoupling the paired bases, each half can serve as a template for re-creating the missing half. Thus if the bases projecting from a single strand follow the sequence A, G, G, C, A, T..., one immediately knows that the complementary bases on the missing strand are T, C, C, G, T, A.... This base-pairing mechanism enables the cell to make accurate copies of the DNA molecule however many times the cell may divide.

When a strand of DNA is taken apart link by link (by treatment with acid or certain enzymes), the phosphate group of the nucleotide may be found attached to carbon No. 3 of the five-carbon deoxyribose sugar. Such a structure is called a 3′-nucleoside monophosphate. We judged, however, that better subunits for purposes of synthesis would be the 5′-nucleoside monophosphates, in which the phosphate linkage is to carbon No. 5 of deoxyribose.

This judgment was based on two lines of evidence. The first had just emerged from an understanding of how the cell itself made nucleotides from glucose, ammonia, carbon dioxide and amino acids. John M. Buchanan of the Massachusetts Institute of Technology had shown that nucleotides containing the bases A and G were naturally synthesized with a 5′ linkage. Our own work had shown the same thing for nucleotides containing T and C. The second line of evidence came from earlier studies my group had conducted at the National Institutes of Health. We had found that certain coenzymes, the simplest molecules formed from two nucleotides, were elaborated from 5′ nucleotide units. For the enzymatic linkage to take place the phosphate of the nucleotide had to be activated by an additional phosphate group [see illustration on next page]. Thus it seemed reasonable that activated 5′ nucleotides (nucleoside 5′ triphosphates) might combine with each other, under the proper enzymatic guidance, to form long chains of nucleic acid.

Our initial attempts at nucleic acid synthesis relied principally on two techniques. The first involved the use of radioactive atoms to label the nucleotide so that we could detect the incorporation of even minute amounts of it into nucleic acid. We sought the enzymatic machinery for synthesizing nucleic acids in the juices of the thymus gland, bone marrow and bacterial cells. Unfortunately such extracts also have a potent capacity for degrading nucleic acids. We added our labeled nucleotides to a pool of nucleic acids and hoped that a few synthesized molecules containing a labeled nucleotide would survive by being mixed into the pool. Even if there were net destruction of the pool of nucleic acids, the synthesis of a few molecules trapped in this pool might still be detected. The second technique exploited the fact that the nucleic acid could be precipitated by making the medium strongly acidic, whereas the nucleotide

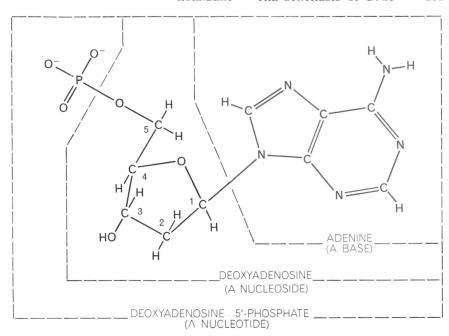

DNA BUILDING BLOCK, the monomer from which DNA polymers are constructed, is termed a nucleotide. There are four nucleotides, one for each of the four bases A, T, G and C. Deoxyadenosine 5′-phosphate, the nucleotide incorporating adenine, is shown here. If the phosphate group is replaced by a hydrogen atom, the structure is called a nucleoside.

precursors remained behind in solution.

Our first experiments with animal-cell extracts were uniformly negative. Therefore we turned to E. coli, which has the virtue of reproducing once every 20 minutes. Here we saw a glimmer. In samples to which we had added a quantity of labeled nucleotides whose radioactive atoms disintegrated at the rate of a million per minute we detected about 50 radioactive disintegrations per minute in the nucleic acid fraction that was precipitated by acid. Although the amount of nucleotide incorporated into nucleic acid was minuscule, it was nonetheless significantly above the level of background "noise." Through this tiny crack we tried to drive a wedge. The hammer was enzyme purification, a technique that had matured during the elucidation of alcoholic fermentation.

DNA Polymerase

In these experiments Uriel Littauer, a Fellow of the Weizmann Institute in Israel, and I observed the incorporation of adenylate (a nucleotide) from ATP into ribonucleic acid (RNA), in which the five-carbon sugar in the backbone of the chain is ribose rather than deoxyribose. Actually the first definitive demonstration of synthesis of an RNA-like molecule in a cell-free system had been achieved in the laboratory of Severo Ochoa in 1955. Working at the New York University School of Medicine, he

and Marianne Grunberg-Manago were investigating an aspect of energy metabolism and made the unexpected observation that one of the reactants, adenosine diphosphate (ADP), had been polymerized by cell juices into a chain of adenylates resembling RNA.

In our first attempts to achieve DNA synthesis in a cell-free system we used the deoxyribonucleoside called deoxythymidine. To Morris E. Friedkin, who was then at Washington University, we are grateful not only for supplying the radioactively labeled compound but also for the knowledge that the compound was readily incorporated into DNA by bone marrow cells and other animal cells. We were hopeful that extracts of E. coli would be able to incoporate deoxythymidine into nucleic acid by converting it first into the 5′ deoxynucleotide and then activating the deoxynucleotide to the triphosphate form. I found this to be the case. In subsequent months Ernest Simms and I were able to prepare separately deoxythymidine 5′-triphosphate and the other deoxynucleoside triphosphates, using enzymes or chemical synthetic routes. (In what follows the various deoxynucleosides in their 5′ triphosphate form will be designated simply by the initial of the base followed by an asterisk. Thus deoxythymidine 5′-triphosphate will be T*.)

In November, 1955, I. Robert Lehman, who is now at Stanford, started on the purification of the enzyme system in *E. coli* extracts that is responsible for converting T* into DNA. We were joined by Maurice J. Bessman some weeks later. Those were eventful days in which the enzyme, now given the name DNA polymerase, was progressively separated from other large molecules. With each step in purification the character of this DNA synthetic reaction became clearer. By June, 1956, when we participated at a conference on the chemical basis of heredity held at Johns Hopkins University, we could report two important facts about DNA synthesis in vitro, although we still lacked the answers to many important questions.

We reported first that preformed DNA had to be present along with DNA polymerase, and that all four of the deoxynucleotides that occur in DNA (A, G, T and C) had to be furnished in the activated triphosphate form. We also reported that DNA from virtually any source—virus, bacterium or animal—could serve with the *E. coli* enzyme. What we still did not know was whether the synthetic DNA was a new molecule or an extension of a preexisting one. There were other questions. Did the synthetic DNA have the same chemical backbone and physical structure as natural DNA? Did it have a chemical composition typical of DNA, in which A equals T and G equals C, and in which, therefore, A plus G equals T plus C? Finally, and crucially: Did the chemical composition of the synthetic DNA reflect the composition of the particular natural DNA used to direct the reaction?

During the next three years these questions and related ones were resolved by the efforts of Julius Adler, Sylvy Kornberg and Steven B. Zimmerman. The synthetic DNA was shown to be a molecule with the chemical structure typical of DNA and the same ratio of A-T pairs to G-C pairs as the particular DNA used to prime, or direct, the reaction [see illustration on page 234]. The relative starting amounts of the four deoxynucleoside triphosphates had no influence whatever on the composition of the new DNA. The composition of the synthetic DNA was determined solely by the composition of the DNA that served as a template. An interesting illustration of this last fact justifies a slight digression.

Howard K. Schachman of the University of California at Berkeley spent his sabbatical year of 1957–1958 with us at Washington University examining the physical properties of the synthetic DNA. It had the high viscosity, the comparatively slow rate of sedimentation and other physical properties typical of natural DNA. The new DNA, like the natural one, was therefore a long, fibrous polymer molecule. Moreover, the longer the mixture of active ingredients was allowed to incubate, the greater the viscosity of the product was; this was direct evidence that the synthetic DNA was continuing to grow in length and in amount. However, we were startled to find one day that viscosity developed in a control test tube that lacked one of the essential triphosphates, G*. To be sure, no reaction was observed during the standard incubation period of one or two hours. On prolonging the incubation for several more hours, however, a viscous substance materialized!

Analysis proved this substance to be a DNA that contained only A and T nucleotides. They were arranged in a perfect alternating sequence. The isolated polymer, named dAT, behaved like any other DNA in directing DNA synthesis: it led to the immediate synthesis of more dAT polymer. Would any G* and C* be polymerized if these nucleotides were present in equal or even far greater amounts than A* and T* in a synthesis directed by dAT polymer? We found no detectable incorporation of G or C under conditions that would have measured the inclusion of even one G for every 100,000 A or T nucleotides polymerized. Thus DNA polymerase rarely, if ever, made the mistake of matching G or C with A or T.

The DNA of a chromosome is a linear array of many genes. Each gene, in turn,

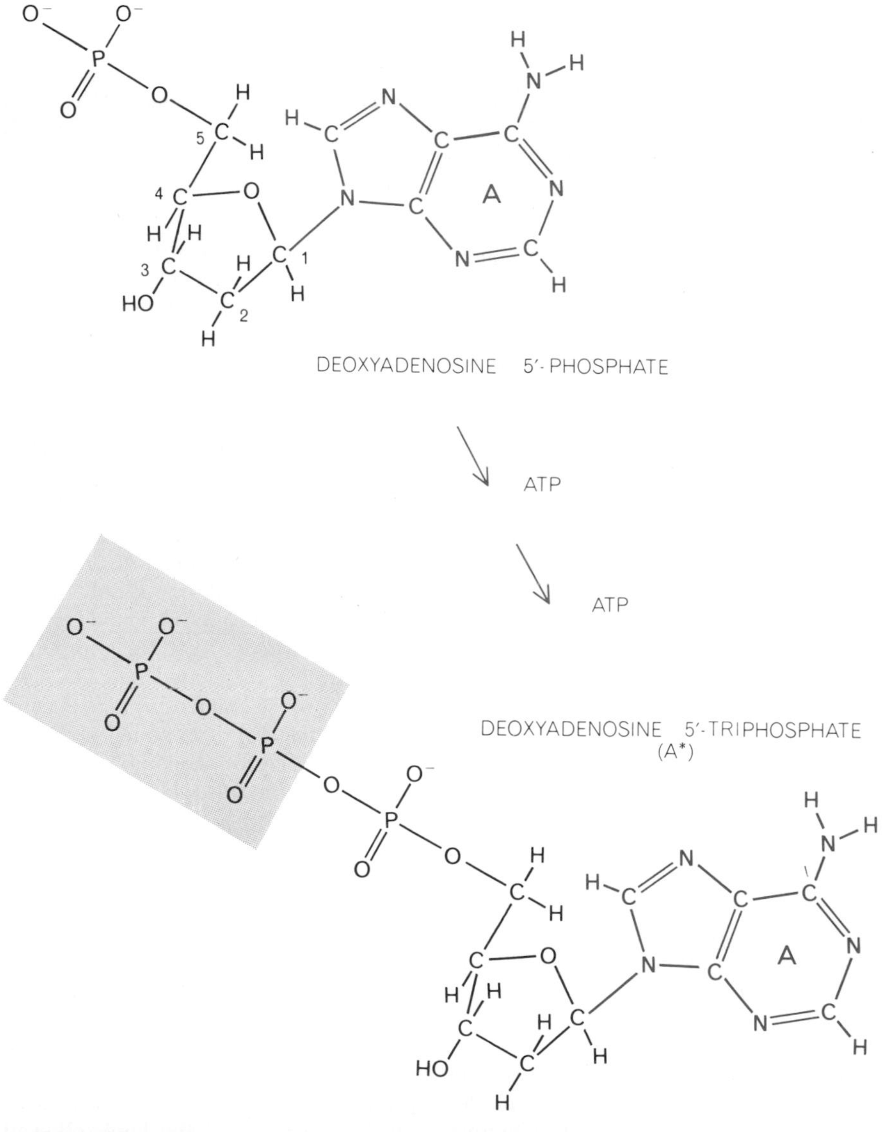

DEOXYADENOSINE 5′-PHOSPHATE

ATP

ATP

DEOXYADENOSINE 5′-TRIPHOSPHATE
(A*)

ACTIVATED BUILDING BLOCK is required when synthesizing DNA on a template of natural DNA with the aid of enzymes. The activated form of the nucleotide containing adenine is deoxyadenosine 5′-triphosphate, symbolized in this article by "A*." It is made from deoxyadenosine 5′-monophosphate by two different enzymes in two steps. Each step involves the donation of a terminal phosphate group from adenosine triphosphate (ATP).

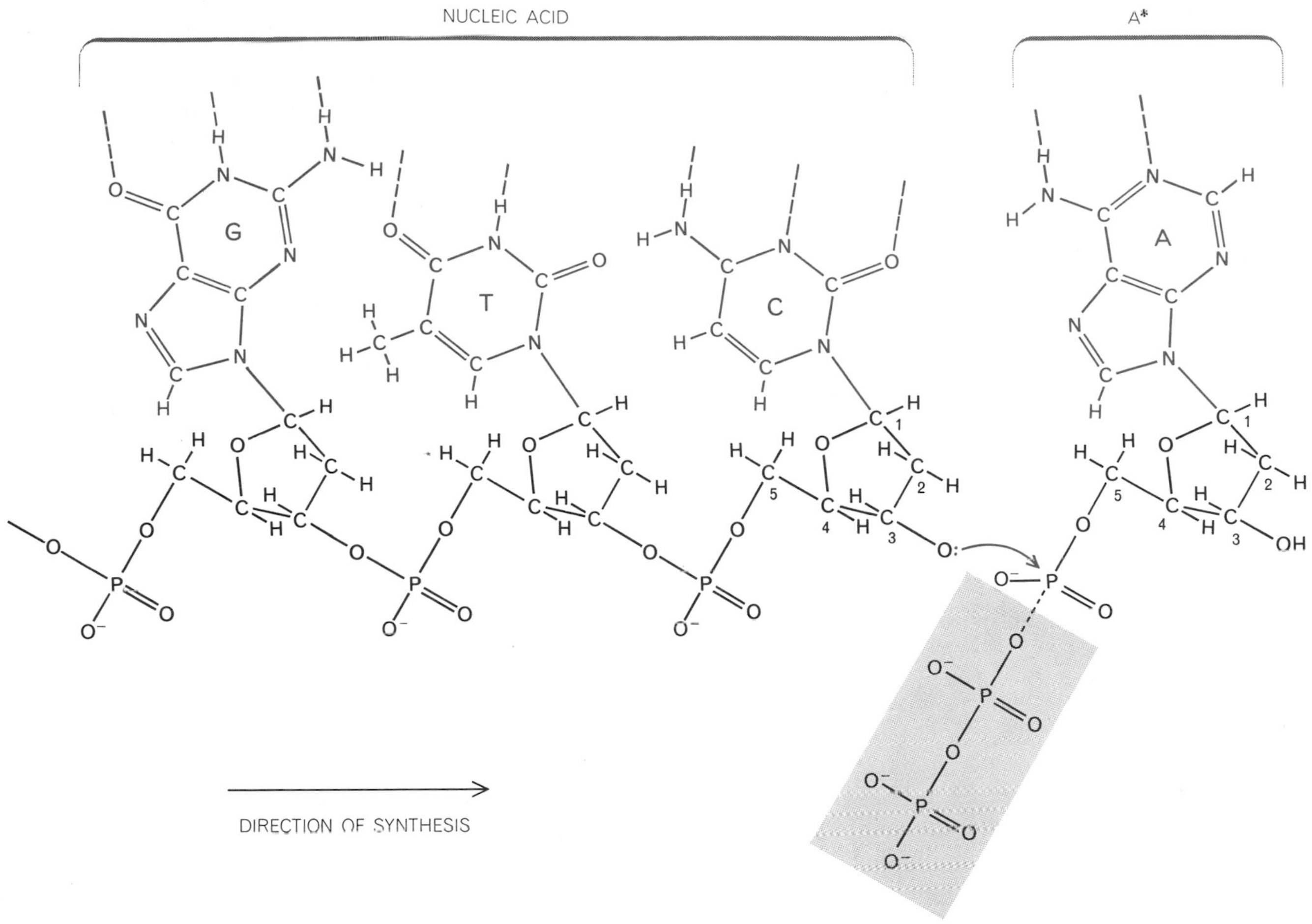

SYNTHESIS OF DNA involves the stepwise addition of activated nucleotides to the growing polymer chain. In this illustration deoxyadenosine 5'-triphosphate (A*) is being coupled through a phosphodiester bond that links the 3' carbon in the deoxyribose portion of the last nucleotide in the growing chain to the 5' carbon in the deoxyribose portion of the newest member of the chain.

is a chain of about 1,000 nucleotides in a precisely defined sequence, which when translated into amino acids spells out a particular protein or enzyme. Does DNA polymerase in its test-tube synthesis of DNA accurately copy the sequential arrangement of nucleotides by base-pairing (A = T, G = C) without errors of mismatching, omission, commission or transposition? Unfortunately techniques are not available for determining the precise sequence of nucleotides of even short DNA chains. Because it is impossible to spell out the base sequence of natural DNA or any copy of it, we have resorted to two other techniques to test the fidelity with which DNA polymerase copies the template DNA. One is "nearest neighbor" analysis. The other is the duplication of genes with demonstrable biological activity.

Nearest-Neighbor Analysis

The nearest-neighbor analysis devised by John Josse, A. Dale Kaiser and myself in 1959 determines the relative frequency with which two nucleotides can end up side by side in a molecule of synthetic DNA. There are 16 possible combinations in all. There are four possible nearest-neighbor sequences of A (AA, AG, AT and AC), four for G (GA, GG, GT and GC) and similarly four for T and four for C. How can the frequency of these dinucleotide sequences be determined in a synthetic DNA chain? The procedure is to use a triphosphate labeled with a radioactive phosphorus atom in conducting the synthesis and to treat the synthesized DNA with a specific enzyme that cleaves the DNA and leaves the radioactive phosphorus atom attached to its nearest neighbor. For example, DNA synthesis is carried out with A* labeled in the innermost phosphate group, the group that will be included in the DNA product. This labeled phosphate group now forms a normal linkage (10^{16} times in a typical experiment!) with the nucleotide next to it in the chain—its nearest neighbor [see

illustration on page 235]. After the synthetic DNA is isolated it is subjected to degradation by an enzyme that cleaves every bond between the 5' carbon of deoxyribose and the phosphate, leaving the radioactive phosphorus atom attached to the neighboring nucleotide rather than to the one (A) to which it had originally been attached. The nucleotides of the degraded DNA are readily separated by electrophoresis or paper chromatography into the four types of which DNA is composed: A, G, T and C. Radioactive assay establishes the radioactive phosphorus content in each of these nucleotides and at once indicates the frequency with which A is next to A, to G, to T and to C.

The entire experiment is repeated, this time with the radioactive label in G* instead of A*. The second experiment yields the frequency of GA, GG, GT and GC dinucleotides. Two more experiments with radioactive T* and C* complete the analysis and establish the 16 possible nearest-neighbor frequencies.

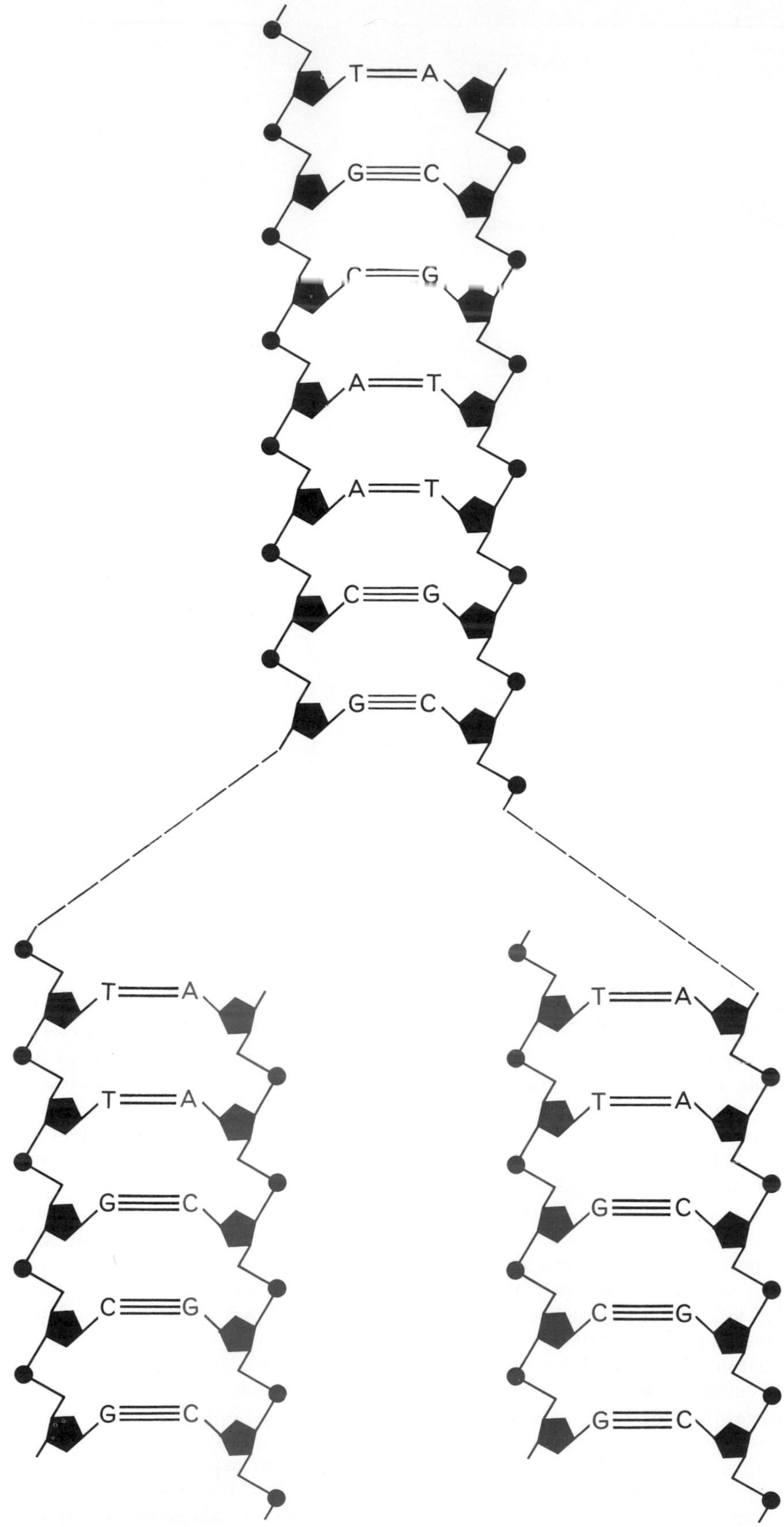

SYNTHESIS OF DUPLEX CHAIN OF DNA yields two hybrid molecules, consisting of a parental strand and a daughter strand, that are identical with each other and with the original duplex molecule. During the replicating process the parental duplex (*black*) separates into two strands, each of which then serves as the template for assembly of a daughter strand (*color*). The pairing of A with T and G with C guarantees faithful reproduction.

Many such experiments were performed with DNA templates obtained from viruses, bacteria, plants and animals. The DNA of each species guided the synthesis of DNA with what proved to be a distinctive assortment of nearest-neighbor frequencies. What is more, when a synthetic DNA was used as a template for a new round of replication, it gave rise to DNA with a nearest-neighbor frequency distribution identical with itself. Among the other insights obtained from these analyses was the recognition of a basic fact about the structure of the double helix. In replication the direction of the DNA chain being synthesized was found to run opposite to that of its template. By inference we can conclude that the chains of the double helix in natural DNA, as surmised by Watson and Crick, must also run in opposite directions.

Even with considerable care the accuracy of nearest-neighbor frequency analysis cannot be better than about 98 percent. Consequently we were still left with major uncertainties as to the precision of copying chains that contain 1,000 nucleotides or more, corresponding to the length of genes. An important question thus remained unanswered: Does DNA that is synthesized on a genetically or biologically active template duplicate the activity of that template?

One way to recognize the biological activity of bacterial DNA is to see if it can carry out "transformation," a process in which DNA from one species of bacteria alters the genetic endowment of a second species. For example, DNA from a strain of *Bacillus subtilis* resistant to streptomycin can be assimilated by a strain susceptible to the antibiotic, whereupon the recipient bacterium and all its descendants carry the trait of resistance to streptomycin. In other words, DNA molecules carrying the genes for a particular characteristic can be identified by their capacity for assimilation into the chromosome of a cell that previously lacked that trait. Yet when DNA was synthesized on a template of DNA that had transforming ability, the synthetic product invariably lacked that ability.

Part of the difficulty in synthesizing biologically active DNA lay in the persistence of trace quantities of nuclease enzymes in our DNA polymerase preparations. Nucleases are enzymes that degrade DNA. The introduction by a nuclease of one break in a long chain of DNA is enough to destroy its genetic activity. Further purification of DNA polymerase was indicated. Efforts over several years by Charles C. Richardson, Thomas Jovin, Paul T. Englund and LeRoy L. Bertsch resulted in a new procedure that was both simple and efficient. Finally, in April, 1967, with the assistance of the personnel and large-scale equipment of the New England Enzyme Center (sponsored by the National Institutes of Health at the Tufts University School of Medicine), we processed 100 kilograms of *E. coli* bacterial paste and obtained about half a gram of pure enzyme, free of the nuclease that puts random breaks in a DNA chain.

Unfortunately even this highly purified DNA polymerase has proved incapable of producing a biologically active DNA from a template of bacterial DNA. The difficulty, we believe, is that the DNA we extract from a bacterium such as *B. subtilis* provides the enzyme with a poor template. A proper template would be the natural chromosome, which is a double-strand loop about one millimeter in circumference. During its isolation from the bacterium the chromosome is broken, probably at random, into 100 or more fragments. The manner in which DNA polymerase and its related enzymes go about the replication of a DNA molecule as large and complex as the *B. subtilis* chromosome is the subject of current study in many laboratories.

The Virus φX174

It occurred to us in 1964 that the problem of synthesizing biologically active DNA might be solved by dealing with a simpler form of DNA that also has genetic activity. This is represented in viruses, such as φX174, whose DNA core is a single-strand loop. This "chromosome" not only is simpler in structure but also it is so small (about two microns in circumference) that it is fairly easy to extract without breakage. We also knew from the work of Robert L. Sinsheimer at the California Institute of Technology that when the DNA of φX174 invades *E. coli*, the first stage of infection involves the "subversion" of one of the host's enzymes to convert the single-strand loop into a double-strand helical loop. Sinsheimer called this first-stage product a "replicative form." Could the host enzyme that copies the viral DNA be the same DNA polymerase we had isolated from *E. coli*?

In undertaking the problem of copying a closed-loop DNA we could foresee some serious obstacles. Would it be possible for DNA polymerase to orient itself and start replication on a DNA

template if the template had no ends? Shashanka Mitra and later Peter Reichard succeeded in finding conditions under which the enzyme, as judged by electron microscope pictures, appeared to copy the single-strand loop. We then wondered if in spite of appearances in the electron micrographs, the DNA of φX174 was really just a simple loop. Perhaps, as had been suggested by other workers, it was really more like a necklace with a clasp, the clasp consisting of substances unrelated to the nucleotides we were supplying. Finally, we were aware from Sinsheimer's work that the DNA of φX174 had to be a completely closed loop in order to be in-

fectious. We knew that our polymerase could only catalyze the synthesis of linear DNA molecules. How could we synthesize a genuinely closed loop? We were still missing either the clasplike component to insert into our product or, if the clasp was a mistaken hypothesis, a new kind of enzyme to close the loop.

Fortunately the missing factor was provided for us by work carried on independently in five different laboratories. The discovery in 1966 of a polynucleotide-joining enzyme was made almost simultaneously by Martin F. Gellert and his co-workers at the National Institutes of Health, by Richardson and Bernard Weiss at the Harvard Medical

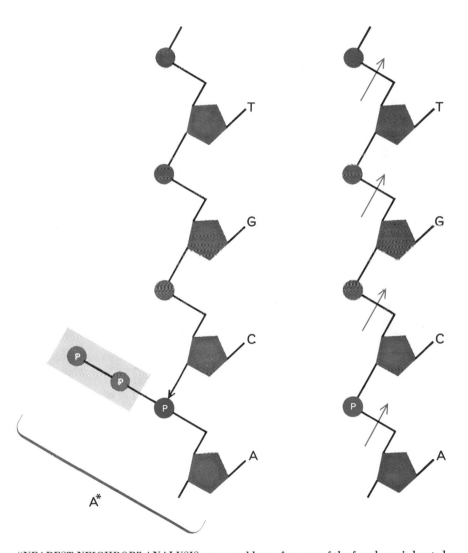

"NEAREST NEIGHBOR" ANALYSIS can reveal how often any of the four bases is located next to any other base in a single strand of synthetic DNA. Thus one can learn how often A is next to A, T, G or C, and so on. A radioactive phosphorus atom (*color*) is placed in the innermost position of one of the activated nucleotides, for example A*. The finished DNA molecule is then treated with an enzyme (*right*) that cleaves the chain between every phosphate and the 5′ carbon of the adjacent deoxyribose. Thus the phosphate is separated from the nucleotide on which it entered the chain and ends up attached to the nearest neighbor instead, C in the above example. The four kinds of nucleotide are separated by paper chromatography and the radioactivity associated with each is measured. The experiment is repeated with radioactive phosphorus linked to the other activated nucleotides.

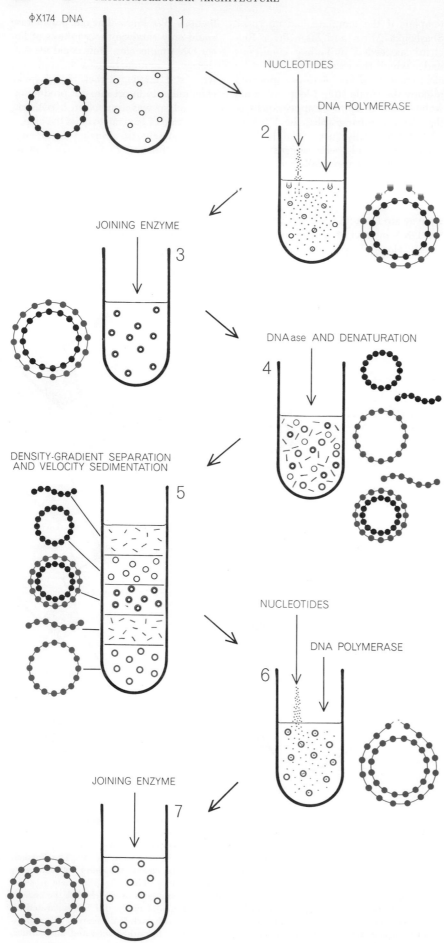

School, by Jerard Hurwitz and his colleagues at the Albert Einstein College of Medicine in New York, by Lehman and Baldomero M. Olivera at Stanford and by Nicholas R. Cozzarelli in my own group. It was the Lehman-Olivera preparation that we now employed in our experiments.

The polynucleotide-joining enzyme has the ability to repair "nicks" in the DNA strand. The nicks occur where there is a break in the sugar-phosphate backbone of one strand of the DNA molecule. The enzyme can repair a break only if all the nucleotides are intact and if what is missing is the covalent bond in the DNA backbone between a sugar and the neighboring phosphate. Provided with the joining enzyme, we were now in a position to find out whether it could work in conjunction with DNA polymerase to synthesize a completely circular and biologically active virus DNA.

By using the DNA of φX174 as a template we gained an important advantage over experiments based on transforming ability. Even if we were successful in synthesizing a DNA with transforming activity, this would still be of relatively limited significance. We could then say only that a restricted section of the DNA—a section as small as a part of a gene—had been assimilated by the recipient cell to replace a comparable section of its chromosome, substituting a proper sequence for a defective or incorrect one. However, Sinsheimer had demonstrated with the DNA of φX174 that a change in even one of its 5,500 nucleotides is sufficient to make the virus noninfective. . Therefore the demonstration of infectivity in a completely synthetic

SYNTHESIS OF φX174 DNA was accomplished by the following steps. Circular single-strand φX174 DNA, tagged with tritium, served as a template (1). Activated nucleotides containing A, G, C and 5′-bromouracil instead of T were added to the template, together with DNA polymerase. One of the activated nucleotides was tagged with radioactive phosphorus. The DNA synthesized on the template was complete but not yet joined in a loop (2). The loop was closed by the joining enzyme (3). Enough nuclease was now added to cut one strand in about half of all the duplex loops (4). This left a mixture of complete duplex loops, template loops, synthetic loops, linear template strands and linear synthetic strands. Since the synthetic strands contained 5′-bromouracil, they were heavier than the template strands and could be separated by centrifugation (5). The synthetic loops were then isolated and used as templates for making wholly synthetic duplex loops (6 and 7).

virus DNA would conclusively prove that we had carried out virtually error-free synthesis of this large number of nucleotides, comprising the five or six genes that carry out the virus's biological function.

In less than a year the test-tube synthesis of ϕX174 DNA was achieved. The steps can be summarized as follows. Template DNA was obtained from ϕX174 and labeled with tritium, the radioactive isotope of hydrogen. Tritium would thereafter provide a continuing label identifying the template. To the template were added DNA polymerase, purified joining enzyme and a cofactor (diphosphopyridine nucleotide), together with A*, T*, G* and C*. One of the nucleoside triphosphates was labeled with radioactive phosphorus. The radioactive phosphorus would thus provide a label for synthetic material analogous to the tritium label for the template. The interaction of the reagents then proceeded until the number of nucleotide units polymerized was exactly equal to the number of nucleotides in the template DNA. This equality was readily determined by comparing the radioactivity from the tritium in the template with the radioactivity from the phosphorus in the nucleotides provided for synthesis.

Such comparison showed that the experiments had progressed to an extent adequate for the formation of complementary loops of synthetic DNA. Complementary loops were designated (−) to distinguish them from the template loop (+). We had to demonstrate that the synthetic (−) loops were really loops. Had the polymerase made a full turn around the template and had the two ends of the chain been united by the joining enzyme? Several physical measurements, including electron microscopy, assured us that our product was a closed loop coiled tightly around the virus-DNA template and that it was identical in size and other details with the replicative form of DNA that appears in the infected cells. We could now exclude the possibility that some clasp material different from the nucleotide-containing compounds we had employed was involved in closing the virus-DNA loop.

The critical questions remaining were whether the synthetic (−) loops had biological activity—that is, infectivity—and whether the synthetic loops could in turn act as templates for the formation of a completely synthetic "duplex" DNA analogous to the replicative forms that were produced naturally inside infected

cells. In order to answer the first of these questions we had to isolate the synthetic DNA strands from the partially synthetic duplexes. For reasons that will be apparent below, we substituted bromouracil, a synthetic but biologically active analogue of thymine, for thymine [see top illustration on page 230]. We then introduced just enough nuclease to produce a single nick in one strand of about half the population of molecules. The duplex loops that had been nicked would release a single linear strand of DNA; these single strands could be separated from their circular companions and from unnicked duplex loops by heating. Thus we were left with a mixture that contained (+) template loops, (−) synthetic loops, (+) template linear forms, (−) synthetic linear forms—all in about equal quantities—and full duplex loops.

It was at this point that the substitution of bromouracil for thymine became useful. Because bromouracil contains a bromine atom in place of the methyl group of thymine, it is heavier than thymine. Therefore a molecule containing bromouracil can be separated from one containing thymine by high-speed centrifugation in a heavy salt solution (the density-gradient technique perfected by Jerome R. Vinograd of Cal Tech). In this system the denser a substance is, the lower in the centrifuge tube it will settle. Thus from top to bottom of the centrifuge tube we obtained fractions containing the light single strands of thymine-containing (+) template DNA, the duplex hybrids of intermediate weight and finally the single-strand synthetic (−) DNA "weighted down" with bromouracil. The reliability of this fractionation was confirmed by three separate peaks of radioactivity corresponding to each of the fractions. We were further reassured by observations that the mean density of each fraction corresponded almost exactly to the mean density of standard samples of virus DNA containing bromouracil or thymine.

Still another physical technique involving density-gradient sedimentation was employed to separate the synthetic linear forms from the synthetic circular forms. The circular forms could then be used in tests of infectivity, by methods previously developed by Sinsheimer to demonstrate the infectivity of circular ϕX174 DNA. We tested our (−) loops by incubating them with E. coli cells whose walls had been removed by the action of the enzyme lysozyme. Infectivity is assayed by the ability of the virus to lyse, or dissolve, these cells when they are "plated" on a nutrient me-

dium. Our synthetic loops showed almost exactly the same patterns of infectivity as their natural counterparts had. Their biological activity was now demonstrated.

One further set of experiments remained in which the (−) synthetic loops were employed as the template to determine if we could produce completely synthetic duplex circular forms analogous to the replicative forms found in cells infected with natural ϕX174 virus. Because the synthetic (−) loops were labeled with radioactive phosphorus, this time we added tritium to one of the nucleotide-containing subunits (C*). The remaining procedures were essentially the same as the ones described above, and we did produce fully synthetic duplex loops of ϕX174. The (+) loops were then separated and were found to be identical in all respects with the (+) loops of natural ϕX174 virus. Their infectivity could also be demonstrated. Sinsheimer had previously shown that, under these assay conditions, a change in a single nucleotide of the virus gave rise to a mutant of markedly decreased infectivity. Therefore the correspondence between the infectivity of our synthetic forms and their natural counterparts attested to the precision of the enzymatic operation.

Future Directions

The total synthesis of infective virus DNA by DNA polymerase with the four deoxynucleoside triphosphates not only demonstrates the capacity of this enzyme to copy a small chromosome (of five or six genes) without error but also shows that this chromosome, at least, is as simple and straightforward as a linear sequence of the standard four deoxynucleotide units. It is a long step to the human chromosome, some 10,000 times larger, yet we are encouraged to extrapolate our current conceptions of nucleotide composition and nucleotide linkage from the tiny ϕX174 chromosome to larger ones.

What are the major directions this research will take? I see at least three immediate and productive paths. One is the exploration of the physical and chemical nature of DNA polymerase in order to understand exactly how it performs its error-free replication of DNA. Without this knowledge of the structure of the enzyme and how it operates under defined conditions in the test tube, our understanding of the intracellular behavior of the enzyme will be incomplete.

A second direction is to clarify the

control of DNA replication in the cell and in the animal. Why is DNA synthesis arrested in a mature liver cell and what sets it in motion 24 hours after part of the liver is removed surgically? What determines the slow rate of DNA replication in adult cells compared with the rate in embryonic or cancer cells? The time is ripe for exploration of the factors that govern the initiation and rate of DNA synthesis in the intact cell and animal. Finally, there are now prospects of applying our knowledge of DNA structure and synthesis directly to human welfare. This is the realm of genetic engineering, and it is our collective responsibility to see that we exploit our great opportunities to improve the quality of human life.

An obvious area for investigation would be the synthesis of the polyoma virus, a virus known to induce a variety of malignant tumors in several species of rodents. Polyoma virus in its infective form is made up of duplex circular DNA and presumably replicates in this form on entering the cell. On the basis of our experience it would appear quite feasible to synthesize polyoma virus DNA. If this synthesis is accomplished, there would seem to be many opportunities for modifying the virus DNA and thus determining where in the chromosome its tumor-producing capacity lies. With this knowledge it might prove possible to modify the virus in order to control its tumor-producing potential.

Our speculations can extend even to large DNA molecules. For example, if a failure in the production of insulin were to be traced to a genetic deficit, then administration of the appropriate synthetic DNA might conceivably provide a cure for diabetes. Of course, a system for delivering the corrective DNA to the cells must be devised. Even this does not seem inconceivable. The extremely interesting work of Stanfield Rogers at the Oak Ridge National Laboratory suggests a possibility. Rogers has shown that the Shope papilloma virus, which is not pathogenic in man, is capable of inducing production of the enzyme arginase in rabbits at the same time that it induces tumors. Rogers found that in the blood of laboratory investigators working with the virus there is a significant reduction of the amino acid arginine, which is destroyed by arginase. This is apparently an expression of enhanced arginase activity. Might it not be possible, then, to use similar nonpathogenic viruses to carry into man pieces of DNA capable of replacing or repairing defective genes?

27

THE RECOGNITION OF DNA IN BACTERIA

SALVADOR E. LURIA

January 1970

Some bacteria have enzyme systems that scan invading DNA molecules injected by viruses and break them unless they are chemically marked at specific recognition sites

The genetic code closely resembles a universal language. As far as anyone knows every word in this language (that is, every triplet of nucleotides) means the same thing for all forms of life: it specifies a particular amino acid. By the same token, organization of a sequence of the words into a sentence, written in the form of a molecule of deoxyribonucleic acid (DNA), carries a similarly unvarying meaning: it specifies the construction of a particular protein. Hence the genetic script, like the script of a book printed in a universal language, is read in the same way by all organisms.

There are situations, however, where the organism examines the DNA script not as a linguist would but as a bibliophile would. A bibliophile may find in the structure of a book's script little signs or marks that identify the printer. It turns out that DNA often has just such identifying marks, and that these decisively affect the behavior of the organisms that recognize them. This unexpected finding has begun a fascinating chapter in the study of molecular biology and evolution.

The story begins with a curious discovery Mary L. Human and I made nearly 20 years ago in experiments with the bacteriophage, or bacterial virus, known as T2 [see "The T2 Mystery," by Salvador E. Luria; SCIENTIFIC AMERICAN Offprint 24]. Ordinarily T2 readily invades and multiplies in the bacterium *Escherichia coli,* but we found that when the virus was grown in a certain *E. coli* strain (called *B/4*), almost all the daughter phages that came out were altered in such a way that they could no longer multiply in the usual *E. coli* hosts of T2. It developed in further experiments that the altered T2 could multiply perfectly well in dysentery bacilli, and this breeding had the effect of transforming the

phage back to full ability to multiply in *E. coli* bacteria.

Since the change T2 had undergone was completely reversible, it was obviously not a genetic mutation; an alteration of a gene or genes would be expected to persist in the progeny that inherited it. What, then, could account for the modification of T2's character? The phenomenon we had observed was not a freak; this was soon shown in other experiments with phages conducted by Giuseppe Bertani in my laboratory at the University of Illinois and independently by Jean J. Weigle at the California Institute of Technology. They found, for example, that when a certain phage called lambda (λ) was grown in the cells of an *E. coli* strain called *C,* only one in 10,000 of the daughter λ phages coming out of *C* could reproduce in a different *E. coli* strain called *K.* The descendants of the few phage particles that did manage to grow in *K* were then able to multiply in *K* cells as well as in *C* cells. A single cycle of growth in *C* cells, however, would restore these phages to the original form: capable of growing in *C* cells but not in *K.* Similarly, it was found that certain other strains of *E. coli* bacteria could restrict the growth of specific phages or modify their form [see *illustrations on pages 240 and 241*]. A given phage could be modified to a series of different forms by shifting it from one host strain to another; each time the daughter phages became adapted to the new host, and the descendants could be returned to the original form by shifting them back to the original host.

The enigma of the changeable phages did not begin to clear up until 1961, when a Swiss investigator, Werner Arber of the University of Geneva, discovered a clue to the reason for the different reception of phages by different strains of

a bacterium. Arber and his colleague Daisy Roulland-Dussoix found that when phage lambda injects its genetic material into an "alien" bacterium (one that will restrict the phage's multiplication), the cell breaks up most of the phage's DNA molecules into small fragments. This happens, for example, when phages grown in *E. coli* of strain *K* invade bacterial cells of strain *B.* The broken DNA of course cannot reproduce new phage. A few of the DNA molecules injected into the *B* bacterium manage, however, to remain intact, and they multiply, producing about 100 daughter phages. These now have the ability to multiply in *B* cells. It turns out that two of them (that is, about 2 percent of modified phages) can also still multiply in *K* bacteria as well. Arber performed a pretty experiment that revealed why this is so [see *top illustration on page 243*].

He used tracer isotopes—heavy nitrogen (^{15}N) and heavy hydrogen (^{2}H, or deuterium)—to label the DNA of phages bred in *K* cells. This was done by growing the bacteria and phages in a medium containing the heavy isotopes. The "heavy" phages were then employed to infect *B* bacteria growing in a medium of ordinary (light) nitrogen and hydrogen. When the new crop of daughter phages emerged from the lysed bacteria, Arber separated the phages by weight by means of a centrifuge that layered the particles (in a cesium salt solution) according to their relative buoyancy, based on their density. Most of the daughter phages proved to be "light," indicating that they were composed of material newly synthesized in the infected bacteria. A few particles, however, contained DNA that was half-heavy; that is, their weight indicated that half of the DNA was made up of new material and the other half of material from the heavy parent phages that had infected the bac-

PROBABILITY OF SUCCESSFUL INFECTION IN

	E. COLI B/4	E. COLI B/4	S. DYSENTERIAE SH
T2 · B	1	1	1
T2 · SH	1	1	1
T2 · B/4 = T*2	10^{-5}	10^{-5}	1
T2 · B/4 · SH	1	1	1

PROBABILITY OF SUCCESSFUL INFECTION IN

	E. COLI K	E. COLI B	E. COLI C
λ · K	1	10^{-4}	1
λ · B	4×10^{-4}	1	1
λ · C	4×10^{-4}	10^{-4}	1

MODIFICATION OF PHAGES can be demonstrated by growing phages in one strain of bacteria and observing how successfully the progeny grow when another strain serves as the host. In these tables the original host strain of T2 or lambda (λ) phages is indicated, in the first column, by a suffix such as *B* or *B/4*, which represents *E. coli* strains *B* or *B/4* respectively. The other columns indicate the probability of successful growth when the phage is obliged to grow again in the same strain or in another one. Fifteen years ago the author and Mary L. Human discovered that when phage T4 was grown in *E. coli B/4*, yielding T2 · *B/4* (originally called T*2), the phage was so altered that it would scarcely multiply in its usual hosts. It would grow freely, however, in dysentery bacilli (*Shigella dysenteriae Sh*). Daughter phages from this cycle, designated T2 · *B/4* · *Sh*, were completely normal.

teria. Since the phage DNA is a typical double-strand molecule, it could be deduced that in these phages one strand of their DNA was light and the other strand was heavy—contributed intact from the parent phage. Arber found that the phage particles containing this hybrid DNA retained the ability to multiply in *K* cells as well as in *B* cells. There was also a small fraction of fully heavy daughter phages (both DNA strands being heavy), and these too proved able to grow in both *K* and *B*.

The result of Arber's experiment made clear that the lambda phage's DNA strands carried some kind of marking that identified specific phages for the bacteria. In a phage generated in *K* cells the DNA somehow acquired a *K* marking, and a phage grown in *B* cells had its DNA marked *B*. When phage DNA that lacked the right marking was injected into a bacterium, the DNA was almost sure to be broken down. The experimental results also said something more: they said that the specific marking on the DNA molecule was attached by a stable, covalent chemical bond, since the mark was retained during the chemical events attending the construction of the daugh-

ter phage. Arber confirmed the stability of the bond by using highly purified phage DNA (instead of the intact phage) to infect bacteria; the DNA retained its specific marking even after going through the chemical purification treatment.

Arber's findings raised a number of interesting questions. What was the nature of the DNA markings? How did a bacterium recognize the absence of the right markings in phage DNA? What mechanism in the bacterium was responsible for destroying DNA with the "wrong" markings? A number of laboratories have looked into these questions and have worked out much of the answer. Actually two sets of answers have come out, one for phages of the lambda type (with *K*, *B* or *C* markings) and one for the T2 and other "T even" phages. Let us look first at the lambda markings.

It was known even before Arber's discoveries that every kind of DNA can occasionally have a seemingly spurious methyl group (CH_3) attached to some of the nucleotides. For example, the methyl group can be tacked on at a certain position in adenine; the base is then called

methylaminopurine, or MAP, instead of simply A. Similarly, the base cytosine (C) can be methylated, so that it becomes methylcytosine (MC). In each case the methyl group is added (with the help of a special enzyme) only after the DNA molecule has been built [*see bottom illustration on page 6*]. And it plays no observable part in the molecule's functioning: MAP behaves just like A, and MC just like C. In short, the methyl group does not alter the genetic spelling.

Could one now suppose the addition does change the style of the script? Is the methyl group perhaps a kind of serif tacked onto the letters? Arber tried the experiment of cutting down the amount of methylation of the phage DNA. He did this by growing *K* bacteria in a medium in which they were deprived of the amino acid methionine, the precursor of the substance that donates methyl groups to DNA. It turned out that phages grown in these bacteria generally did not acquire the *K* marking. Apparently they did not have the serifs (methyl groups) that would identify them as *K*.

Arber found further that the methyl groups serve as markers not just anywhere on the DNA molecule but only at certain strategic spots. This was clearly shown in experiments with a very tiny phage known as fd. Attached to each DNA molecule of this phage are a few MAP groups. Arber and his colleague Urs Kühnlein found that when the fd phage was grown in *K* bacteria, it had two fewer MAP groups than when it was grown in *B* bacteria. The *K*-marked phages failed to grow in *B* cells; these cells broke down the phage DNA.

This suggests that the fd phage's DNA has two sensitive sites, located on adenine bases. If the two sites are methylated (that is, marked MAP), the phage can grow equally well in both *K* and *B* bacteria. If, however, these sites are unmethylated, the *B* bacterium recognizes the phage as being the "wrong" kind and breaks down its DNA. Presumably the recognition is effected by a special enzyme that can split the DNA molecule. The *B* bacterium also has a methylating enzyme that can convert the two critical A bases to MAP (that is, attach a serif) and thus make the phage DNA acceptable so that it multiplies in *B* cells [*see top illustration on page 244*].

This interpretation has now been confirmed by various experiments. Some of the specific enzymes that break the vulnerable DNA sites and some of those that can methylate them have been isolated and identified. For example, Arber's group in Geneva and Matthew S. Meselson and Robert T.-Y. Yuan at

Harvard University have identified the enzymes that break the DNA molecules marked *K* and those that break the DNA molecules marked *B*. As expected from Arber's vulnerable-site hypothesis, it turns out that in each case the enzyme makes just a few breaks in the DNA molecule. Similarly, Arber and his colleagues have isolated a marking enzyme: from bacteria of the *B* strain they extracted the enzyme that can transfer methyl groups from a suitable methyl-donor substance (S-adenosyl methionine) to the *K*-marked (unmethylated) DNA of fd phages.

Thus we see that the bacteria possess a well-defined system for marking and recognizing phages. Shifting from our printing metaphor, we might say that each strain of *E. coli* bacteria stamps its own trademark on the DNA of the phages it produces, just as a factory brands its commercial product.

Work in several laboratories has elicited further details of the bacteria's branding and recognition system. It has been learned that the system in *E. coli* involves three closely linked genes. One directs the synthesis of the DNA-break-

ing enzyme, another controls the synthesis of the methylating enzyme, and the third gene generates a mechanism that is responsible for the recognition of the critical DNA sites that are to be either broken or methylated. It seems likely that this mechanism, or component, is a protein chain that associates itself with both the breaking and the methylating enzymes, and that can recognize the specific sequence of nucleotides that represents the critical DNA sites [*see bottom illustration on page 244*].

The marking and restricting system is widespread among the *E. coli* and related strains of bacteria; it is possessed not only by the *K* and *B* strains but also by others (except for the *C* strain and certain others that have lost the recognition system, perhaps through mutation). Also Arber and his co-workers found that DNA incompatibility is not confined to the case of invasion by a phage. It also applies to exchanges of DNA among the bacteria themselves. When a female *E. coli* cell mates with a male cell carrying the wrong DNA brand (for example, if the female is *B* and the male *K*, or vice versa), the female on receiving the male's DNA will

break it down, just as if it belonged to a phage.

The answer to the T2 mystery turned out to be a different story. The investigation of this problem, carried out in our laboratory at the Massachusetts Institute of Technology by my colleagues Toshio Fukasawa, Costa P. Georgopoulos, Stanley Hattman and Helen R. Revel, demonstrated that the DNA in the T2 phage is not branded in the same way as that in the lambda and fd phages.

The DNA of T2 (and of its even-numbered relatives T4 and T6) carries an oddity: the base cytosine always has a hydroxymethyl group (CH_2OH) attached to it, so that the nucleotide contains hydroxymethylcytosine (HMC) instead of cytosine as its base. The discovery of the HMC base by Gerard R. Wyatt of Yale University and Seymour S. Cohen of the University of Pennsylvania played a major role in phage biochemistry. In most of the HMC nucleotides of these phages one or more sugar molecules (glucose) normally are attached to the hydroxymethyl group [*see illustration on page 245*]. What our laboratory team discovered was that when an *E. coli* bacterium of the *B/4* strain

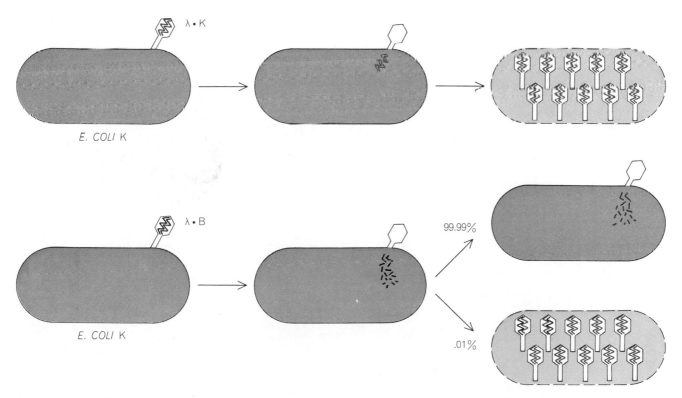

GROWTH RESTRICTION is demonstrated by cells of *E. coli* strain *K*. No restriction occurs when *K* cells are infected by phage $\lambda \cdot K$ (*top*). The deoxyribonucleic acid (DNA) of the phage (*colored zigzag shape*) enters the cell and exploits the cell's chemical machinery to produce about 100 new phage particles, which are released when the cell lyses, or dissolves. Restriction occurs (*bottom*) when *K* cells are infected by $\lambda \cdot B$ particles: lambda phages previously grown on *E. coli* strain *B*. The DNA of the phage (*black zigzag*) enters the cell but is broken down. In about one cell in 10,000, however, the phage DNA manages to multiply and give rise to phage progeny. About 2 percent of the progeny can grow in both *B* and *K* cells because their DNA is a hybrid molecule consisting of one strand of DNA (*black*) from the original $\lambda \cdot B$ and one strand (*color*) newly made inside the *K* cell. The remaining progeny are now modified so that they will grow normally in *K* cells but with a probability of only 10^{-4} in *B* cells.

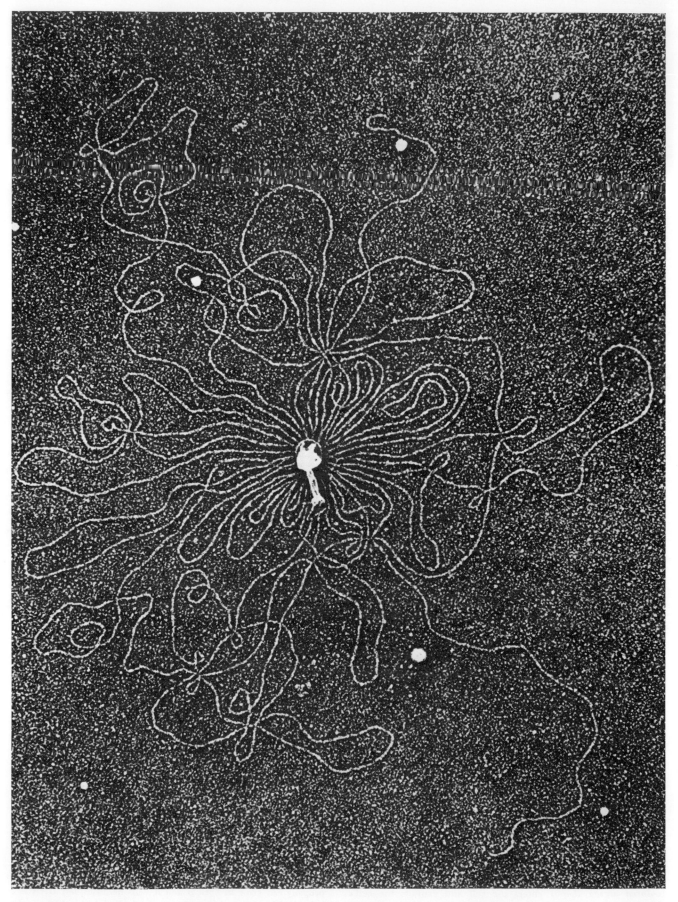

MOLECULE OF BACTERIOPHAGE DNA is shown here as a long, tangled thread after being released from the head of the T2 phage particle. The magnification is 100,000 diameters. The T2 bacteriophage is one of several "T even" phages that infect cells of *Escherichia coli.* When the phage DNA molecule is modified in a certain way, it can be "recognized" as an invader by the cell and destroyed. This electron micrograph was made by Albrecht K. Kleinschmidt of the New York University School of Medicine.

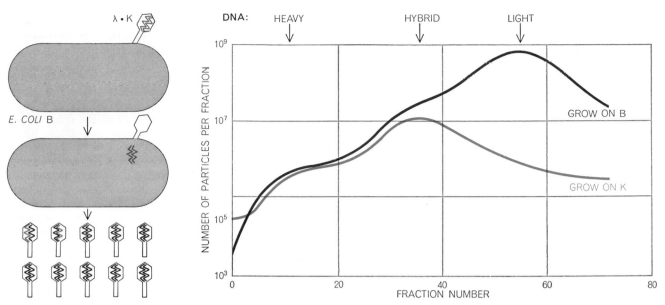

LABELING OF DNA demonstrates that phage particles possessing the ability to grow on both *B* and *K* strains of *E. coli* are predominantly those with hybrid DNA molecules. The experiment was conducted at the University of Geneva by Werner Arber and Daisy Roulland-Dussoix. They infected cells of strain *B* made up of atoms of ordinary weight with phage $\lambda \cdot K$ that had been grown on *E. coli K* cells incorporating heavy hydrogen (deuterium) and heavy nitrogen (^{15}N). Thus the DNA molecules in particles of $\lambda \cdot K$ were labeled with heavy atoms. The progeny produced in the light *B* cells were spun in a centrifuge containing a gradient of cesi-um chloride. The phage particles become distributed in the centrifuge tube according to their weight: heaviest particles at the bottom, lightest at the top. The phage particles in different fractions are then tested for their ability to grow on *B* and *K* cells, with the result plotted at the right. The particles containing hybrid DNA, about 2 percent of the total, grow in both kinds of cell. A small fraction of phage inherits two heavy strands of DNA from its $\lambda \cdot K$ parent and can also grow in both *B* and *K*. For this experiment Arber used a special strain of *E. coli B* that can impress the *B* modification on the phage DNA but lacks the *B* restriction function.

modifies, or brands, a T2 phage, the way it does so is to fail to attach the glucose to the HMC at the sensitive sites in the DNA molecule. In short, it changes the DNA from "sweet" to "sour"!

The soured phage can no longer multiply in *E. coli* cells; the cells detect the absence of glucose at certain critical sites and break down the DNA. Desugared T2 can, however, grow in the cells of dysentery bacilli, and these cells mark the phage DNA by attaching glucose at the necessary sites, thus restoring the phage's ability to grow in *E. coli*.

The special strains of *E. coli* that transform the phage from sweet to sour lack a substance (uridine diphosphoglucose, or UDPG) that is needed to donate glucose to the phage's DNA. This is one way for a T2 phage to acquire the sour branding. The brand can also be put on

5-METHYLCYTOSINE (MC)

6-METHYLAMINOPURINE (MAP)

CYTOSINE

ADENINE

S-ADENOSYL METHIONINE

"MARKED" BASES found in phage DNA that has been modified are 5-methylcytosine (MC) and 6-methylaminopurine (MAP). They are formed from the standard bases cytosine and adenine by the addition of a methyl group (CH_3), supplied by S-adenosyl methionine. Adenine (A) and cytosine (C) supply two of the four "letters" that spell out the genetic message in DNA molecules.

PHAGE	CHARACTERISTIC	PROBABILITY OF GROWTH IN *E. COLI* B
fd · K	2 MAP PER DNA MOLECULE	.001
fd · B	4 MAP PER DNA MOLECULE	1.0
fd m1 · K	3 MAP PER DNA MOLECULE	.03
fd m2 · K	2 MAP PER DNA MOLECULE	1.0

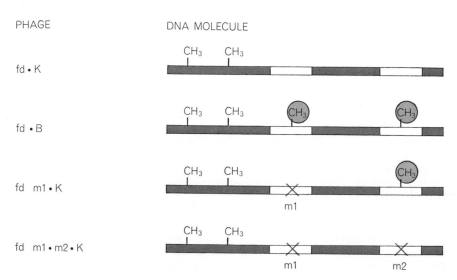

VULNERABLE SITES in the DNA molecule of the phage fd are represented by the white regions in the lower part of the illustration. These sites, or sequences of bases, are broken with a probability of 97 percent by the "restricting" enzyme of *E. coli* B unless an A (adenine) base in that sequence has been converted to MAP by the modifying (that is, methylating) enzyme present in *B* cells. There are also two irrelevant CH_3 groups elsewhere in the fd DNA molecule. The table (*top*) shows for various strains of fd the probability of growth in *E. coli* B. Strains of fd · K with certain mutations, *m*1 and *m*2 (*indicated by X marks*), are insensitive to the restricting enzyme in *E. coli* B even if they are not methylated.

BACTERIAL GENE	FUNCTION	MUTANTS	
r	RESTRICTION MECHANISM ("NUCLEASE" COMPONENT)	r⁻:	RESTRICTING FUNCTION LOST
m	MODIFICATION MECHANISM ("METHYLASE" COMPONENT)	m⁻:	MODIFYING FUNCTION LOST
s	SPECIFICITY (SITE-RECOGNITION COMPONENT)	s⁻:	BOTH RESTRICTING AND MODIFYING FUNCTIONS LOST

GENETIC CONTROL OF RECOGNITION SYSTEM seems to involve three closely linked genes in *E. coli*. Gene *r* makes the restricting enzyme that breaks the phage DNA molecule unless it carries the proper MAP marks. Gene *m* makes the methylating enzymes that provide the marks. Gene *s* makes a component responsible for specific recognition of sites where the marks may or may not be present. Presumably this recognition component is part of both enzymes: the one that can break DNA and the one capable of methylating it.

it, however, by a genetic mutation in the phage itself, that is, by an alteration of the phage genes that normally are responsible for production of the enzymes that attach glucose to the HMC [*see illustration on page 246*]. The *E. coli* bacteria, for their part, possess special genes that endow them with the ability to recognize and destroy sour DNA. These genes produce two enzymes, apparently different, that recognize and attack specific unsweetened sites in the DNA molecule. Experiments have shown that mutation of these two genes eliminates the rejection mechanism, so that unsweetened phages can grow in the mutated *E. coli* cells.

The two enzymes responsible for distinguishing between sweetened and unsweetened DNA have not yet been isolated. All we know about them so far (from genetic analysis) is that they are not the same as the enzymes that detect the difference between the methylated and unmethylated brands on the DNA of lambda and fd phages and on the DNA of different strains of *E. coli* bacteria.

How does the enzyme, in each case, recognize the brand? By what mechanism does the enzyme (protein) molecule find and identify the significant chemical markings on the DNA molecule? This is an intriguing question that applies to many other phenomena in molecular biology. For example, the enzymes that bring about the replication of DNA and those that build RNA molecules using DNA as a template must find the right place on the DNA at which to start the construction of the new molecule. How do they recognize these starting points?

We must suppose that the identifying marker on the DNA molecule in each case is a certain sequence of nucleotides, and that the enzyme establishes recognition by temporary attachment to this sequence—by fit and "feel," so to speak. There are good reasons to believe the identifying sequences are generally rather short; the length of a globular protein would not span many nucleotides in their linear array in the DNA molecule. Arber has suggested that the identifying sequences in branded DNA are probably no more than six to eight nucleotides long. It may soon be possible to work out the exact sequence of bases that constitutes the recognition site for at least one of the restrictive enzymes. This would be a big step forward in the study of protein-DNA interaction.

What kind of search do the enzymes carry on? Are the sensitive sites recognized and attacked only when enzymes

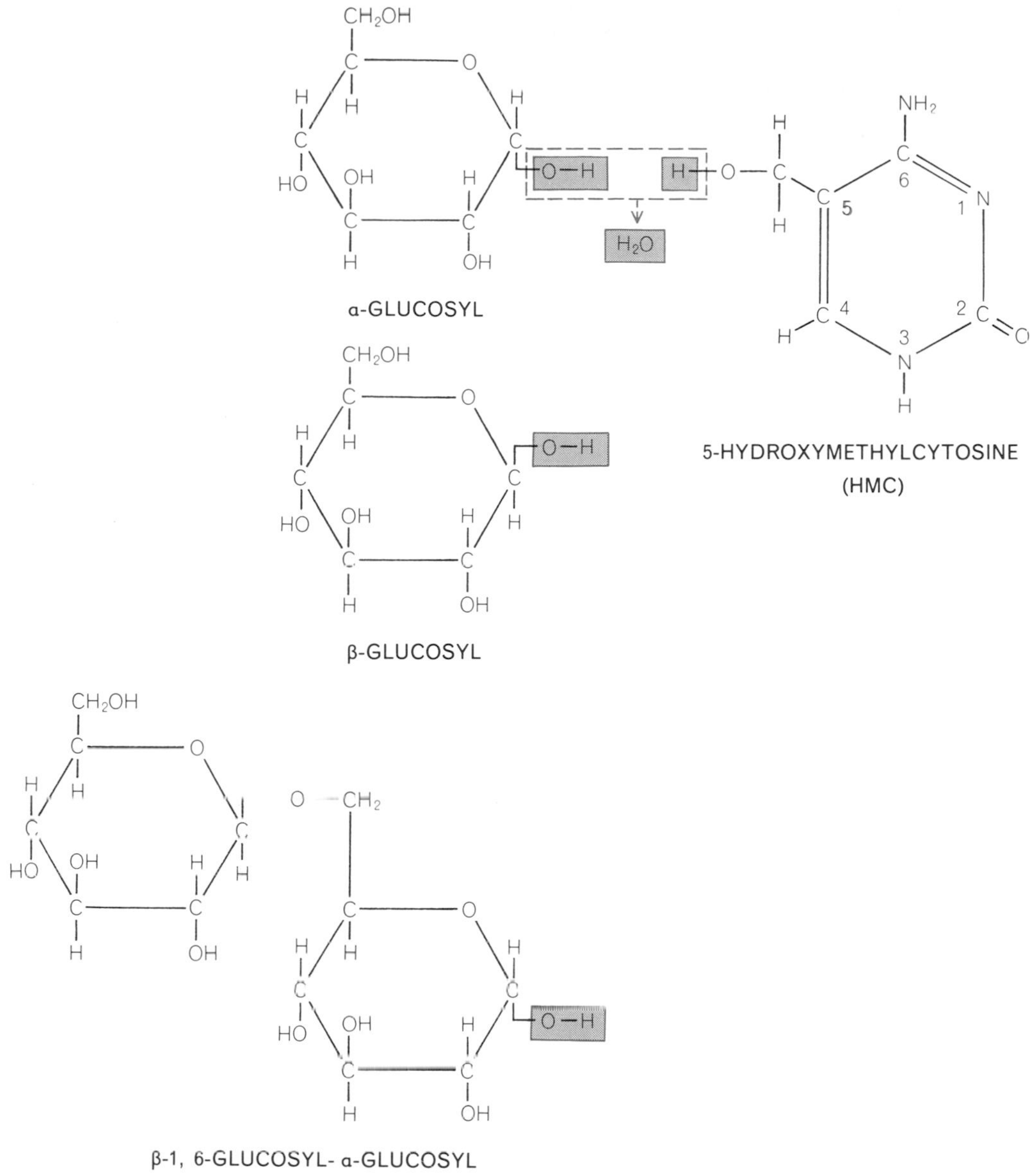

α-GLUCOSYL

β-GLUCOSYL

5-HYDROXYMETHYLCYTOSINE
(HMC)

β-1, 6-GLUCOSYL- α-GLUCOSYL

"SWEET" AND "SOUR" LABELS provide another recognition system that enables *E. coli* cells to discriminate among invading molecules of phage DNA. In this system, which applies to the phages T2, T4 and T6, the base cytosine (C) is replaced by 5-hydroxymethyl-cytosine (HMC), a base formed when a hydroxymethyl group (CH₂OH) replaces hydrogen on the No. 5 carbon atom of cytosine. The normal phage has 70 percent or more of its HMC bases linked to one of the three sugar structures shown at the left. Such DNA molecules can be regarded as sweet. When the sugar units are missing, the DNA is sour and is readily broken down by *E. coli* cells.

happen to fall on them directly? Or do the enzymes browse along the phage DNA molecules, possibly as the latter nose into the bacterium, until they find the telltale markers? A finding by Georg-opoulos and Revel in our laboratory suggests that the second alternative may be correct. They found that even a trace of "sweetness" (that is, glucosylation of comparatively few of the HMC bases, such as occurs in certain mutant T2 phages) was sufficient to protect the DNA of these phages from an enzyme that attacked only one particular site on the molecule and then only when it was unsweetened. It seems reasonable to as-

sume that the small amount of glucose in these phages attaches itself to HMC groups at random, and therefore that the HMC groups at the particular site this enzyme attacks may often be unsweet-ened. We can guess the reason this site usually escapes attack is that the enzyme explores the DNA molecule and stops its exploration or falls off as soon as it en-counters any glucose.

Another puzzling question is posed by the finding that even if only one (ei-ther one) of the two strands of a phage's DNA carries the protective marking, an enzyme will not break the molecule. This indicates that in the case of a vul-

nerable molecule the enzyme scans both strands separately before breaking the molecule. How does it manage to scan the strands individually?

The entire set of phenomena I have described in this article raises a more general question. What is the evolu-tionary significance of the DNA-brand-ing system? What function does it serve for the bacteria? One obvious suggestion is that the system gives a bacterium a defense against certain phages. Evo-lutionary development could, however, have provided the bacteria with more effective defense mechanisms, such as eliminating the surface sites on the bacterial cell to which phages attach themselves. Furthermore, the bacteria that can recognize and destroy branded DNA have no defense against phages that are not branded. It is hard to believe the branding system evolved simply to protect bacteria of one strain against phages coming from another strain. A more plausible speculation is that the system serves primarily to prevent "undesirable" mixing of the bacterial genes (DNA) between the strains of bacteria themselves. Presumably in nature the *E. coli* bacteria do mate or otherwise exchange genetic material. If this is so, the ability of a *B*-strain bacterium to reject *K*-marked DNA (and vice versa) is an effective device for keeping each strain "pure." Thus the branding-and-rejection system facilitates the evolution of bacterial strains in diverging directions, in the same way that isolation mechanisms in cross-fertilization play a role in the evolution of plant and animal species.

The case of sweet and sour DNA seems to tell a clearer and more dramatic story. The T2 phages have the ability to break down the DNA of the bacteria they attack, which contains cytosine instead of HMC. Hence the hydroxymethylation of cytosine in the phages' own DNA may have been an important step in the evolution of their ability to multiply in the bacteria. We can surmise, then, that in reply some strains of *E. coli* by natural selection evolved the ability to break down phage DNA marked with

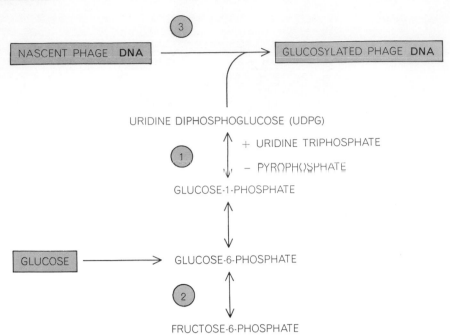

SYNTHESIS OF SWEET DNA requires a series of biosynthetic steps culminating in the addition of glucose units to the nascent DNA of the T-even phages. Mutants of *E. coli* can have blockages in the enzyme systems that carry out either or both of the first two numbered steps. Mutants with a blockage in enzyme No. 1 produce T-even phage with unglucosylated, or sour, DNA (T* phage). Mutants with a block in enzyme No. 2 can make the glucose donor substance, uridine diphosphoglucose, only if glucose is present in the growth medium. The last step, No. 3, is carried out by enzymes not normally present in *E. coli* cells; they are made from instructions coded in the DNA molecules of T-even phages.

HMC. The phage in turn evolved the capacity to tack glucose onto HMC to protect the vulnerable sites of its DNA. Here, then, the evidence of the branded DNA may be unfolding for us a scene in the dynamic acting out of evolution at the most elementary level: the chemistry of the genetic material itself.

28

RNA-DIRECTED DNA SYNTHESIS

HOWARD M. TEMIN

January 1972

The discovery that in certain cancer-causing animal viruses genetic information flows "in reverse"—from RNA to DNA—has important implications for studies of cancer in humans

A major goal of present-day biology is to learn how information is coded in molecular structures and how it is transmitted from molecule to molecule in biological systems. Discovery of the rules governing this transmission is an integral part of understanding how embryonic cells differentiate into the hundreds of distinct types of cell observed in plants and animals and how normal healthy cells become cancerous.

It has now been known for nearly 20 years that the genetic information in all living cells is encoded in molecules of deoxyribonucleic acid (DNA) consisting of two long strands of DNA wound in a double helix. The genetic information for each organism is written in a four-letter alphabet, the "letters" being the four different chemical units called bases. In the normal cell short passages of the genetic message (individual genes) are transcribed from DNA into the closely related single-strand molecule ribonucleic acid (RNA). A length of RNA representing a gene is then translated into a particular protein, a molecule constructed with a 20-letter alphabet, the 20 amino acids. When a cell divides, the information contained in each of the two strands of DNA is replicated, thereby equipping the daughter cell with the full genetic blueprint of the parent.

Francis Crick, one of the codiscoverers of the helical structure of DNA, originally proposed that information can be transferred from nucleic acid to nucleic acid and from nucleic acid to protein, but that "once information has passed into protein it cannot get out again," that is, information cannot be transferred from protein to protein or from protein to nucleic acid. These concepts were simplified into what came to be known as the "central dogma" of molecular biology, which held that information is sequentially transferred from DNA to RNA to protein [see illustration on page 249]. Although Crick's original formulation contained no proscription against a "reverse" flow of information from RNA to DNA, organisms seemed to have no need for such a flow, and many molecular biologists came to believe that if it were discovered, it would violate the central dogma.

I shall describe experiments that originally hinted at a flow of information from RNA to DNA and that since have provided strong evidence that the "reverse" flow of information not only takes place but also accounts for the puzzling behavior of a sizable group of animal viruses whose genetic information is encoded in RNA rather than in DNA. Many of these viruses also produce cancer in animals. Although they have not yet been linked to cancer in man, their ability to transmit information from RNA to DNA inside the living cell makes it attractive to unify two hypotheses of the cause of human cancer that had previously seemed separate: the genetic hypothesis and the viral hypothesis.

There are two broad classes of viruses: viruses whose genome, or complete set of genes, consists of DNA and viruses whose genome consists of RNA. In the cells that they infect the DNA viruses replicate their DNA into new DNA and transmit information from DNA to RNA and thence into protein. Most RNA viruses, such as the viruses that cause poliomyelitis, the common cold and influenza, replicate RNA directly into new copies of RNA and translate information from RNA into protein; no DNA is directly involved in their replication.

In the past few years it has become apparent that a group of viruses, variously called the RNA tumor viruses, the leukoviruses or the rousviruses (after their discoverer, Peyton Rous), replicate by another mode of information transfer. The rousviruses use information transfer from RNA to DNA in addition to the modes of information transfer (DNA to DNA, DNA to RNA and RNA

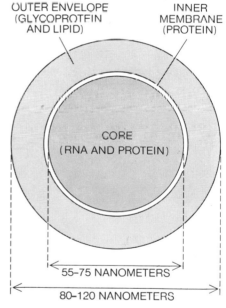

OUTER ENVELOPE (GLYCOPROTEIN AND LIPID)

INNER MEMBRANE (PROTEIN)

CORE (RNA AND PROTEIN)

55–75 NANOMETERS

80–120 NANOMETERS

VIRIONS, or individual particles, of an "RNA-DNA virus," an animal-tumor virus that transfers genetic information from RNA (ribonucleic acid) to DNA (deoxyribonucleic acid) in addition to the normal modes of information transfer used by cells and other viruses, are enlarged about 700,000 diameters in the electron micrograph on following page. The particular RNA-DNA virions shown in thin section in the micrograph cause leukemia in mice; they are similar in structure and function to the Rous sarcoma virions discussed in this article. The electron micrograph was made by N. Sarkar of the Institute for Medical Research in Camden, N.J. A diagram of structure of a virion of this type is given above.

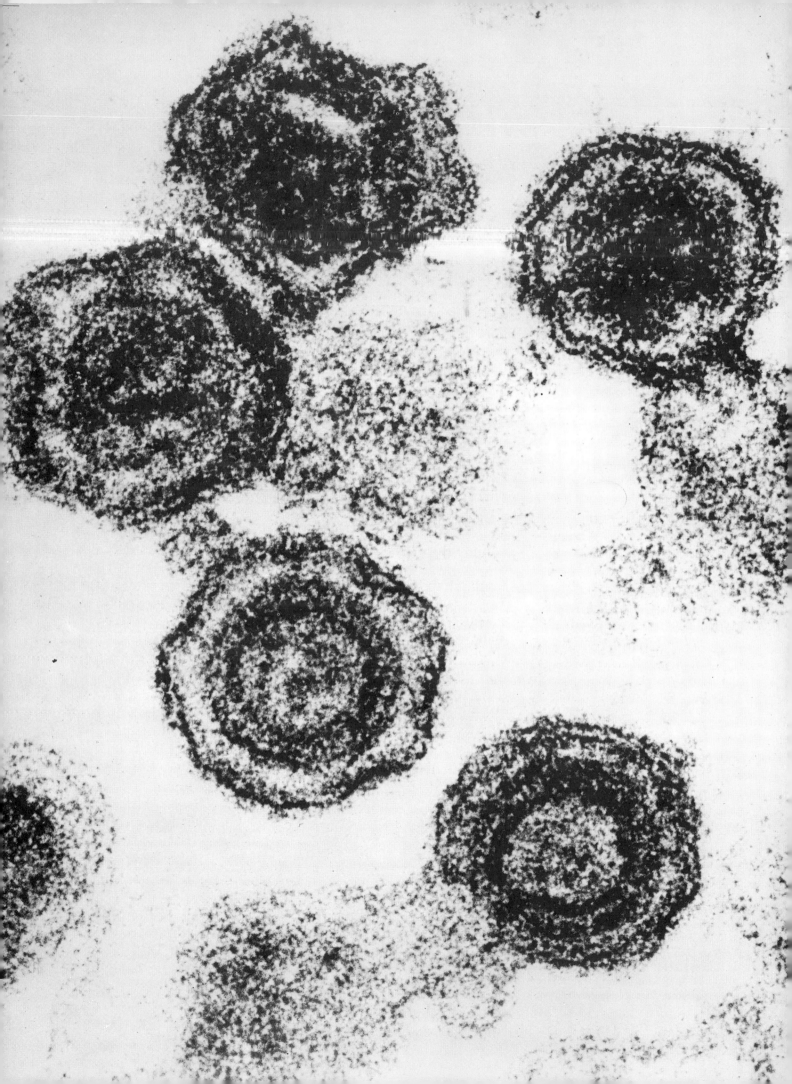

to protein) that are found in cells and in DNA viruses. The rousviruses do not transfer information from RNA to RNA, as other RNA viruses do. The existence of the RNA-to-DNA mode of information transfer in the replication of rousviruses has led some to suggest that there should be three major classes of viruses: DNA viruses, RNA viruses and RNA-DNA viruses [see top illustration on page 254].

The prototype RNA tumor virus, the Rous chicken sarcoma virus, was discovered by Rous 61 years ago at the Rockefeller Institute for Medical Research. An RNA tumor virus had actually been found earlier by V. Ellerman and O. Bang of Copenhagen, but their virus was little studied because it caused leukemia in chickens and was harder to work with than Rous's virus. Rous was studying a transplantable tumor of the barred Plymouth Rock hen. Originally he observed that he could transfer the tumor by the transfer of cells. In 1911 he found that the tumor could also be transferred by means of fluid from which the cells had been filtered. Demonstration that a disease can be transmitted by a cell-free filtrate is commonly accepted as evidence that it is caused by a virus. Descendants of the virus originally discovered by Rous are still being worked on in laboratories all over the world. At the time, however, Rous's discovery was met with disbelief, and after 10 years Rous himself stopped working with the tumor. It was not until nearly 30 years later, when Ludwik Gross of the Veterans Administration Hospital in the Bronx discovered that RNA tumor viruses cause leukemia in mice, that the study of rousviruses became popular.

It is now known that viruses in the same group as the virus originally discovered by Rous, or closely related to it, can cause tumors not only in chickens and mice but also in rats, hamsters, monkeys and many other species of animals. Moreover, viruses of the same group have been isolated from nonmammalian species, including snakes. As yet no bona fide human rousvirus has been discovered. It also appears that some members of this group, for example some of the "associated viruses," do not produce cancer.

In the 1950's, with the beginning of the application of cell-culture methods to animal virology, a tissue-culture assay for the Rous sarcoma virus was developed, first by Robert A. Manaker and Vincent Groupé at Rutgers University and subsequently by Harry Rubin and me at the California Institute of Technology. The assay involves adding suspensions of the virus to sparse cultures of cells taken from the body wall of chicken embryos. The Rous sarcoma virus infects some cells and transforms them into tumor cells. The transformed tumor cells differ in morphology and in growth properties from normal cells and therefore create a focus of altered cells. Assays of the same type have been developed for infections that the Rous sarcoma virus causes in cells taken from turkeys, ducks, quail and rats. Similar assays have also been developed for other transforming rousviruses.

The number of foci of transformed cells is proportional to the number of infectious units of the virus added to the cell culture and provides a rapid and reproducible assay for the Rous sarcoma virus. The use of this assay led to the discovery that the Rous sarcoma virus differs from the other viruses that had been studied up to that time in the way it interacts with the cell. The replication of most viruses is incompatible with cell division; in other words, the virus causes the infected cells to die. Chicken cells infected with the Rous sarcoma virus not only survive but also continue to divide and produce new virus particles [see middle illustration on page 254]. When the Rous sarcoma virus infects rat cells, there is a slightly different interaction of the cell and the virus. The rat cells are transformed into cancer cells, which divide, but the transformed cells do not produce the Rous sarcoma virus even though the genome (DNA) of the virus can be shown to be present. Production of the Rous sarcoma virus can be induced if the transformed rat cells are fused with normal chicken cells.

In the early 1960's the antibiotic actinomycin D was found to be very useful in unraveling the flow of genetic information in cells infected with RNA viruses. The antibiotic inhibits the synthesis of RNA made on a DNA template but not the synthesis of RNA made on an RNA template. The antibiotic therefore stops all RNA synthesis in cells infected by RNA viruses except for RNA specifically related to the viral genome. With this new tool it became easy to determine which RNA's were specific for the viruses.

When I added actinomycin D to cultures of cells producing Rous sarcoma virus, however, I found that the antibiotic inhibited the production of all RNA. One would have expected the replication of RNA on the template of an RNA viral genome to continue without hindrance [see bottom illustration on page 254]. This result was the first direct evidence that the molecular biology of the replication of Rous sarcoma virus was different from that of other RNA viruses. Since that observation was made the inhibition of the replication of rousviruses by actinomycin D has been recognized as one of their defining characteristics. The actinomycin D experiments suggested to me that the Rous sarcoma virus might replicate through a DNA intermediate. This hypothesis is called the DNA provirus hypothesis.

Further experiments, carried out by me and by John P. Bader at the National Cancer Institute, demonstrated that if one inhibits the synthesis of DNA in cells immediately after they have been inoculated with Rous sarcoma virus, one can protect the cells from infection. Here the inhibitors were amethopterin, fluorodeoxyuridine and cytosine arabinoside. These experiments appeared to support the idea that infection requires the synthesis of new viral DNA pro-

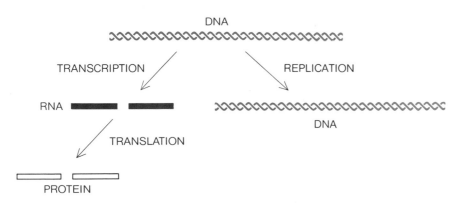

"CENTRAL DOGMA" of molecular biology, originally formulated by Francis Crick, states that within an organism genetic information can be transferred from DNA to DNA or from DNA to RNA to protein, but that it cannot be transferred from protein to protein or from protein to either DNA or RNA. Although a "reverse" flow of genetic information from RNA to DNA was not proscribed in Crick's original formulation, many molecular biologists came to believe that if such a flow were ever discovered, it would violate the central dogma.

duced on an RNA template. This interpretation was not unequivocal, however, because successful production of Rous sarcoma virus requires that the cells divide normally after infection. Therefore the inhibition of DNA synthesis after infection could inhibit production of Rous sarcoma virus not only by blocking possible new viral DNA synthesis but also by preventing normal cell division.

To get around this problem I introduced the idea of infecting cultures of stationary, or nondividing, cells with Rous sarcoma virus. Cells in culture usually require specific factors in blood serum to support their multiplication. If the serum is removed from the medium of the cell cultures, the cells stop dividing. If they are then exposed to Rous sarcoma virus, they become infected but there is no virus production or morphological transformation until serum is added back and the cells divide once again. When such stationary cells are exposed to inhibitors of DNA synthesis, the cells are not killed because they are not making DNA. When the stationary cells are exposed simultaneously to Rous sarcoma virus and to inhibitors of DNA synthesis, the cells are not killed but neither are they infected [see illustration at right].

If one now removes the inhibitor of DNA synthesis and adds serum, enabling the cells to divide once more, one finds that the cells remain free of infection. They do not become transformed and they do not produce virus. These experiments supported the hypothesis that after cells are infected by the Rous sarcoma virus new viral DNA is synthesized at a time different from the cell's normal synthesis of DNA. The new viral DNA is evidently synthesized on a template of viral RNA.

A further extension of this approach to understanding the replication of Rous sarcoma virus was carried out by one of my students, David E. Boettiger, and independently by Piero Balduzzi and Herbert R. Morgan at the University of Rochester School of Medicine and Dentistry. It had been found by others that if 5-bromodeoxyuridine, an analogue of the DNA constituent thymidine, is incorporated into DNA, the DNA becomes sensitized so that it can be inactivated by light. Under the same conditions normal thymidine-containing DNA is not affected by light. Boettiger therefore exposed stationary cells to Rous sarcoma virus in the presence of bromodeoxyuridine and then exposed the cells to light. Although the cells were not killed, the treatment prevented their

being infected by the virus. When serum was again added to enable the cells to divide, they did not become transformed and did not produce virus [see illustration on page 252].

In a related experiment Boettiger showed that the rate of inactivation of the infection by Rous sarcoma virus was dependent on the number of viruses infecting a cell. As he raised the number of viruses infecting each cell, he found that the infection became increasingly resistant to inactivation by light. We interpreted these experiments as showing that each infecting virus makes a new specific DNA, and that the more viruses that infect a cell, the more molecules of new viral DNA that are produced. The experiment seemed to effectively rule out the alternative hypothesis, which was that the infecting virus provokes a new synthesis of some preexisting cellular DNA.

Unfortunately no one has yet been able to unequivocally demonstrate the existence of newly synthesized viral DNA in cells infected with the Rous sarcoma virus. The available techniques are evidently too crude to detect the tiny

amounts of new viral DNA expected to be present. Certain results have been reported, however, with transformed cells. One approach has been to bring DNA from infected cells together with labeled viral RNA to see if single strands of the two molecules would coalesce into a double-strand hybrid molecule. Such hybrids are readily created when the base sequences in the DNA are complementary to the base sequences in the RNA, indicating that both carry the same genetic message and hence that each could arise from the transcription of the other.

The hybridization experiments reported thus far have aroused a great deal of controversy. Although some experiments, notably those of Marcel A. Baluda and Debi P. Nayak of the University of California at Los Angeles, have seemed to demonstrate the presence in infected cells of DNA complementary to viral RNA, the results have not been universally accepted. The finding of an intermediate viral DNA is an essential link in the chain of evidence that is still needed to establish firmly the DNA provirus hypothesis.

Meanwhile strong support for the hy-

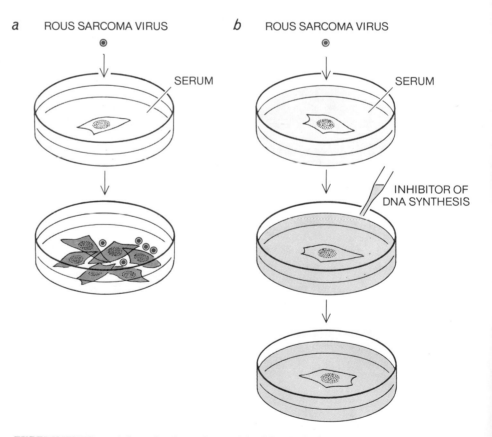

EXPERIMENTS carried out by the author and by John P. Bader at the National Cancer Institute supported the hypothesis that the infection of cells with Rous sarcoma virus requires the synthesis of new viral DNA produced on an RNA template. When the virus is added to cultures of normally dividing cells (a), the cells are transformed into cancer cells, which divide and produce new Rous sarcoma virus. By adding a substance that inhibits the synthesis of DNA in the cells immediately after they have been inoculated with Rous sar-

pothesis has come from experiments of a different kind. In 1969 Satoshi Mizutani, who had written his doctoral thesis on bacterial viruses, came to my laboratory for postdoctoral training. We decided to ask the question: What is the origin of the enzyme (a protein) responsible for forming proviral DNA using the viral RNA as template? When Mizutani exposed stationary cells to Rous sarcoma virus in the presence of inhibitors of protein synthesis, he found that the cells still became infected. We interpreted this experiment to mean that the enzyme that synthesizes DNA from the viral RNA template is already in existence before the infection.

Somewhat earlier other workers had fractionated virions—the actual virus particles as distinct from the forms assumed by the virus inside cells—and had found RNA polymerases, enzymes that catalyze the synthesis of RNA from its building blocks: four different ribonucleoside triphosphates. In 1967 Joseph Kates and B. R. McAuslan of Princeton University and William Munyon, E. Paoletti and J. T. Grace, Jr., of the Roswell Park Institute had found RNA polymer-

ases in a poxvirus, a large DNA virus. Other workers had found another RNA polymerase in a reovirus, a double-strand RNA virus. Therefore we decided to look in the virions of Rous sarcoma virus for a DNA polymerase capable of using the viral RNA as a template. After several months of preliminary experiments we succeeded in showing the existence of a DNA polymerase in purified virions of Rous sarcoma virus.

Before discussing this result I should digress briefly to describe the structure of the Rous sarcoma virus [*see illustration, page 248*]. The virion of the Rous sarcoma virus has a diameter of about 100 nanometers, which makes it larger than the particles of the viruses that cause poliomyelitis and smaller than the particles of the viruses that cause smallpox. The virion of the Rous sarcoma virus consists of a lipid-containing envelope (derived by budding from the cell membrane), an inner membrane and a nucleoid, or core, that contains the viral RNA and certain proteins.

In order to demonstrate that the Rous sarcoma virus contains a polymerase capable of producing DNA on an RNA

template, we first treated the virion with a detergent to disrupt its lipid-containing envelope. We then added to the disrupted virus the four deoxyribonucleoside triphosphates that are the building blocks of DNA. One of the deoxyribonucleoside triphosphates was radioactively labeled.

When the mixture was incubated at 40 degrees Celsius, it incorporated the radioactive label into an acid-insoluble substance that met the usual tests for DNA. The substance was stable in the presence of alkali and the enzyme ribonuclease, treatments that are known to destroy RNA, whereas it was attacked and fragmented by an enzyme that destroys DNA. When we repeated the experiment with disrupted virions pretreated with ribonuclease, an enzyme that destroys RNA, little or no DNA was produced, indicating that intact viral RNA was needed as the template for the synthesis of DNA [*see top illustration on page 253*].

After we had announced these results at the Tenth International Cancer Congress in Houston in May, 1970, we learned that David Baltimore of the

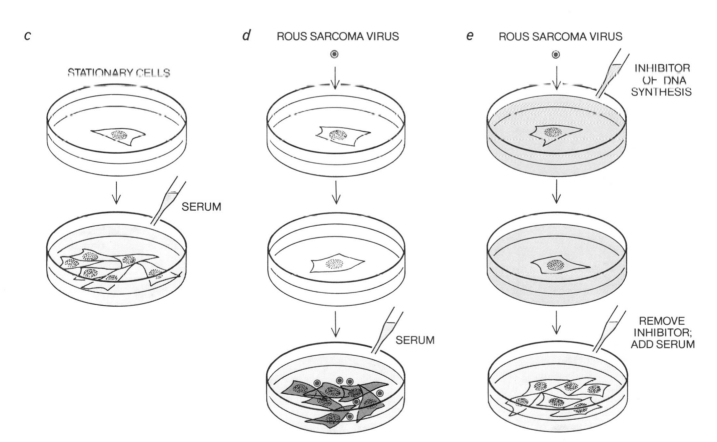

coma virus (*b*), one can protect the cells from infection. In subsequent experiments by the author cultures of stationary, or nondividing, cells were used; when blood serum is added to such cultures (*c*), they divide normally. If such stationary cells are first exposed to Rous sarcoma virus (*d*), however, they become infected but there is no virus production or morphological transformation

until serum is added back and the cells divide once again. When the stationary cells are exposed simultaneously to Rous sarcoma virus and to an inhibitor of DNA synthesis (*e*), the cells are not killed but neither are they infected; when the inhibitor of DNA synthesis is removed and serum is added, cells divide normally, are not infected, do not become transformed and do not produce virus.

Massachusetts Institute of Technology had independently made similar observations with the virion of a mouse leukemia virus. The two papers describing these findings were published together in the June 27, 1970, issue of *Nature*, the British scientific weekly. The two publications stimulated an enormous amount of work whose peak is not yet in sight.

In our early papers we called the new viral enzyme RNA-dependent DNA polymerase because the template was RNA and the product was DNA. Subsequently we and others found that the enzyme could also use DNA as a template for DNA synthesis. We therefore decided to change the word "dependent" to "directed," so that we now refer to the enzyme as RNA-directed DNA polymerase. The revised name makes no statement about the origin of the enzyme or its relation to other DNA polymerases. Independently *Nature* began referring to the enzyme as "reverse

transcriptase," a name that I do not like because of its ambiguity but that has gained wide currency.

All the later studies confirm the original finding that the virions of RNA tumor viruses contain a DNA polymerase system that is activated by treating the virion with a detergent and that is sensitive to ribonuclease. Moreover, the virion enzyme functions only as a DNA polymerase; it will not act as an RNA polymerase. As I have mentioned, however, other unrelated RNA viruses do contain an RNA polymerase.

If the DNA produced by the RNA-directed DNA polymerase is isolated free of protein, the size of its molecule can be estimated by spinning it at high speed in a sucrose gradient in an ultracentrifuge. The molecule is surprisingly small: less than a tenth as long as one would expect a copy of the complete viral RNA to be. The reason for the small size is still elusive. If the isolated DNA product is centrifuged in a cesium sulfate density gradient, which separates RNA from

DNA on the basis of their different densities, one finds that the product has the density of DNA [*see bottom illustration on opposite page*]. Further characterization, for example by treatment with enzymes that specifically attack either single- or double-strand DNA, shows that the product of the DNA polymerase system is a double strand. From such studies one can conclude that the DNA polymerase system of the virion makes short pieces of double-strand DNA.

Many workers have demonstrated that the DNA product of the RNA-directed DNA polymerase system has a base sequence complementary to the viral RNA [*see top illustration on page 255*]. This conclusion is drawn from annealing, or molecular hybridization, experiments. Labeled DNA from the virion polymerase reaction is treated so that the strands of the DNA dissociate. The single-strand DNA is added to unlabeled viral RNA, and the mixture is incubated so that complementary strands can form a hybrid combination. The mixture is then centrifuged in a cesium sulfate density gradient. About half of the product DNA forms a band at a density characteristic of RNA or of hybrid RNA-DNA molecules rather than at a density characteristic of DNA. The test is quite specific and indicates that the DNA polymerase of the virion copies the sequence of the bases of the viral RNA into DNA. This experiment, however, still does not demonstrate that such a copying process takes place in cells infected by Rous sarcoma virus.

The viral DNA polymerase was shown to be present in the core of the virion by the following experiment carried out by George Todaro's group at the National Cancer Institute and by John M. Coffin in my laboratory at the University of Wisconsin. Rous virus virions were treated with a detergent to disrupt the envelope. Then the disrupted virus was centrifuged in a sucrose density gradient. Most of the viral RNA, about 20 percent of the protein and most of the RNA-directed DNA polymerase activity were found to sediment together in "cores," a term given to structures that are denser than whole virions [*see bottom illustration on page 255*]. Further studies showed that with more extensive disruption of the virion the viral DNA polymerase can be freed from the viral RNA and then purified. The purified enzyme is capable of directing the synthesis of DNA on a variety of templates: synthetic and natural DNA, RNA and RNA-DNA hybrids.

The general conclusion from studies in a number of laboratories is that the

FURTHER EXPERIMENTS, carried out by one of the author's students, David E. Boettiger, and independently by Piero Balduzzi and Herbert R. Morgan at the University of Rochester, involved exposing stationary cells to Rous sarcoma virus in the presence of 5-bromodeoxyuridine, an analogue of the DNA constituent thymidine that, when incorporated into DNA, sensitizes the DNA to inactivation by light. As a control some of the treated cells were first not exposed to light (*left*); after serum was added to these cells to enable them to divide they were transformed into cancer cells and began to produce virus. When another culture of treated cells was exposed to light (*right*), the cells were not killed, but the treatment prevented their infection by the virus. When serum was again added to enable these cells to divide, they did not become transformed and did not produce virus.

rousvirus DNA polymerase closely resembles the other DNA polymerases described above that are present in more familiar biological systems and that catalyze the synthesis of DNA on a DNA template. In other words, it is not a unique property of the rousvirus DNA polymerase to be able to use RNA as a template for DNA synthesis. (This was first proposed several years ago by Sylvia Lee Huang and Liebe F. Cavalieri of the Sloan-Kettering Institute.) What is unique so far is the apparent biological role of RNA-directed DNA synthesis in the replication of rousviruses.

Further work in my laboratory has shown that preparations of purified virions of the Rous sarcoma virus contain other enzymes related to DNA replication. The most unusual of them is an enzyme that is named polynucleotide ligase, which repairs breaks in DNA molecules. It is an attractive hypothesis that the function of the ligase is to join the viral DNA to the chromosomal DNA of the host cell, thus integrating the viral genome with the cell genome. After this integration the genetic information of the virus would be replicated with that of the host and passed from the parent cell to the daughter cell. The Rous sarcoma virus virion also contains many other enzymes whose role is completely unknown. We do not know whether they participate in the life cycle of the virus or whether they are merely accidental contaminants picked up in the formation of the virion.

After the first discovery of a DNA polymerase in the virions of RNA tumor viruses, a great many other RNA viruses were examined to see if they contain a similar DNA polymerase system. First it was found that all the viruses previously classified in the RNA tumor virus group contain such an enzyme system. This group of RNA viruses includes both the rousviruses that cause tumors and those that do not cause tumors. Even more interesting, it was found that two types of virus that had not been classified in the same group with RNA tumor viruses also contain a DNA polymerase system. One of these viruses is Visna virus, which causes a slowly developing neurological disease in sheep. After the demonstration of a DNA polymerase in virions of the Visna virus, Kenneth Kaname Takemoto and L. B. Stone at the National Institutes of Health showed that the same virus could cause cancerous transformation of mouse cells in culture. Therefore Visna virus can now be considered a transforming rousvirus. The other type of virus that

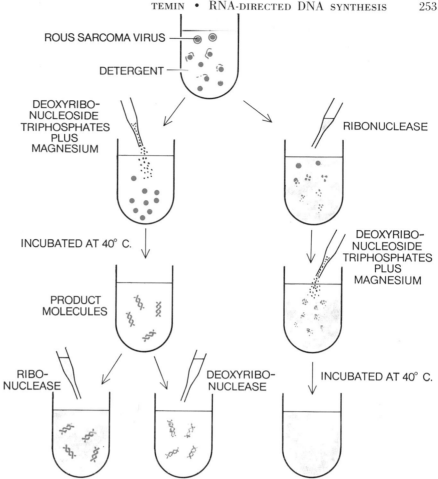

EXISTENCE OF A POLYMERASE capable of producing DNA on an RNA template in RNA tumor viruses was demonstrated by the author and his colleague Satoshi Mizutani (and also independently by David Baltimore of the Massachusetts Institute of Technology). In the experiment conducted by Mizutani and the author purified virions of Rous sarcoma virus were first treated with a detergent to disrupt their lipid-containing envelope. Four deoxyribonucleoside triphosphates, the "building blocks" of DNA, were then added to the disrupted virions. When the mixture was incubated, it incorporated the radioactive label associated with one of the building blocks into an acid-insoluble substance that was stable in the presence of ribonuclease (an enzyme known to destroy RNA), whereas it was fragmented by deoxyribonuclease (an enzyme that destroys DNA). When the experiment was repeated with disrupted virions pretreated with ribonuclease, little or no DNA was produced, indicating that intact viral RNA was needed as template for synthesis of DNA.

has been found to have a DNA polymerase system is the "foamy," or syncytium-forming, viruses. These viruses, isolated from monkeys and cats, have not been connected with any particular disease but are common contaminants of cell cultures. They have not yet been shown to cause tumors or cancerous transformation.

The DNA polymerase present in RNA tumor viruses may not only explain how these viruses produce stable cancerous transformations in the cells they infect but also account for some viral latency,

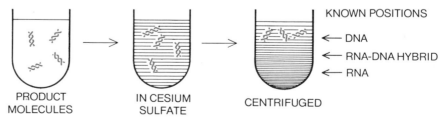

PRODUCT MOLECULES   IN CESIUM SULFATE   CENTRIFUGED

KNOWN POSITIONS
← DNA
← RNA-DNA HYBRID
← RNA

CENTRIFUGATION of the isolated DNA product of the RNA-directed DNA polymerase system in a cesium sulfate density gradient (which separates RNA from DNA on the basis of their different densities) resulted in the finding that the product has the density of DNA. In combination with other findings this result led to the conclusion that the DNA polymerase system of the Rous sarcoma virus virion makes short pieces of double-strand DNA.

254

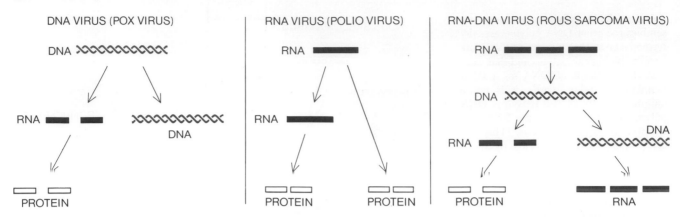

VIRUSES CAN BE GROUPED into three major classes: DNA viruses (*left*), whose genome, or complete set of genes, consists of DNA; RNA viruses (*middle*), whose genome consists of RNA, and RNA-DNA viruses (*right*), the most recently discovered group, whose genome consists alternately of RNA and DNA. A prototype virus in each major class is indicated in parentheses next to the class name. The diagrams illustrate the mode of information transfer that characterizes the replication of viruses in each class.

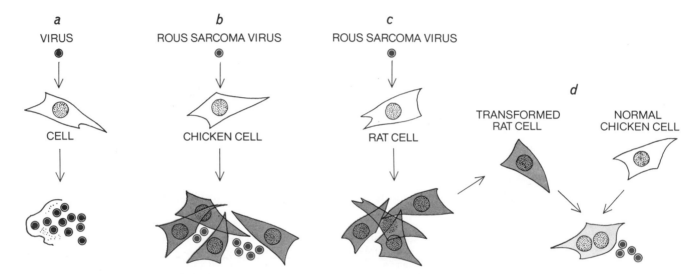

VIRUS-CELL INTERACTION usually leads to the death of the infected cell (*a*), since the replication of most viruses is incompatible with cell division. The Rous sarcoma virus, however, interacts with cells in a different way. Chicken cells infected with the Rous sarcoma virus (*b*) not only survive but also are transformed into cancer cells, which continue to divide and produce new virions. Rat cells infected with the Rous sarcoma virus (*c*) are transformed into cancer cells, which divide but do not produce new virions. By fusing the transformed rat cells with normal chicken cells the production of Rous sarcoma virions can be induced (*d*).

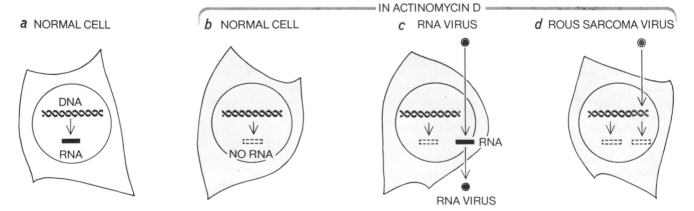

SYNTHESIS OF RNA on a DNA template in normal cells (*a*) is inhibited by the addition of the antibiotic actinomycin D (*b*). Since the antibiotic does not affect the synthesis of RNA made on an RNA template, however, it does not stop RNA synthesis specifically related to the viral genome in cells infected by most RNA viruses (*c*). The finding that actinomycin D inhibited the production of *all* RNA in cells producing Rous sarcoma virus (*d*) was the first direct evidence that the molecular biology of the replication of Rous sarcoma virus was different from that of other RNA viruses. The actinomycin D experiments led the author to propose the DNA provirus hypothesis, which holds that rousviruses such as the Rous sarcoma virus replicate through a DNA intermediate.

the phenomenon in which a virus disappears after infecting an organism only to reappear months or years later. Once an RNA virus has transferred its genetic information to DNA, it would be able to remain latent in a cell and be replicated by the cellular enzyme systems that replicate and repair the cell DNA. After some later activation the virus could appear again as infectious virus particles [*see top illustration on following page*].

About a year ago considerable public excitement was generated by the reported discovery of "RNA-dependent DNA polymerases" in human tumor cells. The general conclusion I would draw now from most of this work has been stated above: All DNA polymerases are capable, under the appropriate conditions, of transcribing information from RNA into DNA. At present we lack generally accepted criteria for determining whether or not such syntheses have any biological role or any relation to rousviruses.

In my laboratory we have taken a slightly different approach to the question of RNA-directed DNA synthesis in cells. We have used detergent activation and ribonuclease sensitivity as criteria in a broad search for DNA polymerase systems in a variety of animal cells. That is, we have looked in cells for a DNA polymerase system similar to viral "cores." Coffin has found such a DNA polymerase system in normal, uninfected rat embryo cells. So far we do not know the full significance of this discovery, but it suggests that ribonuclease-sensitive DNA polymerase systems are present in cells other than tumor cells or virus-infected cells.

For many years I have favored the idea that RNA-directed DNA synthesis may be important in normal cellular processes, particularly those involved in the embryonic differentiation of cells. This idea has been expanded in the form of the protovirus hypothesis [*see bottom illustration on following page*]. The general idea is that in normal cells there are regions of DNA that serve as templates for the synthesis of RNA, and that this RNA serves in turn as a template for the synthesis of DNA that subsequently becomes integrated with the cellular DNA. By this means certain regions of DNA can be amplified. With additional processes that introduce changes in the DNA, the DNA of different cells can be made different. This difference might serve as a means of distinguishing different cells.

What, then, are the general implications of this work for the prevention or treatment of human cancer? We can

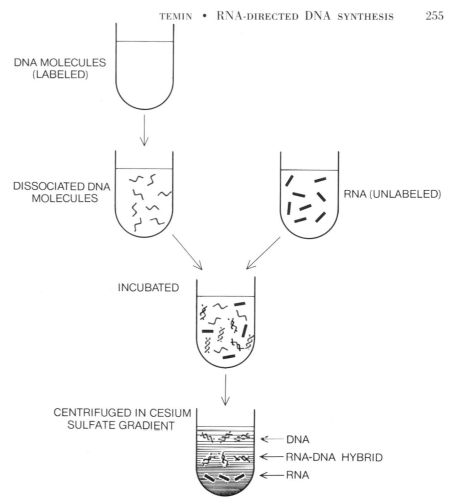

MOLECULAR-HYBRIDIZATION EXPERIMENTS demonstrated that the DNA product of the RNA-directed DNA polymerase system within the virion copies the sequence of bases of the viral RNA into DNA. Labeled DNA from the virion polymerase reaction was first treated so that strands of the DNA dissociated. The single-strand DNA was then added to unlabeled viral RNA, and the mixture was incubated at high temperature so that complementary strands could form a hybrid combination. When the resulting "annealed" mixture was centrifuged in a cesium sulfate density gradient, about half of the product DNA was observed to form a band at a density characteristic of hybrid RNA-DNA molecules.

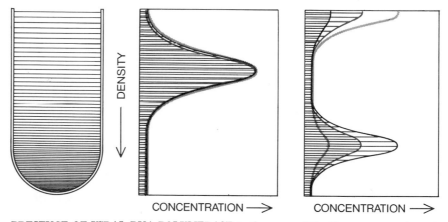

PRESENCE OF VIRAL DNA POLYMERASE in the cores of the Rous sarcoma virus virions was demonstrated by John M. Coffin in the author's laboratory. The curves at center show the density distribution of various radioactively labeled constituents of the whole virions as determined by centrifugation in a sucrose density gradient (*left*). The curves at right show the density distribution of the same constituents determined by centrifugation after the virions were treated with a detergent to disrupt their envelopes. Most of the viral RNA (*black curves*), about 20 percent of the protein (*gray curves*) and most of the RNA-directed DNA polymerase activity (*colored curves*) of the disrupted virions were found to sediment together at a higher density than the corresponding constituents of the whole virions, indicating that these constituents are concentrated in the cores of the virions.

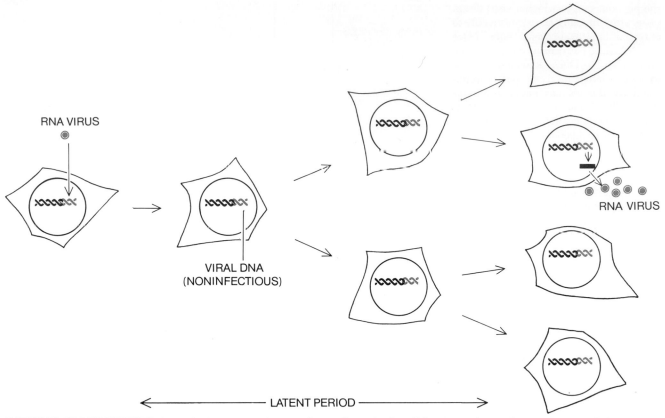

LATENCY OF RNA VIRUSES after infecting an organism may be attributable to the DNA polymerase system present in the cores of such viruses. After transferring its genetic information to DNA in the cell nucleus (*left*), the RNA virus would disappear, remaining latent in the cell by virtue of its replication by the cellular enzyme systems that replicate and repair the cell's DNA (*left center, right center*). Months or years later some form of activation could then cause the infectious RNA virions to appear again (*upper right*).

conclude only that some biological systems utilize a previously undescribed mode of information transfer: from RNA to DNA. It is an interesting coincidence that this new mode of information transfer was first discovered in tumor-causing viruses. We cannot say, however, that RNA-directed DNA synthesis is an exclusive property of such viruses. What the discovery of RNA-directed DNA synthesis does mean is that we now have some simple biochemical tests to determine whether or not newly discovered human viruses are members of the same group as the RNA viruses that produce tumors and cancerous transformations in animal cells, and to look for information related to these viruses in human cancers. We cannot now say that inhibitors of RNA-directed DNA synthesis would have any effect on human cancer. In rousvirus-induced tumors in animals the synthesis of new viral DNA appears to be important only at the initial stage of cancerous transformation, not thereafter.

Probably the most important implication of this discovery for the understanding of cancer in man has been the removal of the dichotomy between viral and genetic theories of the origin of cancer. At a time when genes were thought to consist of DNA alterable only by mutation, and when most of the known cancer-causing animal viruses were of the RNA type, it was hard to imagine common features of genetic and viral theories. Now that we have uncovered evidence that cancer-causing RNA viruses can produce a DNA transcript of the viral RNA, one can readily formulate hypotheses in which elements related to viral RNA are attached to the genome of the cell and transmitted genetically to become activated at some future time and cause "spontaneous" cancer. Experiments designed to test this idea are now in progress in a number of laboratories around the world.

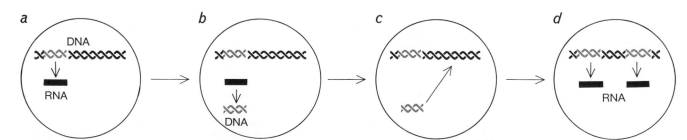

PROTOVIRUS HYPOTHESIS, put forward by the author, embodies the idea that RNA-directed DNA synthesis may be important in normal cellular processes. According to this view, there are regions of DNA in normal cells that serve as templates for the synthesis of RNA (*a*). This RNA serves in turn as a template for the synthesis of DNA (*b*), which later becomes integrated with the cellular DNA (*c*). The amplification of certain regions of DNA resulting from the repetition of the process (*d*) may, in conjunction with additional processes that introduce changes in the DNA, play an important role in the embryonic differentiation of cells.

III

CELLULAR ARCHITECTURE

CELLULAR
ARCHITECTURE

INTRODUCTION

This section includes only three articles, but it introduces the biological membrane, a subject of intense, current interest and research. The membrane not only separates the chemically different regions in a complex biological cell, but also provides the quasi-fluid support on which many of the reactions that are critical to the operation of the cell occur. The first article on the structure of cell membranes describes the history of the development of the lipid bilayer model and introduces the concept of active transport. Other structures for membranes, their physical characteristics, and their fatty-acid content are also considered. The article by Racker on the mitochondrion membrane describes the membrane as a site for controlled and coupled sequences of chemical reactions. Energy production, the Krebs cycle, and membrane-bound enzymes are described. Sharon's article on the bacterial cell wall describes the microscopic and macroscopic structure of this giant molecule and the role of drugs such as penicillin in blocking its formation.

Other articles in *Scientific American* that discuss the architecture of cells include: J. D. Robertson, *The Membrane of the Living Cell* (April 1962, Offprint 151); A. K. Solomon, *Pores in the Cell Membrane* (December 1960, Offprint 76); L. E. and M. R. Hokin, *The Chemistry of Cell Membranes* (October 1965, Offprint 1022); C. de Duve, *The Lysosome* (May 1963, Offprint 156); R. Losick and P. W. Robbins, *The Receptor Site for a Bacterial Virus* (November 1969, Offprint 1161); M. Burnet, *The Structure of the Influenza Virus* (February 1957); R. W. Horne, *The Structure of Viruses* (January 1963, Offprint 147); T. T. Puck, *Single Human Cells In Vitro* (August 1957, Offprint 33); A. K. Solomon, *The State of Water In Red Cells* (February 1971, Offprint 1213); N. K. Wessells, *How Living Cells Change Shape* (October 1971, Offprint 1233); E. Ponder, *The Red Blood Cell* (January 1957); A. A. Moscona, *How Cells Associate* (September 1961, Offprint 95); M. Nomura, *Ribosomes* (October 1969, Offprint 1157); C. Yanofsky, *Gene Structure and Protein Structure* (May 1967, Offprint 1074); N. K. Wessells and W. J. Rutter, *Phases in Cell Differentiation* (March 1969, Offprint 1136); A. O. D. Willows, *Giant Brain Cells in Mollusks* (February 1971, Offprint 1212); D. E. Green, *The Mitochondrion* (January 1964); R. A. Reisfeld and B. D. Kahan, *Markers of Biological Individuality* (June 1972, Offprint 1251).

260

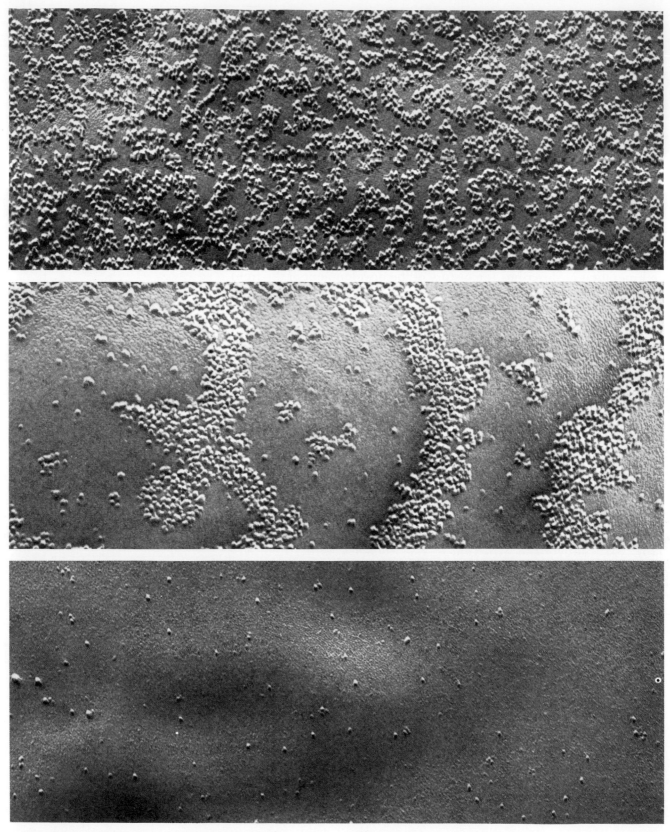

EVIDENCE FOR PROTEINS within the bilayer structure of cell membranes is provided by freeze-etch electron microscopy. A suspension of membranes in water is frozen and then fractured with a sharp blade. The fracture will often split a membrane in the middle along a plane parallel to the surface. After platinum and carbon vapors are deposited along the fracture surface the specimen can be studied in the electron microscope. The micrograph at the top shows many particles 50 to 85 angstroms in diameter embedded in a fractured membrane from rabbit red blood cells. The other two views show how the number of particles is greatly reduced if the membrane is first treated with a proteolytic enzyme that digests 45 percent (*middle*) or 70 percent (*bottom*) of the original membrane protein. The missing particles have presumably been digested by the enzyme. The membrane preparations are enlarged some 95,000 diameters in these micrographs made by L. H. Engstrom in Daniel Branton's laboratory at the University of California at Berkeley.

29

THE STRUCTURE OF CELL MEMBRANES

C. FRED FOX

February 1972

The thin, sturdy envelope of the living cell consists of lipid, phosphate and protein. The proteins act as both gatekeepers and active carriers, determining what passes through the membrane

Every living cell is enclosed by a membrane that serves not only as a sturdy envelope inside which the cell can function but also as a discriminating portal, enabling nutrients and other essential agents to enter and waste products to leave. Called the cytoplasmic membrane, it can also "pump" substances from one side to the other against a "head," that is, it can extract a substance that is in dilute solution on one side and transport it to the opposite side, where the concentration of the substance is many times higher. Thus the cytoplasmic membrane selectively regulates the flux of nutrients and ions between the cell and its external milieu.

The cells of higher organisms have in addition to a cytoplasmic membrane a number of internal membranes that isolate the structures termed organelles, which play various specialized roles. For example, the mitochondria oxidize foodstuffs and provide fuel for the other activities of the cell, and the chloroplasts conduct photosynthesis. Single-cell organisms such as bacteria have only a cytoplasmic membrane, but its structural diversity is sufficient for it to serve some or all of the functions carried out by the membranes of organelles in higher cells. It is clear that any model formulated to describe the structure of membranes must be able to account for an extraordinary range of functions.

Membranes are composed almost entirely of two classes of molecules: proteins and lipids. The proteins serve as enzymes, or biological catalysts, and provide the membrane with its distinctive functional properties. The lipids provide the gross structural properties of the membrane. The simplest lipids found in nature, such as fats and waxes, are insoluble in water. The lipids found in membranes are amphipathic, meaning that one end of the molecule is hydrophobic, or insoluble in water, and the other end hydrophilic, or water-soluble. The hydrophilic region is described as being polar because it is capable of carrying an ionic (electric) charge; the hydrophobic region is nonpolar.

In most membrane lipids the nonpolar region consists of the hydrocarbon chains of fatty acids: hydrocarbon molecules with a carboxyl group (COOH) at one end. In a typical membrane lipid two fatty-acid molecules are chemically bonded through their carboxyl ends to a backbone of glycerol. The glycerol backbone, in turn, is attached to a polar head group consisting of phosphate and other groups, which often carry an ionic charge |*see illustration on next page*|. Phosphate-containing lipids of this type are called phospholipids.

When a suspension of phospholipids in water is subjected to high-energy sound under suitable conditions, the phospholipid molecules cluster together to form closed vesicles: small saclike structures called liposomes. The arrangement of phospholipids in the walls of both liposomes and biological membranes has recently been deduced with the help of X-ray diffraction, which can reveal the distance between repeating groups of atoms. An X-ray diffraction analysis by M. F. Wilkins and his associates at King's College in London indicates that two parallel arrays of the polar-head groups of lipids are separated by a distance of approximately 40 angstroms and that the fatty-acid tails are stacked parallel to one another in arrays of 50 or more phospholipid molecules.

The X-ray data suggest a structure for liposomes and membranes in which the phospholipids are arranged in two parallel layers [*see illustrations on page 263*]. The polar heads are arrayed externally on the bilayer surfaces, and the fatty-acid tails are pointed inward, perpendicular to the plane of the membrane surface. This model of phospholipid structure in membranes is identical with one proposed by James F. Danielli and Hugh Davson in the mid-1930's, when no precise structural data were available. It is also the minimum-energy configuration for a thin film composed of amphipathic molecules, because it maximizes the interaction of the polar groups with water.

Unlike lipids, proteins do not form orderly arrays in membranes, and thus their arrangement cannot be assessed by X-ray diffraction. The absence of order is not surprising. Each particular kind of membrane incorporates a variety of protein molecules that differ widely in molecular weight and in relative numbers; a membrane can incorporate from 10 to 100 times more molecules of one type of protein than of another.

Since little can be learned about the disposition of membrane proteins from a general structural analysis, investigators have chosen instead to study the orientation of one or a few species of the proteins in membranes. In the Danielli-Davson model the proteins are assumed to be entirely external to the lipid bilayer, being attached either to one side of the membrane or to the other. Although information obtained from X-ray diffraction and high-resolution electron microscopy indicates that this is probably true for the bulk of the membrane protein, biochemical studies show that the Danielli-Davson concept is an oversimplification. The evidence for alternative locations has been provided chiefly by Marc Bretscher of the Medical Research Council laboratories in Cambridge and by Theodore L. Steck, G. Franklin and Donald F. H. Wallach of the Harvard Medical School. Their results suggest that certain proteins penetrate the lipid bilayer and that others extend all the way through it.

Bretscher has labeled a major protein of the cytoplasmic membrane of human red blood cells with a radioactive substance that forms chemical bonds with the protein but is unable to penetrate the membrane surface. The protein was labeled in two ways [see illustration on pages 264 and 265]. First, intact red blood cells were exposed to the label so that it became attached only to the portion of the protein that is exposed on the outer surface of the membrane. Second, red blood cells were broken up before the radioactive label was added. Under these conditions the label could attach itself to parts of the protein exposed on the internal surface of the membrane as well as to parts on the external surface.

The two batches of membrane, labeled under the two different conditions, were treated separately to isolate the protein. The purified protein from the two separate samples was degraded into definable fragments by treatment with a proteolytic enzyme: an enzyme that cleaves links in the chain of amino acid units that constitutes a protein. A sample from each batch of fragments was now placed on the corner of a square of filter paper for "fingerprinting" analysis. In this technique the fragments are separated by chromatography in one direction on the paper and by electrophoresis in a direction at right angles to the first. In the chromatographic step each type of fragment is separated from the others because it has a characteristic rate of travel across the paper with respect to the rate at which a solvent travels. In the electrophoretic step the fragments are further separated because they have characteristic rates of travel in an imposed electric field.

Once a separation had been achieved the filter paper was laid on a piece of X-ray film so that the radioactively labeled fragments could reveal themselves by exposing the film. When films from the two batches of fragments were developed, they clearly showed that more labeled fragments were present when both the internal and the external surface of the cell membrane had been exposed to the radioactive label than when the outer surface alone had been exposed. This provides strong evidence that the portion of the protein that gives rise to the additional labeled fragments is on the inner surface of the membrane.

Steck and his colleagues obtained similar results with a procedure in which they prepared two types of closed-membrane vesicle, using as a starting material the membranes from red blood cells. In one type of vesicle preparation (right-side-out vesicles) the outer membrane surface is exposed to the external aqueous environment. In the other type of preparation (inside-out vesicles) the inner surface of the membrane is exposed to the external aqueous environment. When the two types of vesicle are treated with a proteolytic enzyme, only those proteins exposed to the external aqueous environment should be degraded. Steck found that some proteins are susceptible to digestion in both the right-side-out and inside-out vesicles, indicating that these proteins are exposed on both membrane surfaces. Other proteins are susceptible to proteolytic digestion in right-side-out vesicles but not in inside-out vesicles. Such proteins are evidently located exclusively on only one side of the membrane. This information lends credence to the concept of sidedness in membranes. Such sidedness had been suspected for many years because the inner and outer surfaces of cellular membranes are thought to have different biological functions. The development of a technique for preparing vesicles with right-side-out and inside-out configurations should be extremely useful in determining on which side of the membrane a given species of protein resides and thus functions.

Daniel Branton and his associates at the University of California at Berkeley have developed and exploited the technique of freeze-etch electron microscopy to study the internal anatomy of membranes. In freeze-etch microscopy a suspension of membranes in water is frozen rapidly and fractured with a sharp blade. Wherever the membrane surface runs parallel to the plane of fracture much of the membrane will be split along the middle of the lipid bilayer. A thin film of platinum and carbon is then evaporated onto the surface of the fracture. This makes it possible to examine the anatomy of structures in the fracture plane by electron microscopy.

The electron micrographs of the fractured membrane reveal many particles, approximately 50 to 85 angstroms in diameter, on the inner surface of the lipid bilayer. These particles are not observed if the membrane samples are first treated with proteolytic enzymes, indicating that the particles probably consist of protein [see illustration on page 260]. From quantitative estimates of the number of particles revealed by freeze-etching, Branton and his colleagues have suggested that between 10 and 20 percent of the internal volume of many biological membranes is protein.

Somewhere between a fifth and a quarter of all the protein in a cell is physically associated with membranes. Most of the other proteins are dissolved in the aqueous internal environment of the cell. In order to dissolve membrane proteins in aqueous solvents detergents must be added to promote their dispersion. One might therefore expect membrane proteins to differ considerably from soluble proteins in chemical composition. This, however, is not the case.

The amino acids of which proteins are composed can be classified into two groups: polar and nonpolar. S. A. Rosenberg and Guido Guidotti of Harvard University analyzed the amino acid composition of proteins from a number of membranes and found that they contain about the same percentage of polar and nonpolar amino acids as one finds in the soluble proteins of the common colon

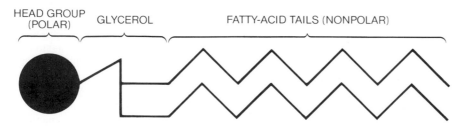

HEAD GROUP (POLAR)   GLYCEROL   FATTY-ACID TAILS (NONPOLAR)

TYPICAL MEMBRANE LIPID is a complex molecular structure, one end of which is hydrophilic, or water-soluble, and the other end hydrophobic. Such a substance is termed amphipathic. The hydrophilic, or polar, region consists of phosphate and other constituents attached to a unit of glycerol. The polar-head group, when in contact with water, often carries an electric charge. The glycerol component forms a bridge to the hydrocarbon tails of two fatty acids that constitute the nonpolar region of the lipid. In this highly schematic diagram the zigzag lines represent hydrocarbon chains; each angle is occupied by a carbon atom and two associated hydrogen atoms. The terminal carbon of each chain is bound to three hydrogen atoms. Phosphate-containing amphipathic lipids are called phospholipids.

bacterium *Escherichia coli.* Thus differences in amino acid composition cannot account for the water-insolubility of membrane proteins.

Studies conducted by L. Spatz and Philipp Strittmatter of the University of Connecticut indicate that the most likely explanation for the water-insolubility of membrane proteins is the arrangement of their amino acids. Spatz and Strittmatter subjected membranes of rabbit liver cells to a mild treatment with a proteolytic enzyme. The treatment released the biologically active portion of the membrane protein: cytochrome b_5. In a separate procedure they solubilized and purified the intact cytochrome b_5 and treated it with the proteolytic enzyme. This treatment also released the water-soluble, biologically active portion of the molecule, together with a number of small degradation products that were insoluble in aqueous solution. The biologically active portion of the molecule, whether obtained from the membrane or from the purified protein, was found to be rich in polar amino acids. The protein fragments that were insoluble in water, on the other hand, were rich in nonpolar amino acids. These observations suggest that many membrane proteins may be amphipathic, having a nonpolar region that is embedded in the part of the membrane containing the nonpolar fatty-acid tails of the phospholipids and a polar region that is exposed on the membrane surface.

We are now ready to ask: How do substances pass through membranes? The nonpolar fatty-acid-tail region of a phospholipid bilayer is physically incompatible with small water-soluble substances, such as metal ions, sugars and amino acids, and thus acts as a barrier through which they cannot flow freely. If one measures the rate at which blood sugar (glucose) passes through the phospholipid-bilayer walls of liposomes, one finds that it is far too low to account for the rate at which glucose penetrates biological membranes. Information of this kind has given rise to the concept that entities termed carriers must be present in biological membranes to facilitate the passage of metal ions and small polar molecules through the barrier presented by the phospholipid bilayer.

Experiments with biological membranes indicate that the hypothetical carriers are highly selective. For example, a carrier that facilitates the transport of glucose through a membrane plays no role in the transport of amino acids or other sugars. An interesting experimental

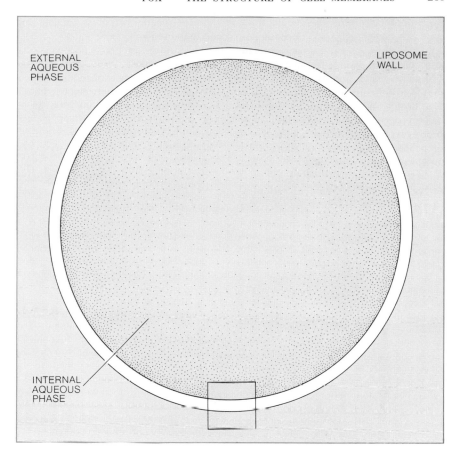

ARTIFICIAL MEMBRANE-ENCLOSED SAC, known as a liposome, is created by subjecting an aqueous suspension of phospholipids to high-energy sound waves. X-ray diffraction shows that the phospholipids in the liposome assume an orderly arrangement resembling what is found in the membranes of actual cells. Area inside the square is enlarged below.

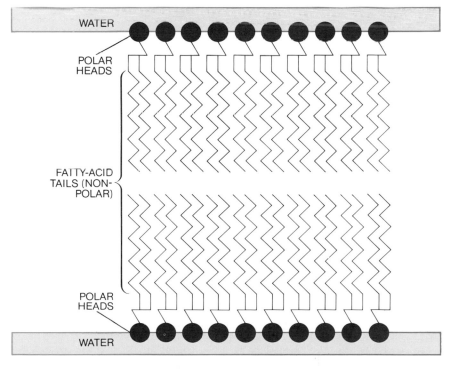

CROSS SECTION OF LIPOSOME WALL shows how the membrane is formed from two layers of lipid molecules. The polar heads of amphipathic lipids face toward the aqueous solution on each side while the nonpolar fatty-acid tails face inward toward one another.

system for measuring selective ion transport was developed by A. D. Bangham, M. M. Standish and J. C. Watkins of the Agricultural Research Council in Cambridge, England, and by J. B. Chappell and A. R. Crofts of the University of Cambridge. As a model carrier they used valinomycin, a nonpolar, fat-soluble antibiotic consisting of a short chain of amino acids (actually 12); such short chains are termed polypeptides to distinguish them from true proteins, which are much larger. Valinomycin combines with phospholipid-bilayer membranes and makes

them permeable to potassium ions but not to sodium ions.

The change in permeability is conveniently studied by measuring the change in electrical resistance across a phospholipid bilayer between two chambers containing a potassium salt in aqueous solution. The experiment is performed by introducing a sample of phospholipid into a small hole between the two chambers. The lipid spontaneously thins out until the chambers are separated by only a thin membrane consisting of a phospholipid bilayer. Electrodes

are then placed in the two chambers to measure the resistance across the membrane.

The resistance across a phospholipid bilayer in the absence of valinomycin is several orders of magnitude higher than the resistance across a typical biological membrane: 10 million ohms centimeter squared compared with between 10 and 10,000. This indicates that phospholipid bilayer membranes are essentially impermeable to small hydrophilic ions. If a small amount of valinomycin (10^{-7} gram per milliliter of salt solution) is intro-

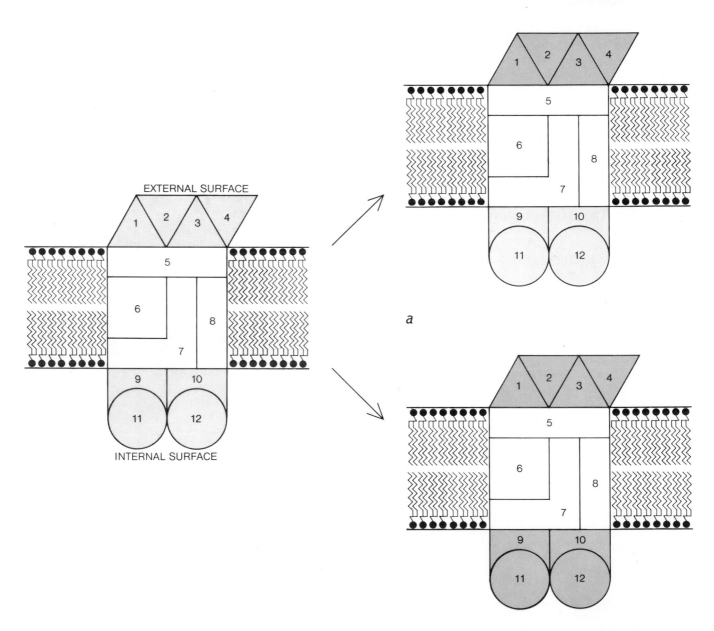

LOCATION OF PROTEINS IN MEMBRANES can be inferred by attaching radioactive labels to the proteins. These diagrams depict an experiment in which a major protein in the membrane of red blood cells was labeled (*a*). When intact cells (*top sequence*) are exposed to the radioactive substance, only the portion of the protein on the outside wall picks up the label (*color*). When the cells are broken before labeling (*bottom sequence*), the radioactive label is able to reach portions of the protein that are exposed to the internal as well as to the external surfaces of the membrane. This can be demonstrated by isolating and purifying the protein labeled under the two conditions. The protein is then broken up into defined fragments (*numbered shapes*) by treating it with a proteolytic enzyme (*b*). Portions of the two batches of fragments are spotted on the corners of filter paper for "fingerprinting" (*c*). This is a

duced into the chambers containing the potassium solution, the resistance falls by five orders of magnitude and the permeability of the phospholipid bilayer to potassium ions rises by a like amount. The permeability of the experimental membrane now essentially duplicates the permeability of biological membranes.

If the experiment is repeated with a sodium chloride solution in the chambers, one finds that the addition of valinomycin causes only a slight change in resistance. Hence valinomycin meets two of the most important criteria for a bio-logical carrier: it enhances permeability and it is highly selective for the transported substance. The question that now arises is: How does valinomycin work?

First of all, valinomycin is nonpolar. Thus it is physically compatible with and can dissolve in the portion of the bilayer that contains the nonpolar fatty-acid tails. Second, valinomycin can evidently diffuse between the two surfaces of the bilayer. S. Krasne, George Eisenman and G. Szabo of the University of California at Los Angeles have shown that the enhancement of potassium-ion transport by valinomycin is interrupted when the bilayer is "frozen" by lowering the temperature. Third, valinomycin must bind potassium ions in such a way that the ionic charge is shielded from the nonpolar region of the membrane. Finally, valinomycin itself must have a selective binding capacity for potassium ions in preference to sodium or other ions.

With valinomycin as a model for carrier-mediated transport, one can postulate three essential steps: recognition of the ion, diffusion of the ion through the membrane, and its release on the other

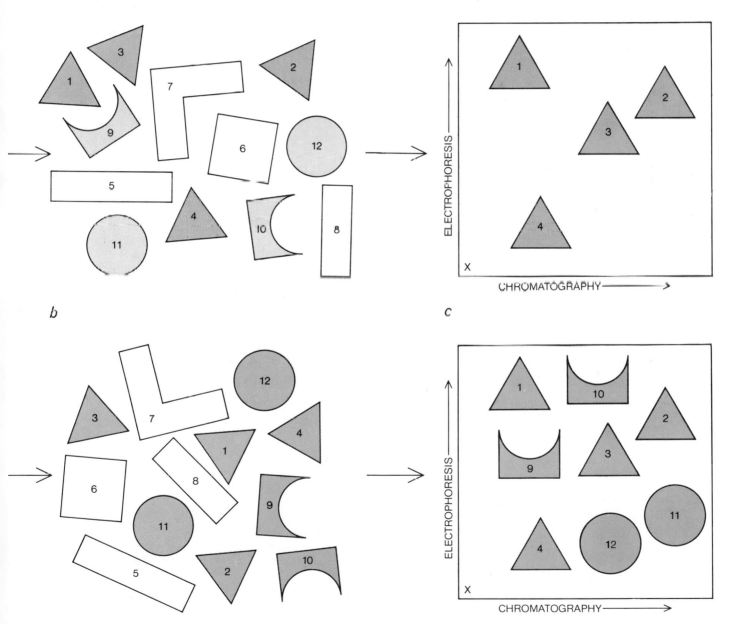

technique that combines chromatography with electrophoresis. By chromatography alone protein fragments would migrate at different rates depending primarily on their solubility in the solvent system. Electrophoresis involves establishing an electric-potential gradient along one axis of the filter paper. Since various fragments have different densities of electric charge they are further separated. A piece of X-ray film is then placed over each sheet of filter paper. Radiation from the labeled fragments exposes the film and reveals where the various fragments have come to rest. A comparison of the X-ray films produced in the parallel experiments shows that more protein fragments are labeled when the red blood cells are broken before labeling and that the additional fragments (9, 10, 11, 12) must represent portions of the original protein that extend through the membrane and penetrate the inner surface.

side. In the first step some part of the valinomycin molecule, embedded in the membrane, "recognizes" the potassium ion as it approaches the surface of the membrane and captures it. In the second step the complex consisting of valinomycin and the potassium ion diffuses through the membrane. Finally, on reaching the opposite surface of the membrane the potassium ion is dissociated from the complex and is released.

The argument to this point can be summarized in a few words. The fundamental structure of biological membranes is a phospholipid bilayer, the phospholipid bilayer is a permeability barrier and carriers are needed to breach it. In addition, the membrane barrier must often be breached in a directional way. In a normally functioning cell hundreds of kinds of small molecule must be present at a higher concentration inside the cell than outside, and many other small molecules must be present at a lower concentration inside the cell than outside. For example, the concentration of potassium ions in human cells is more than 100 times greater than it is in the blood that bathes them. For sodium ions the concentrations are almost exactly reversed. The maintenance of these differences in concentration is absolutely es-

sential; even slight changes can result in death.

Although the model system based on valinomycin provides considerable insight into the function and selectivity of carriers, it sheds no light on the transport mechanism that can pump a substance from a low concentration on one side of the membrane to a higher concentration on the other. Our understanding of concentrative transport (or, as it is usually termed, active transport) owes much to the pioneering effort of Georges Cohen, Howard Rickenberg, Jacques Monod and their associates at the Pasteur Institute in Paris. The Pasteur group studied the transport of milk sugar (lactose) through the cell membrane of the bacterium *Escherichia coli*. Genetic experiments suggested that the carrier for lactose transport was a protein. Studies of the rate of transport revealed that the transport process behaves like a reaction catalyzed by an enzyme, giving further support to the idea that the carrier is a protein. The Pasteur group also found that the lactose-transport system is capable of active transport, producing a lactose concentration 500 times greater inside the cell than outside. The active-transport process depends on the expenditure of metabolic energy; poisons that block energy metabolism destroy

the ability of the cell to concentrate lactose.

A model that accounts for many (but not all) of the properties of the active-transport system that are typified by the lactose system postulates the existence of a carrier protein that can change its shape. The protein is visualized as resembling a revolving door in the membrane wall [*see illustration on opposite page*]. The "door" contains a slot that fits the target substance to be transported. The slot normally faces the cell's external environment. When the target substance enters the slot, the protein changes shape and is thereby enabled to rotate so that the slot faces into the cell. When the target substance has been discharged into the cell, the protein remains with its slot facing inward until the cell expends energy to rotate the protein so that the slot again faces outward.

Working with Eugene P. Kennedy at the Harvard Medical School in 1965, I succeeded in identifying the lactose-transport carrier. We found, as we had expected, that it is a protein with an enzyme-like ability to bind lactose. Since then a number of other transport carriers have been identified, and all turn out to be proteins. The lactose carrier resides in the membrane and is hydrophobic; thus it is physically compatible

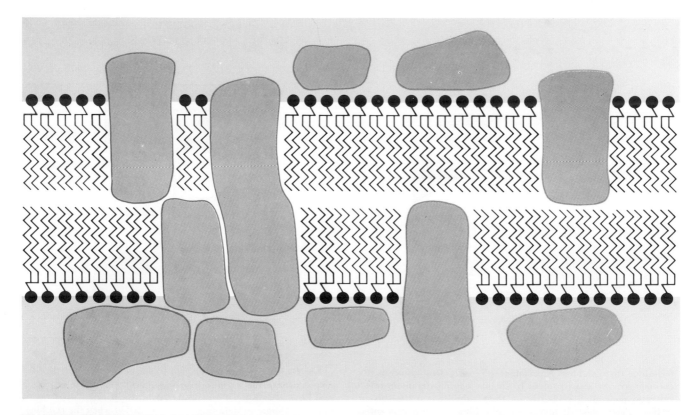

ANATOMY OF BIOLOGICAL MEMBRANE is suggested in this schematic diagram. Phospholipid molecules stacked side by side and back to back provide the basic structure. The gray shapes represent protein molecules. In some cases several proteins (for example the five at the left) are bound into a single functional complex. Proteins can occupy all possible positions with respect to the phospholipid bilayer: they can be entirely outside or inside, they can penetrate either surface or they can extend through the membrane.

with the nonpolar-lipid phase of the membrane.

In 1970 Ron Kaback and his associates at the Roche Institute of Molecular Biology observed that the energy that drives the active transport of lactose and dozens of other low-molecular-weight substances in *E. coli* is directly coupled to the biological oxidation of metabolic intermediates such as D-lactic acid and succinic acid. How energy derived from the oxidation of D-lactic acid can be used to drive active transport is one of the more interesting unsolved problems in membrane biology.

Since transport carriers must be mobile within the membrane in order to move substances from one surface to the other, one might guess that the region of the membrane containing the fatty-acid tails should not have a rigid crystalline structure. X-ray diffraction studies indicate that the fatty acids of membranes in fact do have a "liquid crystalline" structure at physiological temperature, that is, around 37 degrees Celsius. In other words, the fatty acids are not aligned in a rigid crystalline lattice. The techniques of electron paramagnetic resonance and nuclear magnetic resonance can be used to study the flexibility of the fatty-acid side chains in membranes. Several investigators, notably Harden M. McConnell and his associates at Stanford University, have concluded that the fatty acids of membranes are quasi fluid in character.

Membranes incorporate two classes of fatty acids: saturated molecules, in which all the available carbon bonds carry hydrogen atoms, and unsaturated molecules, in which two or more pairs of hydrogen atoms are absent (with the result that two or more pairs of carbon atoms have double bonds). The fluid character of membranes is largely determined by the structure and relative proportion of the unsaturated fatty acids. In phospholipids consisting only of saturated fatty acids the fatty-acid tails are aligned in a rigidly stacked crystalline array at physiological temperatures. In phospholipids consisting of both saturated and unsaturated fatty acids the fatty acids are packed in a less orderly fashion and thus are more fluid. The double bonds of unsaturated fatty acids give rise to a structural deformation that interrupts the ordered stacking necessary for the formation of a rigid crystalline structure [see *illustration on next page*].

My colleagues and I at the University of Chicago (and later at the University of California at Los Angeles) and Peter Overath and his associates at the Uni-

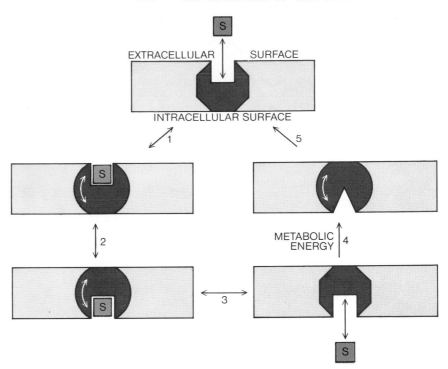

MECHANISM OF "ACTIVE" TRANSPORT may involve a carrier protein (*dark gray*) with the properties of a revolving door. A carrier protein can capture a substance, S, that exists outside the membrane in dilute solution and transport it to the inside of the cell, where the concentration of S is greater than it is outside. When S is bound to the protein, the protein changes shape (*1*), thus enabling it to rotate (*2*). When S becomes detached and enters the cell (*3*), the protein returns to its immobile form. Metabolic energy must be expended (*4*) to alter the protein's shape so that it can rotate and again present its binding site to the cell exterior (*5*). Other protein carriers have the capacity to transport substances from low concentration inside the cell to solutions of higher concentration outside the cell.

versity of Cologne have varied the fatty-acid composition of biological membranes to study the effects of fatty-acid structure on transport. When the membrane lipids are rich in unsaturated fatty acids, transport can proceed at rates up to 20 times faster than it does when the membrane lipids are low in unsaturated fatty acids. These experiments show that normal membrane function depends on the fluidity of the fatty acids.

The temperature at which cells live and grow can have a pronounced effect on the amount of unsaturated fatty acid in their membranes. Bacteria grown at a low temperature have membranes with a greater proportion of unsaturated fatty acid than those grown at a higher temperature. This adjustment in fatty-acid composition is necessary if the membranes are to function normally at low temperature. A similar adjustment can take place in higher organisms. For example, there is a temperature gradient in the legs of the reindeer; the highest temperature is near the body, the lowest is near the hooves. To compensate for this temperature gradient the cells near the hooves have membranes whose lipids are enriched in unsaturated fatty acids.

Although, as we have seen, phospholipids can spontaneously form bilayer films in water, this process only provides a physical rationale as to why the predominant structure in membranes is a phospholipid bilayer. The events leading to the assembly of a biological membrane are far more complex. The cells of higher organisms contain a number of unique membrane structures. They differ widely in lipid composition, and each type of membrane has its own unique complement of proteins. The diversity in protein composition and in the location of proteins within membranes explains the functional diversity of different types of membrane. Rarely does a single species of protein exist in more than one type of membrane.

Since all membrane proteins are synthesized at approximately the same cellular location, what is it that determines that one type of protein will be incorporated only into the cytoplasmic membrane and that another type will turn up only in a mitochondrial membrane? At present this question can be answered only by conjecture tinctured with a few facts. Two general hypotheses for membrane assembly can be offered. One pos-

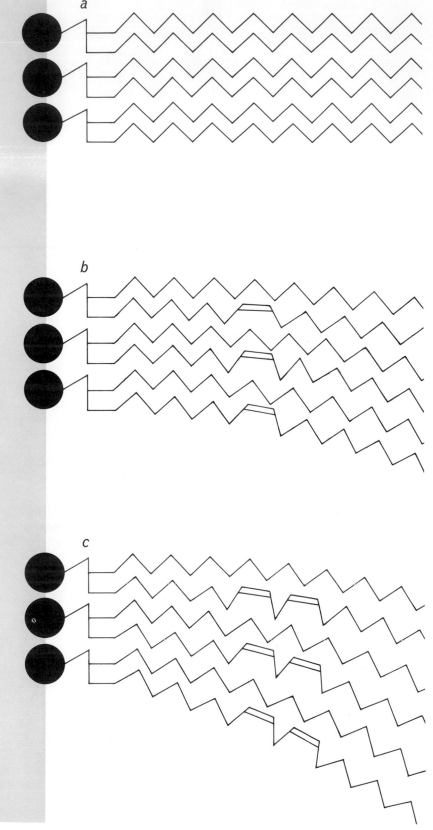

sibility is that new pieces of membrane are made from scratch by a self-assembly mechanism in which all the components of a new piece of membrane come together spontaneously. This new piece could then be inserted into an existing membrane. A second possibility is that newly made proteins are simply inserted at random into a preexisting membrane.

Recent studies in my laboratory at the University of California at Los Angeles and in the laboratories of Philip Siekevitz and George E. Palade at Rockefeller University support the second hypothesis. That is all well and good, but what determines why a given protein is incorporated only into a given kind of membrane? Although this must be answered by conjecture, it is known that many proteins are specifically bound to other proteins in the same membrane. Such protein-protein interactions are not uncommon; many of the functional entities in membranes are complexes of several proteins. Thus the proteins in a membrane may provide a template that is recognized by a newly synthesized protein and that helps to insert the newly synthesized protein into the membrane. In this way old membrane could act as a template for the assembly of new membrane. This might explain why different membranes incorporate different proteins.

Why, then, do different membranes have different lipid compositions? The answers to this question are even more obscure. In general lipids are synthesized within the membrane; the enzymes that catalyze the synthesis are part of the membrane. Some lipids, however, are made in one membrane and then shuttled to another membrane that has no inherent capacity to synthesize them. Since there is an interchange of lipids between various membranes, it seems unlikely that the variations in lipid composition in different membranes can be explained by dissimilarities in the synthetic capacity of a given membrane for a given type of lipid. There are at least two possible ways of accounting for differences in lipid composition. One possibility is that different membranes may destroy different lipids at different rates; another is that the proteins of one species of membrane may selectively bind one type of lipid, whereas the proteins of another species of membrane may bind a different type of lipid. It is obvious from this discussion that concrete evidence on the subject of membrane assembly is scant but that the problems are well defined.

VARIATION IN FATTY-ACID COMPOSITION can disrupt the orderly stacking of phospholipids in a biological membrane. In a lipid layer composed entirely of saturated fatty acids (*a*) the fatty-acid chains contain only single bonds between carbon atoms and thus nest together to form rigid structures. In a lipid layer containing unsaturated fatty acids with one double bond (*b*) the double bonds introduce a deformation that interferes with orderly stacking and makes the fatty-acid region somewhat fluid. When fatty acids with two double bonds are present (*c*), the deformation and the consequent fluidity are greater still.

30

THE MEMBRANE OF THE MITOCHONDRION

EFRAIM RACKER

February 1968

The folded inner membrane of this intracellular body is the site of the major process of energy metabolism in the living cell. It is studied by taking it apart and attempting to put it together again

The seat of oxidative phosphorylation, the process by which most plant and animal cells produce the energy required to sustain life, is the inner membrane of the intracellular particles called mitochondria [*see bottom illustration on page 4*]. Associated with this membrane are the enzymes of oxidative phosphorylation, embedded in a complex matrix that binds them tenaciously in an ordered array. The mechanism of energy production in mitochondria has long defied analysis, since a complex chemical pathway in a living organism cannot really be understood until its intermediate products have been identified and the enzymes that catalyze each step of the process have been individually resolved as soluble components. A decade ago my colleagues and I set out to attack the problem by trying to take the inner membrane of the mitochondrion apart and put it back together. We have been partially successful in the attempt, and along the way we have made some exciting discoveries and developed new methods of studying enzymes bound in membranes.

The universal energy carrier of the cell is adenosine triphosphate (ATP). This molecule functions by transferring its energetic terminal phosphate group to another molecule. In so doing it is converted to adenosine diphosphate (ADP), which in turn can be transformed into ATP by energy-generating systems in the cell. This regeneration of ATP occurs at several stages in the course of the breakdown and oxidation of foodstuffs. Some ATP is formed during glycolysis, a well-understood metabolic pathway utilizing soluble enzymes that break carbohydrates down to simpler compounds.

Most of the ATP is formed, however, during the course of oxidative phosphorylation in mitochondria. Pyruvate, the end product of glycolysis, is delivered to the mitochondria, where it is oxidized to carbon dioxide and water by the enzymes of the Krebs cycle [*see top illustration on pages 273 and 274*]. As hydrogen is removed from the successive intermediate products, it is captured by the coenzyme diphosphopyridine nucleotide (DPN), which contains the vitamin nicotinamide. The electrons of hydrogen are passed along a series of respiratory enzymes, notably yellow flavoproteins and red cytochromes, ultimately combining with protons and oxygen to form water. The energy of this oxidation process is utilized at three sites to regenerate ATP from ADP and inorganic phosphate. Under physiological conditions such "coupling" of oxidation with phosphorylation is compulsory, and respiration takes place only when ADP and phosphate are available, which is to say when ATP is being utilized. This "tight coupling" represents an ingenious control mechanism through which energy production is regulated by the rate of energy consumption.

Two chemicals that affect oxidative phosphorylation serve as tools with which to analyze the process. One is dinitrophenol (DNP), which uncouples oxidation from phosphorylation so that respiration proceeds but produces heat instead of ATP. The other is the antibiotic oligomycin, which acts differently. It interferes with the production of ATP, thereby inhibiting respiration as long as the system is tightly coupled. When dinitrophenol is added, the inhibition by oligomycin is overcome and respiration returns to its original rate, although it produces no ATP.

The enzymes that catalyze electron transport had been isolated and characterized, but only after the disruption of mitochondria with detergents. This process left the oxidation enzymes able to function but damaged the phosphorylation system severely. It was accordingly believed for a long time that the intact mitochondrial structure was essential for oxidative phosphorylation and that the component parts could not be separated without destroying them.

In 1956 the system did begin to yield to fractionation. Independent experiments reported almost simultaneously from the laboratories of Albert L. Lehninger, Henry A. Lardy, David E. Green and W. W. Kielley showed that chemicals such as digitonin and physical methods such as sonic oscillation would break mitochondria into "submitochondrial particles" much smaller than mitochondria and yet able to catalyze oxidative phosphorylation. This accomplishment was an important step forward, and yet there was still no indication that a true resolution—a separation of soluble components—would ever be possible. Such resolution was the task we undertook at the Public Health Research Institute of the City of New York.

In 1957 the first successful resolution of the system of oxidative phosphorylation was achieved when Harvey Penefsky, Maynard E. Pullman and I fragmented beef-heart mitochondria by agitating them with glass beads in a powerful device called a Nossal shaker. We removed the heavier unbroken mitochondria by centrifuging the mixture at low speed and then respun the lighter fraction at high speed [*see top illustration on page 275*]. The resulting sediment—the submitochondrial particles—still contained the respiratory enzymes but could not produce much ATP; the remaining fraction contained a soluble component that was necessary for the coupling of phosphorylation to oxidation. We called this soluble component a coupling factor, F_1. In time various treatments of mitochondria separated other coupling factors that were also required for phosphorylation, and these we called F_2, F_3 and F_4.

The experiments demonstrating the resolution of F_1 were difficult to reproduce. In laboratory jargon our data were "in the right direction"—and that is always a sign of trouble. F_1 as much as doubled phosphorylation, but that was not enough stimulation to provide a reliable assay of its coupling activity. And a reliable assay was required if we were to purify F_1 and characterize it. Now, it was known that mitochondria could catalyze the splitting of ATP into ADP and inorganic phosphorus. In fact, as early as 1945 Lardy, working at the University of Wisconsin, had suggested that this enzymatic ("ATP-ase") activity might be the inverse of some step in oxidative phosphorylation. We discovered that partially purified F_1 did in fact exhibit ATP-ase activity.

Since this was the first time ATP-ase had been extracted as a soluble component from mitochondria, we decided to go after this enzyme. We realized it was a gamble that might not shed any light on oxidative phosphorylation, but the ATP-ase assay was simple and accurate and at least we had had experience in purifying soluble enzymes. We felt that it should not take long to establish whether or not the ATP-ase and the coupling-factor activity were related.

Yet sometimes experience gets in one's way. Working in a "cold room," as one ordinarily does in enzyme research, we found the ATP-ase to be quite unstable (in contrast to the ATP-ase activity of submitochondrial particles, which was quite stable) and we made little progress.

One day we discovered that this enzyme was "cold labile": at 0 degrees centigrade it lost all activity in a few hours, but at room temperature it was stable for days. That was a turning point in our investigations; from then on purification was simple. Furthermore, we had a decisive tool for determining the relationship between ATP-ase and coupling factor. The fact that both activities decayed at the same rate at 0 degrees indicated that the same protein was responsible for both.

Chemical fractionation of F_1 gave us a pure enzyme in good yield. In fact, at first the yield seemed to be too good: often our final preparation had more units of ATP-ase than we had estimated were present in the crude preparation. Moreover, the ratio of coupling activity to ATP-ase activity was not constant during purification. An examination of these discrepancies by Pullman revealed that the crude mitochondrial extract contained a protein that inhibited ATP-ase activity but not coupling activity, and that the removal of this inhibitor during purification explained the unexpected increase in total ATP-ase activity.

The purified F_1 had one puzzling property. Whereas Lardy and his collaborators had shown that both oxidative phosphorylation and the ATP-ase activity of mitochondria were very sensitive to oligomycin, our soluble enzyme was completely insensitive. This apparent discrepancy caused some of our colleagues to challenge the significance of our observations with the soluble enzyme. I had heard that the late Oswald T. Avery had once said: "It doesn't matter if you fall down, as long as you pick up something from the floor when you get up." And so we accepted the challenge and embarked on a project to find

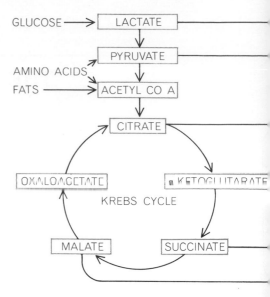

OXIDATIVE PHOSPHORYLATION is the process whereby energy from the oxidation of foodstuffs is harnessed to produce ATP, the energy carrier of the cell. Sugars, fats

out why oligomycin inhibited the enzyme in mitochondrial particles but not the soluble enzyme.

We started with the working hypothesis that there must be a component in mitochondria that confers oligomycin sensitivity on the enzyme. To show this we first had to prepare submitochondrial particles from which all the bound, oligomycin-sensitive ATP-ase had been removed, then add F_1 to them and observe what happened. We were able to eliminate the ATP-ase activity from particles by treating them with urea at 0 degrees, but to our surprise oligomycin-sensitive ATP-ase activity kept reappearing on dilution or aging. It developed that most of the ATP-ase in submitochondrial particles was latent—masked, apparently, by Pullman's inhibitor—and was more resistant to urea than the manifest enzyme was. We had to learn how

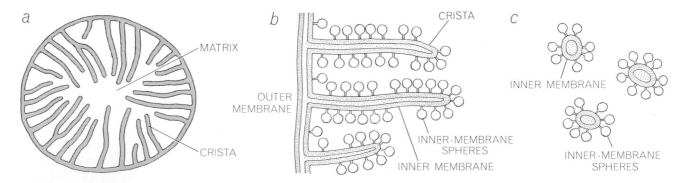

MITOCHONDRION, seen in a schematic cross section (a), has two membranes, each about 60 angstrom units (six millionths of a millimeter) thick. The inner membrane is deeply folded into "cristae" covered with the inner-membrane spheres, each about 85 angstroms in diameter (b). The inner membrane, with its spheres, is the site of oxidative phosphorylation. Mitochondria exposed to sonic oscillation become fragmented into small submitochondrial particles (c), which are still capable of oxidative phosphorylation.

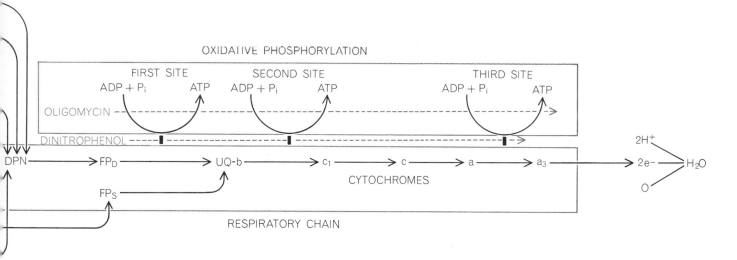

OXIDATIVE PHOSPHORYLATION and proteins are partially metabolized and then, in mitochondria, enter the Krebs cycle, in which they are broken down to carbon dioxide. In the process hydrogen atoms are accepted by the coenzyme diphosphopyridine nucleotide (*DPN*). The chain of respiratory enzymes, including flavoproteins (*FP*) and cytochromes b, c_1, c, a and a_3, catalyze a stepwise transfer of electrons to form water. At three sites phosphorylation is "coupled" to electron transfer. It can be uncoupled by dinitrophenol and inhibited by oligomycin.

to unmask this ATP-ase by removing the inhibitor before urea treatment. We found that if submitochondrial particles were first treated with trypsin, a digestive enzyme, and only then with urea, the resulting "*TU* particles" were depleted of virtually all ATP-ase activity [*see middle illustration on page 273*]. More recently, when we found that the trypsin was damaging the mitochondrial membrane, Lawrence Horstman of our laboratory discovered that the inhibitor could be removed more gently by passing the submitochondrial particles through a column of Sephadex, a molecular sieve that separates small bodies from large ones. When this procedure is followed by treatment with urea, the resulting "*SU* particles" are analogous to *TU* particles but are much more effective in reconstitution experiments.

When we added F_1 to *TU* or *SU* particles, the enzyme was bound to the particles and the ATP-ase activity became not only sensitive to oligomycin but also stable at 0 degrees. Thus our working hypothesis was confirmed: mitochondria contain a component or components that alter the properties of F_1. We have become increasingly aware that this phenomenon is not unusual. Enzymes bound to membranes almost invariably have some properties that are different from those of the same enzymes in solution. Gottfried Schatz of our laboratory suggested the word "allotopy" (from the Greek for "other" and "position") to designate this phenomenon. We observed, furthermore, that the properties not only of the enzyme but also of the

membrane to which it is attached are changed depending on whether they are separate or bound to one another [*see top illustration on page 276*]. An allotopic property of an enzyme can be used to devise a quantitative assay to serve during the purification of the membrane, since one can test successively purer membrane preparations to see if they are still capable of changing the properties of the added enzyme.

The *TU* particles that conferred oligomycin sensitivity on F_1 still contained the entire electron-transport chain, and we went on, with the allotopic property of F_1 as the tool, in an attempt to further resolve this membrane system. One day I subjected *TU* particles to sonic oscillation without including the usual salt buffer. Centrifugation of the resulting mixture at high speed yielded a soluble extract that conferred oligomycin sensitivity on F_1. We called the factor responsible for this property F_0. The discovery seemed even more exciting when the soluble preparation turned out to contain the entire electron-transport chain and even some residual phosphorylating activity: it appeared that we had actually rendered the entire system soluble. Then the addition of salt solution made the preparation turbid, which meant that particles had formed from the soluble system. In other words, in the presence of salt buffer—which must be added in biological experiments to keep the medium constant—F_0 was still particulate. At the time this was disappointing, but the observation led us

into new investigations of the relation between membrane structure and function.

In collaboration with Donald F. Parsons of the University of Toronto Faculty of Medicine and Britton Chance of the University of Pennsylvania School of Medicine, we examined all our membranous preparations of F_0 by negative staining in the electron microscope. We saw, first, that the submitochondrial particles we had started with were similar to those prepared by earlier investigators: sac-shaped structures outlined by a membrane that was covered with the characteristic "inner-membrane spheres" that had been discovered by Humberto Fernandez-Moran of the University of Chicago. The treatment with trypsin caused little change in structure. Subsequent treatment with urea, however, had a dramatic effect: although it left the membrane intact, it removed the inner-membrane spheres [*see bottom illustration on page 273*]. This was unexpected, since David Green had once maintained that these spheres, which he called "elementary particles," represented groups of enzymes of the electron-transport chain [see "The Mitochondrion," by David E. Green; SCIENTIFIC AMERICAN, January, 1964]. We had found, on the contrary, that the *TU* particles (which lacked spheres) contained the entire electron-transport chain!

If the spheres did not contain respiratory enzymes, what did they contain? We calculated that most of the protein removed by urea treatment could be accounted for by the removal of ATP-ase,

272

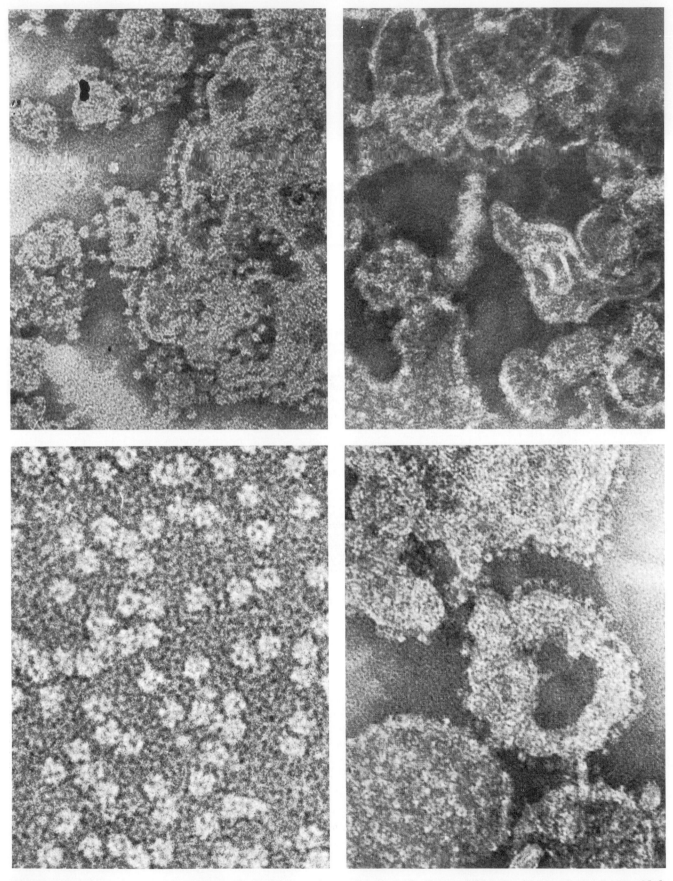

RECONSTITUTION of the mitochondrial membrane begins with submitochondrial particles lined with inner-membrane spheres (*top left*). Treatment with a molecular sieve (Sephadex) and urea produces "SU particles" without spheres (*top right*). When cou-pling factor F_1 (*bottom left*) isolated from mitochondria is added, the characteristic shape of submitochondrial particles is restored (*bottom right*). F_1 spheres are enlarged about 600,000 diameters, other preparations about 300,000 in these electron micrographs.

and so we suspected that the spheres were identical with F_1. We were encouraged in this belief when a preparation of pure F_1 turned out to have the characteristic appearance of the 85-angstrom-unit inner-membrane spheres [*see micrograph at bottom left on page 272*].

One further experiment was needed to identify F_1 unambiguously with the spheres: the reconstitution of a depleted particle by the addition of F_1, resulting in the restoration of the submitochondrial particle's typical shape and function. This was accomplished only recently, after the development of the *SU* particles. The addition of F_1 to these particles yielded a preparation that was indistinguishable in structure from fully functional submitochondrial particles [*see illustration on page 272*] and confirmed that coupling factor F_1 is identical with the inner-membrane spheres.

The morphological reconstitution was not paralleled by restoration of function, however. In an effort to regain oxidative phosphorylation we added three more coupling factors, F_2, F_3 and F_4—proteins that had been obtained from mitochondria by various extraction procedures. *SU* particles reconstituted with all four coupling factors oxidized succinate, a compound of the Krebs cycle, with a high efficiency of ATP synthesis: for each molecule of oxygen consumed, up to 1.8 molecules of ATP were formed. That is very close to the best value—two molecules—that can be achieved with intact mitochondria.

With these experiments one of the aims of our investigation had been achieved: a resolution of soluble components and a reconstitution of structure and function. Another aim has been to get some insight into the mode of action of the coupling factors. How do they fit into the mechanism of oxidative phosphorylation?

There are currently two views of the general nature of that mechanism. One is a chemically oriented hypothesis originally suggested by E. C. Slater of the University of Amsterdam in 1953, in analogy to the mechanism of ATP formation in glycolysis. It proposes that during electron transport high-energy intermediate compounds ($A{\sim}x$, $B{\sim}x$, $C{\sim}x$) are formed at each coupling site, composed of a member of the respiratory chain (A, B, C) and an unknown (x). These compounds are transformed into a common intermediate by interaction with another unknown (y) to form $x{\sim}y$. This intermediate in turn combines with inorganic phosphate to yield $x{\sim}P$, which

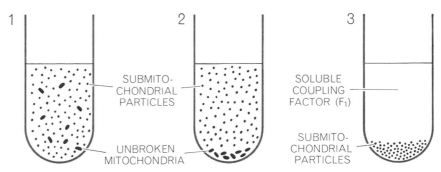

COUPLING FACTOR F_1 is separated by centrifugation. Mitochondria subjected to sonic oscillation (*1*) are centrifuged at low speed to separate particles from intact mitochondria (*2*). Then the particles are spun at high speed. The resulting light fraction (*3*) contains a soluble component (F_1) that is required for ATP production in the course of oxidation.

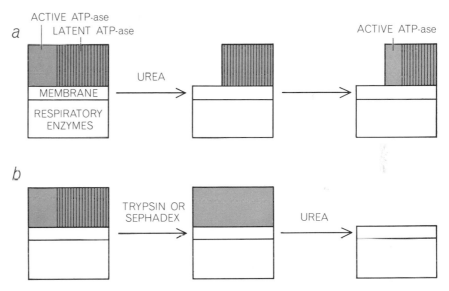

ENZYMATIC ACTIVITY (ATP-ase activity) of soluble F_1 was found resistant to oligomycin, unlike that of the intact membrane. To see if the membrane conferred this sensitivity on soluble F_1, it was first necessary to remove all native F_1 from the membrane. Most of the F_1 is masked by an inhibitor (*bars*), however; treatment with urea removed only exposed F_1, and ATP-ase activity reappeared (*a*). The destruction of inhibitor by trypsin (*b*) exposed the latent F_1 to removal by urea. Later Sephadex was substituted for trypsin.

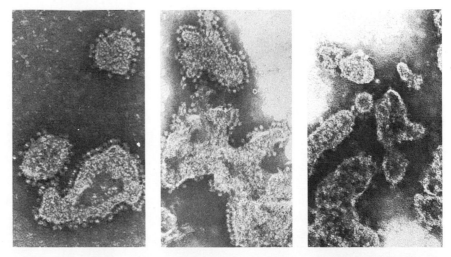

ELECTRON MICROGRAPHS trace the procedure diagrammed in the preceding illustration. The membrane of submitochondrial particles, enlarged about 100,000 diameters, is lined with inner-membrane spheres (*left*). Trypsin has little effect on appearance (*center*). Urea removes the spheres from the particles, leaving "*TU* particles" without spheres (*right*).

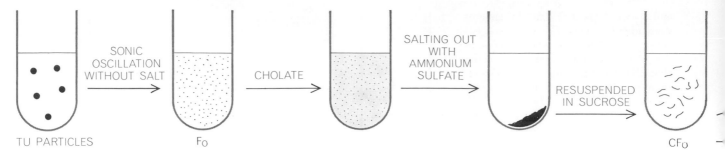

CHOLATE

SALTING OUT
WITH
AMMONIUM
SULFATE

RESUSPENDED
IN SUCROSE

TU PARTICLES F_O CF_O

MEMBRANE OF MITOCHONDRION was isolated as shown here. *TU* particles were subjected to sonic oscillation, producing F_O, which had the capacity to bind F_1. When F_O was dissolved in cho- late and fractionated by "salting out," the colorless precipitate CF_O was obtained. It lacked respiratory enzymes and lipids; added to F_1, it inhibited ATP-ase activity. Addition of phospholipid

ultimately transfers its energetic phos- phate group to ADP to form ATP.

Recently Peter Mitchell of the Glynn Research Ltd. laboratories in England has challenged this chemical hypothesis with some new and provocative ideas. Instead of a high-energy intermediate compound of the respiratory chain, he proposes that an electrical potential de- velops during respiration that provides the energy for ATP production: The pos- itively charged hydrogen ions (protons) are moved to one side of the membrane while the negatively charged electrons are channeled to the other side. The separation of charges is utilized by a complex mechanism to give rise to the high-energy intermediate $x \sim y$, which powers the formation of ATP. At the core of this hypothesis is an ATP-ase

located in the inner membrane.

In some respects the two hypotheses are not much different: both include a high-energy intermediate, $x \sim y$, to gen- erate ATP from ADP and phosphate. In the Mitchell hypothesis, however, $x \sim y$ is formed by means of an electrical mem- brane potential. This requires a much higher integrity of the membrane struc- ture than is required by the chemical hypothesis. Indeed, Mitchell considers that uncouplers such as dinitrophenol act by making the membrane "leaky" to protons, thus preventing a separation of charges. It is apparent, therefore, that further studies of the inner membrane are of utmost importance for the evalua- tion of the two hypotheses.

What is the role of coupling factor according to these two formulations?

Mitchell proposes that F_1, together with F_O, represents the reversible ATP-ase that utilizes the electrical potential to generate ATP. According to the chemi- cal hypothesis, F_1 catalyzes the last step in ATP formation, the "transphosphoryl- ation" from $x \sim P$ to ADP. Indeed, every reaction associated with oxidative phos- phorylation that requires ATP can be shown to be dependent on F_1. June Fes- senden-Raden in our laboratory has pre- pared an antibody against F_1 and has found that these ATP-dependent reac- tions are inhibited by the antibody.

In collaboration with Mrs. Fessenden- Raden, Richard McCarty and Gott- fried Schatz, I have recently found that a coupling factor may have, in addition to a catalytic function that is inhibited by

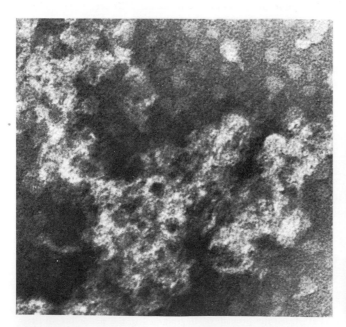

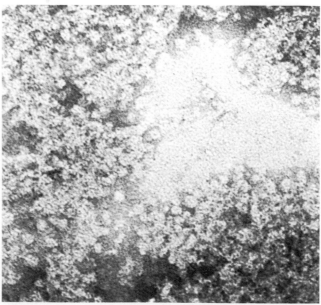

CF_O FRACTION, enlarged about 250,000 diameters in an electron micrograph, appears amorphous (*left*). When F_1 is added to the CF_O preparation, it appears that the F_1 spheres attach themselves to the CF_O, but no distinct structure is seen (*center*). When phospho- lipids are added, the distinct structure that emerges (*right*) re- sembles that of submitochondrial particles. In other words, CF_O

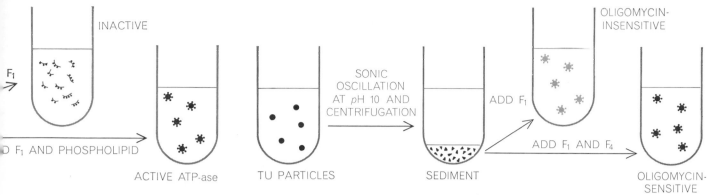

F_1

INACTIVE

ADD F_1 AND PHOSPHOLIPID

ACTIVE ATP-ase

TU PARTICLES

SONIC OSCILLATION AT pH 10 AND CENTRIFUGATION

SEDIMENT

ADD F_1

OLIGOMYCIN-INSENSITIVE

ADD F_1 AND F_4

OLIGOMYCIN-SENSITIVE

restored ATP-ase activity, which was now oligomycin-sensitive. CF_O and phospholipid may thus comprise the membrane proper.

COUPLING FACTOR F_4 is apparently required for oligomycin sensitivity. When *TU* particles are broken down under alkaline conditions, the sedimented particles cannot confer oligomycin sensitivity on F_1. Addition of F_4 in the presence of salt restores this capacity.

its antibody, a second, "structural" function that is not impaired by the antibody. Our first example was the stimulation by F_1 of a reaction that is catalyzed by mitochondria but does not involve ATP. The second example was the observation that in chloroplasts, the energy-generating particles of plant cells, a coupling factor (chloroplast F_1) is required not only for all reactions that involve ADP or ATP but also for a "proton pump" that is driven by light energy without ATP. In contrast to the ATP-dependent reactions, however, this proton pump was not inhibited by an antibody against the chloroplast coupling factor. The factor therefore appears to contribute to the integrity of the chloroplast membrane, which is required for proton transport.

A third example of the "structural"

role of a coupling factor was observed with a preparation of F_1 from yeast mitochondria, which stimulated phosphorylation in beef-heart particles that still contained some residual beef F_1. This stimulation was apparently due to a structural effect of yeast F_1, since it was not inhibited by an antibody against yeast F_1. In beef-heart particles (SU particles) that were completely devoid of native F_1, the yeast factor had no effect; apparently it could not fulfill the catalytic functions of native coupling factor.

We are beginning to suspect that the dual role played by F_1 is representative of a common occurrence in the interaction of enzymes and membranes and is an important expression of the allotopy phenomenon. The contribution of F_1 to the integrity of the mitochondrial mem-

brane and the fact that it is required for the operation of the proton pump of the chloroplast have obvious bearing on the Mitchell hypothesis, and may lead to a clarification of the role played by the membrane in oxidative phosphorylation.

While the work with F_1 was going forward as described above, we therefore also pursued the problem of resolving the inner mitochondrial membrane. F_O had turned out to be a yellow-brown complex of many components including the entire electron-transport chain. By chemical fractionation in the presence of a bile salt, Yasuo Kagawa isolated a virtually colorless fraction (CF_O) with some interesting properties. It lacked respiratory activity, having lost almost all the flavoproteins and cytochromes present in the original submitochondrial particles,

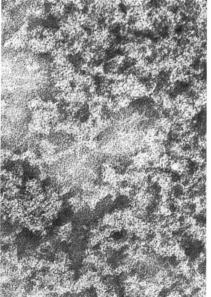

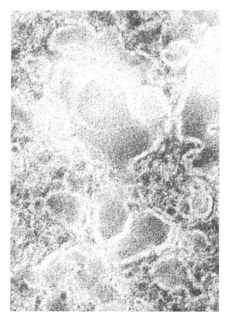

and phospholipid seem to suffice, without respiratory enzymes, to bind F_1 and reconstitute the shape of submitochondrial particles.

F_4 PREPARATION, seen in an electron micrograph at a magnification of 250,000 diameters, appears to be amorphous (*left*). The addition of phospholipid to soluble F_4 yields particles with a sac-shaped structure similar to that of submitochondrial particles (*right*).

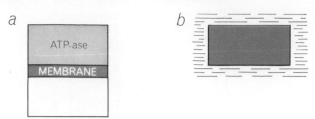

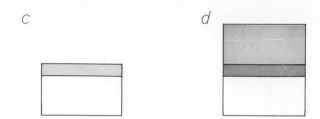

ALLOTOPIC PROPERTIES of purified F_1 and the mitochondrial membrane are indicated. The ATP-ase activity of particles (*a*) was known to be sensitive to oligomycin (*light color*). When soluble F_1 was discovered (*b*), it was found to be resistant to oligomycin (*dark color*), and membrane from which F_1 had been removed (*c*) was sensitive to trypsin (*light gray*). When the enzyme and membrane were bound (*d*), each was changed: the F_1 became sensitive to oligomycin, the membrane resistant to trypsin (*dark gray*).

and it contained only traces of phospholipids, the fat constituents of the membrane.

When F_1 was added to CF_O, the ATP-ase activity of F_1 was almost completely inhibited. The subsequent addition of phospholipid to this inactive complex fully restored ATP-ase activity, which was now sensitive to oligomycin! Equally striking results were observed in the electron microscope. CF_O was amorphous. After the addition of F_1 numerous inner-membrane spheres became attached to CF_O, which remained amorphous. Then, with the addition of phospholipid, the characteristic saclike membranous structures covered with spheres became apparent [*see illustration page 274*]. They could not be distinguished from functional submitochondrial particles, even though they lacked major components of such particles, the respiratory enzymes. These enzymes had always been assumed to be an integral part of the inner membrane—and now they were found not to be present in what appears to be the isolated membrane. How, then, are the respiratory enzymes associated with the membrane? What are the constituents of the inner membrane itself? How are they organized?

To answer some of these questions we proceeded to disrupt the membrane further to see which constituents were necessary for the interaction with F_1. Several years ago Thomas E. Conover and Richard L. Prairie in our laboratory had separated a coupling factor, F_4, that was necessary for phosphorylation in particles obtained by sonic oscillation of mitochondria under highly alkaline conditions. When Kagawa exposed *TU* particles to sonic oscillation under alkaline conditions, after high-speed centrifugation he obtained a sediment ("*TUA* particles") that no longer made F_1 sensitive to oligomycin. The addition of F_4 to *TUA* particles restored oligomycin sensitivity to the complex. Recent experiments by Bernard Bulos in our laboratory at Cornell University revealed that F_1 is bound by *TUA* particles in the presence of salt but is nevertheless not inhibited by oligomycin. On addition of small amounts of a highly purified preparation of F_4, he observed a time-dependent restoration of oligomycin sensitivity, suggesting that an enzymatic process may be taking place. This is the first clue to the mode of action of F_4.

Several years ago Richard S. Criddle, Stephen H. Richardson and their collaborators at the University of Wisconsin isolated an insoluble "structural protein" from mitochondria with the help of detergents and solvents. Our crude preparation of F_1 was similar to that protein in its capacity to combine with some flavoproteins and cytochromes of the respiratory chain but, not having been exposed to damaging chemicals, it remained soluble. In the electron microscope it appeared quite amorphous. When phospholipids were added to soluble F_4, however, a precipitate formed that appeared to be membranous and shaped into sacs [*see illustration at right, page 275*]. We have therefore proposed that F_4 may have a function as an organizational protein in the mitochondrial membrane—a kind of backbone for the association of the respiratory enzymes and the coupling factors involved in the transformation of oxidative energy into ATP.

With this concept as a working hypothesis we have embarked on what promises to be a long and venturesome journey. Taking the membrane-like complex of F_4 and phospholipid as starting material, we are adding isolated soluble flavoproteins and cytochromes of the respiratory chain step by step, checking at each stage to see if some of the allotopic properties of the respiratory chain are restored. In our laboratory Alessandro Bruni and Satoshi Yamashita have constituted sections of the respiratory chain with the appropriate allotopic properties (such as sensitivity to the respiratory poison antimycin). These experiments have given us confidence that we shall eventually achieve a complete reconstitution of the respiratory chain from soluble components. Then we shall turn to the final task: the reconstitution of the system of oxidative phosphorylation from its individual components.

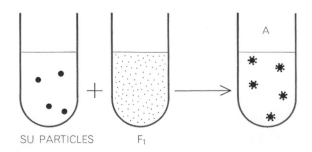

SU PARTICLES

F_1

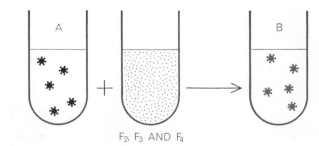

F_2, F_3 AND F_4

SU PARTICLES, from which coupling factor F_1 had been removed, were reconstituted to the shape of submitochondrial particles (*A*) by the addition of F_1 (*see micrographs, page 272*). To restore function as well as structure it was necessary also to add F_2, F_3 and F_4. When this was done, the reconstituted particles (*B*) were capable of generating ATP almost as well as intact mitochondria.

31

THE BACTERIAL CELL WALL

NATHAN SHARON

May 1969

The tough, inflexible envelope that surrounds a bacterium is a single giant molecule made up of amino acid and sugar units. Its formation is blocked by drugs such as penicillin

The pioneer microscopist Anton van Leeuwenhoek, who in the 17th century was the first to observe the one-celled organisms later called bacteria, made sketches showing their shapes. In this way he identified the major classes from which bacteria derive their names: cocci (spherical), bacilli (rod-like) and spirilla (helical). It is evident from van Leeuwenhoek's writings that he realized bacteria have a structure of some kind that holds the organism together and preserves its shape.

This structure is known to us as the bacterial cell wall. It surrounds the fragile plasma membrane of the bacterial cell and protects the membrane and the cytoplasm within from adverse influences in the environment. The rigid wall structure is characteristic of bacteria; the cells of animals are enclosed only by the plasma membrane. Plant cells have a rigid or semirigid envelope, but the structural material of plant cell walls (with the exception of a few seaweeds) is cellulose. The bacterial cell wall is made up of a different material whose composition and structure are far more complex than those of cellulose. Indeed, studies of the bacterial cell wall's molecular architecture have revealed substances and structures found only in bacteria and not in plants and animals.

The composition of the cell-wall material is of particular interest because the cell wall and certain substances that are attached to it are largely responsible for the virulence of bacteria. The symptoms of a broad spectrum of bacterial diseases can be produced by the injection into animals of cell walls rather than whole bacteria. Similarly, animals can be immunized against certain bacteria (for example enteric bacteria, brucellae and pasteurellae) by means of the bacterial cell walls rather than whole killed cells.

The bacterial cell wall became the subject of intensive study about 10 years ago, when it was discovered that certain antibiotics, notably penicillin, kill bacteria by interfering with the synthesis of the cell-wall material. Bacteria that lack cell walls can be grown in the presence of penicillin under certain conditions. Such naked cells, called protoplasts, can also be prepared by means of enzymes, such as lysozyme, that normally kill susceptible bacteria by lysing, or dissolving, their walls. The substances that are stripped from bacteria to expose the na-

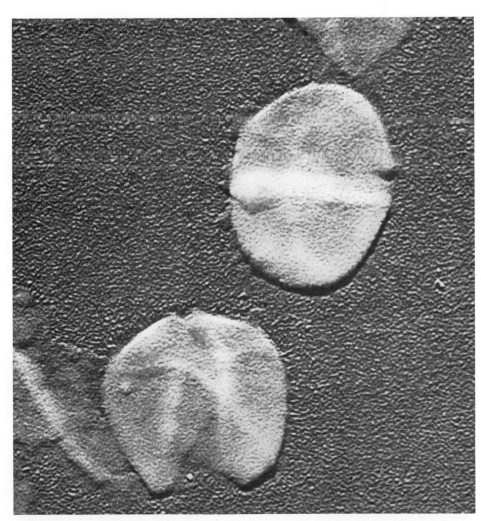

ROUND CELL WALLS were isolated from the spherical bacteria known as *Micrococcus lysodeikticus*. The wall structure is responsible for the characteristic shapes of bacteria.

ked cell have been investigated in order to learn the composition of the cell wall and the process by which bacteria synthesize the wall material. Thanks to these and other studies that I shall discuss, it is now possible to understand the selective action that makes penicillin such an effective antibacterial agent.

The foundations for the investigation of the bacterial cell wall were laid by Milton R. J. Salton at the University of Cambridge in the early 1950's, when he succeeded in isolating cell walls and examining them in the electron microscope. At a large magnification the isolated walls look like empty bags or deflated balloons that have been spread out on a table [see illustrations on pages 277 and 278]. Salton found that, as one might have expected, the cell walls had the same shapes as the bacteria they had formerly housed. The cell walls of cocci are round and the walls of bacilli are elongated.

Although the investigation of these structures has flourished during the past decade, before that time progress was slow. The main difficulty was the dimensions of bacteria. Most bacterial cells are between .5 and one micron wide and between one and five microns long. (A micron is a thousandth of a millimeter.) The investigation of such tiny cells was impeded not only by technical difficulties but also by the widely held premise that cells of such small dimensions must be structureless.

The work that has elucidated the fine structure of bacteria came about as the result both of a radical change in the conception of the living cell and the development of new techniques for its study. An essential tool in investigating bacteria is the electron microscope, whose resolving power is more than 100 times that of the optical microscope. Another important method is paper chromatography, which enables one to examine the chemical composition of cellwall material and conduct experiments on its structure even when only minute quantities of the material are available.

It may appear to be contradictory that whereas bacteria are killed by agents that disrupt their cell walls, the microorganisms can be viable when the cell wall is removed. To clarify this apparent contradiction, let us compare the behavior of bacteria with the behavior of animal cells under certain conditions. Animal cells such as red blood corpuscles will remain intact in a solution that has about the same concentration of salt as the concentration inside the cell (an isotonic solution). On the other hand, if red blood corpuscles are placed in a hypotonic solution (distilled water, for instance) whose osmotic pressure is lower than the pressure of the cell contents, water will enter the cells in accordance with the elementary principles of osmosis. In a short time the corpuscles swell up, and eventually they burst. If the corpuscles are suspended in a hypertonic solution, so that the osmotic pressure is higher than the pressure inside the cell, the corpuscles shrivel up to an unrecog-

ELONGATED CELL WALL retains the shape of the rodlike bacterium *Bacillus licheniformis* from which it was taken. In this electron micrograph and the one on page 277, both of which were made by the author, the cell walls are enlarged 53,000 diameters.

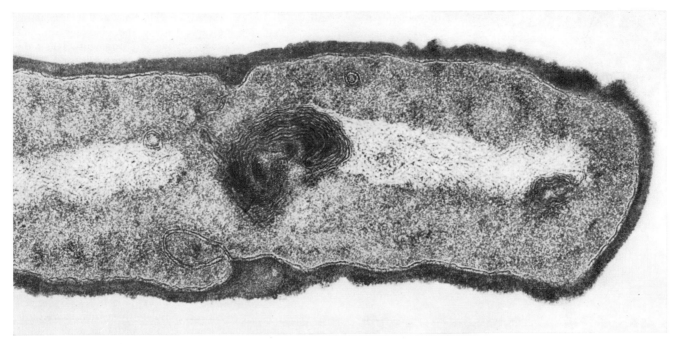

SECTIONAL VIEW of *Bacillus subtilis* shows the cell wall as a dark layer around the edge of the cell. Lining the cell wall is the thin layer of the plasma membrane. Various subcellular particles are visible in the cytoplasm of the bacillus. The bacillus is seen in longitudinal section in this electron micrograph, which was made by Elizabeth H. Leduc, now at Brown University, and Philippe Granboulan at the Institut de Recherches Scientifiques sur le Cancer. The bacillus is enlarged approximately 80,000 diameters.

nizable shape as liquid flows out of them.

In contrast to this behavior, bacteria do not change in their outward appearance when they are in a hypotonic or hypertonic solution. This is because of the tough and inflexible cell wall, whose shape and volume are nearly unchangeable. In bacteria, as in animal cells, the plasma membrane has the property of selective permeability and thus performs the function of "pumping" solutes from the environment into the cell. Indeed, the membrane is so efficient at concentrating metabolites that the osmotic pressure inside the cell can reach 20 atmospheres (300 pounds per square inch). The plasma membrane by itself cannot withstand such pressure; it remains intact only because it transmits the pressure to the cell wall. Hence if the wall of a bacterium is damaged or removed, the membrane bursts.

Weakening or destroying the bacterial cell wall by treating it with lysozyme (or inhibiting the wall's synthesis by means of substances such as penicillin) kills bacteria by causing them to burst. A bacterium that lacks a cell wall will survive, however, if its plasma membrane is preserved intact, which can be done by keeping the microorganism in an isotonic solution. Accordingly one method of preparing the naked protoplasts is to treat bacteria with lysozyme when they are in an isotonic solution. (A relatively concentrated solution of sucrose is used.)

It is interesting that although bacteria have a variety of shapes, their protoplasts are always spherical. The reason is that the weak, flexible plasma membrane assumes the spherical shape of maximum stability. As long as protoplasts remain in an isotonic solution, they take up oxygen and eliminate carbon dioxide, conduct biosynthesis and under special circumstances will even reproduce. In the form of protoplasts, bacteria with flagella retain these whiplike appendages. The flagella do not, however, propel the protoplasts as they do the bacteria, because the solid fulcrum the firm wall structure affords is lacking. Protoplasts from which the cell wall has been entirely removed (some wall-stripping agents do not do a complete job) are not attacked by bacteriophages, the viruses that parasitize bacteria.

If the osmotic pressure of a solution in which protoplasts are suspended is re-

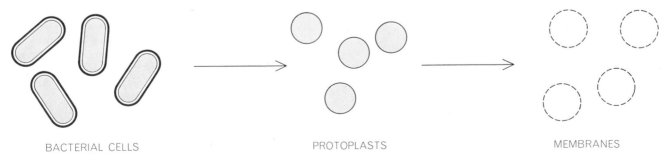

BACTERIAL CELLS   PROTOPLASTS   MEMBRANES

CELL WALLS ARE REMOVED to expose protoplasts (viable naked cells) by treating bacteria with penicillin or with the enzyme lysozyme, which lyses, or dissolves, the wall substance. During the process the bacteria must be kept in a solution that balances the osmotic pressure across the plasma membrane. Bacteria without a wall assume a spherical shape. Plasma membranes of bacteria are prepared by reducing the osmotic pressure of the solution. The protoplasts then swell and burst, releasing their contents.

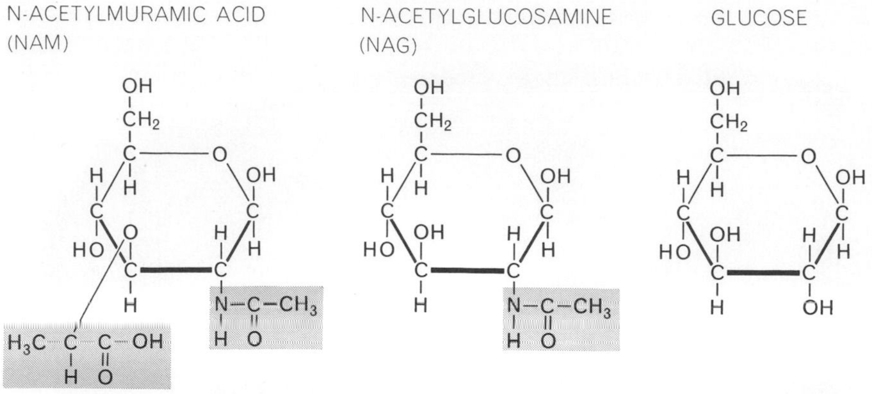

N-ACETYLMURAMIC ACID
(NAM)

N-ACETYLGLUCOSAMINE
(NAG)

GLUCOSE

SACCHARIDE MOLECULES indicate (*color*) where the amino sugars N-acetylmuramic acid (NAM) and N-acetylglucosamine (NAG) differ from glucose. NAM and NAG are components of the glycopeptide that forms the foundation in cell walls of many bacteria. Cellulose, the structural material in plant cell walls, is composed of linked glucose units.

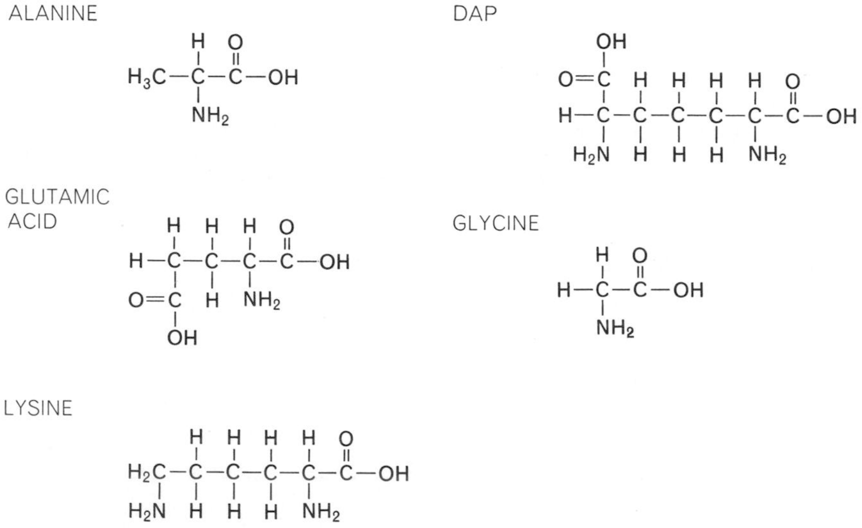

ALANINE

DAP

GLUTAMIC
ACID

GLYCINE

LYSINE

AMINO ACID MOLECULES are the other components of the glycopeptide that gives the cell wall its shape. Alanine and glutamic acid have a configuration different from the one they have in other natural materials (*see illustration below*). In cell walls of many bacteria, lysine is replaced by diaminopimelic acid (DAP). DAP is peculiar to the cell wall.

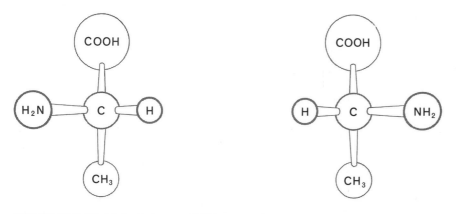

TWO CONFIGURATIONS OF ALANINE are found (in equal amounts) in the glycopeptide of the cell wall. When in solution, one configuration, or stereoisomer, designated *L*-alanine (*left*) rotates plane-polarized light in one direction; its mirror image, *D*-alanine (*right*), rotates polarized light in the opposite direction. *L*-amino acids are normally found in natural materials. Only *D*-glutamic acid appears in cell-wall glycopeptide.

duced, the cells burst and their contents escape, leaving the plasma membrane an empty sac or in fragments. This is the basis of a technique that is used to prepare bacterial plasma membranes, which are now under study in many laboratories.

In investigating the cell wall of bacteria two basic approaches can be followed. In one the wall is studied as part of the intact cell. Thus the examination of very thin slices of bacterium in the electron microscope reveals that the thickness of the cell wall is usually between 15 and 20 millimicrons, which is about 1 percent of the overall thickness of an average bacterial cell. The wall appears to envelop the cell completely; it is also possible to observe changes in the wall during cell division.

The second approach to studying cell walls is to isolate them from bacteria, as one must do in order to analyze their chemical composition. A simple and effective method for the preparation of cell walls is to shake a thick suspension of bacteria and water together with tiny glass beads. The cells are punctured, their contents pour out into the water and the cell walls, which are insoluble, can be readily separated from the denser glass beads and the water-soluble contents of the bacteria. After several washings a white precipitate is obtained that consists solely of cell walls, as can be shown by examination under the electron microscope.

Early investigators were surprised to find that the bacterial wall accounts for 20 to 30 percent of the cell's dry weight. Although the wall is much smaller in volume than the cell cytoplasm, it is mostly solid material whereas the cytoplasm is a relatively dilute solution.

The chemical composition of the cell wall was first analyzed by Salton. His analyses indicated that the wall substance includes almost all the kinds of material that are found in living cells, with the exception of the nucleic acids. He found sugars linked in chains (polysaccharides), amino acids linked in the long chains of proteins and the shorter ones of peptides, and also lipids (fats). As more bacteria were examined, however, it became clear that all cell walls have a few common components: they are two simple sugars and three or four amino acids. These form the "basal structure" of the wall. In the instances where large amounts of other substances were found, it was shown that they came from substances linked to the wall, such as the substances that form the gummy or slimy capsule encasing many bacteria. Treatment with the appropriate solvents (for example salt solutions or organic liquids) was found to remove these substances, leaving an insoluble material that, when it was examined in the electron microscope, still retained the characteristic wall shape. It is therefore this material that provides the cell wall's rigid foundation.

When this material is decomposed by prolonged heating in acid solution, one obtains its constituent saccharides and amino acids. In some bacteria the cell wall is composed almost exclusively of these components. The two sugars are simple ones related to glucose. One is glucosamine, an amino sugar (so named because an amino group—NH_2—is attached to the sugar molecule). Glucosamine is a common constituent of natural polymers. For instance, chitin, the principal structural material of the external skeleton of insects, which among natural polymers is probably second only to cellulose in abundance, is made up solely of glucosamine units. This amino sugar is also found in animal connective tissues and in many protein-carbohydrate compounds. It is usually in the form of N-acetylglucosamine (NAG), and it is in this form in the bacterial cell wall as well.

The other saccharide of the bacterial cell wall is a lactic acid derivative of glucosamine. It was named muramic acid (from *murus*, the Latin word for wall). Muramic acid has been found only in bacteria and microorganisms that resemble bacteria, such as the blue-green algae. Like glucosamine, it is usually in the form of the N-acetyl derivative.

Only a limited number of the 20 amino acids that form protein are found in the cell-wall material. In the walls of many bacteria the amino acids are glutamic acid, glycine, lysine and alanine. Two of the amino acids, however, are found in an "unnatural" configuration, that is, certain of the atoms or groups of atoms in their molecules are arranged differently from the way they are in the amino acids of other natural substances. This configuration, which is designated dextro (D), is a mirror image of the "normal" configuration, levo (L). Glutamic acid appears as a D-stereoisomer in bacterial walls whereas lysine has the usual L configuration. Alanine in the cell wall is found as a D-stereoisomer and also as an L one [see bottom illustration on opposite page]. In many bacteria there is no lysine in the cell wall; instead the wall contains diaminopimelic acid, whose

structure resembles that of lysine. Diaminopimelic acid is peculiar to the bacterial cell wall; it is not found in any protein.

How are the basic constituents of cell-wall material linked to form its rigid structure? In answering this question lysozyme proved a particularly valuable tool. The enzyme was discovered by Sir Alexander Fleming in 1922, six years before his discovery of penicillin. Fleming observed that tears, saliva, nasal mucus and other secretions of the body are capable of dissolving certain bacteria, and he attributed this activity to a cell-lysing enzyme (which he accordingly named lysozyme). He hoped that lysozyme might be useful in fighting pathogenic bacteria. This hope, however, has never been realized.

Lysozyme is also found abundantly in the white of hen's eggs, and it was from this source that lysozyme was isolated and crystallized by other investigators. Fleming found that the bacterium most susceptible to dissolution by lysozyme is *Micrococcus lysodeikticus*. (*Lysodeikticus* means "lysis meter.") This bacterium is still widely used in investigations of lysozyme and enzymes that resemble it. If a tear or a drop of egg white is added to a suspension of *M. lysodeikticus*, in only a minute or so one can see with the unaided eye the turbid suspension clear up. Microscopic examination shows that the cells have simply disappeared.

Until Salton had investigated this phenomenon it was not clear exactly what lysozyme did to the cells, although some information on its action had been gathered by Karl Meyer at Columbia University and by other workers. Salton

STRUCTURAL POLYSACCHARIDES in plant cell walls (*top*), the external skeleton of insects (*middle*) and bacterial cell walls (*bottom*) are similar. The manner in which the chitin and cell-wall polysaccharides differ from cellulose is indicated in color. The polysaccharide of bacterial walls consists of alternating units of NAM and NAG. By cleaving bonds between these units lysozyme dissolves the cell wall. *R* stands for an OH group or a peptide.

showed that lysozyme dissolves the cell wall by cleaving the chemical bonds between its amino sugar subunits.

In investigating the composition of cell walls dissolved by lysozyme, Salton discovered a two-sugar substance composed of acetylglucosamine and acetylmuramic acid. The detailed structure of this disaccharide was worked out in 1963 by Roger W. Jeanloz, Harold M. Flowers and myself in the Laboratory for Carbohydrate Research at the Massachusetts General Hospital. Other investigations have served to clarify the cell wall's architecture, notably those by Jack L. Strominger of the University of Wisconsin School of Medicine, by Jean Marie Ghuysen of the University of Liège, by Howard J. Rogers and H. R. Perkins of the National Institute for Medical Research in London and also those by my colleagues and myself at the Weizmann Institute of Science in Israel. A glycopeptide that appears to represent the fundamental repeating unit of most bacterial cell walls was recently isolated and characterized in our laboratory by David Mirelman.

These investigations have led to the conclusion that the walls of bacteria are made up of two types of polymer, one composed of saccharide subunits and one of amino acid subunits. The saccharide portion of the wall consists of long chains of alternating units of acetylglucosamine and acetylmuramic acid, linked by bridges that include an oxygen atom (glycosidic linkages). This polysaccharide closely resembles the cellulose of plant cell walls and the chitin of insects [see illustration on preceding page].

The cell wall of a bacterium ought to be insoluble by virtue of its polysaccharide alone, as the walls of plants and the external skeletons of insects are. In the bacterial wall, however, there is an additional factor that contributes to the insolubility and also to the strength and toughness of the material. This factor is provided by peptides, the short chains made up of amino acids. The peptide chains, which are cross-linked to one another, are attached to the polysaccharide chains, so that they serve to join them and form a three-dimensional network [see illustration on this page]. The cross-linking of polymers is well known as a means of achieving toughness and mechanical strength in synthetic materials.

The amino acids of the cell wall are linked by peptide bonds (HN–CO), but they are not susceptible to digestion by enzymes, such as trypsin or pepsin, that act to split such bonds (proteolytic enzymes). This resistance may well be re-

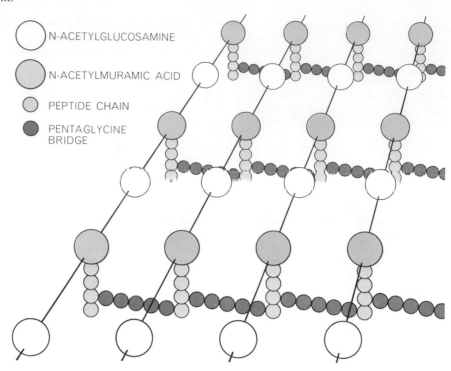

N-ACETYLGLUCOSAMINE

N-ACETYLMURAMIC ACID

PEPTIDE CHAIN

PENTAGLYCINE BRIDGE

STRUCTURE OF CELL-WALL GLYCOPEPTIDE is represented diagrammatically. Attached to the polysaccharide chains are chains of amino acid units (peptides). The peptides, linked to one another by pentaglycine bridges, cross-link the polysaccharide chains. By inhibiting cross-linking, penicillin prevents cell-wall formation in growing bacteria.

lated to the unusual D configuration of the wall components glutamic acid and alanine. Most proteolytic enzymes are unable to split peptide bonds between D-amino acids, and it may be that bacteria are naturally protected from lysis by these enzymes. To be sure, the cell-wall peptides are not totally resistant to all enzymes. If this were so, the world would be full of the cell walls of dead bacteria.

The three-dimensional network of the cell wall can be conceived of in a general way as a gigantic bag-shaped molecule. Such a structure belongs to the general class of glycopeptides; the wall probably is a glycopeptide with a molecular weight of tens of billions. Penicillin, as I have noted, prevents this three-dimensional structure from forming. At what point in the synthesis of the wall glycopeptide does penicillin act?

Fleming himself tried to solve the problem of penicillin's mechanism of action, and the problem has occupied many other workers, particularly during the 1940's, when penicillin came into wide use as a drug. The key to the problem was discovered in 1949 by James T. Park, who is now at the Tufts University Medical School. Park found that when staphylococci are incubated for several hours in a medium containing penicillin, there is an accumulation of substances

that at the time were unknown. The detailed structure of these substances was revealed several years later by Park and by Strominger. They were shown to have a resemblance to the glycopeptide of the cell wall, but they had a low molecular weight. There was also present in these substances in combined form a compound that plays an important role in the biosynthesis of polysaccharides, namely uridine diphosphate.

In 1957 Park and Strominger jointly put forward a hypothesis to account for these findings. The substances associated with uridine diphosphate that accumulate under the influence of penicillin were identified as intermediates in cell-wall synthesis; they result from interference with the synthetic process. This hypothesis explained the well-known fact that penicillin kills only multiplying cells: only at the time of reproduction do bacteria need to synthesize new cell-wall material. It is noteworthy that shortly before Park and Strominger published the results of their work, Joshua Lederberg, who was then at the University of Wisconsin, concluded, on the basis of other evidence, that penicillin kills bacteria by inhibiting the synthesis of their walls.

The hypothesis that the cell wall is penicillin's site of action explained the great susceptibility of bacteria to this drug, which is not toxic to animals. (Only a few micrograms of penicillin are needed to kill staphylococci, but humans are usually not affected by doses 10 million times larger.) This response is to be expected, since penicillin interferes with the synthesis of a structure that is vital for bacteria but is not even present in animals.

During the past decade Park and Strominger's hypothesis has been verified experimentally. The details of the biosynthesis of the intermediate substances and the ways these small water-soluble molecules are converted into the gigantic insoluble glycopeptide of the cell wall have been clarified. Today it is possible to synthesize the uridine diphosphate intermediates from the amino sugars and amino acids of the cell wall with the aid of about 12 specific enzymes that have been extracted from bacteria.

The synthesis of intermediates is only the first step in biosynthesis of the cell wall, and penicillin does not interfere with the process at this stage. In the next stage the uridine-linked intermediate is polymerized together with uridine-linked acetylglucosamine to form the polysaccharide backbone with the aid of enzyme bound to the plasma membrane and a special cofactor (a phosphate derivative of a polyisoprenoid alcohol). Penicillin does not influence this process either. Other antibiotics such as vancomycin and bacitracin do, however, inhibit synthesis at this stage. Quite recently Park and Strominger independently discovered that penicillin inhibits the last stage in synthesis of the cell-wall glycopeptide. This is the synthetic step in which adjacent polysaccharide chains are linked by reaction between their peptide side chains.

Thus on a molecular level the two great discoveries made by Fleming, lysozyme and penicillin, have been joined.

Both substances interact with the cell-wall glycopeptide, although at different sites. The investigation of penicillin and lysozyme has not, however, come to an end. Penicillin's mode of action is being studied in further detail in the search for information that might be useful in combating resistant bacteria, for example through the synthesis of new "tailor-made" antibacterial drugs. Recently there has been an upsurge of interest in lysozyme as the first enzyme whose three-dimensional structure has been precisely determined, so that its properties are understood in atomic detail. It is believed that lysozyme, which we too are studying in terms of its detailed effect on the bacterial cell wall, may offer the key to understanding the secrets of enzyme action.

IV

CHEMICAL BIODYNAMICS

IV

CHEMICAL BIODYNAMICS

INTRODUCTION

The mechanisms by which cells produce and utilize energy are perhaps the most dramatic and fascinating aspect of biochemistry, and in this section we have collected a few of the many available articles on the dynamic function of cells. The first article describes the production of energy in the cell by the Krebs, or tricarboxylic acid, cycle of reactions. The article by Green discusses the enzymatic reactions in the cell — and the analogous reactions *in vitro* — by which long-chain fatty acids are degraded and used. The next two articles describe how proteins are synthesized and degraded. The dynamic translation of DNA into heredity is described in the next two articles: Nirenberg introduces base pairing, the DNA double helix, and protein synthesis, and Fraenkel-Conrat describes the studies on the tobacco mosaic virus. The fascinating subject of chemical paleontology, the discovery and identification of the "chemical fossils" that give clues to the evolutionary development of biopolymer molecules on earth, is described in the next article. The sequence of the next five articles describes the interaction of living systems with light: the first discusses the chemical effects of light, the second the interaction of light with living matter, the third the effect of radiation on nucleic acids, and the last two articles discuss photosynthesis and provide a vivid insight into energy production and use in the cell. These are followed by a description of the mechanism of vision. The article on memory and protein synthesis describes the efforts, which have only just begun, to understand perhaps the most human of man's characteristics, his memory. The next article describes the aging of human cells that appears to occur in tissue-culture studies. The final article recounts the biological role of reactive free radicals — species that have been implicated in the changes produced by radiation and by aging.

We have tried to choose for this volume articles that span an important range of principles and give the student a feeling for the dynamics of any energy production and use in the cell. Important and influential work is summarized in these articles. Four in this section describe research for which Nobel prizes were awarded: article 32, with the prizes awarded to Krebs and to Lipmann; article 36, with the prize to the author; article 42, with the prize to Calvin; and article 44 written by the wife of G. Wald and by one of his collaborators, which describes work both by Wald and by R. S. Mulliken for which they received Nobel prizes.

Many more articles that have been published in *Scientific American* could have been included and were omitted solely for lack of space. Additional articles that are related to topics included here, or that expand on them, are the following: A. L. Lehninger, *How Cells Transform Energy* (September 1961, Offprint 91); D. E. Green, *The Synthesis of Fat* (February 1960, Offprint 67); H. E. Huxley, *The Mechanism of Muscular Contraction* (December 1965, Offprint 1026); M. J. R. Dawkins and D. Hull, *The Production of Heat by Fat* (August 1965, Offprint 1018); J. Changeux, *The Control of Biochemical Reactions* (April 1965, Offprint 1008); F. H. C. Crick, *The Genetic Code* (October 1962, Offprint 123); F. H. C. Crick, *The Genetic Code: III* (October 1966, Offprint 1052); E. H. Davidson, *Hormones and Genes* (June 1965, Offprint 1013); U. W. Goodenough and R. P. Levine, *The Genetic Activity of Mitochondria and Chloroplasts* (November 1970, Offprint 1203); E. Kellenberger, *The Genetic Control of the Shape of a Virus* (December 1966, Offprint 1058); F. Jacob and E. L. Wollman, *Viruses and Genes* (June 1961, Offprint 89); P. C. Hanawalt and R. H. Haynes, *The Repair of DNA* (February 1967, Offprint 1061); P. H. Abelson, *Paleobiochemistry* (July 1956, Offprint 101); J. G. Lawless, C. E. Folsome, and K. A. Kvenvolden, *Organic Matter in Meteorites* (June 1972, Offprint 902); R. Dulbecco, *The Induction of Cancer by Viruses* (April 1967, Offprint 1069); R. Losick and P. W. Robbins, *The Receptor Site for a Bacterial Virus* (November 1969, Offprint 1161); A. Tomasz, *Cellular Factors in Genetic Transformation* (January 1969, Offprint 1130); E. Hadorn, *Transdetermination in Cells* (November 1968, Offprint 1127); G. Feinberg, *Light* (September 1968); G. Wald, *Life and Light* (October 1959, Offprint 61); V. F. Weisskopf, *How Light Interacts with Matter* (September 1968); E. I. Rabinowitch and Govindjee, *The Role of Chlorophyll in Photosynthesis* (July 1965, Offprint 1016); D. I. Arnon, *The Role of Light in Photosynthesis* (November 1960, Offprint 75); W. D. McElroy and H. H. Seliger, *Biological Luminescence* (December 1962, Offprint 141); F. Daniels, Jr., J. C. van der Leun, and B. E. Johnson, *Sunburn* (July 1968); M. Pope, *Electric Currents in Organic Crystals* (January 1967); A. Comfort, *The Life Span of Animals* (August 1961); A. L. Notkins and H. Koprowski, *How the Immune Response to a Virus can Cause Disease* (January 1973); E. Frieden, *The Chemistry of Amphibian Metamorphosis* (November 1963, Offprint 170); F. Verzár, *The Aging of Collagen* (April 1963, Offprint 155).

32

ENERGY TRANSFORMATION IN THE CELL

ALBERT L. LEHNINGER
May 1960

How does the living cell convert the energy of foodstuff into a form that can be utilized and stored? The process involves the sequential action of many enzymes, some of which are part of the cell membranes

A flame and a living cell both burn fuel to yield energy, carbon dioxide and water. The flame, in one step, transforms the chemical energy of the fuel into heat. The cell, in many steps and with little loss to heat, converts this chemical energy into a variety of forms: into the energy of the chemical bonds in the molecules of its own substance, into the mechanical energy of muscle contraction, into the electrical energy of the nerve impulse. In luminescent organisms special cells transform the energy into light.

From the standpoint of thermodynamics the very existence of living things, with their marvelous diversity and complexity of structure and function, is improbable. The laws of thermodynamics say that energy must run "downhill," as in a flame, and that all systems of atoms and molecules must ultimately and inevitably assume the most random configurations with the least energy-content. Continuous "uphill" work is necessary to create and maintain the structure of the cell. It is the capacity to extract energy from its surroundings and to use this energy in an orderly and directed manner that distinguishes the living human organism from the few dollars' (actually $5.66 in today's inflated market) worth of common chemical elements of which it is composed.

The past few years have seen great advances in the investigation of the transformation of energy by the cell. This historic enterprise has engaged the talents of some of the ablest investigators of the century. In its present stage our understanding encompasses not only some of the chemical and physical aspects of the process, but has begun to take in the arrangement of the molecules in the cell that conduct it. Many of the active molecules—the enzymes—have been identified. The intricate chains and cycles of activity by which they extract, trap, exchange and distribute energy have been worked out in sufficient detail to illuminate their principles of operation. And the molecular machinery of these energy-transforming functions has been securely located in the mitochondria, structures found in all cells that burn their fuel in oxygen.

It is, of course, the food intake of the organism that supplies the fuel—sugars, fats and proteins—to the energy-transforming system of the cell. Every student of elementary chemistry learns that a given weight of an organic compound contains a fixed amount of potential energy locked up in the bonds between the atoms of its molecule; for example, the bonds between the carbon, hydrogen and oxygen atoms in the sugar glucose. The energy can be liberated by burning the sugar in oxygen, with the carbon and hydrogen evolving from the flame in the relatively simple, energy-poor molecules of carbon dioxide (CO_2) and water (H_2O). This oxidation yields 690,000 calories per mole of glucose. (A mole is the weight of a substance in grams that is numerically equal to its molecular weight. A mole of glucose weighs 180 grams.) Now it is one of the fundamental principles of thermodynamics that the same total amount of energy is always liberated upon combustion of a given weight of a substance, no matter what the mechanism or pathway of the process. Thus the cellular oxidation of glucose to carbon dioxide and water makes a total of 690,000 calories of energy available to the energy-harnessing activities of cells.

There is an important reason why oxidation in the cell, as contrasted with the uncontrolled combustion that goes on in a flame, must proceed under rigorous control. Living cells are unable to utilize heat in the performance of functions such as muscle contraction, because heat energy can do work only if it flows from a warm region to a cooler one. This is the principle of a heat engine, in which the temperature of the working fluid undergoes a large drop between the combustion chamber and the exhaust. For all practical purposes there is no such temperature differential in the living cell, and the cell cannot function as a heat engine. The cell recovers the energy liberated by the oxidation of foodstuff not primarily as heat, but rather as chemical energy, a form of energy that can do work in a constant-temperature system. To obtain energy in this useful form the cell oxidizes its fuel in a stepwise manner. The agents of this controlled combustion are the enzymes: large molecules that function as catalysts, or promoters of chemical reactions. The cell employs dozens of oxidative enzymes, each specialized to catalyze one reaction in the series that ultimately converts the fuel into carbon dioxide and water.

Investigators have broken down the

MITOCHONDRION, the site of energy transfer in the living cell, is enlarged some 235,000 diameters in this electron micrograph of a rat liver-cell. Cristae, the flattened infoldings of the lining membrane, have been cut at different angles. One lying almost in the plane of the cut forms the wide V at top. Several near the center, cut at right angles, project like fingers from the outer wall. The connections to the wall do not show in those cut at oblique angles. The micrograph was made by Michael L. Watson of the University of Rochester while conducting research under a contract from the Atomic Energy Commission and a grant from the National Institutes of Health.

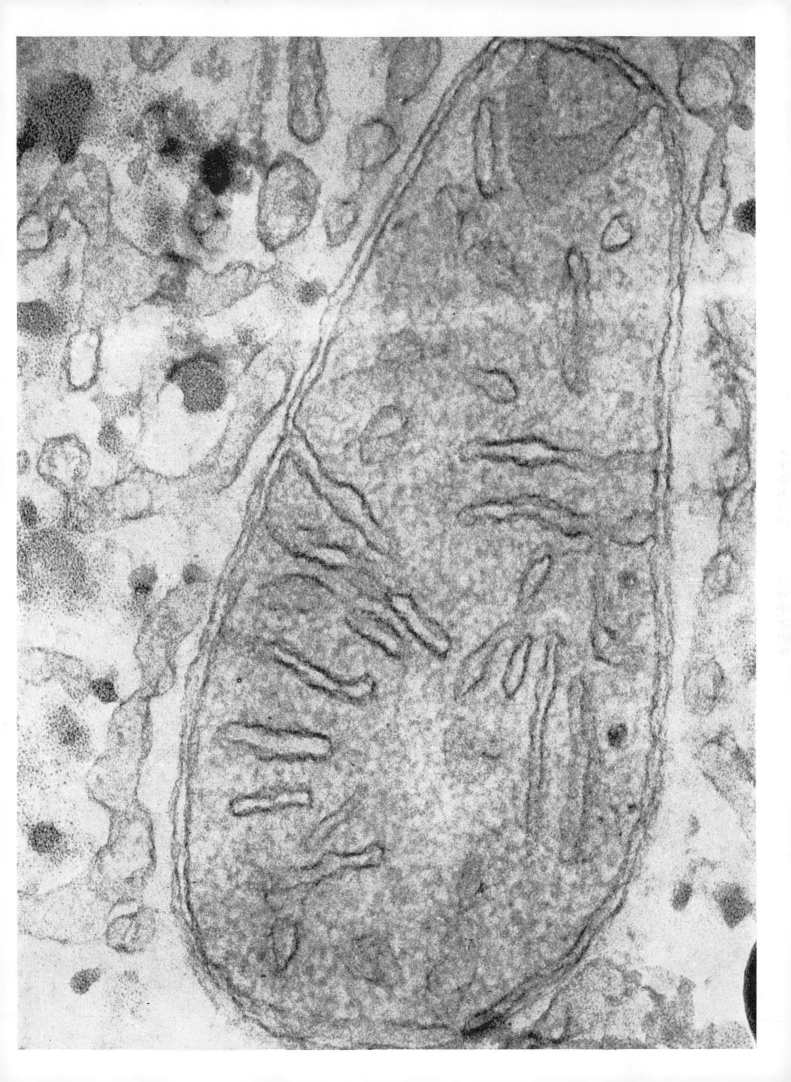

labyrinthine succession of reactions into three major stages. In the first stage enzymes break down the sugar and fat molecules (and protein fragments) into a simpler unit that represents a kind of common denominator of the distinctly different structures of these fuels. In the next two stages other enzymes take this unit apart and oxidize carbon and hydrogen. But the biologically significant product of the whole chain of transactions is energy, not water and carbon dioxide, which are mere waste or exhaust products. As the energy is liberated in the breakdown and oxidation reactions, it is captured in the chemical bonds of a special energy-storing molecule and is delivered thereby to the energy-consuming activities of the cell.

The Stages of Oxidation

In the first-stage breakdown of the glucose molecule, which has six carbon atoms, the enzymes split it into two molecules of pyruvic acid, each of which has three carbon atoms. This conversion is not so simple as it sounds. It involves the sequential action of a dozen specific enzymes [*see top illustration on pages 292 and 293*]. Some 40 years of intensive research went into the resolution of the details of this process and the isolation of the enzymes in pure form.

The intermediate pyruvic acid molecules become the center of activity in the second stage. They are converted to the two-carbon compound acetic acid, in a combined or "activated" form with coenzyme A, a substance that contains pantothenic acid, one of the B vitamins. It is at this point that fats and proteins—broken down to acetic acid by enzyme systems specifically adapted to their structures—also join the common pathway of oxidation. Another set of enzymes acting sequentially and cyclically links up acetic acid with oxalacetic acid, a four-carbon compound, to form citric acid, a six-carbon compound. The second stage is often called the citric acid cycle, after this important intermediate; it is also known as the Krebs cycle in recognition of Sir Hans Krebs of the University of Oxford, who first postulated it in 1937. As the cycle continues, the citric acid undergoes a series of rearrangements and degradations, in the

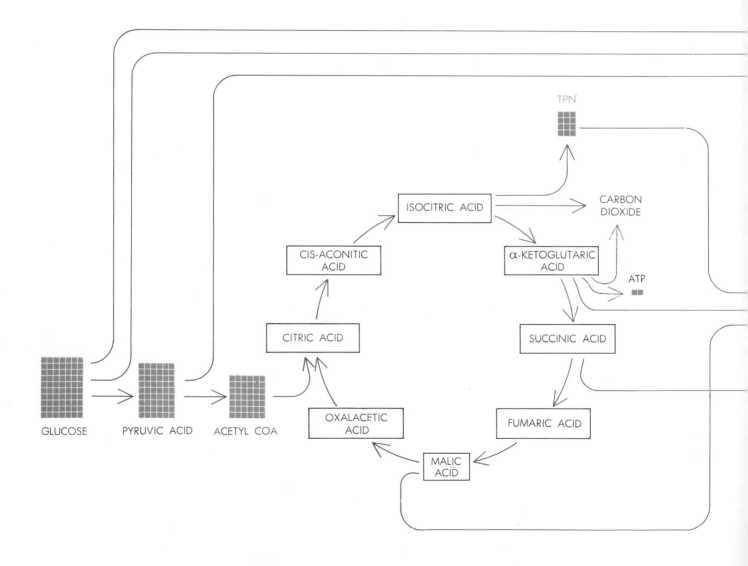

POTENTIAL ENERGY OF GLUCOSE (*far left*) is passed from compound to compound; finally more than 60 per cent is recovered in the form of adenosine triphosphate, or ATP (*top right*). Most of the energy is transferred by the citric acid cycle (*circle*) to en-

course of which oxalacetic acid is regenerated for the next round, and the two carbons from the acetic acid molecule are oxidized to form two molecules of carbon dioxide. Half of the task of oxidation is now completed.

Meanwhile, during the dismemberment of the pyruvic acid molecule in the citric acid cycle, intermediate compounds have picked up the pairs of hydrogen atoms that are attached to carbon atoms. The hydrogens are carried over into the third major multi-enzyme sequence to be combined with oxygen, which in higher animals is brought from the lungs via the bloodstream. This so-called respiratory cycle thus yields water, the second of the two end products of the biological oxidation.

As elementary as the combustion of hydrogen and oxygen may seem, the unraveling of the chain of enzyme activity in the respiratory cycle is the goal of a 50-year campaign of investigation. The contributions of Otto Warburg of Germany and David Keilin of England to this work place them among the major figures in biochemistry. The hydrogen atoms do not by any means enter directly into combination with the oxygen. They or their equivalent electrons, set free when the hydrogen is ionized, travel to this terminus along a chain of hydrogen- and electron-transferring enzyme molecules in the cell. Each of these enzymes possesses a characteristic and specific "active group" that is capable of accepting electrons

from the preceding member of the chain and of passing them along to the next. The chemical nature of the active groups explains why animals must have certain minerals and vitamins in their diet; all the groups contain either a metal, such as iron, or a vitamin, such as riboflavin (vitamin B_2). The lack of any of these essential activators may interrupt the chain and cause faulty or incomplete oxidation of foodstuff in the cell. Not all the links in the chain of enzymes have been identified. Recent work indicates that as many as three additional enzymes may be involved, one containing vitamin K (also essential to the clotting of blood); another containing tocopherol, or vitamin E (also essential to maintenance of muscle tone and to

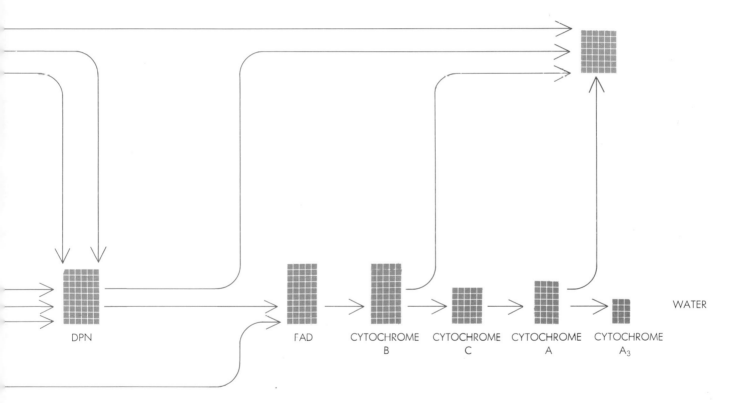

DPN   FAD   CYTOCHROME B   CYTOCHROME C   CYTOCHROME A   CYTOCHROME A₃   WATER

zymes of the respiratory chain (*TPN, DPN, FAD and the cytochromes*), which then pass it to ATP. Small gray squares, each representing approximately 10,000 calories, indicate the portions of the original 690,000 glucose calories that reach various compounds.

reproduction), and a third containing a newly isolated active group called ubiquinone, or coenzyme Q.

The Storage Battery

With the fuel completely oxidized, what has become of the potential energy it contained? This question began to yield to investigation in the late 1930's.

Herman M. Kalckar of Denmark and V. A. Belitser of the U.S.S.R. then independently recognized the significance of a chemical event that occurs along with the oxidation of the fuel. They incubated simple suspensions of ground muscle or kidney with glucose in the presence of oxygen and observed that phosphate ions present in the suspension medium disappeared as the glucose was oxidized.

Further investigation revealed that the phosphate was being incorporated into organic compounds, in particular the compound adenosine triphosphate. Biochemists at once recognized the great significance of this finding. Adenosine triphosphate, now universally known as ATP, had been identified a few years earlier as the energy source in the contraction of muscle. Today it is known

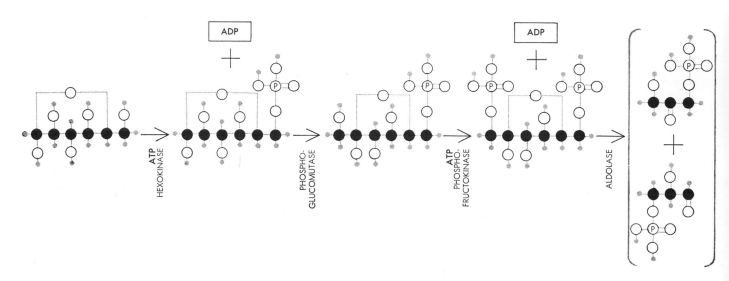

GLYCOLYSIS, the process by which glucose (*first molecule*) is broken down into two molecules of pyruvic acid (*last molecule*), requires the catalytic aid of many enzymes (*light-face type*). Two molecules of ATP are needed to prime the process, but four are

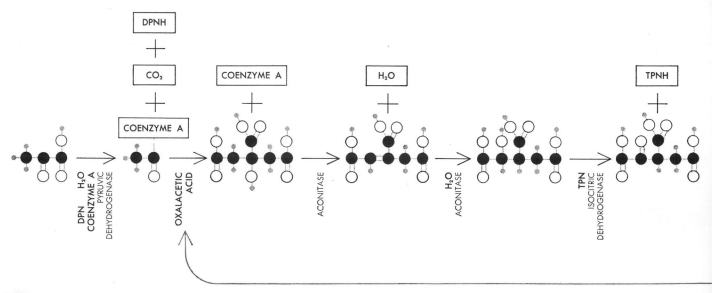

CARBON

HYDROGEN

OXYGEN

PHOSPHORUS

CITRIC ACID CYCLE transfers the energy from pyruvic acid (*first molecule*) to the respiratory enzymes DPN, TPN and FAD by reducing them (*i.e.*, adding hydrogen or electrons to them). Carbon dioxide (CO_2) is released as a waste product. First pyruvic acid is converted to acetyl coenzyme A, an activated form of acetic

that ATP is the universal intracellular carrier of chemical energy.

ATP may be regarded literally as a fully charged storage battery. When the energy of this battery is withdrawn to make muscle contract, for example, the energy-rich ATP molecule transfers its energy to the contracting muscle by losing its terminal phosphate group. ATP thus becomes adenosine diphos-

phate (ADP)—the storage battery in its discharged state. To "recharge" the battery it is obviously necessary to supply a phosphate group plus the energy required to effect the uphill reaction that couples the phosphate to ADP. It was found that ADP as well as free phosphate ions disappear during biological oxidation, and that the two are combined in ATP. Kalckar postulated that this

coupled phosphorylation, often called oxidative phosphorylation, provides the means for converting the energy released by oxidation into a readily usable form. The energy-rich ATP molecule can travel wherever energy is needed in the cell to drive energy-consuming functions, from the contraction of muscle to the synthesis of protein.

This conversion of the energy liber-

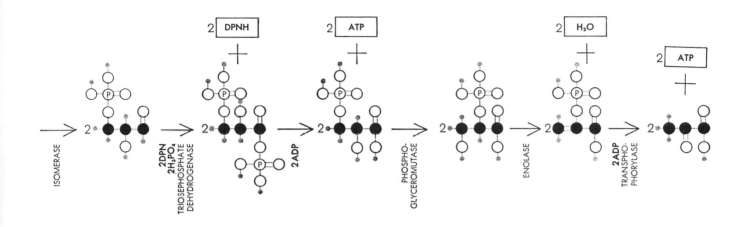

generated, yielding a net gain of two molecules of this energy-rich compound. Energy from glucose is also conserved by the reduction

of the respiratory coenzyme DPN to DPNH (*sixth step*). The glycolytic reactions are reversible with the aid of appropriate enzymes.

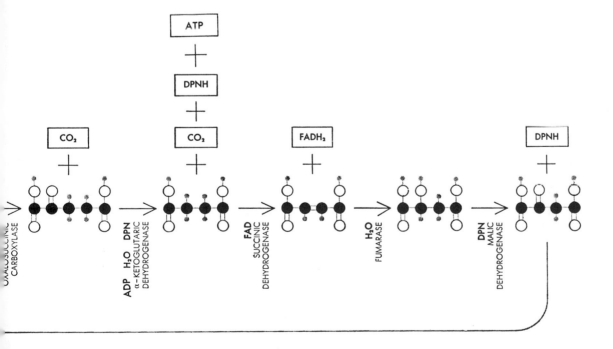

acid (*second molecule*). This reacts with oxalacetic acid to form citric acid (*third molecule*). After a series of rearrangements and oxidations, oxalacetic acid is regenerated (*last molecule*) and can

participate in the cycle again. The substances necessary for each step are named below the arrows (*catalytic enzymes are in light-face type*); side products of the reactions are shown in boxes.

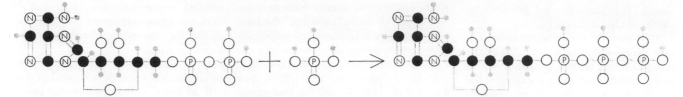

ADENOSINE DIPHOSPHATE, or ADP (*first molecule*), adds the phosphate group from phosphoric acid to generate adenosine tri- phosphate, or ATP (*right of arrow*). Energy is required to forge the high-energy bond (*wavy line*) that links the phosphate groups.

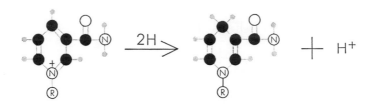

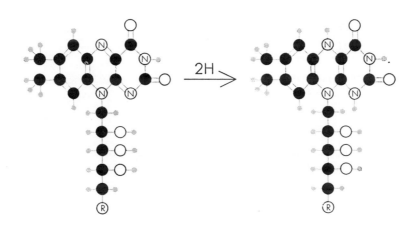

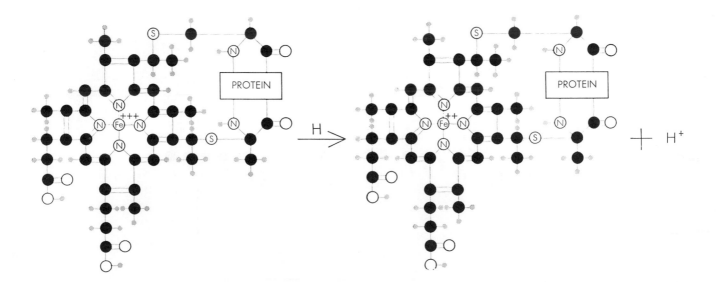

CARBON   HYDROGEN
OXYGEN   NITROGEN
RADICAL   SULFUR
IRON

ACTIVE GROUPS of respiratory enzymes are shown oxidized (*left of arrows*) and reduced (*right*). "R" or "protein" indicates the rest of the molecule. In TPN or DPN (*top*) one hydrogen is joined to a carbon atom; the electron from the other neutralizes the charge on nitrogen. Riboflavin, the active group of FAD (*center*), adds hydrogens to two nitrogen atoms. The active group of cytochromes is heme (*bottom*); an elec- tron from a hydrogen reduces the charge on iron.

ated by the combustion of fuel into the third phosphate bond of ATP proceeds with extraordinary efficiency. For each molecule of glucose completely oxidized to water and carbon dioxide in a tissue preparation, approximately 38 molecules of free phosphate and 38 molecules of ADP combine to form 38 molecules of ATP. In other words the oxidation of each mole of glucose produces 38 moles of ATP. It has been shown that the formation of one mole of ATP from ADP in this reaction as it occurs in the cell requires about 12,000 calories. The formation of 38 moles of ATP therefore requires the input of at least 38 times 12,000 calories, or about 456,000 calories. Since the oxidation of one mole of glucose yields a maximum of 690,000 calories, the recovery of 38 moles of ATP represents a conversion of 66 per cent of the energy. As a comparison, a modern steam-generating plant converts about 30 per cent of its energy input to useful work.

Just how the energy is transferred from the fuel molecules to ATP is a problem that has preoccupied many biochemists over the past 10 years. One early clue to the mechanism of oxidative phosphorylation came from the theoretical calculation of the energy exchanges at each major stage in the oxidation of glucose. Thermodynamics shows, for example, that the first stage in the process—the breakdown of glucose to pyruvic acid—yields little more than 5 per cent of the total energy. From such calculations Belitser predicted over 20 years ago that the combination of hydrogen with oxygen in the third phase—the respiratory cycle—must yield most of the energy. As a matter of fact, the oxidation of one mole of hydrogen to produce one mole of water releases some 52,000 calories. Since the biological oxidation of one mole of glucose reduces 12 atoms of oxygen in the respiratory cycle, the latter must account for 12 times 52,000 calories, or 624,000 calories—90 per cent of the total of 690,000 calories. Conclusive as these calculations seemed to be, it was another dozen years before direct evidence could be adduced to prove that the phosphorylation of ADP is coupled to the respiratory chain. In fact, experimental results seemed if anything to argue against this conclusion.

In 1951 our group, then at the University of Chicago, perfected an experiment that demonstrated unequivocally the presence of supplementary energy-converting enzymes at three points in the respiratory chain. These enzymes har-

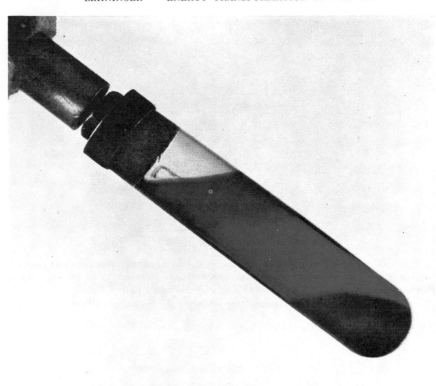

SEPARATION OF MITOCHONDRIA from disrupted cells requires centrifugation at high speed. The tube used for this is made of plastic and has a locking metal cap. The mitochondria are present in the pale middle layer of sediment. The dark layer at the bottom contains cell nuclei; the top layer contains microsomes, the smallest particles of the cell.

CYTOCHROME C in solution changes color visibly when it is oxidized (left) or reduced (right). A sensitive spectrophotometer measures the color differences accurately by registering the transmission of light of various wavelengths through the solution. Special quartz containers of high optical quality, here somewhat enlarged, are used in this instrument.

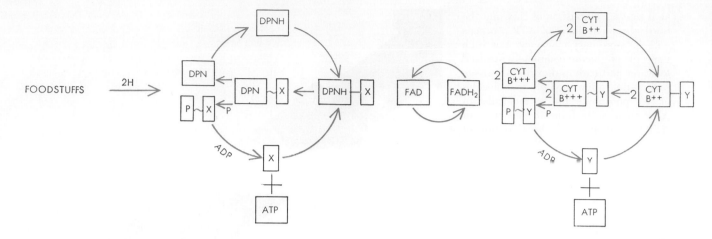

RESPIRATORY ENZYMES transfer energy by a series of cyclic reactions, each set in motion by the one preceding it, like a system of interlocking gears. A pair of hydrogen atoms released in the citric acid cycle reduces one enzyme; this is oxidized again by reducing the next enzyme, and so on. At the end of the chain the hydrogen combines with oxygen to form water. The known carriers in the chain are diphosphopyridine nucleotide (DPN), flavin adenine dinucleotide (FAD), attached to a protein, and four cytochromes. Coupled with the reduction-oxidation cycles of three carriers (DPN, cytochromes B and A) are reactions with unidenti-

ness the energy liberated by the passage of electrons from one link to the next in the chain to phosphorylate ADP to ATP. We found that the passage of each pair of hydrogen atoms or equivalent electrons yields one molecule of ATP at each of three enzymic energy-transfer stations. Since the oxidation of one molecule of glucose sets 12 pairs of hydrogen atoms moving down the chain, the total yield is three times 12, or 36, molecules of ATP. Two additional molecules of ATP are formed in the breakdown of glucose to pyruvic acid. The grand total is then 38 moles of ATP per mole of glucose. These findings fulfilled the prediction from thermodynamic considerations and satisfied the over-all energy balance-sheet of biological oxidation, showing that the respiratory chain is the primary site of energy conversion.

From more recent work we have been able to postulate the probable form of the mechanism by which the energy is coupled at each of the energy-transfer points in the respiratory chain [*see illustration above on these two pages*]. The chain is apparently a series of wheels within wheels, characterized by a cyclic process at each molecule in the chain. Each of these molecules is reduced by the addition of a hydrogen or an electron at one point, and is restored to its original form by oxidation when it delivers the hydrogen or electron to the next point. In three of the cycles there is an intermediate step by which the energy is transferred from the reaction to a coupled reaction that forms

ATP from ADP.

This picture has been modified by the finding that one pair of hydrogen atoms enters the respiratory chain at the middle, and so yields only two molecules of ATP. The deficit is made up, however, by the conversion of ADP to

ATP in one of the reactions of the citric acid cycle. The respiratory chain nonetheless remains the primary site of energy conversion.

The purpose that is served by the stepwise character of the oxidation process in the cell now becomes clear. Na-

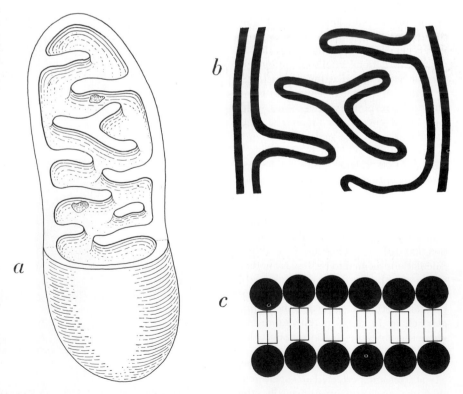

MITOCHONDRIAL STRUCTURE is basically that of a fluid-filled vessel with an involuted wall (*a*). Closer analysis shows that the wall consists of a double membrane (*b*). Each membrane approximates the thickness of a single layer of protein molecules (*spheres at c*),

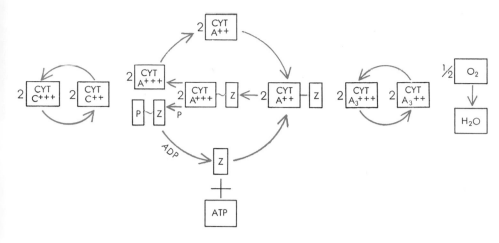

fied enzymes (*designated* X, Y *and* Z) that transfer energy released in the cycles to ATP. The transfer is not fully understood; it is believed to involve the formation of a high-energy bond (*wavy line*) to the transfer enzyme, then combination of this enzyme with a phosphate group (P) and finally the addition of the phosphate to ADP to form ATP. When two hydrogen atoms are passed down the whole chain, they give rise to three molecules of ATP.

Such a program of sequential and cyclic reactions requires that many different enzymes act in proper order and in an integrated, well-controlled way. This suggests that the participating enzymes, perhaps hundreds in number, must have a specific geometric orientation with respect to one another within the cell.

The Geometry of Oxidation

In 1948 Eugene P. Kennedy and I were able to show that enzymes involved in both the citric acid and the respiratory cycles are located in the mitochondria. These tiny oblong or rodlike structures, much smaller than the cell nucleus, occur in the cytoplasm: the extranuclear portion of the cell. A single liver-cell may contain several thousand such bodies; together they may account for about 20 per cent of the total weight of the cell. Earlier in 1948 George H. Hogeboom, Walter C. Schneider and George E. Palade of the Rockefeller Institute had perfected a method for isolating mitochondria intact and in large quantities by spinning down cell extracts in the ultracentrifuge.

When we incubated mitochondria with pyruvic acid and other intermediates of the citric acid cycle in the presence of oxygen, we found that all of the complex reactions of the citric acid and respiratory cycle proceeded at a high rate and in an orderly manner. On the other hand, we found that nuclei and other cell structures were incapable of conducting the oxidation process. We also discovered that the mitochondria carry out the vital energy-recovery process of oxidative phosphorylation, generating ATP from ADP. The mitochondria are thus the "power plants" of the cell.

ture usually chooses simple ways to do things, and it would be simpler to accomplish the combustion in one step. The many-membered respiratory chain serves, however, to break up or quantize the 52,000 calories liberated by the oxidation of each pair of hydrogens into

three smaller packets. Each of these packets contains the approximate amount of energy, namely 12,000 calories, required to phosphorylate ADP to ATP. The process thus achieves efficient conversion of energy in terms of the energy currency of the cell.

These bodies are so small that they are barely recognizable as oblongs or rods when they are viewed in the light microscope. Yet recent advances in instrumentation have made it possible to sketch a molecular description of the mitochondrion and to discern at least dimly the spatial arrangement of the many enzymes concerned with biological oxidation.

The first approach was to look at the ultrastructure of the mitochondrion under the electron microscope. In 1952 Fritiof S. Sjöstrand in Sweden and Palade in New York began to apply a newly perfected means of obtaining ultrathin sections of tissue to study mitochondria. Their pictures of thin sections cut at different angles through

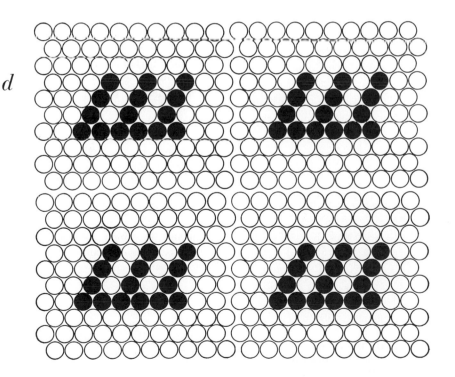

and the space between them equals the thickness of a double layer of fat molecules (*prongs*). The respiratory-chain enzymes (*black spheres at d*) form part of the membranes; they are evidently arranged in sets distributed at regular intervals in the membranes' protein layers.

single mitochondria demonstrated that the mitochondrion is not just an amorphous blob of protoplasm, but rather a highly organized structure with much fine detail—almost a cell within the cell. It consists of an outer enclosing membrane separated by a thin space from an inner membrane which at intervals apparently folds inward to form the so-called cristae. The semicompartmented space inside is filled with a semifluid "matrix."

From electron micrographs it is possible to estimate the dimensions of these structures. The membranes have a thickness of from 60 to 70 angstrom units. (An angstrom unit is a hundred millionth of a centimeter.) The space between the membranes and across the cristae measures about 60 angstroms. Mitochondria of all cells, regardless of the tissue or the species, have the same structural plan, in accord with their similarity of function.

The constant thickness of the membrane has significance from another point of view. It happens to approximate the dimension of a single protein molecule plus a lipid (fat) molecule. Chemical analysis shows that the membrane consists of about 65 per cent protein and 35 per cent lipid. These findings suggest that the membranes are arranged in a sandwich: The single layer of protein molecules that forms the outer membrane is apparently lined with oriented lipid molecules abutting a similar layer of lipid molecules on the outer surface of the layer of protein that forms the inner membrane.

When mitochondria are subjected to intense sound waves or to chemical agents such as detergents, the membranes break up, and the internal matrix escapes. The insoluble membrane fragments can easily be separated by centrifugation from the soluble matrix material, and the two fractions can then be analyzed separately. By these procedures we have found that the matrix contains most of the enzymes of the citric acid cycle, and that the enzymes of the respiratory cycle turn up exclusively in the membrane fragments.

Britton Chance of the University of Pennsylvania has employed spectroscopic techniques to study the respiratory enzymes in intact mitochondria. Each of the respiratory enzymes, having a characteristic active group, possesses a distinctive "color" and spectrum. Chance has succeeded in establishing not only the number of molecules of each type in the mitochondria but also the sequence in which the electrons move from one carrier to another in the respiratory chain.

The Enzymes Assembled

Such information, supplemented by our chemical studies, leads me to believe that the respiratory-chain enzymes are organized in assemblies or sets containing only one molecule of each of the enzymes. One such set would be made up of perhaps eight different molecules. Attached laterally to these would probably be six (perhaps nine) other catalytically

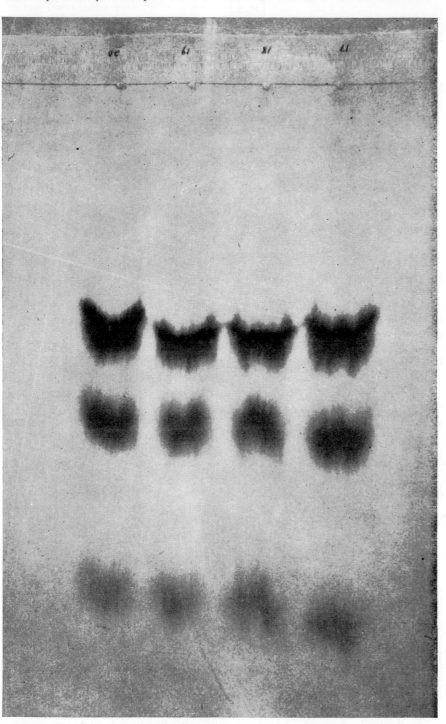

CHROMATOGRAM separates ATP (*top row of spots*) from ADP (*second row*) and AMP (*bottom*). These substances are hard to separate by other methods. In this experiment samples from a flask containing mitochondrial fragments incubated with ADP and DPNH were placed on the numbered marks. A solvent, allowed to flow downward from the top edge of the paper by capillary action, then carried the compounds with it at varying speeds. The paper was later viewed in ultraviolet radiation, under which the compounds show up at characteristic sites as dark spots against the fluorescent paper. ATP spots confirm that fragments formed ATP from ADP. Spots can be cut out and ATP dissolved for testing.

active protein molecules, the function of which is to carry out the coupled formation of ATP. A complete assembly would thus contain 15 or more active protein molecules arranged in close geometric array. An individual liver mitochondrion might contain several thousand such units. Since the total mass and the protein content of the mitochondrial membrane are known, it is possible to calculate that these assemblies comprise as much as 40 per cent of its substance. This calculation, along with our recent finding that the assemblies are evenly distributed in the membrane, show that the membrane is not an inert wall or container but an active molecular machine. The highly ordered arrangement of the specialized enzyme molecules determines the organization and programming of the enzymatic activity of the living cell.

Mitochondria in intact cells have been observed to swell and shrink, apparently by the uptake of water from the cytoplasm. This activity may serve mainly to move water and other substances through the cell. In this connection we have made the interesting discovery that the membrane itself changes its dimensions in the course of its activity. Like a sheet of muscle tissue, it can relax or contract. We have found that this change in dimension is related to the concentration of ATP; the membrane contracts when the concentration is high and relaxes when it is low. This suggests that the rate of oxidation (and of energy recovery in the mitochondria) may be regulated by the local concentrations of ATP and ADP, which occur in inverse relationship to each other. Overproduction of ATP may thus automatically throttle down this mechano-chemical system and gear its rate of power production to the demands of the cell. This same mechano-chemical system may also be responsible for "pumping" water and for the remarkable motility of mitochondria in some cells.

The integration of chemistry and geometry in the structure of the power plant of the cell poses new challenges to the investigator. It sets as the supreme goal not only the duplication of the catalytically active enzyme assemblies by proper linkage of the individual energy-transferring enzyme molecules, but also the reconstitution of the detailed structure of the mitochondrial membrane.

WARBURG APPARATUS is used in biochemical experiments to test for oxidative activity. Flasks containing the mitochondria and a solution of pyruvate are incubated inside the drumlike water bath. Each flask is connected to the top of one arm of a U-tube, containing colored fluid. As oxygen is used up by the test material, the change in gas pressure forces the fluid to rise in one arm and fall in the other. The rate and amount of oxygen consumption can be calculated by taking readings of the levels at various times.

33

THE METABOLISM OF FATS

DAVID E. GREEN
January 1954

Together with the metabolism of sugar it provides the bulk of animal energy. Its key is the oxidation of fatty acids, a full knowledge of which has been obtained only this year

Animals generate nearly all their energy by oxidizing sugars and fats. Biochemists have known for some time how the body oxidizes sugars, but only within the past year have they filled in an equally detailed picture of how it oxidizes fats. It had taken some 50 years of intensive work in laboratories all over the world to complete the picture. Last month the Nobel prize committee recognized the importance of phases of that work when it awarded the 1953 prize in physiology and medicine to two men who had made key discoveries. The his-

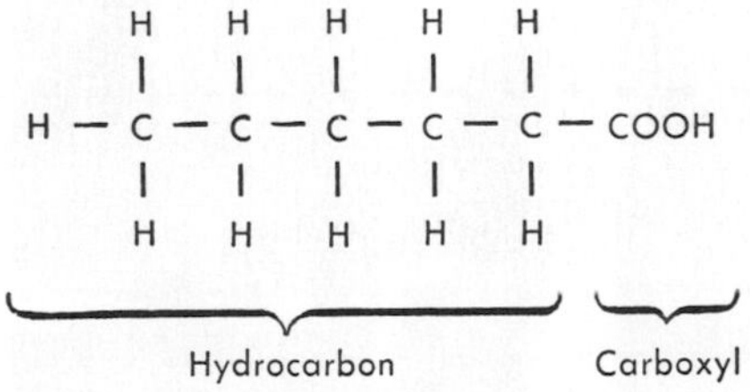

FATTY ACID is composed of a hydrocarbon chain with a carboxyl group at one end.

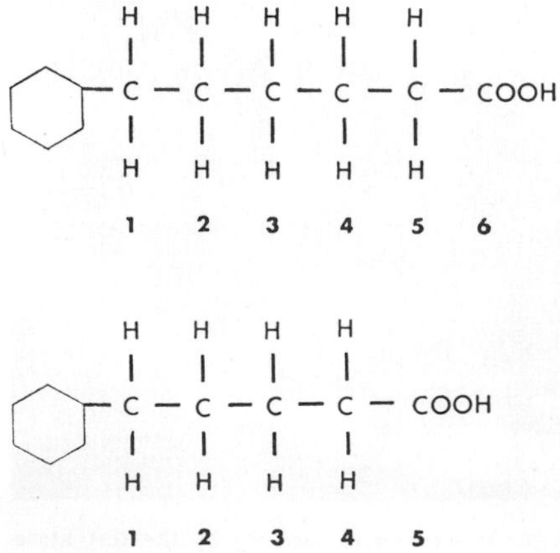

PHENYL FATTY ACIDS have even or odd number of carbons attached to benzene ring.

tory of the research on oxidation of fatty acids is one of the truly adventurous episodes in biochemistry.

The substances called fats consist of a combination of a fatty acid with an alcohol, such as glycerol. The part of this combination that the body burns for fuel is the fatty acid. It is a hydrocarbon—the kind of stuff man has been burning for fuel ever since he lit the first wax candle. A fatty acid is a hydrocarbon with a carboxyl group (COOH) attached at one end of the molecule [*see diagram at the left*]. In the common fatty acids the hydrocarbon chain is usually 16 to 20 carbon atoms long.

The process nature has developed for burning fatty acids in the body is roundabout, complex and under beautiful control. The only way a chemist can reconstruct it is from deductions based on examination of the combustion products, and the main problem has been to catch these products at the successive stages of the combustion.

In 1904 the German biochemist Franz Knoop opened the door to an understanding of fatty acid oxidation with a brilliantly thought-out experiment. He conceived the idea of attaching a fatty acid to a more obdurate substance, as one would fasten a piece of cheese to a wooden block, and then examining the products when the animal body so to speak "chopped off" (oxidized) successive slices of the "cheese." The block he used was the benzene ring. Attachment of a fatty acid to this ring forms what organic chemists call a phenyl fatty acid. Knoop synthesized two kinds of phenyl fatty acids—one with an even number of carbon atoms in the hydrocarbon chain, the other with an odd number. Then he fed them to experimental animals for oxidation and analyzed the animals' urine to find out how they had "de-

graded" (chopped down) the fatty acids.

As he had hoped, the odd-numbered and even-numbered phenyl fatty acids yielded different end products. The even-numbered chain was reduced to one carbon atom plus the carboxyl group, i.e., two carbon atoms in all. This is phenyl acetic acid. The odd-numbered chain was degraded to just the carboxyl group, attached directly to the benzene ring, that is, benzoic acid. From these results Knoop, and independently the chemist Henry Dakin, concluded that fatty acids were chopped down two carbon atoms at a time. Each chop (oxidation) removed two carbon atoms, including the one in the carboxyl group, and then a new carboxyl group was formed at the cut end of the chain. The successive cleavages are shown in the series of diagrams on page 244. Chemists called the carbon atom that was cut from the chain the alpha atom and the one from which it was separated the beta atom, and the process became known as beta oxidation. Knoop and Dakin visualized each oxidation as taking place in four steps, shown in the formulas at the top of page 245.

Knoop's theory hit the mark almost exactly, but it took half a century to prove the theory correct and to identify the intermediate products formed in the four steps of beta oxidation. The difficulty was that only the final product appeared in the urine, and the intermediate stages could not be isolated. The number of investigators who beat their heads against this difficulty is legion. But occasionally an ingenious experimenter had a flash of inspiration, and one inspiration led to another.

In 1906 the German chemist Gustave Embden made the first break in the wall by studying fatty acid metabolism in a

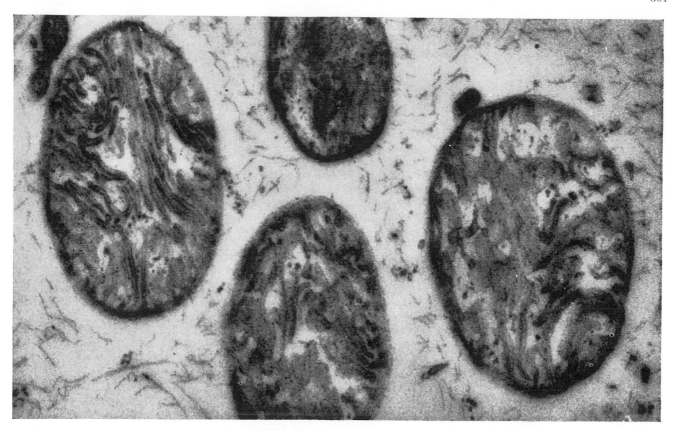

MITOCHONDRIA are particles where fats are oxidized. The four mitochondria in this electron micrograph of rat muscle are enlarged 75,000 diameters. The micrographs on this page were made by George E. Palade of the Rockefeller Institute for Medical Research.

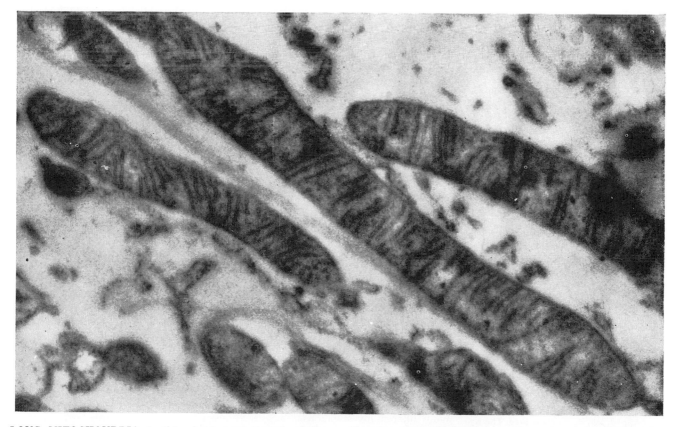

LONG MITOCHONDRIA in this electron micrograph of the proximal convoluted tubule of a rat kidney are enlarged 41,000 diameters. The mitochondria are shown in cross section by impregnating the tissue with plastic and cutting it with a glass knife.

PHENYL ACETIC ACID is the end product of the oxidation of a phenyl fatty acid that has an even number of carbon atoms.

BENZOIC ACID is the end product of the similar oxidation of a phenyl fatty acid that possesses an odd number of carbon atoms.

EVEN-NUMBERED phenyl fatty acid is chopped down (*vertical lines*) two carbon atoms at a time, leaving phenyl acetic acid.

ODD-NUMBERED phenyl fatty acid is similarly chopped, leaving benzoic acid. The hydrocarbon hydrogens are omitted for clarity.

single organ instead of the whole animal. He separated the liver from an animal and kept it functioning by pumping nutrients into it through a closed circulation system. When he introduced fatty acids into the entering veins, he found that the isolated liver did indeed oxidize many kinds of fatty acids. But the only product he could recognize in the outgoing blood was diacetic acid—a combination of two molecules of acetic acid. This was one of the end products, not an intermediate. Yet Embden had begun to blaze the right path, though it took many years for investigators to follow it up. That path was to simplify the experimental conditions. In 1935 the English biochemist J. H. Quastel went a step further by examining fatty acid oxidation in thin slices of tissue, rather than in a whole organ. Although he obtained valuable information on the extent of combustion of fatty acids by various tissues, he too failed to isolate any intermediate products.

Then in 1939 Luis Leloir and J. M. Muñoz of Argentina reported an epoch-making discovery. They crushed liver cells and found that tiny granules from the cell, after being separated from the debris, were able to carry out fatty acid oxidation as effectively as the intact cell. No outsider can really appreciate what this meant to investigators: at last they were freed from the shackles of working with complex biological systems and could probe fatty acid oxidation at the molecular level.

At first the results were disappointing, because the oxidation products of the granules were essentially identical with those obtained in the more complex systems. But studies of the granules eventually opened a completely new approach. The cell granules, later identified as mitochondria, were found to house hundreds of enzymes which catalyzed fatty acid oxidation and other related processes. Most important, experiments with the mitochondrial system carried out in my laboratory disclosed that the fatty acids were not oxidized as such; they had to be converted to something else first. No oxidation took place unless an oxidizable substance called a "sparker" was added to the mitochondrial suspension. However, it soon became apparent that the oxidation of the fatty acid was sparked not by the oxidation of the sparker itself but by some event which accompanied this oxidation. The tracking down of that mysterious event involved the piecing together of clues from many different investigations.

One of the clues came from the dis-

covery of the so-called citric acid or Krebs cycle by Hans A. Krebs of England (who for his discovery was awarded half of the 1953 Nobel prize in physiology). This cycle has to do with the oxidation of pyruvic acid (a breakdown product of sugar) into carbon dioxide and water, and it takes place in five separate oxidative steps. At each step the oxidation of the intermediate product is accompanied by the simultaneous conversion of inorganic phosphate into adenosine triphosphate (ATP)—that famous substance which triggers so many chemical reactions in the cell.

Now the mysterious role of the sparker in fatty acid oxidation began to unravel. Any one of the five oxidative steps in the citric acid cycle can spark fatty acid oxidation; thus the sparker could be any of the four substances formed from pyruvic acid during the operation of the cycle. It was apparent that the oxidation of the sparker led to the formation of ATP, and ATP in turn converted the fatty acid to something else. What that something was, and how it took part in the oxidative chopping down of the fatty acid, became the next important question to be answered. And here the chief clues stemmed from the work of Fritz A. Lipmann of Harvard University (winner of the other half of the 1953 Nobel physiology prize).

In 1945 Lipmann discovered in animal tissues a substance which was essential for the utilization of acetic acid in the body. He named it coenzyme A: a coenzyme is much smaller than an enzyme (about the relative size of the moon compared to the earth), is not a protein and is usually resistant to breakdown by heat. Lipmann and his group proceeded to purify and analyze coenzyme A. It was found to be made of three main building stones: (1) pantothenic acid (one of the B vitamins), (2) a phosphate related to ATP and (3) thioethanolamine (discovered by Esmond E. Snell of Wisconsin).

Lipmann recognized that the acetic acid participating in cell reactions is not the original form but a more reactive derivative. He and his colleagues found many signs pointing to the likelihood that acetic acid interacted with coenzyme A to form acetyl coenzyme A. But he was not able to isolate that product. It remained for Feodor Lynen of Munich to recognize what made coenzyme A tick. That knowledge provided him with the needed clue which enabled him to announce in 1951 success in isolating acetyl coenzyme A from yeast.

There was a striking parallel to the

case of the unknown active form of the fatty acids, but it was not immediately recognized as such. The first to draw the parallel were the microbiologists Horace A. Barker and Earl Stadtman at the University of California. They had been studying a microorganism which showed the remarkable capacity to synthesize fatty acids in the absence of oxygen in a medium containing ethyl alcohol as the sole source of carbon. In effect this organism was carrying out fatty acid oxidation in reverse. Barker and Stadtman discovered that the organism needed coenzyme A for this process, and they surmised that the active form of fatty acid for which investigators had been searching was a derivative of coenzyme A, just as the active form of acetic acid was acetyl coenzyme A. In short, the x in the fatty acid equation turned out to be coenzyme A.

Biochemists were now able to get on with working out the details of fatty acid oxidation. This called for a radical change in strategy. First ways and means had to be found for preparing the various fatty acid derivatives that were oxidized. Then the enzymes which catalyzed the successive reactions had to be isolated one by one. The investigation was blocked at the outset by the scarcity of purified coenzyme A; it was available only in milligram amounts. A group of investigators at the University of Wisconsin solved this difficulty by recognizing that yeast was a rich and convenient source of the coenzyme, that the substance could readily be concentrated in charcoal chromatographic columns and that the coenzyme could then be purified by precipitation of its copper salt with a copper salt of glutathione. This made it possible to prepare coenzyme A in gram quantities instead of milligrams.

My colleagues Henry Mahler and Saleh Wakil then found a way to carry out the next step: synthesis of the needed fatty acid derivatives. They isolated from beef liver an enzyme which in the presence of ATP could convert various fatty acids into their corresponding coenzyme A derivatives. The process was uncomfortably expensive, but it could supply all the derivatives required. Then came the problem of identifying the enzyme that catalyzed each reaction. Our group at Wisconsin, following up an earlier observation by George Drysdale and Henry Lardy, who had worked only with rats, developed a technique for preparing mitochondria, in which the enzymes are housed, from slaughterhouse animals. After isolating the en-

Oxidation

$$RCH_2CH_2COOH \longrightarrow RCH = CHCOOH$$

- -

Addition of water

$$RCH = CHCOOH \longrightarrow RCHOHCH_2COOH$$

- -

Oxidation

$$RCHOHCH_2COOH \longrightarrow RCOCH_2COOH$$

- -

Addition of water and splitting

$$RCOCH_2COOH \longrightarrow RCOOH + CH_2COOH$$

CLASSICAL SCHEME of fatty-acid oxidation was called beta oxidation. The carbon atom next to the carboxyl group is labeled alpha; the next carbon atom to the left, beta. The remainder of the molecule is indicated by R. The double lines represent double chemical bonds.

Acyl CoA dehydrogenase

$$RCH_2CH_2COS\ CoA \longrightarrow RCH = CHCOS\ CoA$$

- -

Hydrase

$$RCH = CHCOS\ CoA + H_2O \longrightarrow RCHOHCH_2COS\ CoA$$

- -

Beta-hydroxy acyl CoA dehydrogenase

$$RCHOHCH_2COS\ CoA \longrightarrow RCOCH_2COS\ CoA$$

- -

Cleavage enzyme

$$RCOCH_2COS\ CoA + CoA \longrightarrow RCOS\ CoA + CH_3COS\ CoA$$

MODERN SCHEME of fatty acid oxidation supposes much the same four steps as the classical scheme. The principal difference is due to the role of coenzyme A, denoted by the symbol CoA. Above each of the four reactions is the name of the enzyme known to catalyze it.

COENZYME A is essential to the oxidation of fatty acids. The upper part of its molecule consists of thioethanolamine (NH to SH) and pantothenic acid. The lower part, joined to the upper by a phosphate group (P), is adenylic acid. Another phosphate is near the bottom.

zymes from the mitochondria, the Wisconsin group systematically tested them and found, as might have been expected, that there was a separate enzyme for each of the four stages in the stepwise oxidation of fatty acids. The four enzymes and the reaction products have now been identified [*see illustration at bottom of preceding page*].

Lynen in Germany, collaborating with Severo Ochoa of New York University, reached the same objectives as the Wisconsin group by a completely different stratagem. Lacking their large-scale supplies of coenzyme A and the enzyme needed to prepare coenzyme A derivatives, he hit upon the ingenious expedient of offering synthetic derivatives for the oxidation-catalyzing enzymes to act upon. He used fatty acid derivatives of thioethanolamine, which is one of the building blocks of coenzyme A and which can be synthesized in the laboratory without too much difficulty. Some of the enzymes of the fatty acid oxidizing system did act upon these synthetic derivatives, and Lynen thus was able to purify those particular enzymes. It is not often that enzymes can be fooled in this way, but Lynen's bait did the trick.

In retrospect it is now easy to see why it was so difficult to isolate the intermediate products of fatty acid oxidation. The attachment of a fatty acid to coenzyme A is like the kiss of death. Once attached, the fatty acid has no alternative but to go through the entire cycle of chemical change. Only at the end of the cycle is it possible for the end product (diacetic acid) to let go of the coenzyme. Nature has seen to it that release from the coenzyme takes place at the end of the ride and not before. This restriction makes sense from the standpoint of the cell. Since ATP has to foot the bill for making a coenzyme A derivative of the fatty acid, and since ATP is a most valuable asset which the cell can ill afford to squander, nature appears to have taken special precautions to prevent the breakdown of coenzyme A derivatives during fatty acid oxidation.

Perhaps you are wondering how the cell gets any energy profit from the oxidation of fatty acids if ATP has to be used to initiate the process. The answer is that the chain of oxidation reactions returns as many as 100 molecules of ATP for each one invested. For each penny on deposit the cell gets back a dollar.

Why all this fuss about finding out the enzymatic details of the steps in fatty acid oxidation? Biochemists are in no position at present to say what fruits may come from it, but the study of mechanisms has in general been one of the most rewarding pursuits of biochemistry. The knowledge of the mechanism of fatty acid oxidation will permit us to probe deeply into areas which hitherto have been impenetrable. Some of the nuggets of information which it has uncovered thus far are that three B vitamins (flavin, niacin and pantothenic acid) are involved in the oxidation process, that ATP sparks it by energizing the conversion of a fatty acid to its coenzyme A derivative, that copper teams up with flavin to assist the enzyme which carries out the first oxidative step. By way of illustration of how one finding leads to another, this last discovery has led our group to the recognition that molybdenum and iron serve as functional groups in other flavin-containing enzymes not involved in fatty acid oxidation.

From the height the enzyme chemist has scaled in elucidating the oxidation of fatty acids he can see other eminences which may be approachable by similar tactics. The reverse of this oxidation process, *i.e.*, the synthesis of long-chain fatty acids, proceeds in somewhat the same way as the synthesis of cholesterol and other steroids, of plant carotenoids (*e.g.*, vitamin A) and of rubber. All these synthetic processes appear to depend upon the combination of the same fundamental units. They differ only in detail. In fatty acid synthesis carbon atoms join together in a linear arrangement like beads on a string. In carotenoid synthesis the only difference is that side spurs of carbon atoms are attached at regular intervals along the string. In cholesterol synthesis the string loops around and forms a series of rings. From what has been learned about the enzymes participating in fatty acid oxidation there is good reason to believe that it will be possible to carry out these synthetic processes artificially with isolated enzymes within the next five to 10 years.

Our new knowledge about fatty acid oxidation may also eventually help to explain some mysteries of diabetes. Many diabetics are unable to oxidize fats completely: their urine contains abnormal amounts of products of partial oxidation. Furthermore, the amount of fat in their tissues declines to a very low level. Injections of insulin enable these patients to carry the oxidation to completion and increase remarkably their capacity to synthesize and deposit fat. It seems altogether likely that a block in the enzyme systems involved in the synthesis of fat plays a substantial part in diabetes. In this connection it should be pointed out that the same enzymes which bring about fatty acid oxidation in animal tissues can be made to work backward and under appropriate conditions synthesize fatty acids.

34

HOW PROTEINS START

BRIAN F. C. CLARK AND KJELD A. MARCKER

January 1968

*The chain of amino acid units that constitutes a protein
molecule begins to grow when a variant of one of the
standard amino acids is delivered to the site of synthesis
by a specific transfer agent*

Over the past 15 years a tremendous amount of information has been amassed on how the living cell makes protein molecules. Step by step investigators in laboratories all over the world are clarifying the architecture of specific proteins, the nature of the genetic material that incorporates the instructions for building them, the code in which the instructions are written and the processes that translate the instructions into the work of construction. With the information now available experimenters have already synthesized a number of protein-like molecules from cell-free materials, and the day seems not far off when we shall be able to describe, and perhaps control, every step in the making of a protein

How is the building of a protein initiated? Until recently this question seemed to create no special problems. Given a supply of the amino acids from which a protein is made, the cell assembles them into a polypeptide chain that grows into a protein molecule, and it did not appear that the cell used any special machinery to start the construction of the chain. We have now learned, however, that the cell does indeed possess a starting mechanism. With the discovery of this mechanism it has become possible to study in detail the first step in the production of a protein molecule.

In order to discuss this new development we must first review the general features of protein synthesis by the cell. Proteins are made up of some 20 varieties of amino acid. A protein molecule consists of a long chain of amino acid units, typically from 100 to 500 or more of them, linked together in a specific sequence. The instructions for the particular order in each protein (the cell manufactures hundreds of different proteins) reside in the chainlike molecule of deoxyribonucleic acid (DNA). The DNA molecule consists of units called nucleotides; each nucleotide contains a side group of atoms called a base, and the sequence of bases along the DNA chain specifies the sequence for amino acids in the protein. There are four different bases in DNA: adenine (A), guanine (C), thymine (T) and cytosine (C). A "triplet" (a sequence of three bases) constitutes the "codon" that specifies a particular amino acid. The four bases taken three at a time in various sequences provide 64 possible codons; thus the four-letter language of DNA provides a vocabulary that is more than sufficient to designate the 20 amino acids. (In fact, some amino acids can be indicated by more than one codon.)

DNA does not guide the construction of the protein directly. Its message is first transcribed into the daughter molecule called messenger ribonucleic acid (mRNA). Messenger RNA also has four bases; three of them (A, G and C) are the same as in DNA, but the fourth, taking the place of thymine, is uracil (U). The RNA molecule is generated from DNA by a coupling process based on the fact that U couples to A and G couples to C. Thus during the transcription of DNA into RNA the four bases A, G, T and C in DNA give rise respectively to U, C, A and G in RNA [*see illustration on following page*].

The coded message is then read off the messenger RNA and translated into the construction of a protein molecule. This process takes place on the cell particles known as ribosomes, and it requires the assistance of smaller RNA molecules called transfer RNA (tRNA) that bring amino acids to the indicated sites. Each transfer RNA is specific for a particular amino acid, to which it attaches itself with the aid of an enzyme. It possesses an "anticodon" corresponding to a particular codon on the messenger RNA molecule. The ribosome moves along the messenger RNA molecule, reading off each codon in succession, and in this way it mediates the placement of the appropriate amino acids as they are delivered. As the amino acids join the chain they are linked together through peptide bonds formed by means of enzymes.

The decipherment of the genetic code for protein synthesis began in 1961 when Marshall W. Nirenberg and J. Heinrich Matthaei of the National Institutes of Health synthesized a simplified form of messenger RNA, composed of just one type of nucleotide, and found that it could generate the formation of a protein-like chain molecule made up of one variety of amino acid. Their artificial messenger RNA was the polynucleotide called "poly-U," containing uracil as the base. When it was added to a mixture of amino acids, extracts from cells of the bacterium *Escherichia coli* and energy-supplying compounds, it caused the synthesis of a polypeptide chain composed of the amino acid phenylalanine. Thus the poly-U codon (UUU) was found to specify phenylalanine.

This breakthrough quickly led to the identification of the codons for a number of other amino acids by means of the same device: using synthetic forms of messenger RNA. The experiments suggested that the initiation of synthesis of a protein was a perfectly straightforward matter. It appeared that the first codon in the messenger RNA chain simply called forth the delivery and placement of the specified amino acid and that no special starting signal was required. In 1964, however, Frederick Sanger and one of the authors of this article (Marcker) discovered a peculiar form of an amino acid, in combination with its transfer RNA, that threw entirely new light on the situation.

Using extracts from the *E. coli* bacterium, we were studying the chemical characteristics of the combination of the amino acid methionine with its specific tRNA. In the course of this study we decided to investigate the breakdown of the compound by pancreatic ribonuclease, an enzyme known to split RNA chains at certain specific bonds [*see top illustration on page 308*]. In order to facilitate identification of the products we labeled the methionine in advance with radioactive sulfur, and after treatment of the methionine-tRNA compound with

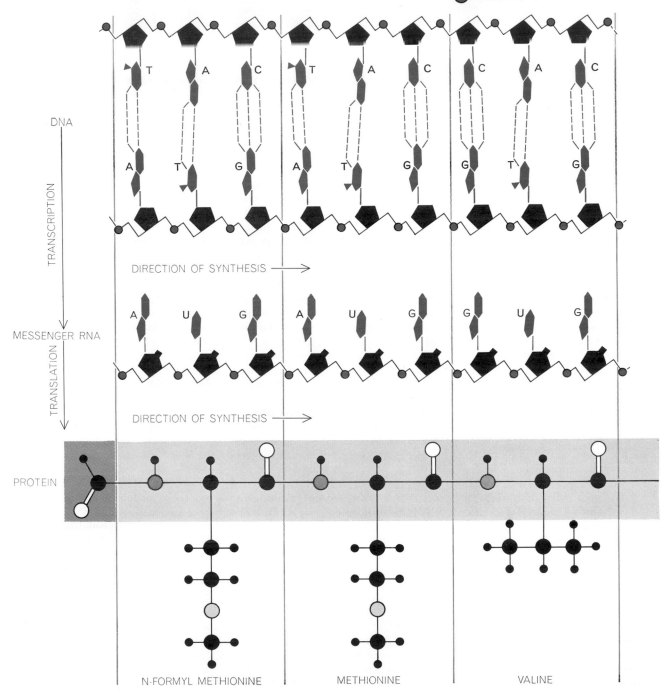

CARBON   A ADENINE

OXYGEN   T THYMINE

HYDROGEN   G GUANINE

NITROGEN   C CYTOSINE

SULFUR   U URACIL

PHOSPHORUS

DNA

TRANSCRIPTION

DIRECTION OF SYNTHESIS →

MESSENGER RNA

TRANSLATION

DIRECTION OF SYNTHESIS →

PROTEIN

N-FORMYL METHIONINE   METHIONINE   VALINE

TRANSMISSION OF GENETIC INFORMATION takes place in two main steps. First the linear code specifying a particular protein is transcribed from deoxyribonucleic acid (DNA) into messenger ribonucleic acid (RNA). The code letters in DNA are the four bases adenine (A), thymine (T), guanine (G) and cytosine (C). Hydrogen bonds (*broken lines*) between the complementary bases A–T and G–C hold the two strands of the DNA molecule together. The strands, which run antiparallel, consist of alternating units of deoxyribose sugar (*pentagons*) and phosphate (PO₃H). The code letters in messenger RNA duplicate those attached to one strand of the DNA except that uracil (U) replaces thymine. In RNA the sugar is ribose. In the second step of the process messenger RNA is translated into protein. The code letters in RNA are read in triplets, or codons, each of which specifies one (or sometimes more) of the 20 amino acids that form protein molecules. It has now been found that the codon AUG can specify a modification of methionine known as formyl methionine, which signals the start of a protein chain. Inside the chain AUG specifies ordinary methionine. The codon GUG, which codes for valine inside the chain, can also specify formyl methionine and initiate chain synthesis.

the enzyme we separated the products by means of electrophoresis, the technique that segregates electrically charged molecules according to their charge, size and shape. As was to be expected, one of the products was the compound known as methionyl-adenosine, a combination of methionine with the terminal adenosine portion of the tRNA molecule. But we also found, to our surprise, that the products included a considerable amount of a formylated variety of this compound, that is, a variation in which a formyl group (CHO) replaced a hydrogen atom in the amino group (NH_2) of the molecule. It turned out that this was by no means an artifact of the treatment to which the original compound had been subjected; growing cells proved to contain a high proportion of formylated methionine tRNA.

It was immediately evident that formylated methionine must occupy a special position in the protein molecule. The attachment of the formyl group to the amino group would prevent the amino group from forming a peptide bond [see illustration at right]. Consequently the formylated amino acid must be an end unit in the protein molecule. Since an amino group forms the "front" end of protein molecules when they are being assembled, formylated methionine must constitute the initial unit of the molecule.

We were able to separate the methionine tRNA of E. coli into two distinct species, and found that only one can be formylated. The formylatable species constitutes about 70 percent of the bacterium's methionine tRNA [see bottom illustration on following page]. Recent work in our laboratory at the Medical Research Council in Cambridge has established that the compound is formylated (at methionine's amino group) only after the amino acid has become attached to the tRNA molecule. The donor of the formyl group is 10-formyl tetrahydrofolic acid, and the reaction is catalyzed by a specific enzyme that acts exclusively on the combination of methionine with the formylatable species of tRNA.

Our laboratory and others have proceeded to analyze the initiation of protein formation by several experimental techniques. We began by testing a number of different synthetic messenger RNA's for their ability to bring about synthesis of a polypeptide incorporating methionine. Only two of the synthetic polynucleotides we tried proved to be capable of doing this. One contained the bases uracil, adenine and guanine (poly-UAG); the other had only uracil and guanine (poly-UG). We found that in a

mixture of amino acids and other cell-free materials where only the formylatable species of methionine tRNA was present, either poly-UAG or poly-UG would cause the synthesis of a polypeptide with methionine in the starting position—and only in that position. Surprisingly, this was true even when no formyl group was attached to the methionine-tRNA compound. We had to conclude that the formylatable version of the tRNA for methionine possessed a special adaptation that helped it to function as a polypeptide-chain initiator.

A thorough search was made for formylated varieties of other tRNA's: that is, of tRNA's for amino acids other than methionine. None were found. This raised an interesting question. In the proteins produced by E. coli cells the amino acid at the "front" end of the protein molecule is not always methionine; often it is alanine or serine. These amino acids are never found to be formylated. How, then, does either of them become the initial member of the protein chain?

Experiments with natural messenger RNA's (rather than synthetic polynucleotides) have suggested an explanation. Jerry Adams and Mario Capecchi, work-

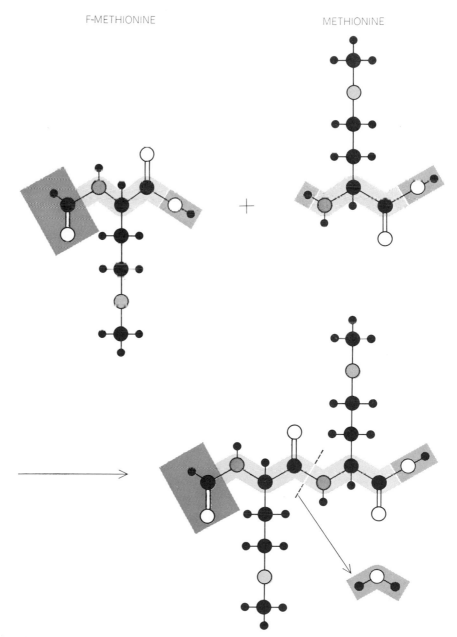

F-METHIONINE   METHIONINE

FORMYL METHIONINE, abbreviated F-Met, has a formyl group (CHO) where methionine (Met) has a hydrogen atom as part of a terminal amino (NH_2) group. When an amino acid enters a protein chain, one of the hydrogens from the amino end of one molecule combines with an OH group from the carboxyl (COOH) end of another molecule to form a molecule of water. The two molecules are then linked by a peptide bond. The formyl group prevents this reaction, hence F-Met can appear only at the beginning of a protein chain.

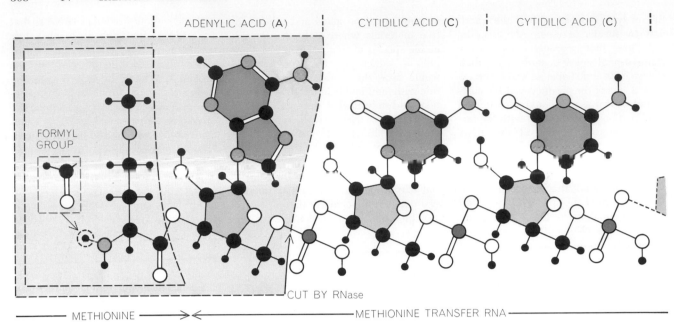

ADENYLIC ACID (A)   CYTIDILIC ACID (C)   CYTIDILIC ACID (C)

FORMYL GROUP

CUT BY RNase

——— METHIONINE ———>< ——————— METHIONINE TRANSFER RNA ———————

TRANSFER OF AMINO ACID to the site of protein synthesis is accomplished by molecules of transfer RNA (tRNA). There is at least one species of transfer RNA for each amino acid. All transfer RNA molecules contain the base sequence CCA at the terminal that holds the amino acid. Such a terminal is diagrammed here and shown coupled to methionine. Methionine that subsequently can be converted to formyl methionine is transferred by a different tRNA. When treated with the enzyme ribonuclease (RNase), the final base (adenine) and its coupled amino acid are split off from the rest of the transfer RNA. The fragment is called an aminoacyl adenosine.

ing in the laboratory of James D. Watson at Harvard University, and Norton D. Zinder and his collaborators at Rockefeller University have used messenger RNA's extracted from bacterial viruses. These RNA's direct the synthesis of the proteins that form the coat of the virus. The experimenters in Watson's and Zin-der's laboratories found that when such an RNA was added to cell-free materials in the test tube, formylated methionine turned up at the starting end of the coat proteins that were synthesized. This was most surprising, because normally in living systems the initial amino acid of the viruses' coat protein is alanine. A signifi-cant clue was found, however, in the fact that the coat proteins synthesized in the cell-free systems invariably had an ala-nine in the second position, following the formyl methionine. From this it seems reasonable to deduce that in living sys-tems, as in the cell-free system, the for-mation of the protein starts with formyl

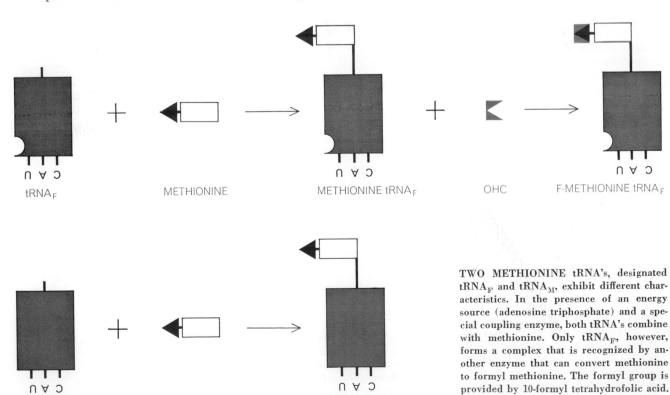

tRNA_F   METHIONINE   METHIONINE tRNA_F   OHC   F-METHIONINE tRNA_F

tRNA_M   METHIONINE   METHIONINE tRNA_M

TWO METHIONINE tRNA's, designated tRNA_F and tRNA_M, exhibit different char-acteristics. In the presence of an energy source (adenosine triphosphate) and a spe-cial coupling enzyme, both tRNA's combine with methionine. Only tRNA_F, however, forms a complex that is recognized by an-other enzyme that can convert methionine to formyl methionine. The formyl group is provided by 10-formyl tetrahydrofolic acid. The complex that results is F-Met-tRNA_F.

methionine, and that the bacterial cells supply an enzyme that chops off the formyl methionine later, leaving alanine in the first position.

Experiments with *E. coli* RNA in our laboratory and others have produced similar results. Messenger RNA extracted from these bacteria, like that extracted from bacterial viruses, causes cell-free systems to synthesize proteins with formyl methionine in the first position. On the other hand, the proteins extracted from living *E. coli* cells usually have unformylated methionine or alanine or serine in the lead position. It therefore seems likely that the living cells remove the formyl group from methionine or split off the entire formyl methionine unit after synthesis of the protein chain has got under way. The significance of the frequent appearance of alanine and serine at the front end of *E. coli* proteins is not clear; no satisfactory explanation has yet been found for the cell's selection of alanine and serine to follow formyl methionine. At all events, what does seem plausible now is that in *E. coli* the synthesis of all proteins starts with formyl methionine as the first unit.

How does the messenger RNA convey the message calling for formyl methionine as the starting unit? Does it use a special codon addressed specifically to the formylatable variety of methionine tRNA? We tested various codons for their ability to bring about the delivery of formyl methionine to the protein-synthesizing ribosomes. A codon for methionine was already known: it is AUG. We found that AUG was "read" by both varieties of methionine tRNA—the formylatable and the unformylatable. Either variety of tRNA delivered and bound methionine to the ribosome in response to AUG. We found that the formylatable tRNA (but not the other variety) also

recognized and responded to another codon: GUG.

These findings were consistent with our earlier observation that either poly-UAG or poly-UG could effect the incorporation of methionine into a polypeptide in a cell-free system. Poly-UAG, of course, can contain the codons AUG and GUG, depending on the sequence in which the bases happen to be arranged in this polynucleotide; poly-UG provides the codon GUG. That both AUG and GUG can initiate the synthesis of a methionine polypeptide was confirmed and clearly spelled out in detail by experiments in the laboratory of H. Gobind Khorana at the University of Wisconsin. Using synthetic messenger RNA's in which the bases were arranged only in these triplet sequences (AUG and GUG), Khorana's group showed that both codons led to the formation of a chain with formyl methionine in the starting position. AUG also placed methionine in internal positions in the chain, but GUG, which can code only for the formylatable version of the tRNA, incorporated methionine only at the starting end [see illustration below].

Investigators in the laboratories of Severo Ochoa at New York University and Paul M. Doty at Harvard obtained the same results. They also noted that both codons possess a certain versatility as signals, depending on their location in the messenger RNA. Located at or near the beginning of the messenger RNA chain, the AUG triplet is recognized by the formylatable variety of tRNA and leads to the placement of formylated methionine at the starting end of the polypeptide; farther on in the messenger RNA chain the same triplet is recognized by unformylatable tRNA and causes the placement of unformylated methionine in the internal part of the polypeptide. In short, at the "front" end

tRNA	CODONS
MET-tRNA$_M$	AUG
MET-tRNA$_F$	AUG
F-MET-tRNA$_F$	GUG

CODON ASSIGNMENTS show the bases in messenger RNA that cause the two Met-tRNA's to deliver methionine or formyl methionine for insertion in a protein chain.

of the RNA message the AUG codon says to the cell's synthesizing machinery, "Start the formation of a protein"; when it is located internally in the message, AUG simply says, "Place a methionine here." Similarly, the codon GUG was found to have two possible meanings: located at the beginning of the message, it orders the initiation of a protein with formylated methionine; in an internal position in the message it is the code word for placement not of methionine but of the amino acid valine.

How is it that each of these codons signifies a starting signal in one position and has a different meaning in another? Obviously this question will have to be answered in order to clarify the language of the protein-starting mechanism. Indeed, we cannot be sure that a codon in itself constitutes the entire message for the initiation of a protein. The signaling mechanism may be more complex than one might assume from the findings developed so far. Those findings are based almost entirely on work done with artificial messenger RNA's, and it is possible that the messages they provide are only approximations—meaningful enough to stimulate the cell machinery but not the full story.

When we consider how important the

SYNTHETIC MESSENGER	SOURCE OF METHIONINE		POSITION OF METHIONINE IN POLYPEPTIDE		CODONS USED
	MET-tRNA$_M$	MET-tRNA$_F$ / F-MET-tRNA$_F$	INTERNAL	N-TERMINAL	
RANDOM POLY-UG	−	+	−	+	GUG
RANDOM POLY-AUG	+	+	+	+	AUG, GUG
POLY-(UG)$_n$	−	+	−	+	GUG
POLY-(AUG)$_n$	+	+	+	+	AUG

INCORPORATION OF METHIONINE in protein-like chains has been studied with synthetic messenger RNA's and the two species of methionine transfer RNA: tRNA$_M$ and tRNA$_F$. The plus sign indicates combinations that lead to incorporation. In random poly-UG and random poly-AUG the bases can occur in any sequence, but presumably the only effective sequences are GUG and AUG. Poly-(UG)$_n$ and poly-(AUG)$_n$ are synthetic chains of RNA consisting of 30 or more repetitions of the base sequences indicated.

codons AUG and GUG are in initiating the synthesis of polypeptides, it is certainly odd that a synthetic messenger RNA such as poly-U, which of course cannot supply those codons, nevertheless manages to cause the ribosomes to produce a polypeptide. We can only conclude that they do so by mistake, so to speak, that is, by acting in a way not entirely specified by the available information. (It is ironic that the genetic code was broken because artificial systems were able to make the right kind of mistake!) Are there circumstances that tend to assist these systems in accomplishing proper mistakes? One influential factor has been found. It is the concentration of magnesium in the cell-free system of building materials. A high magnesium concentration makes it possible for many kinds of synthetic messenger RNA to generate polypeptides; when the magnesium concentration is lowered, only the RNA's that contain AUG or GUG succeed in doing so. What magnesium may have to do with polypeptide initiation is still unclear.

Let us come back to the placement of the initial methionine as the normal first step in the construction of a protein. We have noted that the methionine-tRNA complex that places the amino acid in the initial position does not necessarily contain a formyl group. Evidently under conditions of a relatively high concentration of magnesium the formyl group of itself plays no essential role in the installation of the amino acid. What seems to be important is the character of the tRNA: only the formylatable variety of methionine tRNA can initiate the synthesis, and it can do so even when it is not formylated. What, then, are the specific properties that account for its role as an initiator?

A reasonable supposition is that this variety of tRNA has a special shape or configuration that helps it to fit into a particular site on the ribosome. As a matter of fact there is evidence that ribosomes possess two kinds of site for the attachment of tRNA's. One kind, called an amino acid site, simply receives and positions the tRNA when it arrives with its amino acid; the other kind, called a peptide site, holds the tRNA while a peptide bond is formed between its amino acid and an adjacent neighbor [see illustration at left]. It is therefore plausible to suppose that the formylatable variety of the tRNA for methionine may have a shape that helps it to fit into a peptide site on the ribosome and thus be in a position to start the linking together of amino acids.

Evidence in support of this hypothesis has been obtained in our laboratory by Mark S. Bretscher and one of us (Marcker) and by Philip Leder and his associates at the National Institutes of Health in experiments using the antibiotic puromycin. The structure of puromycin is similar to that of the end of a tRNA molecule that attaches to an amino acid [see illustration on next page]. Because it has an NH_2 group, puromycin can form a peptide bond with an amino acid, but since it lacks the free carboxyl (COOH) group of a normal amino acid it cannot form a second peptide bond. Thus it cannot participate in chain elongation. Various experiments indicate that puromycin will add on to—and terminate—a growing polypeptide chain only when the tRNA holding the chain is bound in the peptide site.

In other experiments it has been found that the formylatable variety of methionine tRNA, when bound to a ribosome, will combine with puromycin; the unformylatable variety of the tRNA, on the other hand, will not react with puromycin. The experimental results therefore indicate that there are indeed two kinds of ribosomal site or state: one where a

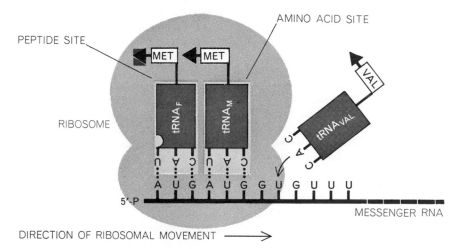

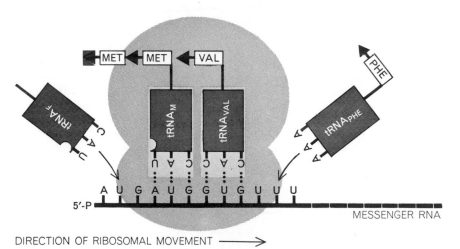

PROTEIN SYNTHESIS takes place on cellular particles called ribosomes, which travel along the "instruction tape" of messenger RNA, reading off the genetic message. The ribosome evidently has two sites for accommodating molecules of transfer RNA: a peptide site and an amino acid site. It appears that the structure of tRNA$_F$ enables it to go directly to the peptide site, thereby initiating the protein chain. This special structure is symbolized by a notch in tRNA$_F$. Other tRNA's may acquire the configuration needed for the peptide site after first occupying the amino acid site. In step 1 (top) the codon AUG at the front end (5′-phosphate end) of messenger RNA pairs with the anticodon CAU that is believed to exist on tRNA$_F$, which delivers a molecule of formyl methionine to start the protein chain. The codon AUG in the second position is paired with the CAU anticodon of tRNA$_M$, which delivers a molecule of ordinary methionine. In step 2 (bottom) the tRNA$_F$ molecule has moved away and the peptide site has been occupied by tRNA$_M$, which is now coupled to the growing protein chain. Valine transfer RNA has moved into the amino acid site.

peptide bond cannot be formed between the peptide chain and puromycin and one where it can. Most likely the latter is the peptide site. Furthermore, the experiments have strengthened the suspicion that the formylatable tRNA possesses a unique structure that somehow helps it to move into the peptide site on a ribosome. Apparently the structure of the formylatable tRNA has been particularly tailor-made for its function as a chain initiator.

The question therefore arises: What is the precise role of the formyl group? If the formyl group per se has nothing to do with placing methionine in the starting position, what function does it have? Our earlier experiments, in which we used a relatively high magnesium concentration, suggested that the formyl group is involved somehow in the formation of the first peptide bond, which launches the building of the polypeptide chain. When the methionine tRNA complex is formylated, synthesis of the polypeptide proceeds much faster than when it is not. This effect can be ascribed to the fact that the presence of the formyl group somehow facilitates the entry into the peptide site. It still remains to be determined just how the formyl group helps to promote such an effect.

Further light has been shed on the problem of protein-chain initiation in the past year by the work of several laboratories, including our own. Special protein agents, still poorly defined, have been implicated together with a cofactor in the formation of the initiation complex on the ribosome. When these new components are present and the supply of magnesium is low, the formyl group is necessary if the formylatable methionine tRNA is to be attached to the ribosomal peptide site by a messenger. Quite recently the cofactor has been identified as being a nucleotide derivative: guanosine triphosphate. Hence we are coming to the view that the conditions prevailing within the living cell are approached by these low-magnesium conditions, where there is strict specificity for forming the initiation complex and for unambiguous polypeptide formation. In our present state of knowledge, however, it is still unclear how these new components help to ensure the placement of the formylated methionine tRNA in the peptide site on the ribosome.

The specific findings concerning the initiation of protein synthesis that we have discussed in this article apply only to bacterial cells. So far no such form of tRNA (containing the formyl group or any other blocking agent) has been

found in the cells of mammals. Accordingly the mechanism of protein-chain initiation is possibly different in mammalian cells from the mechanism discussed here. The process of polypeptide initiation in the cells of higher organisms is currently under study in several laboratories.

Meanwhile the investigation of the *E. coli* system is being pursued with experiments that promise to yield further insights. The way in which the vaguely characterized protein agents and guanosine triphosphate are involved in the initiation of a polypeptide chain is being explored. Much work is under way on analyzing the sequence of nucleotides in natural messenger RNA's, with a view to determining whether or not AUG or GUG constitutes a complete coding signal for protein initiation. We are searching for differences between the formylatable and unformylatable varieties of methionine tRNA, in their nucleotide sequences and in their three-dimensional structures, that may throw light on their respective interactions with the ribosomes.

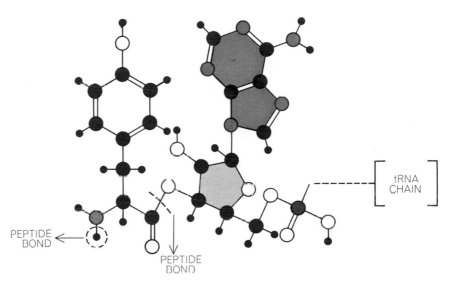

TYROSINE tRNA

PUROMYCIN

PROTEIN-CHAIN TERMINATION can be induced by adding puromycin, an antibiotic, to a protein-synthesizing system. The structure of puromycin closely resembles the structure formed by the amino acid tyrosine and the terminal base of tRNA. Colored disks mark the atomic differences. Tyrosine can be inserted in a protein chain because it can form two peptide bonds. Puromycin can form only one peptide bond because the —CONH— linkage (*inside broken line*) is less reactive than the —COO— linkage in tyrosine tRNA.

35

PROTEIN-DIGESTING ENZYMES

HANS NEURATH

December 1964

These enzymes, which are themselves proteins, play an essential role in the process of digestion. The study of their molecular structure provides clues as to how all enzymes produce their catalytic effects

A problem that fascinates biochemists is how the living cell performs with such speed and precision a multitude of chemical reactions that otherwise occur with immeasurable slowness at the same temperature and pressure. It has long been known that the secret is to be found in the remarkable catalysts created by the living cell. A catalyst is defined as a substance that promotes a chemical reaction without itself being used up in the process. The biological catalysts that mediate all the chemical reactions necessary for life are called enzymes. All enzymes are proteins, although some work in concert with the simpler inorganic or organic compounds known as coenzymes. Each enzyme is tailored to accelerate one specific type of chemical reaction, and since the cell must carry out thousands of different reactions it must have the services of thousands of different enzymes.

One important class of enzymes has the task of degrading, or digesting, other proteins. Because the process is also termed lysis (from the Greek word for "loosing"), these enzymes are called proteolytic. Present in all living organisms, they degrade cellular proteins as part of the cell's metabolic cycle. They also take apart the protein molecules that the organism ingests as food and make the subunits available for constructing the new proteins the organism needs for its own sustenance.

The molecules of proteins are complex structures composed of hundreds or thousands of amino acid subunits of about 20 different kinds. These subunits are linked together in various proportions—exact for each protein—to form polypeptide chains. The term "peptide" refers to the bond that is formed when two amino acids link together and in the process release a molecule of water. The portion of the amino acid molecule that remains in the polypeptide chain after the loss of the atoms in water is termed an amino acid residue.

Proteolytic enzymes have the ability to restore the elements of water by a process of hydrolysis that involves only the protein, the proteolytic enzyme and the water in which both are dissolved. Protein hydrolysis occurs when the peptide bond, which links a carbon atom on one amino acid residue with a nitrogen atom on the next, is broken. Simultaneously the hydroxyl ion (OH^-) of a water molecule is added to the carbon end of the broken protein chain and the remaining hydrogen ion (H^+) is added to the nitrogen end [*see top illustration on page 314*].

Like all other chemical reactions, protein hydrolysis involves an exchange of electrons between certain atoms of the reacting molecules. In the absence of a catalyst this exchange occurs so slowly that it cannot be measured. It can be accelerated by the addition of acids, which increase the supply of hydrogen ions, or of bases, which provide hydroxyl ions. The acids and bases act as true catalysts: they are not consumed in the process. If a protein is boiled in a concentrated acid, it will decompose completely into free amino acids. Such conditions would obviously destroy a living cell. Proteolytic enzymes produce the same result even faster and with no harm to the organism. And whereas hydrogen ions act indiscriminately on all proteins and on all peptide linkages in any protein, proteolytic enzymes are specific, acting only on certain kinds of linkage.

Since proteolytic enzymes are themselves proteins, their action is directed toward the same compounds from which they are made. What, then, differentiates a protein that is a proteolytic enzyme from the protein the enzyme digests? How does a proteolytic enzyme exert its catalytic function without destroying itself or the cell in which it originates? An answer to these fundamental questions would do much to clarify how all enzymes perform their functions. In the 30 years since a proteolytic enzyme (trypsin) was first isolated and crystallized (by Moses Kunitz of the Rockefeller Institute) proteolytic enzymes have served as prototypes in studies relating protein structure to enzyme function.

Because of their role in one of man's most important functions, that of nourishing himself, the proteolytic enzymes of the digestive tract have a long history of investigation, surpassed perhaps only by the study of the enzymes involved in the fermentation of alcohol by yeast (from which the term "enzyme," meaning "in yeast," is derived). Among the digestive enzymes, those secreted by the pancreas—trypsin, chymotrypsin and the carboxypeptidases—are the most completely analyzed and best understood. Hence they will serve to illustrate what is now known about the specificity, structure and mode of action of all proteolytic enzymes.

The proteolytic enzymes of the pancreas are synthesized in the form of precursors called zymogens, and they are stored in the pancreas in intracellular bodies known as zymogen granules. In the zymogen form proteolytic enzymes are inactive and so are prevented from exerting their destructive action on the protein components of the tissue in which they originate. Activation takes place after the zymogens are secreted into the small intestine, where a certain enzyme makes small but important changes in the structure of the zymogen molecule. Such changes will be considered in more detail later in this article.

As I have indicated, proteolytic en-

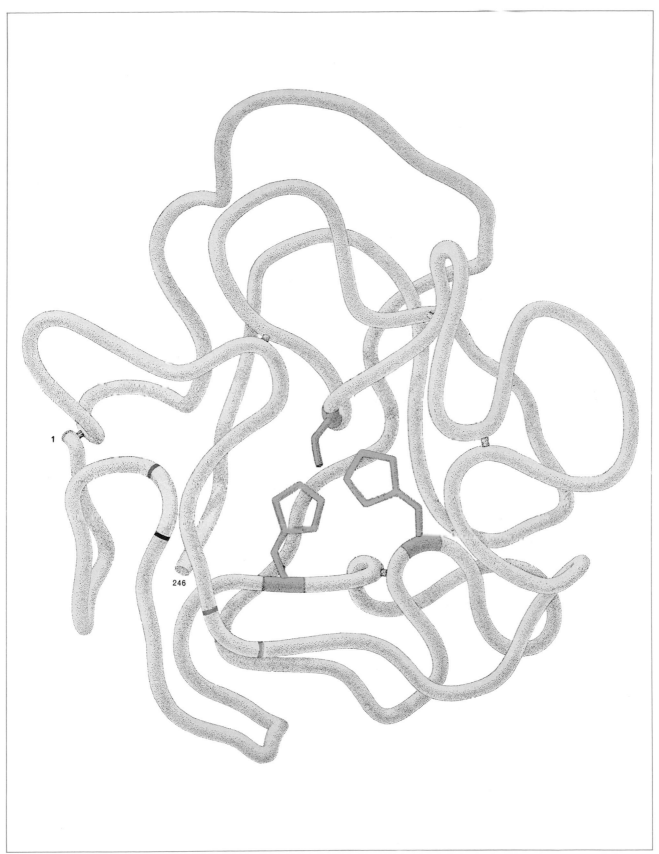

CHYMOTRYPSINOGEN MOLECULE, the inactive precursor of the protein-digesting enzyme chymotrypsin, may look something like this. This drawing, based on a model built by the author, shows the hypothetical route of the molecule's central chain composed of 246 amino acid subunits, often referred to as residues. The model takes into account many of the known chemical features of this enzyme secreted by the pancreas. The three-dimen-sional conformation is dictated, in part, by sulfur-sulfur linkages, or disulfide bridges, which tie the molecular chain together at five points. The molecule becomes an active enzyme when split by trypsin at the point indicated by the black ring. Three gray rings show secondary cleavage points. The active site of the molecule is believed to lie in a small region that includes two residues of histidine and a residue of serine, all three shown in dark color.

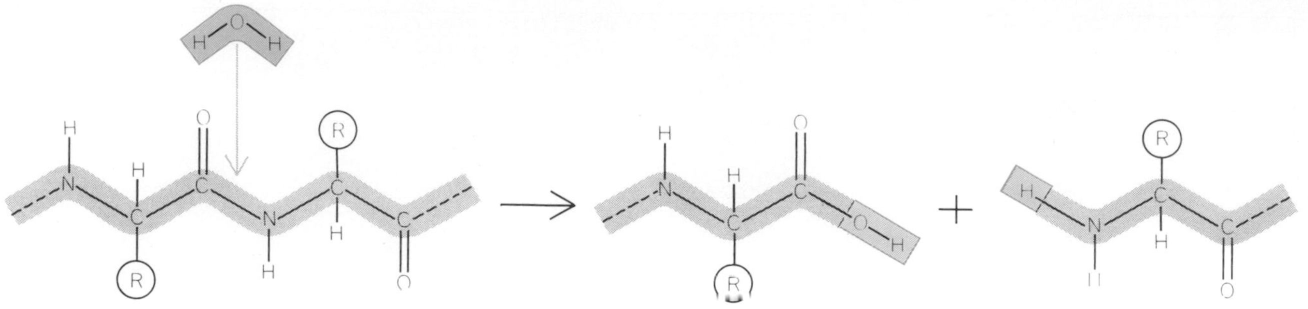

CLEAVAGE OF A "PEPTIDE" BOND is the fundamental step in the degradation, or digestion, of a protein molecule. A peptide bond is the carbon-nitrogen linkage formed when two amino acids are united with the simultaneous release of a molecule of water.

The reverse process, peptide hydrolysis, is shown here. Without a catalyst, hydrolysis is immeasurably slow. The letter "R" represents any of the various side groups found in the 20 common amino acids from which the polypeptide chains of proteins are assembled.

zymes are highly selective in their catalytic effect. An enzyme such as trypsin will hydrolyze certain bonds in a protein molecule but will have no measurable effect on certain others. The explanation is that, although all peptide bonds are chemically similar, their immediate environment is modified by the chemical nature of the amino acid residues on either side of a given bond. Each of the proteolytic enzymes requires a specific chemical environment in order to catalyze the hydrolysis of a peptide bond.

Trypsin, for example, hydrolyzes only peptide bonds whose carbonyl group (C=O) is contributed by an amino acid (e.g., arginine or lysine) that has a positively charged side group. Chymotrypsin acts preferentially on peptide bonds whose carbonyl group is

adjacent to an amino acid (e.g., tyrosine or phenylalanine) that has a six-carbon ring in its side group. In contrast, carboxypeptidases exclusively hydrolyze the last peptide bond in a polypeptide chain. Carboxypeptidase A acts preferentially on peptide bonds adjacent to terminal amino acid residues with a six-carbon-ring side group, and carboxypeptidase B on those adjacent to terminal residues (e.g., lysine or arginine) whose side groups end in an amino group (NH_2). An illustration of the specificities of these enzymes toward a hypothetical polypeptide is shown at the bottom of these two pages.

Broadly speaking, trypsin and chymotrypsin cleave peptide bonds that occupy internal positions in the polypeptide chain, and hence they are referred to as "endopeptidases." Carboxypepti-

dases, which cleave the outermost peptide bonds, are called exopeptidases. As we shall see, it happens that the two endopeptidases, trypsin and chymotrypsin, operate by a common mechanism that is different from the one involved in the action of carboxypeptidases.

A typical protein molecule containing several hundred amino acid residues in a linear sequence will have many peptide bonds that can be attacked by an enzyme such as trypsin. The enzyme can hydrolyze the chain at every peptide bond adjacent to an arginine or a lysine residue, giving rise to polypeptide fragments with arginine or lysine residues at the ends. If the same protein is exposed to the action of a carboxypeptidase rather than trypsin, the terminal peptide bonds will be hydrolyzed sequentially, giving rise to free

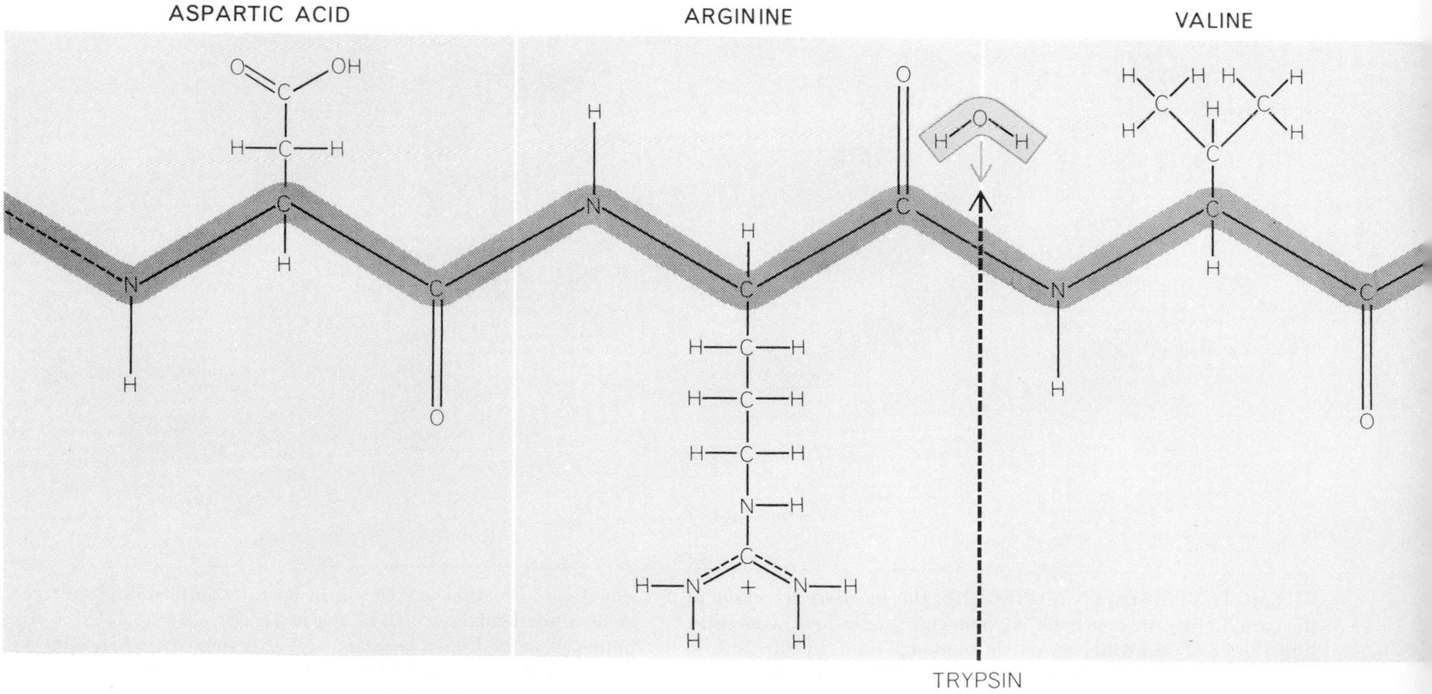

SPECIFICITY OF PEPTIDE BOND CLEAVAGE is an important characteristic of enzyme action. Three of the principal "proteo-

lytic," or protein-digesting, enzymes are trypsin, chymotrypsin and carboxypeptidase A. Each will split only peptide bonds that are in

amino acids one at a time until a peptide bond is reached that does not meet the specific requirements of the attacking enzyme. Through the combined action of endopeptidases and exopeptidases a protein can be digested into fragments of various lengths and ultimately degraded into free amino acids. This process is physiologically significant: it provides both free amino acids and larger fragments for absorption through the intestinal wall. The selective hydrolysis of proteins by proteolytic enzymes has also been an important experimental tool for the determination of the amino acid sequence of proteins [see the article "The Chemical Structure of Proteins," by William H. Stein and Stanford Moore, beginning on page 147].

A major step toward understanding the specificity of proteolytic enzymes was taken some 25 years ago at the Rockefeller Institute, when Max Bergmann, Joseph S. Fruton (now at the Yale University School of Medicine) and their associates found that relatively simple synthetic compounds with only one or two peptide bonds were also susceptible to hydrolysis by proteolytic enzymes. They found, for instance, that trypsin rapidly hydrolyzes a compound designated N-acyl argininamide, whose molecular structure is derived from the amino acid arginine [see upper illustration on next page]. Molecules of this kind have an amide (a nitrogen-containing group) where arginine has a carboxyl group (COOH); at the other end they have an acyl group (a derivative of an organic acid, such as acetic acid) where arginine has its primary amino group. The hydrolysis of such molecules by trypsin yields N-acyl arginine and ammonia.

Several years later my associates George Schwert, Seymour Kaufman, John Snoke and I, then working at the Duke University School of Medicine, were able to narrow down still further the minimum structural elements required in a molecule that could be hydrolyzed by trypsin. We found that when the nitrogen of the amide group in N-acyl argininamide is replaced by oxygen, the resulting esters are hydrolyzed even more rapidly than the original amide compound. These synthetic compounds, like all other substances on which enzymes act, are commonly called substrates.

The recognition that substrates of such relatively simple and specified properties are attacked in the same way as complex proteins has helped considerably in clarifying how proteolytic enzymes work. It was immediately evident that the large size of the natural substrate molecule does not play a unique part in the action of proteolytic enzymes. Furthermore, synthetic substrate molecules with a limited number of reactive groups have provided a tool for testing the contribution of each structural element of the substrate to the specificity of the enzyme, just as a set of keys can be used to determine the precision and uniqueness of a lock. A great many investigations of this kind, extending the work of Bergmann and his associates, have been conducted in numerous laboratories, particularly those of Emil L. Smith at the University of Utah College of Medicine, of Carl Niemann at the California Institute of Technology and in our own laboratory at the University of Washington. These studies have provided a rich catalogue of information relating the structure of synthetic molecules to their interaction with proteolytic enzymes.

This catalogue contains a special section listing a group of compounds that react with both trypsin and chymotrypsin as if they had little regard for the different specificities of the two enzymes. These compounds are highly reactive esters, formed by the coupling of an "aromatic" alcohol (one having a six-carbon ring) with certain organic acids. Let us therefore consider at this point some basic principles of the mechanism of action of these enzymes.

Each atom has only a limited number of electrons that can be shared with surrounding atoms to form chemical bonds. When additional electrons are offered such an atom, it can accept them only by letting go of electrons already being used, causing that bond to rupture. In the hydrolysis of polypeptides, bond breakage is usually initiated by a nucleophile, a compound with electrons to spare [see lower illus-

TYROSINE   ALANINE   PHENYLALANINE

CHYMOTRYPSIN   CARBOXYPEPTIDASE A

a particular chemical environment. Thus trypsin splits the peptide bond to the right of an arginine residue; chymotrypsin splits the bond to the right of a tyrosine residue; carboxypeptidase A splits a terminal peptide bond to the left of a phenylalanine residue.

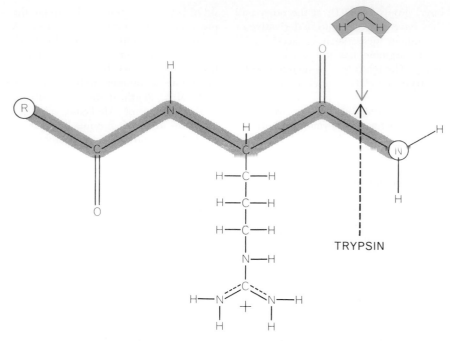

TRYPSIN

CLEAVAGE OF SYNTHETIC SUBSTRATES, compounds that are acted on by enzymes, can provide important information about enzyme mechanisms. The molecule shown here, *N*-acyl argininamide, is a derivative of the amino acid arginine and one of the simplest compounds cleaved by trypsin. The acyl group (R·CO) replaces a hydrogen atom normally present in arginine; the amino group (NH$_2$) at right replaces the hydroxyl group (OH).

CLEAVAGE MECHANISM is believed to require a nucleophile, a reactive group with electrons to spare. The nucleophile represented here by :Y has two lone electrons in its outer shell. The nucleophile reacts with the substrate (R·CO·X) to form an unstable intermediate in which each of the carbon atom's four bonds is linked to another atom or group. When the C·X bond breaks, X is expelled. X is a nitrogen atom in polypeptides or an oxygen atom in esters. An ester is produced when an acid combines with an alcohol.

tration above]. As the electrons are accepted by the carbon atom of the peptide bond, the link between the carbon and the nitrogen is broken and the amide group is detached. An enzyme such as chymotrypsin is believed to promote this reaction because certain groups on the surface of its molecule are effective electron donors.

It is clear, however, that many other features of an enzyme are involved in the process. To begin with, the enzyme binds the substrate to itself and thereby brings the reacting groups of both substrate and enzyme in close contact with each other. The enzyme's reacting groups include effective nucleophiles, such as the five-atom imidazole ring of the amino acid histidine. Other enzyme groups tend to stabilize the intermediate products formed during hydrolysis. One of these is the short but reactive side group of the amino acid serine,

which can transiently capture the acid portion of the substrate, forming an ester [*see illustration on opposite page*].

The operation of these factors can be illustrated by the hydrolysis of the synthetic substrate *p*-nitrophenylacetate, which is the acetic acid ester of the aromatic alcohol nitrophenol. That this substrate can be hydrolyzed both by trypsin and by chymotrypsin was shown by B. S. Hartley and B. A. Kilby of the University of Cambridge. The hydrolysis begins when the enzyme encounters the substrate in solution and the two molecules form what is described as an enzyme-substrate complex. In the second step the ester is split into its acidic and basic constituents. The former, or acetyl portion, combines with the enzyme, giving rise to an intermediate called the acyl enzyme; the latter is released as the alcohol nitrophenol. In the third step the acyl enzyme is hydro-

lyzed by water, which regenerates free enzyme and liberates the acyl group as the corresponding acid (in this case acetic acid).

The first step happens so fast that it cannot be separated from the second; the third step, however, takes place more slowly. The initial burst of activity representing the first two steps can be followed by measuring the liberation of nitrophenol; the measurements are made easier by the fact that this alcohol is yellow. The third step can be followed by measuring the release of the acid. Since it is the slowest step, it governs the overall rate of the process. The mechanism of each of these steps has been studied intensively, notably by Thomas C. Bruice, first at Cornell University and more recently at the University of California at Santa Barbara, and by Myron L. Bender at Northwestern University. A particular effort has been made to identify the specific groups on the enzyme surface that participate in each of the postulated steps.

So far two kinds of group have been clearly implicated in the proteolytic activity of trypsin and chymotrypsin. The reaction with *p*-nitrophenylacetate can be stopped at the end of the second step, for instance, by carrying out the reaction in a mildly acidic solution. When the acyl enzyme is removed and analyzed, it is found that the acyl fraction is invariably bound to the hydroxyl group of a serine residue of the enzyme. Although trypsin and chymotrypsin each contain some 30 serine residues per molecule, only one of them is capable of capturing the acyl group during reaction with synthetic substrates. The location of this active serine has been identified within the structure of each of the two enzymes. Because serine is involved in their reactivity, trypsin, chymotrypsin and several other enzymes that function in the same way have been given the name "serine proteases."

The other group that has been definitely implicated as a participant in the enzymatic reaction is the imidazole ring of a histidine residue, which contains three carbon atoms and two nitrogen atoms. It is an effective nucleophile because it can donate two electrons from one of the two nitrogen atoms of the ring structure. Therefore it is believed that the histidine residue partakes in the second step of the reaction by initiating the attack on the peptide bond or the equivalent bond in an ester, and possibly also in the third step by initiating the hydrolysis of the acyl enzyme.

Chymotrypsin contains two histidine

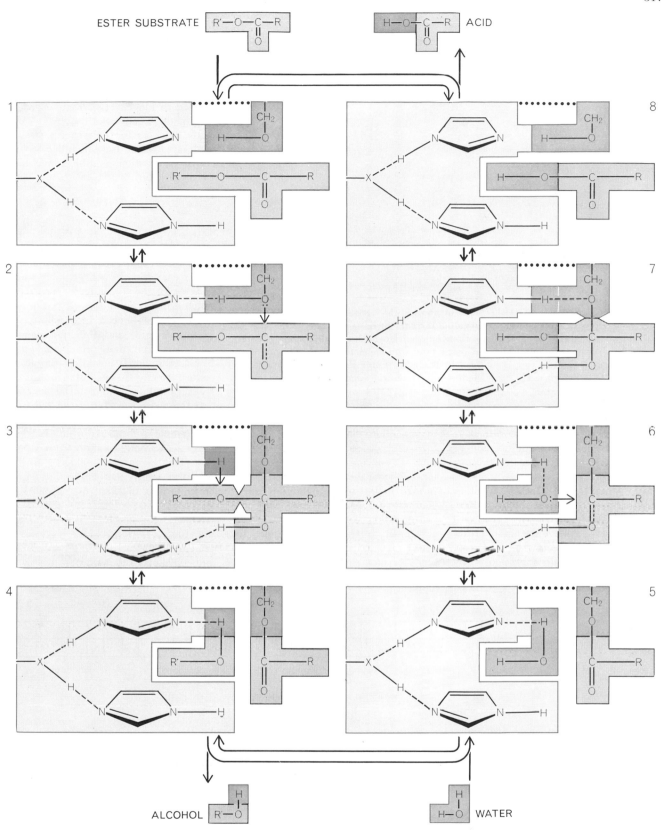

ESTER SUBSTRATE R'—O—C—R

H—O—C—R ACID

ALCOHOL R'—O

H—O WATER

HYDROLYSIS OF ENZYME SUBSTRATE by the active site in chymotrypsin or trypsin may take place as shown here. The active site contains the functional groups of two histidine residues (*light colored area containing two imidazole rings*) and of a serine residue (*dark color*). The chemical nature of the group X is not established. In Step 1 the ester enters the active region. In Step 2 the OH group of the serine residue begins its nucleophilic attack. In Step 3 an unstable intermediate is formed. In Step 4 the ester is cleaved and an acyl enzyme is formed. The alcohol created by the cleavage of the ester is released and is replaced by a molecule of water (*Step 5*). In Step 6 a second nucleophilic attack is launched by the OH group of water. In Step 7 another unstable intermediate is formed. And in Step 8 the acid fraction of the ester is released and the free enzyme is restored to its original active condition. The scheme shown here is based in part on concepts proposed by Myron L. Bender and F. J. Kézdy of Northwestern University.

the terms suggest that only a portion of the enzyme molecule is involved in enzyme action. The active region is only the relatively small area that comes in direct contact with the substrate. Considerable effort is currently being expended to identify the reactive groups in the active centers of enzymes. Much of the evidence for the existence of such centers has come from investigations of enzyme inhibition, a subject to which I shall now turn.

Enzyme Inhibition

It is to be expected that the functioning of an enzyme will be impaired if any of the groups required for binding or for catalysis are not available to interact with the substrate. Certain naturally occurring inhibitors of trypsin and chymotrypsin have a high affinity for these enzymes, and they have molecules so large that presumably they block off much if not all of the active center. Among these substances are the trypsin inhibitor found in the pancreas and the trypsin inhibitors isolated from the soybean. Other inhibitors structurally resemble a normal substrate but lack the reactive bond in the required position. These substrate analogues are believed to occupy the binding site on the enzyme, thus blocking the normal substrate.

One of the earliest observations implicating a serine residue as a component of the active center of trypsin and chymotrypsin was the discovery

SYNTHETIC ENZYME INHIBITORS can abolish the activity of chymotrypsin and trypsin by reacting with one of the histidine residues in the active site of the enzyme. The compound shown here, *p*-toluenesulfonyl-phenylalanyl-chloromethylketone (TPCK), resembles chymotrypsin substrates and reacts only with chymotrypsin. If the side chain containing the six-carbon ring is replaced by the straight side chain found in lysine, the compound (then known as TLCK) resembles trypsin substrates and reacts only with trypsin.

residues and trypsin three of them. As we shall see, there is reason to believe that not one but two histidine residues are in fact cooperating in this process. One of them pushes the electrons toward the bond being attacked; the other pulls electrons from the opposite side. In this reaction scheme only the second and third steps involve the actual breaking of chemical bonds. The first step, on the other hand, does not necessarily

lead to a reaction. This step is dependent on the affinity of the enzyme for the substrate and hence is an expression of enzyme specificity, as distinguished from the catalytic function performed in the second and third steps. The groups on the enzyme that contribute either to specificity or to catalysis constitute what is often termed the active site, or active center. The existence of such a region is hypothetical;

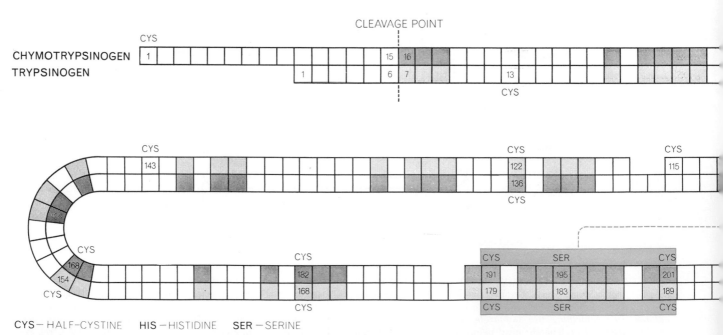

CYS — HALF-CYSTINE   HIS — HISTIDINE   SER — SERINE

LINEAR SEQUENCES of the amino acid residues of chymotrypsinogen and trypsinogen, the inactive precursor of trypsin, have been established except for one short stretch (84–87) in trypsinogen. When the two sequences are aligned as shown here, it is found that about 40 percent of the positions in the two chains are occupied by the same amino acid residues. These regions of

by A. K. Balls of Purdue University and Eugene F. Jansen of the U.S. Department of Agriculture that the two enzymes are inhibited by a nerve gas, diisopropylfluorophosphate (DFP). Two years earlier Abraham Mazur and Oscar Bodansky of the Medical Division of the Chemical Warfare Service at Edgewood Arsenal had shown that DFP similarly inhibits acetylcholine esterase, an enzyme involved in the conduction of nerve impulses. When the inhibited chymotrypsin was subsequently broken down and analyzed, a phosphorus-containing fragment of DFP was invariably found to be attached to a specific serine residue—the same residue that later was found to be acylated during reaction with *p*-nitrophenylacetate. Thus the chemical modification of a single residue—serine—completely abolished enzyme function.

A clear demonstration that a histidine residue is another constituent of the active site has recently been provided by Elliott N. Shaw and his co-workers at Tulane University. These investigators synthesized compounds that in structure simulate specific substrates but contain, instead of the bond normally hydrolyzed, a group that reacts irreversibly with a histidine residue. For example, the compound *p*-toluenesulfonyl - phenylalanyl - chloromethylketone (TPCK) resembles substrates for chymotrypsin [*see top illustration on opposite page*]. It reacts with a nucleophile in the active center, specifically a histidine residue, completely inhibiting

the enzyme. The compound known as TLCK is analogous, except that it simulates specific substrates for trypsin and therefore reacts with a histidine residue in the active center of trypsin only. In addition to serine and histidine, other specific residues seem to be important for the catalytic function of trypsin and chymotrypsin, but their role is still to be clarified.

Enzyme Structure

The relation between the structure of the entire enzyme molecule and its active site is not fully understood. In most cases, however, the maintenance of the three-dimensional structure of the entire molecule is necessary for maintaining the conformation of the active center. Hence many processes that change the conformation of the protein are accompanied by a loss of enzymatic activity. Trypsin, chymotrypsin and carboxypeptidases contain some 200 to 300 amino acid residues per molecule, equivalent to between 3,000 and 5,000 atoms. The active site involves only a small fraction of the residues, perhaps not more than 20. In order to appreciate the magnitude of the problem of characterizing the active site, let us consider what is now known about the chemical composition and structure of these enzymes.

The chemical composition of a compound is usually expressed by the number of atoms of which it is composed. On this basis glucose has the composi-

tion $C_6H_{12}O_6$. Chymotrypsinogen—the inactive precursor, or zymogen, of chymotrypsin—has the elemental composition $C_{1,130}H_{1,782}O_{356}N_{308}S_{12}$. This formula tells us virtually nothing, however, about the arrangement of the atoms in the molecule. For such giant protein molecules the composition is usually expressed in terms of the constituent amino acid residues rather than the atoms. Thus the composition of chymotrypsinogen, which contains 246 amino acid residues, can be written: alanine$_{22}$ arginine$_4$ aspartic acid$_9$ asparagine$_{14}$ half-cystine$_{10}$ glutamic acid$_5$ glutamine$_{10}$ glycine$_{23}$ histidine$_2$ isoleucine$_{10}$ leucine$_{19}$ lysine$_{14}$ methionine$_2$ phenylalanine$_6$ proline$_9$ serine$_{29}$ threonine$_{23}$ tryptophan$_8$ tyrosine$_4$ valine$_{23}$.

But even this representation is not adequate to describe the properties of a protein, any more than the formula $C_6H_{12}O_6$ genuinely describes the chemical properties of glucose. A more meaningful representation of the structure of proteins is given in terms of their amino acid sequence, that is, the linear arrangement of the amino acid residues. Such an analysis is a formidable undertaking; until recently it had been accomplished for only one enzyme—ribonuclease, a pancreatic enzyme whose molecule is made up of 124 amino acid residues.

The amino acid sequence of both trypsinogen (the precursor of trypsin) and chymotrypsinogen have been announced within the past year. The trypsinogen molecule is composed of a

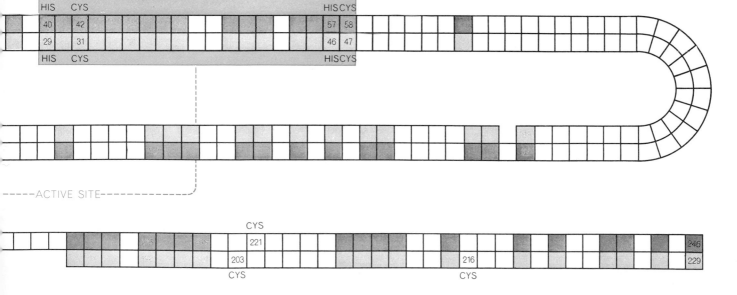

homology are indicated by filled squares. The longest homologous sequences involve areas believed to be part of the active sites of the two enzymes. The 10 half-cystine residues in chymotrypsinogen link up to form five disulfide bridges and the 12 half-cystines in trypsinogen form six disulfide bridges. The effect on the two-dimensional geometry of the two enzymes is shown on the next two pages.

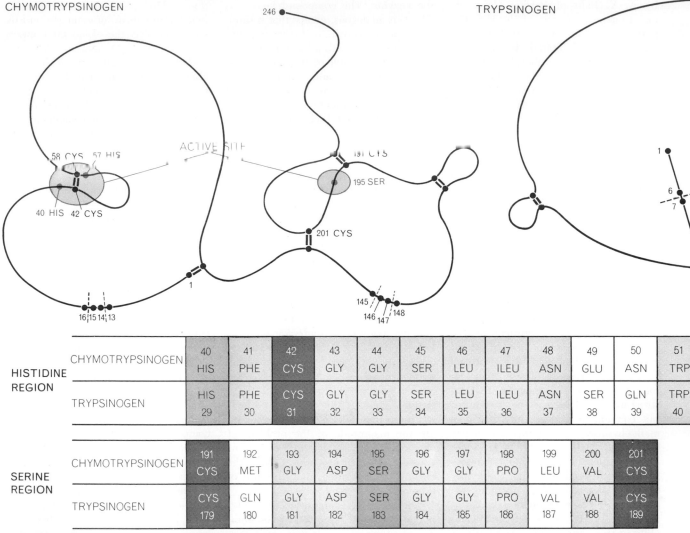

CHYMOTRYPSINOGEN   TRYPSINOGEN

HISTIDINE REGION	CHYMOTRYPSINOGEN	40 HIS	41 PHE	42 CYS	43 GLY	44 GLY	45 SER	46 LEU	47 ILEU	48 ASN	49 GLU	50 ASN	51 TRP
	TRYPSINOGEN	HIS 29	PHE 30	CYS 31	GLY 32	GLY 33	SER 34	LEU 35	ILEU 36	ASN 37	SER 38	GLN 39	TRP 40

SERINE REGION	CHYMOTRYPSINOGEN	191 CYS	192 MET	193 GLY	194 ASP	195 SER	196 GLY	197 GLY	198 PRO	199 LEU	200 VAL	201 CYS
	TRYPSINOGEN	CYS 179	GLN 180	GLY 181	ASP 182	SER 183	GLY 184	GLY 185	PRO 186	VAL 187	VAL 188	CYS 189

ACTIVE SITES in chymotrypsinogen and trypsinogen, labeled "histidine region" and "serine region," are almost identical. In each homologous region there are two half-cystine residues (*dark gray*) that form disulfide bridges in the actual molecule and serve to lock the active histidine and serine residues in a fixed position. The disulfide bridges are shown as short double bonds in the schematic two-dimensional representations of the two enzymes. In trypsinogen the heavy bonds indicate the three disulfide bridges whose location is known; the lighter bonds show possible bridge locations. When cleaved between residues 15 and 16, chymotrypsinogen

single polypeptide chain of 229 amino acid residues, or 17 fewer than chymotrypsinogen. The amino acid sequence of chymotrypsinogen was established by Hartley at Cambridge and independently by B. Keil and F. Šorm of the Czechoslovak Academy of Science in Prague. The amino acid sequence of trypsinogen was subsequently reported from our laboratory by K. A. Walsh and his co-workers, and a partial sequence was published by the Prague group. In each laboratory some 15 man-years of work have gone into the sequence analysis of each of these two proteins.

Although the amino acid sequence does not reveal the three-dimensional structure of a protein, it provides much information of value to the enzyme chemist. For example, it permits the identification of any residue that participates directly or indirectly in enzyme catalysis. In the case of the zymogens chymotrypsinogen and trypsinogen it also discloses the location of the peptide bonds in the molecule that are cleaved during activation. Chymotrypsinogen is converted to chymotrypsin by cleavage of the bond between the 15th and 16th amino acid residues in the chymotrypsinogen chain. The activation of trypsinogen is accomplished by a cleavage between its sixth and seventh residues. Subsidiary cleavage points in chymotrypsinogen will be mentioned later.

An important feature of linear polypeptide chains is that they are often folded and cross-linked at one or more points. Such cross-links, or bridges, can have an important role in establishing a protein's three-dimensional structure. A cross-link is formed when two sulfur atoms protruding at different points on the polypeptide chain join in a disulfide

(—S—S—) bond. The individual sulfur atoms enter the chain appended to cysteine residues, but when two such residues are linked by a disulfide bond, the combination entity is called a cystine residue. The chymotrypsinogen molecule has five cystine (or 10 half-cystine) residues, which serve to tie the molecule together at five different points. The location of these disulfide bridges is shown at the upper left in the illustration above. The trypsinogen molecule has 12 half-cystine residues and therefore six bridges, but only three have been precisely located.

The Active Centers

When the amino acid sequences of chymotrypsinogen and trypsinogen were unraveled, biochemists immediately looked for the location of the histidine

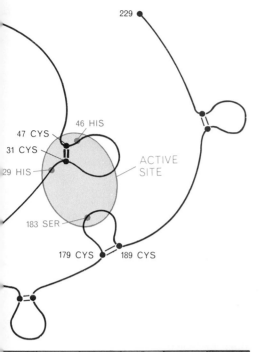

53	54	55	56	57	58
VAL	THR	ALA	ALA	HIS	CYS
VAL	SER	ALA	ALA	HIS	CYS
42	43	44	45	46	47

ALA	ALANINE	**LEU**	LEUCINE
ASN	ASPARAGINE	**MET**	METHIONINE
ASP	ASPARTIC ACID	**PHE**	PHENYLALANINE
CYS	HALF-CYSTINE	**PRO**	PROLINE
GLN	GLUTAMINE	**SER**	SERINE
GLU	GLUTAMIC ACID	**THR**	THREONINE
GLY	GLYCINE	**TRP**	TRYPTOPHAN
HIS	HISTIDINE	**VAL**	VALINE
ILEU	ISOLEUCINE		

is converted to the active enzyme chymotrypsin. Cleavage of trypsinogen between residues 6 and 7 yields active trypsin. (The lighter broken lines in chymotrypsinogen indicate secondary cleavage points.)

and serine residues that had been identified with the active centers of the two molecules. In chymotrypsinogen the active histidines are at positions 40 and 57; in trypsinogen, the shorter molecule, they are at positions 29 and 46. In both molecules the spacing between the two histidines is exactly the same. In both there are neighboring cystine residues that tie the polypeptide chain together in such a way that the two histidine residues are brought close together. Thus chymotrypsinogen has a cystine bridge between positions 42 and 58, which gives rise to a loop in the molecular chain. Similarly, trypsinogen has a cystine bridge between positions 31 and 47, forming another loop. Brought into proximity at the entrance to these loops, the two histidines can cooperate in acting as electron donors in protein hydrolysis.

Where in the two structures are the reactive serines? In chymotrypsinogen the reactive serine is at position 195 and in trypsinogen it is at position 183. Again note the similarity in location. And although in each molecule there is a large linear distance between the two histidines and the reactive serine, there is good reason to believe that the three residues actually lie close together in the active region of the three-dimensional molecule.

The resemblance in chemical structure between chymotrypsinogen and trypsinogen goes even further. When the amino acid sequences of the two molecules are compared side by side, after a common starting point has been chosen, it is found that about 40 percent of the amino acid residues are the same in both molecules [*see illustration on pages 254 and 255*]. Moreover, the most striking similarity in sequence occurs in the neighborhood of the histidine residues and the reactive serine [*see illustration at left*].

This marked similarity in structure of two enzymes that operate by the same mechanism but differ in specificity suggests that the two may have evolved from a common archetype. In the course of evolution those elements of the structure that are necessary for enzyme function would have remained unchanged. A similar conservation of chemical character, together with preservation of biological function, had previously been observed in the amino acid sequence of the cytochrome and hemoglobin molecules found in various species of animals [see "The Hemoglobin Molecule," by M. F. Perutz; SCIENTIFIC AMERICAN Offprint 196].

The three-dimensional structure of chymotrypsinogen and trypsinogen is still unknown. The method of X-ray crystallography, which has been so successful in determining the three-dimensional structure of hemoglobin and its close relative myoglobin, has proved to be much more difficult to apply to the proteolytic enzymes. The data available so far for chymotrypsinogen and chymotrypsin reveal something that has been facetiously described as "complex but unintelligible." In the absence of direct evidence we have attempted to construct a hypothetical model of chymotrypsinogen that incorporates certain principles we believe to be correct [*see illustration on page 313*]. The model shows, for instance, the close juxtaposition of the two histidines and the active serine, the restrictions imposed by the disulfide bonds, the tendency of water-soluble amino acid residues to be on the

outside of the molecule and water-insoluble residues to be on the inside, and the necessity for the peptide bonds cleaved during activation to be readily accessible. How many of the model's features are correct only future research will tell.

The zymogen precursors of proteolytic enzymes are manufactured within the cells of the digestive organs but are converted into active enzymes after they have been secreted by the cells. In all known cases the conversion involves the splitting of a specific peptide bond in the zymogen molecule. Representative examples of zymogen activation, in addition to the activation of chymotrypsinogen and trypsinogen, are the conversion of pepsinogen to pepsin and of procarboxypeptidase to carboxypeptidase. The phenomenon is not limited to proteolytic enzymes; for instance, many of the proteins involved in the process of blood clotting require similar activation, such as the conversion of prothrombin to thrombin, of plasminogen to plasmin and of fibrinogen to fibrin.

Zymogen Activation

The key enzyme in the activation of pancreatic zymogens is enterokinase, a proteolytic enzyme secreted by the mucous membranes of the intestine. Its prime function is to convert trypsinogen to trypsin; trypsin then becomes the key for the activation of all other pancreatic zymogens. Enterokinase is not, however, the only enzyme that can activate trypsinogen; the activation can be accomplished by trypsin itself and by several other enzymes, some of them found in bacteria. As far as is known, all these enzymes function in the same way: they cleave a specific peptide bond in the polypeptide chain of trypsinogen. In the trypsinogen obtained from cattle the cleavage occurs between the sixth and seventh amino acid residues of the chain. Since each molecule of trypsin released can activate another molecule of trypsinogen, it is apparent that the activation of trypsinogen is a self-accelerating process. Chymotrypsinogen similarly becomes an active enzyme when it is cleaved by trypsin. The cleavage takes place near one end of the chain (between residues 15 and 16) but no fragment is released because residue 1 is tied to the remainder of the molecule by a disulfide bond.

In these activation reactions the hydrolytic effect of trypsin comes to a halt after the first peptide bond has been cleaved. Presumably all other peptide bonds in the zymogen molecule that could have been cleaved are so located within the structure of the molecule as to remain completely resistant to the action of the enzyme. In the activation of chymotrypsinogen, hydrolysis can go a bit further because the chymotrypsin formed acts as a proteolytic enzyme on itself and can cleave three additional peptide bonds; they lie between positions 13 and 14, 145 and 146, and 147 and 148. All the resulting fragments are enzymatically active. But the bond between positions 15 and 16 must first be split by trypsin.

All known zymogens of the pancreas follow the same pattern of activation when they become converted to the active form. The activation requires the cleavage of a peptide bond adjacent to an arginine or a lysine residue near the beginning of the polypeptide chain of the zymogen. Some zymogens, however, must undergo additional transformations before they become activated. For example, the zymogen procarboxypeptidase A exists in the pancreatic juice as an aggregate of three large molecular subunits. Only one of these, subunit I, can be regarded as the immediate precursor of the enzyme carboxypeptidase A. Subunit II is the precursor of an enzyme that is similar in specificity to chymotrypsin but differs in composition from the chymotrypsin we have been discussing. The nature and role of subunit III is unknown. When trypsin is added to the three aggregated subunits, subunit I is converted to carboxypeptidase A only after subunit II has first become activated.

Why and how does the cleavage of a single and specific peptide bond in the zymogen give rise to an active enzyme? At present one can offer only a partial explanation, based largely on reasoning and only indirectly supported by experimental facts. I have mentioned previously that the action of an enzyme seems to involve two steps: the specific binding of the substrate and the subsequent bond-cleaving. Singly or in combination these steps require the proper configuration of groups on the enzyme molecule: the binding site and the catalytic site. The absence of either site or both in the zymogen would preclude enzyme function. Experiments have shown that chymotrypsinogen can bind substrates or inhibitors for chymotrypsin in much the same way that chymotrypsin itself does. These observations suggest that the binding site exists in the zymogen and hence that the catalytic site is not functional unless and until the specific peptide bond is cleaved during activation.

This cleavage is believed to change the conformation of the protein molecule so as to bring the groups involved in the bond-cleaving mechanism into the proper spatial relation. In the case of trypsin and chymotrypsin these groups are the reactive serine residue and the two histidine residues. Since the binding site and the catalytic site together form the active center, however, they must occupy adjacent or overlapping areas. Inasmuch as one of them is already present in the zymogen, it follows that the conformational changes involved in zymogen activation must be localized within a very small region of the molecule. The relation between these two sites is perhaps analogous to the relation of the back and the seat of a folding chair: the back can be regarded as determining the specificity and the seat the function. Obviously it is only after the chair is unfolded that it becomes functional.

Although this article has drawn almost exclusively on trypsin and chymotrypsin to illustrate the relation between chemical structure and enzymatic function, it would be wrong to conclude that all proteolytic enzymes utilize the same mechanism for the hydrolysis of peptide bonds. The range of structures essential for catalytic function is not limited to the cooperation of serine and histidine residues. For instance, the proteolytic enzymes of a group found in plants require a free thiol group (SH) for catalytic activity. These include papain, a proteolytic enzyme extracted from papaya juice, and ficin, an enzyme obtained from the fig tree. Papain, commercially used as a meat tenderizer, has a polypeptide chain of some 180 amino acid residues; the chain of ficin is probably somewhat longer.

Other proteolytic enzymes, such as pepsin and certain enzymes found in bacteria, are active only in slightly acid solution. Some aminopeptidases and the carboxypeptidases require the participation of specific metals for enzymatic function. (The former are exopeptidases that cleave only the first bond in a polypeptide chain; the latter are exopeptidases that cleave only the last bond.) The aminopeptidases usually require as coenzymes the ions of such metals as manganese and magnesium. The carboxypeptidases usually contain in their active center a firmly bound atom of a metal such as zinc.

Carboxypeptidase A exemplifies the behavior of these metal-containing enzymes. It has a single chain of some 300 amino acid residues; their complete sequence is still unknown. Yet the mode of action of this enzyme and the structure of its active site are remarkably well understood. Bert L. Vallee and his associates at the Harvard Medical School have shown that the natural enzyme contains an atom of zinc that is believed to be anchored to a thiol group somewhere in the chain and to an amino group of the first residue in the chain. When zinc is present, the enzyme will hydrolyze both polypeptides and their corresponding esters. When the zinc is removed, the protein completely loses its enzymatic activity.

If zinc is replaced by another metal, the specificity of the enzyme is altered. For example, when zinc is replaced by cadmium, the enzyme no longer hydrolyzes peptides but hydrolyzes esters even more effectively than the zinc enzyme does. When zinc is replaced by copper, the enzyme again becomes in-

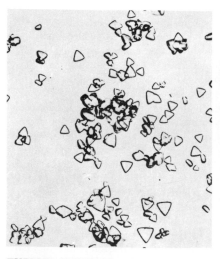
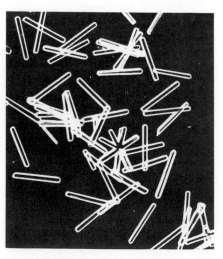

ENZYME CRYSTALS of bovine trypsinogen are at left and of trypsin at right. It can be seen that activation of the former produces a striking change in crystal structure. The two photomicrographs were made in 1935 by Moses Kunitz of the Rockefeller Institute.

active. Vallee and his associates have shown that it is possible not only to modify the activity of carboxypeptidase by replacement of the metal but also to manipulate the activity and specificity of the enzyme still further by chemical alteration of the active center. Their work has shown that at least five groups must participate in the enzyme's proteolytic activity: the metal, the two amino acid residues to which the metal is bound and two tyrosine residues that, when chemically modified, produce a change in the enzyme's specificity.

The Task Ahead

The foregoing examples were selected to illustrate the range and variety of structures encountered in the search for a chemical explanation of the function and mechanism of proteolytic enzymes.

We have only recently determined the linear structure of a few hydrolytic enzymes. We still do not know how they are folded into three-dimensional structures, but we are by no means sure that a three-dimensional model of the molecule will tell us how it functions as an enzyme. Scientific inquiries do not always proceed in an orderly and systematic manner, and perhaps we shall not have to wait until the last detail of the structure of these enzymes is known to understand their function.

The most exciting problem is the elucidation of the structure of the active center. One attempts to do this by "mapping" those groups on the enzyme molecule that are directly involved in enzyme function. The mapping process essentially consists of testing the effect of chemical modification of specific groups in abolishing or inducing enzyme activity. Yet there is no assurance that the notion of an active center is a physical reality and that an isolated structure that reproduced the groupings believed to constitute this site, in the proper spatial relations, would display full enzymatic function. Nature is seldom wasteful, and there is a real possibility that the entire enzyme molecule may be necessary for it to exhibit its full range of function.

Although much remains to be learned about proteolytic enzymes, enough is now known about them to enable investigators to use them to test various hypotheses of the way all enzymes function. It would be fitting if some of the earliest known enzymes, serving one of the most fundamental needs of man, should enable us to understand the mode of action of enzymes in general.

36

THE GENETIC CODE: II

MARSHALL W. NIRENBERG
March 1963

A sequel to F. H. C. Crick's article of October 1962 which discussed how the hereditary material embodies the code for the manufacture of proteins. The nature of the code has now been further elucidated

Just 10 years ago James D. Watson and Francis H. C. Crick proposed the now familiar model for the structure of DNA (deoxyribonucleic acid), for which they, together with Maurice H. F. Wilkins, received a Nobel prize last year. DNA is the giant helical molecule that embodies the genetic code of all living organisms. In the October 1962 issue of *Scientific American* F. H. C. Crick described the general nature of this code. By ingenious experiments with bacterial viruses he and his colleagues established that the "letters" in the code are read off in simple sequence and that "words" in the code most probably consist of groups of three letters. The code letters in the DNA molecule are the four bases, or chemical subunits, adenine, guanine, cytosine and thymine, respectively denoted A, G, C and T.

This article describes how various combinations of these bases, or code letters, provide the specific biochemical information used by the cell in the construction of proteins: giant molecules assembled from 20 common kinds of amino acids. Each amino acid subunit is directed to its proper site in the protein chain by a sequence of code letters in the DNA molecule (or molecules) that each organism inherits from its ancestors. It is this DNA that is shaped by evolution. Organisms compete with each other for survival; occasional random changes in their information content, carried by DNA, are sometimes advantageous in this competition. In this way organisms slowly become enriched with instructions facilitating their survival.

The exact number of proteins required for the functioning of a typical living cell is not known, but it runs to many hundreds. The great majority, if not all, of the proteins act as enzymes, or biological catalysts, which direct the hundreds of different chemical reactions that go on simultaneously within each cell. A typical protein is a molecular chain containing about 200 amino acid subunits linked together in a specific sequence. Each protein usually contains all or most of the 20 different kinds of amino acids. The code for each protein is carried by a single gene, which in turn is a particular region on the linear DNA molecule. To describe a protein containing 200 amino acid subunits a gene must contain at least 200 code words, represented by a sequence of perhaps 600 bases. No one yet knows the complete base sequence for a single gene. Viruses, the smallest structures containing the blueprints for their own replication, may contain from a few to several hundred genes. Bacteria may contain 1,000 genes; a human cell may contain a million. The human genes are not strung together in

EXPERIMENT BEGINS when cells of the colon bacillus are ground in a mortar with finely divided aluminum oxide. "Sap" released from ruptured cells still synthesizes protein.

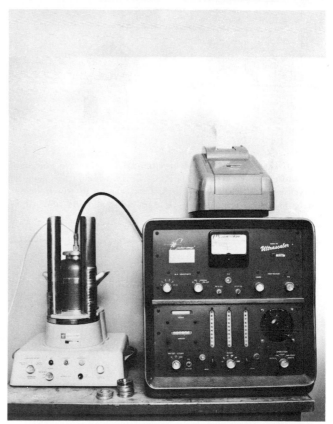

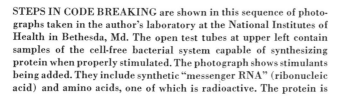

STEPS IN CODE BREAKING are shown in this sequence of photographs taken in the author's laboratory at the National Institutes of Health in Bethesda, Md. The open test tubes at upper left contain samples of the cell-free bacterial system capable of synthesizing protein when properly stimulated. The photograph shows stimulants being added. They include synthetic "messenger RNA" (ribonucleic acid) and amino acids, one of which is radioactive. The protein is produced when the samples are incubated 10 to 90 minutes. At upper right the protein is precipitated by the addition of trichloroacetic acid (TCA). At lower left the precipitate is transferred to filter-paper disks, which will be placed in carriers called planchettes. At lower right the planchettes are stacked in a radiation counting unit. Radiation measurement indicates how well a given sample of messenger RNA has directed amino acids into protein.

BASES

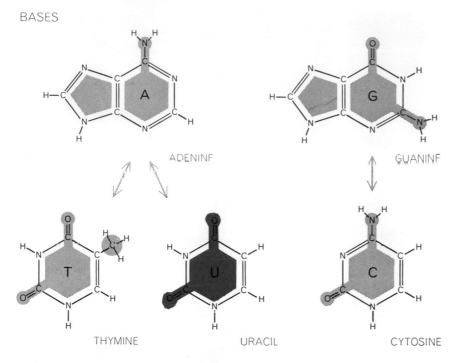

CHAIN COMPONENTS

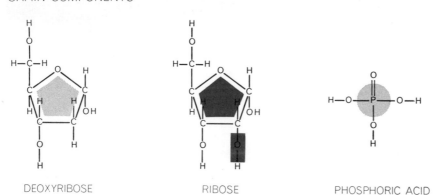

DEOXYRIBOSE   RIBOSE   PHOSPHORIC ACID

COMPONENTS OF DNA (deoxyribonucleic acid) are four bases adenine, guanine, thymine and cytosine (symbolized A, G, T, C), which act as code letters. Other components, deoxyribose and phosphoric acid, form chains to which bases attach (*see below*). In closely related RNA, uracil (U) replaces thymine and ribose replaces deoxyribose.

DNA STRUCTURE

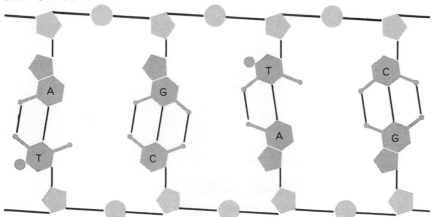

DNA MOLECULE resembles a chain ladder (actually twisted into a helix) in which pairs of bases join two linear chains constructed from deoxyribose and phosphate subunits. The bases invariably pair so that A links to T and G to C. The genetic code is the sequence of bases as read down one side of the ladder. The deoxyribose-phosphate linkages in the two linear chains run in opposite directions. DNA molecules contain thousands of base pairs.

one long chain but must be divided among at least 46 DNA molecules. The minimum number is set by the number of human chromosomes (46), which collectively carry the hereditary material. In fact, each chromosome apparently carries not one or two but several copies of the same genetic message. If it were possible to assemble the DNA in a single human cell into one continuous thread, it would be about a yard long. This three-foot set of instructions for each individual is produced by the fusion of egg and sperm at conception and must be precisely replicated billions of times as the embryo develops.

The bottom illustration at left shows how the bases in DNA form the cross links connecting two helical strands composed of alternating units of deoxyribose (a simple sugar) and phosphate. The bases are attached to the sugar units and always occur in complementary pairs: A joined to T, and G joined to C. As a result one strand of the DNA molecule, with its associated bases, can serve as the template for creating a second strand that has a complementary set of bases. The faithful replication of genes during cell division evidently depends on such a copying mechanism.

The coding problem centers around the question: How can a four-letter alphabet (the bases A, G, C and T) specify a 20-word dictionary corresponding to the 20 amino acids? In 1954 the theoretical physicist George Gamow, now at the University of Colorado, pointed out that the code words in such a dictionary would have to contain at least three bases. It is obvious that only four code words can be formed if the words are only one letter in length. With two letters 4×4, or 16, code words can be formed. And with three letters $4 \times 4 \times 4$, or 64, code words become available—more than enough to handle the 20-word amino acid dictionary [*see top illustration on page* 332]. Subsequently many suggestions were made as to the nature of the genetic code, but extensive experimental knowledge of the code has been obtained only within the past 18 months.

The Genetic Messenger

It was recognized soon after the formulation of the Watson-Crick model of DNA that DNA itself might not be directly involved in the synthesis of protein, and that a template of RNA (ribonucleic acid) might be an intermediate in the process. Protein synthesis is conducted by cellular particles called ribosomes, which are about half protein and half RNA (ribosomal RNA). Several years ago Jacques Monod and François Jacob of the Pasteur Institute in Paris coined the term "messenger RNA" to describe the template RNA that carried genetic messages from DNA to the ribosomes.

A few years ago evidence for the enzymatic synthesis of RNA complementary to DNA was found by Jerard Hurwitz of the New York University School of Medicine, by Samuel Weiss of the University of Chicago, by Audrey Stevens of St. Louis University and their respective collaborators [see "Messenger RNA," by Jerard Hurwitz and J. J. Furth; SCIENTIFIC AMERICAN Offprint 119]. These groups, and others, showed that an enzyme, RNA polymerase, catalyzes the synthesis of strands of RNA on the pattern of strands of DNA.

RNA is similar to DNA except that RNA contains the sugar ribose instead of deoxyribose and the base uracil instead of thymine. When RNA is being formed on a DNA template, uracil appears in the RNA chain wherever adenine appears at the complementary site on the DNA chain. One fraction of the RNA formed by this process is messenger RNA; it directs the synthesis of protein. Messenger RNA leaves the nucleus of the cell and attaches to the ribosomes. The sequence of bases in the messenger RNA specifies the amino acid sequence in the protein to be synthesized.

The amino acids are transported to the proper sites on the messenger RNA by still another form of RNA called transfer RNA. Each cell contains a specific activating enzyme that attaches a specific amino acid to its particular transfer RNA. Moreover, cells evidently contain more than one kind of transfer RNA capable of recognizing a given amino acid. The significance of this fact will become apparent later. Although direct recognition of messenger RNA code words by transfer RNA molecules has not been demonstrated, it is clear that these molecules perform at least part of the job of placing amino acids in the proper position in the protein chain. When the amino acids arrive at the proper site in the chain, they are linked to each other by enzymic processes that are only partly understood. The linking is accomplished by the formation of a peptide bond: a chemical bond created when a molecule of water is removed from two adjacent molecules of amino acid. The process requires a transfer enzyme, at least one other enzyme and a cofactor: guanosine triphosphate. It appears that amino acid subunits are bonded into the growing protein chain one at a time, starting at the end of the chain carrying an amino group (NH_2)

and proceeding toward the end that terminates with a carboxyl group (COOH).

The process of protein synthesis can be studied conveniently in cell-free extracts of the colon bacillus (*Escherichia coli*). The bacteria grow rapidly in suitable nutrients and are harvested by sedimenting them out of suspension with a centrifuge. The cells are gently broken open by grinding them with finely powdered alumina [see illustration on page 324]; this releases the cell sap, containing DNA, messenger RNA, ribosomes, enzymes and other components. Such extracts are called cell-free systems, and when they are fortified with energy-rich substances (chiefly adenosine triphosphate), they readily incorporate amino acids into protein. The incorporation process can be followed by using amino acids containing carbon 14, a radioactive isotope of carbon.

Optimal conditions for protein synthesis in bacterial cell-free systems were determined by workers in many laboratories, notably Alfred Tissières of Harvard University, Marvin Lamborg and Paul C. Zamecnik of the Massachusetts General Hospital, G. David Novelli of the Oak Ridge National Laboratory and Sol Spiegelman of the University of Illinois. When we began our work at the National Institutes of Health, our

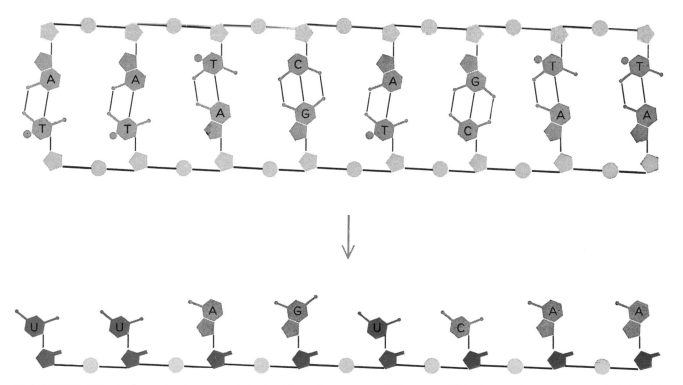

MESSENGER RNA is the molecular agent that transcribes the genetic code from DNA and carries it to the sites in the cell (the ribosomes) where protein synthesis takes place. The letters in messenger RNA are complementary to those in one strand of the DNA molecule. In this example UUAGUCAA is complementary to AATCAGTT. The exact mechanism of transcription is not known.

progress was slow because we had to prepare fresh enzyme extracts for each experiment. Later my colleague J. Heinrich Matthaei and I found a way to stabilize the extracts so that they could be stored for many weeks without appreciable loss of activity.

Normally the proteins produced in such extracts are those specified by the cell's own DNA. If one could establish the base sequence in one of the cell's genes—or part of a gene—and correlate it with the amino acid sequence in the protein coded by that gene, one would

be able to translate the genetic code. Although the amino acid sequence is known for a number of proteins, no one has yet determined the base sequence of a gene, hence the correlation cannot be performed.

The study of cell-free protein syn-

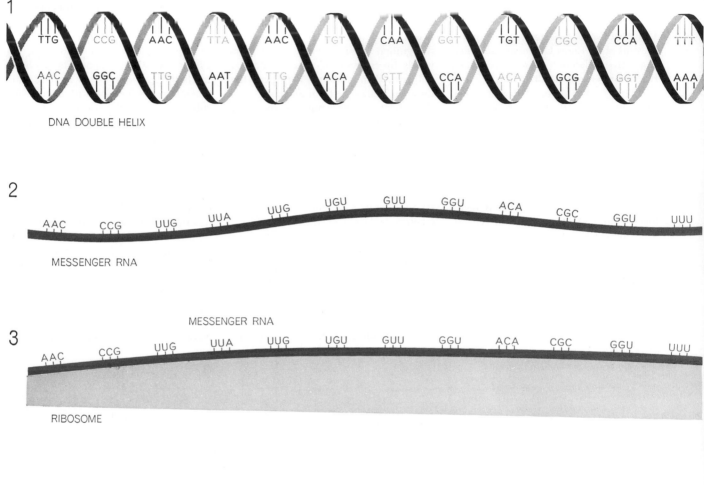

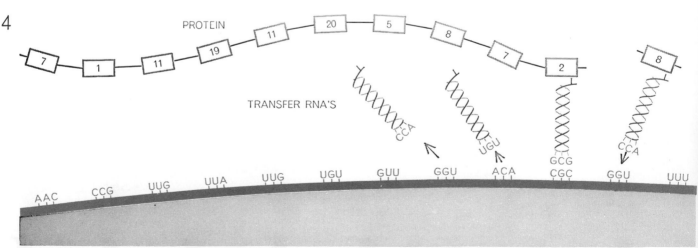

SYNTHESIS OF PROTEIN begins with the genetic code embodied in DNA (1). The code is transcribed into messenger RNA (2). In the diagram it is assumed that the message has been derived from the DNA strand bearing dark letters. The messenger RNA finds its way to a ribosome (3), the site of protein synthesis. Amino acids, indicated by numbered rectangles, are carried to proper sites on the messenger RNA by molecules of transfer RNA (*see illustration on opposite page*). Bases are actually equidistant, not

thesis provided an indirect approach to the coding problem. Tissières, Novelli and Bention Nisman, then at the Pasteur Institute, had reported that protein synthesis could be halted in cell-free extracts by adding deoxyribonuclease, or DNAase, an enzyme that specifically de-

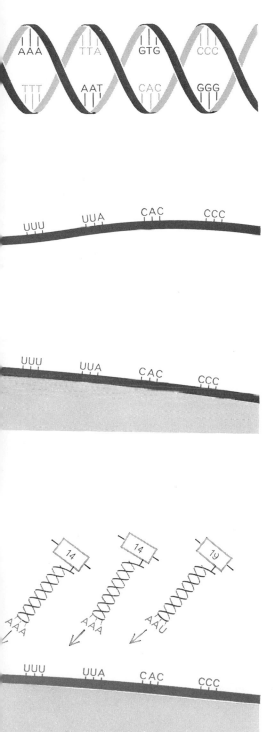

stroys DNA. Matthaei and I also observed this effect and studied its characteristics. It seemed probable that protein synthesis stopped after the messenger RNA had been depleted. When we added crude fractions of messenger RNA to such extracts, we found that they stimulated protein synthesis. The development of this cell-free assay for messenger RNA provided the rationale for all our subsequent work.

We obtained RNA fractions from various natural sources, including viruses, and found that many of them were highly active in directing protein synthesis in the cell-free system of the colon bacillus. The ribosomes of the colon bacillus were found to accept RNA "blueprints" obtained from foreign organisms, including viruses. It should be emphasized that only minute amounts of protein were synthesized in these experiments.

It occurred to us that synthetic RNA containing only one or two bases might direct the synthesis of simple proteins containing only a few amino acids. Synthetic RNA molecules can be prepared with the aid of an enzyme, polynucleotide phosphorylase, found in 1955 by Marianne Grunberg-Manago and Severo Ochoa of the New York University School of Medicine. Unlike RNA polymerase, this enzyme does not follow the pattern of DNA. Instead it forms RNA polymers by linking bases together in random order.

A synthetic RNA polymer containing only uracil (called polyuridylic acid, or poly-U) was prepared and added to the active cell-free system together with mixtures of the 20 amino acids. In each mixture one of the amino acids contained radioactive carbon 14; the other 19 amino acids were nonradioactive. In this way one could determine the particular amino acid directed into protein by poly-U.

It proved to be the amino acid phenylalanine. This provided evidence that the RNA code word for phenylalanine was a sequence of U's contained in poly-U. The code word for another amino acid, proline, was found to be a sequence of C's in polycytidylic acid, or poly-C. Thus a cell-free system capable of synthesizing protein under the direction of chemically defined preparations of RNA provided a simple means for translating the genetic code.

The Code-Word Dictionary

Ochoa and his collaborators and our group at the National Institutes of

grouped in triplets, and mechanism of recognition between transfer RNA and messenger RNA is hypothetical. Linkage of amino acid subunits creates a protein molecule.

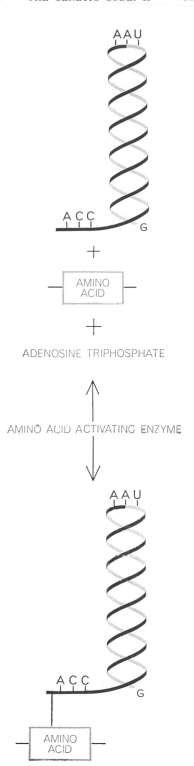

TRANSFER RNA is a special helical form of RNA that transports amino acids to their proper site in the protein chain. There is at least one transfer RNA for each of the 20 common amino acids. All, however, seem to carry the bases ACC where the amino acids attach and G at the opposite end. The attachment requires a specific enzyme and energy supplied by adenosine triphosphate. Unpaired bases in transfer RNA (AAU in the example) may provide the means by which the transfer RNA "recognizes" the place to deposit its amino acid package.

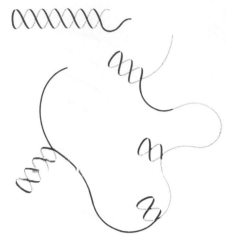

RNA STRUCTURE can take various forms. Transfer RNA (*top*) seems to be a fairly short double helix (probably less perfect than shown) that is closed at one end. Some RNA molecules contain a mixture of coiled and uncoiled regions (*bottom*).

Health, working independently, have now synthesized and tested polymers containing all possible combinations of the four RNA bases A, G, C and U. In the initial experiments only RNA polymers containing U were assayed, but recently many non-U polymers with high template activity have been found by M. Bretscher and Grunberg-Manago of the University of Cambridge, and also by Oliver W. Jones and me. All the results so far are summarized in the table at the bottom of pages 332 and 333. It lists the

RNA polymers containing the minimum number of bases capable of stimulating protein formation. The inclusion of another base in a polymer usually enables it to code for additional amino acids.

With only two kinds of base it is possible to make six varieties of RNA polymer: poly-AC, poly-AG, poly-AU, poly-CG, poly-CU and poly-GU. If the ratio of the bases is adjusted with care, each variety can be shown to code with great specificity for different sets of amino acids. The relative amount of one amino acid directed into protein compared with another depends on the ratio of bases in the RNA. Assuming a random sequence of bases in the RNA, the theoretical probabilities of finding particular sequences of two, three or more bases can be calculated easily if the base ratio is known. For example, if poly-UC contains 70 per cent U and 30 per cent C, the probability of the occurrence of the triplet sequence UUU is .7 × .7 × .7, or .34. That is, 34 per cent of the triplets in the polymer are expected to be UUU. The probability of obtaining the sequence UUC is .7 × .7 × .3, or .147. Thus 14.7 per cent of the triplets in such a polymer are probably UUC. This type of calculation, however, assumes randomness, and it is not certain that all the actual polymers are truly random.

It had been predicted by Gamow, Crick and others that for each amino acid there might be more than one code word, since there are 64 possible triplets and

only 20 amino acids. A code with multiple words for each object coded is termed degenerate. Our experiments show that the genetic code is indeed degenerate. Leucine, for example, is coded by RNA polymers containing U alone, or U and A, or U and C, or U and G.

It must be emphasized that degeneracy of this sort does not imply lack of specificity in the construction of proteins. It means, rather, that a specific amino acid can be directed to the proper site in a protein chain by more than one code word. Presumably this flexibility of coding is advantageous to the cell in ways not yet fully understood.

A molecular explanation of degeneracy has been provided recently in a striking manner. It has been known that some organisms contain more than one species of transfer RNA capable of recognizing a given amino acid. The colon bacillus, for example, contains two readily distinguishable species that transfer leucine. Bernard Weisblum and Seymour Benzer of Purdue University and Robert W. Holley of Cornell University separated the two leucine-transfer species and tested them in cell-free systems. They found that one of the species recognizes poly-UC but not poly-UG. The other species recognizes poly-UG but not poly-UC [*see top illustration on page 331*]. Although the number of transfer RNA species per cell is unknown, it is possible that each species corresponds to a different code word.

There is, however, the possibility of

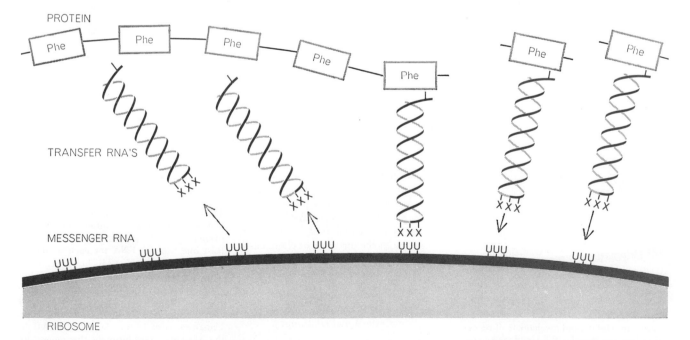

FIRST BREAK IN GENETIC CODE was the discovery that a synthetic messenger RNA containing only uracil (poly-U) directed the manufacture of a synthetic protein containing only one amino acid, phenylalanine (*Phe*). The finding was made by the author and J. Heinrich Matthaei. The X's in transfer RNA signify that the bases that respond to code words in messenger RNA are not known.

real ambiguity in protein synthesis. This would occur if one code word were to direct two or more kinds of amino acid into protein. So far only one such ambiguity has been found. Poly-U directs small amounts of leucine as well as phenylalanine into protein. The ratio of the two amino acids incorporated is about 20 or 30 molecules of phenylalanine to one of leucine. In the absence of phenylalanine, poly-U codes for leucine about half as well as it does for phenylalanine. The molecular basis of this ambiguity is not known. Nor is it known if the dual coding occurs in living systems as well as in cell-free systems.

Base sequences that do not encode for any amino acid are termed "nonsense words." This term may be misleading, for such sequences, if they exist, might have meaning to the cell. For example, they might indicate the beginning or end of a portion of the genetic message. An indirect estimate of the frequency of nonsense words can be obtained by comparing the efficiency of random RNA preparations with that of natural messenger RNA. We have found that many of the synthetic polymers containing four, three or two kinds of base are as efficient in stimulating protein synthesis as natural polymers are. This high efficiency, together with high coding specificity, suggests that relatively few base sequences are nonsense words.

In his article in *Scientific American* F. H. C. Crick presented the arguments for believing that the coding ratio is either three or a multiple of three. Recently we have determined the relative amounts of different amino acids directed into protein by synthetic RNA preparations of known base ratios, and the evidence suggests that some code words almost surely contain three bases. Yet, as the table at the bottom of the next two pages shows, 18 of the 20 amino acids can be coded by words containing only two different bases. The exceptions are aspartic acid and methionine, which seem to require some combination of U, G and A. (Some uncertainty still exists about the code words for these amino acids, because even poly-UGA directs very little aspartic acid or methionine into protein.) If the entire code indeed consists of triplets, it is possible that correct coding is achieved, in some instances, when only two out of the three bases read are recognized. Such imperfect recognition might occur more often with synthetic RNA polymers containing only one or two bases than it does with natural messenger RNA, which always contains a mixture of all four. The results obtained with synthetic RNA may dem-

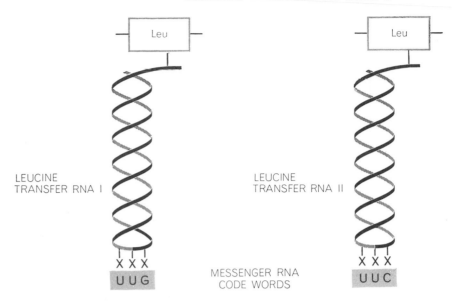

TWO KINDS OF TRANSFER RNA have been found, each capable of transporting leucine (*Leu*). One kind (*left*) recognizes the code word UUG; the other (*right*) recognizes UUC.

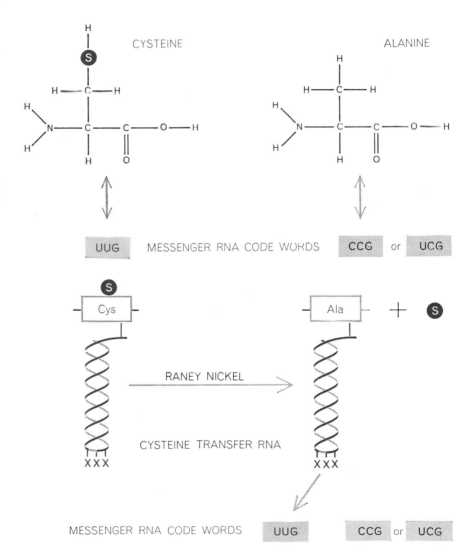

INGENIOUS EXPERIMENT showed that code-word recognition depends on the specificity of transfer RNA, not on the structure of the amino acid being transported. Cysteine is coded by UUG, alanine by CCG or UCG. Cysteine was hooked to its specific transfer RNA and sulfur was removed by a catalyst (Raney nickel). With sulfur removed from the molecule, cysteine became alanine. It was still directed into protein, however, as if it were cysteine.

SINGLET CODE (4 WORDS)	DOUBLET CODE (16 WORDS)				TRIPLET CODE (64 WORDS)			
					AAA	AAG	AAC	AAU
					AGA	AGG	AGC	AGU
					ACA	ACG	ACC	ACU
					AUA	AUG	AUC	AUU
					GAA	GAG	GAC	GAU
					GGA	GGG	GGC	GGU
A	AA	AG	AC	AU	GCA	GCG	GCC	GCU
G	GA	GG	GC	GU	GUA	GUG	GUC	GUU
C	CA	CG	CC	CU	CAA	CAG	CAC	CAU
U	UA	UG	UC	UU	CGA	CGG	CGC	CGU
					CCA	CCG	CCC	CCU
					CUA	CUG	CUC	CUU
					UAA	UAG	UAC	UAU
					UGA	UGG	UGC	UGU
					UCA	UCG	UCC	UCU
					UUA	UUG	UUC	UUU

CODE-LETTER COMBINATIONS increase sharply with the length of the code word. Since at least 20 code words are needed to identify the 20 common amino acids, the minimum code length is a sequence of three letters, assuming that all words are the same length.

onstrate the coding potential of the cell; that is, it may reveal code words that function routinely in the living cell and potential words that would be recognized if appropriate mutations were to occur in the cellular DNA. The table on page 12 summarizes the code-word dictionary on the assumption that all code words are triplets.

The Universality of the Code

Does each plant or animal species have its own genetic code, or is the same genetic language used by all species on this planet? Preliminary evidence suggests that the code is essentially universal and that even species at opposite ends of the evolutionary scale use much the same code. For instance, a number of laboratories in the U.S. and England have recently reported that synthetic RNA polymers code the same way in mammalian cell-free systems as they do in the bacterial system. The base compositions of mammalian code words corresponding to about six amino acids have been determined so far. It nevertheless seems probable that some differences may be found in the future. Since certain amino acids are coded by multiple words, it is not unlikely that one species may use one word and another species a different one.

An indirect check on the validity of code words obtained in cell-free systems can be made by studying natural proteins that differ in amino acid composition at only one point in the protein chain. For example, the hemoglobin of an individual suffering from "sickle cell" anemia differs from normal hemoglobin in that it has valine at one point in the chain instead of glutamic acid. Another abnormal hemoglobin has lysine at the same point. One might be able to show, by examining the code-word dictionary, that these three amino acids—glutamic acid, valine and lysine—have similar code words. One could then infer that the two

abnormal hemoglobins came into being as a result of a mutation that substituted a single base for another in the gene that controls the production of hemoglobin. As a matter of fact, the code-word dictionary shows that the code words are similar enough for this to have happened. One of the code groups for glutamic acid is AGU. Substitution of a U for A produces UGU, the code group for valine. Substitution of an A for a U yields AGA, one of the code groups for lysine. Similar analyses have been made for other proteins in which amino acid substitutions are known, and in most cases the substitutions can be explained by alteration of a single base in code-word triplets. Presumably more code words will be found in the future and the correlation between genetic base sequences and amino acid sequences can be made with greater assurance.

The Nature of Messenger RNA

Does each molecule of messenger RNA function only once or many times in directing the synthesis of protein? The question has proved difficult because most of the poly-U in the experimental system is degraded before it is able to function as a messenger. We have found, nevertheless, that only about 1.5 U's in poly-U are required to direct the incorporation of one molecule of phenylalanine into protein. And George Spyrides and Fritz A. Lipmann of the Rockefeller Institute have reported that only about .75 U's are required per molecule of amino acid in their studies. If the coding is done by triplets, three U's would be required if the messenger functioned only once. Evidently each poly-U molecule directs the synthesis of more than one long-chain molecule of polyphenylalanine. Similar results have been obtained in intact cells. Cyrus Levinthal and his associates at the Massachusetts Institute

	U	A	C	G
AMINO ACIDS CODED	PHENYLALANINE	LYSINE	PROLINE●	
	LEUCINE■			

■ POLY U CODES PREFERENTIALLY FOR PHENYLALANINE
● REPORTED BY ONLY ONE LABORATORY; STILL TO BE CONFIRMED
▲ REQUIRES ONLY FIRST OF TWO BASES LISTED
△ REQUIRES ONLY SECOND OF TWO BASES LISTED

SPECIFICITY OF CODING is shown in this table, which lists 18 amino acids that can be coded by synthetic RNA polymers containing no more than one or two kinds of base. The only amino acids that seem to require more than two bases for coding are aspartic acid and methionine, which need U, A and G. The relative amounts of amino acids directed into

of Technology inhibited messenger RNA synthesis in living bacteria with the antibiotic actinomycin and found that each messenger RNA molecule present at the time messenger synthesis was turned off directed the synthesis of 10 to 20 molecules of protein.

We have observed that two factors in addition to base sequence have a profound effect on the activity of messenger RNA: the length of the RNA chain and its over-all structure. Poly-U molecules that contain more than 100 U's are much more active than molecules with fewer than 50. Robert G. Martin and Bruce Ames of the National Institutes of Health have found that chains of poly-U containing 450 to 700 U's are optimal for directing protein synthesis.

There is still much to be learned about the effect of structure on RNA function. Unlike DNA, RNA molecules are usually single-stranded. Frequently, however, one part of the RNA molecule loops back and forms hydrogen bonds with another portion of the same molecule. The extent of such internal pairing is influenced by the base sequence in the molecule. When poly-U is in solution, it usually has little secondary structure; that is, it consists of a simple chain with few, if any, loops or knots. Other types of RNA molecules display a considerable amount of secondary structure [see top illustration on page 330].

We have found that such a secondary structure interferes with the activity of messenger RNA. When solutions of poly-U and poly-A are mixed, they form double-strand (U-A) and triple-strand (U-A-U) helices, which are completely inactive in directing the synthesis of polyphenylalanine. In collaboration with Maxine F. Singer of the National Institutes of Health we have shown that poly-UG containing a high degree of ordered secondary structure (possibly due to G-G hydrogen-bonding) is unable to code for amino acids.

It is conceivable that natural messenger RNA contains at intervals short regions of secondary structure resembling knots in a rope. These regions might signify the beginning or the end of a protein. Alternative hypotheses suggest that the beginning and end are indicated by particular base sequences in the genetic message. In any case it seems probable that the secondary structure assumed by different types of RNA will be found to have great influence on their biological function.

The Reading Mechanism

Still not completely understood is the manner in which a given amino acid finds its way to the proper site in a protein chain. Although transfer RNA was found to be required for the synthesis of polyphenylalanine, the possibility remained that the amino acid rather than the transfer RNA recognized the code word embodied in the poly-U messenger RNA.

To distinguish between these alternative possibilities, a brilliant experiment was performed jointly by François Chapeville and Lipmann of the Rockefeller Institute, Günter von Ehrenstein of Johns Hopkins University and three Purdue workers: Benzer, Weisblum and William J. Ray, Jr. One amino acid, cysteine, is directed into protein by poly-UG. Alanine, which is identical with cysteine except that it lacks a sulfur atom, is directed into protein by poly-CG or poly-UCG. Cysteine is transported by one species of transfer RNA and alanine by another. Chapeville and his associates enzymatically attached cysteine, labeled with carbon 14, to its particular type of transfer RNA. They then exposed the molecular complex to a nickel catalyst, called Raney nickel, that removed the sulfur from cysteine and converted it

to alanine—without detaching it from cysteine-transfer RNA. Now they could ask: Will the labeled alanine be coded as if it were alanine or cysteine? They found it was coded by poly-UG, just as if it were cysteine [see bottom illustration on page 331]. This experiment shows that an amino acid loses its identity after combining with transfer RNA and is carried willy-nilly to the code word recognized by the transfer RNA.

The secondary structure of transfer RNA itself has been clarified further this past year by workers at King's College of the University of London. From X-ray evidence they have deduced that transfer RNA consists of a double helix very much like the secondary structure found in DNA. One difference is that the transfer RNA molecule is folded back on itself, like a hairpin that has been twisted around its long axis. The molecule seems to contain a number of unpaired bases; it is possible that these provide the means for recognizing specific code words in messenger RNA [see illustration at right on page 329].

There is still considerable mystery about the way messenger RNA attaches to ribosomes and the part that ribosomes play in protein synthesis. It has been known for some time that colon bacillus ribosomes are composed of at least two types of subunit and that under certain conditions they form aggregates consisting of two subunits (dimers) and four subunits (tetramers). In collaboration with Samuel Barondes, we found that the addition of poly-U to reaction mixtures initiated further ribosome aggregation. In early experiments only tetramers or still larger aggregates supported the synthesis of polyphenylalanine. Spyrides and Lipmann have shown that poly-U makes only certain "active" ribosomes aggregate and that the remaining monomers and dimers do not support polyphenylalanine syn-

BASES PRESENT IN SYNTHETIC RNA

UA	UC	UG	AC	AG	CG
PHENYLALANINE ▲	PHENYLALANINE ▲	PHENYLALANINE ▲	LYSINE ▲	LYSINE ▲	PROLINE ▲
LYSINE △	PROLINE △	LEUCINE	PROLINE △	GLUTAMIC ACID	ARGININE ●
TYROSINE	LEUCINE	VALINE	HISTIDINE	ARGININE ●	ALANINE ●
LEUCINE	SERINE	CYSTEINE	ASPARAGINE	GLUTAMINE ●	
ISOLEUCINE		TRYPTOPHAN	GLUTAMINE	GLYCINE ●	
ASPARAGINE ●		GLYCINE	THREONINE		

protein by RNA polymers containing two bases depend on the base ratios. When the polymers contain a third and fourth base, additional kinds of amino acids are incorporated into protein. Thus the activity of poly-UCG (an RNA polymer containing U, C and

G) resembles that of poly-UC plus poly-UG. Poly-G has not been found to code for any amino acid. Future work will undoubtedly yield data that will necessitate revisions in the table. An RNA-code-word dictionary derived from the table appears on page 334.

AMINO ACID	RNA CODE WORDS			
ALANINE	CCG	UCG ■		
ARGININE	CGC	AGA	UCG ■	
ASPARAGINE	ACA	AUA		
ASPARTIC ACID	GUA			
CYSTEINE	UUG △			
GLUTAMIC ACID	GAA	AGU ■		
GLUTAMINE	ACA	AGA	AGU ■	
GLYCINE	UGG	AGG		
HISTIDINE	ACC			
ISOLEUCINE	UAU	UAA		
LEUCINE	UUG	UUC	UUA	UUU □
LYSINE	AAA	AAG ●	AAU ●	
METHIONINE	UGA ■			
PHENYLALANINE	UUU			
PROLINE	CCC	CCU ▲	CCA ▲	CCG ▲
SERINE	UCU	UCC	UCG	
THREONINE	CAC	CAA		
TRYPTOPHAN	GGU			
TYROSINE	AUU			
VALINE	UGU			

△ UNCERTAIN WHETHER CODE IS UUG OR GGU

■ NEED FOR U UNCERTAIN

□ CODES PREFERENTIALLY FOR PHENYLALANINE

● NEED FOR G AND U UNCERTAIN

▲ NEED FOR U A G UNCERTAIN

GENETIC-CODE DICTIONARY lists the code words that correspond to each of the 20 common amino acids, assuming that all the words are triplets. The sequences of the letters in the code words have not been established, hence the order shown is arbitrary. Although half of the amino acids have more than one code word, it is believed that each triplet codes uniquely for a particular amino acid. Thus various combinations of AAC presumably code for asparagine, glutamine and threonine. Only one exception has been found to this presumed rule. The triplet UUU codes for phenylalanine and, less effectively, for leucine.

thesis.

A possibly related phenomenon has been observed in living cells by Alexander Rich and his associates at the Massachusetts Institute of Technology. They find that in reticulocytes obtained from rabbit blood, protein synthesis seems to be carried out predominantly by aggregates of five ribosomes, which may be held together by a single thread of messenger RNA. They have named the aggregate a polysome.

Many compelling problems still lie ahead. One is to establish the actual sequence of bases in code words. At present the code resembles an anagram. We know the letters but not the order of most words.

Another intriguing question is whether in living cells the double strand of DNA serves as a template for the production of a single strand of messenger RNA, or whether each strand of DNA serves as a template for the production of two different, complementary strands of RNA. If the latter occurs—and available evidence suggests that it does—the function of each strand must be elucidated.

Ultimately one hopes that cell-free systems will shed light on genetic control mechanisms. Such mechanisms, still undiscovered, permit the selective retrieval of genetic information. Two cells may contain identical sets of genes, but certain genes may be turned on in one cell and off in another in highly specific fashion. With cell-free systems the powerful tools of enzymology can be brought to bear on these and other problems, with the promise that the molecular understanding of genetics will continue to advance rapidly in the near future.

37

THE GENETIC CODE OF A VIRUS

HEINZ FRAENKEL-CONRAT
October 1964

The tobacco mosaic virus consists of hereditary material and a single protein. Artificial changes in the hereditary material elucidate how it directs the synthesis of the three-dimensional molecule of protein

A few years ago it seemed that the virus that causes the mosaic disease of tobacco plants might serve as a Rosetta stone for deciphering the genetic code. The sequence of amino acid subunits in the protein that forms the coat of the tobacco mosaic virus was almost completely established. The ribonucleic acid (RNA) of the virus was believed to carry the coded information needed for the construction of this one protein. It was hoped that it would not be too difficult to work out the sequence of nucleotide subunits in the RNA; the code could then be deduced directly by matching up the sequence of amino acids in the protein with the sequence of groups of nucleotides, or "words," in the RNA molecule. Such a translation would go a long way toward disclosing how the chainlike molecules of RNA and deoxyribonucleic acid (DNA) are able to direct the construction of three-dimensional living cells.

Unhappily (or happily, as some prefer to look at it) nature does not make things easy by providing simple Rosetta stones. The RNA molecule of the tobacco mosaic virus turned out to be an enormously long chain composed of 6,400 nucleotide subunits. Furthermore, the molecule bears various messages: it carries directions for synthesizing not only the coat protein but also other proteins, that is, certain enzymes. Thus it has become apparent that decipherment of its language will be a complicated task. This article will describe some current attacks on the problem and the progress that has been made.

The recent advances in chemical genetics have been recounted in several articles in *Scientific American* [see "The Genetic Code," by F. H. C. Crick, Offprint 123, and "The Genetic Code: II," by Marshall W. Nirenberg in this book,

page 324]. I need review here only a few of the principal features of the chemical machinery for the reproduction of viruses. The tobacco mosaic virus consists of a long strand of RNA wrapped in a coat of protein [*see illustration on this page*]. The 6,400 nucleotide subunits of the RNA are of four kinds: guanine (G), cytosine (C), adenine (A) and uracil (U). After the viral RNA has invaded a tobacco leaf cell, it reproduces by acting as a template for the formation of complementary chains. Each guanine in the original chain combines with a cytosine from the intracellular environment and each cytosine with a guanine; similarly, each adenine combines with a uracil and each uracil with an adenine. Each of the subunits includes a ribose group and a phosphate group; the subunits are linked by bridges of phosphate. When a duplicate chain has been formed on the template, it peels off and is ready to combine with coat protein to form a new virus particle. The process requires the catalytic assistance of at least one enzyme—more probably two. It used to be supposed that the host cell supplied these enzymes ready-made, but it has now been established that the viral RNA directs the synthesis of the enzymes, using the cell's amino acids as the building material.

The coat protein of the tobacco mosaic virus has 158 amino acid subunits. If we suppose the RNA code word, or "codon," for each amino acid consists of three nucleotides, then a chain of 474 nucleotides would suffice to provide the information for synthesizing the coat protein. Adding the message required for synthesizing an enzyme (presumably a larger molecule than the coat protein) could raise the requirement to a chain of about 1,500 nucleotides. The fact that the RNA molecule of the virus is four times longer suggests that it probably directs the synthesis of more than one molecule of enzyme or other protein.

To decipher the code embodied in the virus's RNA we are confronted, then, with a molecule 6,400 nucleotides long carrying a series of different messages along its length. A beginning toward analysis of the structure of this molecule has been made by chopping off the nucleotides at the ends of the chain one by one with enzymes and alkalies [*see illustration on page 338*]. The RNA chain is so long, however, that this method is not likely to get us very far toward determining the full structure of the molecule or deciphering its code.

The specific topic of this article is

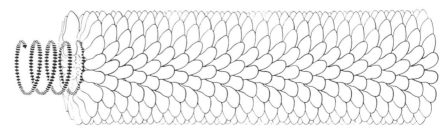

TOBACCO MOSAIC VIRUS has a coat of protein molecules (*the radially arranged white structures*) surrounding a strand of ribonucleic acid (RNA), represented by the black helix.

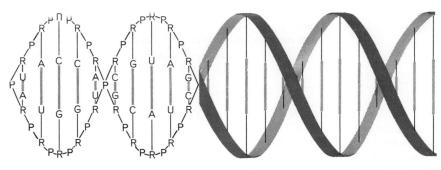

RIBONUCLEIC ACID is represented as a chain of repeating units of ribose sugar and phosphate. Extending from each ribose is a base: guanine, cytosine, adenine or uracil.

VIRAL RNA replicates by forming a double helix. The tobacco mosaic virus helix would consist of 6,400 subunits of guanine, cytosine, adenine and uracil (G, C, A and U).

CYTOSINE → NITROUS ACID → URACIL

ADENINE → NITROUS ACID → HYPOXANTHINE

GUANINE → NITROUS ACID → XANTHINE

CHEMICAL MUTATION occurs when nitrous acid causes a replacement of the amino group (NH_2) in cytosine and adenine with the hydroxyl group (OH). The respective products of such mutation, uracil and hypoxanthine, appear at top and middle right. Guanine does not carry its amino group at a corresponding site on the molecule. Since the colored parts of the molecules are those interacting during replication or in double-strand molecules such as helical RNA, no mutant RNA results from the mutation of guanine to xanthine.

another approach to the problem, which, although more roundabout, has yielded some highly rewarding results. This method, pursued in our laboratory at the University of California at Berkeley and in several others, consists in making slight changes in the chemical structure of the RNA and then observing what effect these changes have on its genetic activity. Roughly speaking, the strategy is analogous to changing a letter or two in a verbal message to see how it changes the sense of the message.

Of the various reagents used to change the RNA, by far the most useful has proved to be nitrous acid. A particle of tobacco mosaic virus treated with this chemical is often so changed that it produces different disease symptoms in the tobacco plant it infects. Its new properties are transmitted to its progeny. Frequently the behavior of the altered strain of virus resembles that of a known natural strain—a known mutant—of the virus. Clearly in such cases the change in the treated virus represents a genuine mutation.

A study of the reaction between nitrous acid and the nucleotides of RNA shows how the mutation is brought about. Nitrous acid causes the replacement of the amino group (NH_2) in a nucleotide with a hydroxyl group (OH). In the case of cytosine this results in transformation of the cytosine to uracil [see bottom illustration on this page]. The conversion of one, two or three cytosines in the RNA chain into uracils may well convert the virus to a viable new strain that produces somewhat different disease symptoms. In the case of adenine the deamination by nitrous acid changes the adenine to hypoxanthine, a nucleotide that is not normally present in RNA but that is like guanine in part of its structure and therefore can combine with cytosine. Because adenine is thus converted to a base that resembles guanine (combining with cytosine instead of uracil), this change may sometimes result in a mutant virus. No mutation results, however, when guanine is deaminated to xanthine, because xanthine behaves like guanine itself; that is, it links up with cytosine as guanine does.

It has been established, then, that a localized change in the nucleotide composition of the viral RNA can produce a noticeable change in the activity of the virus. This circumstance does not provide a means for locating the changed nucleotides in the RNA's of different strains of tobacco mosaic virus. In a molecule with 6,400 nucleotides, comprising between 1,100 and 1,800 of

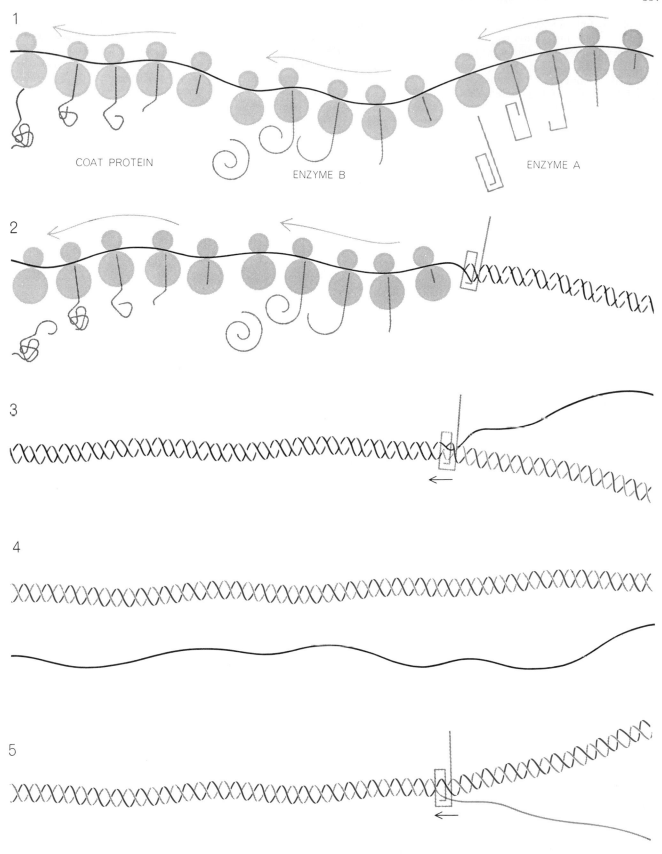

WITHIN THE CELL viral RNA (*black strand at top*) attaches to ribosomes (*double balls*) that travel along the strand synthesizing peptide chains of various proteins. Products of the process might include the material of the virus coat and two enzymes, represented by differently shaped structures emerging in the first step. In the second step the single RNA strand binds to an enzyme that catalyzes production of a complementary strand (*dark colored line*). This double strand RNA is shown uncurling (*step 3*), again under the influence of an enzyme. It makes a new strand (*light colored line in step 3*) and releases the original one. The first complementary strand then releases its progeny strand (*step 4*) and makes another (*gray curve in step 5*) under the influence of an enzyme. A released strand either can make protein as in step 1, or undergo a process of replication as in steps 2 through 4.

each of the four nucleotide types, the problem of detecting a difference of just a few nucleotides between one strain and another is beyond any present analytical technique. We therefore turned to studies of the proteins synthesized by artificially altered RNA.

The protein coat wrapped around the

RNA core of the tobacco mosaic virus consists of nearly 2,200 molecules. Thus any change in the RNA that is reflected in the construction of the protein is amplified many times, and the change in the structure of the protein should be comparatively easy to detect. Moreover, a change in just one of the 16

different kinds of amino acid composing the protein amounts to a change of 5 percent or more in the total amino acid composition of the molecule. This composition was known to vary considerably even among natural strains of tobacco mosaic virus, and it could be assumed that such variations in the

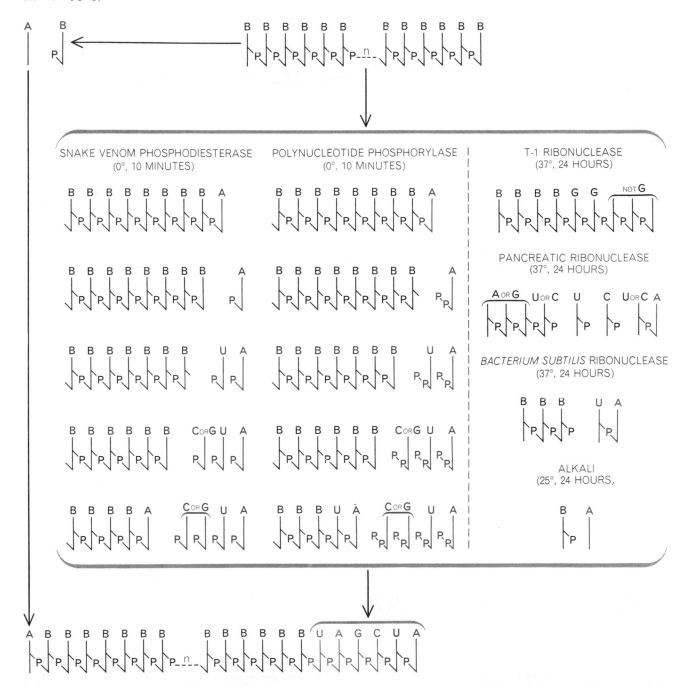

END GROUPS removed from RNA provide clues to its structure. RNA appears at top right as a chain of bases (B), phosphates (P) and ribose (*symbolized by vertical lines*), the approximate length of which, 6,400 subunits, is denoted by n (for number of nucleotides). The method of removing an end group is given above the evidence thus gathered. The certain presence of guanine, cytosine, adenine or uracil at a given site is shown in dark color. If experiments only narrow the possibilities, light color is used. The effects of an enzyme from snake venom, phosphodiesterase, are shown in five stages, during which end groups are successively broken off. The effect of polynucleotide phosphorylase is similarly represented. In the RNA at bottom the identified nucleotides are dark-colored.

chemically modified strains would be easily and accurately measurable. Systematic programs of analysis of the proteins in mutant tobacco mosaic viruses were undertaken in our laboratory by two visiting Japanese workers, Akira Tsugita and Gunku Funatsu, and at the University of Tübingen by Hans Wittmann.

Most of the studies undertaken in our laboratory made use of virus mutants that produce unusual lesions in tobacco plants of the variety *Nicotiana sylvestris*. The common tobacco mosaic virus causes a discoloration and a distortion of leaves that spreads over the entire plant. A typical mutant we have investigated causes only local, walled-off lesions at the sites of inoculation. Altogether some 200 chemically induced mutants have been studied in the two laboratories. In general both laboratories agree in their findings concerning the changes in the amino acids of these mutants' coats, but they differ in some respects, on which I shall comment later.

Of the 200 mutants, about 120 apparently still had the same coat protein as they had had before mutation. At least the protein's overall amino acid composition was unchanged. Although it is possible that changes of the amino acids within the molecule might have been masked by alterations in one direction offsetting those in another, we can safely dismiss this explanation as highly unlikely, in view of the very tiny probability that such precisely balancing changes would take place in more than half of the mutants. Of the approximately 80 mutants that did show a change in the coat protein, nearly all were altered in only one, two or occasionally three amino acids. (There were a few that differed radically—in as many as 30 amino acids—from the common strain of the tobacco mosaic virus; possibly these were not artificial mutants but uncommon strains that happened to be present in the inoculated material.)

Certain patterns showed up in the mutations. All the changes were one-way; for example, there were many cases of conversion of the amino acid serine to the amino acid phenylalanine but not a single instance of transformation of phenylalanine to serine. Most interesting was the fact that of the 272 possible conversions of one amino acid to another (among the 16 present in the virus) only 21 actually occurred, and of these transformations only 14 showed up more than once [*see illustration on next two pages*]. Let us now consider

LEAVES from the tobacco plant *Nicotiana sylvestris* are compared. At top is a leaf free from infection. Below it is a leaf generally discolored by infection with tobacco mosaic virus. At bottom is a leaf with local lesions, an effect caused by some 200 mutant strains.

what interpretations we can extract from the results.

In their work on the genetic code Marshall W. Nirenberg and his associates at the National Institutes of Health had connected certain nucleotide combinations, or code words, with specific amino acids (as Nirenberg explained in the *Scientific American* article to which I have already referred). He had found that a synthetic RNA consisting only of uracils caused just the amino acid phenylalanine to form a chain, from which he concluded that the code word for incorporating phenylalanine into a protein was UUU. Similarly, an RNA-like molecule composed only of cytosine (thus constituting the codon CCC) carried specific instructions for polymerizing the amino acid proline. A combination of two parts of

uracil with one part of cytosine, forming a codon containing two U's and one C in some unknown order, directed the polymerization of leucine; a switch in the nucleotide proportions to one part of uracil and two parts of cytosine, that is, to a codon containing one U and two C's, favored the polymerization of serine.

Applying these findings to our virus mutants, we found that the two corresponded remarkably well. Let us say that the treatment of the virus with nitrous acid deaminated one of the cytosines in its RNA and thereby changed it to uracil. This might alter a CCC codon to one containing a U and two C's, which would result in the replacement of a proline by a leucine in the protein coat of the virus. In the same way a CUU sequence would be changed to UUU, which would lead to

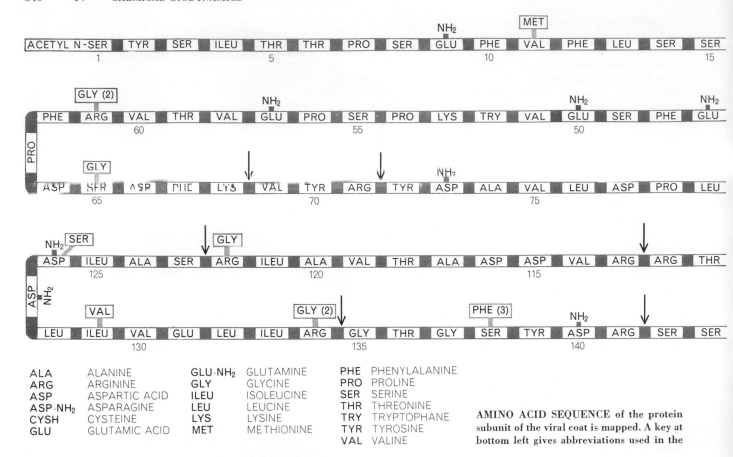

ALA	ALANINE	GLU-NH₂	GLUTAMINE	PHE	PHENYLALANINE
ARG	ARGININE	GLY	GLYCINE	PRO	PROLINE
ASP	ASPARTIC ACID	ILEU	ISOLEUCINE	SER	SERINE
ASP-NH₂	ASPARAGINE	LEU	LEUCINE	THR	THREONINE
CYSH	CYSTEINE	LYS	LYSINE	TRY	TRYPTOPHANE
GLU	GLUTAMIC ACID	MET	METHIONINE	TYR	TYROSINE
				VAL	VALINE

AMINO ACID SEQUENCE of the protein subunit of the viral coat is mapped. A key at bottom left gives abbreviations used in the

the replacement of a serine in the protein by a phenylalanine. As we have noted, in our mutants there was often a change from serine to phenylalanine or from proline to leucine, but it was never the other way around.

Studying the protein-building effects of various nucleotide combinations, Nirenberg and others have steadily enlarged the codon dictionary. Unfortunately the dictionary has grown in ambiguity as it has grown in size. Some amino acids apparently can be coded by as many as five different codons, or nucleotide triplets; leucine, for example, has been found to be represented by five codons and serine by four. This indicates that the code is highly ambiguous, or "degenerate." The reasons for this ambiguity in the genetic language remain obscure and certainly hide complexities still not understood. At all events, Wittmann has proposed a scheme of step-by-step transformations of the codons for four amino acids that illustrates the degeneracy of the code and may provide a way to determine the sequence of the nucleotides in each codon, which is not yet known [*see top illustration on page 343*].

The table at the bottom of page 10 summarizes all the amino acid changes in artificial mutants that have been observed in our laboratory and in Witt-

mann's at Tübingen. It includes cases in which the RNA nucleotides were altered by the attachment of a bromine atom or a methyl group instead of by deamination. The summary shows that all the amino acid transformations that occurred more than once in our laboratory can be accounted for either by a conversion of cytosine to uracil or by a change of adenine to guanine by way of hypoxanthine. But it also raises some puzzling questions. What about those cases, particularly some of the transformations obtained repeatedly by Wittmann, that cannot be explained by such conversions? By what chemical mechanism can methylation or bromination give rise to the same amino acid replacements, even though they do not affect the cytosine or adenine as deamination does? These questions are still unanswered.

Pursuing another line of investigation, Nirenberg initiated an intriguing experiment in collaboration with our laboratory. His system for exploring the coding effects of RNA uses a medium containing extracts from cells of the bacterium *Escherichia coli*. It contains ribosomes, enzymes, adenosine triphosphate (as the energy source) and amino acids attached to "transfer" RNA—in short, all the apparatus needed for the synthesis of proteins or polypeptides with the ex-

ception of the genetic material itself. The addition to this system of any type of RNA or DNA, even a synthetic RNA of the simplest kind, will bring about the linking of amino acids into chains in a sequence specified by the added RNA. Would the tobacco mosaic virus RNA induce the system to synthesize the tobacco mosaic virus coat protein?

The experiment was undertaken, and our first interpretation of the results was guardedly optimistic. This interpretation, however, was later corrected; there was no evidence that the system produced any identifiable tobacco mosaic virus protein. On the other hand, Daniel Nathans and his colleagues at the Rockefeller Institute found that the RNA of a virus that attacks *E. coli* would cause a cell-free extract of the bacterium to produce the coat protein of that virus. The most plausible explanation for this discrepancy seems to be that the code is too ambiguous, or degenerate, to carry over from a plant to a bacterial system. The tobacco plant cells and the *E. coli* system may preferentially use different codons to represent a given amino acid, and therefore a message may become garbled when it is transferred from one system to the other. It is as if the genetic code, although universal in principle, contained varying dialects, the cells of different species

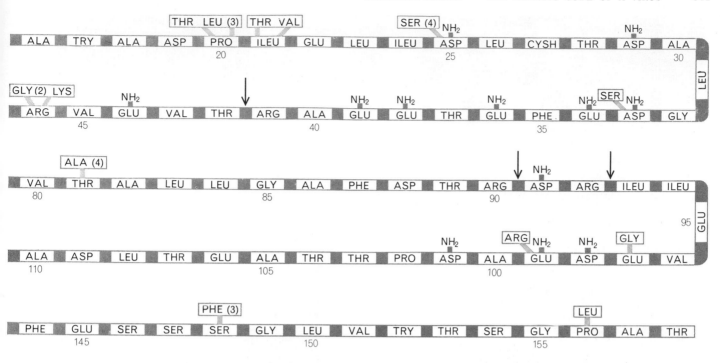

map. When the protein is decomposed by the enzyme trypsin, it fragments at sites marked by arrows. Replacements of amino acids occur in chemically induced mutants as shown in rectangles above sites at which they appear, with numbers indicating instances of observation. The protein coat of a virus particle has nearly 2,200 subunits. Its RNA, now under study, contains three times as many.

using different versions of the general language.

Apart from what our virus mutants may reveal about the genetic code, they interest us profoundly for another reason. The experiments in controlled alteration of their protein coats promise to help unravel the three-dimensional structure of the protein. The elucidation of protein structure is widely recognized as a central problem in biology. Next to the machinery of heredity, the three-dimensional structure of proteins holds perhaps the most important key to all the processes of life. Molecular structure determines the activities of enzymes in catalyzing biochemical reactions, of antibodies in precipitating foreign substances, of protein hormones and other specific proteins in regulating metabolism; in sum, it is a prime factor accounting for the properties and functions of all proteins from those in the coat of a virus to those in the cells of the brain. Each protein is characterized by specific internal bonds that maintain its three-dimensional form and by special surfaces that selectively bind to it certain ions, simple molecules or other proteins.

The spatial organization of a few proteins has been worked out by studying the patterns of X rays diffracted by crystals of the proteins in a dry state.

Such analyses do not necessarily show what form the proteins take in their natural condition in water solution, nor do they throw light on the nature of the internal bonds that maintain the protein molecule's shape. These questions are being investigated indirectly in many laboratories by gentle chemical methods probing the reactivity of specific protein molecules at various points in the molecular chain. Along this line our virus mutants have provided helpful information.

The protein coat of the tobacco mosaic virus performs certain definite functions in protecting the integrity and promoting the infectivity of the virus, and we have studied these functions in detail in our laboratory [see "Rebuilding a Virus," by Heinz Fraenkel-Conrat; SCIENTIFIC AMERICAN Offprint 9]. One can assume that the protein coat of the common strain of the virus, as it has evolved by natural selection, is highly efficient, and that any mutation is likely to reduce the virus's viability. We were therefore interested in seeing just how and to what extent each chemically induced change in the amino acid sequence would affect the virus.

As I have mentioned, almost the entire sequence of the 158 amino acids in this protein was known. It was also known that, when one attacked the protein coat of the common tobacco mosaic virus with an enzyme that removes amino acids from the carboxyl (COOH) end of a protein chain, it was able to chop off only a single amino acid, threonine, at the very end of the chain. Surprisingly, it turned out that this amputation (removing a total of 2,200 threonine units from the 2,200 protein molecules forming the coat of the virus) did not markedly affect the biological properties of the virus. Our very first mutant, however, showed a dramatic increase in vulnerability. As it happened, this mutation had replaced a proline near the end of the chain (No. 156 in the sequence) with leucine. The change made the protein much more susceptible to digestion by the enzyme. The enzyme was now able to clip three amino acids off the protein in the virus (and many more than three when it attacked the protein alone, stripped away from the virus). The three-amino-acid amputation made the virus distinctly less viable. This showed clearly that a single mutation, producing only an apparently minor change in the protein, could greatly reduce the virus's chances of survival.

Later studies have indicated that some RNA mutations render the RNA incapable even of forming the protein

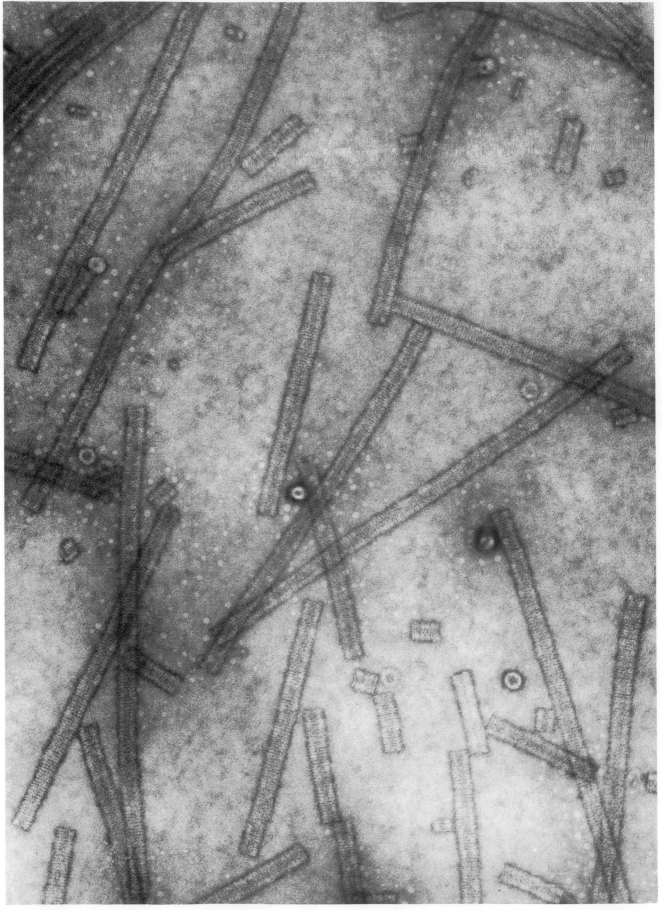

RODLIKE STRANDS of the tobacco mosaic virus are magnified 750,000 times in this electron micrograph made by H. L. Nixon of the Rothamsted Experimental Station in England. Only the larger particles are full-length viruses, capable of infecting cells.

coat. Most often, however, the mutant proteins that were examined showed exchanges of amino acids only near the ends of the chain, where the alteration might do the least damage to the functioning of the protein.

In contrast to the chain ends and certain inner parts around the middle of the chain, which also showed exchanges, there is a certain segment (between No. 108 and No. 122 in the sequence) that strongly resists attack. This part of the chain is the same in all natural strains of the virus and remains unchanged in all the mutants that have been investigated. Its stability suggests that it constitutes a portion of the molecule that is particularly important for the proper folding of the chain. In this segment there is a pair of arginines close to a pair of aspartic acid units, and it may be that these amino acid pairs play a role in the folding.

It should be recognized that the frequently recurring exchanges are probably not the result of mutations occurring preferentially at these sites, but are due to natural selection. It appears certain that these are sites where exchanges cause the least harm to the function of the protein of forming a protective shell for the RNA. Thus the exchanges greatly predominate among those mutants that are viable enough to be isolated in amounts sufficient for chemical study. A change of a serine to a phenylalanine elsewhere in the molecule than at positions No. 138 or No. 148 presumably renders the protein nonfunctional.

By means of chemical probing and genetic mutation the entire protein molecule is being explored for clues to its three-dimensional structure. Such clues include the distances between specific groups in the chain and the chemical reactivity of the various parts of the chain. It can be deduced, for instance, that units in the chain that resist reaction with applied reagents are likely to be inside folds where they are tied up in internal bonding. One such probe has shown that the tyrosine at position No. 72 in the sequence is remarkably recalcitrant to reaction with any chemical applied to the intact virus; the tyrosine at position No. 139, on the other hand, readily reacts with iodine.

Gradually, through genetic and chemical soundings of this kind, we hope to build up a complete picture of the protein bonding and structure that give the tobacco mosaic virus its extraordinary architectural perfection and stability.

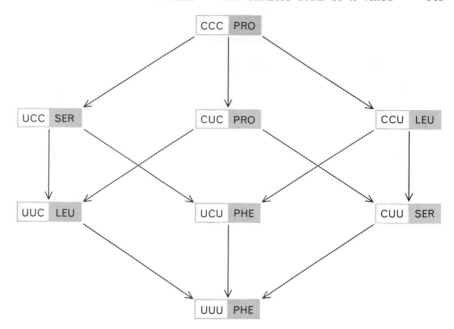

DEAMINATION that changes cytosine to uracil is represented by this octet arrangement of single steps from the triplet CCC to UUU. Amino acid equivalents given at right of the "codons" represent often-observed exchanges in mutants induced by deamination.

NATURE OF EXCHANGES		DEAMINATION	BROMINATION	METHYLATION	POSSIBLE CODON EXCHANGES	
ASP	GLY	(2)			AGC	GGC
	ALA	(4)				
ASP·NH₂	SER	4 (2)	3		ACA	GCA
THR	ALA	2			ACA	GCA
	ILEU	(8)			CAA	UAA
	MET	(3)				
SER	PHE	4 (4)	2	2	CUU	UUU
	GLY		1		ACG	GCG
	LEU	(2)				
GLU	GLY	1 (1)			AUG	GUG
GLU·NH₂	ARG	1				
	VAL	(2)				
PRO	LEU	3 (1)	4	4	CUC	UUC
	SER	(3)			CCU	UCU
VAL	MET	1				
ILEU	VAL	2 (3)			AUU	GUU
	THR		1			
	MET	(1)			AUA	GUA
LEU	PHE	(1)			CUU	UUU
ARG	GLY	3	3	1	AGA	GGA
	LYS	1	1			

REPLACEMENTS in chemically induced mutants of the tobacco mosaic virus are charted by frequency of observation. At left is the amino acid exchange taking place; in middle, the process that induced it; at right, possible codon exchanges for various mutants. The figures listed parenthetically were obtained by Hans Wittmann of the University of Tübingen.

38

CHEMICAL FOSSILS

GEOFFREY EGLINTON AND MELVIN CALVIN
January 1967

Certain rocks as much as three billion years old have been found to contain organic compounds. What these compounds are and how they may have originated in living matter is under active study

If you ask a child to draw a dinosaur, the chances are that he will produce a recognizable picture of such a creature. His familiarity with an animal that lived 150 million years ago can of course be traced to the intensive studies of paleontologists, who have been able to reconstruct the skeletons of extinct animals from fossilized bones preserved in ancient sediments. Recent chemical research now shows that minute quantities of organic compounds—remnants of the original carbon-containing chemical constituents of the soft parts of the animal—are still present in some fossils and in ancient sediments of all ages, including some measured in billions of years. As a result of this finding organic chemists and geologists have joined in a search for "chemical fossils": organic molecules that have survived unchanged or little altered from their original structure, when they were part of organisms long since vanished.

This kind of search does not require the presence of the usual kind of fossil—a shape or an actual hard form in the rock. The fossil molecules can be extracted and identified even when the organism has completely disintegrated and the organic molecules have diffused into the surrounding material. In fact, the term "biological marker" is now being applied to organic substances that show pronounced resistance to chemical change and whose molecular structure gives a strong indication that they could have been created in significant amounts only by biological processes.

One might liken such resistant compounds to the hard parts of organisms that ordinarily persist after the soft parts have decayed. For example, hydrocarbons, the compounds consisting only of carbon and hydrogen, are comparatively resistant to chemical and biological attack. Unfortunately many other biologically important molecules such as nucleic acids, proteins and polysaccharides contain many bonds that hydrolyze, or cleave, readily; hence these molecules rapidly decompose after an organism dies. Nevertheless, several groups of workers have reported finding constituents of proteins (amino acids and peptide chains) and even proteins themselves in special well-protected sites, such as between the thin sheets of crystal in fossil shells and bones [see "Paleobiochemistry," by Philip H. Abelson; SCIENTIFIC AMERICAN Offprint 101].

Where complete destruction of the organism has taken place one cannot, of course, visualize its original shape from the nature of the chemical fossils it has left behind. One may, however, be able to infer the biological class, or perhaps even the species, of organism that gave rise to them. At present such deductions must be extremely tentative because they involve considerable uncertainty. Although the chemistry of living organisms is known in broad outline, biochemists even today have identified the principal constituents of only a few small groups of living things. Studies in comparative biochemistry or chemotaxonomy are thus an essential parallel to organic geochemistry. A second uncertainty involves the question of whether or not the biochemistry of ancient organisms was generally the same as the biochemistry of present-day organisms. Finally, little is known of the chemical changes wrought in organic substances when they are entombed for long periods of time in rock or a fossil matrix.

In our work at the University of California at Berkeley and at the University of Glasgow we have gone on the assumption that the best approach to the study of chemical fossils is to analyze geological materials that have had a relatively simple biological and geological history.

The search for suitable sediments requires a close collaboration between the geologist and the chemist. The results obtained so far augur well for the future.

Organic chemistry made its first major impact on the earth sciences in 1936, when the German chemist Alfred Treibs isolated metal-containing porphyrins from numerous crude oils and shales. Certain porphyrins are important biological pigments; two of the best-known are chlorophyll, the green pigment of plants, and heme, the red pigment of the blood. Treibs deduced that the oils were biological in origin and could not have been subjected to high temperatures, since that would have decomposed some of the porphyrins in them. It is only during the past decade, however, that techniques have been available for the rapid isolation and identification of organic substances present in small amounts in oils and ancient sediments. Further refinements and new methods will be required for detailed study of the tiny amounts of organic substances found in some rocks. The effort should be worthwhile, because such techniques for the detection and definition of the specific architecture of organic molecules should not only tell us much more about the origin of life on the earth but also help us to establish whether or not life has developed on other planets. Furthermore, chemical fossils present the organic chemist with a new range of organic compounds to study and may offer the geologist a new tool for determining the environment of the earth in various geological epochs and the conditions subsequently experienced by the sediments laid down in those epochs.

If one could obtain the fossil molecules from a single species of organism, one would be able to make a direct correlation between present-day biochemistry and organic geochemistry. For example, one could directly compare the lipids, or fatty compounds, isolated from a living organism with the lipids of its fossil ancestor. Unfortunately the fossil lipids and other fossil compounds found in sediments almost always represent the chemical debris from many organisms.

The deposition of a compressible fine-grained sediment containing mineral particles and disseminated organic matter takes place in an aquatic environment in which the organic content can be partially preserved; an example would be the bottom of a lake or a delta. The organic matter makes up something less than 1 percent of many ancient sediments. The small portion of this carbon-containing material that is soluble in organic solvents represents a part of the original lipid content, more or less modified, of the organisms that lived and died while the sediment was being deposited.

The organic content presumably consists of varying proportions of the components of organisms—terrestrial as well as aquatic—that have undergone chemical transformation while the sediment was being laid down and compressed. Typical transformations are reduction, which has the effect of removing oxygen from molecules and adding hydrogen, and decarboxylation, which removes the carboxyl radical (COOH). In addition, it appears that a variety of reactive unsaturated compounds (compounds having available chemical bonds) combine to form an insoluble amorphous material known as kerogen. Other chemical changes that occur with the passage of time are related to the temperature to which the rock is heated by geologic processes. Thus many petroleum chemists and geologists believe petroleum is created by progressive degradation, brought about by heat, of the organic matter that is finely disseminated throughout the original sediment. The organic matter that comes closest in structure to the chains and rings of carbon atoms found in the hydrocarbons of petroleum is the matter present in the lipid fraction of organisms. Another potential source of petroleum hydrocarbons is kerogen itself, presumably formed from a wide variety of organic molecules; it gives off a range of straight-chain, branched-chain and ring-containing hydrocarbons when it is strongly heated in the laboratory. One would also like to know more about the role of bacteria in the early steps of sediment formation. In the upper layers of most newly formed sediments there is strong bacterial activity, which must surely re-

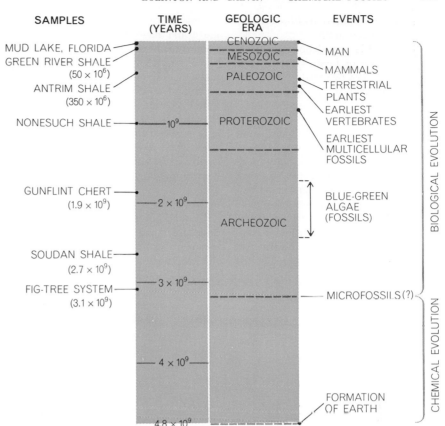

GEOLOGICAL TIME SCALE shows the age of some intensively studied sedimentary rocks (*left*) and the sequence of major steps in the evolution of life (*right*). The stage for biological evolution was set by chemical evolution, but the period of transition is not known.

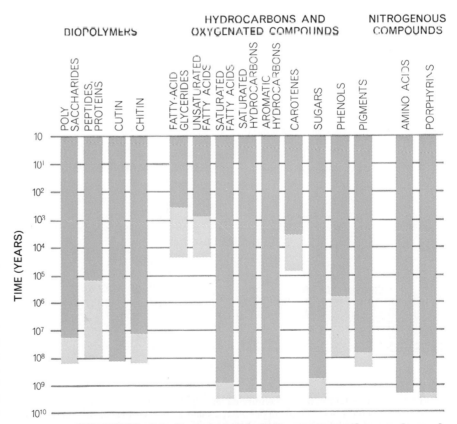

ORGANIC COMPOUNDS originally synthesized by living organisms and more or less modified have now been found in many ancient rocks that began as sediments. The dark bars indicate reasonably reliable identification; the light bars, unconfirmed reports. Cutin and chitin are substances present respectively in the outer structures of plants and of insects.

sult in extensive alteration of the initially deposited organic matter.

In this article we shall concentrate on the isolation of fossil hydrocarbons. The methods must be capable of dealing with the tiny quantity of material available in most rocks. Our general procedure is as follows.

After cutting off the outer surface of a rock specimen to remove gross contaminants, we clean the remaining block with solvents and pulverize it. We then place the powder in solvents such as benzene and methanol to extract the organic material. Before this step we sometimes dissolve the silicate and carbonate minerals of the rock with hydrofluoric and hydrochloric acids. We separate the organic extract so obtained into acidic, basic and neutral fractions. The compounds in these fractions are converted, when necessary, into derivatives that make them suitable for separation by the technique of chromatography. For the initial separations we use column chromatography, in which a sample in solution is passed through a column packed with alumina or silica, Depending on their nature, compounds in the sample pass through the column at different speeds and can be collected in fractions as they emerge.

In subsequent stages of the analysis finer fractionations are achieved by means of gas-liquid chromatography. In this variation of the technique, the sample is vaporized into a stream of light gas, usually helium, and brought in contact with a liquid that tends to trap the compounds in the sample in varying degree. The liquid can be supported on an inorganic powder, such as diatomaceous earth, or coated on the inside of a capillary tube. Since the compounds are alternately trapped in the liquid medium and released by the passing stream of gas they progress through the column at varying speeds, with the result that they are separated into distinct fractions as they emerge from the tube. The temperature of the column is raised steadily as the separation proceeds, in order to drive off the more strongly trapped compounds.

The initial chromatographic separation is adjusted to produce fractions that consist of a single class of compound, for example the class of saturated hydrocarbons known as alkanes. Alkane molecules may consist either of straight chains of carbon atoms or of chains that include branches and rings. These subclasses can be separated with the help of molecular sieves: inorganic substances, commonly alumino-silicates, that have a fine honeycomb structure. We use a sieve whose mesh is about five angstrom units, or about a thousandth of the wavelength of green light. Straight-chain alkanes, which resemble smooth flexible rods about 4.5 angstroms in diameter, can enter the sieve and are trapped. Chains with branches and rings are too big to enter and so are held back. The straight-chain alkanes can be liberated from the sieve for further analysis by dissolving the sieve in hydrofluoric acid. Other families of molecules can be trapped in special crystalline forms of urea and thiourea, whose crystal lattices provide cavities with diameters of five angstroms and seven angstroms respectively.

The families of molecules isolated in this way are again passed through gas chromatographic columns that separate the molecular species within the family. For example, a typical chromatogram of straight-chain alkanes will show that molecules of increasing chain length emerge from the column in a regularly spaced sequence that parallels their increasing boiling points, thus producing a series of evenly spaced peaks. Although the species of molecule in a particular peak can often be identified tentatively on the basis of the peak's position, a more precise identification is usually desired. To obtain it one must collect the tiny amount of substance that produced the peak—often measured in micrograms— and examine it by one or more analytical methods such as ultraviolet and infrared

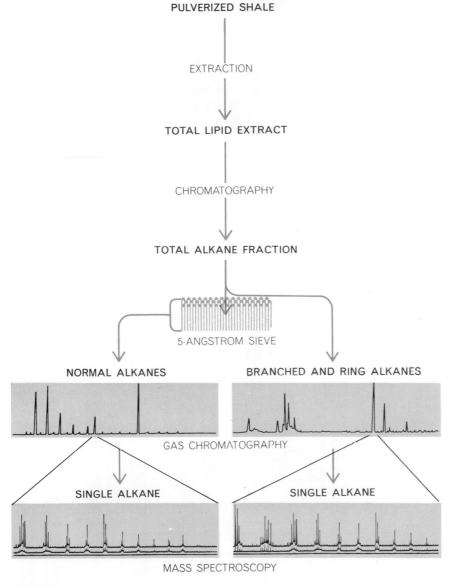

PULVERIZED SHALE

EXTRACTION

TOTAL LIPID EXTRACT

CHROMATOGRAPHY

TOTAL ALKANE FRACTION

5-ANGSTROM SIEVE

NORMAL ALKANES   BRANCHED AND RING ALKANES

GAS CHROMATOGRAPHY

SINGLE ALKANE   SINGLE ALKANE

MASS SPECTROSCOPY

ANALYTICAL PROCEDURE for identifying chemical fossil begins with the extraction of alkane hydrocarbons from a sample of pulverized shale. In normal alkanes the carbon atoms are arranged in a straight chain. (Typical alkanes are illustrated on page 348.) Molecular sieves are used to separate straight-chain alkanes from alkanes with branched chains and rings. The two broad classes are then further fractionated. Compounds responsible for individual peaks in the chromatogram are identified by mass spectrometry and other methods.

spectroscopy, mass spectrometry or nuclear magnetic resonance. In one case X-ray crystallography is being used to arrive at the structure of a fossil molecule.

A new and useful apparatus is one that combines gas chromatography and mass spectrometry [*see illustration at right*]. The separated components emerge from the chromatograph and pass directly into the ionizing chamber of the mass spectrometer, where they are broken into submolecular fragments whose abundance distribution with respect to mass is unique for each component. These various analytical procedures enable us to establish a precise structure and relative concentration for each organic compound that can be extracted from a sample of rock.

How is it that such comparatively simple substances as the alkanes should be worthy of geochemical study? There are several good reasons. Alkanes are generally prominent components of the soluble lipid fraction of sediments. They survive geologic time and geologic conditions well because the carbon-hydrogen and carbon-carbon bonds are strong and resist reaction with water. In addition, alkane molecules can provide more information than the simplicity of their constitution might suggest; even a relatively small number of carbon and hydrogen atoms can be joined in a large number of ways. For example, a saturated hydrocarbon consisting of 19 carbon atoms and 40 hydrogen atoms could exist in some 100,000 different structural forms that are not readily interconvertible. Analysis of ancient sediments has already shown that in some cases they contain alkanes clearly related to the long-chain carbon compounds of the lipids in present-day organisms [*see illustration on next page*]. Generally one finds a series of compounds of similar structure, such as the normal, or straight-chain, alkanes (called *n*-alkanes); the compounds extracted from sediments usually contain up to 35 carbon atoms. Alkanes isolated from sediments may have been buried as such or formed by the reduction of substances containing oxygen.

The more complicated the structure of the molecule, the more valuable it is likely to be for geochemical purposes: its information content is greater. Good examples are the alkanes with branches and rings, such as phytane and cholestane. It is unlikely that these complex alkanes could be built up from small subunits by processes other than biological ones, at least in the proportions found. Hence we are encouraged to look

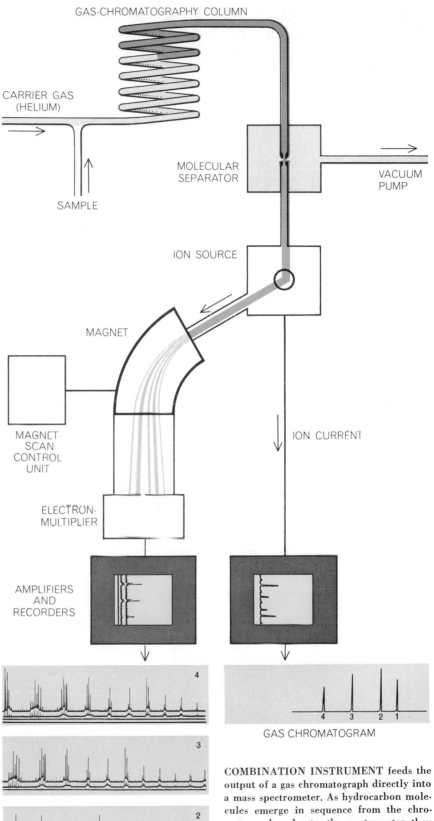

GAS CHROMATOGRAPHY COLUMN

CARRIER GAS (HELIUM)

SAMPLE

MOLECULAR SEPARATOR

VACUUM PUMP

ION SOURCE

MAGNET

MAGNET SCAN CONTROL UNIT

ION CURRENT

ELECTRON-MULTIPLIER

AMPLIFIERS AND RECORDERS

GAS CHROMATOGRAM

MASS SPECTRA

COMBINATION INSTRUMENT feeds the output of a gas chromatograph directly into a mass spectrometer. As hydrocarbon molecules emerge in sequence from the chromatograph and enter the spectrometer, they are ionized, or broken into charged fragments. The size of the ionization current is proportional to the amount of material present at each instant and can be converted into a chromatogram. In the spectrometer the charged fragments are directed through a magnetic field, which separates them according to mass. Each species of molecule produces a unique mass-distribution pattern.

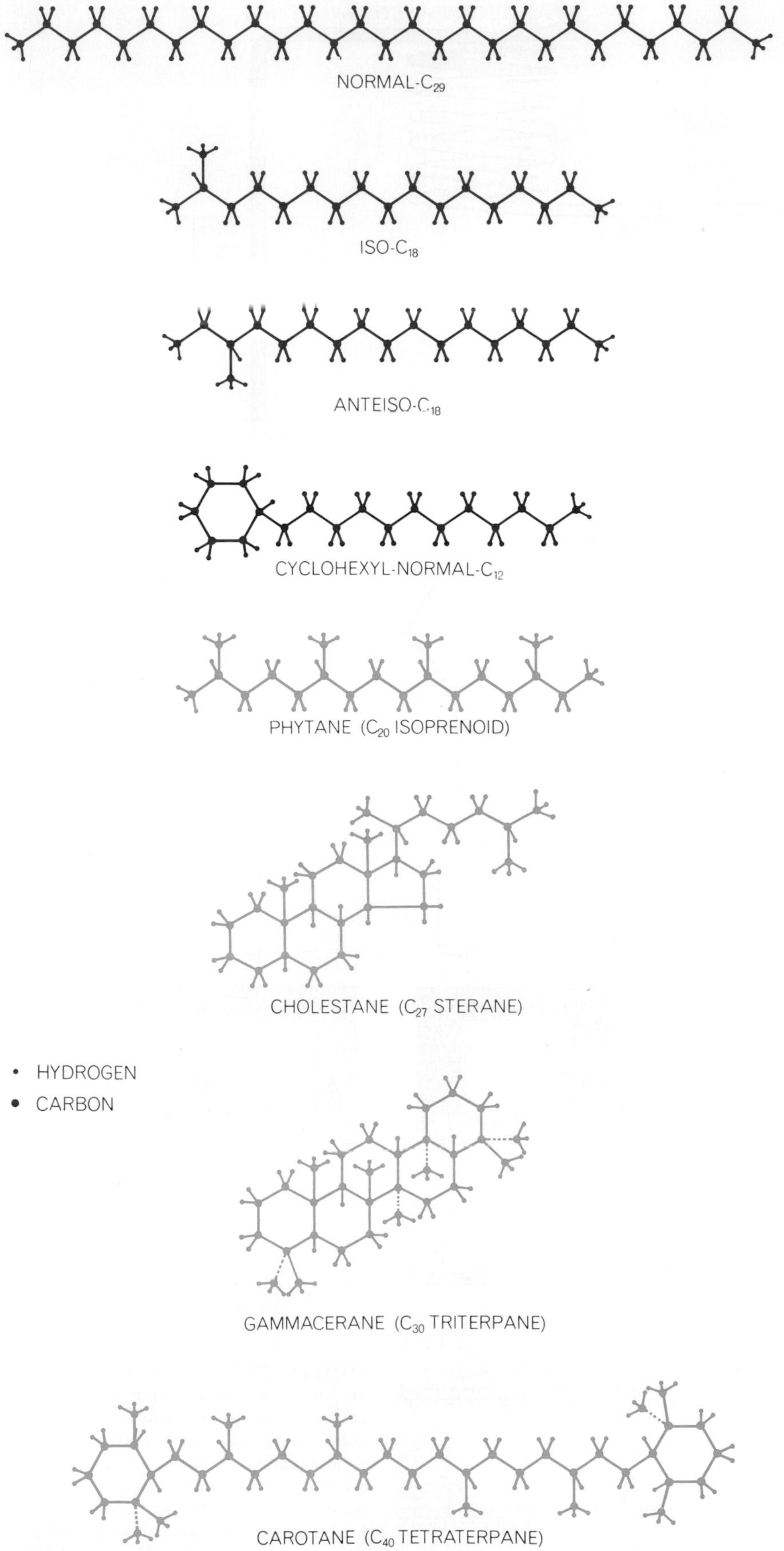

NORMAL-C$_{29}$

ISO-C$_{18}$

ANTEISO-C$_{18}$

CYCLOHEXYL-NORMAL-C$_{12}$

PHYTANE (C$_{20}$ ISOPRENOID)

CHOLESTANE (C$_{27}$ STERANE)

• HYDROGEN
• CARBON

GAMMACERANE (C$_{30}$ TRITERPANE)

CAROTANE (C$_{40}$ TETRATERPANE)

ALKANE HYDROCARBON MOLECULES can take various forms: straight chains (which are actually zigzag chains), branched chains and ring structures. Those depicted here have been found in crude oils and shales. The molecules shown in color are so closely related to well-known biological molecules that they are particularly useful in bespeaking the existence of ancient life. The broken lines indicate side chains that are directed into the page.

for biological precursors with appropriate preexisting carbon skeletons.

In conducting this kind of search one makes the assumption, at least at the outset, that the overall biochemistry of past organisms was similar to that of present-day organisms. When lipid fractions are isolated directly from modern biological sources, they are generally found to contain a range of hydrocarbons, fatty acids, alkanols, esters and so on. The mixture is diverse but by no means random. The molecules present in such fractions have structures that reflect the chemical reaction pathways systematically followed in biological organisms. There are only a few types of biological molecule wherein long chains of carbon atoms are linked together; two examples are the straight-chain lipids, the end groups of which may include oxygen atoms, and the lipids known as isoprenoids.

The straight-chain lipids are produced by what is called the polyacetate pathway [see illustration on opposite page]. This pathway leads to a series of fatty acids with an even number of carbon atoms; the odd-numbered molecules are missing. One also finds in nature straight-chain alcohols (n-alkanols) that likewise have an even number of carbon atoms, which is to be expected if they are formed by simple reduction of the corresponding fatty acids. In contrast, the straight-chain hydrocarbons (n-alkanes) contain an odd number of carbon atoms. Such a series would be produced by the decarboxylation of the fatty acids.

The second type of lipid, the isoprenoids, have branched chains consisting of five-carbon units assembled in a regular order [see illustration on page 350]. Because these units are assembled in head-to-tail fashion the side-chain methyl groups (CH$_3$) are attached to every fifth carbon atom. (Tail-to-tail addition occurs less frequently but accounts for several important natural compounds, for example beta-carotene.) When the isoprenoid skeleton is found in a naturally occurring molecule, it is reasonable to assume that the compound has been formed by this particular biological pathway.

Chlorophyll is possibly the most widely distributed molecule with an isoprenoid chain; therefore it must make some contribution to the organic matter in sediments. Its fate under conditions of geological sedimentation is not known, but it may decompose into only two or three large fragments [see illustration on page 351]. The molecule of chlorophyll a consists of a system of interconnected rings and a phytyl side chain, which is an isoprenoid. When chlorophyll is decomposed, it seems likely that the phytyl chain is split off and converted to phytane (which has the same number of carbon atoms) and pristane (which is shorter by one carbon atom). When both of these branched alkanes are found in a sediment, one has reasonable presumptive evidence that chlorophyll was once present. The chlorophyll ring system very likely gives rise to the metal-containing porphyrins that are found in many crude oils and sediments.

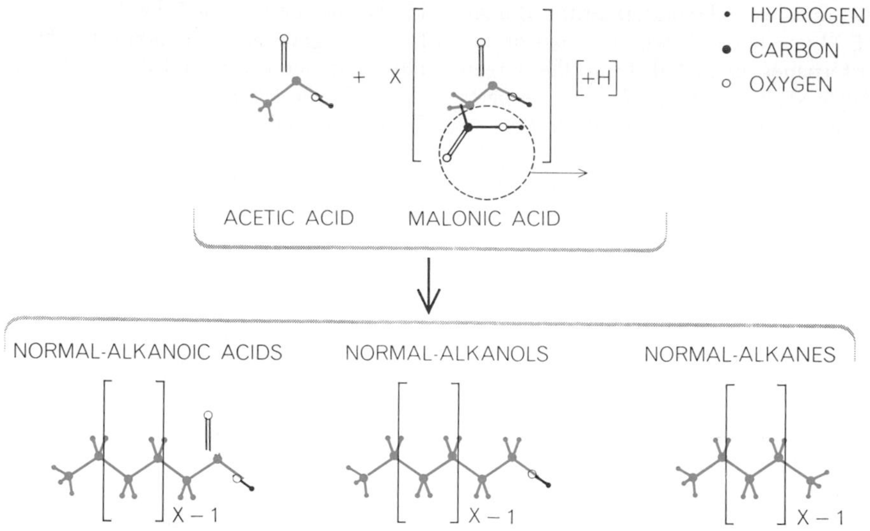

Phytane and pristane may actually enter the sediments directly. Max Blumer of the Woods Hole Oceanographic Institution showed in 1965 that certain species of animal plankton that eat the plant plankton containing chlorophyll store quite large quantities of pristane and related hydrocarbons. The animal plankton act in turn as a food supply for bigger marine animals, thereby accounting for the large quantities of pristane in the liver of the shark and other fishes.

An indirect source for the isoprenoid alkanes could be the lipids found in the outer membrane of certain bacteria that live only in strong salt solutions, an environment that might be found where ancient seas were evaporating. Morris Kates of the National Research Council of Canada has shown that a phytyl containing lipid (diphytyl phospholipid) is common to bacteria with the highest salt requirement but not to the other bacteria examined so far.

This last example brings out the point that in spite of the overall oneness of present-day biochemistry, organisms do differ in the compounds they make. They also synthesize the same compounds in different proportions. These differences are making it possible to classify living species on a chemotaxonomic, or chemical, basis rather than on a morphological, or shape, basis. Eventually it may be possible to extend chemical classification to ancient organisms, creating a discipline that could be called paleochemotaxonomy.

Our study of chemical fossils began in 1961, when we decided to probe the sedimentary rocks of the Precambrian period in a search for the earliest signs of life. This vast period of time, some four billion years, encompasses the beginnings of life on this planet and its early development to the stage of organisms consisting of more than one cell [see illustrations on page 345]. We hoped that our study would complement the efforts being made by a number of workers, including one of us (Calvin), to imitate in the laboratory the chemical evolution that must have preceded the appearance of life on earth. We also saw the possibility that our work could be adapted to the study of meteorites and of rocks obtained from the moon or nearby planets. Thus it even includes the possibility of uncovering exotic and alien biochemistries. The exploration of the ancient rocks of the earth provides a testing ground for the method and the concepts involved.

We chose the alkanes because one might expect them to resist fairly high temperatures and chemical attack for long periods of time. Moreover, J. G. Bendoraitis of the Socony Oil Company, Warren G. Meinschein of the Esso Research Laboratory and others had already identified individual long-chain alkanes, including a range of isoprenoid types, in certain crude oils. Even more encouraging, J. J. Cummins and W. E. Robinson of the U.S. Bureau of Mines had just made a preliminary announcement of their isolation of phytane, pristane and other isoprenoids from a relatively young sedimentary rock: the Green River shale of Colorado, Utah and Wyoming. Thus the alkanes seemed to offer the biological markers we were seeking. Robinson generously provided our laboratory with samples of the Green River shale, which was deposited some 50 million years ago and constitutes the major oil-shale reserve of the U.S.

The Green River shale, which is the remains of large Eocene lakes in a rather stable environment, contains a considerable fraction (.6 percent) of alkanes. Using the molecular-sieve technique, we split the total alkane fraction into alkanes with straight chains and those with branched chains and rings and ran the resulting fractions through the gas chromatograph [see illustration at top left on page 352]. The straight-chain alkanes exhibit a marked dominance of molecules containing an odd number of carbon atoms, which is to be expected for straight-chain hydrocarbons from a biological source. The other fraction shows a series of prominent sharp peaks; we conclusively identified them as isoprenoids, confirming the results of Cummins and Robinson. The large proportion of phytane, the hydrocarbon corresponding to the entire side chain of chlorophyll, is particularly noteworthy. The oxygenated counterparts of the steranes and triterpanes (27 to 30 carbon atoms) and the high-molecular-weight n-alkanes (29 to 31 carbons) are typical constituents of the waxy covering of the leaves and pollen of land plants, leading to the inference that such plants made major contributions to the organic matter deposited in the Green River sediments.

Although the gross chemical structure (number of rings and side chains) of the steranes and triterpanes was established in this work, it was only recently that the precise structure of one of these hydrocarbons was conclusively established. E. V. Whitehead and his associates in the British Petroleum Company and Robin-

· HYDROGEN
● CARBON
○ OXYGEN

ACETIC ACID   MALONIC ACID

NORMAL-ALKANOIC ACIDS   NORMAL-ALKANOLS   NORMAL-ALKANES

X − 1   X − 1   X − 1

STRAIGHT-CHAIN LIPIDS are created in living organisms from simple two-carbon and three-carbon compounds: acetate and malonate, shown here as their acids. The complex biological process, which involves coenzyme A, is depicted schematically. The fatty acids (n-alkanoic acids) and fatty alcohols (n-alkanols) produced in this way have an even number of carbon atoms. The removal of carbon dioxide from the fatty acids, the net effect of decarboxylation, would give rise to a series of n-alkanes with an odd number of carbon atoms.

son and his collaborators in the Bureau of Mines have shown that one of the triterpanes extracted from the Green River shale is identical in all respects with gammacerane [*see illustration on page 208*]. Conceivably it is produced by the reduction of a compound known as gammaceran-3-beta-ol, which was recently isolated from the common protozoon *Tetrahymena pyriformis.* Other derivatives of gammacerane are rather widely distributed in the plant kingdom.

At our laboratory in Glasgow, Sister Mary T. J. Murphy and Andrew McCormick recently identified several steranes and triterpanes and also the tetraterpane called perhydro-beta-carotene, or carotane [*see top illustration on page 354*]. Presumably carotane is derived by reduction from beta-carotene, an important red pigment of plants. A similar reduction process could convert the familiar biological compound cholesterol into cholestane, one of the steranes found in the Green River shale [*see same illustration on page 354*]. The mechanism and sedimentary site of such geochemical reduction processes is an important problem awaiting attack.

W. H. Bradley of the U.S. Geological Survey has sought a contemporary counterpart of the richly organic ooze that presumably gave rise to the Green River shale. So far he has located only four lakes, two in the U.S. and two in Africa, that seem to be reasonable candidates. One of them, Mud Lake in Florida, is now being studied closely. A dense belt of vegetation surrounding the lake filters out all the sand and silt that might otherwise be washed into it from the land. As a result the main source of sedimentary material is the prolific growth of microscopic algae. The lake bottom uniformly consists of a grayish-green ooze about three feet deep. The bottom of the ooze was deposited about 2,300 years ago, according to dating by the carbon-14 technique.

Microscopic examination of the ooze shows that it consists mainly of minute fecal pellets, made up almost exclusively of the cell walls of blue-green algae. Some pollen grains are also present. Decay is surprisingly slow in spite of the ooze's high content of bound oxygen and the temperatures characteristic of Florida. Chemical analyses in several laboratories, reported this past November at a meeting of the Geological Society of America, indicate that there is indeed considerable correspondence between the lipids of the Mud Lake ooze and those of the Green River shale. Eugene McCarthy of the University of California at Berkeley has also found beta-carotene in samples of Mud Lake ooze that are about 1,100 years old. The high oxygen content of the Mud Lake ooze seems inconsistent, however, with the dominance of oxygen-poor compounds in the Green River shale. The long-term geological mechanisms that account for the loss of oxygen may have to be sought in sediments older than those in Mud Lake.

Sediments much older than the Green River shale have now been examined by our groups in Berkeley and Glasgow, and by workers in other universities and in oil-industry laboratories. We find that the hydrocarbon fractions in these more ancient samples are usually more complex than those of the Green River shale; the gas chromatograms of the older samples tend to show a number of partially resolved peaks centered around a single maximum. One of the older sediments we have studied is the Antrim shale of Michigan. A black shale probably 350 million years old, it resembles other shales of the Chattanooga type that underlie many thousands of square miles of the eastern U.S. Unlike the Green River shale, the straight-chain alkane fraction of the Antrim shale shows little or no predominance of an odd number of carbon atoms over an even number [*see middle illustration of three at top of pages 352 and 353*]. The alkanes with branched chains and rings, however, continue to be rich in isoprenoids.

The fact that alkanes with an odd number of carbon atoms are not predominant in the Antrim shale and sediments of comparable antiquity may be owing to the slow cracking by heat of carbon chains both in the alkane component and in the kerogen component. The effect can be partially reproduced in the laboratory by heating a sample of the Green River shale for many hours above 300 degrees centigrade. After such treatment the straight-chain alkanes show a reduced dominance of odd-carbon molecules and the branched-chain-and-ring fraction is more complex.

The billion-year-old shale from the Nonesuch formation at White Pine, Mich., exemplifies how geological, geochemical and micropaleontological techniques can be brought to bear on the problem of detecting ancient life. With the aid of the electron microscope Elso S. Barghoorn and J. William Schopf of Harvard University have detected in the Nonesuch shale "disaggregated particles of condensed spheroidal organic matter." In collaboration with Meinschein the Harvard workers have also found evidence that the Nonesuch shale contains isoprenoid alkanes, steranes and porphyrins. Independently we have analyzed the Nonesuch shale and found that it contains pristane and phytane, in addition to iso-alkanes, anteiso-alkanes and cyclohexyl alkanes.

Barghoorn and S. A. Tyler have also detected microfossils in the Gunflint chert of Ontario, which is 1.9 billion years old, almost twice the age of the

ACETIC ACID (3 UNITS)   MEVALONIC ACID   ISOPENTENYL PYROPHOSPHATE

DIMER, $(C_5)_2$

TRIMER, $(C_5)_3$

TETRAMER, $(C_5)_4$

POLYMER $(C_5)_n$

BRANCHED-CHAIN LIPIDS are produced in living organisms by an enzymatically controlled process, also depicted schematically. In this process three acetate units link up to form a six-carbon compound (mevalonic acid), which subsequently loses a carbon atom and is combined with a high-energy phosphate. "Head to tail" assembly of the five-carbon subunits produces branched-chain molecules that are referred to as isoprenoid structures.

Nonesuch shale. They have reported that the morphology of the Gunflint microfossils "is similar to that of the existing primitive filamentous blue-green algae."

One of the oldest Precambrian sediments yet analyzed is the Soudan shale of Minnesota, which was formed about 2.7 billion years ago. Although its total hydrocarbon content is only .05 percent, we have found that it contains a mixture of straight-chain alkanes and branched-chain-and-ring alkanes not unlike those present in the much younger Antrim shale [*see third illustration of three at top of next two pages*]. In the branched-chain-and-ring fraction we have identified pristane and phytane. Steranes and triterpanes also seem to be present, but we have not yet established their precise three-dimensional structure. Preston E. Cloud of the University of California at Los Angeles has reported that the Soudan shale contains microstructures resembling bacteria or blue-green algae, but he is not satisfied that the evidence is conclusive.

A few reports are now available on the most ancient rocks yet examined: sediments from the Fig Tree system of Swaziland in Africa, some 3.1 billion years old. An appreciable fraction of the alkane component of these rocks consists of isoprenoid molecules. If one assumes that isoprenoids are chemical vestiges of chlorophyll, one is obliged to conclude that living organisms appeared on the earth only about 1.7 billion years after the earth was formed (an estimated 4.8 billion years ago).

Before reaching this conclusion, however, one would like to be sure that the isoprenoids found in ancient sediments have the precise carbon skeleton of the biological molecules from which they are presumed to be derived. So far no sample of pristane or phytane—the isoprenoids that may be derived from the phytyl side chain of chlorophyll—has been shown to duplicate the precise three-dimensional structure of a pure reference sample. Vigorous efforts are being made to clinch the identification.

Assuming that one can firmly establish the presence of biologically structured isoprenoid alkanes in a sediment, further questions remain. The most serious one is: Were the hydrocarbons or their precursors deposited when the sediment was formed or did they seep in later? This question is not easily answered. A sample can be contaminated at any point up to—and after—the time it reaches the laboratory bench. Fossil fuels, lubricants and waxes are omnipresent, and laboratory solvents contain

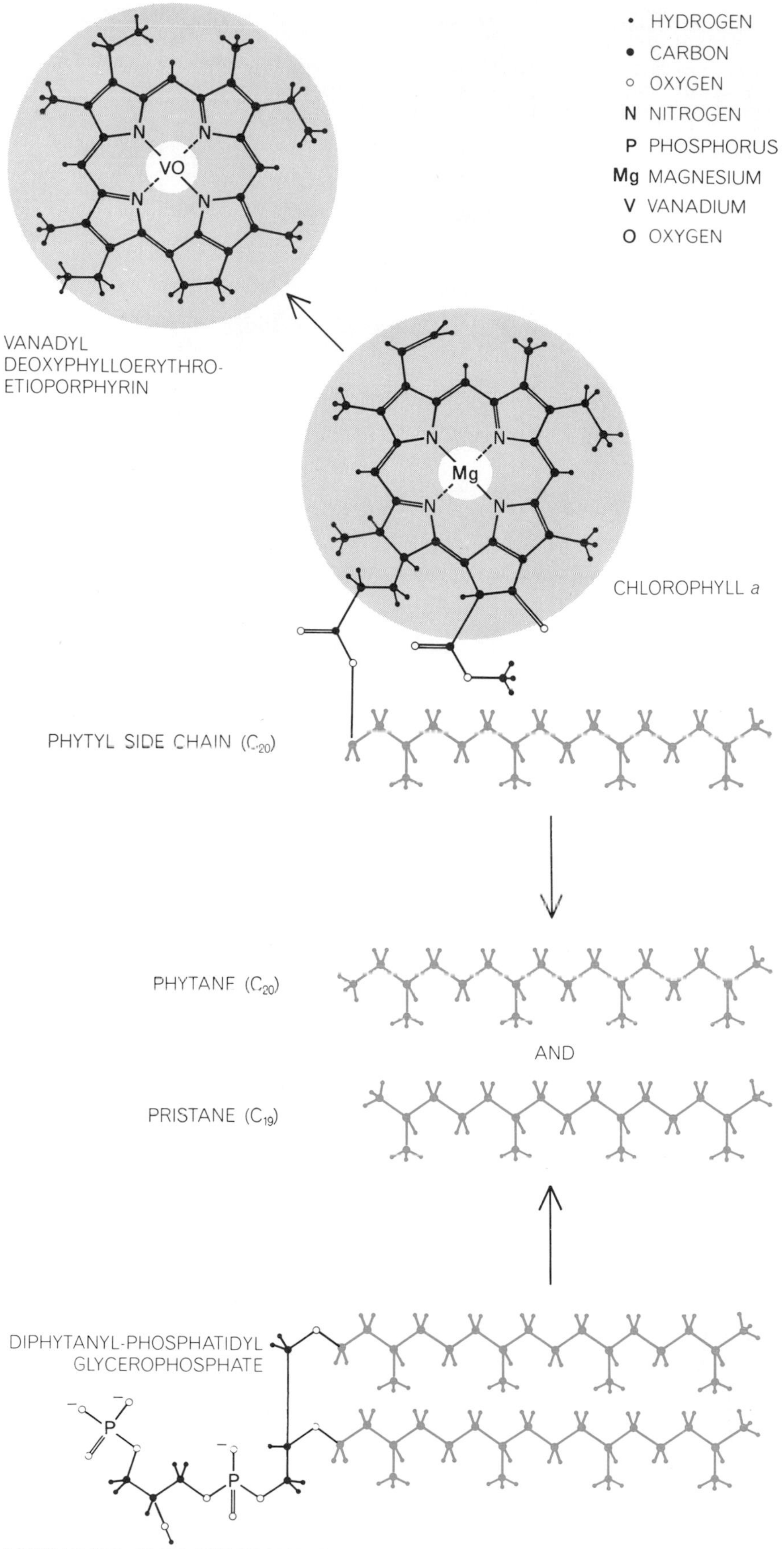

• HYDROGEN
● CARBON
○ OXYGEN
N NITROGEN
P PHOSPHORUS
Mg MAGNESIUM
V VANADIUM
O OXYGEN

VANADYL DEOXYPHYLLOERYTHRO-ETIOPORPHYRIN

CHLOROPHYLL *a*

PHYTYL SIDE CHAIN (C_{20})

PHYTANE (C_{20})

AND

PRISTANE (C_{19})

DIPHYTANYL-PHOSPHATIDYL-GLYCEROPHOSPHATE

DEGRADATION OF CHLOROPHYLL *A*, the green pigment in plants, may give rise to two kinds of isoprenoid molecules, phytane and pristane, that have been identified in many ancient sediments. It also seems likely that phytane and pristane can be derived from the isoprenoid side chains of a phosphate-containing lipid (*bottom structure*) that is a major constituent of salt-loving bacteria. The porphyrin ring of chlorophyll *a* is the probable source of vanadyl porphyrin (*upper left*) that is widely found in crude oils and shales.

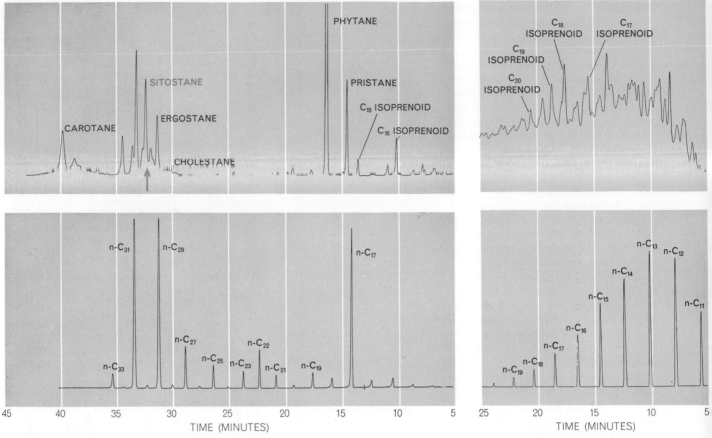

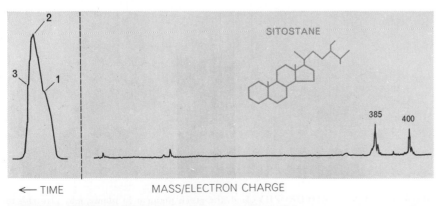

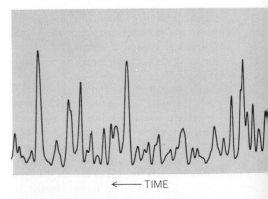

HYDROCARBONS IN YOUNG SEDIMENT, the 50-million-year-old Green River shale, produced these chromatograms. Alkanes with branched chains and rings appear in the top curve, normal alkanes in the bottom curve. The alkanes in individual peaks were identified by mass spectrometry and other methods. Such alkanes as phytane and pristane and the predominance of normal alkanes with an odd number of carbon atoms affirm that the hydrocarbons are biological in origin. The bimodal distribution of the curves is also significant.

OLDER SEDIMENTS are represented by the Antrim shale (*left*), which is 350 million years old, and by the Soudan shale (*right*), which is 2.7 billion years old. Alkanes with branched chains and rings again are shown in the top curves, normal alkanes

tiny amounts of pristane and phytane unless they are specially purified.

One way to determine whether or not rock hydrocarbons are indigenous is to measure the ratio of the isotopes carbon 13 and carbon 12 in the sample. (The ratio is expressed as the excess of carbon 13 in parts per thousand compared with

the isotope ratio in a standard: a sample of a fossil animal known as a belemnite.) The principle behind the test is that photosynthetic organisms discriminate against carbon 13 in preference to carbon 12. Although we have few clues to the abundance of the two isotopes throughout the earth's history, we can

at least test various hydrocarbon fractions in a given sample to see if they have the same isotope ratio. As a simple assumption, one would expect to find the same ratio in the soluble organic fraction as in the insoluble kerogen fraction, which could not have seeped into the rock as kerogen.

IDENTIFICATION OF SITOSTANE in the Green River shale was accomplished by "trapping" the alkanes that produced a major peak in the chromatogram (*colored arrow at top left on this page*) and passing them through the chromatograph-mass spectrometer. As the chromatograph drew the curve at the left, three scans were made with the spectrometer. Scan 1 (*partially shown at right*) is identical with the scan produced by pure sitostane.

NINETEEN-CARBON ISOPRENOID was identified in the Antrim shale by using a co-injection technique together with a high-resolution gas chromatograph. These two high-resolution curves, each taken from a

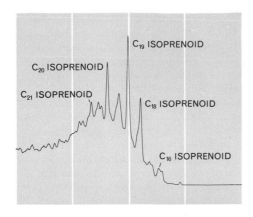

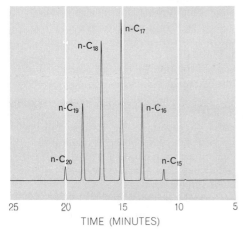

in the bottom curves. These chromatograms lack a pronounced bimodal distribution and the normal alkanes do not show a predominance of molecules with an odd number of carbon atoms. Nevertheless, the prevalence of isoprenoids argues for a biological origin.

Philip H. Abelson and Thomas C. Hoering of the Carnegie Institution of Washington have made such measurements on sediments of various geological ages and have found that the isotope ratios for soluble and insoluble fractions in most samples agree reasonably well. In some of the oldest samples, however,

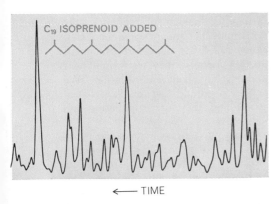

much longer trace, show the change in height of a specific peak when a small amount of pure 19-carbon isoprenoid was coinjected with the sample. Other peaks can be similarly identified by coinjecting known alkanes.

there are inconsistencies. In the Soudan shale, for example, the soluble hydrocarbons have an isotope ratio expressed as −25 parts per thousand compared with −34 parts per thousand for the kerogen. (In younger sediments and in present-day marine organisms the ratio is about midway between these two values: −29 parts per thousand.) The isotope divergence shown by hydrocarbons in the Soudan shale may indicate that the soluble hydrocarbons and the kerogen originated at different times. But since nothing is known of the mechanism of kerogen formation or of the alterations that take place in organic matter generally, the divergence cannot be regarded as unequivocal evidence of separate origin.

On the other hand, there is some reason to suspect that the isoprenoids did indeed seep into the Soudan shale sometime after the sediments had been laid down. The Soudan formation shows evidence of having been subjected to temperatures as high as 400 degrees C. The isoprenoid hydrocarbons pristane and phytane would not survive such conditions for very long. But since the exact date, extent and duration of the heating of the Soudan shale are not known, one can only speculate about whether the isoprenoids were indigenous and survived or seeped in later. In any event, they could not have seeped in much later because the sediment became compacted and relatively impervious within a few tens of millions of years.

A still more fundamental issue is whether or not isoprenoid molecules and others whose architecture follows that of known biological substances could have been formed by nonbiological processes. We and others are studying the kinds and concentrations of hydrocarbons produced by both biological and nonbiological sources. Isoprene itself, the hydrocarbon whose polymer constitutes natural rubber, is easily prepared in the laboratory, but no one has been able to demonstrate that isoprenoids can be formed nonbiologically under geologically plausible conditions. Using a computer approach, Margaret O. Dayhoff of the National Biomedical Research Foundation and Edward Anders of the University of Chicago and their colleagues have concluded that under certain restricted conditions isoprene should be one of the products of their hypothetical reactions. But this remains to be demonstrated in the laboratory.

It is well known, of course, that complex mixtures of straight-chain, branched-chain and even ring hydrocarbons can readily be synthesized in the

laboratory from simple starting materials. For example, the Fischer-Tropsch process, used by the Germans as a source of synthetic fuel in World War II, produces a mixture of saturated hydrocarbons from carbon monoxide and water. The reaction requires a catalyst (usually nickel, cobalt or iron), a pressure of about 100 atmospheres and a temperature of from 200 to 350 degrees C. The hydrocarbons formed by this process, and several others that have been studied, generally show a smooth distribution of saturated hydrocarbons. Many of them have straight chains but lack the special characteristics (such as the predominance of chains with an odd number of carbons) found in the similar hydrocarbons present in many sediments. Isoprenoid alkanes, if they are formed at all, cannot be detected.

Paul C. Marx of the Aerospace Corporation has made the ingenious suggestion that isoprenoids may be produced by the hydrogenation of graphite. In the layered structure of graphite the carbon atoms are held in hexagonal arrays by carbon-carbon bonds. Marx has pointed out that if the bonds were broken in certain ways during hydrogenation, an isoprenoid structure might result. Again a laboratory demonstration is needed to support the hypothesis. What seems certain, however, is that nonbiological syntheses are extremely unlikely to produce those specific isoprenoid patterns found in the products of living cells.

Another dimension is added to this discussion by the proposal, made from time to time by geologists, that certain hydrocarbon deposits are nonbiological in origin. Two alleged examples of such a deposit are a mineral oil found enclosed in a quartz mineral at the Abbott mercury mine in California and a bitumen-like material called thucolite found in an ancient nonsedimentary rock in Ontario. Samples of both materials have been analyzed in our laboratory at Berkeley. The Abbott oil contains a significant isoprenoid fraction and probably constitutes an oil extracted and brought up from somewhat older sediments of normal biological origin. The thucolite consists chiefly of carbon from which only a tiny hydrocarbon fraction can be extracted. Our analysis shows, however, that the fraction contains trace amounts of pristane and phytane. Recognizing the hazards of contamination, we are repeating the analysis, but on the basis of our preliminary findings we suspect that the thucolite sample represents an oil of biological origin that has been almost completely carbonized. We are aware, of course, that one runs the risk of invoking circular arguments in such discussions. Do isoprenoids demonstrate biological origin (as we and others are suggesting) or does the presence of isoprenoids in such unlikely substances indicate that they were formed nonbiologically? The debate may not be quickly settled.

There is little doubt, in any case, that organic compounds of considerable variety and complexity must have accumulated on the primitive earth during the prolonged period of nonbiological chemical development—the period of chemical evolution. With the appearance of the first living organisms biological evolution took command and presumably the "food stock" of nonbiological compounds was rapidly altered. If the changeover was abrupt on a geological time scale, one would expect to find evidence of it in the chemical composition of sediments whose age happens to bracket the period of transition. Such a discontinuity would make an intensely exciting find for organic geochemistry. The transition from chemical to biological evolution must have occurred earlier than three billion years ago. As yet, however, no criteria have been established for distinguishing between the two types of evolutionary process.

We suggest that an important distinction should exist between the kinds of molecules formed by the two processes. In the period of chemical evolution autocatalysis must have been one of the dominant mechanisms for creating large molecules. An autocatalytic system is one in which a particular substance promotes the formation of more of itself. In biological evolution, on the other hand, two different molecular systems are involved: an information-bearing system based on nucleic acids and a catalytic system based on proteins. The former directs the synthesis of the latter. A major problem, subject to laboratory experiment, is visualizing how the two systems originated and were linked.

The role of lipids in the transition may have been important. Today lipids form an important part of the membranes of all living cells. A. I. Oparin, the Russian investigator who was among the first to discuss in detail the chemical origin of life, has suggested that an essential step in the transition from chemical to biological evolution may have been the formation of membranes around droplets, which could then serve as "reaction vessels." Such self-assembling membranes might well have required lipid constituents for their function, which would be to allow some compounds to enter and leave the "cell" more readily than others. These membranes might have been formed nonbiologically by the polymerization of simple two-carbon and three-carbon units. According to this line of reasoning, the compounds that are now prominent constituents of living things are prominent precisely because they were prominent products of chemical evolution. We scarcely need add that this is a controversial and therefore stimulating hypothesis.

What one can say with some confidence is that autocatalysis alone seems unlikely to have been capable of producing the distribution pattern of hydrocarbons observed in ancient Precambrian rocks, even when some allowance is made for subsequent reactions over the course of geologic time. That it could have produced compounds of the observed type is undoubtedly possible, but

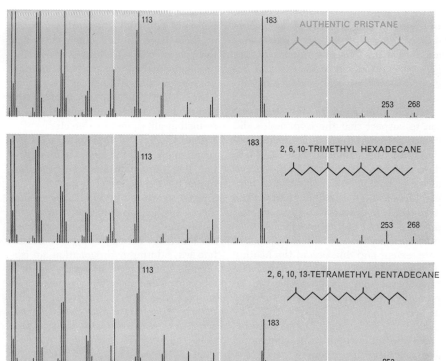

SIMILARITY OF MASS SPECTRA makes it difficult to distinguish the 19-carbon isoprenoid pristane from two of its many isomers (molecules with the same number of carbon and hydrogen atoms). The three records shown here are replotted from the actual tracings produced by pure compounds. When the sample contains impurities, as is normally the case, the difficulty of identifying authentic pristane by mass spectrometry is even greater.

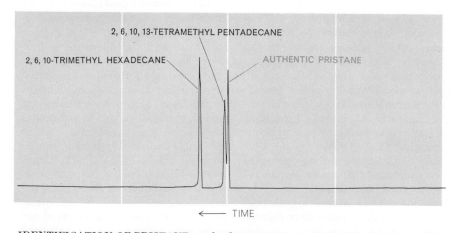

IDENTIFICATION OF PRISTANE can be done more successfully with the aid of a high-resolution gas chromatograph. When pure pristane and the isomers shown in the illustration above are fed into such an instrument, they produce three distinct peaks. This curve and the mass spectra were made by Eugene McCarthy of the University of California at Berkeley. He also made the isoprenoid study shown at the bottom of preceding two pages.

it seems to us that the observed pattern could not have arisen without the operation of those molecular systems we now recognize as the basis of living things. Eventually it should be possible to find in the geological record certain molecular fossils that will mark the boundary between chemical and biological evolution.

Another and more immediate goal for the organic geochemist is to attempt to trace on the molecular level the direction of biological evolution. For such a study one would like to have access to the actual nucleic acids and proteins synthesized by ancient organisms, but these are as yet unavailable (except perhaps in rare instances). We must therefore turn to the geochemically stable compounds, such as the hydrocarbons and oxygenated compounds that must have derived from the operation of the more perishable molecular systems. These "secondary metabolites," as we have referred to them, can be regarded as the signatures of the molecular systems that synthesized them or their close relatives.

It follows that the carbon skeletons found in the secondary metabolites of present-day organisms are the outcome of evolutionary selection. Thus it should be possible for the organic geochemist to arrange in a rough order of evolutionary sequence the carbon skeletons found in various sediments. There are some indications that this may be feasible. G. A. D. Haslewood of Guy's Hospital Medical School in London has proposed that the bile alcohols and bile acids found in present-day vertebrates can be arranged in an evolutionary sequence: the bile acids of the most primitive organisms contain molecules nearest chemically to cholesterol, their supposed biosynthetic precursor.

Within a few years the organic geochemist will be presented with a piece of the moon and asked to describe its organic contents. The results of this analysis will be awaited with immense curiosity. Will we find that the moon is a barren rock or will we discover traces of organic compounds—some perhaps resembling the complex carbon skeletons we had thought could be produced only by living systems? During the 1970's and 1980's we can expect to receive reports from robot sampling and analytical instruments landed on Mars, Venus and perhaps Jupiter. Whatever the results and their possible indications of alien forms of life, we shall be very eager to learn what carbon compounds are present elsewhere in the solar system.

TWO ALKANES IN GREEN RIVER SHALE, cholestane and carotane, probably have been derived from two well-known biological substances: cholesterol and beta-carotene. The former is closely related to the steroid hormones; the latter is a red pigment widely distributed in plants. These two natural substances can be converted to their alkane form by reduction: a process that adds hydrogen at the site of double bonds and removes oxygen.

EFFECT OF HEATING ALKANES is to produce a smoothly descending series of products (normal alkenes) if the starting material is a straight-chain molecule such as n-octadecane. (The term "alkene" denotes a hydrocarbon with one carbon-carbon double bond.) If, however, the starting material is an isoprenoid such as pristane, heating it to 600 degrees centigrade for .6 second produces an irregular series of alkenes because of the branched chain. Such degradation processes may take place in deeply buried sediments. These findings were made by R. T. Holman and his co-workers at the Hormel Institute in Minneapolis, Minn.

39

THE CHEMICAL EFFECTS OF LIGHT

GERALD OSTER

September 1968

*Visible light triggers few chemical reactions (except
in living cells), but the photons of ultraviolet radiation
readily break chemical bonds and produce short-lived
molecular fragments with unusual properties*

Our everyday world endures because most substances, organic as
well as inorganic, are stable in
the presence of visible light. Only a few
complex molecules produced by living
organisms have the specific property of
responding to light in such a way as to
initiate or participate in chemical reactions ["How Light Interacts with Living
Matter," by S. B. Hendricks, beginning on page 365]. Outside of living systems only a few kinds of molecules
are sufficiently activated by visible light
to be of interest to the photochemist.

The number of reactive molecules increases sharply, however, if the wavelength of the radiant energy is shifted
slightly into the ultraviolet part of the
spectrum. To the photochemist that is
where the action is. Thus he is primarily
concerned with chemical events that
are triggered by ultraviolet radiation in
the range between 180 and 400 nanometers. These events usually happen so
swiftly that ingenious techniques have
had to be devised to follow the molecular transformations that take place. It
is now routine, for example, to identify
molecular species that exist for less than
a millisecond. Species with lifetimes
measured in microseconds are being
studied, and new techniques using laser
pulses are pushing into the realm where
lifetimes can be measured in nanosec-
onds and perhaps even picoseconds.

The photochemist is interested in such
short-lived species not simply for their
own sake but because he suspects that
many, if not most, chemical reactions
proceed by way of short-lived interme-
diaries. Only by following chemical reactions step by step in fine detail can he
develop plausible models of how chemical reactions proceed in general. From
such studies it is often only a short step
to the development of chemical proc-

esses and products of practical value.

When a quantum of light is absorbed
by a molecule, one of the electrons of
the molecule is raised to some higher
excited state. The excited molecule is
then in an unstable condition and will
try to rid itself of this excess energy
by one means or another. Usually the
electronic excitation is converted into vibrational energy (vibration of the atoms
of the molecule), which is then passed on
to the surroundings as heat. Such is the
case, for example, with a tar roof on a
sunny day. An alternative pathway is for
the excited molecule to fluoresce, that is,
to emit radiation whose wavelength is
slightly longer than that of the exciting
radiation. The bluish appearance of quinine water in the sunlight is an example
of fluorescence; the excitation is produced by the invisible ultraviolet radiation of the sun.

The third way an electronically excited molecule can rid itself of energy is
the one of principal interest to the photochemist: the excited molecule can undergo a chemical transformation. It is the
task of the photochemist to determine
the nature of the products made, the
amount of product made per quantum
absorbed (the quantum yield) and how
these results depend on the concentrations of the starting materials. His next
step is to combine these data with the
known spectroscopic and thermodynamic properties of the molecules involved
to make a coherent picture. It must be
admitted, however, that only the simplest photochemical reactions are understood in detail.

There is also a fourth way an excited
molecule can dissipate its energy: the
molecule may be torn apart. This is
called photolysis. As might be expected,
photolysis occurs only if the energy of
the absorbed quantum exceeds the en-

ergy of the chemical bonds that hold the
molecule together. The energy required
to photolyse most simple molecules corresponds to light that lies in the ultraviolet region [*see illustration on page 359*].
For example, the chlorine molecule is
colored and thus absorbs light in the
visible range (at 425 nanometers), but it
has a low quantum yield of photolysis
when exposed to visible light. When it is
exposed to ultraviolet radiation at 330
nanometers, on the other hand, the
quantum yield is close to unity: each
quantum of radiation absorbed ruptures
one molecule.

Albert Einstein proposed in 1905 that
one quantum of absorbed light leads
to the photolysis of one molecule, but it
required the development of quantum
mechanics in the late 1920's to explain
why the quantum yield should depend
on the wavelength of the exciting light.
James Franck and Edward U. Condon,
who carefully analyzed molecular excitation, pointed out that when a molecule
makes a transition from a ground state to
an electronically excited state, the transition takes place so rapidly that the interatomic distances in the molecule do not
have time to change. The reason is that
the time required for transition is much
shorter than the period of vibration of
the atoms in the molecule.

To understand what happens when a
molecule is excited by light it will be
helpful to refer to the illustration on
the following page. The lower curve represents the potential energy of a vibrating diatomic molecule in the ground
state. The upper curve represents the
potential energy of the excited molecule,
which is also vibrating. The horizontal
lines in the lower portion of each curve
indicate the energy of discrete vibrational levels. If the interatomic distance

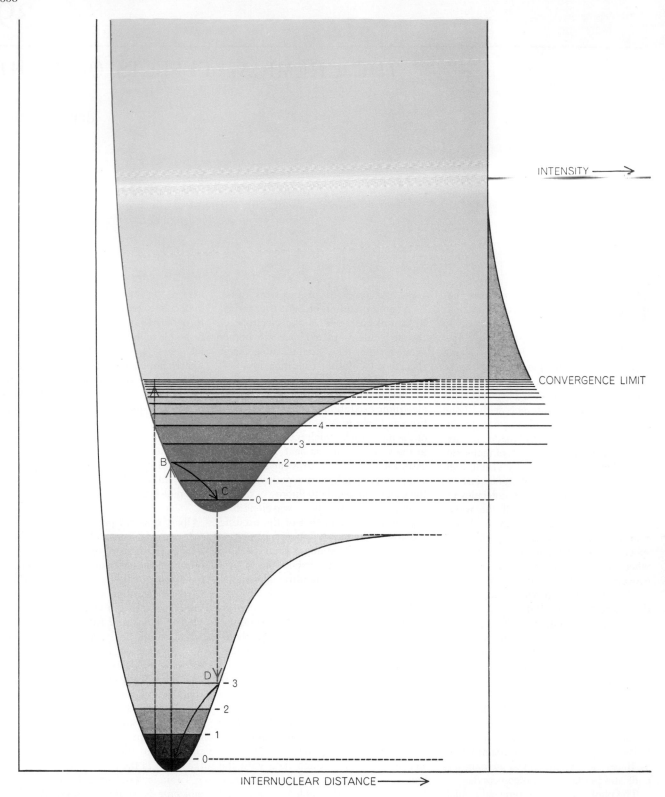

INTENSITY ⟶

CONVERGENCE LIMIT

4

3

2

1

0

D

3

2

1

A

0

INTERNUCLEAR DISTANCE ⟶

RESPONSE OF SIMPLE MOLECULES TO PHOTONS can be followed with the help of potential-energy curves. The lower curve represents the potential energy of a typical diatomic molecule in the ground state; the upper curve represents its potential energy in the first excited electronic state. Because the two atoms of the molecule are constantly vibrating, thus changing the distance between atomic nuclei, the molecule can occupy different but discrete energy levels (*horizontal lines*) within each electronic state. The molecule in the lowest ground state can be dissociated, or photolysed, if it absorbs a photon with an energy equal to or greater than ΔE_1. This is the energy required to carry the molecule to or beyond the "convergence limit." The length of the horizontal lines at the right below that limit represents the probability of transition from the ground electronic state to a particular vibrational level in the excited electronic state. Thus a photon with an energy of ΔE_2 will raise the molecule to the second level (B) of that state. There it will vibrate, ultimately lose energy to surrounding molecules and fall to C. It can now emit a photon with somewhat less energy than ΔE_2 and fall to D. This is called fluorescence. After losing vibrational energy molecule will return to A.

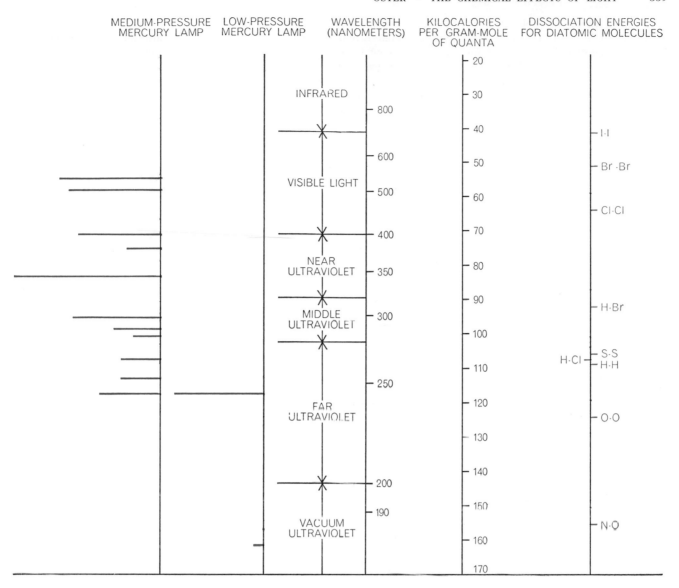

MEDIUM-PRESSURE MERCURY LAMP	LOW-PRESSURE MERCURY LAMP	WAVELENGTH (NANOMETERS)	KILOCALORIES PER GRAM-MOLE OF QUANTA	DISSOCIATION ENERGIES FOR DIATOMIC MOLECULES

DISSOCIATION ENERGIES of most common diatomic molecules are so high that the energy can be supplied only by radiation of ultraviolet wavelengths. The principal exceptions are molecules of chlorine, bromine and iodine, all of which are strongly colored, indicating that they absorb light. The energy carried by a quantum of radiation, or photon, is directly proportional to its frequency, or inversely proportional to its wavelength. There are 6.06×10^{23} photons in a gram-mole of quanta. This is the number required to dissociate a gram-mole of diatomic molecules (6.06×10^{23} molecules) if the quantum yield is unity. A gram-mole is the weight in grams equal to the molecular weight of a molecule, thus a gram-mole of oxygen (O_2) is 32 grams. The principal emission wavelengths of two commonly used types of mercury lamp are identified at the left. Lengths of the bars are proportional to intensity.

becomes large enough in the ground state, the molecule can come apart without ever entering the excited state. The curve for the excited state is displaced to the right of the curve for the ground state, indicating that the average interatomic distance (the minimum in each curve) is somewhat greater in the excited state than it is in the ground state. That is, the excited molecule is somewhat "looser."

The molecule can pass from the ground state to one of the levels of the excited state by absorbing radiation whose photon energy is equal to the energy difference between the ground state and one of the levels of the excited state. Provided that the quantum of radiation is not too energetic the molecule will remain intact and continue to vibrate. After a brief interval it will emit a quantum of fluorescent radiation and drop back to the ground state. Because the emission occurs when the excited molecule is at the lowest vibrational level, the emitted energy is less than the absorbed energy, hence the wavelength of the fluorescent radiation is greater than that of the absorbed radiation.

When the absorbed radiation exceeds a certain threshold value, the molecule comes apart; it is photolysed. At this point the absorption spectrum, shown at the right side of the illustration, becomes continuous, because the molecule is no longer vibrating at discrete energy levels. As long as the molecule is intact only discrete wavelengths of light can be absorbed.

It is possible for the excited state to pass to the ground state without releasing a quantum of radiation, in which case the electronic energy is dissipated as heat. Franck and Condon explained that this was accomplished by an overlapping, or crossing, of the two potential-energy curves, so that the excited molecule slides over, so to speak, to the ground state, leaving the molecule in an abnormally high state of vibration. This vibrational energy is then readily transferred to surrounding molecules.

As far as life on the earth is concerned, the most important photolytic reaction in nature is the one that creates a cano-

py of ozone in the upper atmosphere. Ozone is a faintly bluish gas whose molecules consist of three atoms of oxygen; ordinary oxygen molecules contain two atoms. Ozone absorbs broadly in the middle- and far-ultraviolet regions with a maximum at 255 nanometers. Fortu-

nately ozone filters out just those wavelengths that are fatal to living organisms.

Ozone production begins with the photolysis of oxygen molecules (O_2), which occurs when oxygen strongly absorbs ultraviolet radiation with a wavelength of 190 nanometers. The oxygen atoms released by photolysis may simply recombine or they may react with other oxygen molecules to produce ozone (O_3). When ozone, in turn, absorbs ultraviolet radiation from the sun, it is either photolysed (yielding O_2 and O) or it contributes to the heating of the atmosphere. A dynamic equilibrium is reached in which ozone photolysis balances ozone synthesis.

Early in this century physical chemists were presented with a photolytic puzzle. It was observed that when pure chlorine and hydrogen are exposed to ultraviolet radiation, the quantum yield approaches one million, that is, nearly a million molecules of hydrogen chloride (HCl) are produced for each quantum of radiation absorbed. This seemed to contradict Einstein's postulate that the quantum yield should be unity. In 1912 Max Bodenstein explained the puzzle by proposing that a chain reaction is involved [see upper illustration at left].

The chain reaction proceeds by means of two reactions, following the initial photolysis of chlorine (Cl_2). The first reaction, which involves the breaking of the fairly strong H-H bond, creates a small energy deficit. The second reaction, which involves the breaking of the weaker Cl-Cl bond, makes up the deficit with energy to spare. Breaking the H-H bond requires 104 kilocalories per gram-mole (the equivalent in grams of the molecular weight of the reactants, in this case H_2). Breaking the Cl-Cl bond requires only 58 kilocalories per gram-mole. In both of the reactions that break these bonds HCl is produced, yielding 103 kilocalories per gram-mole. Consequently the first reaction has a deficit of one kilocalorie per gram-mole and the second a surplus of 45 (103 − 58) kilocalories per gram-mole. The two reactions together provide a net of 44 kilocalories per gram-mole. Thus the chain reaction is fueled, once ultraviolet radiation provides the initial breaking of Cl-Cl bonds.

The chain continues until two chlorine atoms happen to encounter each other to form chlorine molecules. This takes place mainly at the walls of the reaction vessel, which can dissipate some of the excess electronic excitation energy of the chlorine atoms and allow chlorine mole-

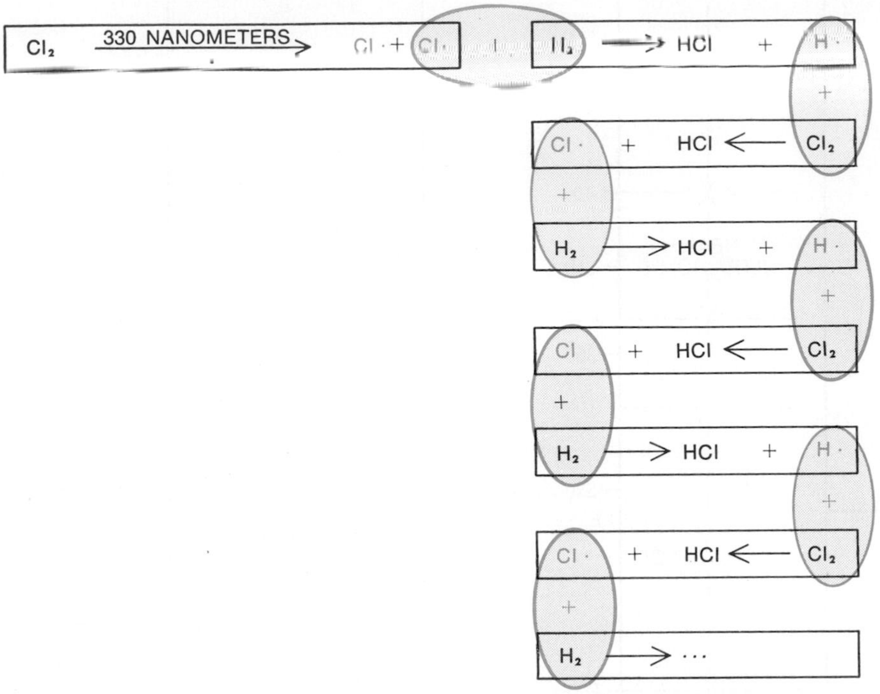

CHAIN REACTION is produced when pure chlorine and hydrogen are exposed to ultraviolet radiation. A wavelength of 330 nanometers is particularly effective. Such radiation is energetic enough to dissociate chlorine molecules, which requires only 58 kilocalories per gram-mole, but it is too weak to dissociate hydrogen molecules, which requires 104 kilocalories per gram-mole. The formation of HCl in the subsequent reactions provides 103 kilocalories per gram-mole. Since 104 kilocalories are needed for breaking the H-H bond, the reaction of atomic chlorine (Cl·) and H_2 involves a net deficit of one kilocalorie per gram-mole. However, the next reaction in the chain, involving H· and Cl_2, provides a surplus of 45 kilocalories (103 − 58). This energy surplus keeps the chain reaction going.

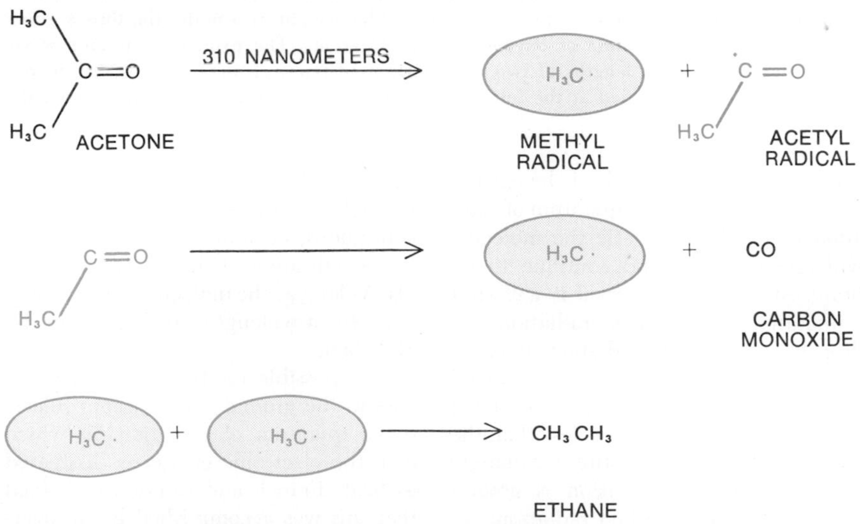

PHOTOLYSIS OF ACETONE, which yields primarily ethane and carbon monoxide, is a much studied photochemical reaction. It was finally understood by postulating the existence of short-lived free radicals, fragments that contain unsatisfied valence electrons.

cules to form. The free atoms may also be removed by impurities in the system.

Bromine molecules will likewise undergo a photochemical reaction with hydrogen to yield hydrogen bromide. The quantum yield is lower than in the chlorine-hydrogen reaction because atomic bromine reacts less vigorously with hydrogen than atomic chlorine does. Bromine atoms react readily, however, with olefins (linear or branched hydrocarbon molecules that contain one double bond). Each double bond is replaced by two bromine atoms. This is the basis of the industrial photobromination of hydrocarbons. Bromination can also be carried out by heating the reactants in the presence of a catalyst, but the product itself may be decomposed by such treatment. The advantage of the photochemical process is that the products formed are not affected by ultraviolet radiation.

An important industrial photochlorination process has been developed by the B. F. Goodrich Company. There it was discovered that when polyvinyl chloride is exposed to chlorine in the presence of ultraviolet radiation, the resulting plastic withstands a heat-distortion temperature 50 degrees Celsius higher than the untreated plastic does. As a result this inexpensive plastic can now be used as piping for hot-water plumbing systems.

A much studied photolytic reaction is one involving acetone (C_2H_6CO). When it is exposed to ultraviolet radiation, acetone gives rise to ethane (C_2H_6) with a quantum yield near unity, together with carbon monoxide and a variety of minor products, depending on the wavelength of excitation. The results can be explained by schemes that involve free radicals—fragments of molecules that have unsatisfied valence electrons. Photolysis of acetone produces the methyl radical (CH_3) and the acetyl radical (CH_3CO). Two methyl radicals combine to form ethane [see lower illustration on page 360].

W. A. Noyes, Jr., of the University of Rochester and others assumed the existence of these free radicals in order to explain the end products of the photolysis. Because the lifetime of free radicals may be only a ten-thousandth of a second, they cannot be isolated for study. Since the end of World War II, however, the technique of flash spectroscopy has been developed for recording their existence during their brief lifetime.

Flash spectroscopy was devised at the University of Cambridge by R. G. W. Norrish and his student George Porter, who is now director of the Royal Institution. They designed an apparatus [see illustration, page 362] in which a sample is illuminated with an intense burst of ultraviolet to create the photolytic products. A small fraction of a second later weaker light is beamed into the reaction chamber; at the far end of the chamber the light enters a spectrograph, which records whatever wavelengths have not been absorbed. The absorbed wavelengths provide clues to the nature of the short-lived species produced by photolysis. In 1967 Norrish and Porter shared the Nobel prize in chemistry with Manfred Eigen of the University of Göttingen, who had also developed techniques for studying fast reactions.

Flash spectroscopy has greatly increased chemists' knowledge about the "triplet state," an excited state that involves the pairs of electrons that form chemical bonds in organic molecules. Normally the spins of the paired electrons are antiparallel, or opposite to each other. When exposed to ultraviolet radiation, the molecules are raised to the first excited state and then undergo a nonradiative transition to an intermediate state in which the spins of two electrons in the same state are parallel to each other. This is the triplet state. If it is again exposed to ultraviolet or visible radiation, the triplet state exhibits its own absorption spectrum, which lies at a longer wavelength than the absorption spectrum of the normal ground state, or state of lowest energy [see illustration at left].

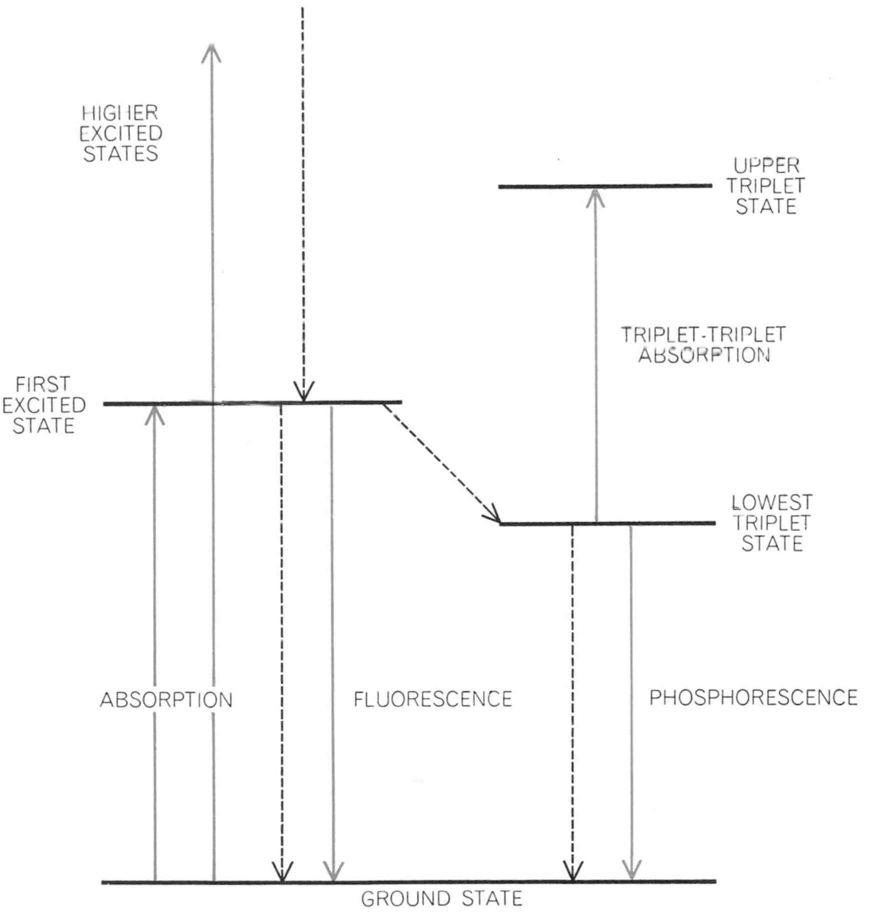

TRIPLET STATE has become an important concept in understanding the photochemical reactions of many organic molecules. Like all molecules, they can be raised to an excited state by absorption of radiation. They can also return to the ground state by normal fluorescence: reemission of a photon. Alternatively, they can drop to the triplet state without emission of radiation. (Broken lines indicate nonradiative transitions.) The existence of this state can be inferred from the wavelength of the radiation it is then able to absorb in passing to a higher triplet state. The triplet state arises when the spins of paired electrons point in the same direction rather than in the opposite direction, as they ordinarily do.

The concept of the triplet state in organic molecules is due mainly to the work of G. N. Lewis and his collaborators at the University of California at Berkeley in the late 1930's and early 1940's. These workers found that when dyes (notably fluorescein) are dissolved in a rigid medium such as glass and are exposed to a strong light, the dyes change color. When the light is removed, the dyes revert to their normal color after a second or so. This general phenomenon is called photochromism. Lewis deduced the existence of the triplet state and ascribed its fairly long duration to the time required for the parallel-spin electrons to become uncoupled and to revert to their normal antiparallel arrangement.

In 1959 Porter and M. W. Windsor used flash spectroscopy to search for the triplet state in the spectra of organic molecules in ordinary fluid solvents. They were almost immediately successful. They found that under such conditions the triplet state has a lifetime of about a millisecond.

In his Nobel prize lecture Porter said: "Any discussion of mechanism in organic photochemistry immediately involves the triplet state, and questions about this state are most directly answered by means of flash photolysis. It is now known that many of the most important photochemical reactions in solution, such as those of ketones and quinones, proceed almost exclusively via the triplet state, and the properties of this state therefore become of prime importance."

While studying the photochemistry of dyes in solution, my student Albert H. Adelman and I, working at the Polytechnic Institute of Brooklyn, demonstrated that the chemically reactive

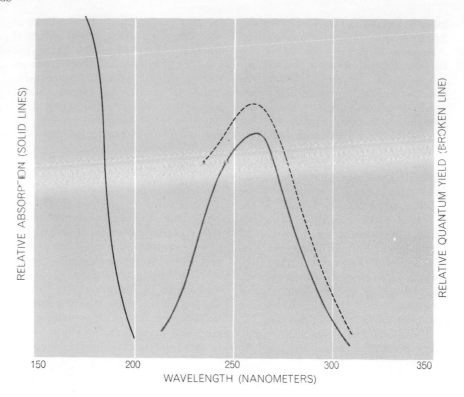

ABSORPTION SPECTRUM OF OZONE (*solid curve in color*) peaks at about 250 nanometers in the ultraviolet. As a happy consequence, the canopy of ozone in the upper atmosphere removes the portion of the sun's radiation that would be most harmful to life. The biocidal effectiveness of ultraviolet radiation is shown by the broken black line. The solid black curve is the absorption spectrum of molecular oxygen. For reasons not well understood, ultraviolet radiation of 200 nanometers does not penetrate the atmosphere.

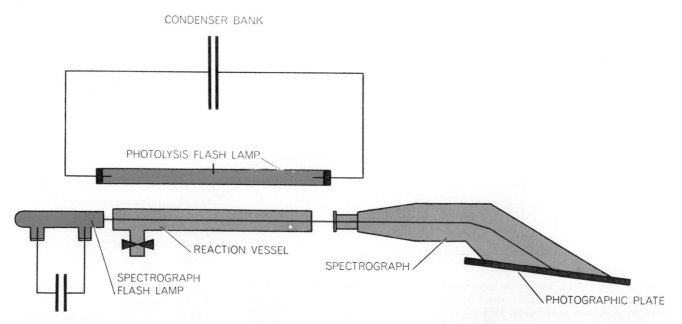

APPARATUS FOR FLASH PHOTOLYSIS was devised by R. G. W. Norrish and George Porter at the University of Cambridge. With it they discovered the short-lived triplet state that follows the photolysis of various kinds of molecules, organic as well as inorganic. The initial dissociation is triggered by the photolysis flash lamp, which produces an intense burst of ultraviolet radiation. A millisecond or less later another flash lamp sends a beam of ultraviolet radiation through the reaction vessel. Free radicals in the triplet state absorb various wavelengths ("triplet-triplet" absorption) and the resulting spectrum is recorded by the spectrograph.

species is the triplet state of the dye. Specifically, when certain dyes are excited by light in the presence of electron-donating substances, the dyes are rapidly changed into the colorless ("reduced") form. Our studies showed that the reactive state of the dye—the triplet state—has a lifetime of about a tenth of a millisecond. The dye is now a powerful reducing agent and will donate electrons to other substances, with the dye being returned to its oxidized state [see illustration on page 364]. In other words, the dye is a photosensitizer for chemical reductions; visible light provides the energy for getting the reaction started.

In the course of these studies I discovered that free radicals are created when dyes are photoreduced. The free radicals make their presence known by causing vinyl monomers to link up into polymers. The use of free radicals for bringing about polymerization of monomers is well known in industry. It occurred to me that adding suitable dyes to monomer solutions would provide the basis for a new kind of photography. In such a solution the concentration of free radicals would be proportional to the intensity of the visible light and thus the degree of polymerization would be controlled by light. It has turned out that very accurate three-dimensional topographical maps can be produced in plastic by this method.

The use of dyes as photosensitizing agents is, of course, fundamental to photography. In 1873 Hermann Wilhelm Vogel found that by adding dyes to silver halide emulsions he could make photographic plates that were sensitive to visible light. At first such plates responded only to light at the blue end of the spectrum. Later new dyes were found that extended the sensitivity farther and farther toward the red end of the spectrum, making possible panchromatic emulsions. Photographic firms continue to synthesize new dyes in a search for sensitizers that will act efficiently in the infrared part of the spectrum. The nature of the action of sensitizers in silver halide photography is still obscure, nearly 100 years after the effect was first demonstrated. The effect seems to depend on the state of aggregation of the dye absorbed to the silver halide crystals.

The reverse of photoreduction—photo-oxidation—can also be mediated by dyes, as we have found in our laboratory. Here again the reactive species of the dye is the dye in the triplet state. We have found that the only dyes that will serve as sensitizers for photooxidation are those that can be reduced in the presence of light.

The oxidized dye—the dye peroxide—is a powerful oxidizing agent. In the process of oxidizing other substances the dye is regenerated [see illustration on page 364]. My student Judith S. Bellin and I have demonstrated this phenomenon, and we have employed dye-sensitized photooxidation to inactivate some biological systems. These systems include viruses, DNA and ascites tumor cells. That dyes are visible-light sensitizers for biological inactivation was first demonstrated in 1900 by O. Raab, who observed that a dye that did not kill a culture of protozoa did so when the culture was placed near a window.

The inactivation that results from dye sensitization is different from the inactivation that results when biological sys-

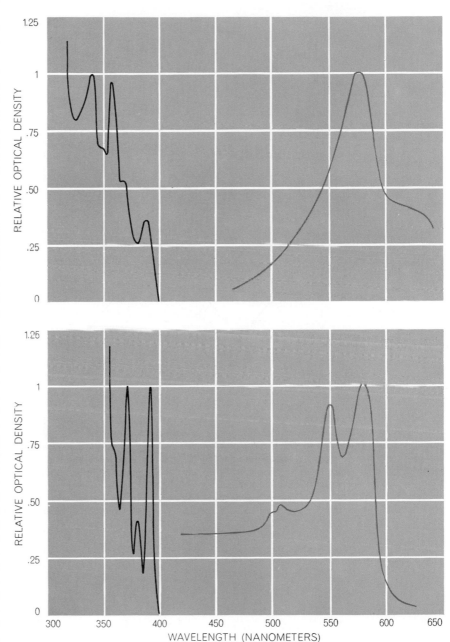

TRIPLET-TRIPLET ABSORPTION OF VISIBLE LIGHT has been observed in the author's laboratory at the Polytechnic Institute of Brooklyn. His equipment sends a beam of ultraviolet radiation into samples embedded in a plastic matrix in one direction and visible light at right angles to the ultraviolet radiation. The visible absorption spectra are then recorded in the presence of ultraviolet radiation. The black curves at the left in these two examples show the absorption of the electronic ground state. The colored curves at the right show the absorption of visible wavelengths that raises the excited molecule from the lowest triplet state to upper triplet states. The top spectra were produced by chrysene, the lower spectra by 1,2,5,6-dibenzanthracene. Both are aromatic coal-tar hydrocarbons.

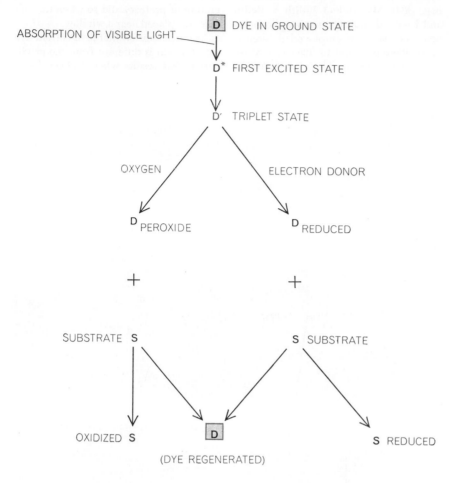

ABSORPTION OF VISIBLE LIGHT ⎯⎯ [D] DYE IN GROUND STATE

D* FIRST EXCITED STATE

D' TRIPLET STATE

OXYGEN   ELECTRON DONOR

D PEROXIDE   D REDUCED

+   +

SUBSTRATE S   S SUBSTRATE

OXIDIZED S   [D]   S REDUCED
(DYE REGENERATED)

UNUSUAL PROPERTIES OF TRIPLET STATE have been explored by the author. Certain dyes in the triplet state can act either as strong oxidizing or as strong reducing agents, depending on the conditions to which the triplet state itself is exposed. In the presence of a substance that donates electrons (i.e., a reducing agent), the dye is reduced and can then donate electrons to some other substance (*substrate S*). In the presence of an oxidizing agent, the dye becomes highly oxidized and can then oxidize, or remove electrons from, a substrate. In both cases the dye is regenerated and returns to its normal state. The author's studies show that the reactive state of the dye lives only about .1 millisecond.

tems are exposed to ultraviolet radiation. Here the inactivation often seems to result from the production of dimers: the cross-linking of two identical or similar chemical subunits. Photodimerization is implicated, for example, in the bactericidal action of ultraviolet radiation. It has long been known that the bactericidal action spectrum (the extent of killing as a function of wavelength) closely parallels the absorption spectrum of DNA, the genetic material. If dried-down films of DNA are irradiated with ultraviolet, they become cross-linked. According to one view the cross-linking occurs by means of the dimerization of thymine, one of the constituent groups of DNA.

Although this may well be the mode of action of ultraviolet radiation, my own feeling is that insufficient consideration has been given to the photolysis of the disulfide bonds of the proteins in bacteria. This bond is readily cleaved by ultraviolet radiation and has an absorption spectrum resembling that of DNA. Disulfide bonds are vital in maintaining the structure and activity of proteins; their destruction by ultraviolet radiation could also account for the death of bacteria.

In using dyes as sensitizers for initiating chemical reactions we are taking our first tentative steps into a realm where nature has learned to work with consummate finesse. Carbon dioxide and water are completely stable in the presence of visible light. Inside the leaves of plants, however, the green dye chlorophyll, when acted on by light, mediates a sequence of chemical reactions that dissociates carbon dioxide and water and reassembles their constituents into sugars and starches. A dream of photochemists is to find a dye, or sensitizer, that will bring about the same reactions in a nonliving system. There is reason to hope that such a system could be a good deal simpler than a living cell.

40

HOW LIGHT INTERACTS WITH LIVING MATTER

STERLING B. HENDRICKS

September 1968

*Light activates three key processes of life:
photosynthesis, vision and photoperiodism
(the response of plants and animals to the cycle
of night and day). Such activation is mediated
by specific pigments*

Life is believed to have arisen in a primordial broth formed by sunlight acting on simple molecules at the surface of the cooling earth. It could have been sustained by the broth for aeons, but eventually, with the arrival of photosynthesis, some living things came to use sunlight more directly. So it remains today, with photosynthesis by plants serving to capture sunlight for the energy needs of all forms of life.

As various kinds of animals evolved, the ones that were best able to sense their surroundings were favored to survive. Because light acts over considerable distances it is well suited to sensing. To exploit light the animals needed some kind of detector: a tissue, an eyespot or an eye. The detector had to be coupled to a responding system: a ganglion or a brain. Signals from the system controlled locomotion toward food or away from danger.

Photosynthesis and vision do not exhaust the potential of the luminous environment. Both plants and animals have evolved mechanisms to respond to the changing daily cycle of light and dark. It is this photoperiodism that provides the seasonal schedule for, among other things, the flowering of plants, the pupation of insects and the nesting of birds.

To understand these phenomena one must ask how light acts in life. Part of the answer is very simple: it acts by exciting certain absorbing molecules. What happens to the molecules in the course of absorption is more difficult to describe, but many details of the processes are now reasonably well known. On the other hand, our ideas about how the molecular events are coupled to the responses of plants and animals are still quite tentative.

In discussing the present state of knowledge about light and life I shall treat vision first because this phenomenon has some features in common with both photosynthesis and photoperiodism. In all three processes light acts through absorption by a small, colored molecule—a chromophore—that is associated with a large molecule of protein. In the case of vision the light-sensitive molecules are responsible for the pink and purplish color of the retina. In the retina of the human eye there are some 100 million thin rod-shaped cells and five million slightly cone-shaped ones. Each is connected through a synapse, or junction, to a nerve fiber leading to the brain. Electron micrographs show that the outer end of both rods and cones is packed with thin membranous sacs, and with these sacs are associated the light-absorbing chromophores. (Vision and photosynthesis share this association of a chromophore with a membrane.) Excitation of the chromophore by light causes some kind of change in the membrane, and this change gives rise to a signal in the nerve fiber.

In vision the nature of the receiving chromophore and the manner of its excitation by light are well understood. Both have much in common with light reception in photoperiodism. As George Wald of Harvard University established, the receiving chromophore is vitamin-A aldehyde (in structural terms 11-*cis* retinal). The chromophore is found in association with a protein, opsin. The opsins are fatty proteins; thus they have an affinity for the sac membranes, which consist largely of lipid—that is, fatty—material. There are four types of opsin, one in the rods and three in the cones. Combined with 11-*cis* retinal, they respectively form rhodopsin and three kinds of iodopsin. On excitation by light all four opsins change in the same way.

In vision, photosynthesis and photoperiodism alike the chromophore molecule is notable for its alternating single and double chemical bonds. Known to chemists as conjugated systems, molecules of this kind are structurally quite stable because the groups of atoms attached by double bonds cannot rotate around the bonds. Each conjugated system, if it is adequately extended, has a rather low energy state that can be excited by visible light. When the system is excited, its double-bond character is somewhat relaxed, so that a *cis* configuration can change to a *trans* one [*see top illustration on page 367*]. This ability to change form is a key element in vision and photoperiodism. In photosynthesis, however, no change of form takes place because the change is constrained by the ring structures of the chlorophyll chromophores.

The effects of light in vision and photoperiodism are determined by measur-

LIGHT-SENSITIVE PIGMENT that triggers photoperiodic responses in plants is shown in its two states in the photograph on the following page. Called phytochrome, the pigment, which is seen here in a .2 percent solution, is instrumental in a number of seasonal occurrences such as plants flowering and seeds germinating. In one state (*left*) phytochrome is excitable by far-red light, in the other (*right*) by red light. Alternating exposures to these colors change the pigment from one state to the other and back again. The phytochrome shown here was extracted from oat seedlings in the laboratory of F. E. Mumford and E. L. Jenner in the Central Research Department of E. I. du Pont de Nemours & Co. It is contained in square quartz cells designed for studies of light absorption. The faint numerals near the top indicates the length of the light path through the cell: 1.000 centimeter.

ing the molecular changes produced by light excitation. Some of these changes are very rapid: they may occur in less than a millionth of a second. Changes as fast as this can be followed only if they are excited in an even shorter time, for instance by a very brief but intense flash of light. The measurement of the change also must be made quite rapidly. The method employed is flash excitation at room temperature or lower, followed by photoanalysis—the technique for which George Porter and R. G. W. Norrish shared (with Manfred Eigen) the 1967 Nobel prize in chemistry [see "The Chemical Effects of Light," by Gerald Oster, beginning on page 357]. Low temperatures slow down the molecular changes and make them more amenable to measurement.

When the vision chromophore is excited by light, it changes from the cis to the trans form. The result is the conversion of rhodopsin into prelumirhodopsin with an all-trans chromophore. The production of this single change is the one and only role that light plays in vision. The change is followed by several rapid shifts in the structure of the opsin and also changes in the relation of the chromophore to the opsin. To judge by the time it takes for a retinal-cell signal to arrive at a nerve ending, the signal is induced by the shifts that take place in the first thousandth of a second.

The course of the molecular changes can be traced by studying rhodopsin in solution. Prelumirhodopsin, identifiable by its maximum absorption of light at a wavelength of 543 nanometers, can be held at temperatures below −140 degrees Celsius and reversibly changed back into rhodopsin. When the prelumirhodopsin is warmed to −40 degrees C., it is converted into lumirhodopsin. The same conversion probably occurs at body temperature but much more rapidly. This change and subsequent ones, including the formation of metarhodopsins, involve shifts in the molecular configuration of opsin. Among vertebrates the changes finally lead to the dissociation of the chromophore from the protein. The released all-trans retinal then has to be reduced to the alcohol form and oxidized back to the aldehyde form to regenerate 11-cis retinal. Once the cis retinal is regenerated it spontaneously recombines with opsin to form rhodopsin again.

Analysis of the changes in electric potential that occur simultaneously with these molecular changes shows that a potential appears within 25 millionths of

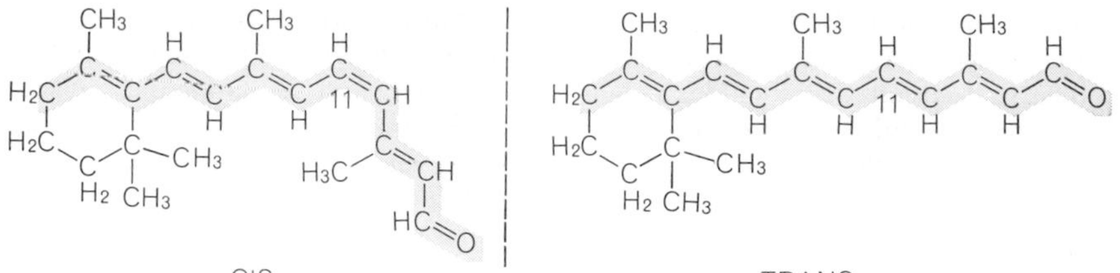

VISION depends on a light-sensitive chromophore molecule, 11-cis retinal (left), which has alternating single and double bonds (color). When light excites the molecule, its configuration changes from cis to the trans form (right), thereby setting in train a series of complex changes in the structure of the proteins with which the retinal chromophore is associated.

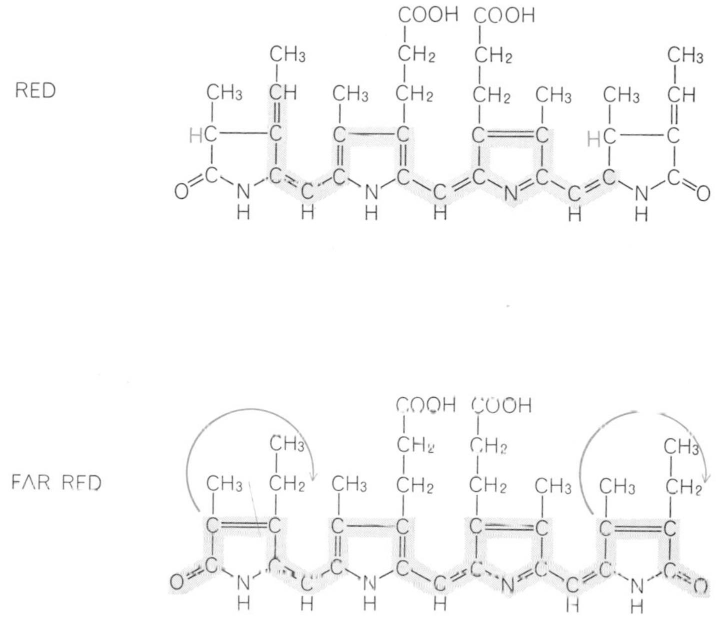

PHOTOPERIODISM in plants depends on phytochrome, another molecule that is sensitive to light. Like the retinal molecule, phytochrome has alternating single and double bonds (color). When excited by light, it changes from a configuration sensitive to red light (top) to one sensitive to far-red light, probably because two hydrogen atoms shift (bottom).

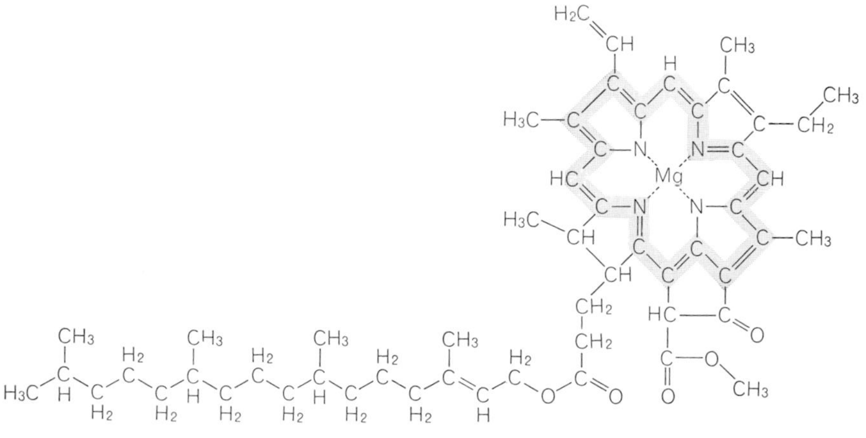

PHOTOSYNTHESIS depends on the light-sensitive chromophore molecules of several kinds of chlorophylls that have differing side groups. The molecule shown here is chlorophyll a. Like the vision and photoperiodism chromophores, chlorophyll molecules include singly and doubly bonded atoms, but these form a closed loop within the chlorophyllin portion of the molecule (color). When excited by light, chlorophylls forward the energy they receive to centers where it induces chemical changes (see illustrations on page 373).

a second after a flash of light. The potential is positive with respect to the cornea in a circuit that includes neighboring tissues and the retina. The positive potential is followed in a thousandth of a second by a growing signal of the opposite sign. These events take place during the period when prelumirhodopsin and lumirhodopsin are present. The first potential probably accompanies the change of rhodopsin to prelumirhodopsin. The second depends markedly on the temperature at some distance from the place of light action, probably in the outer membrane of the rod or cone. Currently there is much interest in the possible identification of these changes in

potential as early steps in the eventual excitation of the nerve fiber. Another view is that nerve excitation is associated with the transitions involving metarhodopsin I and metarhodopsin II [see upper illustration below].

Color vision depends on the three opsins, each found in a different cone cell. Their absorption spectra have been measured, and curves were found with peaks at wavelengths of 450, 525 and 555 nanometers (respectively in the blue, green and yellow regions of the visible spectrum). Activation by light leads to the same sequence of molecular changes described for rhodopsin. The singularity of the nerve associations with the rod and

cone cells preserves the retinal detail, or register, in the transmission of the visual signal; the differences in absorption among the three kinds of cone retain the color pattern of the image.

The responses of plants to variations in the length of day and night involve light-induced molecular changes that closely parallel those involved in vision. Because photoperiodic responses are not as well known as visual ones I shall present some illustrative examples. Chrysanthemums and many other plants flower in response to the increasing length of the nights as fall approaches. If the long nights are experimentally inter-

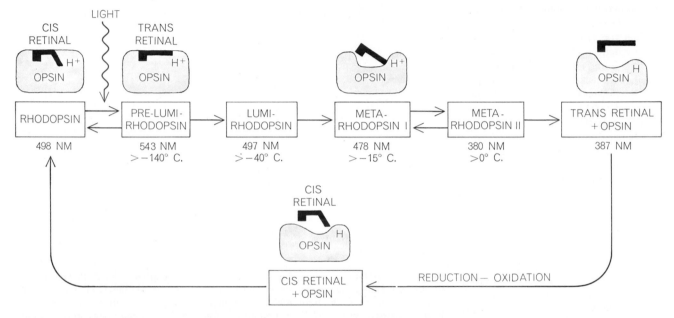

CHANGES IN RHODOPSIN, the visual pigment contained in the rod cells of the vertebrate eye, can be traced in the laboratory at low temperatures. Four successive forms of the pigment appear (rectangles) as the retinal molecule (black) first attains its trans configuration and then dissociates from the protein opsin. When it becomes cis retinal again, it recombines and completes the cycle.

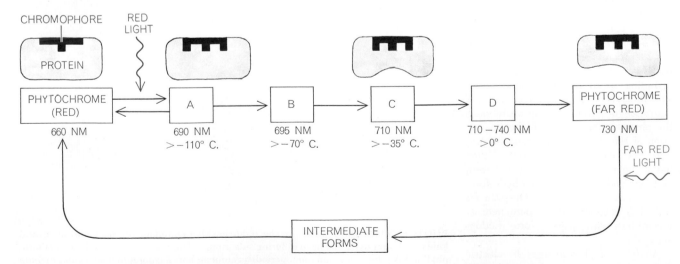

CHANGES IN PHYTOCHROME, the plant-photoperiodism chromophore, can also be traced at low temperature. As with rhodopsin, intermediate forms with characteristic light-absorption peaks appear before the initial red-absorbing form of the pigment is turned into the far-red-absorbing form. Unlike retinal, the chromophore (black) remains associated with its protein throughout the cycle.

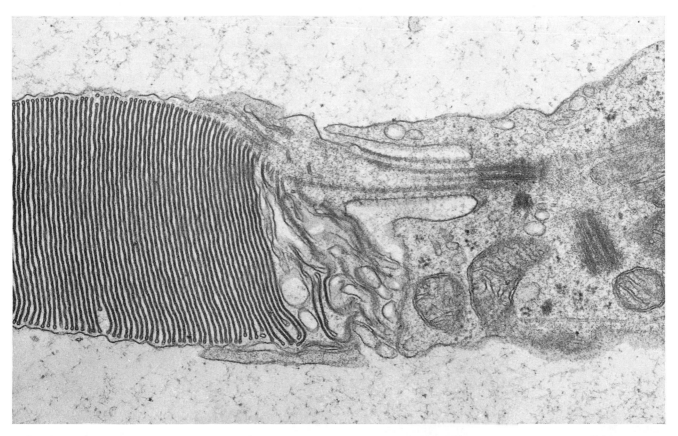

PART OF HUMAN ROD is magnified 44,000 times in electron micrograph. The outer segment of the rod (*left*) is filled with the membrane of the lamellae. The inner segment (*right*) is less complex in structure. This micrograph was also made by Kuwabara.

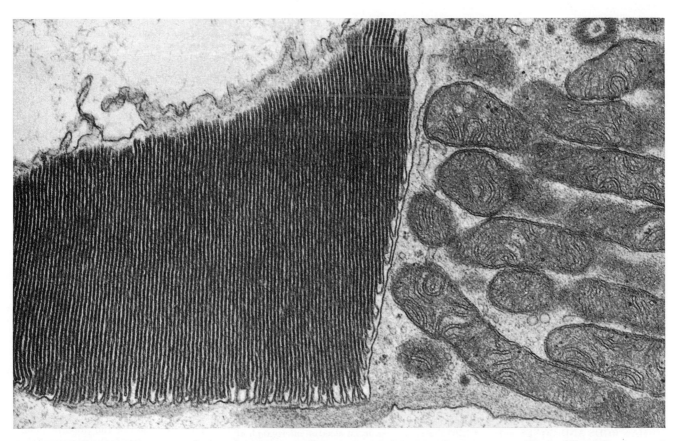

PART OF HUMAN CONE is magnified 44,000 times in another micrograph by Kuwabara that emphasizes the area connecting the structure's inner and outer segments. The lamellae differ from rod lamellae in being "packaged," some singly and some in groups.

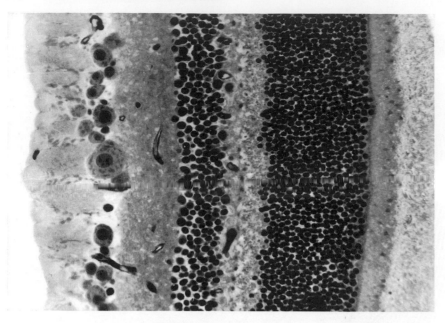

CAT RETINA is seen in cross section, enlarged 670 times. The nerve-fiber layer (*left*) is the part of the retina that lies in contact with the eye's vitreous body. The entering light must penetrate this and five additional layers of retinal tissue before reaching rods and cones (*right*). The micrograph was made by A. J. Ladman of the University of New Mexico.

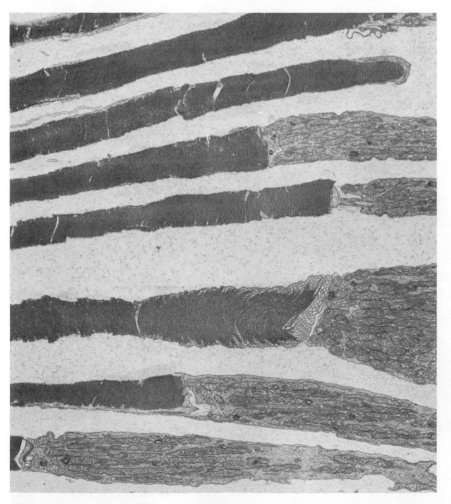

HUMAN RODS AND CONES, enlarged 7,200 times, are seen in an electron micrograph of retina. The visual pigments are concentrated in platelike layers of membrane called lamellae. The micrograph was made by Toichiro Kuwabara of the Harvard Medical School.

rupted by exposing the plants to short periods of light near midnight, the plants will not flower. Red light with an absorption maximum at a wavelength of 660 nanometers is most effective in preventing flower formation. Thus we anticipate that the light-receiving pigment in the plant is blue—the complementary color to the absorbed red. If shortly after exposure to red light the plants are exposed to light near the limit of vision in the far red (700 nanometers), they will flower.

The ornamental plant kalanchoe clearly illustrates the reversible response. The red light evidently converts the photoreversible pigment to a far-red-absorbing form. This changes the plant from the flowering state to the nonflowering one. The far-red light returns the pigment to its red-absorbing form, which enables flowering to proceed. Control of flowering by length of night is a very important factor in determining what varieties of soybean, wheat and other commercial crops are best suited for being grown in various latitudes with different periods of light and darkness.

Many kinds of seeds will germinate only if the photoreversible pigment has been activated. The seeds of some pine and lettuce species, for example, will not germinate in the laboratory unless they are briefly exposed to red light (or, to be sure, light containing red light). If the red-light activation of the seeds is followed by a short exposure to far-red light before the seeds are returned to darkness, the seeds remain dormant. The activation-reversal cycle can be repeated many times; germination or continued dormancy depends on the last exposure in the sequence.

The requirement of light for seed germination is a major cause of the persistence of weeds in cultivated crops. A seed that is dormant when it first falls to the ground is usually covered by soil in the course of the winter. As the seed lies buried the pigment that controls its germination changes into the red-absorbing form; now the seed will not germinate until it is again exposed to sunlight by cultivation or some other disturbance of the soil. When it is exposed, the sunlight converts some of the red-absorbing pigment back to the far-red-absorbing form, and germination begins. Seeds of one common weed, lamb's-quarters, are known to have lain buried for 1,700 years and then to have germinated on exposure to light.

The activation of the photoreversible pigment also controls the growth of trees and many common flowering plants. If such plants are to continue growing, they must have long periods of daylight. As the days become short growth stops and the plants' buds go into a dormant state that protects them against the low temperatures of winter.

The photoreversible pigment of plants has been named phytochrome. It is invisible in plant tissue because of its low concentration. It was isolated by methods widely used in the preparation of enzymes and other proteins. The pigment is indeed blue [see illustration on page 366]. Its photoreversibility is exactly what was expected on the basis of plant responses to light.

The chemical structure of the phytochrome molecule shows that it is related to the greenish-yellow pigments of human bile and the blue pigments of blue-green algae. The molecule comprises an open group of atoms that is closely related to the rings in the chlorophyll molecule. It has two side groups that can change from the cis form to the trans when they are excited by light. A more probable excitation change, however, is a shift in the position of the molecule's hydrogen atoms.

The changes in the phytochrome molecule following excitation by a flash of light are similar to those in rhodopsin. The first excitation response takes place in a few millionths of a second and gives rise to a form of the molecule that is analogous to prelumirhodopsin. The change stops at this point if the temperature is below −110 degrees C. At these low temperatures the molecule can be reconverted into its initial red-absorbing form by the action of light. At temperatures higher than −110 degrees several more intermediate phytochromes are formed before the final far-red-absorbing molecule appears. These intermediate stages also involve alterations in the molecular form of the protein associated with phytochrome, just as there are alterations in the form of opsin, the protein of rhodopsin. In its final form phytochrome differs from rhodopsin in that the molecule of phytochrome remains linked to the protein rather than being dissociated from it. Far-red light will reverse the process and convert the final form of phytochrome back to its initial red-absorbing form, although a different series of intermediate molecular forms is involved.

Flowering, seed germination and most other plant responses follow slowly on the excitation of phytochrome. Unlike vision, in which the response follows the rapid appearance of intermediate

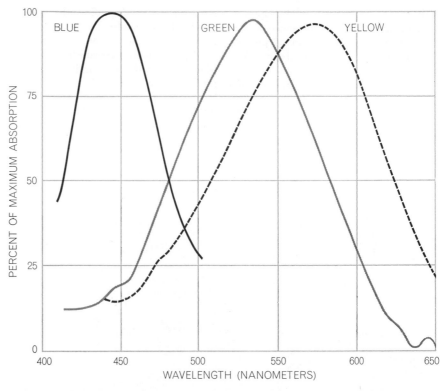

COLOR PERCEPTION in humans arises from the combination of retinal with three dissimilar opsins in the cones of the retina. The three different iodopsin pigments formed thereby absorb the greatest amount of visible light at three different wavelengths. The differences between the signals from each group of cones reflect the color pattern of the image.

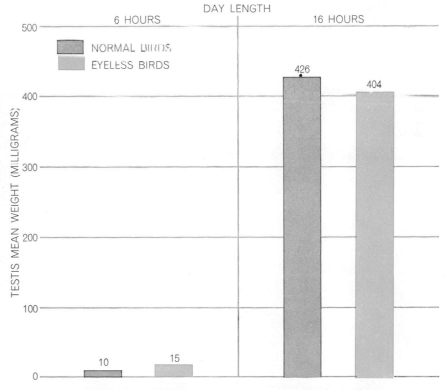

PHOTOPERIODISM IN SPARROWS has been shown to involve some light receptor other than the eye. The testis weight of both eyeless and normal sparrows remained low when their cages were lighted to simulate short days and long nights over a two-month period (left). When eyeless and normal birds instead underwent two months of long days and short nights, their testis weight showed a nearly identical increase (right). The experiment was conducted by Michael Menaker and Henry Keatts of the University of Texas.

molecules, the photoperiodic response of plants depends on the presence of the final, far-red-absorbing form of phytochrome. Little is known about how the far-red-absorbing molecule does its work. One view is that it regulates enzyme production by controlling the genetic material in cell nuclei. Another view is that the molecule's lipid solubility results in its being attached to membranes in the cell, such as the cell wall and the membrane of the nucleus. Changes in the form of the phytochrome molecule would then affect the permeability of the membranes and therefore the functioning of the cell.

The continuous exposure of plants to blue and far-red wavelengths in the visible spectrum opposes the action of the far-red-absorbing form of the phytochrome molecule. It may be that excitation by far-red light causes a continuous displacement of the far-red-absorbing molecules from cell membranes. Continuous excitation of this kind is what happens, for example, during the long light periods that so markedly influence the growth of Douglas firs. If the trees are exposed to 12-hour days and 12-hour nights, they remain dormant. If the length of the day increases, however, they grow continuously.

Photoperiodism is not confined to the plant kingdom: animals also respond to changes in the length of the day. The migration and reproduction of many birds, the activity cycles of numerous mammals and the diapause (suspended animation) of insects are controlled in

LONG LIGHT PERIODS markedly influence the growth of Douglas fir. When exposed to short days, or days and nights of equal length, the tree will remain dormant (*left*). Excitation by additional light produces continuous growth. One tree (*center*) received an hour of dim illumination during its 12-hour night; the other (*right*) had its 12-hour day extended by eight hours of dim light.

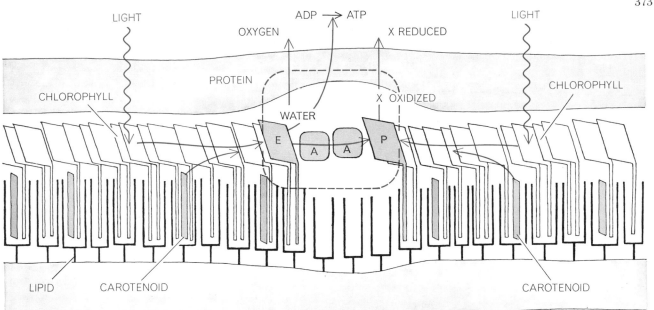

PLANT PHOTORECEPTORS contain molecular arrays, shown here in schematic form. Within the chloroplast, chlorophyll molecules are held together both by their mutual attraction and by the affinity of each molecule's phytol "tail" for lipids and its main body for proteins. Other molecules of pigment, such as carotenoids, are also embedded in the array. For each 500 or so chlorophyll molecules is found a specialized energy-transfer center, comprising two energy sinks, E and P (color, center), linked by a system for transferring electrons, represented here by units labeled A. The electron-cascade system by means of which this array of molecules turns energy from light (colored inward arrows) into chemical energy (outward arrows) is seen in detail in the illustration below.

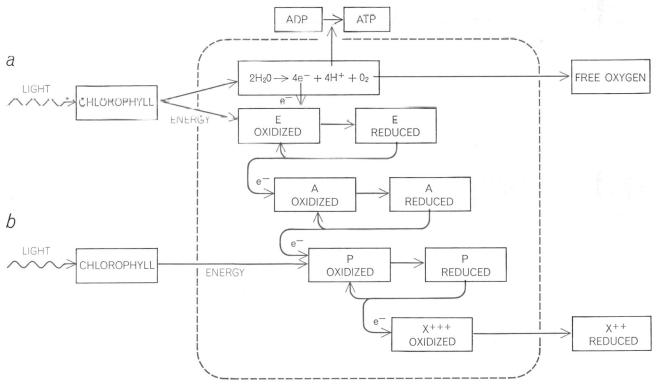

ELECTRON CASCADE that is responsible for the main action of photosynthesis, the use of energy from light to reduce carbon dioxide to sugar, is shown schematically here. Within a zone that contains two energy sinks, E and P, water and its components (top rectangle) are in a state of equilibrium until several chlorophyll molecules pass the energy received from light (a) to the first energy sink, E. This event starts the cascade; E, driven to a higher state of excitation by the energy it receives, seizes an electron (colored arrow) from a water component. Next E falls back to its lower state; the seized electron is released and cascades onward via the transfer system A. It arrives at the second energy sink, P, soon after that sink has received energy from other light-excited chlorophyll molecules (b). The cascade ends when P falls back to its lower state, passing both electron and energy to X, an electron-rich compound. The event energizes X sufficiently to let it power the carbon dioxide reduction process (arrow, lower right); this is the main photosynthetic action. Two other events, however, are also consequences of the cascade. Hydrogen ions (top arrow) provide the energy gradient needed to transform adenosine diphosphate (ADP) into adenosine triphosphate (ATP). Similarly, the ion neutralized by the electron loss that initiated the cascade joins with other components to form water, thereby freeing oxygen (arrow, upper right).

this way. These examples of photoperiodism (and some less clear-cut responses in man) depend on the action of several hormones working in sequence. Such sequences of hormone action can have a regular rhythm. They provide a basis for the circadian (meaning about one day) rhythms of "biological clocks." The 24-hour cycle of such clocks is established by light.

The diapause of insects illustrates one form of the interplay of hormone action and light. Some silkworms and the larva of the codling moth, for example, go into a dormant form when the days are short. In this state, which helps the insects to survive the winter, the release of a hormone from a group of cells in the central part of the brain is suspended. The unreleased hormone is the first in a series that leads to a final hormone, ecdysone, that controls the metamorphosis of the pupa into an adult moth. When the brain cells of the dormant pupa are exposed to light for long days, the brain hormone is released and triggers the metamorphosis. Ecdysone injected into a resting pupa brings on metamorphosis even when the days are short.

Here the sole action of the light is the release of a brain hormone. The pigment in the brain cells that absorbs the light has not yet been identified, but current work in the U.S. Department of Agriculture on the response to light by codling moths in diapause promises an answer. The blue-green part of the spectrum (between 500 and 560 nanometers), and probably shorter wavelengths as well, appears to be most effective for breaking diapause. The pigment is possibly of the porphyrin type, with a central structure resembling the ring system of chlorophyll.

Man's dependence on a biological clock is apparent in the unease he feels when the relation of his circadian rhythm to the actual cycle of day and night is quickly disrupted, as it is when he travels by air for distances measured in many degrees of longitude. His hormonal controls are disturbed or out of phase. Deer mice and other small mammals also display cyclic periods of activity such as running that seem to be regulated by light.

Involved at an early stage in the release of the hormones that trigger activity cycles is the region of the brain known as the hypothalamus. Whether this region contains a pigment receptor for the small amount of light that might penetrate the skull or whether it is stimulated by a signal from the eye, or the region of the eye, is unknown. The hypothalamus also controls the pituitary, hormones from which affect the reproductive organs, the cortex of the adrenal gland and other target organs. At present, however, the existence of a pigment responsible for vertebrate photoperiodism, its physical location and the nature of its action on the molecular level remain to be established.

By exciting a chromophore light acts as a trigger both in vision and in photoperiodism, initiating processes that depend for their energy on the organism's own metabolism. In the third major area of light's interaction with living matter— photosynthesis—the opposite is true: the energy of light is utilized to manufacture the fuels that support life. For this to happen there must be (1) a system to receive the light, (2) an arrangement to transfer energy between molecules and (3) some means of coupling light energy to chemical change. Chlorophyll molecules (or rather the molecules of several chlorophylls that differ in their side groups) constitute the principal receiving system. An electron in the chlorophyll molecule is excited from its normal energy level to a higher level by the impact of visible light. The excited electron reverts to its normal state in less than a hundred-millionth of a second. This reversion might be accomplished by the reemission of visible light, but that, of course, would not advance the photosynthetic process. Instead the reversion proceeds in several steps during which the energy necessary for photosynthesis is transferred along a chain of molecules. A small part of the energy released by the reversion is reemitted as light of a longer wavelength, and hence of lower energy, than the light that was absorbed. (This is the dark red light characteristically emitted by chlorophyll when it is excited to fluorescence.) The remainder of the energy is transferred by way of other chlorophyll molecules to an ultimate recipient, a molecule that receives the energy and effects the chemi-

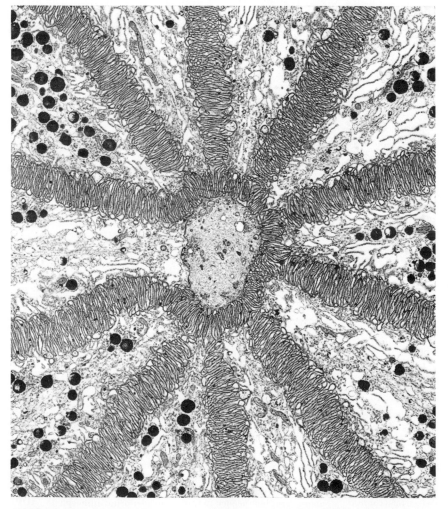

LIGHT-RECEPTOR ORGAN of an invertebrate is enlarged 8,000 times in an electron micrograph. Called an ommatidium, it is one of about 1,000 such units that comprise a horseshoe crab's compound eye. The spokelike arrays and the central ring are photosensitive. William H. Miller of the Yale University School of Medicine made the micrograph.

cal synthesis. For these energy transfers to be efficient the molecules in the chain must meet two criteria. First, they must be physically close together. Second, there must be a close match between the amount of energy available from the donor molecule and the amount acceptable by the recipient.

The plastids of plant cells, the microscopic bodies that contain the chlorophyll pigments, are made up of layered structures known as lamellae that have a high content of protein and lipid [*see top illustration on page 373*]. The chlorophyll molecule has one end (the phytol end) that is soluble in lipid and a main body (the chlorophyllin end) that has an affinity for protein. These affinities give rise to a structural system in which the chlorophyll molecules are closely packed.

The lamellae also contain other molecules with conjugated bonds. These include carotenoids that are similar in structural arrangement to retinal, and phycocyanin, which has a chromophore closely related to the chromophore of phytochrome. These accessory molecules also absorb radiant energy and transfer it to the chlorophyll molecules.

The accumulated energy is finally transferred from the chlorophyll molecules to a relatively few molecules that act as energy-trapping "sinks." In each lamella there is about one sink for every 500 chlorophyll molecules. This small number, whereas it effects a desirable parsimony in the systems required for the chemical steps of photosynthesis, constitutes a bottleneck insofar as energy transfer is concerned. When light reaches a level of intensity about a fifth the intensity of full sunlight, energy arrives at the sinks faster than it can be utilized. Saturation at this level of intensity is nonetheless a good compromise because the average plant leaf is somewhat shaded and seldom receives energy much above the one-fifth level.

The energy accumulated by the sink molecules is ultimately applied to split water molecules into hydrogen and oxygen and to yield an electron-rich compound, which here I shall call "X," that acts as a final electron-acceptor. An oxidized material (that is, one that has given up electrons) is formed as a waste product. In green plants this material is oxygen. It is as a result of this aspect of photosynthesis that the earth's atmosphere contains the oxygen essential to all animal life.

Measured in terms of its products, the effectiveness of the photosynthetic chemical system decreases as the wavelength of the light being absorbed becomes longer. Absorption in the far-red region of the visible spectrum can be made effective, however, if supplementary light of shorter wavelength is also present. This suggests that two steps are involved in electron transfer rather than one; perhaps two energy sinks work together in some kind of booster action. The processes associated with each type of sink are a subject of current investigation, as is the manner in which electron flow might be coordinated between the two steps.

The electron-transport system, as it is now conceived, can be represented schematically [*see illustrations on page 373*]. Trapping centers are indicated as points *E* and *P*. An electron is thought to be transferred from *P* to *X* by one act of light absorption (*b*): the electron loss leaves *P* oxidized, whereupon *X*, the electron-acceptor, becomes an electron-rich, or reduced, compound. In close order a second act of light absorption (*a*) transfers an electron from water to point *E*, leaving *E* electron-rich and leading eventually to free oxygen as the oxidized substance in green plants. The scheme is completed by electron transfer from reduced *E* to oxidized *P*. Functioning of the electron transport steps from water to *X* again requires close association of the necessary parts in the lipid-rich lamellae of the plastids.

The bare skeleton of the scheme serves the purpose of exposition as far as the "photo" part of photosynthesis is concerned, but it leaves much to be told about the "synthesis" part. "Synthesis" implies an output that can be used in a life process. Oxygen, although it is an

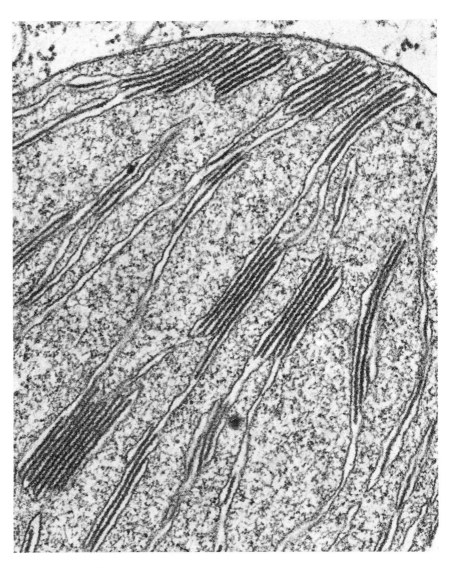

PART OF PHOTOSYNTHETIC ORGAN, a chloroplast of the alga *Nitella*, is seen enlarged 133,000 times in an electron micrograph made by Myron C. Ledbetter of the Brookhaven National Laboratory. Chlorophylls and other pigments involved in the process of photosynthesis are associated with the many lamellae scattered throughout the chloroplast.

oxidized waste product of the scheme, eventually closes back on the electron-rich compound X through the process of respiration. X is an immediately useful product for reactions outside the lamellae. It serves as the chief energy-transferring agent in the reduction of carbon dioxide to sugar. A further reaction transforms X through intermediates, along with carbon dioxide and water, into electron-poor, or oxidized, X and phosphorus-containing sugars. The reaction needs more energy than can be supplied by X alone. This energy, as well as the needed phosphate, comes from adenosine triphosphate (ATP). ATP is formed by the removal of water from adenosine diphosphate (ADP) and the addition of a phosphate molecule. Energy for the transformation of X is available when the added phosphate of ATP is transferred to some other molecule or is split from ATP by water.

Returning to our illustration of the scheme, it appears that the energy difference in the electron transfer from E to P is adequate to make ATP. This until recently was thought to be the most likely way at least part of a plant's supply of ATP was produced. A view that is now being vigorously debated suggests that hydrogen ions appear inside the lamella in the electron transfer that follows light absorption. The enhanced acidity with respect to the outside of the retaining membrane that the ions inside the membrane provide would give an energy gradient adequate for the formation of ATP. With regard to the first action of light (b), electrons excited by that event can also be transferred through X back to the starting point P with coupling to ATP along the way—a process known as cyclic phosphorylation.

This broad outline of energy transfer in photosynthesis has been developed chiefly during the past 10 years. There is still much to be learned about the molecular details of oxygen liberation, the formation of ATP and the coordination of electron flow in various parts of the process. New discoveries may well alter some of today's concepts of photosynthesis at the most basic level. The situation is much the same with regard to our present understanding of vision and photoperiodism. Examination of the immediate changes after light absorption has proved to be a more fruitful realm of study than the search for the ensuing steps that lead to the responses of sight, growth and biological rhythm.

41

ULTRAVIOLET RADIATION AND NUCLEIC ACID

R. A. DEERING
December 1962

The damaging effects of ultraviolet on living things have long been known. Now they are being explained in terms of specific changes in molecules of the genetic material

Ever since the discovery in 1877 that ultraviolet radiation can kill bacteria, workers in several disciplines have been studying the effects of the radiation on living things. Its actions have turned out to be many and varied. Ultraviolet can temporarily delay cell division and can also delay the synthesis of certain substances by cells; it can change the way in which substances pass across the membranes of the cell; it can cause abnormalities in chromosomes; it can produce mutations. Obviously it is a potent tool for the study of living cells, and it has been extensively employed by experimenters. If its exact modes of action at the molecular level were fully understood, the tool would be even sharper and more useful. This article reports the considerable progress that has been made in the past few years toward understanding the biophysical and biochemical role of ultraviolet.

Most of the recent work has concentrated on the interaction of ultraviolet radiation and the molecule of the genetic material deoxyribonucleic acid (DNA), and that is what I shall discuss. There is no doubt that many of the effects of ultraviolet are exerted solely or chiefly by means of changes in DNA. The fact that DNA strongly absorbs ultraviolet,

and that its absorption spectrum resembles the ultraviolet "action spectrum" for many biological changes (that is, the biological effectiveness of various wavelengths), show that this must be true. Therefore DNA is the logical starting point in the investigation of the biological activity of ultraviolet radiation.

This radiation falls between visible light and X rays in the spectrum of electromagnetic waves, ranging in wavelength from about 4,000 to a few hundred angstrom units. (An angstrom unit is one hundred-millionth of a centimeter.) The important wavelengths for the biologist are those between 2,000 and 3,000 angstroms. The sun is a powerful emitter of ultraviolet, but a layer of ozone in the upper atmosphere absorbs most of the radiation below 2,900 angstroms. Were it not for the ozone, sunlight would damage or kill every exposed cell on earth.

In the laboratory, working with monochromatic ultraviolet radiation at various wavelengths, investigators have established that the region most potent in its effects on living things is near 2,600 angstroms. When DNA was isolated, it was found to absorb most strongly at just these wavelengths. In the past five years workers in several laboratories

have begun to discover what happens to the DNA molecule when it absorbs ultraviolet energy.

Natural DNA, as the readers of this magazine are well aware, normally consists of a double-strand helix. The helices proper—the twin "backbones" of the molecule—consist of an alternation of sugar (deoxyribose) and phosphate groups. Attached to each of the sugars is one of four nitrogenous "bases," generally adenine, guanine, thymine and cytosine. The bases on the two backbones are joined in pairs by hydrogen bonds, the adenine on one chain always being paired with thymine on the other, and the guanine with cytosine. The hydrogen bonds that join the base pairs are weaker than ordinary chemical bonds. Simply heating double-strand DNA breaks the bonds and partially or completely separates the two backbones into two strands of "denatured" DNA.

Ultraviolet radiation falling on DNA is absorbed primarily by the bases, which exhibit about the same absorption peak at 2,600 angstroms as the whole DNA molecule does. This being the case, the first approach was to study the effects of ultraviolet radiation on the isolated bases. It soon turned out that thymine and cytosine, which belong to the class

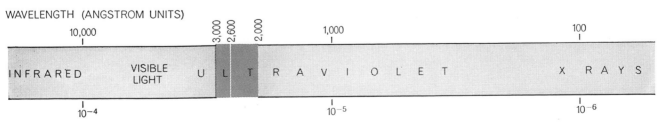

WAVELENGTH (ANGSTROM UNITS)

10,000   3,000 2,600 2,000   1,000   100

INFRARED   VISIBLE LIGHT   U L T R A V I O L E T   X R A Y S

10^{-4} 10^{-5} 10^{-6}

WAVELENGTH (CENTIMETERS)

ULTRAVIOLET portion of the electromagnetic spectrum lies between visible light and X rays. The wavelengths between 2,000 and 3,000 angstrom units are of primary biological importance. DNA, the genetic material, absorbs most strongly at 2,600 angstroms.

of substances called pyrimidines, are far more sensitive to ultraviolet than are adenine and guanine, which are purines. About one in every 100 quanta of ultraviolet energy absorbed by pyrimidines alters the molecules; for purines the ratio is one in 10,000. (In general only a few of the quanta absorbed by a molecule will be effective in producing permanent changes.) The search was therefore narrowed to the pyrimidines.

The first effect to be discovered was that ultraviolet acts on cytosine molecules or the cytosine units of DNA in water solution, adding a water molecule across a double bond [see middle illustration on page 302]. Heating the altered cytosine, even to the temperatures required for biological growth, or acidifying it, partly reverses the reaction. Therefore the hydration of cytosine did not seem likely to be of major biological importance.

For some years, however, this hydration was the only sensitive, ultraviolet-induced change in the bases that could be detected. Heavy doses of radiation did produce complex rearrangements, but these doses were far in excess of the smallest ones known to have biological effects. About three years ago a breakthrough in the photochemistry of DNA came when R. Beukers, J. Ijlstra and W. Berends of the Technological University of Delft in Holland and Shih Yi Wang, now of Johns Hopkins University, discovered that although in a liquid solution thymine is not particularly sensitive to ultraviolet, in a concentrated, frozen aqueous solution it is extremely sensitive. It developed that irradiation of the frozen solution causes thymine molecules to combine and form two-molecule chains, or dimers. As in the case of the cytosine conversion, a double bond changes to a single, and new bonds between carbon atoms link the two thymines [see bottom illustration on page 380]. Unlike the altered cytosine, the thymine dimer is stable to heat and acid. But when the solution is melted, irradiation can convert the dimer back into the two original thymine molecules. What the freezing does is to hold the thymines close together in a crystalline or semicrystalline configuration, making it possible for the dimer bonds to form between two neighboring thymines when they absorb ultraviolet. It seemed likely that such a conversion would also occur in DNA, where thymine units are sometimes adjacent to each other on a helical strand and are held in relatively fixed positions. In 1960 Adolf Wacker and his associates at the University of Frankfort found thymine dimers in DNA extracted from irradiated bacteria.

In order to get more complete information on the formation and splitting of thymine dimers in polymer chains such as DNA, Richard B. Setlow and I carried out experiments on some model polymers at the Oak Ridge National Laboratory. Similar experiments were performed independently at the California Institute of Technology by Harold Johns and his collaborators. The compounds we used were short polymers—in effect short single strands of DNA in which all the bases were thymine. Some of our test molecules contained only two backbone units and two thymines; others had 12 or more. Since the sugar-phosphate backbone holds the thymines in fairly close proximity, we anticipated that ultraviolet radiation should form dimers between adjacent thymines in a chain even in a liquid solution. And we expected that once the dimers had formed they would be subject to breakage by ultraviolet, as were the isolated thymine dimers. When thymine loses a double bond in changing to a dimer, it also loses its ability to absorb light at 2,600 angstroms. Therefore measuring the change in 2,600-angstrom absorption gives an indication of the ratio between thymine monomers and thymine dimers in the solution.

When we irradiated our polymers, dimers were in fact produced. Since the rate of formation did not vary with thymine concentration, we concluded

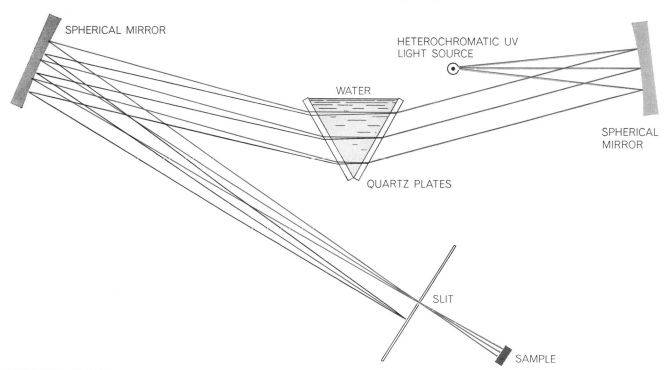

MONOCHROMATOR provides ultraviolet light of a single wavelength for experiments. Light of mixed wavelengths is rendered parallel by a spherical mirror biand passes through a quartz-and-water prism. (Glass would not transmit the desired wavelengths.) The prism splits the light into many components of different wavelengths, only two of which are indicated here, and the beams are refocused by a second mirror. The sample to be irradiated is positioned behind a slit that excludes all but the desired wavelength.

that they were formed within, rather than between, individual polymers. Dimers also broke up into monomers, but not at the same rate at which they formed. The process is analogous to a reversible chemical reaction in which forward and backward reactions proceed at different rates, with an equilibrium eventually being reached between the reactants and the products. We found that for every wavelength of ultraviolet there is at high doses an equilibrium between the number of dimers being formed and the number being broken [see bottom left illustration page 381]. At each wavelength and intensity there is a certain rate for dimer formation and a different one for breakage; the equilibrium level is determined by the relative rates of the forward and backward reactions. At 2,800 angstroms the equilibrium state is on the dimer side: most of the thymines are dimerized. At 2,400 angstroms the opposite is true: most of the thymines are monomers. The relative number of monomers and dimers in the polymer solutions can be controlled by changing the wavelength of the incident ultraviolet.

When the data from a number of experiments are plotted [see bottom right illustration page 381], the resulting curves show the ability of each wavelength to make and break dimers in these model polymers. The curves approximately parallel the absorption spectra of the monomer and dimer respectively, indicating that it is difference in absorption capacity that accounts for the different action of various wavelengths. The "quantum yield," or number of molecules altered by each quantum, does not change greatly with wavelength; for dimer formation in the polymers containing only thymine it is of the order of .01 and for breakage it is near 1.

The next step was to relate molecular changes to changes in the properties of DNA and in its biological activity. Julius Marmur and Lawrence Grossman of Brandeis University have shown recently that when double-strand DNA is exposed to ultraviolet, the two strands become more strongly linked, apparently by chemical bonds rather than by the original weak hydrogen bonds. Marmur and Grossman believe the strong link is the result of interchain dimerization, that is, the formation of dimers between thymine units on opposite strands of the double helix.

At Oak Ridge, Frederick J. Bollum and Setlow found that ultraviolet can induce dimer linkages between adjacent thymine units in single-strand DNA.

They suspect that the same thing can happen between adjacent thymines in natural DNA, but in this case some of the hydrogen bonds in a local region may have to be broken before dimerization is possible. Marmur and Grossman have shown that irradiation does indeed disrupt hydrogen bonding between strands of natural DNA.

Another effect of ultraviolet on isolated DNA that has been clearly identified is a breaking of the sugar-phosphate backbone, but this occurs only at uninterestingly high doses. Among the sensitive reactions only the cytosine and thymine conversions are understood well enough for their biological implications to be assessed. There are surely other important effects, but they remain to be discovered.

Although the biological significance of the cytosine hydration has generally been discounted because it reverses at body temperature or lower, the reversal may be slower in intact DNA than in the isolated base. There is no direct evidence that the hydration product would be detrimental to the biological activity of DNA, but it might affect the hydrogen bonding in a segment of the helix and thereby give rise to the broken bonds observed by Marmur and Grossman.

The formation of thymine dimers should in theory be of great biological significance. When DNA makes a replica of itself, according to the widely accepted hypothesis, the hydrogen bonds break and a new complementary chain forms along each of the old strands. A dimer cross link between strands would interrupt the separation, blocking replication. Dimers between adjacent thymines on the same strand would interfere with proper pairing of the bases. Normally an adenine should come into position opposite each thymine on the parent strand. The joining of two adjacent thymines would probably change matters enough to impair the proper incorporation of adenine; replication might stop short at

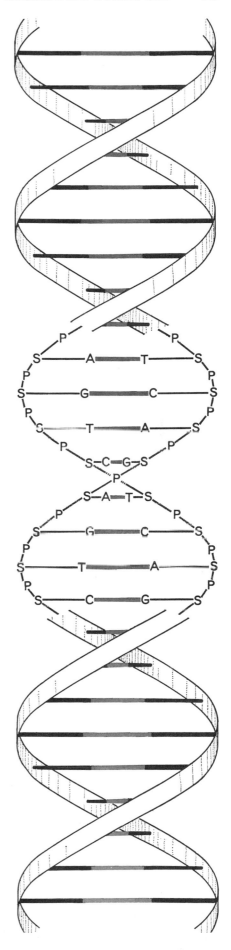

DNA MOLECULE is a double helix, diagramed here schematically. (One strand is actually displaced along the axis of the helix with regard to the other.) The backbone strands are composed of alternating sugar (S) and phosphate (P) groups. Attached to each sugar is one of four bases, usually adenine (A), guanine (G), thymine (T) and cytosine (C). Hydrogen bonds (gray) between bases link the strands. Adenine is always paired with thymine, guanine with cytosine. Genetic information is provided by the sequence of bases along a strand.

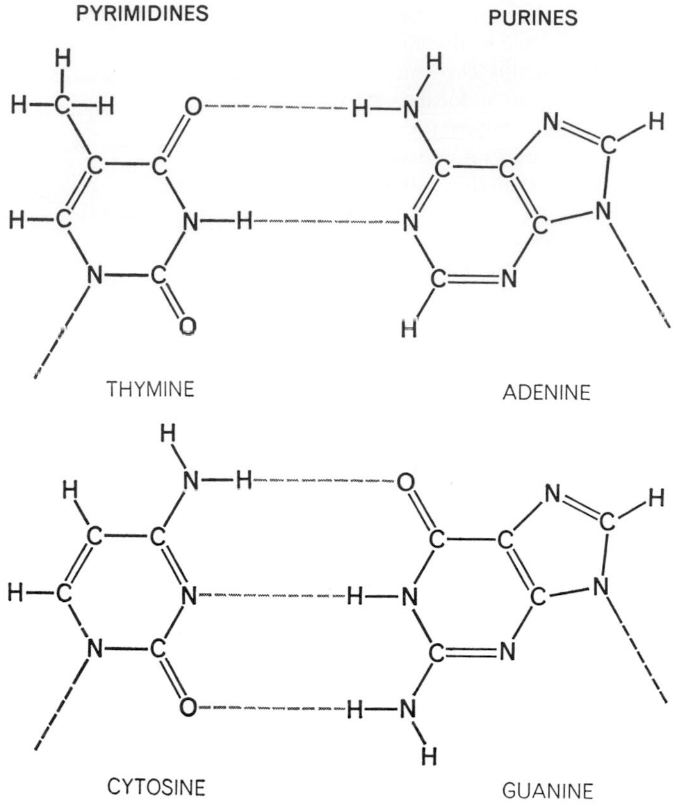

PYRIMIDINES   PURINES

THYMINE   ADENINE

CYTOSINE   GUANINE

FOUR BASES are diagramed as they are paired in DNA. Adenine and guanine, the larger molecules, are purines; thymine and cytosine are pyrimidines. The broken black lines show points of attachment to sugar groups; the broken gray lines are interchain hydrogen bonds.

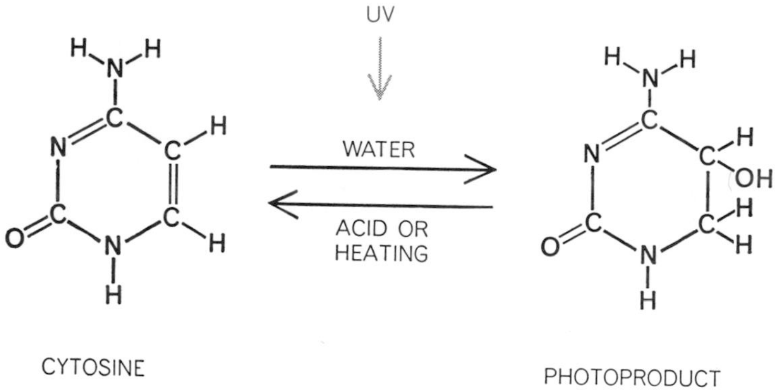

CYTOSINE   PHOTOPRODUCT

CYTOSINE in a water solution is altered by irradiation with ultraviolet. A water molecule is added across the double bond between two carbon atoms, the double bond changing to a single bond. When the cytosine solution is heated or acidified, the process is reversed.

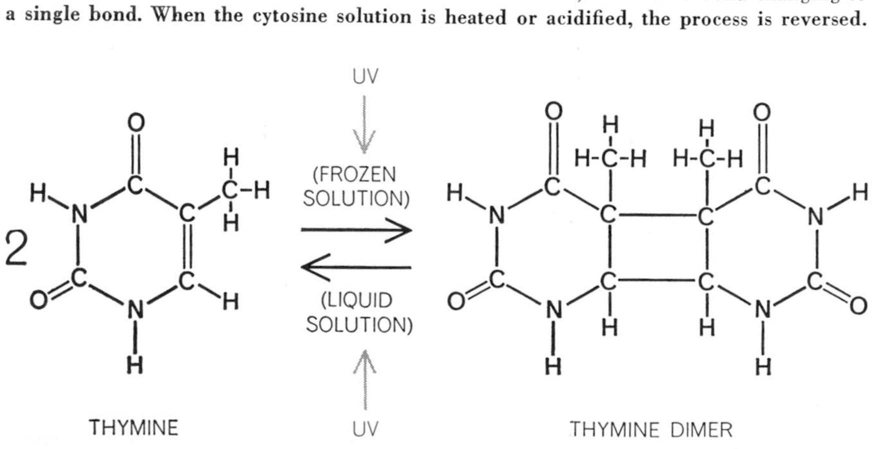

THYMINE   THYMINE DIMER

THYMINE in a frozen solution undergoes the reaction shown here when it is irradiated. The double bond between carbon atoms changes to single and two thymines are linked in a double molecule, or dimer. When the solution is melted, irradiation breaks the dimer.

that point or might proceed incorrectly, with an altered base sequence on the newly formed chain. On subsequent replication this altered strand would replicate itself, producing a molecule with the wrong base sequence in both strands —in other words, a mutated gene.

Recent work at Oak Ridge has provided direct experimental proof that thymine dimerization is one of the important ways in which the biological activity of DNA is altered by ultraviolet. Setlow and Bollum studied the ability of irradiated single strands of DNA to serve as a template in the manufacture of new DNA in a variety of cell-free test-tube preparations. Irradiation at 2,800 angstroms cut down the priming ability of DNA, the reduction being proportional to the adenine-thymine content of the various preparations. Subsequent irradiation at 2,400 angstroms partially restored template activity. Presumably irradiation at 2,800 angstroms formed dimers between adjacent thymines on the template DNA, blocking or slowing down the normal synthesis of new DNA strands. Irradiation at 2,400 angstroms evidently broke some of the dimers, partially restoring template activity.

In another series of experiments Setlow and his wife Jane K. Setlow worked with a form of DNA called "transforming principle," studying its ability to carry specific bits of genetic information from one cell to another. The measure of the biological activity of the DNA in this case was its effectiveness in transforming a given trait in the new cell. The Setlows found that irradiation at 2,800 angstroms destroyed the transforming ability of a portion of the DNA molecules. Again, when the irradiated DNA was exposed to 2,400-angstrom radiation, some of its molecules regained their transforming ability. The experimenters could account quantitatively for their results by assuming that about 50 per cent of the inactivation of the transforming DNA was due to thymine dimerization. They do not know what changes account for the rest.

When some types of cells that have been damaged by ultraviolet are exposed to ordinary blue light, a great deal of the damage is reversed; even bacteria that appear to have been killed are revived [see "Revival by Light," by Albert Kelner; SCIENTIFIC AMERICAN, May, 1951]. Claud S. Rupert and his associates at Johns Hopkins University had shown that this photoreactivation takes place through light-mediated enzyme reactions, but the details were not known. Recently Daniel L. Wulff and

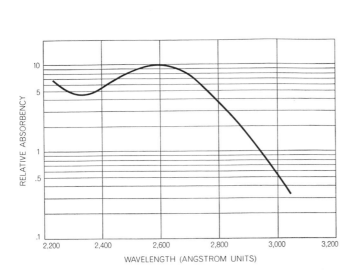

ABSORPTION SPECTRUM of DNA shows its ability to absorb ultraviolet radiation of various wavelengths. The peak is at 2,600 angstrom units, the wavelength known to be most harmful to cells.

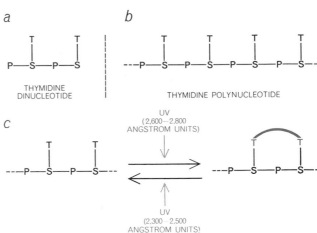

MODEL POLYMERS exposed to ultraviolet were synthetic all-thymine DNA strands composed of either two nucleotide (phosphate-sugar-base) units (*a*) or 12 or more units (*b*). Irradiation both formed and broke dimers between adjacent thymines (*c*). Irradiation with high doses of the longer wavelengths led to an equilibrium condition in which most of the thymines were dimerized; exposure to shorter wavelengths tended to break the dimers.

Rupert have identified one mechanism of reactivation: a particular enzyme preparation, in the presence of blue light, can break up to 90 per cent of the thymine dimers in irradiated DNA. Marmur and Grossman have also shown that the enzyme system can break the ultraviolet-induced cross links between two DNA strands, thereby strengthening the idea that these links result from interchain thymine dimers

Some bacteria apparently can produce enzymes that repair ultraviolet-damaged DNA without the need for visible light. The extreme resistance to ultraviolet displayed by certain bacteria may come from an ability to produce large amounts of these repair enzymes.

To sum up, it is clear that ultraviolet can change DNA in specific ways and can partially reverse those changes. Moreover, both the forward and backward alterations are reflected in DNA function. There is considerable evidence that much of the damage that ultraviolet radiation inflicts on cells and viruses is caused directly by its effects on DNA.

In the case of viruses this may be the whole story. When they infect a cell to

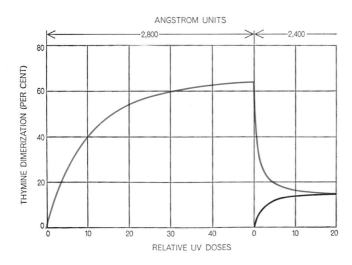

DEGREE OF DIMERIZATION varies with wavelength and dose. Increasing the exposure at 2,800 angstrom units increases the proportion of thymine units that are dimerized until an equilibrium state is attained with 65 per cent of the thymine units as dimers (*colored curve*). Irradiation at 2,400 angstrom units of the same sample, or of a different sample (*black curve*), results in a new equilibrium level with about 17 per cent of the thymines dimerized.

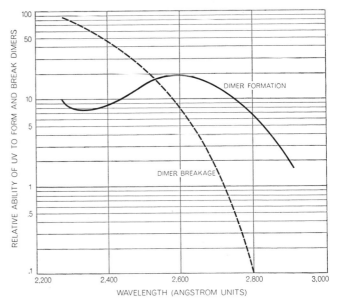

VARIOUS WAVELENGTHS differ in their ability to form and break dimers. Those over 2,540 angstrom units are more able to form dimers; the reverse is true of the shorter wavelengths.

replicate themselves, only their nucleic acid is injected into the cell. Therefore damage to DNA must be directly reflected in the ability of the virus to take over the cell's metabolism and multiply. In cells, however, one cannot assume that the only important effects of ultraviolet are those involving DNA. The radiation is absorbed by proteins, by ribonucleic acid (RNA) outside the nucleus and by other substances that play a part in cell metabolism, and it presumably changes their structures too. The task of identifying all the ultraviolet reactions in living organisms has been well begun, but there is much to learn.

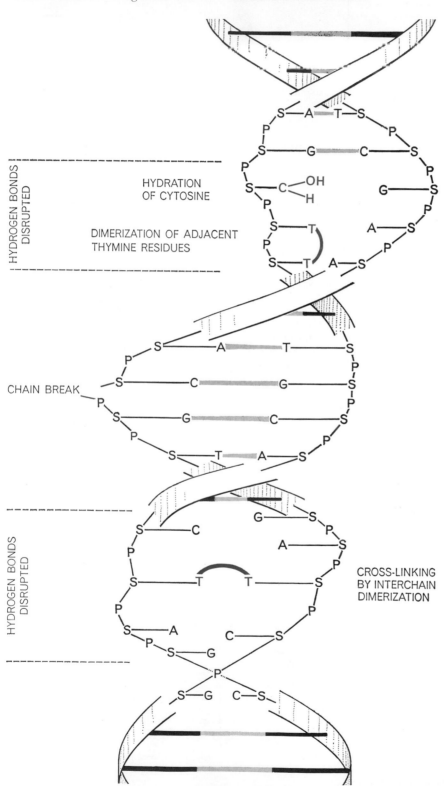

ALTERATIONS IN DNA caused by ultraviolet irradiation are diagramed here. The formation of thymine dimers is the change most likely to do damage to living cells and viruses.

42

THE PATH OF CARBON IN PHOTOSYNTHESIS

J. A. BASSHAM

June 1962

The carbon atoms of the carbon dioxide assimilated by plants traverse an intricate cycle of chemical reactions. The immediate products are not only carbohydrates but also amino acids, fats and other compounds

The processes of life consist ultimately of the synthesis and breakdown of carbon compounds. Because a carbon atom can bind four other atoms to itself at a time and is thereby able to link up with other atoms—especially other carbon atoms—in chains and rings, carbon lends itself to the construction of a virtually endless variety of molecules. These molecules derive their physical characteristics and chemical activity not only from their composition but also from their size and intricacy of structure. The rich variety of life suggests in turn that living cells have gone far in the elaboration of such compounds and the processes that make and unmake them. All these processes depend in the end on a first one. This is the process of photosynthesis, which takes carbon and several other common elements from the environment and builds them into the substances of life.

The plant finds most of these elements already bonded to oxygen in oxides such as carbon dioxide (CO_2), water (H_2O), nitrate (NO_3^-) and sulfate (SO_4^{--}). Before the plant can bind the elements other than oxygen together as organic compounds, it must remove some of the excess oxygen as oxygen gas (O_2), and this accomplishment takes a large amount of energy.

In the simplest terms photosynthesis is the process by which green plants trap the energy of sunlight by using that energy to break strong bonds between oxygen and other elements, while forming weaker bonds between the other elements and forcing oxygen atoms to pair as oxygen gas. For example, to make the sugar glucose ($C_6H_{12}O_6$) the plant must split out six molecules of oxygen in order to combine the carbon and half the oxygen of six carbon dioxide molecules with the hydrogen of six water molecules.

The glucose and other organic compounds taken up in the chemical machinery of the plants and the animals that live on plants serve both as fuel and as the raw materials for the synthesis of higher organic compounds. That considerable solar energy is bound by photosynthesis becomes apparent when wood or coal is burned. In living cells the controlled combustion of respiration extracts this energy to power the other processes of life. Both kinds of combustion take oxygen from the air and break down organic compounds to carbon dioxide and water again. In its end result photosynthesis can be defined as the opposite of respiration. Together these complementary processes drive the cyclic flow of matter and the noncyclic flow of energy through the living world [*see illustration below*].

From such generalizations about the effect and function of photosynthesis in nature it is a long step to the explanation of how photosynthesis works. Yet much of the explanation is now complete. The work has been greatly facilitated by the earlier and more nearly complete resolution of the chemistry of respiration. The two processes, it turns out, are in some ways complementary on the molecular scale, just as they are on the grand scale in the biosphere. Each involves some 20 to 30 discrete reactions and as many intermediate compounds; half a dozen of these reactions and their intermediates are common to both photosynthesis and respiration. Only the first few steps in photosynthesis are driven directly by light. The energy of light is trapped in the bonds of a few specific compounds. These energy carriers deliver the energy in discrete units to the steps of synthesis that follow. The same or closely similar carriers perform corresponding operations in respiration, picking up energy from the stepwise dismemberment of the fuel molecule and

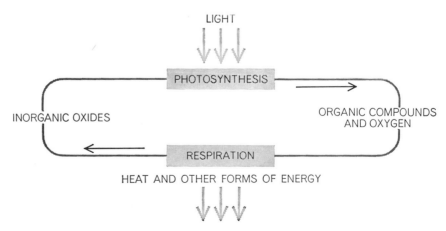

PHOTOSYNTHESIS AND RESPIRATION are the complementary processes that drive the cyclic flow of matter and the noncyclic flow of energy through the biosphere. Photosynthesis uses light energy to convert inorganic oxides to oxygen and organic compounds such as glucose. In respiration of plants and animals oxygen reacts with these compounds to produce the inorganic oxides carbon dioxide and water as well as biologically useful energy.

delivering it to the energy-consuming processes of the cell. Although the first, energy-trapping stage in photosynthesis remains to be clarified, it is now possible to trace the path of carbon from the very first step in which a single atom of carbon is captured in the bonds of an evanescent intermediate compound.

The term "carbohydrate" recalls the deduction of early 19th-century investi-gators that photosynthesis made glucose directly by combining atoms of carbon with molecules of water, as the formula for glucose suggests. In line with this idea it was thought that the oxygen transpired by green leaves came from the splitting of carbon dioxide. The progress of chemistry, however, failed to disclose any processes that would accom-plish these results so simply. Accumulat-ing evidence to the contrary became convincing some 30 years ago, when C. B. van Niel of Stanford University discovered that certain bacteria produce organic compounds by a process of photosynthesis similar to that in plants but with one important difference. These bacteria use hydrogen sulfide (H_2S) in-stead of water and liberate elemental sulfur instead of gaseous oxygen. The

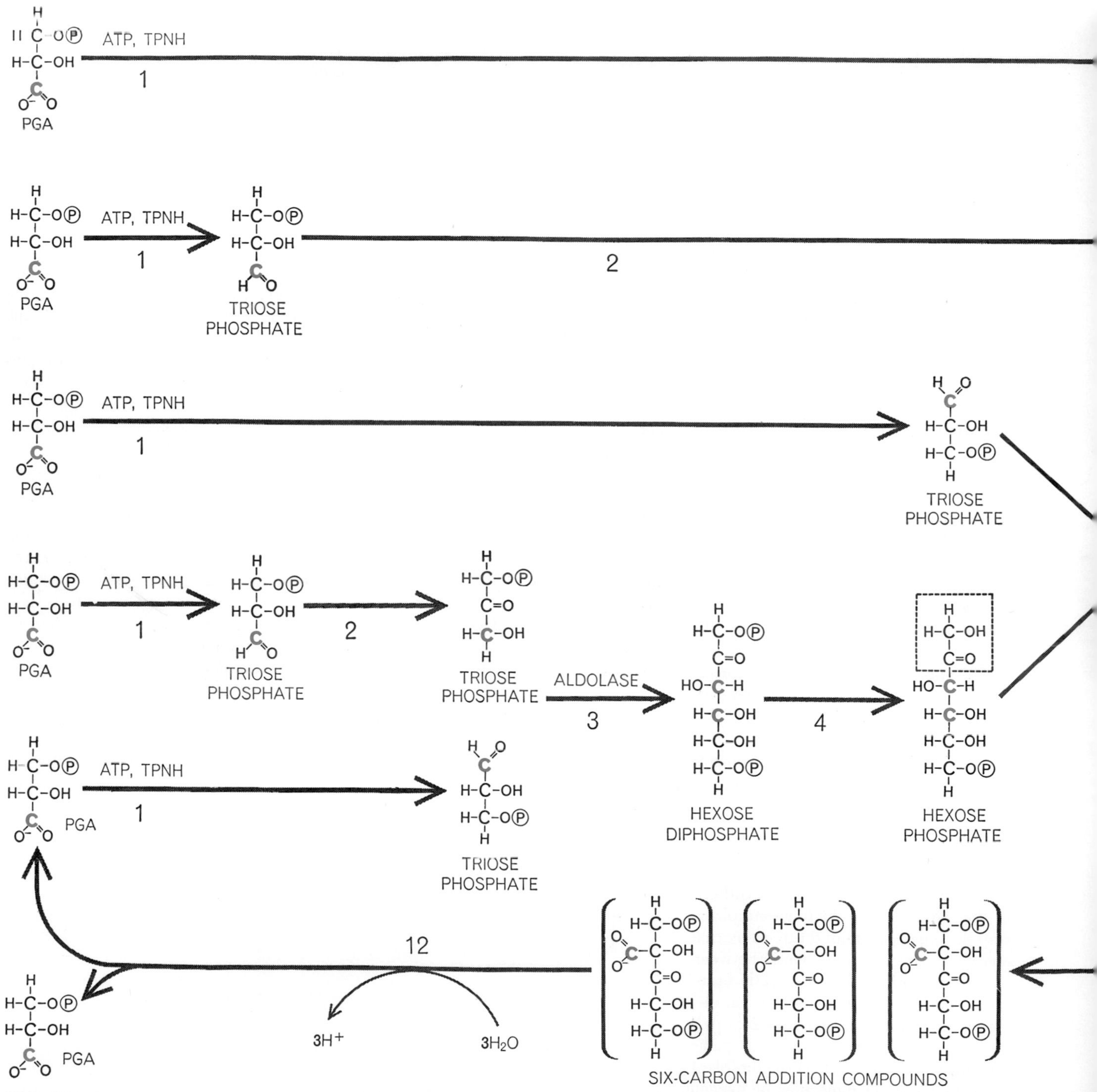

PATH OF RADIOACTIVE CARBON (*color*) was determined from experiments described in the text. Five molecules of PGA, the first stable intermediate product to appear, are reduced (*1*) by cofactors ATP and TPNH to five triose phosphate molecules. A circled *P* represents a phosphate group (—HPO_3^-). Two of these are converted to a different type of triose phosphate (*2*); the subse-quent condensation of one of each kind of triose into hexose diphos-phate (*3*) is mediated by the enzyme aldolase. Hexose diphosphate then loses a phosphate group (*4*). Transketolase, another enzyme, removes two carbons from the hexose and adds them to a triose

otherwise complete similarity of the two processes strongly suggested that the oxygen evolved by green plants must come from the splitting of water.

The Capture of Light

Photosynthesis could now be described in terms of familiar chemistry. The splitting of water would be accomplished by the process of oxidation (which means the removal of hydrogen atoms from a molecule), with oxygen gas as the product of the reaction. The free hydrogen atoms would then be available to carry through the equally familiar and opposite process of reduction (which means the addition of hydrogen atoms to a molecule). By the addition of hydrogen atoms (or electrons plus hydrogen ions) the carbon dioxide would be reduced to an organic compound.

It is during the first, energy-converting stage of photosynthesis that the water molecule is split. Initially the energy of light impinging on the plant cell is transformed into the chemical potential energy of electrons excited from their normal orbits in molecules of the green pigment chlorophyll and other plant pig-

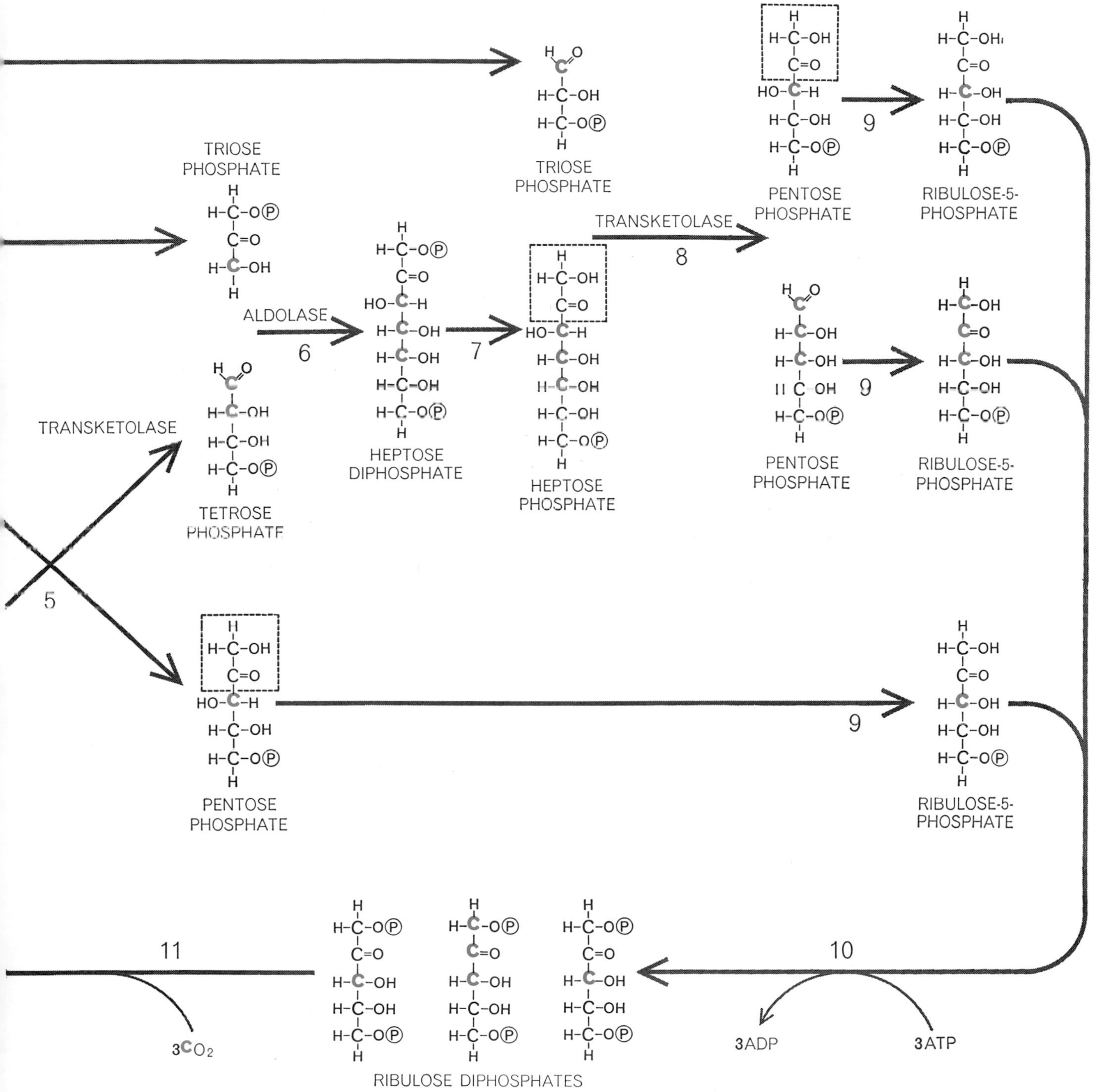

TRIOSE PHOSPHATE

TRIOSE PHOSPHATE

PENTOSE PHOSPHATE

RIBULOSE-5-PHOSPHATE

TRANSKETOLASE 8

ALDOLASE 6

7

HEPTOSE DIPHOSPHATE

HEPTOSE PHOSPHATE

PENTOSE PHOSPHATE

RIBULOSE-5-PHOSPHATE

9

TRANSKETOLASE

TETROSE PHOSPHATE

5

PENTOSE PHOSPHATE

9

RIBULOSE-5-PHOSPHATE

11

10

3 CO₂

3ADP

3ATP

RIBULOSE DIPHOSPHATES

phosphate (5), making one tetrose and one pentose phosphate. The tetrose condenses with a triose (6) to form a heptose diphosphate, which then loses a phosphate group (7). Transketolase removes two carbons from the heptose and adds them to a triose (8), making two more pentose phosphates for a total of three; these are converted to ribulose-5-phosphate (9), then to ribulose diphosphate (10). Addition of three carbon dioxide molecules (11) produces three unstable compounds (*brackets indicate unknown structure*) that begin the cycle again. Addition of three water molecules (12) results in six PGA molecules for a net gain of one in the cycle.

ments. A large part of this energy eventually goes into the splitting of water as electrons and hydrogen ions are transferred from water to the substance triphosphopyridine nucleotide (TPN⁺), which is thereupon reduced to the form designated TPNH. The TPNH thus becomes not only a carrier of energy but also the bearer of electrons for the subsequent reduction of carbon dioxide. Along with the movement of electrons from water to TPNH, some energy goes to charging the energy-carrying molecule adenosine triphosphate (ATP), specifically by promoting the attachment of a third phosphate group ($-OPO_3^{--}$) to adenosine diphosphate (ADP), the discharged form of the carrier. Both ATP and TPNH belong to the family of compounds known as cofactors or coenzymes, which work with enzymes in the catalysis of chemical reactions. ATP, the universal currency of energy transactions in the cell, plays a significant role in respiration as well as in photosynthesis.

Needless to say, the manufacture of each of these cofactors involves an intricate cycle of reactions [see "The Role of Light in Photosynthesis," by Daniel I. Arnon; SCIENTIFIC AMERICAN Offprint 75]. Although the cycles are not yet fully understood, it is enough for the purpose of the present discussion to know that ATP and TPNH, or closely similar compounds, furnish the energy for the second stage of photosynthesis, during which the carbon atom of carbon dioxide is reduced and joined to a hydrogen atom and a carbon atom in place of an oxygen.

The process of reduction goes forward in small steps. Each reaction brings about some change in a carbon compound until the starting material is at last transformed to the final product. For each reaction there is therefore an intermediate compound. Since every life process involves a more or less extended series of intermediates, cells typically contain a large number of intermediates. Many of them turn up in two or more pathways leading to different end products. The tracing of the path of carbon in photosynthesis required first of all a technique for identifying the intermediates proper to it and for establishing their sequence along the path.

Samuel Ruben and Martin D. Kamen, then at the University of California, met this need some 20 years ago by their discovery of the radioactive isotope of carbon with a mass number of 14. This isotope has a conveniently long half life of more than 5,000 years; over the time period of an experiment, therefore, carbon 14 has an effectively constant radioactivity. Ruben and his colleagues recognized at once the potential usefulness of this isotope as a label for the identification of compounds in biological processes. They prepared carbon dioxide in which the carbon atoms were carbon 14. When they exposed green plants to an atmosphere containing this gas instead of normal carbon dioxide ($C^{12}O_2$), the plants took up the $C^{14}O_2$ and made compounds from it. The presence of the carbon 14 in these compounds could be detected by various devices, such as the Geiger-Müller counter, and by radioautography on X-ray film. Unfortunately this work was cut short by the war and by Ruben's death in a laboratory accident.

In 1946 Melvin Calvin organized a new group at the Lawrence Radiation Laboratory of the University of California with the primary objective of tracing the path of carbon in photosynthesis with $C^{14}O_2$ as one of its principal tools. Starting as a graduate student in 1947, I had the good fortune to participate in this work with Calvin, Andrew A. Benson and others.

The early experiments were quite simply contrived. We used leafy plants and often just the leaves of plants. After allowing a leaf to photosynthesize for a given length of time in an atmosphere of $C^{14}O_2$ in a closed chamber, we would bring biochemical activity to a halt by immersing the leaf in alcohol. With the enzymes inactivated, the reactions converting one intermediate compound into another would stop, and the pattern of labeling would be "frozen" at that point. We soon discovered, however, that photosynthesis proceeds too rapidly for completely reliable observation by such a procedure. With a few seconds' delay in the penetration of alcohol into the cell, for example, the labeling pattern would be disarrayed and no longer representative of the stage at which we tried to halt the photosynthesis.

Since rapid and precisely timed killing of the plant is important, we adopted single-celled algae—*Chlorella pyrenoidosa* and *Scenedesmus obliquus*—as the subject for many of our experiments. In both species the plant consists of a single cell so small that it can be seen only with a microscope. Alcohol can quickly penetrate the cell wall and deactivate the enzymes. The algae offer another advantage: they can be grown in continuous cultures, assuring us a supply of material with highly constant properties.

An experimental sample is taken from the culture in a thin-walled, transparent closed vessel. Illuminated through the walls of the vessel and supplied with a stream of ordinary carbon dioxide, which is bubbled through the suspension, the algae photosynthesize at the normal rate. We shut off the supply of carbon dioxide and inject a solution of

REDUCTION OF PGA to triose phosphate requires both ATP and TPNH. At top ATP gives up its terminal phosphate group to PGA to produce phosphoryl-3-PGA. At bottom TPNH donates a hydrogen atom and an electron (*broken circle and arrow*), thereby displacing a phosphate group and forming triose phosphate. The second step is in reality more complex than shown here and involves other cofactors in addition to TPNH.

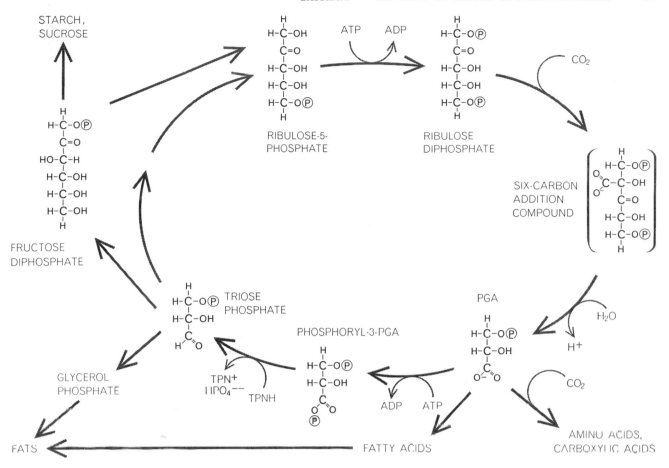

END PRODUCTS OF PHOTOSYNTHESIS are not limited to carbohydrates (e.g., sucrose and starch), as first thought, but include, among other things, fatty acids, fats, carboxylic acids and amino acids. Carbon cycle shown here is highly simplified; it involves at least 12 discrete reactions. Moreover, the steps from PGA to fatty acids and to amino and carboxylic acids have not been indicated.

radioactive bicarbonate ion (carbon dioxide dissolved in our algae culture medium is mostly converted into bicarbonate ion). After a few seconds or minutes the cells are killed. We then extract the soluble radioactive compounds from the plant material and analyze them.

The Reduction of CO_2

Calvin and his colleagues soon found that the carbon 14 label was distributed among several classes of biochemical compounds, including not only sugars but also amino acids: the subunits of proteins. As the exposure time was reduced to a few seconds, the first stable intermediate product of photosynthesis was found to be the three-carbon compound 3-phosphoglyceric acid (PGA).

The next step was to determine which of the three carbon atoms in the first generation of PGA molecules synthesized in the presence of radioactive carbon dioxide bears the carbon 14 label. We first removed from PGA the phosphate group [see *illustration on opposite page*] and then diluted the free glyceric acid with glyceric acid containing the stable carbon 12 isotope in order to have enough material for analysis by ordinary chemical methods. Treatment with reagents that severed the bonds between the carbons produced three different products, one from each carbon atom. By measuring the radioactivity of each of the products we were able to identify the labeled carbon.

In PGA from plants that had been exposed to labeled carbon dioxide for only five seconds we found that virtually all the carbon 14 was located in the carboxyl atom, the carbon at one end of the chain that is bound to two oxygens. This was not surprising because the carboxyl group most nearly resembles carbon dioxide. The carbon is bound to the oxygens by three bonds, however, instead of four; the fourth bond now ties it to the middle carbon in the PGA chain. The transfer of this bond from one of the oxygens to a carbon constitutes the first step in the reduction of the carbon dioxide. This was evidence also that the re-

duction is accomplished by some sort of carboxylation reaction, a reaction in which carbon dioxide is added to some organic compound. Ultimately, of course, the two other carbons of PGA must come from carbon dioxide. But it was some time before investigation disclosed the specific compound that picks up the carbon dioxide and the cyclic pathway that makes this carbon dioxide acceptor from PGA.

The discovery of the pathway intermediates was made much easier by a then comparatively new technique: two-dimensional paper chromatography, developed by the British chemists A. J. P. Martin and R. L. M. Synge. Closely similar compounds can be distinguished in this procedure by slight differences in their relative solubility in an organic solvent and in water. The extract from the plant is dropped on a sheet of filter paper near one corner. An edge of the paper adjacent to the corner is immersed in a trough containing an organic solvent; the paper is held taut by a weight and the whole assembly is placed in a water-saturated atmosphere in a vapor-tight box. The solvent traveling through the paper by capillarity dissolves the compounds and carries them along with it. As they move along in the solvent, however, the compounds tend to distribute themselves between the solvent and the water absorbed by the fibers of the paper. In general the more soluble the compound is in water compared with the organic solvent, the slower it travels. If the compound is also absorbed to some extent by the cellulose fibers, its movement will be even slower. As a result the compounds are distributed in a row in one dimension. Depending on the solubility of the compounds and the nature of the solvent used, some compounds may still overlap one another. Repetition of the procedure, with a different solvent traveling at right angles

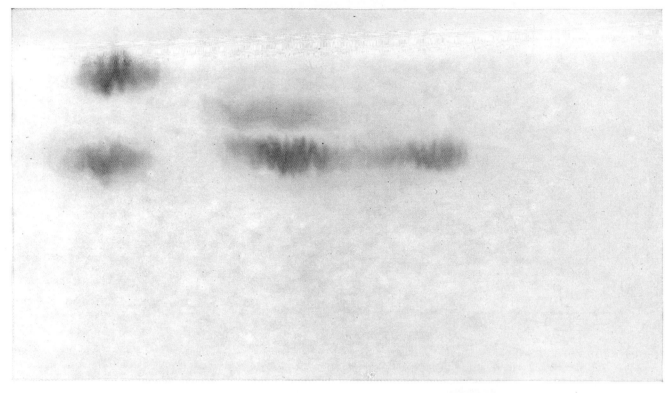

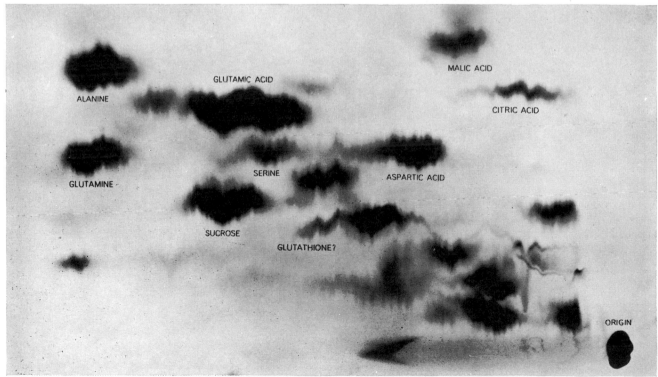

CHROMATOGRAM AND RADIOAUTOGRAPH used to corroborate the identity of amino acids produced by photosynthesizing algae appear at top and bottom respectively. The method of identifying such substances is described in the text. Areas of the radioautograph corresponding to colored areas in the chromatogram are alanine, glutamine, glutamic acid, serine and aspartic acid.

to the direction of the first run, will usually separate these compounds in a second dimension.

Since most of the compounds are colorless, special techniques are needed to locate them on the paper. Those that are radioactive will locate themselves, however, if the chromatographic paper is placed in contact with a sheet of X-ray film for a few days. The resulting radioautograph will show as many as 20 or 30 radioactive compounds in the substances extracted from algae exposed to carbon 14 for only 30 seconds [*see illustration at right*]. Clearly the synthetic apparatus of the plant works rapidly.

Chromatographs and Radioautographs

In order to identify these compounds we prepared a chromatographic map by running samples of many known compounds through the same chromatographic system and recording the locations at which we found them on the paper. The locations can be made visible in these cases by spraying the paper with a mist of some chemical that is known to react with the compound to produce a colored spot. Comparison of the radioautograph of an unknown compound with the map yields a first clue to its identity. This can be corroborated by washing the radioactive compound out of the paper with water and mixing it with a larger sample of the suspected authentic substance. The mixture is applied to a new piece of filter paper and chromatographed. With enough of the authentic material to yield a colored spot, comparison with a radioautograph of the same paper now shows whether or not the radioactive and the authentic material really coincide. The possibility that the authentic material and the radioactive material are still not the same can be tested by using different solvent systems in the preparation of the chromatograph and by other means.

Over the years these procedures have established the identity of a great many of the intermediate and end products of photosynthesis. Some of the sugar phosphates labeled by carbon 14 proved to be well-known derivatives of triose (three-carbon) and hexose (six-carbon) sugars. Others were discovered for the first time among the intermediates produced by our algae. Benson showed that among these are a seven-carbon sugar phosphate and also five-carbon phosphates, including in particular ribulose-1,5-diphosphate.

The rapid building of carbon 14 into the more familiar triose and hexose phosphates suggested certain biochemi-

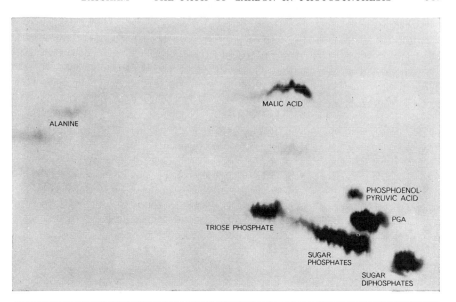

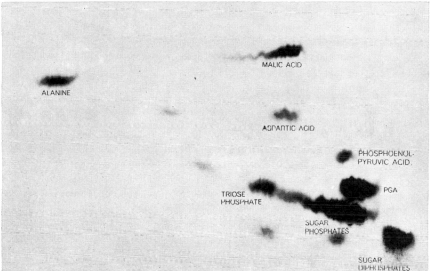

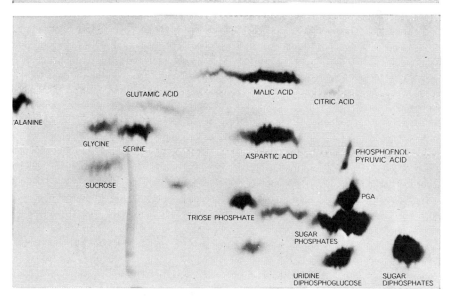

THREE RADIOAUTOGRAPHS reveal the compounds containing radioactive carbon that were produced by *Chlorella* algae during five (*top*), 10 (*middle*) and 30 seconds (*bottom*) of photosynthesis. Alanine, the first amino acid to appear in the process, shows up very faintly at first (*top*); glycine, serine, glutamic acid and aspartic acid appear as photosynthesis progresses. These radioautographs were made at the author's laboratory by exposing X-ray-sensitive film to chromatograms of compounds extracted from three samples of algae.

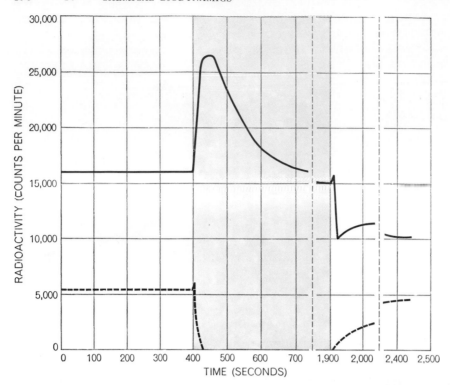

EFFECT OF SUDDEN DARKNESS on PGA (*solid curve*) and ribulose diphosphate (*broken curve*) is shown here. The conversion of ribulose diphosphate to PGA continues after the light is turned off (*colored area*), so that the ribulose concentration drops to zero. The concentration of PGA, which is no longer reduced to triose phosphate by ATP and TPNH, increases momentarily before it is used up in the production of other compounds.

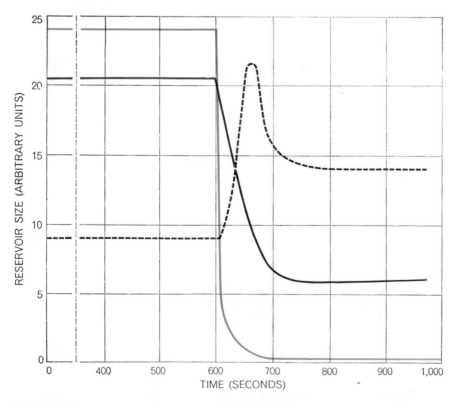

SUDDEN DEPLETION OF CARBON DIOXIDE (*colored curve*) slows the carboxylation of ribulose diphosphate to PGA. Because the light remains on after depletion, ribulose diphosphate (*broken curve*) continues to be formed and its concentration rises. PGA (*black curve*) is still reduced to triose phosphate, so that its concentration drops. "Reservoir size" refers to the average size of the "pool" of any one compound per unit quantity of algae.

cal pathways already established in studies of respiration. It seemed likely that PGA might be linked to these phosphates by the reverse of a sequence of respiratory reactions first mapped many years ago by the German chemists Otto Meyerhof, Gustav Embden and Jakob Parnas. In the respiratory pathway hexose phosphate is split into two molecules of triose phosphate, with the split occurring between the two carbon atoms in the middle of the chain. The triose phosphate is then oxidized to give PGA. The electrons from this energy-yielding operation are picked up by diphosphopyridine nucleotide (DPN+), which is thereupon reduced to DPNH. The DPN+ is a close relative of the TPN+ that turns up in photosynthesis. In addition this oxidation yields enough energy to make a molecule of ATP from ADP and phosphate ion.

In the reverse pathway of these reactions in photosynthesis, Calvin concluded, the plant uses the cofactors ATP and TPNH, made earlier by the transformation of the energy of light, to bring about the reduction of PGA to triose phosphate. In the first step the terminal phosphate group of ATP is transferred to the carboxyl group of PGA to form a "carboxyl phosphate" (really an acyl phosphate). Some of the chemical potential energy that was stored in ATP is now stored in the acyl phosphate, making the new intermediate compound highly reactive. It is now ready for reduction by TPNH. This reducing agent donates two electrons to the reactive intermediate. One carbon-oxygen bond is thereby severed and the oxygen atom, carried off with the phosphate group, is replaced by a hydrogen atom. In this way the carboxyl carbon atom is reduced to an aldehyde carbon atom; that is, it now has two bonds to oxygen instead of three and one bond each to carbon and hydrogen [*see illustration on page 386*]. This is the point in the cycle at which the major portion of the solar energy captured in the first stage of photosynthesis is applied to the reduction of carbon.

The Unstable Intermediate

The next development in the plotting of the carbon pathway came from a series of experiments first performed in our laboratory by Peter Massini. He hoped to see which intermediates would be most strongly increased or decreased in concentration by turning off the light and allowing the synthetic process to go on for a while in the dark. In order to establish the concentration of the various intermediates when the reaction proceeds in the light, he bubbled radioactive carbon dioxide through the culture for more than half an hour. At the end of this period every intermediate was as highly radioactive as the incoming carbon dioxide. The radioactivity from each compound therefore gave a measure of the concentration of the compound. He then turned off the light and after a few seconds took another sample of algae in which he measured the relative concentration of compounds by the same technique. Comparison with the compounds sampled in the light showed that the concentration of PGA was greatly increased. This finding could be readily explained: turning off the light stopped the production of the ATP and TPNH required to reduce PGA to triose phosphate.

Of the sugar phosphates present, only one, the five-carbon ribulose diphosphate, was found to have changed significantly; its concentration dropped to zero. Because the PGA had simultaneously increased in concentration, it was apparent that ribulose diphosphate was consumed in the production of PGA. This finding was of great significance because it indicated for the first time that

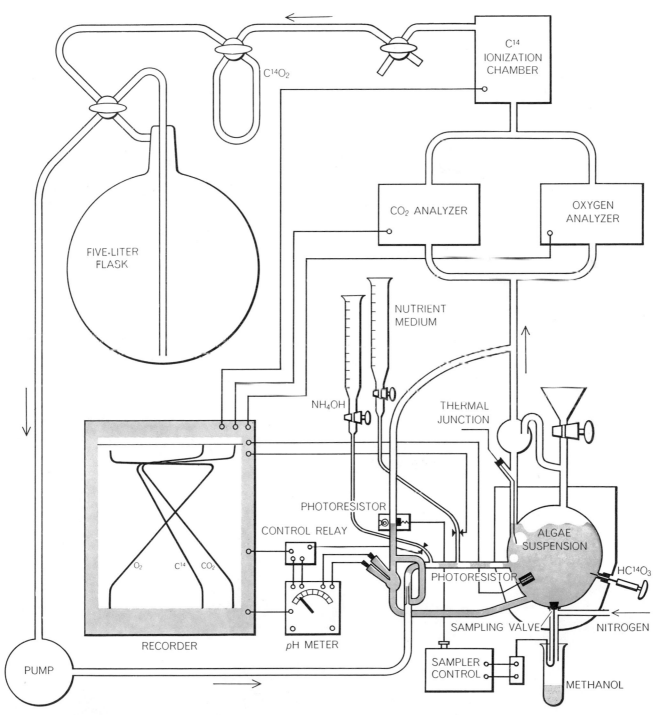

STEADY-STATE APPARATUS permits experimental control and study of photosynthesis. The algae are suspended in nutrient in a transparent vessel (*lower right*). A gas pump circulates a mixture of air, ordinary carbon dioxide and labeled carbon dioxide (when needed) to the vessel, where it bubbles through the suspensions. Labeled carbon can also be added in the form of bicarbonate ($HC^{14}O_3$). Measurements of the oxygen, carbon dioxide and labeled carbon levels in the gas are recorded continuously. The *p*H is maintained at a constant value by means of the *p*H meter. The sampler control allows removal of samples into the test tube.

ribulose diphosphate is the intermediate to which carbon dioxide is attached by the carboxylation reaction.

For this reaction ribulose diphosphate is prepared by an earlier reaction that goes on in the light and in which ATP donates its terminal phosphate group to ribulose monophosphate. The more reactive diphosphate molecule now adds one molecule of carbon dioxide by carboxylation. The details of this reaction remain obscure because the resulting six-carbon intermediate is so unstable that we have not been able to detect it

by our methods of analysis. As its first stable product this sequence of events yields two three-carbon PGA molecules.

Massini's experimental results were confirmed by a parallel experiment devised by Alex Wilson, then a graduate student in our laboratory. Instead of turning out the light Wilson shut off the supply of carbon dioxide. In this situation one might expect to find an increase in the concentration of the compound that is consumed in the carboxylation reaction; ribulose diphosphate showed such an increase. Correspondingly, one

would look for a decrease in the concentration of the product of this reaction; PGA did in fact decrease in concentration.

The first steps along the path were thus established. The photosynthesizing plant starts with ribulose monophosphate and converts it to ribulose diphosphate, using chemical potential energy trapped from the light in the terminal phosphate bond of ATP. Carbon dioxide is joined to this compound, and the resulting six-carbon intermediate splits to two molecules of PGA. With energy and

EXPERIMENT to determine the path of carbon in photosynthesis is outlined. After removal of an algae sample from its culture tube, the sample is placed in a transparent vessel (*top left*). At start of experiment labeled bicarbonate is injected into the vessel (*second from*

top left). A sample is then removed by pressing a button on the sampler control (*third from top left*); alcohol in the test tube kills the algae. The sample is concentrated by evaporation in a special flask to which a vacuum has been applied (*top right*)

electrons supplied by ATP and TPNH, PGA is reduced to triose phosphate. In the next step, it was apparent, two triose phosphates must be joined end to end in the reverse of a familiar respiratory pathway to form a hexose phosphate. The pathway from hexose to pentose phosphate remained to be uncovered.

We continued the carbon-by-carbon dissection and analysis of these chains by the methods that had earlier shown the carbon 14 in PGA to be located first in the carboxyl carbon. In the hexose molecules we had found the labeled carbon concentrated in the two middle carbons, just where it should be if two triose molecules made from PGA were linked together by their labeled ends. We also took apart the seven-carbon and five-carbon sugar phosphates to establish the position of the carbon 14 atoms in their chains. As the result of these degradations we were able to show that the overall economy of the photosynthetic process starts with five three-carbon PGA's, variously transforms them through three-, six-, four- and seven-carbon phosphate intermediates and returns three five-carbon ribulose diphosphates to the starting point [*see illustrations on pages 218 and 219*]. From the carboxylation of these three chains and their immediate bisection, the cycle at last yields six PGA molecules. The net result, therefore, is the conversion of three carbon dioxide molecules to one PGA molecule.

The Calvin Cycle

With these steps filled in, the carbon reduction cycle in photosynthesis, called the Calvin cycle, was complete. The in-

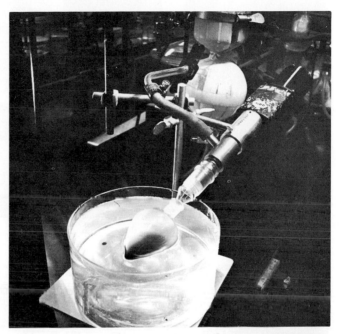

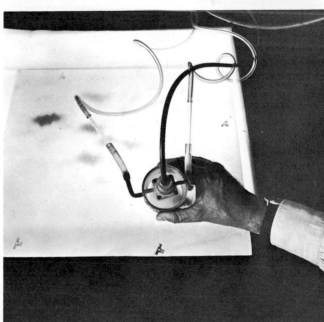

and the extract applied to chromatogram paper (*bottom left*). The paper is placed in a trough of chromatographic solvent (*second from bottom left*), which diffuses through the paper; after eight hours the paper is turned at right angles and the process is repeated. A radioautograph is made by exposing X-ray-sensitive film to the chromatogram (*third from bottom left*). The radioactivity of the compounds in the chromatogram is then measured (*bottom right*), using the radioautograph as a guide to their location.

THREE CULTURE TUBES are part of the continuous-culture apparatus used by the Bio-Organic Chemistry Group at the University of California to grow green algae under constant conditions in an aqueous medium. Two tubes are empty; the third contains algae.

"LOLLIPOP" is a thin, transparent vessel to which algae are transferred from culture tubes. Carbon dioxide containing radioactive carbon is bubbled through the algae suspension in experiments to determine the path taken by carbon in the photosynthetic process.

termediates formed in the cycle depart from it on various pathways to be converted to the end products of photosynthesis. From triose phosphate, for example, one sequence of reactions leads to the six-carbon sugar glucose and the large family of carbohydrates.

Because the cycle had been established primarily by experiments with algae and the leaves of a few higher plants, it was important to see whether or not the cycle prevailed throughout the plant kingdom. Calvin and Louisa and Richard Norris carried out experiments with a wide variety of photosynthetic organisms. In every case, although they found variation in the amounts of particular intermediates formed, the pattern was qualitatively the same.

It also had to be shown that the pathway we had traced out is quantitatively the most important route of carbon reduction in photosynthesis. To this end

Martha Kirk and I undertook an intensive study of the kinetics of the flow of carbon in photosynthesis. Our study has helped to solve other general problems, particularly the question of how carbon enters into the pathways leading to the synthesis of proteins and fats. The biological materials for this work are supplied by an algae culture system under automatic feedback control. In this apparatus we are able to maintain the photosynthetic process in a steady state, with nutrients supplied at a constant rate and with temperature, density, salinity and acidity held within narrow limits.

At the start of a run we inject radioactive bicarbonate ion into the culture medium along with radioactive carbon dioxide gas and so bring the ratio of carbon 14 to carbon 12 immediately to its final level in both the gas and the liquid phase. An automatic recorder measures the rate at which carbon is absorbed by the photosynthesizing cells. We take samples every few seconds and kill the cells immediately by immersing them in alcohol. After we have chromatographed the photosynthetic intermediates and measured their radioactivity we then plot the appearance of labeled carbon

in each of these compounds as a function of time.

By the end of three to five minutes, our records show, all the stable intermediates of the cycle are saturated with carbon 14. Taking the total amount of carbon thus fixed in compounds and comparing it with the rate of uptake of carbon in the culture, we found that the cycle accounts for more than 70 per cent of the total carbon fixed by the algae. A small but significant amount is also taken up by the addition of carbon dioxide to a three-carbon compound, phosphoenolpyruvic acid, to give four-carbon compounds.

From the earliest work with carbon 14 in our laboratory, it had been apparent that carbon dioxide finds its way rather quickly into products other than carbohydrates in the photosynthesizing plant. This was at variance with traditional ideas about photosynthesis that regarded carbohydrates as the sole organic products of the process. It was important to ask, therefore, whether fats and amino acids could be formed directly from the cycle as products of its intermediates or whether these noncarbohydrates were synthesized only from the carbohydrate end products of photosynthesis. Our kinetic studies show that certain amino acids must indeed be formed from the intermediates and must therefore be regarded as true products of photosynthesis. The amino acid alanine, for example, shows up labeled by carbon 14 at least as rapidly as any carbohydrate; it would be labeled with carbon 14 much more slowly if it were made from carbohydrate, since the carbohydrate would have to be labeled first. We have been able to show that more than 30 per cent of the carbon taken up by the algae in our steady-state system is incorporated directly into amino acids. There is some evidence that fats may also be formed as products of the cycle.

The discovery that plants make these other compounds as direct products of photosynthesis lends new interest and importance to the chloroplast, the subcellular compartment of green cells that contains pigments and the rest of the photosynthetic apparatus. It has been known for some time that this highly structured organelle is responsible for the absorption of light, the splitting of water and the formation of the cofactors for carbon reduction. More recent studies have shown that it is the site of the entire carbon-reduction cycle. Now the chloroplast emerges as a complete photosynthetic factory for the production of just about everything necessary to the plant's growth and function.

1 HEXOSE AND HEPTOSE MONOPHOSPHATES

2 PGA

3 SUCROSE

4 HEXOSE AND HEPTOSE DIPHOSPHATES

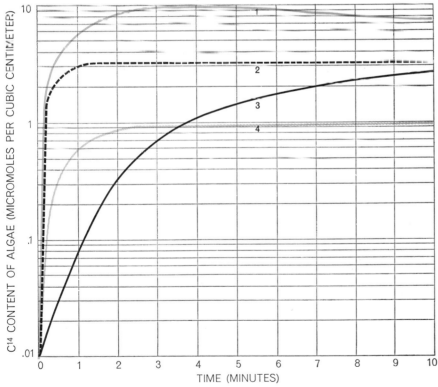

CALVIN CYCLE (*see illustration on pages 384 and 385*) was shown to be the most important route of carbon reduction in photosynthesis in studies by the author and Martha Kird. As seen here, all stable intermediates of the cycle become saturated with labeled carbon within three to five minutes. Comparison with the rate of carbon uptake in the algae culture showed that the cycle accounts for more than 70 per cent of the carbon fixed in compounds.

43

THE MECHANISM OF PHOTOSYNTHESIS

R. P. LEVINE
December 1969

Light of two wavelengths is required to activate two photochemical systems. Together they provide the electrons, protons and energy-rich molecules needed to convert carbon dioxide and water into food

Light interacts with living organisms in such processes as vision, bioluminescence and photosynthesis, but it is apparently only during photosynthesis that light energy is converted into useful forms of chemical energy. That energy, in turn, is used to build up complex molecules—notably carbohydrates—that animals require as food. Photosynthetic organisms also provide most of the oxygen in the atmosphere, and the evolution of animals was certainly dependent on the existence of oxygen-evolving microorganisms in the primitive seas.

The study of photosynthesis spans the disciplines of photophysics, photochemistry, biochemistry and physiology. In recent decades such studies have revealed many remarkable aspects of the photosynthetic process. The simple equation that summarizes the process has been known since the 19th century: water (H_2O) plus carbon dioxide (CO_2) yields some form of carbohydrate (represented by CH_2O) and oxygen (O_2). The reaction is driven by light energy. Photons are first absorbed by chlorophyll and other photosynthetic pigments. The red and blue-green algae, for example, contain pigments called phycobilins in addition to chlorophyll [*see illustration on page 398*].

Once the light energy is absorbed, it is used for two purposes. First, it is used to generate what a chemist calls "reducing power." Reduction involves the addition of electrons or the removal of protons, or both. Molecules that are rich in reducing power can transfer electrons to more oxidized molecules. The reducing agent produced by photosynthesis is NADPH, the reduced form of nicotinamide adenine dinucleotide diphosphate (NADP). Second, the light energy becomes converted into the energy-rich phosphate compound adenosine triphosphate (ATP). ATP and NADPH have certain structural elements in common [*see illustrations on page 5*]. Both are needed to reduce CO_2, a relatively oxidized molecule, into carbohydrate. The overall balance sheet for photosynthesis shows that three molecules of ATP and two molecules of NADPH are required for each molecule of CO_2 reduced.

In most algae and in higher plants photosynthesis occurs in the intricate, membrane-filled structure known as the chloroplast [*see illustration on opposite page*]. Within the chloroplast light energy is trapped and the rapid photophysical and photochemical reactions take place that generate the ATP and NADPH to be used in the more leisurely biochemical process of reducing carbon dioxide to carbohydrate. Once ATP and NADPH have been formed they are released into the nonmembranous, or soluble, phase of the chloroplast; there the fixation of carbon dioxide can proceed in the absence of light with the assistance of a number of soluble enzymes.

The Absorption of Light Energy

When chlorophyll or one of the other photosynthetic pigments absorbs photons, the pigment passes from its lowest energy state, or ground state, to a higher energy state. The excited state is not stable, and the pigment can return to the ground state within 10^{-9} second. If in that brief period the energy is not used for the generation of ATP or NADPH, it can be dissipated as fluorescent light. Since 100 percent efficiency is never achieved in biological systems, fluorescence is always observed during photosynthesis; the fluorescence is at a longer —hence less energetic—wavelength than the wavelength originally absorbed. As we shall see, fluorescence has been useful to the investigator of photosynthesis.

Each of the photosynthetic pigments has its characteristic absorption spectrum: it absorbs more or less light at different wavelengths depending on its molecular structure. For example, the two chlorophylls designated *a* and *b* have major and distinctive absorption bands in the blue and red regions of the spectrum [*see top illustration on page 400*]. When studied in isolation outside the chloroplast, each pigment also has a characteristic fluorescence spectrum. Inside the chloroplast, however, chlorophyll *b* fluorescence is never detected, even when the incident light is of a wavelength known to be absorbed by that chlorophyll. Similarly, fluorescence is never observed from the carotenoid pigments and the phycobilins. Only one of the pigments fluoresces naturally inside the chloroplast: chlorophyll *a*.

This surprising phenomenon has now been explained, largely through the work of Louis N. M. Duysens of the Netherlands. He showed that chlorophyll *b*, the carotenoids and the phycobilins do not participate directly in photosynthesis but rather act only as "antennas" to help gather light energy. When they absorb energy and become excited, they transfer their excitation energy to chlorophyll *a*. Only chlorophyll *a* is actively involved in the subsequent reactions of photosynthesis; when its energy cannot be used for photosynthesis, it dissipates its excitation energy as fluorescence.

The mechanism for the transfer of excitation energy between pigment molecules in photosynthesis is not clearly understood, but a process called inductive resonance is one possibility. If an excited molecule is close enough to an unexcited one (say within 30 angstroms), it can dissipate its energy by inducing an excited state in the neighboring molecule. In this way energy can pass from chlorophyll *b* to chlorophyll *a*. The reverse process is not possible, however, because to become excited chlorophyll *b* requires more excitation energy than

an excited chlorophyll *a* can provide.

Ultimately the energy of excitation reaches a photosynthetic reaction center, where it is transferred to a special long-wavelength form of chlorophyll *a*. Because this pigment absorbs at a longer wavelength, and hence at a lower ener-

gy, than the surrounding pigment molecule, it can be considered a kind of energy sink. The transfer of excitation energy from an excited molecule of normal chlorophyll *a* to such a special chlorophyll *a* molecule probably takes place within 10^{-12} second, which is 1,000 times

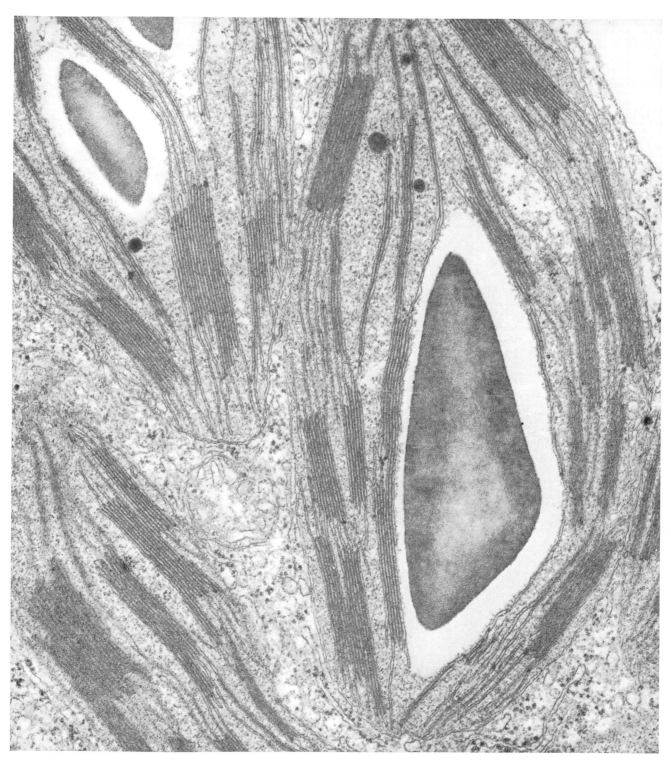

SITE OF PHOTOSYNTHESIS is the organelle known as a chloroplast, present in the cells of all higher plants and most algae. This electron micrograph made by Peter Hepler of Harvard University shows portions of three chloroplasts. Photosynthesis takes place inside the dark membranes that lie in long parallel bundles. The large triangular object in the chloroplast at the right is a kernel of starch, produced by photosynthesis. The chloroplast at the upper left contains two kernels. The magnification is 58,000 diameters.

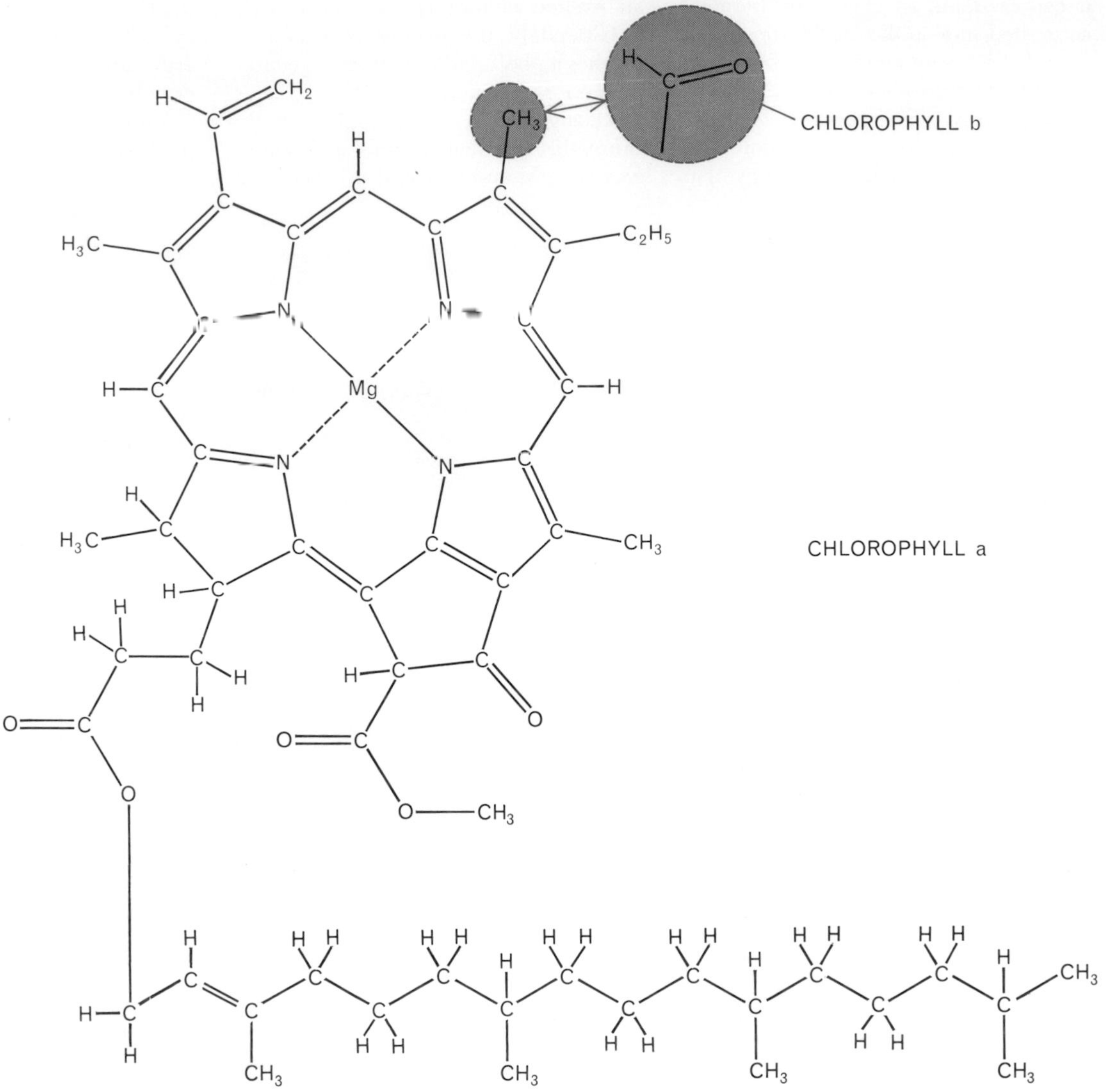

CHLOROPHYLL b

CHLOROPHYLL a

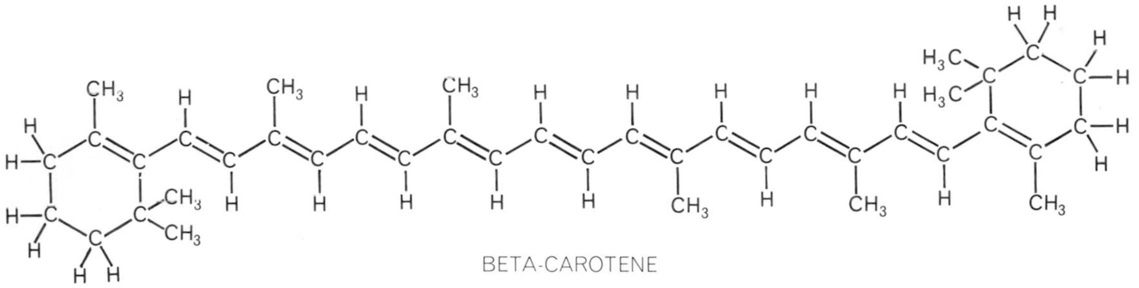

BETA-CAROTENE

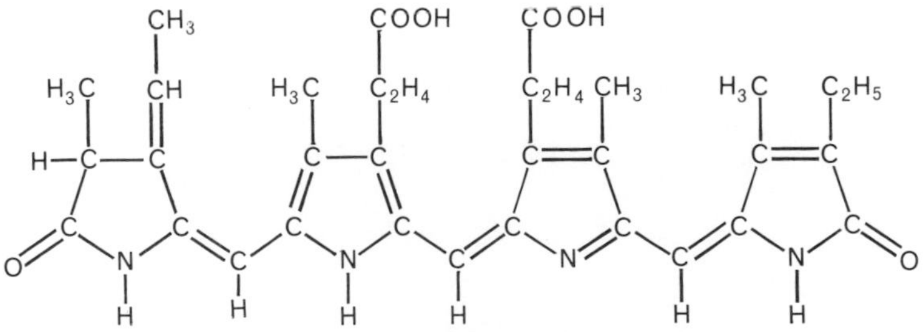

PHYCOCYANOBILIN

PHOTOSYNTHETIC PIGMENTS have the ability to capture photons and convert their energy into molecular excitation energy. Chlorophyll *a* is the pigment found in algae and in the leaves of higher plants. Chlorophyll *b* has the same structure except that a –CHO group replaces a –CH$_3$ group in one corner of the porphyrin ring. Beta-carotene is another photosynthetic pigment present in many higher plants. Red and blue-green algae contain still a third class of pigments, the phycobilins, of which phycocyanobilin is one.

faster than the time taken for the "waste" energy of chlorophyll *a* to emerge as fluorescence. Thus there is ample time for an excited chlorophyll molecule to disperse its energy in a chemically useful way.

The First Chemical Steps

Once light energy has been relayed to the special chlorophyll *a* molecule, the energy sink in the reaction center, the chemistry begins: the excitation energy must be used to form an oxidant and a reductant. The oxidant must be capable of oxidizing water, that is, capable of splitting the water molecule into free oxygen, protons and electrons. (Actually two molecules of H_2O are split into one molecule of O_2 plus four electrons and four protons.) The reductant must accept the reducing equivalents (electrons and protons) that arise from the oxidation of water. Ultimately these equivalents will be used in the reduction of carbon dioxide. The oxidant and the reductant must be formed within the very short lifetime of the excited state of chlorophyll *a*. How this comes about constitutes one of the biggest gaps in our knowledge of photosynthesis, and a great deal of what follows must be conjecture.

One simple way to visualize the initiation of the first chemical steps of photosynthesis is to imagine the existence of an electron-donor molecule *D* and an electron-acceptor molecule *A*. The donor in the oxidized form (D^+) will oxidize water and the acceptor in reduced form (A^-) will ultimately transfer its reducing equivalents to NADP, converting it to NADPH.

A simple model of the primary reaction sequence involving the donor, the acceptor and an excited molecule of chlorophyll *a* is shown in the top illustration on page 401. In this sequence the chlorophyll *a* in the reaction center is raised to an excited state by receiving excitation energy from surrounding pigment molecules. Each reaction-center chlorophyll molecule is in close association with the donor and acceptor molecules in the membrane of the chloroplast. When the chlorophyll returns to the ground state, the release of the excitation energy is sufficient to extract an electron from the donor molecule, thereby oxidizing it to D^+, and to transfer this electron to the acceptor, thus reducing it to A^-. Such charge-transfer processes are known to operate in nonbiological systems where the organic molecules involved have properties similar to those involved in photosynthesis.

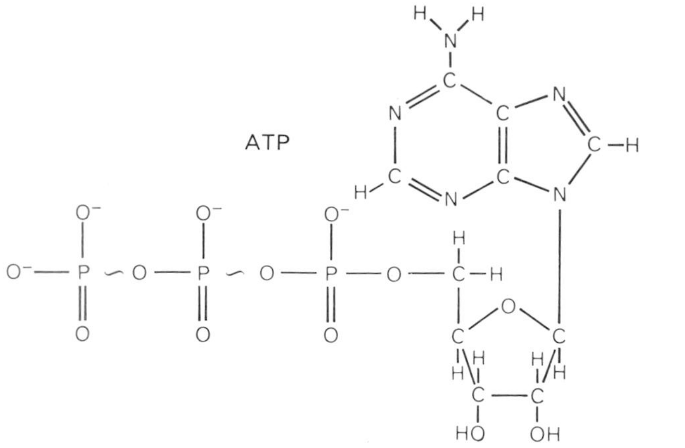

ADENOSINE TRIPHOSPHATE, or ATP, is produced from adenosine diphosphate (ADP), with the energy collected by the photosynthetic pigments. The wavy lines are links to energy-rich phosphate groups. If the last group (*color*) is removed, ATP becomes ADP. In the process ATP supplies energy for converting carbon dioxide into carbohydrates.

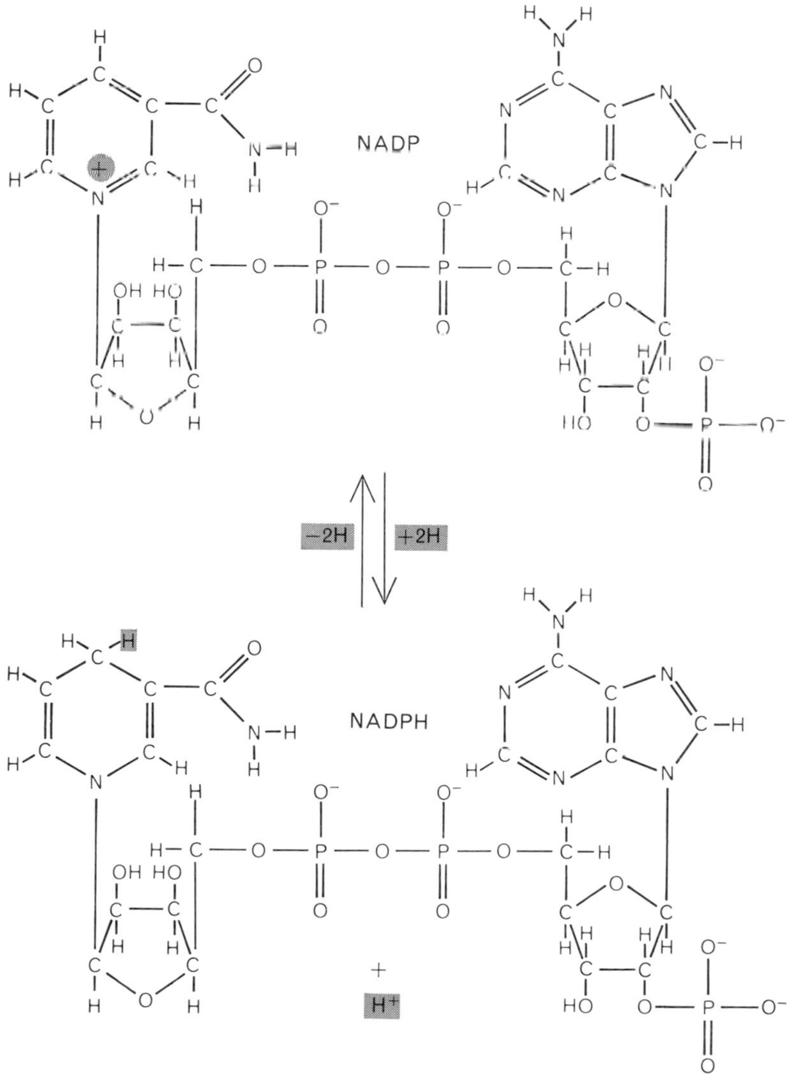

NICOTINAMIDE ADENINE DINUCLEOTIDE DIPHOSPHATE, or NADP, is reduced to NADPH during photosynthesis. NADP becomes NADPH by the addition of two hydrogen atoms. One binds directly to the molecule while the other loses its electron and is released as a proton (H^+). NADPH supplies "reducing power" for fixation of carbon dioxide.

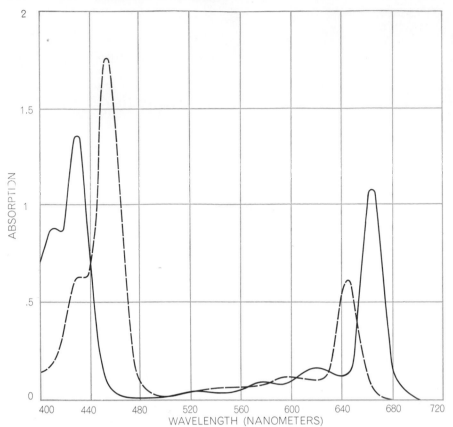

ABSORPTION SPECTRA show that chlorophyll *a* (*solid line*) and chlorophyll *b* (*broken line*) strongly absorb blue and far-red light. The green, yellow and orange wavelengths lying between the peaks are reflected and give both pigments their familiar green color.

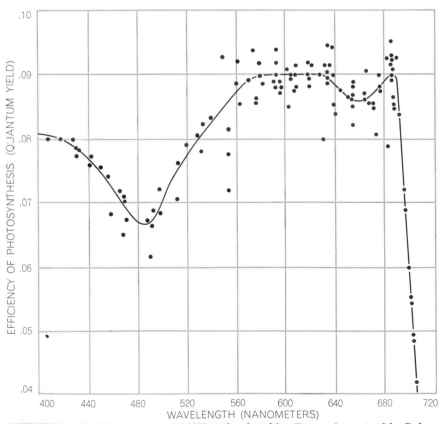

EFFICIENCY OF LIGHT ABSORPTION in the alga chlorella was determined by Robert Emerson of the University of Illinois. Efficiency of photosynthesis falls off sharply in the far-red region beyond 680 nanometers even though chlorophyll *a* still absorbs at that wavelength. If light of shorter wavelength is added to far-red light, efficiency rises sharply.

Regardless of how the charge transfer may be accomplished in the chloroplast, its net effect is to separate oxidizing and reducing equivalents. Very little is known about how the hypothetical D^+ participates in the oxidation of water with the concomitant evolution of oxygen. Much more is known about the transfer of reducing equivalents from the hypothetical A^- to NADP. With this step the mechanism of photosynthesis moves from a high-speed photochemical phase to a slower biochemical phase in which electrons are transported through a series of reactions, ultimately to yield NADPH and ATP.

The Biochemical Phase

Our understanding of the biochemical phase of photosynthesis owes much to investigations showing that two light reactions (and not one, as has been tacitly assumed so far) take place in the photosynthetic process used by algae and higher plants. Two experiments set the stage for this discovery. The first provided measurements of the rate of photosynthesis at different wavelengths of light over the range absorbed by the photosynthetic pigments. The result is a curve showing how the quantum efficiency of the process varies at different wavelengths [*see bottom illustration at left*]. The curve reveals a curious fact: in the far-red region, beyond a wavelength of 680 nanometers, the efficiency of photosynthesis falls rapidly to zero even though the pigments still absorb light.

This surprising result led to the second set of experiments, reported in 1956 by Robert Emerson and his colleagues at the University of Illinois. They found that although photosynthesis is very inefficient at wavelengths greater than 680 nanometers, it can be enhanced by adding light of a shorter wavelength, 650 nanometers for example. Moreover, the rate of photosynthesis in the presence of both wavelengths is greater than the sum of the rates obtained when the two wavelengths are supplied separately. This phenomenon, now known as the Emerson enhancement effect, can be explained if photosynthesis is assumed to require two light-driven reactions, both of which can be driven by light of less than 680 nanometers but only one by light of longer wavelength.

These two sets of experiments marked the beginning of an exciting period in the effort to understand the mechanism of photosynthesis. They gave rise to a provocative hypothesis and to several revealing lines of research. The hypothesis was provided by Robert Hill and Fay Bendall of the University of Cambridge. They proposed a scheme showing how electrons could be transported along a biochemical chain in which two separate reactions are triggered by light. Before describing the Hill-Bendall scheme I should briefly touch on some characteristics of electron transport.

One must distinguish first between a transfer in which electrons go *with* an electrochemical gradient (the easy direction) and a transfer in which the electrons go *against* that gradient (the hard direction). Electron-donor and electron-acceptor molecules can be characterized by the quantity called oxidation-reduction potential, which can be positive or negative and is usually expressed in volts. Electrons can be transferred from donors that have a more negative potential to acceptors that have a more positive potential without any input of energy. In fact, when electron transfer takes place along this gradient, energy is released; the greater the gap in potential between donor and acceptor, the greater the yield of energy. To transfer electrons against the electrochemical gradient, on the other hand, requires an input of energy. The greater the gap between the donor and the acceptor, the greater the energy required.

In the mitochondria of both plant and animal cells energy, in the form of ATP, is generated as a consequence of electron transport down an electrochemical gradient between a series of electron-donors and -acceptors called cytochromes. At least some of the electron-transport steps in photosynthesis, however, must go against an electrochemical gradient because the oxidation-reduction potential of water (the primary electron-donor) is +.8 volt whereas that of NADP (the terminal acceptor) is −.3 volt.

The Two Photochemical Systems

Hill and his co-workers had earlier identified and characterized a number of cytochromes found in the chloroplast; Hill and Bendall saw that ATP might be generated in photosynthesis if advantage were taken of the difference in oxidation-reduction potential between two of these cytochromes. One of them, a *b*-type cytochrome, has a potential close to zero and the other, a *c*-type cytochrome, has a potential of about +.35 volt. In the Hill-Bendall scheme, therefore, two light reactions provide the energy to go *against* the electrochemical

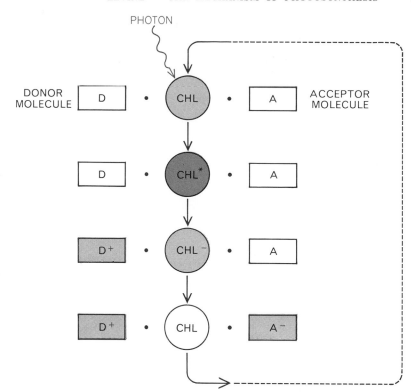

FIRST CHEMICAL STEPS in photosynthesis involve an electron-donor molecule (*D*) and an electron-acceptor molecule (*A*) in close association with a special chlorophyll (*Chl*) in the reaction center. An incoming photon can raise the chlorophyll to an excited state (*Chl**). When the excited chlorophyll returns to the ground state in less than 10^{-9} second, the energy released extracts an electron from *D*, oxidizing it to D^+, and transfers the electron to *A*, reducing it to A^-. Later D^+ oxidizes water and A^- reduces NADP to NADPH.

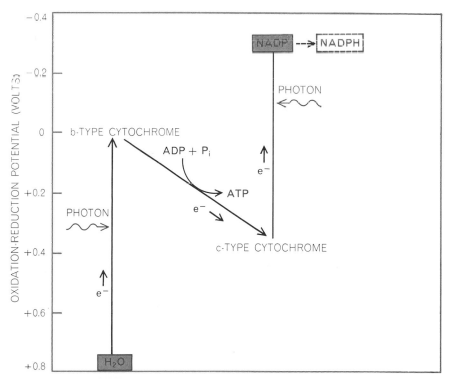

HILL-BENDALL MODEL of electron transport in photosynthesis suggests how electrons (e^-) removed from water can be boosted against an electrochemical gradient, finally reaching NADP. With the aid of protons, also provided by water, NADP is converted to NADPH. A change in an upward direction requires an input of energy, supplied by photons; a change in a downward direction yields energy. ADP is presumably converted to ATP on the downslope between the two cytochromes, which act as electron-acceptors and -donors.

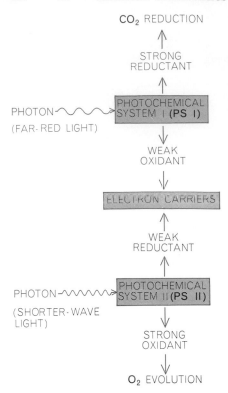

CO₂ REDUCTION
↑
STRONG
REDUCTANT
↑
PHOTON∿∿∿ PHOTOCHEMICAL SYSTEM I (PS I)
(FAR-RED LIGHT)
↓
WEAK
OXIDANT
↓
ELECTRON CARRIERS
↓
WEAK
REDUCTANT
↑
PHOTON∿∿∿ PHOTOCHEMICAL SYSTEM II (PS II)
(SHORTER-WAVE LIGHT)
↓
STRONG
OXIDANT
↓
O₂ EVOLUTION

PS I AND PS II, two photochemical systems, cooperate in the fixation of carbon dioxide in algae and higher plants. Each system has its own reaction center containing a photosynthetic pigment. The pigment in PS I is a species of chlorophyll *a* known as P-700 because its maximum absorption is at a wavelength of 700 nanometers. The strong oxidant of PS II is able to oxidize water. The strong reductant of PS I has the power to reduce NADP to NADPH. The reactions driven by the photochemical systems are shown in the equations at the bottom of the page.

gradient while electron transport between the two cytochromes goes *with* the gradient [*see bottom illustration on preceding page*].

The Hill-Bendall formulation indicated that the two light reactions occur in two different photochemical systems [*see illustration above*]. Each system has a reaction center within which an oxi-

dant and a reductant are formed. Photochemical system II (PS II) sensitizes a reaction that results in the oxidation of water and in the formation of a weak reductant. The chlorophyll in the reaction center of PS II has not yet been identified, but it is presumed to be some form of chlorophyll *a*. Photochemical system I (PS I) sensitizes a reaction that yields a weak oxidant and a strong reductant. The chlorophyll in the reaction center of PS I has been identified as a species of chlorophyll *a* whose absorption peak is at 700 nanometers and is therefore known as P-700. The two photochemical systems are linked in series by electron-carriers, so that the weak reductant produced in PS II is oxidized by the weak oxidant produced in PS I.

Duysens and his co-workers provided some of the early evidence for this model of two photochemical systems acting in series when they showed that the *c*-type cytochrome in the chloroplast is reduced by the shorter-wavelength light absorbed by PS II and oxidized by the longer-wavelength light absorbed by PS I. Such "antagonistic" effects indicate that the cytochrome lies in the path of electron flow between the two systems. Other investigators have since demonstrated similar antagonistic effects on the *b*-type cytochrome of the chloroplast and on P-700.

Additional evidence for the series model has involved the use of the potent weed killer DCMU (dichlorophenyldimethyl urea), which owes its effectiveness to its ability to inhibit the flow of electrons from water to NADP. In its presence both the *c*-type and the *b*-type cytochromes can be oxidized by PS I but they cannot be reduced by PS II. One can therefore assume that the DCMU acts at a site somewhere between PS II and the cytochromes [*see illustration on page 9*]. The photoreduction of NADP is thus blocked by DCMU, but it can be restored if an artificial electron-donor (such as a reduced

indophenol dye) is introduced into the chloroplast. In the presence of the dye NADP can once again be photoreduced but now light absorbed by PS I alone is sufficient. The effect of DCMU indicates the existence of two light reactions coupled by a system of electron-donors and -acceptors.

Electron Path from PS II

The portion of the photosynthetic electron-transport chain that carries electrons from water to PS II is known as the oxidizing "side" of PS II. As mentioned above, the oxidation of water is effected by the oxidized form of a hypothetical donor molecule, D^+. Experimental evidence for electron transport between water and PS II has recently been provided by Takashi Yamashita and Warren L. Butler of the University of California at San Diego, but the nature of the electron-carrier (or carriers) involved, and its relation to D, has not yet been determined.

The reducing "side" of PS II is the portion of the electron-transport chain between PS II and its electron-acceptor A. The fluorescence properties of the chloroplast have provided information on the nature of A. If chloroplasts are irradiated with short-wavelength light, the yield of fluorescence is high, but if they are illuminated with the longer wavelength of light that can be absorbed by PS I, the fluorescence yield decreases. From these observations Duysens and his associates have inferred that when A is reduced by PS II, fluorescence is high, but when it is oxidized by PS I, fluorescence is low. They called the acceptor component Q rather than A, Q standing for quencher of fluorescence. Q in the oxidized form quenches fluorescence, whereas Q in the reduced form does not and therefore the fluorescence yield increases. The yield increases even more in the presence of DCMU, suggesting that the weed killer acts at a site between Q and PS I in the photosynthetic electron-transport chain.

The chemical nature of Q has not been determined with certainty. Norman I. Bishop of Oregon State University has obtained evidence suggesting that it may be a compound known as plastoquinone. Regardless of its chemical nature, Q is probably the electron-acceptor of PS II.

Electron Path from Q to P-700

Proceeding along the electron-transport chain from PS II to PS I, one finds

$$2\,H_2O \longrightarrow O_2 + 4e^- + 4H^+$$

$$2NADP + 4e^- + 2H^+ \longrightarrow 2NADPH$$

$$2H^+ + 2NADPH + CO_2 \longrightarrow 2NADP + H_2O + CH_2O$$

$$\text{NET:} \quad CO_2 + H_2O \longrightarrow CH_2O + O_2$$

CHEMISTRY OF PHOTOSYNTHESIS is summarized by these four equations. Water is oxidized, releasing free oxygen, electrons and protons. The electrons and two protons reduce NADP to NADPH. NADPH plus two protons and carbon dioxide yield NADP, water and carbohydrate (CH_2O). Thus carbon dioxide and water yield carbohydrate and oxygen.

that there is a cytochrome on the downhill slope from Q to PS I. This is a b-type cytochrome. It is followed by a c-type cytochrome. The reduction of both cytochromes is sensitized by PS II; their oxidation is sensitized by PS I. As mentioned above, this differential oxidation and reduction pattern localizes the cytochromes between the two photochemical systems.

Between the b-type and the c-type cytochromes there is at least one more component, which is not yet identified. Evidence for its existence comes from experiments that Donald Gorman and I have conducted in the Biological Laboratories of Harvard University, using mutant strains of a unicellular green alga that had lost the capacity to carry out normal photosynthetic electron transport. Of the mutant strains, one lacked the b-type cytochrome, another lacked the c-type cytochrome and a third lacked an unknown component.

The third mutant strain proved to possess both cytochromes (b-type and c-type), but when it was illuminated with long-wavelength light of the kind absorbed by PS I, only the c-type cytochrome was oxidized. When light of the kind absorbed by PS II was used, only the b-type cytochrome was reduced. Since the first kind of light normally oxidizes both cytochromes and the second kind of light normally reduces both cytochromes, it was clear that some component was missing from the mutant strain that ordinarily acts as an electron donor and -acceptor between the two cytochromes. For want of a specific identification we have designated it M.

Another component in the electron-transport chain is the copper-containing protein plastocyanin. Although this protein can act as an electron-acceptor and -donor, its role is not clearly understood. At present some experiments indicate that plastocyanin lies between the c-type cytochrome and PS I, and other experiments give equally convincing evidence that it lies on the uphill side of the c-type cytochrome.

We now come to P-700, the chlorophyll that absorbs far-red light in the reaction center of PS I. Its discoverers, Bessel Kok of the Research Institute for Advanced Study in Baltimore and George E. Hoch of the University of Rochester, showed that P-700 is oxidized by light absorbed by PS I and reduced by light absorbed by PS II. On being oxidized it transfers its electron to its electron-acceptor. It can then be reduced again by electrons coming from water and passed along the transport

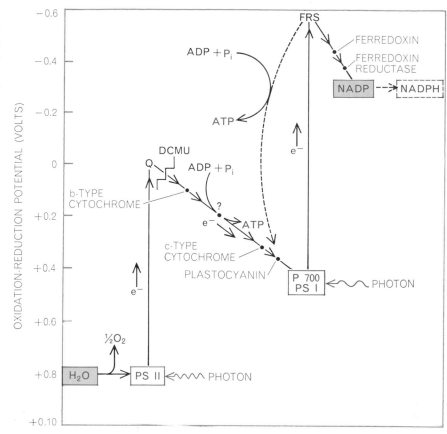

ELECTRON-TRANSPORT CHAIN in photosynthesis follows the Hill-Bendall model (*see bottom illustration on page 401*). It shows how electrons removed from water are passed along by various acceptors and donors until they finally reach NADP and participate in its reduction to NADPH. Along the chain, at one or more places not yet clearly identified, energy is extracted to form ATP from ADP and inorganic phosphate (P_i). The energy for boosting electrons against the electrochemical gradient is supplied by photons that excite chlorophyll molecules in the two photochemical systems, PS I and PS II. Electrons leaving PS II are evidently accepted directly by a substance called Q (for quencher of fluorescence) and electrons leaving PS I by the ferredoxin-reducing substance (FRS). The downhill path from Q to PS I contains a number of donors and acceptors, which are discussed in the text. The path that transports electrons from FRS to NADP appears to be less complicated; the end result is NADPH. Many details of the electron-transport chain have been clarified with the help of a weed killer called DCMU, which interrupts electron flow to the right of Q.

chain from PS II. The immediate electron-donor to P-700 is either the c-type cytochrome or plastocyanin or both.

The Path between P-700 and NADP

Moving along the electron-transport chain, we see that the electron donated by PS I, which has a potential of about +.45 volt, must travel against a large electrochemical gradient before it can reach NADP, which has a potential of −.3 volt. The electron-acceptor of PS I has been much debated ever since it was proposed a few years ago that the acceptor might be ferredoxin, an electron-acceptor and -donor molecule whose negative potential (−.43 volt) is even higher than that of NADP. It was clear that if light energy could boost an electron against the potential gradient from

P-700 to ferredoxin, the final step to NADP would be downhill. The difficulty was that several investigators found that PS I can boost electrons against potentials even more negative than that of ferredoxin. It did not seem economical of nature to provide a greater boosting capacity than is actually required, and this cast doubt on ferredoxin's being the primary acceptor of electrons from PS I.

Recently Charles Yocum and Anthony San Pietro of Indiana University and Achim Trebst of the University of Göttingen have discovered what is apparently the true acceptor. A substance with a potential of about −.6 volt, it has been given the tentative name ferredoxin-reducing substance, or FRS. Its chemical nature is now under study. Its absorption spectrum suggests that it will turn out to have a complex structure consisting of more than one molecular species.

We have now nearly reached the end of the photosynthetic electron-transport chain. The FRS transfers its electron to ferredoxin, and NADP is reduced to NADPH in the presence of an enzyme called ferredoxin-NADP reductase.

ATP Formation and CO₂ Fixation

Much current research is focused on how the production of ATP is coupled to photosynthetic electron transport. Theoretically, as Hill and Bendall originally suggested, sufficient energy is available in the downhill flow of electrons between the b-type and the c-type cytochromes to phosphorylate a molecule of ADP, converting it into a molecule of ATP. And indeed there is evidence for a site of ATP formation between the two cytochromes. There is also evidence for a cyclic flow of electrons around PS I (most likely involving FRS), and ATP formation is coupled to this electron flow.

In spite of extensive investigation there is uncertainty regarding the mechanism of the coupling of ATP formation and electron flow not only in the chloroplast but also in mitochondria. To review current opinions regarding the mechanism would require an article in itself.

We have now reached the last phase of the photosynthetic process: the reduction of carbon dioxide to carbohydrate. Much of our knowledge of this final phase is due to the work of Melvin Calvin, James A. Bassham and Andrew A. Benson of the University of California [see the article "The Path of Carbon in Photosynthesis," by J. A. Bassham beginning on page 383.] In this cycle one molecule of ribulose diphosphate and one molecule of carbon dioxide react, with the aid of suitable enzymes, to form two molecules of phosphoglyceric acid (PGA). The two molecules of PGA are converted to two molecules of glyceraldehyde phosphate in a reaction that requires two molecules of NADPH and two of ATP. One other step requires ATP (the production of ribulose diphosphate from the monophosphate), so that the overall requirement is three molecules of ATP and two of NADPH for each molecule of carbon dioxide reduced to carbohydrate [see illustration below]. This sequence is thought to represent the pathway of carbon dioxide fixation in higher plants, algae and photosynthetic bacteria.

Quite recently, however, M. D. Hatch and C. R. Slack in Australia have shown that there is a different kind of pathway in certain species of tropical grasses. The first step of CO₂ fixation in these grasses involves the carboxylation of phosphopyruvic acid (rather than of ribulose diphosphate), yielding oxaloacetic acid, which then serves as a precursor of PGA.

We have now followed the mechanism of photosynthesis from the initial trapping of the electromagnetic energy of light, through the conversion of energy into chemical energy, then through the electron-transport steps that lead to the generation of NADPH and ATP and finally to the terminal events of carbon dioxide fixation. We have seen that some parts of the process are much better understood than others. The most enigmatic part is the one associated with events at photochemical system II. The means by which four electrons and four protons are extracted from water with the concomitant evolution of a molecule of oxygen is one of the most fascinating problems still to be solved.

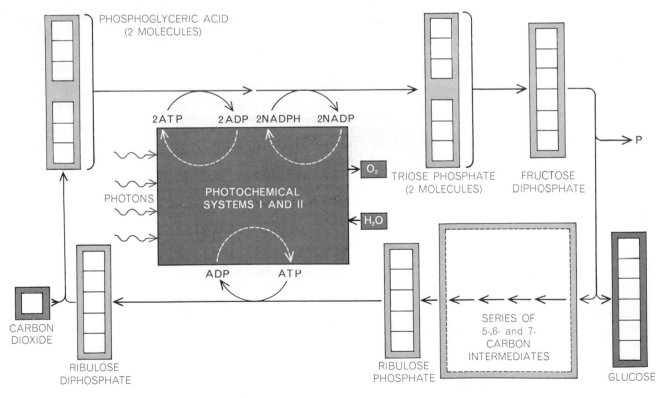

FIXATION OF CARBON DIOXIDE is achieved by a cycle of chemical reactions powered by photons that are trapped by the two photochemical systems. These systems package part of the energy in the form of ATP and remove electrons and protons from water, releasing oxygen. The electrons and protons enter the cycle in the form of NADPH. Two molecules of NADPH and three of ATP are required to fix one molecule of carbon dioxide, shown entering the cycle at the lower left. In the cycle each white square represents a carbon atom. The carbon atoms from CO₂ can be incorporated into a variety of compounds and removed at various points in the cycle. Here six atoms of carbon supplied by CO₂ are shown leaving the cycle as glucose, $C_6H_{12}O_6$, a simple carbohydrate.

44

MOLECULAR ISOMERS IN VISION

RUTH HUBBARD AND ALLEN KROPF

June 1967

Certain organic compounds can exist in two or more forms that have the same chemical composition but different molecular architecture. One of them is the basis for vision throughout the animal kingdom.

Molecular biology, which today is so often associated with very large molecules such as the nucleic acids and proteins, actually embraces the entire effort to describe the structure and function of living organisms in molecular terms. We are coming to see how the manifold activities of the living cell depend on interactions among molecules of thousands of different sizes and shapes, and we can speculate on how evolutionary processes have selected each molecule for its particular functional properties. The significance of precise molecular architecture has become a central theme of molecular biology.

One of the more recent observations is that biological molecules are not static structures but, in a number of well-established cases, change shape in response to outside influences. As an example, the molecule of hemoglobin has one shape when it is carrying oxygen from the lungs to cells elsewhere in the body and a slightly different shape when it is returning to the lungs without oxygen [see "The Hemoglobin Molecule," by M. F. Perutz; SCIENTIFIC AMERICAN Offprint 196]. A somewhat similar changeability in the molecule of lysozyme, which breaks down the walls of certain bacterial cells, is described in the article beginning on page 170 by David C. Phillips of the University of Oxford. In this article we shall describe some of the simplest changes in shape that can take place in much smaller organic molecules and show how change of this type provides the basis of vision throughout the animal kingdom.

A "Childish Fantasy"

The notion that molecules of the same atomic composition might have different spatial arrangements is less than 100 years old. It dates back to a paper titled "*Sur les formules de structure dans l'espace*," written in 1874 by Jacobus Henricus van't Hoff, then an obscure chemist at the Veterinary College of Utrecht. At that time it was still respectable to doubt the existence of atoms; to speak of the three-dimensional arrangement of atoms in molecules was a speculative leap of great audacity. Van't Hoff's paper provoked Hermann Kolbe, one of the most eminent organic chemists of his day, to publish a withering denunciation.

"Not long ago," Kolbe wrote in 1877, "I expressed the view that the lack of general education and of thorough training in chemistry of quite a few professors of chemistry was one of the causes of the deterioration of chemical research in Germany. . . . Will anyone to whom my worries may seem exaggerated please read, if he can, a recent memoir by a Herr van't Hoff on 'The Arrangements of Atoms in Space,' a document crammed to the hilt with the outpourings of a childish fantasy. This Dr. J. H. van't Hoff, employed by the Veterinary College at Utrecht, has, so it seems, no taste for accurate chemical research. He finds it more convenient to mount his Pegasus (evidently taken from the stables of the Veterinary College) and to announce how, on his daring flight to Mount Parnassus, he saw the atoms arranged in space."

Van't Hoff's "childish fantasy" was put forth independently by the French chemist Jules Achille le Bel and was soon championed by a number of leading chemists. In spite of Kolbe's opinion, evidence in support of the three-dimensional configuration of molecules rapidly accumulated. In 1900 van't Hoff was named the first recipient of the Nobel prize in chemistry.

Even before van't Hoff's paper of 1874 chemists had begun using the concept of the valence bond, commonly represented by a line connecting two atoms. It was not unnatural, therefore, to associate the valence bond concept with the idea that atoms were arranged precisely in space. The simplest hydrocarbon, methane (CH_4), would then be represented as a regular tetrahedron with a hydrogen atom at each vertex joined by a single valence bond to a carbon atom at the center of the structure [see illustration on page 407].

The valence bond remained an elusive concept, however, until G. N. Lewis postulated in 1916 that a common type of bond—the covalent bond—was formed when two atoms shared two electrons. "When two atoms of hydrogen join to form the diatomic molecule," he wrote, "each furnishes one electron of the pair which constitutes the bond. Representing each valence electron by a dot, we may therefore write as the graphical formula of hydrogen H : H." He visualized this bond to be "that 'hook and eye,' which is part of the creed of the organic chemist." To explain why electrons should tend to pair in this manner, Lewis could offer nothing beyond an intuitive principle that he called "the rule of two."

The rule of two entered the physicist's description of the atom when Wolfgang Pauli put forward the exclusion principle in 1923. This states that electrons in atoms and molecules are found in "orbitals" that can accommodate at most two electrons. Since electrons can be regarded as minuscule spinning negative charges, and thus as tiny electromagnets, the two electrons in each orbital must be spinning in opposite directions.

Let us return, however, to some of the chemical observations that gave rise to van't Hoff's ideas of stereochemistry late in the 19th century. Chemists were confronted by a series of puzzling observations best exemplified by two simple compounds: maleic acid and fumaric acid [see illustrations on page 408]. Here were two distinct, chemically pure substances, each with four atoms of carbon, four of hydrogen and four of oxygen ($C_4H_4O_4$). It was known, moreover, that the connections between atoms in the two molecules were exactly the same and that the two central carbon atoms in each molecule were connected by a double bond. Yet the two compounds were indisputably different. Whereas crystals of maleic acid melted at 128 degrees centigrade, crystals of fumaric acid did not melt until heated to about 290 degrees C. Furthermore, maleic acid was

about 100 times more soluble in water and 10 times stronger as an acid than fumaric acid. When maleic acid was heated in a vacuum, it gave off water vapor and became a new substance, maleic anhydride, which readily recombined with water and reverted to maleic acid. Fumaric acid underwent no such reaction. On the other hand, if either compound was heated in the presence of hydrogen, it was transformed into the identical compound succinic acid ($C_4H_6O_4$), which contains two more hydrogen atoms per molecule than maleic or fumaric acid.

It was known that the four carbon atoms in maleic and fumaric acids form a chain. The only way to explain the differences between the compounds is to assume that the two halves of a molecule

that are connected by a double bond are not free to rotate with respect to each other. Thus the form of the $C_4H_4O_4$ molecule in which the two terminal COOH groups lie on the same side of the double bond (maleic acid) is not identical with the form in which the COOH groups lie on opposite sides (fumaric acid). Molecules that assume distinct shapes in this way are called geometrical or cis-trans isomers of one another. "Cis" is from the Latin meaning on the same side; "trans" means on opposite sides. Therefore maleic acid is the cis isomer of the $C_4H_4O_4$ molecule and fumaric acid is the trans isomer. When the double carbon-carbon bond of either isomer is reduced to a single bond by the addition of two more hydrogen atoms, the two halves of the molecule are free

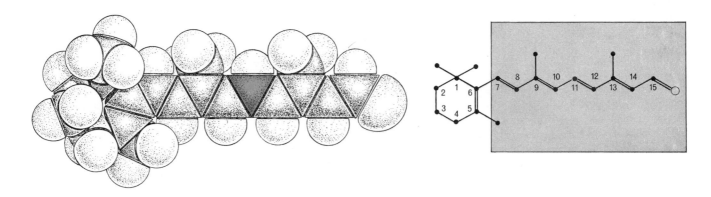

ALL-*TRANS* RETINAL

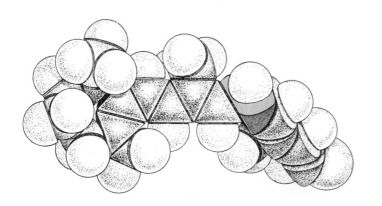

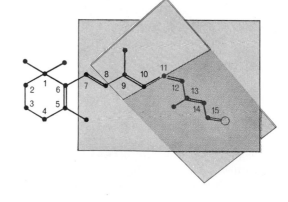

11-*CIS* RETINAL

FUNDAMENTAL MOLECULE OF VISION is retinal ($C_{20}H_{28}O$), also known as retinene, which combines with proteins called opsins to form visual pigments. Because the nine-member carbon chain in retinal contains an alternating sequence of single and double bonds, it can assume a variety of bent forms. Each distinct form is termed an isomer. Two isomers of retinal are depicted here. In the models (*left*) carbon atoms are dark, except carbon No. 11, which is shown in color; hydrogen atoms are light. The large atom attached to carbon No. 15 is oxygen. In the structural formulas (*right*) hydrogen atoms are omitted. The parts of each isomer that lie in a plane are marked by background panels. When tightly bound to opsin, retinal is in the bent and twisted form known as 11-*cis*. When struck by light, it straightens out into the all-*trans* configuration. This simple photochemical event provides the basis for vision.

to rotate with respect to each other and a single compound results: succinic acid.

Van't Hoff proposed that when two carbon atoms are joined by a single bond they can be regarded as the centers of two tetrahedrons that meet apex to apex, thus allowing the two bodies to rotate freely. To represent a double bond, he visualized the two tetrahedrons as being joined edge to edge so that they were no longer free to rotate. Apart from minor modifications his proposals have stood up extremely well.

Electrons in Orbitals

Van't Hoff's explanation, of course, was a purely formal one and provided no real insight into *why* a double bond prevents the parts of a molecule it joins from rotating with respect to each other. This was not understood for another 50 years, when the development of wave mechanics by Erwin Schrödinger set the stage for one of the most productive periods in theoretical chemistry. With Schrödinger's wave equation to guide them, chemists and physicists could compute the orbitals around atoms where pairs of electrons could be found. The valence bonds, which chemists had been drawing as lines for almost a century, now took on physical reality in the form of pairs of electrons confined to orbitals that were generally located in the regions where the valence lines had been drawn.

The first molecule to be analyzed successfully by the new wave mechanics was hydrogen (H_2). Walter Heitler and Fritz London applied Schrödinger's prescription and obtained the first profound insight into the nature of chemical bonding. Their results define the region in space most likely to be frequented by the pair of electrons associated with the two hydrogen atoms in the hydrogen molecule. The region resembles a peanut, each end of which contains a proton, or hydrogen nucleus [*see top illustration on page 409*].

The phrase "most likely to be frequented" must be used because, as Max Born convincingly argued in the late 1920's, the best one can do in the new era of quantum mechanics is to calculate the probability of finding electrons in certain regions; all hope of placing them in fixed orbits must be abandoned. The new methods were quickly applied to many kinds of molecule, including some with double bonds.

One of the most fruitful methods for describing doubly bonded molecules— the molecular-orbital method—was devised by Robert S. Mulliken of the Uni-

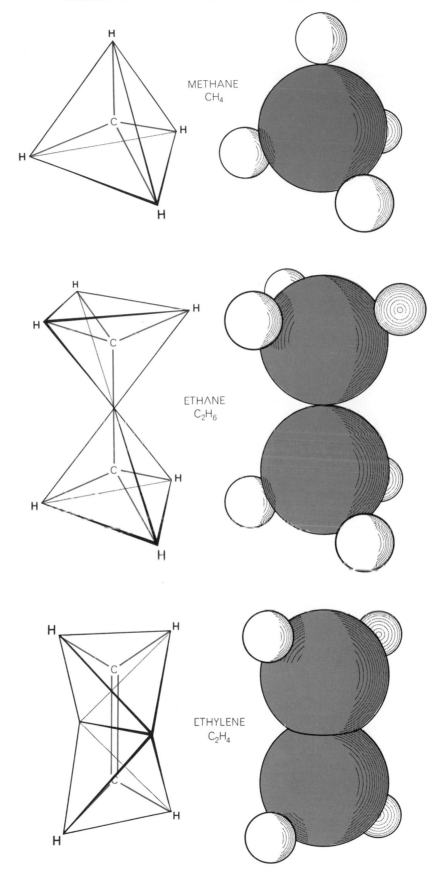

METHANE
CH_4

ETHANE
C_2H_6

ETHYLENE
C_2H_4

SIMPLEST HYDROCARBON MOLECULES are methane, ethane and ethylene. In methane the carbon atom lies at the center of a tetrahedron that has a hydrogen atom at each apex. The models at right show relative diameters of carbon and hydrogen. Ethane can be visualized as two tetrahedrons joined apex to apex. Ethylene, the simplest hydrocarbon that has a carbon-carbon double bond, can be visualized as two tetrahedrons joined edge to edge. The C=C bond in ethylene is about 15 percent shorter than the C—C bond in ethane.

versity of Chicago, who last year received the Nobel prize in chemistry. In Mulliken's concept a double bond can be visualized as three peanut-shaped regions [*see bottom illustration on next page*]. The central peanut, the "sigma" orbital, encloses the nuclei of the two adjacent atoms, as in the hydrogen molecule. The other two peanuts, which jointly form the "pi" orbital, lie along each side of the sigma orbital. The implication of this model is that in forming the sigma orbital the two electrons occupy a common volume, whereas in forming the pi orbital both electrons tend to occupy the two separate volumes simultaneously.

One can say that the sigma bond connects the atoms like an axle that joins two wheels but leaves them free to rotate separately. The pi bond ties the

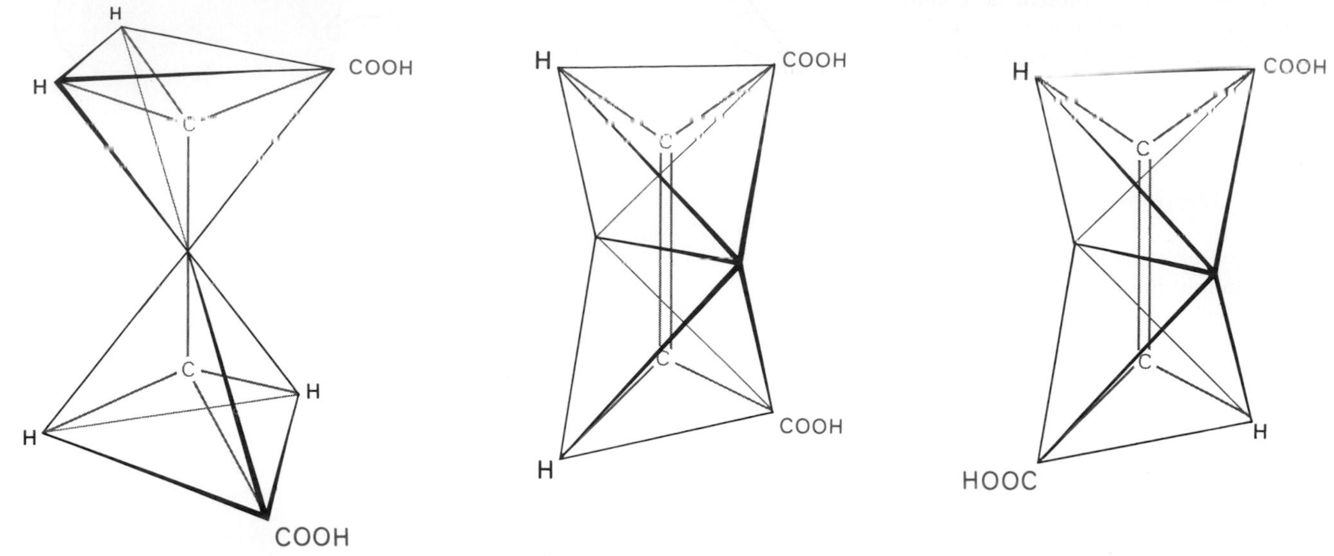

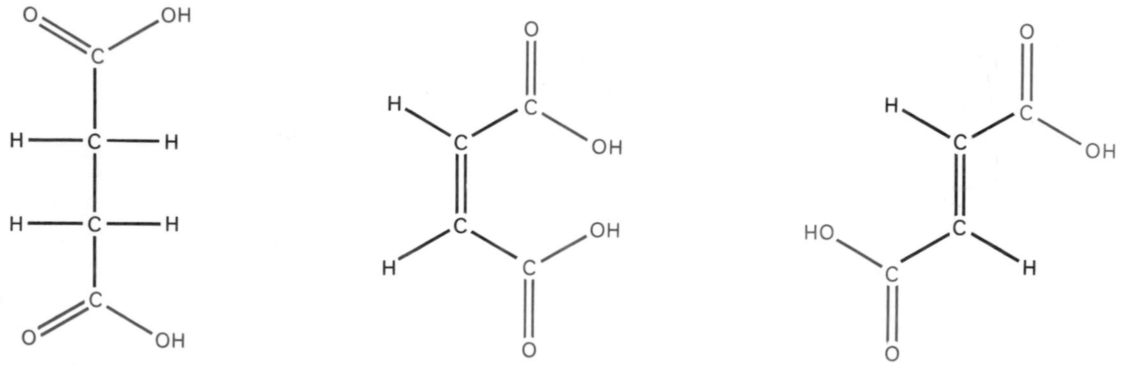

SUCCINIC ACID

MALEIC ACID

FUMARIC ACID

MOLECULAR PUZZLE was presented to chemists of the 19th century by maleic acid and fumaric acid, which have the same formula, $C_4H_4O_4$, and can be converted to succinic acid by the addition of two hydrogen atoms. Nevertheless, maleic and fumaric acids have very different properties. In 1874 Jacobus Henricus van't Hoff suggested that the central pair of carbon atoms in the three acids could be visualized as occupying the center of tetrahedrons that were joined edge to edge in the case of maleic and fumaric acids and apex to apex in the case of succinic acid. Thus the spatial relations of the two carboxyl (COOH) groups would be rigidly fixed in maleic and fumaric acids but not in succinic acid, because in the latter molecule the tetrahedrons would be free to rotate.

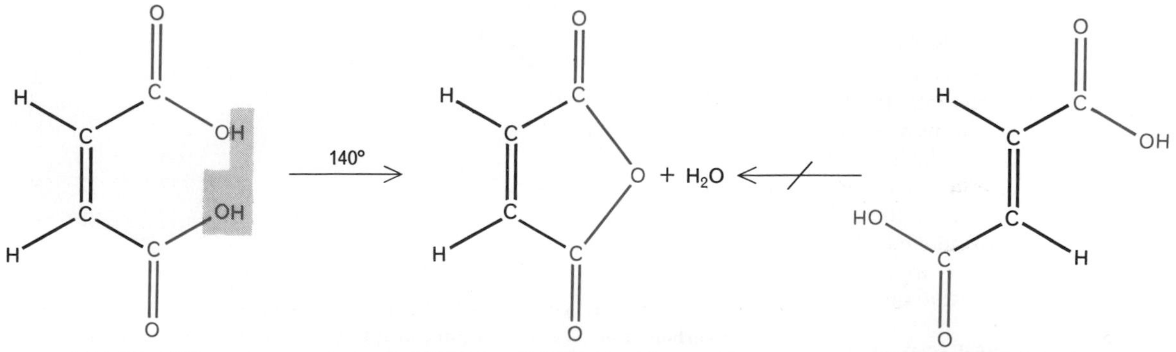

MALEIC ACID

MALEIC ANHYDRIDE + WATER

FUMARIC ACID

ONE CONSEQUENCE OF ISOMERISM is that maleic acid readily loses a molecule of water when heated, yielding maleic anhydride. Fumaric acid does not undergo this reaction because its carboxyl groups are held apart at opposite ends of the molecule.

two wheels together so that they must rotate as a unit. It also forces the two halves of the molecule to lie in the same plane, exactly as if two tetrahedrons were cemented edge to edge. In this way the molecular-orbital description of bonding provides a quantum-mechanical explanation of *cis-trans* isomerism.

The single bond joining two atoms, such as the carbon atoms of the two methyl groups in ethane (CH_3–CH_3), is a sigma bond, which leaves the groups attached to the two carbons free to rotate with respect to each other. Actually the two methyl groups in ethane are known to have a preferred configuration, so that they are not completely freewheeling. Nonetheless, at ordinary temperatures enough energy is available to make 360-degree rotations so frequent that derivatives of ethane (in which one hydrogen in each methyl group is replaced by a different kind of atom) do not form *cis-trans* isomers. There are exceptions, however, if the groups of atoms that replace hydrogen are so bulky that they collide and prevent rotation. In general, therefore, *cis-trans* isomerism is confined to molecules incorporating double bonds.

Electrons Delocalized

So much for molecules that have one double bond. What is the situation when a molecule has two or more double bonds? Specifically, what stereochemical behavior can be expected when single and double bonds alternate to form what is called a conjugated system?

The simplest conjugated system is found in 1,3-butadiene, a major ingredient in the manufacture of synthetic rubber, which can be written C_4H_6 or CH_2=CH–CH=CH_2. The designation "1,3" indicates that the double bonds originate at the first and third carbon atoms. Some of the properties of the biologically more interesting conjugated molecules are exhibited by butadiene. From the foregoing discussion one might expect that the second and third carbon atoms in the molecule would be free to rotate around the sigma bond connecting them. In actuality the rotation is not free: all the atoms in butadiene tend to lie in a plane.

It can also be shown that the energy content of each double bond in butadiene differs significantly from the energy content of the one double bond in the closely related compound 1-butene (CH_2=CH–CH_2–CH_3). The energy released in changing the two double bonds

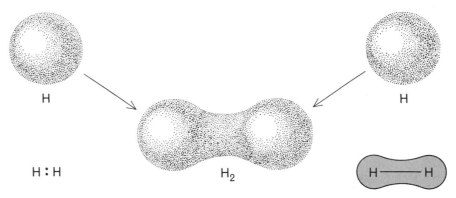

HYDROGEN MOLECULE is formed when two atoms of hydrogen (*H*) are joined by a chemical bond. The bond is created by the pairing of two electrons, one from each atom, which must have opposite magnetic properties if the atoms are to attract each other. The position of the electrons as they orbit around the hydrogen nuclei cannot be precisely known but can be represented by an "orbital," a fuzzy region in which the electrons spend most of their time. Known as a sigma orbital, it can be stylized as at lower right. The formula for the hydrogen molecule can be written as at lower left; the dots indicate electrons.

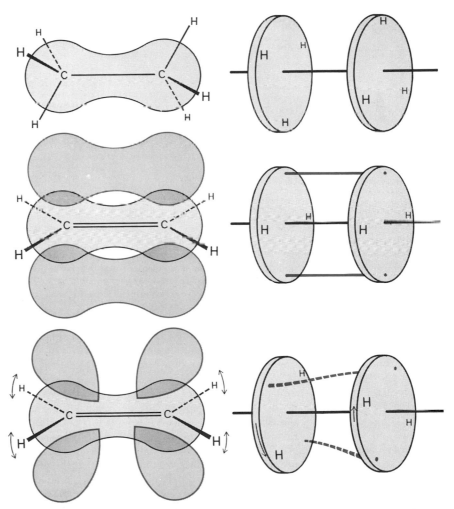

MOLECULAR ORBITALS help to explain why molecules held together by single bonds differ from molecules with double bonds. For example, the two carbon atoms in ethane are joined by two electrons in a sigma orbital, similar to the orbital in the hydrogen molecule. The two ends of the molecule, like wheels joined by a simple axle, are able to rotate. The two carbon atoms in ethylene (*middle*) are joined by two additional electrons in a "pi" orbital (*color*), as well as by two electrons in a sigma orbital. The four hydrogen atoms in ethylene are held in a plane perpendicular to the plane of the orbitals. The effect is as if two wheels were held together by two rigid rods in addition to an axle. When ethylene is in an "excited" state (*bottom*), one of the pi electrons occupies the four-lobed orbital. This lessens the rigidity of the double bond and gives it more of the character of a single bond.

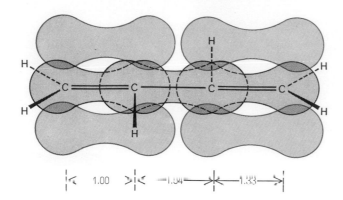
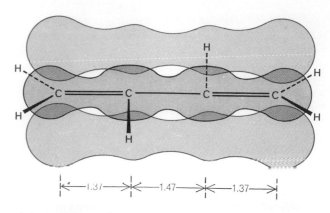

DELOCALIZED ORBITALS are found in "conjugated" systems: molecules in which single and double bonds alternate. The simplest conjugated molecule is 1,3-butadiene (C_4H_6). If its pi orbitals (*color*) were simply confined to the double bonds, as in ethylene, its orbital structure and carbon-carbon distances (in angstrom units) would be as shown at the left. Even in the lowest energy state, however, the pi electrons tend to spread across the entire molecule (*right*). As a result double bonds are lengthened and the single bond is shortened, making each type of bond more like the other. As a consequence the entire molecule is planar, or flat.

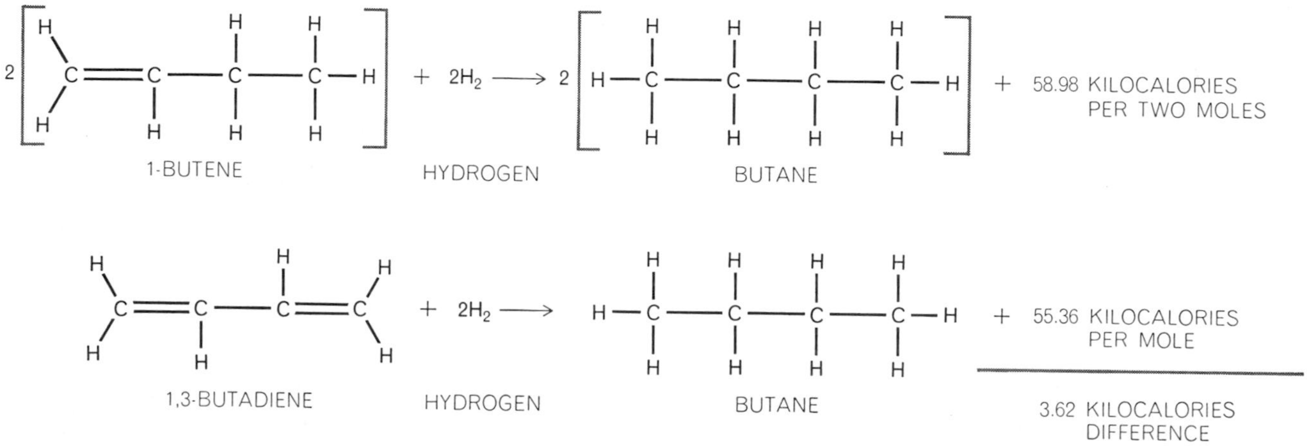

EVIDENCE FOR BOND MODIFICATION in a conjugated molecule can be obtained by measuring the energy released when double bonds are converted to single bonds by adding hydrogen. Hydrogenation of the double bond in 1-butene, which is not a conjugated molecule, yields 29.49 kilocalories for every mole of reactant. A mole is a weight in grams equal to the molecular weight of a substance: 56 for butene and 54 for butadiene. Hydrogenation of two moles of butene, hence the hydrogenation of twice as many double bonds, would therefore yield 58.98 kilocalories. Hydrogenation of the same number of double bonds in butadiene (present in a single mole) yields only 55.36 kilocalories. The difference is 3.62 kilocalories per mole for the two bonds, or 1.81 kilocalories for each double bond. The lesser energy in the butadiene double bonds indicates that they are more stable than the double bond in 1-butene.

of butadiene into the single bonds of butane ($CH_3-CH_2-CH_2-CH_3$) is about 55,400 calories per mole of butane formed. (A mole is a weight in grams equal to the molecular weight of the molecule: 58 for butane, 54 for butadiene.) The energy released in converting 1-butene, which has only one double bond, into butane is about 29,500 calories per mole. When expressed in terms of equivalent numbers of double bonds hydrogenated, the latter reaction yields some 1,800 calories more than the former [*see lower illustration above*]. The greater energy release means that the double bond in 1-butene is more reactive than either of those in 1,3-butadiene.

The added stability of the bonds in 1,3-butadiene was not unexpected; the same kind of result had been obtained for benzene, whose famous ring structure is formed by six carbon atoms connected alternately by single and double bonds. One can picture the extra energy of stabilization as arising from the tendency of electrons in the pi orbitals to leak out and become delocalized. Indeed, the phenomenon is called delocalization. The pi orbitals spread over larger portions of conjugated molecules than one might have thought, so that the properties of delocalized systems can no longer be described in terms of the properties of the double and single bonds as they are usually drawn. In order to represent the pi orbitals of 1,3-butadiene more accurately one must stretch them across all four carbon atoms of the molecule [*see upper illustration above*]. The stretching helps to explain why butadiene is not completely free to rotate around the central single bond: the bond has some of the characteristics of a double bond.

The altered character of butadiene's central carbon-carbon bond has been confirmed by X-ray-diffraction studies of butadiene. Whereas the usual carbon-carbon bond lengths are about 1.54 angstrom units for a single bond and 1.33 angstroms for a double bond, the length of the central single bond in 1,3-butadiene is only 1.47 angstroms. (An angstrom is 10^{-8} centimeter.) Linus Pauling, who did much to clarify the nature of the chemical bond, has estimated that the observed shortening of the central carbon-carbon bond of butadiene implies that it has about 15 percent of the double-bond character. One consequence of this is that the molecular configuration of butadiene tends to remain planar, or flat.

The tendency toward planarity in

conjugated systems was clearly demonstrated by the Scottish X-ray crystallographer J. M. Robertson and his colleagues in the mid-1930's. They compared the configurations of dibenzyl and *trans* stilbene, both of which contain two benzene rings joined by two carbon atoms [*see illustration at right*]. The difference between the two molecules is that in dibenzyl the two carbons are joined by a single bond, whereas in stilbene they are joined by a double bond. Robertson showed that the rings in dibenzyl are essentially at right angles to the connecting carbon-carbon bridge. In the *trans* form of stilbene the rings and the bridge lie in a plane, and the single bonds that join the rings to the two carbons in the bridge are foreshortened from the normal single-bond length to about 1.44 angstroms. In the *cis* form of stilbene the two rings cannot lie in a plane because they bump into each other.

Light-sensitive Molecules

We turn now to *cis-trans* isomerism in the family of molecules we have worked with most directly, the carotenoids and their near relatives, retinal (also known as retinene) and vitamin A. These molecules are built up from units of isoprene, which is like 1,3-butadiene in every respect but one: at the second carbon of isoprene a methyl group (CH_3) replaces the hydrogen atom present in butadiene. Natural rubber is polyisoprene, a long conjugated chain of carbon atoms with a methyl group attached to every fourth carbon.

The compound known as beta-carotene, which is responsible for the color of carrots, consists of an 18-carbon conjugated chain terminated at both ends by a six-member carbon ring, each of which adds another double bond to the conjugated system. The molecule has 40 carbon atoms in all and is presumably assembled from eight isoprene units [*see illustration on page 412*].

Until about 15 years ago the *cis-trans* isomers of the carotenoids entered biology in only one rather trivial way: in determining the color of tomatoes. Laszlo T. Zechmeister and his collaborators at the California Institute of Technology found in the early 1940's that normal red tomatoes contain the carotenoid lycopene in the all-*trans* configuration. (Lycopene differs from beta-carotene only in that the six carbon atoms at each end of the molecule do not close to form rings.) The yellow mutant known as the tangerine tomato contains

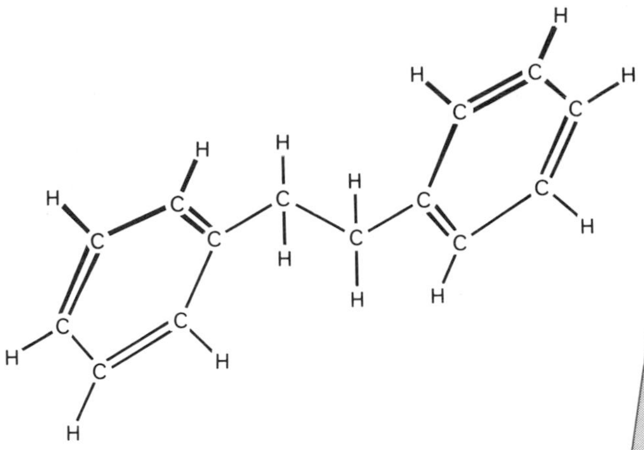

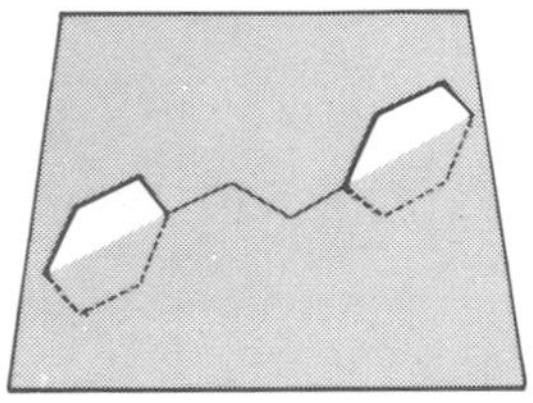

DIBENZYL

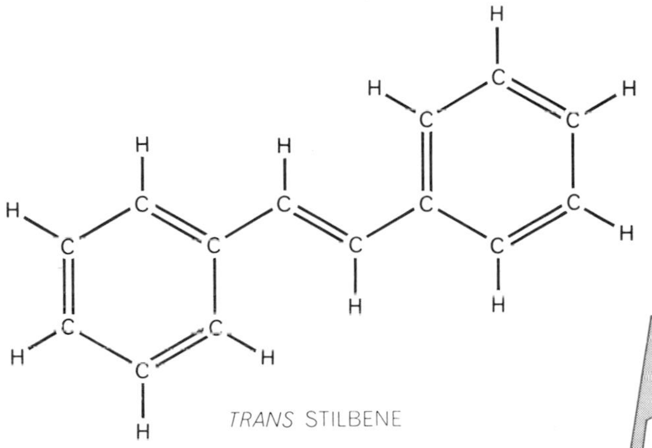

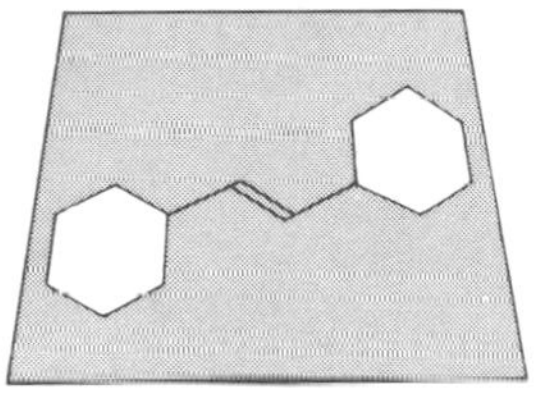

TRANS STILBENE

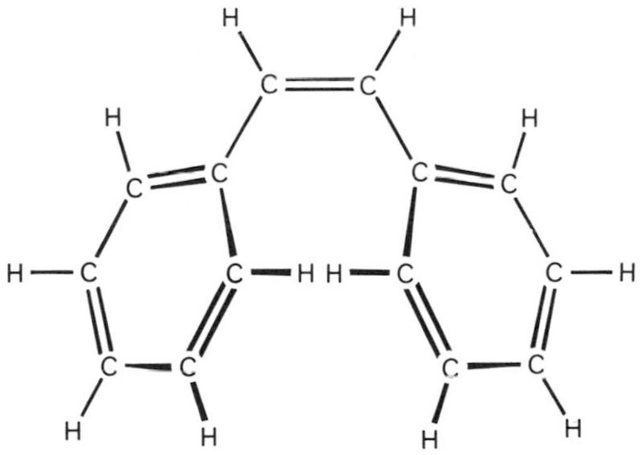

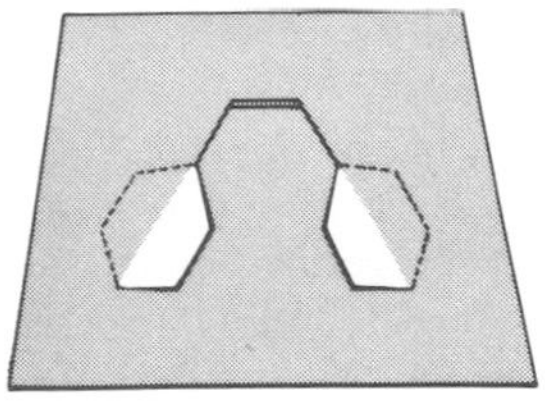

CIS STILBENE

PREFERENCE FOR FLATNESS in conjugated systems is exhibited by molecules of dibenzyl and two isomers of stilbene. The latter have a double bond in the carbon-carbon bridge linking the two benzene rings, whereas dibenzyl has a single bond. In dibenzyl the two rings are practically at right angles to the plane of the bridge. In *trans* stilbene all the atoms lie essentially in a plane. In *cis* stilbene, as can be demonstrated with molecular models, the two rings interfere with each other and thus cannot lie flat. The twisting of the rings has been established by X-ray studies of a related compound, *cis* azobenzene, in which the two rings are joined through a doubly bonded nitrogen ($N=N$) bridge.

ALL-*TRANS* CAROTENE

ALL-*TRANS* LYCOPENE

TWO NATURAL CAROTENOIDS are examples of highly conjugated systems. Like other carotenoids, they are built up from units of isoprene (C_5H_8), also known as 2-methyl butadiene. In these diagrams hydrogen atoms are omitted so that the carbon skeletons can be seen more clearly. Both molecules contain 40 carbon atoms and are symmetrical around the central carbon-carbon double bond, numbered 15–15'. Beta-carotene gives carrots their characteristic orange color. *Trans* lycopene is responsible for the red color of tomatoes.

a yellow *cis* isomer of lycopene called prolycopene. Zechmeister, who contributed more than anyone else to our present understanding of carotenoid chemistry, liked to demonstrate how a yellow solution of prolycopene, extracted from tangerine tomatoes, could be converted into a brilliant orange solution of all-*trans* lycopene simply by adding a trace of iodine in the presence of a strong light.

The discovery that the *cis-trans* isomerism of a carotenoid plays a crucial role in biology was made in the laboratory of George Wald of Harvard University, where it was found that the *cis-trans* isomerism of retinal is intrinsic to the way in which visual pigments react to light. The discovery of these pigments is usually attributed to the German physiologist Franz Boll.

The Chemistry of Vision

In 1877, the same year that Kolbe was ridiculing van't Hoff's work on stereochemistry, Boll noted that a frog's retina, when removed from the eye, was initially bright red but bleached as he watched it, becoming first yellow and finally colorless. Subsequently Boll observed that in a live frog the red color of the retina could be bleached by a strong light and would slowly return if the animal was put in a dark chamber. Recognizing that the bleachable substance must somehow be connected with the frog's ability to perceive light, Boll named it "erythropsin" or "Sehrot" (visual red). Before long Willy Kühne of Heidelberg found the red pigment in the retinal rod cells of many animals and renamed it "rhodopsin" or "Sehpurpur"

(visual purple), which it has been called ever since. Kühne also named the yellow product of bleaching "Sehgelb" (visual yellow) and the white product "Sehweiss" (visual white).

The chemistry of the rhodopsin system remained largely descriptive until 1933, when Wald, then a postdoctoral fellow working in Otto Warburg's laboratory in Berlin and Paul Karrer's laboratory in Zurich, demonstrated that the eye contains vitamin A. Wald showed that the vitamin appears when rhodopsin is bleached by light—the physiological process known as light adaptation—and disappears when rhodopsin is resynthesized during dark adaptation [see "Night Blindness," by John E. Dowling; SCIENTIFIC AMERICAN Offprint 1053]. He found that rhodopsin consists of a colorless protein (later named opsin) that carries as its chromophore, or color bearer, an unknown yellow carotenoid that he called retinene. Wald went on to show that the bleaching of rhodopsin to visual yellow corresponds to the liberation of retinene from its attachment to opsin, and that the fading of visual yellow to visual white represents the conversion of retinene to vitamin A. During dark adaptation rhodopsin is resynthesized from these precursors.

The chemical relation between retinene and vitamin A was elucidated in 1944 by R. A. Morton of the University of Liverpool. He showed that retinene is formed when vitamin A, an alcohol, is converted to an aldehyde, a change that involves the removal of two atoms of hydrogen from the terminal carbon atom of the molecule. As a result of Morton's finding the name retinene was recently changed to retinal.

In 1952 one of us (Hubbard), then a graduate student in Wald's laboratory, demonstrated that only the 11-*cis* isomer of retinal can serve as the chromophore of rhodopsin. This has since been confirmed for all the visual pigments whose chromophores have been examined. These pigments found in both the rod and cone cells of the eye contain various opsins, which combine either with retinal (strictly speaking retinal₁) or with a slightly modified form of retinal known as retinal₂. One other isomer of retinal, the 9-*cis* isomer, also combines with opsins to form light-sensitive pigments, but they are readily distinguishable from the visual pigments in their properties and have never been found to occur naturally. They have been called isopigments.

In 1959 we showed that the only thing light does in vision is to change the shape of the retinal chromophore by isomerizing it from the 11-*cis* to the all-*trans* configuration [*see illustration on page 406*]. Everything else—further chemical changes, nerve excitation, perception of light, behavioral responses—are consequences of this single photochemical act.

The change in the shape of the chromophore alters its relation to opsin and ushers in a sequence of changes in the mutual interactions of the chromophore and opsin, which is observed as a sequence of color changes. In vertebrates the all-*trans* isomer of retinal and opsin are incompatible and come apart. In some invertebrates, such as the squid, the octopus and the lobster, a metastable state is reached in which the all-*trans* chromophore remains bound to opsin.

Until the structure of opsin is established there is no way to know just how 11-*cis* retinal is bound to the opsin molecule. In the 1950's F. D. Collins, G. A. J. Pitt and others in Morton's laboratory showed that in cattle rhodopsin the aldehyde (C=O) group of 11-*cis* retinal forms what is called a Schiff's base with an amino (NH₂) group in the opsin molecule. Recently Deric Bownds in Wald's laboratory has found that the amino group belongs to lysine, one of the amino acid units in the opsin molecule, and has identified the amino acids in its immediate vicinity. There is little doubt that 11-*cis* retinal also has secondary points of attachment to opsin; otherwise it would be hard to explain why only the 11-*cis* isomer serves as the chromophore in visual pigments. Light changes the shape of the chromophore and thus alters its spatial relation to opsin. This leads, in turn, to changes in the shape of the opsin molecule [*see lower illustration on page 412*]. The details of these changes, however, are still obscure.

How Molecules Twist

Let us examine somewhat more closely the various isomers of retinal. The six known isomers are illustrated at the left: the all-*trans* isomer and five *cis* isomers of one kind or another. Experiments with models, together with other evidence, show that four of the six isomers are essentially planar. The two that are not are the 11-*cis* isomer and the 11,13-*dicis* isomer. In these isomers there is considerable steric hindrance, or intramolecular crowding, between the hydrogen atom on carbon No. 10 (C_{10}) and the methyl group attached to C_{13}. Thus the double bond that joins C_{11} and C_{12} cannot be rotated by 180 degrees from the *trans* to a planar *cis* configuration. In the 11-*cis* isomers the tail of the molecule from C_{11} through C_{15} is therefore twisted out of the plane formed by the rest of the molecule.

This twisted geometry introduces two configurations, called enantiomers, that are mirror images of each other; if the molecule could be viewed from the ring end, one form would be twisted to the left and the other to the right. It is possible that opsin may combine selectively with only one enantiomer.

As Pauling had predicted in the 1930's, the steric hindrance that necessitates the twist in the 11-*cis* isomer makes it less stable than the all-*trans* or the 9-*cis* and 13-*cis* forms. We have recently found, for example, that the 11-*cis* form contains about 1,500 calories more "free energy" per mole than the *trans* form. One has to put in about 25,000 calories per mole, however, to rotate the molecule from one form to the other. This amount of energy, which is much more than a molecule is likely to acquire through chance collisions with its neighbors, is known as the activation energy: the energy required to surmount the barrier that separates the *cis* and *trans* states.

This raises an important point. How can two parts of a molecule be rotated around a double bond? The interconversion of *cis* and *trans* isomeric forms to another requires gross departures from flatness. How can this be accomplished? Here we must introduce the concept of the excited state. One can think of molecules as existing in two kinds of state: a "ground," or stable, state of relatively low internal energy and various less stable states of higher energy—

the excited states. Molecules are raised from the ground state into one or another excited state by a sudden influx of energy, which can be in the form of heat or light. They return to the ground state by giving up their excess energy, usually as heat but occasionally as light, as in fluorescence or phosphorescence.

The orbital diagrams we have described apply to molecules in the ground state. When molecules are in an excited state, their electrons have more energy and therefore occupy different orbitals. Quantum-mechanical calculations show that an excited pi electron divides its time between the two ends of a double bond [*see bottom illustration on page 409*]. The net effect is to make the double bond in an excited molecule more like a single bond and less like a double bond. In a conjugated molecule, in which pi electrons are already delocalized, the changes in bond character are not uniform throughout the conjugated system but depend on the nature of the excitation and the structure of the molecule that is excited.

When one tries to isomerize carotenoids in the laboratory, it is usually helpful to add catalysts such as bromine or iodine. (The reader will recall that Zechmeister used iodine in his demonstrations.) Heat and light favor the existence of excited states. Bromine and iodine probably function by dissociating into atomic bromine and iodine, a process that is also favored by light. A bromine or iodine atom adds fleetingly to the double bond and converts it into a single bond, which is then momentarily free to rotate until the bromine or iodine atom has departed. The actual lifetime of the singly bonded form can be very brief indeed: the time required for one rotation around a carbon-carbon single bond is only about 10^{-12} second.

The Sensitivity of Eyes

It may seem remarkable that all animal visual systems so far studied depend on the photoisomerization of retinal for light detection. Three main branches of the animal kingdom—the mollusks, the arthropods and the vertebrates—have evolved types of eyes that differ profoundly in their anatomy. It seems that various anatomical (that is, optical) arrangements will do; apparently the photochemistry, once it had evolved, was universally accepted. Presumably the visual pigments of all animals must within narrow limits be equally sensitive to light, otherwise the more light-sensitive animals would eventually

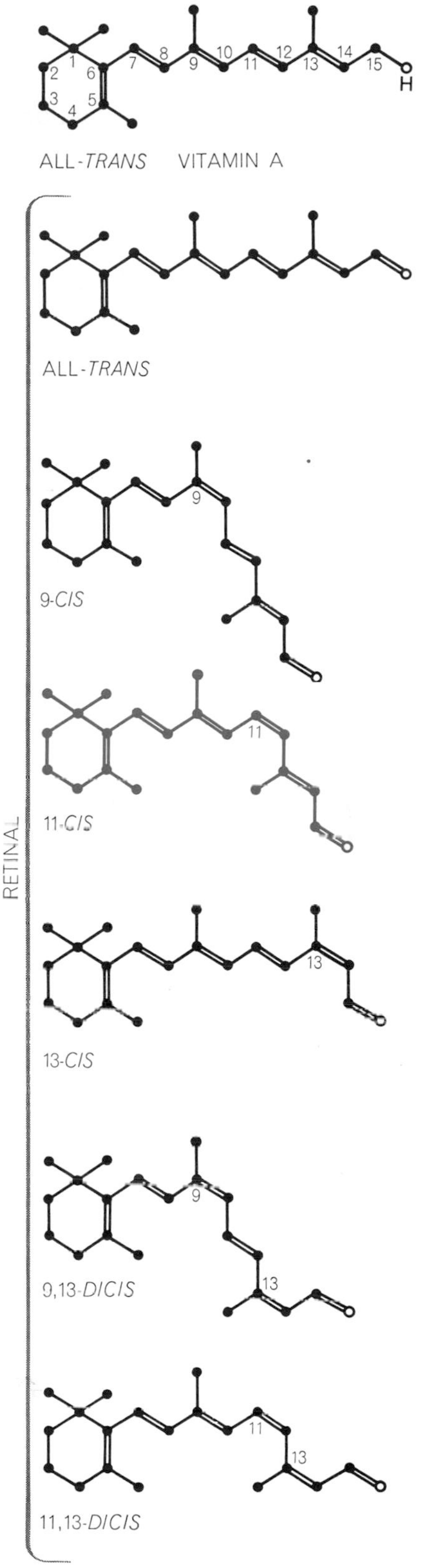

ALL-*TRANS* VITAMIN A

ALL-*TRANS*

9-*CIS*

11-*CIS*

13-*CIS*

9,13-*DICIS*

11,13-*DICIS*

RETINAL

SIX ISOMERS OF RETINAL are represented in skeleton form below the structure of all-*trans* vitamin A. Hydrogen atoms are omitted, except for the H in the hydroxyl group of vitamin A. If that H and one other on the final carbon are removed, all-*trans* retinal results. This isomer and 11-*cis* retinal, which combines with opsin to form rhodopsin, are the isomers involved in vision.

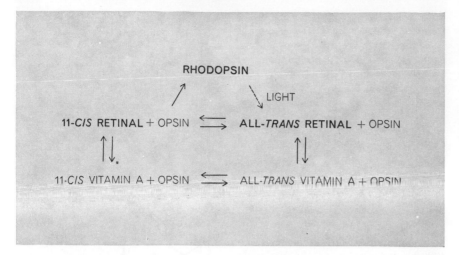

RHODOPSIN

11-*CIS* RETINAL + OPSIN ⇌ ALL-*TRANS* RETINAL + OPSIN

11-*CIS* VITAMIN A + OPSIN ⇌ ALL-*TRANS* VITAMIN A + OPSIN

PHOTOCHEMICAL EVENTS IN VISION involve the protein opsin and isomers of retinal and its derivative, vitamin A. Opsin joined to 11-*cis* retinal forms rhodopsin. When struck by light, the 11-*cis* chromophore is converted to an all-*trans* configuration and subsequently all-*trans* retinal becomes detached from opsin. With the addition of two hydrogen atoms, all-*trans* retinal is converted to all-*trans* vitamin A. Within the eye this isomer must be converted to 11-*cis* vitamin A, thence to 11-*cis* retinal, which recombines with opsin to form rhodopsin.

replace those whose eyes were less sensitive.

How sensitive to light is the animal retina? In a series of experiments conducted about 1940 Selig Hecht and his collaborators at Columbia University showed that the dark-adapted human eye will detect a very brief flash of light when only five quanta of light are absorbed by five rod cells. From this Hecht concluded that a single quantum is enough to trigger the discharge of a dark-adapted rod cell in the retina.

It is therefore essential that the quantum efficiency of the initial photochemical event be close to unity. In other words, virtually every quantum of light absorbed by a molecule of rhodopsin must isomerize the 11-*cis* chromophore to the all-*trans* configuration. It was shown many years ago by the British workers H. J. A. Dartnall, C. F. Goodeve and R. J. Lythgoe that an absorbed quantum has about a 60 percent chance of bleaching frog rhodopsin. One of us (Kropf) has found a similar quantum efficiency for the isomerization of the 11-*cis* retinal chromophore of cattle rhodopsin. Our work also shows that 11-*cis* retinal is more photosensitive than either the 9-*cis* or the all-*trans* isomers when they are attached as chromophores

to opsin, and this may be the reason why the geometrically hindered and therefore comparatively unstable 11-*cis* isomer has evolved into the chromophore of the visual pigments.

We have also recently measured the quantum efficiency of the photoisomerization of retinal and several closely related carotenoids in solution. Retinal turns out to be considerably more photosensitive than any of them and nearly as photosensitive as rhodopsin.

Although all animal eyes seem to employ 11-*cis* retinal as their light-sensitive agent, there are slight variations in the opsins that combine with retinal, just as there are variations in other proteins, such as hemoglobin, from species to species. Within the next few years we may learn the complete amino acid sequence of one of the opsins, and thereafter we should be able to compare such sequences for two or more species. It may be many years, however, before X-ray crystallographers have established the complete three-dimensional structure of an opsin molecule and are able to describe the site that binds it to retinal. One can conjecture that the binding site will be quite similar in the various opsins, even those from animals of different phyla, but there may be surprises in store. Whatever the precise details, it is clear that evolution has produced a remarkably efficient system for translating the absorption of light into the language of biochemistry—a language whose vocabulary and syntax are built on the various ways proteins interact with one another and with smaller molecules in their environment.

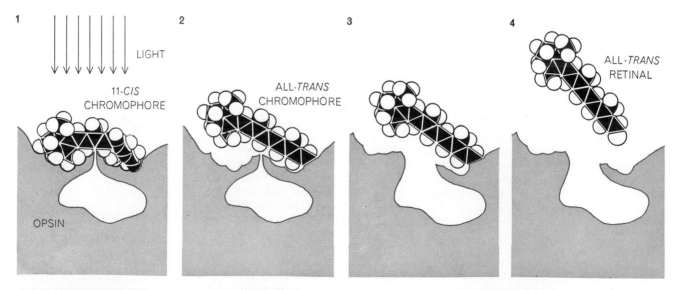

MOLECULAR EVENTS IN VISION can be inferred from the known changes in the configuration of 11-*cis* retinal after the absorption of light. In these schematic diagrams the twisted isomer is shown attached to its binding site in the much larger protein molecule of opsin (*1*). After absorbing light the 11-*cis* chromophore straightens into the all-*trans* isomer (*2*). Presumably a change in the shape of opsin (*3*) facilitates the release of all-*trans* retinal (*4*). The configuration of the binding site in opsin is not yet known.

45

MEMORY AND PROTEIN SYNTHESIS

BERNARD W. AGRANOFF

June 1967

If a goldfish is trained to perform a simple task and shortly thereafter a substance that blocks the manufacture of protein is injected into its skull, it forgets what it has been taught

What is the mechanism of memory? The question has not yet been answered, but the kind of evidence needed to answer it has slowly been accumulating. One important fact that has emerged is that there are two types of memory: short-term and long-term. To put it another way, the process of learning is different from the process of memory-storage; what is learned must somehow be fixed or consolidated before it can be remembered. For example, people who have received shock treatment in the course of psychiatric care report that they cannot remember experiences they had immediately before the treatment. It is as though the shock treatment had disrupted the process of consolidating their memory of the experiences.

In our laboratory at the University of Michigan we have demonstrated that there is a connection between the consolidation of memory and the manufacture of protein in the brain. Our experimental animal is the common goldfish (*Carassius auratus*). Basically what we do is train a large number of goldfish to perform a simple task and at various times before, during and after the training inject into their skulls a substance that interferes with the synthesis of protein. Then we observe the effect of the injections on the goldfish's performance.

Why seek a connection between memory and protein synthesis? For one thing, enzymes are proteins, and enzymes catalyze all the chemical reactions of life. It would seem reasonable to expect that memory, like all other life processes, is dependent on enzyme-catalyzed reactions. What is perhaps more to the point, the manufacture of new enzymes is characteristic of long-term changes in living organisms, such as growth and the differentiation of cells in the embryo. And long-term memory is by definition a long-term change.

The investigation of a connection between memory and protein synthesis is made possible by the profound advances in knowledge of protein synthesis that have come in the past 10 years. A molecule of protein is made from 20 differ-

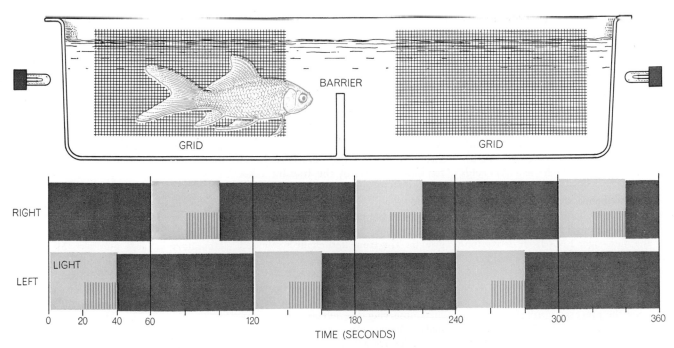

TRAINING TANK the author used was designed so that goldfish learned to swim from the light end to the dark end. A learning trial began with the illumination of the left end of the tank (*chart at bottom*), followed after a pause by mild electric shocks (*colored vertical lines*) from grids at that end. At first a fish would swim over the central barrier in response to shock; then increasingly the fish came to respond to light cue alone as sequence of light, shock and darkness was alternately repeated at each end of the tank.

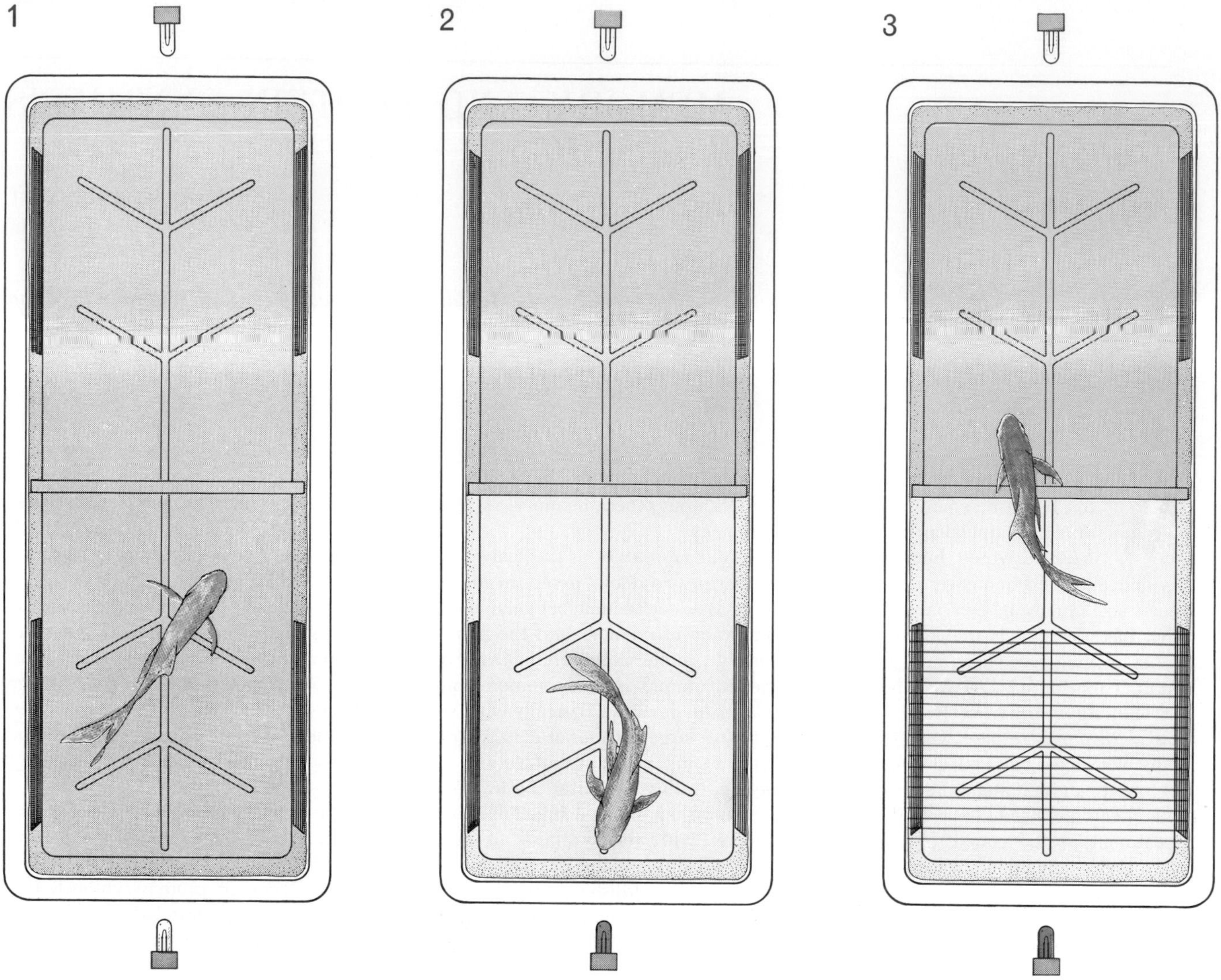

1   2   3

GOLDFISH LEARN in successive trials to solve the problem the shuttle box presents. Following 20 seconds of darkness (*1*) the end of the box where the fish is swimming is lighted for an equal period of time (*2*). The fish fails to respond, swimming over the barrier

ent kinds of amino acid molecule, strung together in a polypeptide chain. The stringing is done in the small bodies in the living cell called ribosomes. Each amino acid molecule is brought to the ribosome by a molecule of transfer RNA, a form of ribonucleic acid. The instructions according to which the amino acids are linked in a specific sequence are brought to the ribosome by another form of ribonucleic acid: messenger RNA. These instructions have been transcribed by the messenger RNA from deoxyribonucleic acid (DNA), the cell's central library of information.

With this much knowledge of protein synthesis one can begin to think of examining the process by interfering with it in selective ways. Such interference can be accomplished with antibiotics. Whereas some substances that interfere with the machinery of the cell, such as cyanide, are quite general in their ef-

fects, antibiotics can be highly selective. Indeed, some of them block only one step in cellular metabolism. As an example, the antibiotic puromycin simply stops the growth of the polypeptide chain in the ribosome. This it does by virtue of the fact that its molecule resembles one end of the transfer RNA molecule with an amino acid attached to it. Accordingly the puromycin molecule is joined to the growing end of the polypeptide chain and blocks its further growth. The truncated chain is released attached to the puromycin molecule.

Numerous workers have had the idea of using agents such as puromycin to block protein synthesis in animals and then observing the effects on the animals' behavior. Among them have been C. Wesley Dingman II and M. B. Sporn of the National Institutes of Health, who injected 8-azaguanine into rats; T. J. Chamberlain, G. H. Rothschild and

Ralph W. Gerard of the University of Michigan, who administered the same substance to rats, and Josefa B. Flexner, Louis B. Flexner and Eliot Stellar of the University of Pennsylvania, who injected puromycin into mice. Such experiments encouraged us to try our hand with the goldfish.

We chose the goldfish for our experiments because it is readily available and can be accommodated in the laboratory in large numbers. Moreover, a simple and automatically controlled training task for goldfish had already been developed by M. E. Bitterman of Bryn Mawr College. One might wonder if a fish has such a thing as long-term memory; in the opinion of numerous psychologists and anglers there can be no doubt of it.

Our training apparatus is called a shuttle box. It is an oblong plastic tank

4   5   6

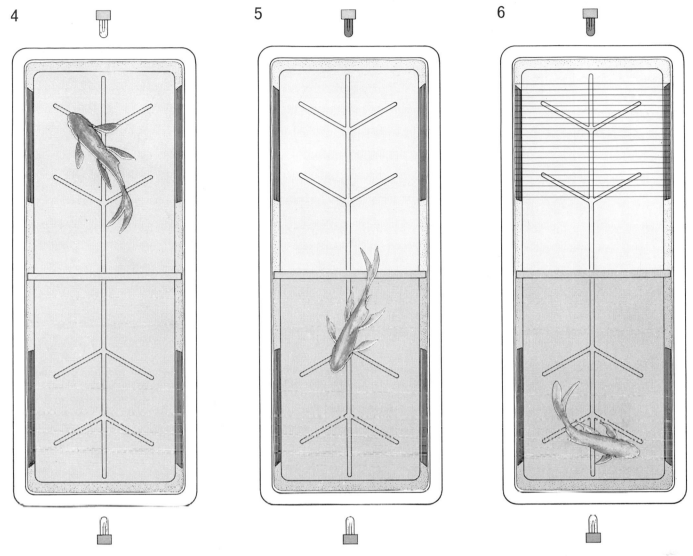

only after the shock period heralded by light has begun (3). When the same events are repeated at the other end of the box (4, 5 and 6), the fish shown here succeeds in crossing the barrier during the 20 seconds of light that precede the period of intermittent shock.

divided into two compartments by a barrier that comes to within an inch of the water surface [see illustration on page 415]. At each end of the box is a light that can be turned on and off. On opposite sides of each compartment are grids by means of which the fish can be given a mild electric shock through the water.

The task to be learned by the fish is that when it is in one compartment and the light goes on at that end of the box, it should swim over the barrier into the other compartment. In our initial experiments we left the fish in the dark for five minutes and then gave it five one-minute trials. Each trial consisted in (1) turning on the light at the fish's end of the box, (2) 20 seconds later intermittently turning on the shocking grids and (3) 20 seconds after that turning off both the shocking grids and the light. If the fish crossed the barrier into the other

compartment during the first 20 seconds, it *avoided* the shock; if it crossed the barrier during the second 20 seconds, it *escaped* the shock.

An untrained goldfish almost always escaped the shock, that is, it swam across the barrier only when the shock began. Whether the fish escaped the shock or avoided it, it crossed the barrier into the other compartment. Then, after 20 seconds of darkness, the light at that end was turned on to start the second trial. Thus the fish shuttled back and forth with each trial. If a fish failed to either avoid or escape, it missed the next trial. Such missed trials were rare and generally came only at the beginning of training.

In these experiments the goldfish went through five consecutive cycles of five minutes of darkness followed by five training trials; accordingly they received a total of 20 trials in 40 minutes. They

were then placed in individual "home" tanks—plastic tanks that are slightly smaller than the shuttle boxes—and kept there for three days. On the third day they were returned to the shuttle box, where they were given 10 more trials in 20 minutes.

The fish readily learned to move from one compartment to the other when the light went on, thereby avoiding the shock. Untrained fish avoided the shock in about 20 percent of the first 10 trials and continued to improve with further trials. If they were allowed to perform the task day after day, the curve of learning flattened out at about 80 percent correct responses.

What was even more significant for our experiments was what happened when we changed the interval between the first cycle of trials and the second, that is, between the 20th and the 21st of the 30 trials. If the second cycle was

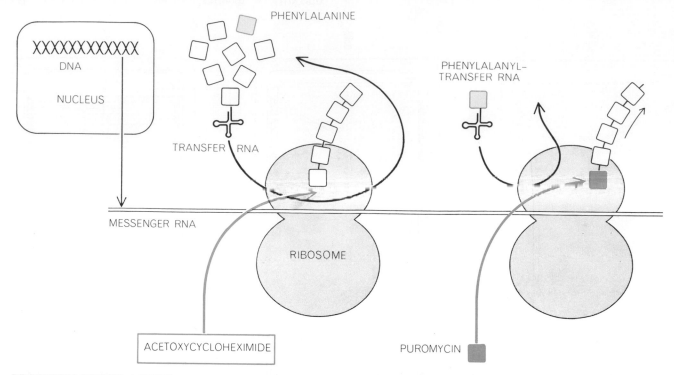

PROTEIN-BLOCKING AGENTS can interrupt the formation of molecules at the ribosome, where the amino acid units of protein are linked according to instructions embodied in messenger ribonucleic acid (mRNA). One agent, acetoxycycloheximide, interferes with the bonding mechanism that links amino acids brought to the ribosome by transfer RNA (tRNA). Puromycin, another agent, resembles the combination of tRNA and the amino acid phenylalanine. Thus it is taken into chain and prematurely halts its growth.

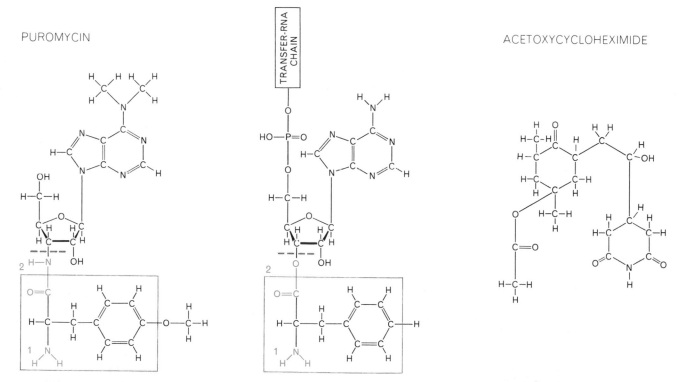

MOLECULAR DIAGRAMS show the resemblance between puromycin and the combination phenylalanyl-tRNA. In both cases the portion of the molecule below the broken line is incorporated into a growing protein molecule, joining at the free amino group (*1*). But in puromycin the CONH group (*2*), unlike the corresponding group (COO) of phenylalanyl, will not accept another amino acid and the chain is broken. Acetoxycycloheximide does not resemble amino acid but slows rate at which the chain forms.

begun a full month after the first, the fish performed as well as they did on the third day. If the second cycle was begun on the day after the first, the fish performed equally well, as one would expect. In short, the fish had perfect memory of their training.

We found that we could predict the training scores of groups of fish on the third day on the basis of their scores on the first day. This made it easier for us to determine the effect of antibiotics on the fish's memory: we could compare the training scores of fish receiving antibiotics with the predicted scores. Since we conducted these initial experiments we have made several improvements in our procedure. We now record the escapes and avoidances automatically with photodetectors, and we have arranged matters so that a fish does not miss a trial if it fails to escape. We have altered the trial sequence and the time interval between the turning on of the light and the turning on of the shocking grid. The results obtained with these improved procedures are essentially the same as our earlier ones.

The principal antibiotic we use in our experiments is puromycin, whose effect on protein synthesis was described earlier. We inject the drug directly into the skull of the goldfish with a hypodermic syringe. A thin needle easily penetrates the skull; 10 microliters of solution is then injected over the fish's brain (not into it). In an early series of experiments we injected 170 micrograms of puromycin in that amount of solution at various stages in our training procedures.

We found that if the puromycin was injected immediately after training, memory of the training was obliterated. If the same amount of the drug was injected an hour after training, on the other hand, memory was unaffected. Injection 30 minutes after training produced an intermediate effect. Reducing the amount of puromycin caused a smaller loss of memory.

After the injection the fish seemed to swim normally. We were therefore encouraged to test whether or not puromycin interferes with the changes that occur in the brain as the fish is being trained. This we did by injecting the fish before their initial training. We found that they learned the task at a normal rate, that is, their improvement during the first 20 trials was normal. Fish tested three days later, however, showed a profound loss of memory. This indicated to us that puromycin did not block the short-term memory demonstrated during

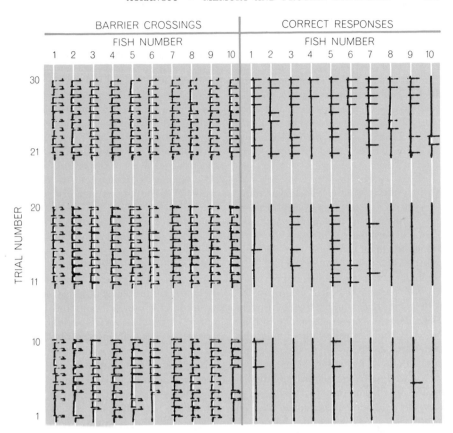

TRACE FROM RECORDER shows the performances of 10 goldfish in 30 trials. Each horizontal row represents a trial, beginning at the bottom with trial 1. A blip (*left side*) indicates that a fish either escaped or avoided the shock; a dash in the same row (*right*) signifies an avoidance, that is, a correct response for the trial. These fish learned at the normal rate.

learning but did interfere with the consolidation of long-term memory. And since an injection an hour after training has no effect on long-term memory, whereas an injection immediately after training obliterates it, it appears that consolidation can take place within an hour.

One observation puzzled us. The animals had received their initial training during a 40-minute period, 20 minutes of which was spent in the dark. Puromycin could erase all memory of this training; none of the memory was consolidated. Yet the experiment in which we injected puromycin 30 minutes after training had shown that more than half of the memory was consolidated during that period. How was it that no memory at all was consolidated at least toward the end of the 40-minute training period? To be sure, the fish that had been injected 30 minutes after the training period had been removed from the shuttle boxes and placed in their home tanks. But what was different about the time spent in the shuttle box and the time spent in the home tank that memory could be consolidated in the home tank

but could not be in the shuttle box?

Roger E. Davis of our laboratory undertook further experiments to clarify the phenomenon. He found that fish that were allowed to remain in the shuttle box for several hours after training and were then returned to their home tank showed no loss of memory when they were tested four days later. On the other hand, fish that were allowed to remain in the shuttle box for the same length of time and were then injected with puromycin and returned to their home tank had a marked memory loss! In other words, the fish in the first group did not consolidate memory of their training until after they had been placed in their home tank. It appears that simply being in the shuttle box prevents the fixation of memory. Subsequent studies have led us to the idea that memory fixation is blocked when the organism is in an environment associated with a high level of stimulation. This effect indicates that the formation of memory is environment-dependent, just as the consolidation of memory is time-dependent.

We conclude from all these experiments that long-term memory of training in the goldfish is formed by a puromy-cin-sensitive step that begins after training and requires that the animal be removed from the training environment. The initial acquisition of information by the fish is puromycin-insensitive and is a qualitatively different process. But what does the action of puromycin on memory formation have to do with its known biochemical effect: the inhibition of protein synthesis?

We undertook to establish that puromycin blocks protein synthesis in the goldfish brain under the conditions of our experiments. This we did in the following manner. First we injected puromycin into the skull of the fish. Next we injected into the abdominal cavity of the fish leucine that had been labeled with tritium, or radioactive hydrogen. Now, leucine is an amino acid, and if labeled leucine is injected into a goldfish's abdominal cavity, it will be incorporated into whatever protein is being synthesized throughout the goldfish's body. By measuring the amount of labeled leucine incorporated into protein after, say, 30 minutes, one can determine the rate of protein synthesis during that time.

We compared the amount of labeled leucine incorporated into protein in goldfish that had received an injection of puromycin with the amount incorporated in fish that had received either no injection or an injection of inactive salt solution. We found that protein synthesis in the brain of fish that had been injected with puromycin was deeply inhibited. The effects of different doses of puromycin and the length of time it took the drug to act did not, however, closely correspond to what we had observed in our experiments involving the behavioral performance of the goldfish. In retrospect this result is not surprising. Various experiments, including our own, had shown that the rate of memory consolidation can be altered by changes in the conditions of training. Moreover, the rate of leucine incorporation can be affected by complex physiological factors.

Another way to check whether or not puromycin exerts its effects on memory by inhibiting protein synthesis would be to perform the memory experiments with a second drug known to inhibit such synthesis. Then if puromycin blocks long-term memory by some other mechanism, the second drug would have no effect on memory. It would be even better if the second drug did not resemble puromycin in molecular structure, so that its effect on protein synthesis would not be the same as puromycin's. Such a drug exists in acetoxycycloheximide. Where puromycin blocks the growth of the polypeptide chain by taking the place of an amino acid, acetoxycyclo-heximide simply slows down the rate at which the amino acids are linked together. We found that a small amount of this drug (.1 microgram, or one 1,700th the weight of the amount of puromycin we had been using) produced a measurable memory deficit in goldfish. Moreover, it commensurately inhibited the synthesis of protein in the goldfish brain.

These experiments suggest that protein synthesis is required for the consolidation of memory, but they are not conclusive. Louis Flexner and his colleagues have found that puromycin can interfere with memory in mice. On the other hand, they find that acetoxycyclohex-imide has no such effect. They conclude that protein is required for the expression of memory but that experience acts not on protein synthesis directly but on messenger RNA. The conditions of their experiments and the fact that they are working with a different animal do not allow any ready comparison with our experiments.

Our studies of the goldfish have led us to view learning and memory as a form of biological development. One may think of the brain of an animal as being completely "wired" by heredity; all possible pathways are present, but not all are "soldered." It may be that in short-term memory, pathways are selected rapidly but impermanently. In that case protein synthesis would not be required, which may explain why puromycin has no effect on short-term memory. If the consolidation of memory calls for more permanent connections among pathways, it seems reasonable that protein synthesis would be involved. The formation of such connections, of course, would be blocked by puromycin and acetoxycycloheximide.

Another possibility is that the drugs block not the formation of permanent pathways but the transmission of a signal to fix what has just been learned. There is some evidence for this notion in what happens to people who suffer damage to certain parts of the brain (the mammillary bodies and the hippocampus). They retain older memories and are capable of new learning, but they cannot form new long-term memories. Experiments with animals also provide some evidence for a "fix" signal. We are currently doing experiments in the hope of determining which of these hypotheses best fits the effects of puromycin and acetoxycycloheximide on memory in the goldfish.

Quite apart from our own work, it has been suggested by others that it is possible to transfer patterns of behavior

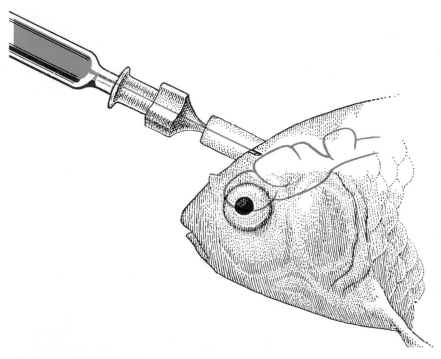

ANTIBIOTIC WAS INJECTED through the thin skull of a goldfish and over rather than into the brain. The antibiotic was puromycin, which inhibits protein synthesis. Following its injection the fish were able to swim normally. They could then be tested for memory loss.

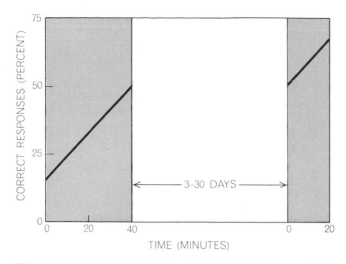

NORMAL LEARNING RATE of goldfish in 30 shuttle-box trials is shown by the black curve. Whether the last 10 trials were given three days after the first 20 (the regular procedure) or as much as a month later, fish demonstrated the same rate of improvement.

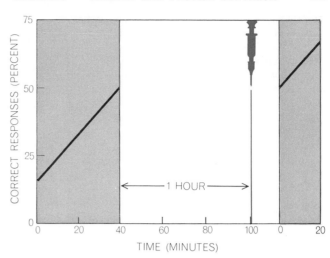

INJECTION WITH PUROMYCIN one hour after completion of 20 learning trials did not disrupt memory. Goldfish given the antibiotic at this point scored as well as those in the control group in the sequence of 10 trials that followed three days afterward.

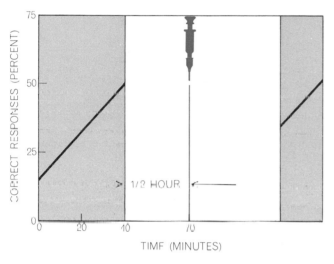

INJECTION HALF AN HOUR AFTER the first 20 trials cut the level of correct responses to half the level without such injection.

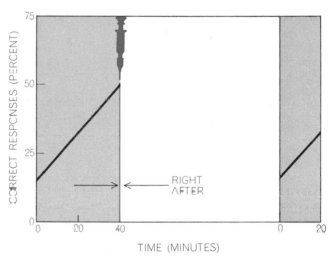

INJECTION IMMEDIATELY AFTER the first 20 trials erased all memory of training. The fish scored at the untrained level.

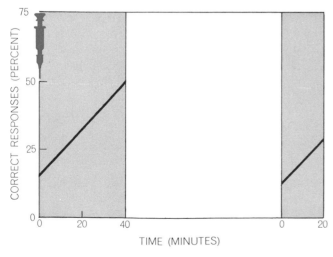

INJECTION PRIOR TO TRAINING did not affect the rate at which goldfish learned to solve the shuttle-box problem. But puromycin given at this point did suppress the formation of long-term memory, as shown by the drop in the scores three days afterward.

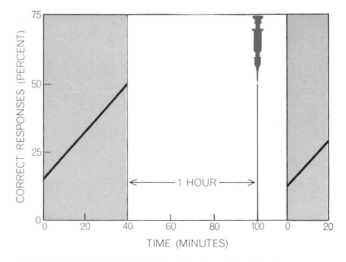

ENVIRONMENTAL FACTOR in the formation of lasting memory was seen when fish remained in training (instead of "home") tanks during the fixation period. Under these conditions fixation did not occur. Puromycin given at end of period still erased memory.

from one animal to another (even to an animal of a different species) by injecting RNA or protein from the brain of a trained animal into the brain (or even the abdominal cavity) of an untrained one. If such transfers of behavior patterns can actually be accomplished, they imply that memory resides in molecules of RNA or protein. Nothing we have learned with the goldfish argues for or against the possibility that a behavior pattern is stored in such a molecule.

It can be observed, however, that there is no precedent in biology for such storage. What could be required would be a kind of somatic mutation: a change in the cell's store of information that would give rise to a protein with a new sequence of amino acids. It seems unlikely that such a process could operate at the speed required for learning.

It might also be that learning and memory involve the formation of short segments of RNA or protein that somehow label an individual brain cell. Richard Santen of our laboratory has calculated (on the basis of DNA content) the number of cells in the brain of a rat: it comes to 500 million. With this figure one can calculate further that a polypeptide chain of seven amino acids, arranged in every possible sequence, could provide each cell in the rat's brain with two unique markers.

The concept that each nerve cell has its own chemical marker is supported by experiments on the regeneration of the optic nerve performed by Roger W. Sperry of the California Institute of Technology. If the optic nerve of a frog is cut and the two ends of the nerve are put back together rotated 180 degrees with respect to each other, the severed fibers of the nerve link up with the same

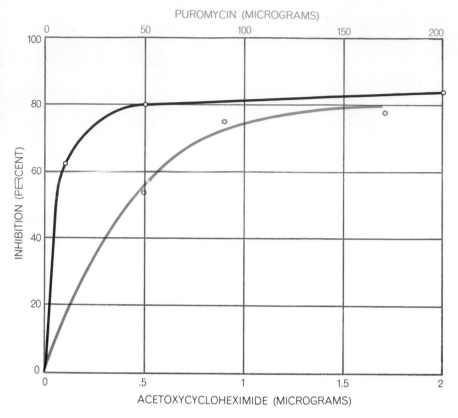

SLOWED PROTEIN SYNTHESIS in the brain of goldfish is induced both by acetoxycycloheximide (*black line*) and the antibiotic puromycin (*colored line*), agents that block the fixation of memory. The author tested the effect on goldfish of various quantities of the two drugs; acetoxycycloheximide was found several hundred times more potent than puromycin.

fiber as before. This of course suggests that each fiber has a unique marker that in the course of regeneration enables it to recognize its mate.

Is it possible, then, that a cell "turned on" by the learning process manufactures a chemical marker? And could such a process give rise to a substance that, when it is injected into another animal, finds its way to the exact location where

it can effectuate memory? Thus far the evidence put forward in support of such ideas has not been impressive. In this exciting period of discovery in brain research clear-cut experiments are more important than theories. Certain long-term memories held by investigators in this area may be more of a hindrance than a help in exploring all its possibilities.

HUMAN CELLS AND AGING

LEONARD HAYFLICK

March 1968

When normal cells are grown outside the body, their capacity to survive dwindles after a period of time. This deterioration may well represent aging and an ultimate limit to the span of life

The common impression that modern medicine has lengthened the human life-span is not supported by either vital statistics or biological evidence. To be sure, the 20th-century advances in control of infectious diseases and of certain other causes of death have improved the longevity of the human population as a whole. These accomplishments in medicine and public health, however, have merely extended the *average* life expectancy by allowing more people to reach the upper limit, which for the general run of mankind still seems to be approximately the Biblical fourscore years. Aging and a limited life-span apparently are characteristic of all animals that stop growing after reaching a fixed, mature size. In the case of man, after the age of 30 there is a steady, inexorable increase in the probability of death from one cause or another; the probability doubles about every eight years as one grows older. This general probability is such that even if the major causes of death in old age—heart disease, stroke and cancer—were eliminated, the average life expectancy would not be lengthened by much more than 10 years. It would then be about 80 years instead of the expectancy of about 70 years that now prevails in advanced countries.

Could man's life-span be extended—or is there an inescapable aging mechanism that restricts human longevity to the present apparent limit? Until recently few biologists ventured to attempt to explore the basic processes of aging; obviously the subject does not easily lend itself to detailed study. It is now receiving considerable attention, however, in a number of laboratories [see "The Physiology of Aging," by Nathan W. Shock; SCIENTIFIC AMERICAN, January, 1962]. In this article I shall discuss some new findings at the cellular level.

No doubt many mechanisms are involved in the aging of the body. At the cell level at least three aging processes are under investigation. One is a possible decline in the functional efficiency of nondividing, highly specialized cells, such as nerve and muscle cells. Another is the progressive stiffening with age of the structural protein collagen, which constitutes more than a third of all the body protein and serves as the general binding substance of the skin, muscular and vascular systems [see "The Aging of Collagen," by Frederic Verzár; SCIENTIFIC AMERICAN Offprint 155]. In our own laboratory at the Wistar Institute we have addressed ourselves to a third question: the limitation on cell division. Our studies have focused particularly on the structural cells called fibroblasts, which produce collagen and fibrin. These cells, like certain other "blast" cells, go on dividing in the adult body. We set out to determine whether human fibroblasts in a cell culture could divide indefinitely or had only a finite capacity for doing so.

Alexis Carrel's famous experiments more than a generation ago suggested that animal cells per se (that is, cells removed from the body's regulatory mechanisms) might be immortal. He apparently succeeded in keeping chick fibroblasts growing and multiplying in glass vessels for more than 30 years—a great deal longer than a chicken's life expectancy. Later experimenters reported similar successes with embryonic cells from laboratory mice. It has since been learned, however, through improved techniques and a better understanding of cell cultures, that the conclusions drawn from those early experiments were erroneous.

In the case of chick fibroblasts it has been repeatedly demonstrated that, if care is taken not to add any living cells to the initial population in the glass vessel, the cell colony will not survive long. The early cultures, including Carrel's, were fed a crude extract taken from chick embryos, and it is now believed these feedings must have contained some living chick cells. That is to say, in all probability the reason the cultures continued to grow indefinitely was that new, viable fibroblasts were introduced into the culture at each feeding.

Restudy of the experiments in culturing mouse cells has brought to light a highly interesting fact. It has been found that when normal cells from a laboratory mouse are cultured in a glass vessel, they frequently undergo a spontaneous transformation that enables them to divide and multiply indefinitely. This type of transformation takes place regularly in cultures of mouse cells but only rarely in cultures of the fibroblasts of man or other animals. These transformed cell populations have several abnormal properties, but they are truly immortal: many of the mouse-derived cultures have survived for decades. Similarly, the famous line of transformed human cells called HeLa, originally derived from cervical tissue in 1952 by George O. Gey of the Johns Hopkins University School of Medicine, is still growing and multiplying in glass cultures.

On microscopic examination the transformed cells show themselves to be indeed abnormal. Instead of the normal number of 46 chromosomes in a human diploid cell, the "mixoploid" HeLa cells may have anywhere from 50 to 350 chromosomes per cell. They differ from normal chromosomes in size and shape and also stain differently. Moreover, they often behave like cancer cells: inoculated into a suitable laboratory animal, they can grow as tumors. This property of transformed cells has become an important tool in investigations of the genesis of cancer. Although the spontaneous transformation of human cells is rare, investigators of cancer are making use of the recent discovery that normal human cells can be routinely transformed into cancer cells by exposing them to the monkey virus known as SV-40.

A crucial consideration for the relevance of cells in culture to aging, of course, is that they are normal cells. Our interest is therefore directed not to abnormal cells but to the observation that normal cells do not divide indefinitely. Whereas a population of transformed cells will proliferate and survive for decades in cell culture, no one has succeeded in perpetuating a culture of normal animal cells. The same is true of cells implanted in a living animal. Transformed cells will go on growing indefinitely in a series of tissue transplants from animal to animal, but normal cells will not. Peter L. Krohn of the University of Birmingham has shown, for example, that normal mouse skin can survive only a limited number of serial grafts from one mouse to another in the same inbred strain.

Over the past seven years in our laboratory Paul S. Moorhead and I have been studying cell cultures of normal human fibroblasts. Unlike highly specialized cells, fibroblasts (which serve as the structural bricks for most body

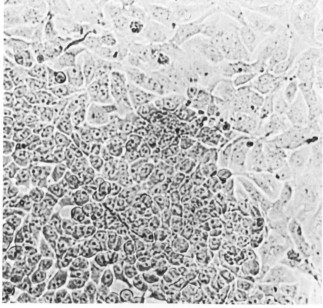

CULTIVATION OF HUMAN CELLS in the author's laboratory begins with the breakdown of lung tissue into separate cells. This is accomplished by means of the digestive enzyme

tissues) will grow and multiply in a nutrient medium in glass bottles. We have used lung tissue as our principal source of the cells. We break down the tissue into separated cells by means of the digestive enzyme trypsin, then remove the trypsin by centrifugation and seed the cells in a bottle containing a suitable growing medium. After a few days of incubation at 99 degrees Fahrenheit the

fibroblasts have spread out on the glass surface and begun to divide. In a week or so they cover the entire available surface with a layer one cell deep. Since they will proliferate only in a single layer, we strip off the layer that has covered the bottle surface, again separate the cells with trypsin and plant half of the cells in each of two new bottles with fresh medium. In three or four days the

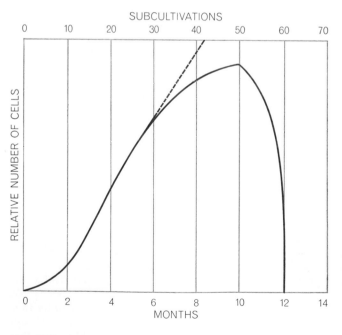

LIFETIME OF HUMAN CELLS was determined by allowing a population to multiply until it had doubled in size. After a culture of cells from embryonic tissue had grown to a particular point, it was divided in two (see illustration at top of these two pages). Cell division ceased after about 50 such subcultivations had doubled. It is possible at any time (although it is rare) for a spontaneous change to occur after which the cells multiply indefinitely (broken line).

TRANSFORMED HUMAN CELLS are distinguished by their morphology and reaction to staining. The darker amnion cells, forming a large island at lower left, have undergone a spontaneous transformation. They will continue to divide after the neighboring cells have died. Transformed, or "mixoploid," cells have more chromosomes than diploid cells do; they are utilized in cancer research. The magnification of this photomicrograph is about 180 diameters.

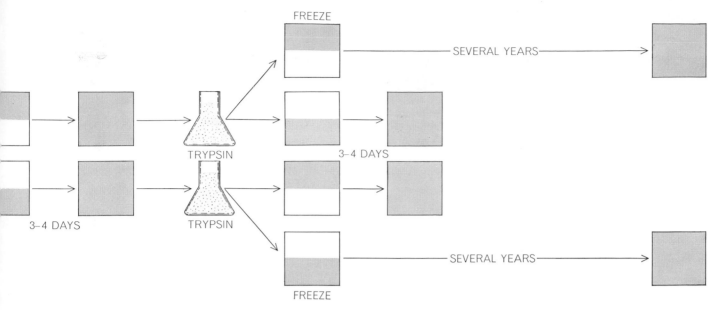

trypsin. After the cells, seeded in a bottle, have multiplied to cover its surface, they are again treated with trypsin, then divided into two halves and replanted. Most cells thus grown are placed in cold storage. Thawed and planted years later, they resume division.

inoculated fibroblasts grow over all the available surface in each bottle, thus doubling the number of cells taken from the original bottle. As this procedure is repeated at four-day intervals, the fibroblasts continue to proliferate in new bottles, doubling in number each time.

We found that fibroblasts taken from four-month-old human embryos doubled in this way about 50 times (the limit ranged between 40 and 60 doublings). After reaching this limit of capacity for division the cell population died. It could therefore be concluded that human fibroblasts derived from embryonic tissue and grown in cell culture have a finite lifetime amounting to approximately 50 population doublings (which in our culture covered a span of six to eight months).

Further study reinforced this conclusion. It turned out, for example, that if cell division was interrupted and then resumed, the total number of population doublings was not altered. In our experiments we did not, of course, double the number of bottles at each step; after only 10 doubling passages we would have had 1,024 bottles, and 50 doublings of the original seeding of fibroblasts could have produced about 20 million tons of cells! To keep the yield within reasonable bounds we set aside most of the cells from the subcultivations and put them in cold storage. We found they could be kept in suspended animation for apparently unlimited periods; even after six years in storage they proved to be capable of resuming division when they were thawed and placed in a culture medium. They "remembered" the doubling level they had reached before stor-

age and completed the course from that point. For example, cells that had been stored at the 30th doubling went on to divide about 20 more times.

The geometric rate of increase of the cells in culture has made it possible to provide essentially unlimited supplies of the cells for experiments. Samples of one of our strains of normal human fibroblasts (WI-38), "banked" after the eighth doubling in liquid nitrogen at 190 degrees below zero centigrade, have been distributed to hundreds of research laboratories around the world. The stored cells presumably would be available for study far in the future. For instance, well-protected capsules containing frozen cells might be buried in Antarctica or deposited in orbit in the cold of outer space for retrieval many generations hence. Investigators might then be able to use them to study, among other things, whether or not time had brought about any evolutionary change in the aging of man or other animals at the cellular level.

We found further confirmation of the finite lifetime of human fibroblasts when we cultured such cells from adult donors. Samples of lung fibroblasts taken at the time of death from eight adults, ranging in age from 20 to 87, underwent 14 to 29 doublings in cell culture afterward. The number of doublings in these tests did not show a clear correlation with the age of the donor, but presumably this was because our method of measuring the doubling of cell populations in bottles is not sufficiently precise to disclose such a correlation in detail. Our current experiments do suggest,

however, that consistent differences can be found between broad age groups. It appears that fibroblasts from human embryos will divide in cell culture 50 ± 10 times, those from persons between birth and the age of about 20 will divide 30 ± 10 times and those from donors over 20 will divide 20 ± 10 times.

We have tested fibroblasts from several human embryonic tissues besides lung tissue and found that they too are limited to a total of about 50 divisions in culture. The doubling lifetimes of fibroblasts from animals other than man have also been studied. As one would expect, the cells of the shorter-lived vertebrates show less capacity for division. For example, normal fibroblasts from embryos of chickens, rats, mice, hamsters and guinea pigs usually double no more than 15 times in cell culture, and cells that have been taken from adults of the same species undergo considerably fewer than 15 divisions.

Early in our experiments it became evident that we had to examine the possibility that a lethal factor in the culture medium or a defect in the culture technique might be responsible for the limitation of division and ultimate death of the cells. Did the cells stop dividing because of some lack in the nutrient mixture or the presence of contaminating microorganisms such as viruses? We explored these possibilities by various experiments, one of which consisted in culturing a mixture of normal fibroblasts taken from male and female donors. Female cells can be distinguished from male cells either by the presence of special chromatin bodies (found only in fe-

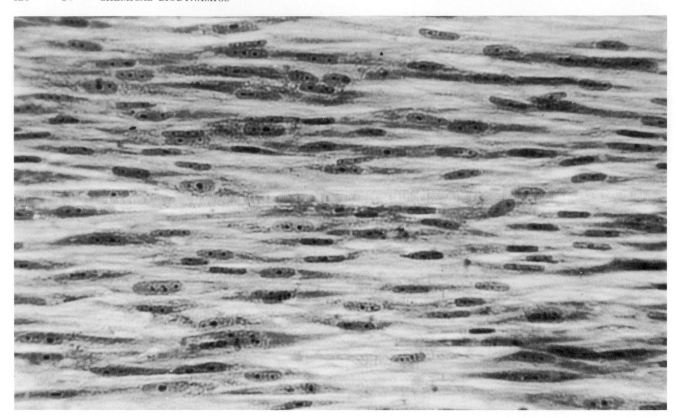

GROWING HUMAN CELLS cover the surface of the glass vessel in which they were planted. They form a layer one cell deep. Under proper conditions this normal cell population will continue to proliferate; it will not, except in unusual cases, multiply indefinitely.

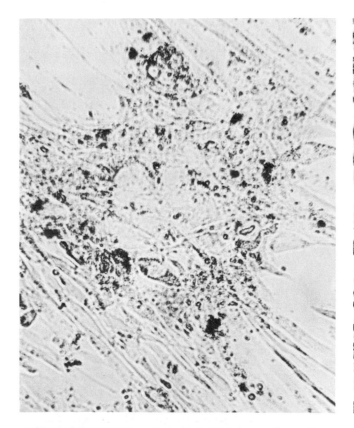

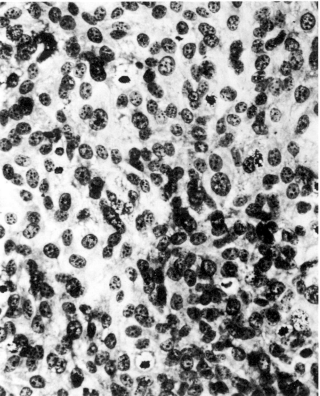

AGED HUMAN CELLS are irregular in appearance; they no longer divide. The aging of such normal cell populations is apparently due to an intrinsic process, not a deficiency in growing conditions. The cells were taken from lung tissue and grown in glass. This photomicrograph and the one at top were made by the author.

IMMORTAL HUMAN CELLS appear in this photomicrograph made by Fred C. Jensen of the Wistar Institute. The cells, once normal like those at top, have been treated with the monkey virus SV-40; thus transformed, they can apparently multiply indefinitely. The magnification in these photomicrographs is 300 diameters.

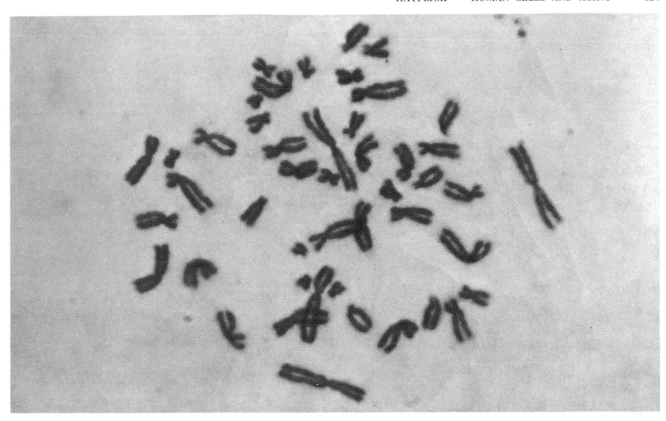

HUMAN CHROMOSOMES are seen magnified 3,000 diameters in a normal diploid cell grown in culture. The chromosomes assume this compact form during mitotic cell division. When grown in culture, human cells display a limited capacity to divide; their finite longevity may be related to the finite span of human life. The chromosomes in the picture are stained with a dye called Giemsa.

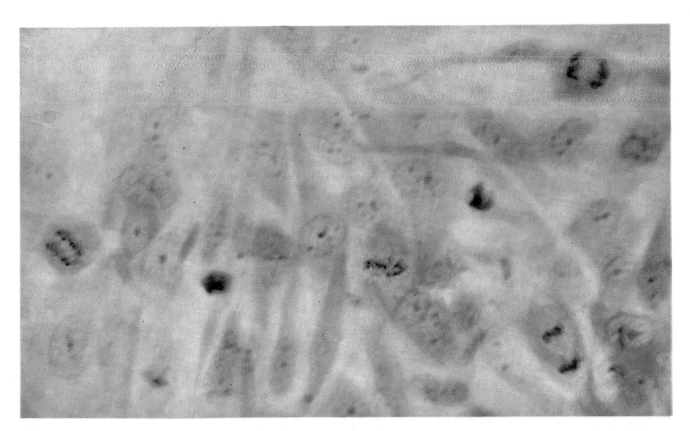

STAGES OF CELL DIVISION appear in this photomicrograph of human cells. The chromosomes are seen as dark clusters. Those in the nucleus at the center have not begun to divide. At left is a nucleus in which the chromosome cluster has separated into two parts; at right, top and bottom, are two nuclei in a later stage of the division cycle. The culture is stained with aceto-orcein and green counterstain. The magnification is 750 diameters. Both photomicrographs were made by Paul S. Moorhead of the Wistar Institute.

male cells) or by the visible difference between the *XX* female sex chromosomes and the *XY* male chromosomes. As a consequence we were able to use these cell "markers" as a label for following the progeny of the respective original parents.

We seeded a bottle with a certain number of fibroblasts from a male population that had undergone 40 doublings and with an equal number of fibroblasts from a female population that had gone through only 10 doublings. If some inadequacy of the culture or the accumulation of a killing factor in the medium were the primary cause of cell death in our cultures, then in this experiment the number of doublings in culture should have been the same for both the male and the female cells after we had mixed them together; there is no reason to suppose that a nutritional inadequacy or a lethal factor such as a virus would act preferentially on the cells of a particular sex. Actually the male and female cells composing the mixture, presumably because of their difference in age, proved to have sharply different survival rates. Most of the male population, having already undergone 40 doublings before the mixture was made, died off after 10 more doublings of the mixed culture, whereas the "young" female population of cells in the same mixed culture, with only 10 previous doublings, went on dividing for many more than the male. After 25 doublings all the male cells had disappeared and the culture contained only female cells. These results appear to confirm in an unambiguous way that the life-span of fibroblast populations in our cell cultures is determined by intrinsic aging of the cells rather than by external agencies.

Does this aging result from depletion or dilution of the cells' own chemical resources? We considered the possibility that the eventual death of the cells might be attributable to the exhaustion of some essential metabolite the cells could not synthesize from the culture medium. If that were the case, however, the original store of this substance must be very large indeed to enable the cells to multiply for 50 generations. Simple mathematics showed that in order to provide at least one molecule of the hypothetical substance for each cell by the 50th doubling, the original cell would have to have at least three times its known weight even if the substance in question were the lightest element (hydrogen) and the original parent cell were composed entirely of that element!

By isolating individual cells and developing clones (colonies) from them we have been able to establish that each human fibroblast from an embryo is capable of giving rise to about 50 doubling generations in cell culture. As the cell proliferation proceeds there is a gradual decline in the capacity for reproduction. Investigators in several laboratories have found that in successive generations of the multiplying cells a larger and larger fraction of the progeny becomes incapable of dividing until, by the 45th or 50th doubling generation, the entire population has lost this ability.

Curiously, as the population of human fibroblasts approaches the end of its lifetime, aberrations often crop up in the chromosomes. Chromosome aberrations and cell-division peculiarities related to age have also been observed in human leucocytes and in the liver tissue of living mice. The question of whether or not cell abnormalities are a common accompaniment of human aging remains a moot point, however; there is no clear clinical or laboratory evidence on the matter.

Our information on the aging of cells so far is limited to what we can observe in cell cultures in glassware. It has not yet been established that fibroblasts behave the same way in living animals as they do in an artificial culture. However, in man several organs lose weight after middle age, and this is directly attributable to cell loss. The human brain weighs considerably less in old age than it does in the middle years, the kidney also shows a large reduction in nephrons accompanying cell loss and the number of taste buds per papilla of the tongue drops from an average of 245 in young adults to 88 in the aged. We cannot be certain that fibroblasts stop dividing or divide at a lower rate as an animal ages or that the bodily signs of aging can be explained on that basis. It is well known, however, that certain cell systems in animals do stop dividing and die in the normal course of development. Familiar examples are the larval tissues of insects, the tail and gills of the tadpole and some embryonic kidney tissues (the pronephros and mesonephros) in the higher vertebrates.

Reviewing these phenomena, John W. Saunders, Jr., of the School of Veterinary Medicine of the University of Pennsylvania has suggested that the death of cells resulting in the demise of specific tissues is a normal, programmed event in the development of multicellular animals. By the same reasoning we can surmise that the aging and finite lifetime of normal cells constitute a programmed mechanism that sets an overall limit on an organism's length of life. This would suggest that, even if we were able to checkmate all the incidental causes of human aging, human beings would inevitably still succumb to the ultimate failure of the normal cells to divide or function.

In this connection it is interesting to consider what engineers call the "mean time to failure" in the lifetime of machines. Every machine embodies built-in obsolescence (intentional or unintentional) in the sense that its useful lifetime is limited and more or less predictable from consideration of the durability of its parts. By the repair or replacement of elements of a machine as they fail its lifetime can be extended, but barring total replacement of all the elements eventual "death" of the machine as a functioning system is inevitable.

What might determine the "mean time to failure" of an animal organism? I suggest that animal aging may result from deterioration of the genetic program that orchestrates the development of cells. As time goes on, the DNA of dividing cells may become clouded with an accumulation of copying errors (analogous to the "noise" that develops in the serial copying of a photograph). The coding and decoding system that governs the replication of DNA operates with a high degree of accuracy, but the accuracy is not absolute. Moreover, there is some experimental evidence that, as Leslie E. Orgel of the Salk Institute for Biological Studies has suggested, certain enzymes involved in the transcription of information from DNA for the synthesis of proteins may deteriorate with age. At all events, since the ability of cells to divide or to function is controlled by the inherited information-containing molecules, it seems likely that some inherent degeneration of these molecules may hold the key to the aging and eventual death of cells.

Pursuing the machine analogy, we might surmise that man is endowed with a longer life-span than other mammals because human cells have evolved a more effective system for correcting or repairing errors as they arise. Such an evolution would account for the generally progressive lengthening of the fixed life-span from the lower to the higher animals; presumably the march of evolution has developed improvements in the cells' error-repairing mechanisms. It is clear, however, that even in man this system is far from perfect. In the idiom of computer engineers we might say that man, like all other animals, has a "mean time to failure" because his normal cells eventually run out of accurate program and capacity for repair.

47

FREE RADICALS IN BIOLOGICAL SYSTEMS

WILLIAM A. PRYOR

August 1970

Short-lived and highly reactive, free radicals are essential intermediates in many chemical processes. In living systems radicals play important roles in radiation damage and in aging

In early 1969 an exquisite creature named Vanessa showed a mouse called Mimi to Simon Templar, hero of the television program "The Saint." Vanessa explained that the mouse's life expectancy had been increased by more than 40 percent as the result of an experiment conducted by Vanessa's father: he had fed Mimi a diet including "butylated hydroxytoluene," or BHT. The episode was actually based on a scientific report: Denham Harman of the University of Nebraska has found that BHT and several other chemicals of widely varying types appear to increase the average life-span of laboratory animals [see "Science and the Citizen," SCIENTIFIC AMERICAN, March, 1969]. The most obvious similarity among these life-lengthening chemicals is that they all interfere with or entirely stop the reactions of ephemeral chemical entities called free radicals.

A free radical is a chemical compound that has an odd number of electrons and is therefore generally highly reactive and unstable and cannot be isolated by ordinary methods. In contrast, most chemical compounds have an even number of electrons and are stable. As we shall see, chemical bonds are made up of a pair of electrons, and the high reactivity of free radicals stems directly from the fact that they have an odd electron. Any species with an odd electron seeks another odd-electron species, and the two proceed to pair their odd electrons to unite and form a bond.

Free radicals are known to be key intermediates in many laboratory, industrial and biochemical processes. Most of the reactions of oxygen involve free radicals, including the slow degradation of organic materials in air, burning and the drying of paints. Many plastics are made by processes that involve free radicals as transient intermediates. Radicals can also be detected in most animal and plant cells, and it is clear that radical chemistry plays a vital role in life processes. Radicals appear to be involved in the production of at least some types of cancer, and the concentration of radicals is different in cancerous cells from what it is in normal cells. Some of the reactions that mediate respiration by living organisms involve radicals. Fats in foods become rancid on oxidation, and so inhibitors of radical reactions, such as BHT, are added to lengthen the storage life of foods. Radiation damage to living systems occurs partly through free radicals. Finally, aging itself has been postulated to involve random destructive reactions by radicals present in the body.

To understand the great reactivity of free radicals it is necessary to understand first why electrons pair to form the normal two-electron bond. Electrons have a magnetic moment and are small magnets; as such they can be considered to have the property termed spin, which can be either "pointing up" or "pointing down" for any electron. Chemical compounds are assigned a "multiplicity" depending on the arrangement of the spins of their electrons. If a molecule has an even number of electrons entirely arranged in pairs with opposed spins, the molecular species is said to be a singlet. If a molecule has an odd number of electrons, then the odd electron must have an unpaired spin; the molecule is called a doublet and is a radical. If a molecule has an even number of electrons but the electrons of one pair have parallel spins, the species is said to be a triplet. Bonds are formed only between electrons that are paired and have antiparallel spins; clearly, then, most molecules will be singlets.

When an organic molecule containing a normal electron-pair bond is heated above a certain temperature (for example, when a hydrocarbon molecule such as is found in petroleum is heated to a temperature between 700 and 800 degrees Celsius), the weakest bond in the compound breaks and the two fragments fly apart. In this process of bond thermolysis, or thermal bond scission, the two electrons in the bond divide, one going with each fragment. The process can be symbolized as $A:B \rightarrow A\cdot + B\cdot$, where AB is an ordinary molecule and the two dots represent the two electrons in the A-B bond that hold the molecule together. A process such as this one, in which the bonding pair of electrons divides symmetrically as the bond breaks, is called homolysis, or homolytic bond scission. The two electrons in the A-B bond must be paired and have antiparallel spins, and so the odd electrons in A· and B· initially must have antiparallel spins. If these two radicals come together again, they can re-form the A-B

METHANE is a simple molecule consisting of a carbon atom bonded to four hydrogen atoms (*left*). The methyl free radical (*center*) lacks one hydrogen. A chemical bond consists of two paired electrons; the methyl radical has an odd electron (*dot*) and so it is highly reactive. Two methyl radicals can combine to form an ethane molecule (*right*).

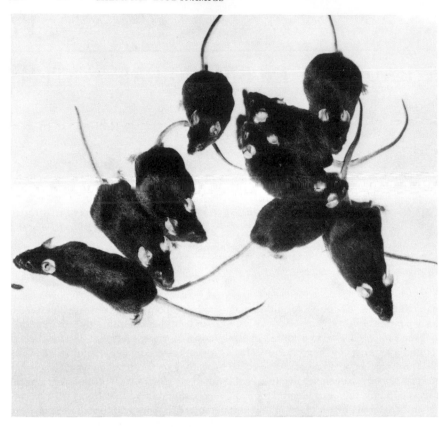

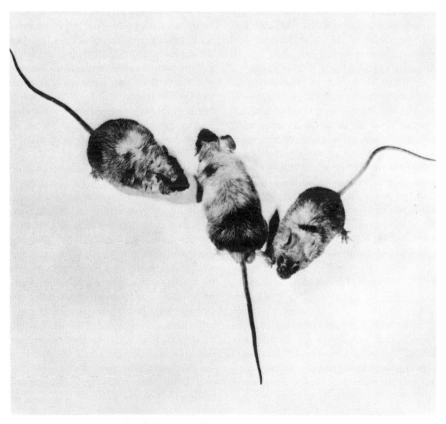

FREE RADICALS appear to be implicated in the process of aging and in damage to tissues from radiation. The connection between radiation and aging is demonstrated in these photographs, made by Howard J. Curtis of the Brookhaven National Laboratory, of two groups of 14-month-old mice. Originally there were nine mice in each group. One group received a large but nonlethal dose of radiation as young adults. The untreated mice are healthy (*top*). Of the irradiated mice only three survive (*bottom*) and they are senile and gray.

bond. This bond-making process is extremely fast for almost all radicals, so that radicals exist only in very low concentrations. As soon as their concentrations build up, A• and B• collide with each other more frequently, and stable A-B molecules are produced.

Suppose, however, that the A• and B• radicals have not come from the same A-B molecule. If a random pair of A• and B• radicals collide in solution, the spins of the odd electrons in the two radicals are randomly oriented, and the chance that a pair of radicals that have odd electrons with antiparallel spins will collide will be only one in four. This is the case because the triplet state is three times as probable as the singlet, and three in four of the collisions lead to a triplet multiplicity, which cannot form a bond.

There are two possibilities for these triplet pairs: either one of the electrons could "flip" its spin to the other direction, converting the triplet to a singlet pair so that the A-B bond could form, or diffusion could occur faster than this conversion, and the two radicals would simply separate without reacting. Experimentally it is found that triplet radical pairs usually diffuse apart and do not combine to form a bond. Diffusive separation of a pair of radicals occurs in solvents of ordinary viscosity in a time period of only about 10^{-10} second; apparently the triplet-to-singlet conversion requires more time than that. In summary, an A-B molecule will be re-formed every time A• and B• radicals with antiparallel spins encounter each other in solution. However, A• and B• radicals that do not have correlated spins will re-form A-B only one time in four; three in four of the encounters form triplet pairs of radicals, which cannot form bonds.

Since radicals are extremely reactive substances that normally exist only in very dilute solutions, it is not surprising that most practical radical reactions are chain reactions. Radicals, formed in an initiation phase, react in a cyclic propagation sequence in which a product molecule is produced at the same time that another radical is formed to carry on the chain. The cycle is ended by a termination phase, in which radicals recombine. For example, the reaction of a hydrocarbon with chlorine has a chain length of 1,000 or more: every primary radical produced eventually leads to the formation of 1,000 or more molecules of product.

The chain reaction can be initiated by irradiation with light or X rays, by simply heating the system to a high temperature or by the use of an "initiator," a compound with an unusually weak bond that breaks to form radicals at a convenient temperature. The cracking of petroleum is initiated by the scission of one of the carbon-carbon bonds in the petroleum hydrocarbon molecules themselves. A simple hydrocarbon such as ethane has a carbon-carbon bond with a bond strength of 85 kilocalories per mole. The rate of breaking of such a bond becomes appreciable only at temperatures near 700 degrees C. The compounds called peroxides, on the other hand, contain the oxygen-oxygen bond, which is unusually weak. For example, the bond-dissociation energy of the central bond in hydrogen peroxide, HO—OH, is 48 kilocalories per mole. Different organic peroxides dissociate at temperatures ranging from 50 to 200 degrees, so that one can generally find a commercially available peroxide initiator that decomposes to produce radicals at a convenient rate at any desired temperature.

The decomposition of an initiator is not always as simple a process in solution as it is in the gas phase. In the gas phase the A· and B· fragments fly apart and each A-B molecule yields two radicals. In solution, however, it is often observed that each initiator does not produce the two free-radical fragments expected. The reason is that when an initiator undergoes bond scission in solution, the A· and B· fragments are held together very briefly by the "cage" of surrounding solvent molecules. The two fragments strike these solvent molecules as they try to separate and are reflected back toward each other. Consider the decomposition of an azo compound of the type R—N=N—R, which decomposes by splitting out a nitrogen molecule, N≡N, and forming two R· radicals. Since these two radicals have opposed spins and can immediately couple, the formation of the stable molecule R—R in the cage can compete with the diffusion apart of the two R· radicals. This reduces the number of free R· radicals that can initiate chemical reactions, and azo initiators therefore range from about 50 to 100 percent in efficiency.

One of the striking features of radical reactions is that only two general types of propagation reaction are commonly observed: atom abstraction and addition. In the abstraction reaction a free radical attacks another molecule to pull off an atom with one valence electron, usually a hydrogen atom. This reaction can be symbolized as M· + RH → MH + R·, where M· is any free radical and

INITIATION

a Cl—Cl $\xrightarrow{\text{LIGHT}}$ 2 Cl·

b INITIATOR $\xrightarrow{\text{HEAT}}$ R·

R· + Cl₂ $\longrightarrow$ R—Cl + Cl·

PROPAGATION

c Cl· + CH₄ $\longrightarrow$ HCl + ·CH₃

d ·CH₃ + Cl₂ $\longrightarrow$ Cl—CH₃ + Cl·

SUM CH₄ + Cl₂ $\longrightarrow$ Cl—CH₃ + HCl

TERMINATION

2 Cl· $\longrightarrow$ Cl₂

Cl· + CH₃· $\longrightarrow$ Cl—CH₃

2CH₃· $\longrightarrow$ CH₃—CH₃

SIMPLE CHAIN PROCESS, the chlorination of methane, begins with an initiation step in which chlorine atoms (free radicals) are produced (a) by light, which dissociates chlorine molecules, or (b) by other free radicals (R·) provided by the decomposition of a chemical initiator. Two propagation steps (c, d) make up the chain sequence, during which the number of radicals is conserved; the steps can be summed to give the overall chemical change. The chain reaction can be terminated by the coupling of any two radicals. A propagation reaction such as c or d, involving the transfer of an atom from a molecule to a radical, is an atom abstraction, one of the two common types of free-radical propagation reactions.

RH is any hydrogen-containing molecule. Notice that in this reaction, as in all propagation reactions of radicals, the number of radicals is not reduced; one radical is used and another is made.

In the addition reaction a radical adds to a material that contains a double bond: M· + CH₂=CHR → M—CH₂—CHR. Notice that two of the electrons in the double bond unpair; one joins with the odd electron of the free radical to form a new bond and one becomes localized on the adjacent carbon atom to form a new radical center. Materials such as polyethylene, polyvinyl chloride and polystyrene are produced by processes that involve this reaction. A monomer

such as ethylene, vinyl chloride or styrene is mixed with an initiator and the mixture is heated to a temperature at which the initiator decomposes to form free radicals [see bottom illustration at right].

A dramatic advance in the study of radical reactions was made in 1945 with the invention of an instrument that detects and identifies radicals by the magnetic properties of their odd electron. In this technique, called electron-spin resonance, or ESR, a strong magnetic field is applied to the sample and the energy absorption is measured when the odd electrons flip their spins from being aligned in the same direction as the field to being aligned in the opposite direction. The

INITIATION

CH₂=CH₂ $\xrightarrow{\text{HOT AIR AND HIGH PRESSURE}}$ ROO·

ROO· + CH₂=CH₂ ⟶ ROO—CH₂—·CH₂

PROPAGATION

ROO—(CH₂—CH₂)ₙ—CH₂—·CH₂ + CH₂=CH₂ ⟶

ROO—(CH₂—CH₂)ₙ—CH₂—CH₂—CH₂—·CH₂

TERMINATION

ROO—(CH₂—CH₂)ₙ—CH₂—·CH₂ + ROO· ⟶

ROO—(CH₂—CH₂)ₙ—CH₂—CH₂—OOR

ADDITION REACTION is another common free-radical propagation reaction. In it a radical adds to a compound containing a double bond. For example, polyethylene is made from ethylene by using air as an initiator. The air oxidizes the ethylene to produce peroxide radicals (ROO·), which initiate the polymerization (molecular chain-building) process.

amplitude of the resulting energy peaks and the field strengths at which they occur give information on the concentration of radicals and their nature. Often, particularly if the radicals can be studied either in the liquid phase or in a single crystal, the detailed structure of the radical can be identified from its ESR spectrum. This technique has had great impact on the study of radicals in biochemical systems and in tissues, since the radicals can be studied even when the chemistry of the system is incompletely understood.

A living organism is a machine that requires vast amounts of energy for the chemical and physical work it must perform. Organisms obtain their energy by the oxidation of biological materials—by burning food as fuel. The burning does not proceed in the random and inefficient way it would in a furnace; rather, enzymes act as catalysts and the oxidations take place in a controlled sequence of small steps in which more nearly the maximum obtainable energy is liberated.

Oxidation is defined as the loss of electrons; the reverse process, reduction, is the gaining of electrons. Oxidation can proceed through loss of electrons from a substance either in pairs or one at a time. For example, a reaction in which a substance is oxidized by the overall loss of two electrons to form a stable product might occur in one step, in which both electrons are transferred, or in two steps, by the transfer of one electron at a time with a free radical as a transient intermediate. If this intermediate were very unstable, it would rapidly lose the second electron to form the ultimate product and would exist only at extremely low concentrations and be very difficult or even impossible to detect. The difference between oxidations that involve radicals and proceed one electron at a time and those that proceed in two-electron steps is therefore not always obvious just from the products that are formed.

In the early 1930's Leonor Michaelis of the Rockefeller Institute initiated a series of investigations to prove that biological oxidations might involve free radicals. In 1946 he published his highly provocative statement that "all oxidations of organic molecules, although they are bivalent, proceed in two successive univalent [one-electron] steps, the intermediate being a free radical." We now know that this theory is incorrect: there are two-electron oxidations in biochemistry that proceed pairwise and do not involve radicals as intermediates. Nev-

ertheless, Michaelis' views inspired research that is continuing today, providing insight into the nature of the processes by which living organisms obtain energy. Studies by numerous workers in the 1940's and 1950's indicated that an intermediate could sometimes be identified in enzymatic oxidation-reduction reactions of biological molecules, but in spite of Michaelis' conviction there was no proof that the intermediate was a free radical. Starting in 1954, however, and using the newly developed techniques of ESR, Barry Commoner and George E. Pake of Washington University, Helmut Beinert of the University of Wisconsin, Anders Ehrenberg of the Nobel Institute of Sweden and others were able to show that a paramagnetic intermediate can be detected in some enzyme-substrate systems. Commoner in 1956, Melvin Calvin in 1957 and later other investigators showed that ESR signals are produced in plant systems during photosynthesis.

Early ESR instruments were insensitive. They could detect radicals only in biological materials that had been freeze-dried, killing most living systems, and so it was difficult to correlate concentrations of radicals with biological activity. The possibility existed that the radicals being detected were artifacts not directly involved in the biochemical reactions under investigation. Because water absorbs microwave energy very near the frequency used for most commercial ESR instruments, it was particularly difficult to study biological samples in aqueous solution in their natural state.

Around 1957 various workers developed new ESR methods that made it possible to study aqueous samples, and as the ESR instruments have im-

proved so has the observed correlation between radical concentration and biological activity. It now appears that radicals are important intermediates in a number of biological processes. In some systems a correlation between biological activity and the concentration of radicals can be established, in others the structure of the radicals can be identified and in still others the rate of reaction of the radicals can be followed. To date, however, all three of these things have been done for very few systems. Two areas in which ESR techniques have been successfully applied are enzymatic oxidations and the mechanisms by which radiation damages organic materials.

Enzymatic oxidations usually proceed through the removal of electrons from the substrate by an enzyme and their transfer to a coenzyme. These are steps in an elaborate electron-transport system that carries the electrons from the substrate to oxygen and effects the reduction of oxygen to water, and they occur in the cigar-shaped organelles of the cell called mitochondria. It is probable that all living cells contain some radicals, but ESR signals can be detected only in certain cells. In general stronger signals are detected in cells with high concentrations of mitochondria. The greatest success in correlating ESR signals with biological activity has come from studies of the flavin coenzymes and coenzyme Q. Several workers have shown that relatively stable radicals are formed in these enzyme systems and that radicals are directly involved as intermediates in the oxidations.

Although most enzymes do not transfer electrons from the substrate directly to oxygen, the oxidases and some of the flavins are able to do so. They generally

PEROXIDE	STRUCTURE	TEMPERATURE (DEGREES CELSIUS)
t-BUTYL HYDROPEROXIDE	$(CH_3)_3CO-OH$	172
CUMENE HYDROPEROXIDE	$C_6H_5C(CH_3)_2O-OH$	158
t-BUTYL PEROXIDE	$(CH_3)_3CO-OC(CH_3)_3$	127
t-BUTYL PERBENZOATE	$C_6H_5CO_2-OC(CH_3)_3$	105
t-BUTYL PERACETATE	$CH_3CO_2-OC(CH_3)_3$	100
BENZOYL PEROXIDE	$C_6H_5CO_2-O_2CC_6H_5$	76
ACETYL PEROXIDE	$CH_3CO_2-O_2CCH_3$	67
t-BUTYL PHENYLPERACETATE	$C_6H_5CH_2CO_2-OC(CH_3)_3$	66
t-BUTYL TRIPHENYLPERACETATE	$(C_6H_5)_3CCO_2-OC(CH_3)_3$	11

INITIATORS can be selected from among a number of organic compounds that dissociate to produce free radicals at a wide range of temperatures. Most such compounds are peroxides, which contain the readily broken oxygen-oxygen bond. The table lists some peroxide initiators and gives the temperature at which each of them has a half-life of 10 hours.

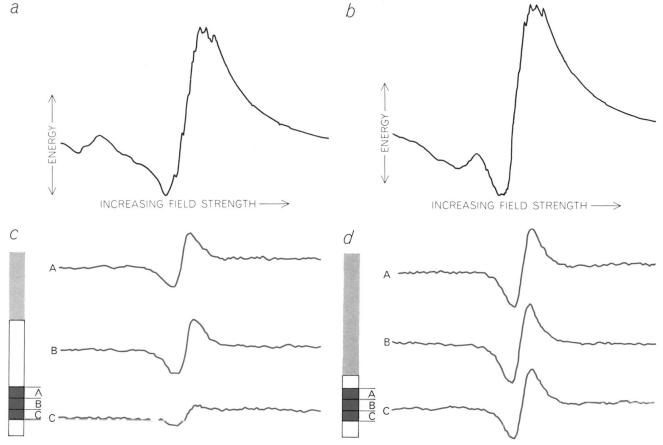

ELECTRON-SPIN RESONANCE (ESR) detects unpaired electrons and thus measures the number and properties of free radicals in a sample placed in a magnetic field. An ESR curve traces the changes in energy associated with the flipping of electron spins as the magnetic field varies; the amplitude of a peak and its location are characteristic of the concentrations and the chemical sites of free radicals. These ESR spectra from the Japan Electron Optics Company are those of the tobacco in an unsmoked cigarette (*a*) and in a smoked butt (*b*) and of various levels (*A, B and C*) of the filters of half-smoked (*c*) and nine-tenths-smoked (*d*) cigarettes.

reduce oxygen to hydrogen peroxide, which is subsequently reduced to water by another class of enzymes called catalases or peroxidases. Hydrogen peroxide is produced in all cells through the reduction of oxygen, and the peroxidase enzymes are important in keeping peroxide concentrations down to levels that are not damaging to the cell. Lawrence H. Piette of the University of Hawaii and others have shown that peroxidase-substrate systems give ESR signals, and the activity of the enzyme has been correlated with the strength of the signal.

The processes by which radiation interacts with organic materials are quite complex, and we shall consider only the simplest scheme here. Ionizing radiation can consist of electromagnetic radiation such as very-short-wavelength light or X rays, or of highly energetic particles such as high-speed electrons or alpha particles. Radiation causes ionization by knocking electrons away from the molecules to which they belong, thus producing free electrons and positively charged ions, or cation radicals. The ions and

electrons may recombine to produce neutral molecules in an excited state, which can undergo homolysis to produce free radicals; radicals can also be produced by reactions between the cation radicals and the neutral molecules.

Cells are from 60 to 80 percent water and when a plant or an animal is irradiated, most of the energy is deposited in the aqueous phase; less often will a primary ionization occur in an organic biomolecule. A portion of the damage to living systems therefore results from reactive particles that are formed in the water phase and diffuse to an organic molecule in the cell, causing secondary reactions. The chief radical species produced in the radiolysis of water and implicated in radiation damage are solvated electrons (electrons associated with water molecules), hydrogen atoms and hydroxyl radicals (HO·).

Cells are extremely sensitive to radiation. Calculations show that radiation that destroys perhaps only one molecule in 100 million can have profound biological consequences and can even kill

the cell. The most reasonable explanation of this biological magnification is that the energy from radiation can be transferred to critical polymer molecules in the nucleus of the cell. Aside from certain effects on cellular membranes, which will be discussed later, it is probable that the cell nucleus and its chromosomes are critical in determining radiation sensitivity. About half of the dry mass of the nucleus consists of chromosomes and of this about half is deoxyribonucleic acid (DNA) and about half is associated protein material. Two types of reactions are therefore particularly important in terms of the damage they can cause. The first is the reaction of radicals from water with DNA and the second is their reaction with proteins.

Each strand of the double helix of DNA is a chain of subunits called nucleotides, arranged in an ordered sequence to spell out genetic instructions according to the genetic code. The distinctive element of each nucleotide is one of four nitrogenous bases: thymine,

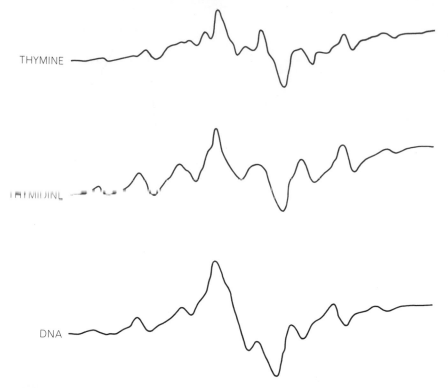

THYMINE

THYMIDINE

DNA

THYMINE, one of the four bases in DNA, is sensitive to radiation, probably at least in part because hydrogen atoms add to thymine to form free radicals, which can react to alter the DNA. The top ESR spectrum, from an experiment conducted by Anders Ehrenberg and his colleagues, shows the pattern typical of the radical, produced when thymine was subjected to gamma radiation. The other two spectra show that the same radical is present in thymidine (a part of the DNA molecule that contains thymine) and in a purified DNA sample.

adenine, guanine and cytosine. All three of the radicals produced in water radiolysis react with these bases at a high rate. For example, hydrogen atoms add to thymine to produce a radical that can be identified by its characteristic ESR spectrum. The same spectrum can be observed when hydrogen atoms react with the thymine nucleotide, thymidylic acid, or when either solid DNA or an aqueous solution of DNA is irradiated [see top illustration]. The thymine radicals are not stable and they react further to alter the DNA molecule and interfere with the coding of genetic information. Such a change could be nonlethal, since the cell has mechanisms for excising and repairing damage to its DNA, but it would be expected to often alter a gene in such a way that it would kill the cell.

Dov Elad and his co-workers at the Weizmann Institute of Science have re-

cently shown that certain organic compounds can add to the bases in DNA. These reactions proceed in the same way whether they are induced by light, high-energy radiation or normal free-radical initiators, and they unquestionably involve free radicals as intermediates.

Another reaction that may cause mutations has been identified by H. J. Rhaese of the National Institutes of Health. He has shown that hydrogen peroxide reacts with adenine and modifies its structure. This reaction occurs when adenine is irradiated with X rays or when it is simply treated with hydrogen peroxide and ferric ions, a mixture known to produce hydroxyl radicals. Although it has not yet been established that this slight modification of adenine causes mutations in organisms, there is some evidence that it may contribute to the weak mutagenic effect of X rays. For example, there appears to be a higher incidence of chromosome aberrations in cells with low concentrations of catalase and consequently with higher than normal concentrations of hydrogen peroxide.

Proteins, which play an important role in the chemistry of all plant and animal cells, are made up of many amino acid molecules joined together in a peptide chain. Proteins are held in a particular configuration by molecular forces, and the shape of each protein is critical to its exact functioning. One amino acid, cysteine, contains a sulfur-hydrogen group, and two of these groups can be converted to produce a disulfide bond between two cysteine residues in the protein chain. This S—S link helps to hold the protein molecules in their active configuration. (A macroscopic example of this is the permanent-waving of hair. Hair is made of the protein keratin, which contains disulfide cross-links that help the hair to hold its shape. In permanent-waving these bonds are first reduced to S—H groups; then the molecules are arranged in the desired conformation by rolling the hair on rods and the S—S bonds are then remade by oxidation with peroxide.)

In 1955 Walter Gordy of Duke University examined the ESR spectra of irradiated proteins and proposed that two types of radical are produced. One gives a two-peak spectrum that varies slightly from one protein to another. It is now known that this spectrum results from an odd electron localized on a carbon atom of the peptide-chain backbone. The second type of radical gives an ESR absorption at the low-field end of the spectrum and is easy to identify

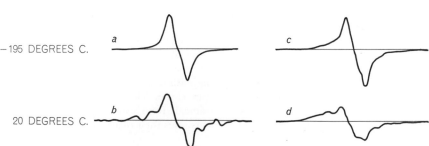

SPERM HEADS   SPERM HEADS PLUS MEA

−195 DEGREES C.   a   c

20 DEGREES C.   b   d

EFFECTS of temperature and of a sulfur-containing drug on radical formation were demonstrated by Peter Alexander. Fish-sperm heads, which are rich in DNA, were cooled to −195 degrees Celsius and irradiated. The frozen sample shows a typical low-temperature spectrum (a). When the sample is warmed, the DNA spectrum appears as the radicals become mobile (b). Addition of MEA, containing thiol (S—H) groups, has little effect on the frozen sample (c). When the sample is warmed, the radicals transfer from DNA to thiols and a new plateau characteristic of sulfur radicals appears at the left in the spectrum (d).

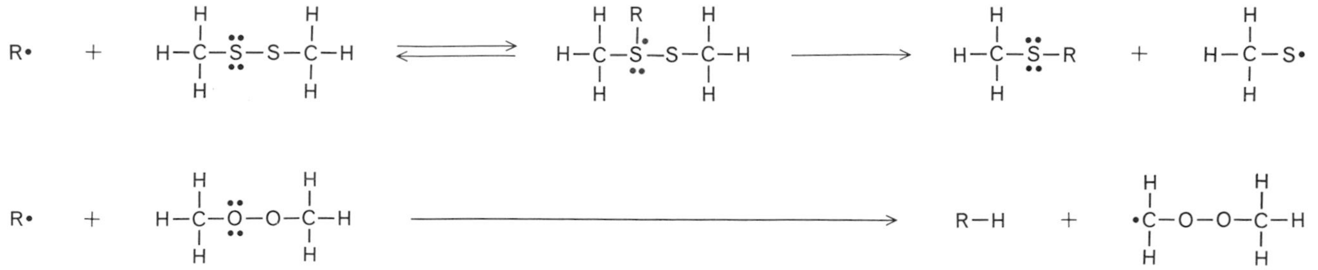

$$R\cdot \;+\; H_3C{-}\overset{..}{\underset{..}{S}}{-}S{-}CH_3 \;\rightleftharpoons\; H_3C{-}\overset{R}{\underset{..}{S}}{-}S{-}CH_3 \;\longrightarrow\; H_3C{-}\overset{..}{\underset{..}{S}}{-}R \;+\; H_3C{-}S\cdot$$

$$R\cdot \;+\; H_3C{-}\overset{..}{\underset{..}{O}}{-}O{-}CH_3 \;\longrightarrow\; R{-}H \;+\; \cdot C H_2{-}O{-}O{-}CH_3$$

SULFUR COMPOUNDS protect against radiation because they re-act readily with radicals, which attack the sulfur-sulfur bond of disulfides faster than they attack the oxygen-oxygen bond of per-oxides, according to the author and his colleagues. It may be that an attack by a radical on methyl disulfide, for example (*top*), can form a relatively stable intermediate because the sulfur atom can accommodate nine electrons (*color*). An attack by a radical on methyl peroxide (*bottom*), on the other hand, cannot involve this path, since the oxygen atom can only accommodate eight electrons. As a result the reaction proceeds by hydrogen abstraction instead.

as a sulfur radical, RS•; it is observed in enzymes that contain S—H or disulfide groups and in mixtures of enzymes with added sulfur compounds. Thormod Hen-riksen in Norway, Peter Alexander and M. G. Ormerod in England and Harold C. Box and several other workers in this country have shown that irradiation of proteins at low temperatures produces ESR signals of the carbon-radical type, but that heating the material gradually converts at least some of these carbon radicals to sulfur radicals. Clearly pro-tein molecules possess a mechanism for transferring the site of the destruction from one part of the molecule to another or to nearby molecules.

Although the sulfur-sulfur bond is criti-cal in determining the shape of many proteins, it is cleaved remarkably easily. My research group at Louisiana State University has studied the cleavage of the disulfide bond by radicals and we believe it may be faster than analogous reactions of oxygen compounds because of the ability of sulfur to react with radi-cals to form a relatively stable intermedi-ate [*see illustration above*]. There is evi-dence from other laboratories that some of the reactions of ions with sulfur-sulfur bonds may also proceed through a re-lated intermediate complex.

Recently several workers in this coun-try, Gabriel Stein in Israel and H. Jung in Germany have shown that the hydro-gen atom attacks the sulfur-sulfur bond so efficiently that each hydrogen atom produced in a model system leads to the inactivation of an entire enzyme mole-cule. Since enzymes are among the largest molecules known and the hydro-gen atom is the smallest atom, this is a bit like killing an elephant with a BB pellet. When a cell is irradiated, very few of the hydrogen atoms from the aqueous phase are likely to collide with an enzyme molecule, but apparently those that do can cause the total destruc-tion of the biological activity of the enzyme.

Since sulfur groups react readily with radicals, it is not surprising that sulfur compounds act as drugs that protect against radiation. There are a number of mechanisms by which a molecule might protect a cell from radiation damage. For example, compounds containing S—H groups (thiols) can protect impor-tant biological molecules through a re-pair process. A number of workers have shown that radiation can result in the removal of a hydrogen atom from a bio-logical polymer, leaving a radical. The thiol can then repair this damage by a hydrogen-transfer reaction, donating a

hydrogen atom to the polymer radical and creating a less lethal thiyl radical, R—S•. These reactions are thought to be partly responsible for the significant ra-diation-protection activity of thiols. Most compounds containing S—H or S—S bonds act as protective agents in labora-tory systems, but in the body problems of solubility, diffusion and toxicity arise and so only certain sulfur compounds are effective radiation-protection drugs.

It is interesting that the potent radia-tion protection drug beta-mercapto-ethylamine (MEA), or cysteamine, also has been found by Harman to be effec-tive in lengthening the mean life-span of mice. There are radicals in the body and it is clear that they can damage biochem-ical systems in cells; radiation involves radical reactions and also effects aging. These facts have suggested to many in-vestigators that aging itself must be at least partly due to damage caused by radical reactions within the body.

In an organism such as man, aging can be expected to result from many chemi-cal reactions, and there is no reason to anticipate a single cause or mechanism. On the contrary, it is likely that several different and perhaps complex mecha-nisms make significant contributions to the total changes that occur in our bodies

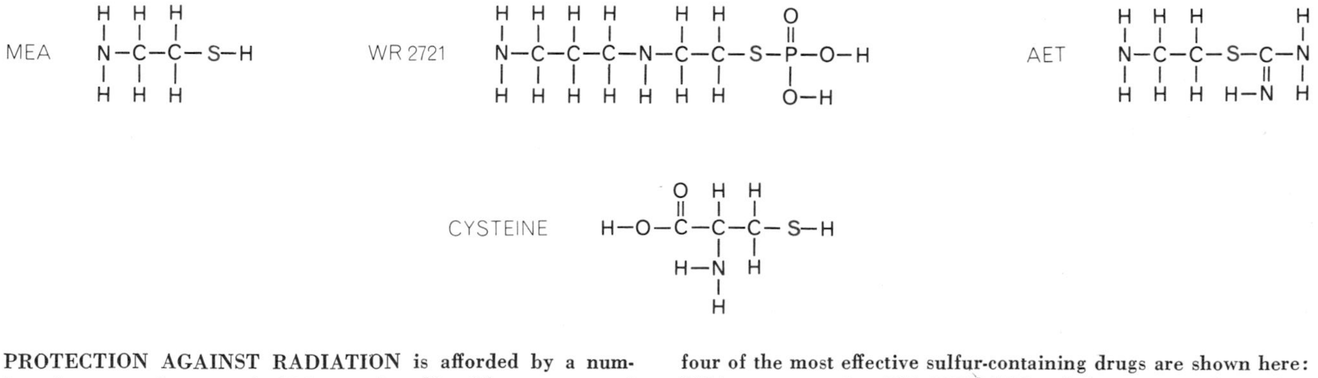

MEA

$$N H_2{-}C H_2{-}C H_2{-}S{-}H$$

WR 2721

$$N H_2{-}C H_2{-}C H_2{-}C H_2{-}N H{-}C H_2{-}C H_2{-}S{-}\overset{O}{\underset{O-H}{P}}{-}O{-}H$$

AET

$$N H_2{-}C H_2{-}C H_2{-}S{-}C(=NH){-}N H_2$$

CYSTEINE

$$H{-}O{-}\overset{O}{C}{-}\underset{NH_2}{C H}{-}C H_2{-}S{-}H$$

PROTECTION AGAINST RADIATION is afforded by a num-ber of compounds, most of which contain sulfur. The structures of four of the most effective sulfur-containing drugs are shown here: MEA, or cysteamine, WR 2721, AET and the amino acid cysteine.

COLLAGEN, a protein connective tissue, becomes less flexible with age because its fibers become cross-linked. In the speculative scheme shown here the cross-linking is caused by free radicals. Carbon atoms, which for the sake of simplicity are shown here with two hydrogen atoms (*left*), are attacked by radicals; they lose a hydrogen atom, giving rise to radical centers on the peptide chain. Pairs of radical centers could couple with each other, producing cross-links that bind the fibers to one another, stiffening them.

with time. There is increasing evidence that at least some of these mechanisms involve the reactions of free radicals.

One theory suggests that aging may be partly due to changes in the connective tissues collagen, elastin and reticulin. These are the structural materials of the body, largely protein in composition, that give tissues shape, plasticity, resilience and elasticity. The biological role of collagen depends on its high plasticity and its ability to bear stress and maintain shape and form; it is present throughout the body but it occurs in particularly high concentrations in flexible organs such as the lungs, blood vessels, skin and muscles. With age, collagen fibers become denser, stiffer, thicker and less plastic. They also become insoluble in organic solvents, indicating that they have become cross-linked by chemical bonds. It is not unreasonable to suggest that some of this cross-linking could result from free-radical reactions.

Most collagen occurs in organs with high concentrations of blood serum, in which radicals are known to be present, and ESR studies show that free radicals attack proteins, producing radical centers on the protein chains. It might be expected that the subsequent combination of these protein radicals could sometimes lead to cross-linking between the protein collagen fibers, and thus to stiffness and increased density [see illustration on this page]. It may not be the free radicals themselves that cause cross-linking but rather some product of their reactions. It is known that under most conditions radiation does not cross-link collagen but rather stimulates collagen synthesis at an increased rate, producing some of the same symptoms as cross-linking. This is one of the ways in which radiation mimics the effects of aging, as William F. Forbes of the University of Waterloo in Canada has pointed out, rather than truly accelerating natural aging.

A theory favored by Howard J. Curtis of the Brookhaven National Laboratory and others suggests that aging in mammals results from mutations in the animals' somatic cells (the body cells, as distinguished from the germ cells involved in reproduction). The theory is conceptually quite simple: DNA directs the synthesis of ribonucleic acid (RNA), which in turn directs the synthesis of all the proteins produced by the cell. If errors gradually accumulate in the genetic information of the DNA or RNA, the cell begins to malfunction and may die. As in other aging theories, the concept is simple but the proof is not. The DNA molecule is enormously complex and there is no direct chemical evidence that it changes with time. Furthermore, it is not possible to measure mutations in somatic cells directly and indirect techniques must be found. In support of the mutation theory, it has been shown that certain radiation effects on chromosome aberrations and on life expectancy are similar, that short-lived animals develop chromosome aberrations faster than long-lived ones and that mouse strains with a longer lifespan also have a lower sensitivity to radiation. There remain several serious difficulties with this theory, however, and it is not universally accepted.

A theory that aging is due at least in part to the peroxidation of lipids explicitly implicates radical reactions. The lipids, which include the fats, constitute the most concentrated source of energy available to the organism. They are oxidized in the cell, chiefly in the mitochondria, in a series of reactions that normally proceed in carefully controlled enzyme-regulated steps. Like all organic materials, however, lipids can also react with oxygen in a nonenzymatic, free-radical pathway. Those that contain reactive hydrogen atoms called allylic hydrogens are particularly prone to undergo the radical reaction. This type of hydrogen occurs in the polyunsaturated fatty acids, which account for about 13 percent of the caloric intake of the human diet.

The peroxidation of lipids is a typical free-radical chain process involving both hydrogen transfer reactions and addition reactions [see illustration below]. The importance of lipid peroxidation to aging rests on the belief that damage to cells accumulated over the lifetime of the organism gradually reduces the efficiency with which the cell carries out its functions. For example, A. L. Tappel of the University of California at Davis has shown that when enzymes are present in systems in which fatty acids are

ABSTRACTION $P—H$ $\xrightarrow{ROO·}$ $P·$

ADDITION $P·$ + O_2 $\longrightarrow$ $POO·$

ABSTRACTION $P—H$ + $POO·$ $\longrightarrow$ $P·$ + $POOH$

PEROXIDATION OF LIPIDS involves both abstraction and addition. A lipid polymer (*P—H*) first loses a hydrogen atom through abstraction by a peroxide radical, forming a lipid radical (*P·*). Then the lipid adds a molecule of oxygen to form a peroxidized radical (*POO·*), which in turn abstracts a hydrogen, leaving a radical (*P·*) to carry on the chain.

being oxidized, the biological activity of the enzymes is destroyed. The oxygen does not react with the enzymes; rather the enzymes are attacked either by radicals produced through the interaction of the lipids and oxygen or by nonradical molecular products from the oxidation of the fats, or by both. Furthermore, there are numerous similarities between the damage to enzymes produced by lipid peroxidation and the damage from radiation.

The body provides a natural lipid antioxidant in vitamin E, and the products derived from vitamin E show that it functions at least partly as a free-radical inhibitor that is sacrificially oxidized to protect the lipids. It is striking that the effects of diets deficient in vitamin E resemble certain effects of both radiation damage and aging. In all three cases there is evidence for structural damage to various cellular membranes.

The membranes in a cell are the partitions that compartmentalize reaction systems so that they do not interfere with one another. Membranes also appear to contain many of the specific receptor sites for the binding of hormones and drugs. Moreover, they must be permeable to particular chemical species at precisely defined rates. They are therefore highly structured and extremely sensitive components, and any deterioration of a cell's membranes must seriously affect its ability to function and could be responsible for some of the effects of cellular aging. Membranes consist of lipids and proteins in varying proportions; the membranes of mitochondria, for example, are about 27 percent lipid and 73 percent protein. Research in a number of laboratories makes it clear that both radiation and free radicals produce significant structural deterioration in membranes and that peroxidation of lipids leads to products that cause cross-linking, reduced permeability and structural decay in membranes.

A particularly clear case of free-radical damage to membranes has been demonstrated for human red blood cells by Edward M. Kosower of the State University of New York at Stony Brook and his wife Nechama S. Kosower of the Albert Einstein College of Medicine. They used a drug that reacts with glutathione, a thiol in blood cells, inhibiting its protective action. The same drug also produces an intermediate that, under the controlled conditions of their experiments, reacts with oxygen in the cell membrane to form radicals; the radicals cause lipid autoxidation, which destroys the viability of the cell wall and bursts

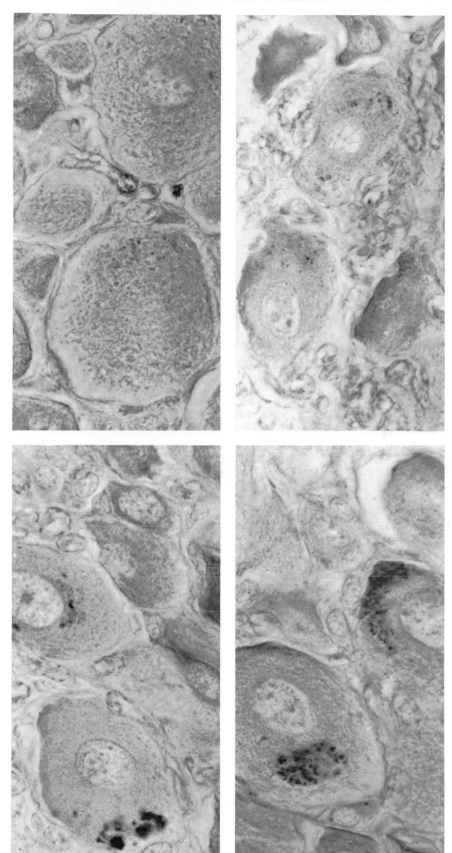

AGE PIGMENTS accumulate in cells with age and there is evidence that they are produced in part by radical reactions. Photomicrographs made by Thaddeus Samorajski and his colleagues at the Cleveland Psychiatric Institute compare nerve cells from the dorsal root ganglia of mice four (*top left*), eight (*top right*), 20 (*bottom left*) and 30 (*bottom right*) months old. The increase in size and concentration of the dark granules is clearly seen.

the cell. This hemolysis occurs even when only a small fraction of the lipid in the cell wall is damaged.

A link between lipid peroxidation and aging is found in the chemistry of the age pigments. These materials, a kind of metabolic debris of the cell, are fluorescent brown compounds that accumulate slowly in cells that are not regularly replaced, such as the nondividing cells in the heart, nervous system and lungs. Age pigments are about 60 percent protein, 25 percent lipid and 15 percent carbohydrate. Several observations indicate that radical reactions are probably important in the production of age pigments. The lipid fragments in them appear to be peroxidized, their proteins are cross-linked in ways suggestive of the cross-links found in proteins exposed to peroxidizing lipids and

Tappel has shown that lipid peroxidation of some cellular organelles gives materials that have the characteristic fluorescent spectra of the age pigments.

The color-bearing components of these age pigments appear in many cases to belong to the melanin class of compounds, which are also responsible for the color of the pigments of skin, eyes and hair. Melanins consist of large polymeric networks that have been known for more than 15 years to contain free-radical centers that can give rise to ESR signals. Donald C. Borg of Brookhaven has shown recently that some age pigments from various organs produce the type of ESR signals characteristic of melanins, indicating that age pigments also contain free radicals of the melanin type.

Bernard L. Strehler of the University of Southern California and others have

shown that age pigments increase almost linearly with age in human heart tissue, for example, and it is clear that these materials are related to aging. Whether they play an important role in the process or are merely harmless byproducts is unknown. The occurrence of peroxidized lipid fragments in these age pigments does prove, however, that lipids are attacked by radicals and are peroxidized in living systems. Furthermore, the fact that lipids are attacked by peroxidizing radicals confirms that reactive free radicals actually are present in living tissue, an assumption that underlies many of the free-radical theories of aging. In this area, as in so many others, it is apparent that the study of radical reactions will continue to provide important new lines of research for chemists and biologists.

I BIOLOGICAL REGULATORS

1. The Carbon Cycle

GEOGRAPHIC VARIATIONS IN PRODUCTIVITY. J. H. Ryther in *The Sea: Ideas and Observations on Progress in the Study of the Seas, Vol. 2: The Composition of Sea-Water—Comparative and Descriptive Oceanography*, edited by M. N. Hill. Interscience, 1963.

THE INFLUENCE OF ORGANISMS ON THE COMPOSITION OF SEA-WATER. A. C. Redfield, B. H. Ketchum, and F. A. Richards in *The Sea: Ideas and Observations on Progress in the Study of the Seas, Vol. 2: The Composition of Sea-Water—Comparative and Descriptive Oceanography*, edited by M. N. Hill. Interscience, 1963.

THE ROLE OF VEGETATION IN THE CARBON DIOXIDE CONTENT OF THE ATMOSPHERE. Helmut Lieth in *Journal of Geophysical Research*, Vol. 68, No. 13, pages 3887–3898; July 1, 1963.

GROSS-ATMOSPHERIC CIRCULATION AS DEDUCED FROM RADIOACTIVE TRACERS. Bert Bolin in *Research in Geophysics, Vol. 2: Solid Earth and Interface Phenomena*, edited by Hugh Odishaw. The M.I.T. Press, 1964.

PHOTOSYNTHESIS. E. Rabinowitch and Govindjee. John Wiley & Sons, 1969.

IS CARBON DIOXIDE FROM FOSSIL FUEL CHANGING MAN'S ENVIRONMENT? Charles D. Keeling in *Proceedings of the American Philosophical Society*, Vol. 114, No. 1, pages 10–17; February 16, 1970.

THE NITROGEN CYCLE. C. C. Delwiche in *Scientific American*, Vol. 223, pages 136–146; September, 1970.

THE OXYGEN CYCLE. Preston Cloud and Aharon Gibor in *Scientific American*, Vol. 223, pages 110–123; September, 1970.

2. The Chemical Elements of Life

TRACE ELEMENTS IN BIOCHEMISTRY. H. J. M. Bowen. Academic Press, 1966.

CONTROL OF ENVIRONMENTAL CONDITIONS IN TRACE ELEMENT RESEARCH: AN EXPERIMENTAL APPROACH TO UNRECOGNIZED TRACE ELEMENT REQUIREMENTS. Klaus Schwarz in *Trace Element Metabolism in Animals*, edited by C. F. Mills. E. & S. Livingstone, 1970.

THE PROTEINS: METALLOPROTEINS, Vol. 5. Edited by Bert L. Vallee and Warren E. C. Wacker. Academic Press, 1970.

CERULOPLASMIN: A LINK BETWEEN COPPER AND IRON METABOLISM. Earl Frieden in *Bioinorganic Chemistry*, Advances in Chemistry Series 100. American Chemical Society, 1971.

TRACE ELEMENTS IN HUMAN AND ANIMAL NUTRITION. E. J. Underwood. Academic Press, 1971.

MINERAL NUTRITION OF PLANTS: PRINCIPLES AND PERSPECTIVES. E. Epstein. John Wiley & Sons, 1972.

3. The Shapes of Organic Molecules

SYMPOSIUM: THREE-DIMENSIONAL CHEMISTRY in *Journal of Chemical Education*, Vol. 41, No. 2, pages 65–85; February, 1964.

CONFORMATIONAL ANALYSIS. Ernest L. Eliel, Norman L. Allinger, Stephen J. Angyal, and George A. Morrison. Interscience, 1965.

CONFORMATION THEORY. Michael Hanack. Academic Press, 1965.

CONFORMATIONAL ANALYSIS. Joseph B. Dence in *Chemistry*, Vol. 43, No. 6, pages 6–10; June, 1970.

4. Steroids

STEROIDS. Louis F. Fieser and Mary Fieser. Van Nostrand-Reinhold, 1959.

STEROIDS: STEROLS AND BILE ACIDS. C. J. W. Brooks in *Rodd's Chemistry of Carbon Compounds*, Vol. 2, 2nd edition, edited by S. Coffey. Elsevier, 1970.

THE BIOSYNTHESIS OF STEROIDS. L. J. Mulheirn and P. J. Ramm in *Quarterly Review, Chemical Society of London*, pages 259–291; 1972.

5. Alkaloids

AN INTRODUCTION TO THE ALKALOIDS. G. A. Swan. Wiley-Interscience, 1967.

THE BIOCHEMISTRY OF ALKALOIDS. T. Robinson. Springer-Verlag, 1968.

CHEMISTRY OF THE ALKALOIDS. Edited by S. W. Pelletier. Van Nostrand-Reinhold, 1970.

A HANDBOOK OF ALKALOIDS AND ALKALOID-CONTAINING PLANTS. R. F. Raffauf. Wiley-Interscience, 1970.

6. The Hormones of the Hypothalamus

NEURAL CONTROL OF THE PITUITARY GLAND. G. W. Harris. Williams & Wilkins, 1955.

CHARACTERIZATION OF OVINE HYPOTHALAMIC HYPOPHYSIOTROPIC TSH-RELEASING FACTOR. Roger Burgus, Thomas F. Dunn, Dominic Desiderio, Darrell N. Ward, Wylie Vale, and Roger Guillemin in Nature, Vol. 226, No. 5243, pages 321–325; April 25, 1970.

THE HYPOTHALAMUS: PROCEEDINGS OF THE WORKSHOP CONFERENCE ON INTEGRATION OF ENDOCRINE AND NON ENDOCRINE MECHANISMS IN THE HYPOTHALAMUS. Edited by L. Martini, M. Motta, and F. Fraschini. Academic Press, 1970.

STRUCTURE OF THE PORCINE LH- AND FSH-RELEASING HORMONE, 1: THE PROPOSED AMINO ACID SEQUENCE. H. Matsuo, Y. Baba, R. M. G. Nair, A. Arimura, and A. V. Schally in Biochemical and Biophysical Research Communications, Vol. 43, No. 6, pages 1334–1339; June 18, 1971.

SYNTHETIC LUTEINIZING HORMONE-RELEASING FACTOR: A POTENT STIMULATOR OF GONADOTROPIN RELEASE IN MAN. S. S. C. Yen, R. Rebar, G. VandenBerg, F. Naftolin, Y. Ehara, S. Engblom, K. J. Ryan, K. Benirschke, J. Rivier, M. Amoss, and R. Guillemin in The Journal of Clinical Endocrinology and Metabolism, Vol. 34, No. 6, pages 1108–1111; June, 1972.

SYNTHETIC POLYPEPTIDE ANTAGONISTS OF THE HYPOTHALAMIC LUTEINIZING HORMONE RELEASING FACTOR. Wylie Vale, Geoffrey Grant, Jean Rivier, Michael Monahan, Max Amoss, Richard Blackwell, Roger Burgus, and Roger Guillemin in Science, Vol. 176, No. 4037, pages 933–934; May 26, 1972.

7. Cyclic AMP

CYCLIC ADENOSINE MONOPHOSPHATE IN BACTERIA. Ira Pastan and Robert Perlman in Science, Vol. 169, No. 3943, pages 339–344; July 24, 1970.

CYCLIC AMP. G. Alan Robison, Reginald W. Butcher, and Earl W. Sutherland. Academic Press, 1971.

CYCLIC AMP AND CELL FUNCTIONS. Edited by G. Alan Robison, Gabriel G. Nahas, and Lubos Triner in Annals of the New York Academy of Sciences, Vol. 185; December 3, 1971.

8. Prostaglandins

PROSTAGLANDINS. J. W. Hinman in Annual Review of Pharmacology, Vol. 12, pages 161–178; 1972.

PROSTAGLANDINS. E. W. Horton in Monographs on Endocrinology, Vol. 7. Springer-Verlag, 1972.

PROSTAGLANDINS. T. Oesterling, W. Morozowich, and T. J. Roseman in Journal of Pharmaceutical Sciences, Vol. 61, page 1861; 1972.

PROSTAGLANDINS. J. R. Weeks in Annual Review of Pharmacology, Vol. 12, pages 317–336; 1972.

THE PROSTAGLANDINS: CLINICAL APPLICATIONS IN HUMAN REPRODUCTION. Edited by E. M. Southern. Futura, 1972.

INTERNATIONAL CONFERENCE ON PROSTAGLANDINS, Vienna, September 25–28, 1972, in Advances in Biosciences, Vol. 9. Pergamon Press-Vieweg, 1973.

THE PROSTAGLANDINS, Vol. 1. Edited by P. W. Ramwell. Plenum Press, 1973.

9. Analgesic Drugs

ANALGETICS. Edited by George de Stevens. Academic Press, 1965.

SYNTHETIC ANALGESICS, 2A: MORPHINANS, by J. Hellerbach, O. Schnider, H. Besendorf, and B. Pellmont; 2B: BENZOMORPHANS, by Nathan B. Eddy and Everette L. May. Pergamon Press, 1966.

NARCOTIC ANALGESICS and DRUG ADDICTION AND DRUG ABUSE in The Pharmacological Basis of Therapeutics, 4th edition, edited by Louis S. Goodman and Alfred Gilman, Chapters 15 and 16. Macmillan, 1970.

NARCOTIC DRUGS. Edited by Davis H. Clouet. Plenum Press, 1971.

10. Barbiturates

SEDATIVES AND HYPNOTICS. M. T. Bush in Physiological Pharmacology, Vol. 1, edited by W. S. Root and F. G. Hoffman. Academic Press, 1963.

CLINICAL PHARMACOLOGY OF THE BARBITUATES. J. W. Dundee in International Anesthesiology Clinics, Vol. 7, pages 3–29; 1969.

HYPNOTICS AND SEDATIVES in The Pharmacological Basis of Therapeutics, 4th edition, edited by Louis S. Goodman and Alfred Gilman, Chapter 9. Macmillan, 1970.

11. The Hallucinogenic Drugs

THE CLINICAL PHARMACOLOGY OF THE HALLUCINOGENS. Erik Jacobsen in Clinical Pharmacology and Therapeutics, Vol. 4, No. 4, pages 480–504; July–August, 1963.

LYSERGIC ACID DIETHYLAMIDE (LSD-25) AND EGO FUNCTIONS. G. D. Klee in Archives of General Psychiatry, Vol. 8, No. 5, pages 461–474; May, 1963.

PROLONGED ADVERSE REACTIONS TO LYSERGIC ACID DIETHYLAMIDE. S. Cohen and K. S. Ditman in Archives of General Psychiatry, Vol. 8, No. 5, pages 475–480; May, 1963.

THE PSYCHOTOMIMETIC DRUGS: AN OVERVIEW. Jonathan O. Cole and Martin M. Katz in The Journal of the American Medical Association, Vol. 187, No. 10, pages 758–761; March, 1964.

12. The Stereochemical Theory of Odor

THE SCIENCE OF SMELL. Robert H. Wright. Allen & Unwin, 1964.

THE CHEMICAL SENSES. R. W. Moncrieff. Leonard Hill, 1967.

MOLECULAR SHAPE AND ODOUR: PATTERN ANALYSIS BY "PAPA". John E. Amoore, Guido Palmieri, and Enzo Wanke in *Nature*, Vol. 216, No. 5120, pages 1084–1087; December 16, 1967.

SPECIFIC ANOSMIA: A CLUE TO THE OLFACTORY CODE. John E. Amoore in *Nature*, Vol. 214, No. 5093, pages 1095–1098; June 10, 1967.

MOLECULAR BASIS OF ODOR. John E. Amoore. Charles C Thomas, 1970.

13. L-asparagine and Leukemia

L-ASPARAGINASE THERAPY FOR LEUKEMIA AND OTHER MALIGNANT NEOPLASMS. J. M. Hill, J. Roberts, E. Loeb, A. Khan, A. MacLellan, and R. W. Hill in *Journal of the American Medical Association*, Vol. 202, pages 882–888; 1967.

INHIBITION OF LEUKEMIAS IN MAN BY L-ASPARA-GINASE. H. F. Oettgen, L. J. Old, E. A. Boyse, H. A. Campbell, F. S. Philips, B. D. Clarkson, L. Tallal, R. D. Leeper, M. K. Schwarts, and J. H. Kim in *Cancer Research*, Vol. 27, pages 2619–2631; 1967.

L-ASPARAGINASE, SYNTHETASE ACTIVITY OF MOUSE LEUKEMIAS. B. Horowitz, B. K. Madras, A. Meister, L. J. Old, E. A. Boyse, and E. Stockert in *Science*, Vol. 160, pages 533–535; 1968.

EXPERIMENTAL AND CLINICAL EFFECTS OF L-ASPARA-GINASE. Edited by E. Grundmann and H. F. Oettgen in *Recent Results in Cancer Research*, Vol. 33. Springer-Verlag, 1970.

14. The Induction of Interferon

PERSPECTIVES IN THE CONTROL OF VIRAL DISEASES INCLUDING CANCER. M. R. Hilleman in *Immunity in Viral and Rickettsial Diseases*, edited by Alexander Kohn and Marcus A. Kingberg. Plenum Press.

CLINICAL STUDIES OF INDUCTION OF INTERFERON BY POLYINOSINIC POLYCYTIDYLIC ACID (The Gustav Stern Symposium, Chapter 12). D. A. Hill, S. Baron, H. B. Levy, J. Bellanti, C. E. Buckler, G. Cannellos, P. Carbone, R. M. Chanock, V. DeVita, M. A. Guggenheim, E. Homan, A. Z. Kapikian, R. L. Kirsch-stein, J. Mills, J. C. Perkins, J. E. Van Kirks, and M. Worthington in *Perspectives in Virology, 7: From Molecules to Man*, edited by Morris Pollard. Academic Press, 1971.

DOUBLE-STRANDED RNA'S IN RELATION TO INTER-FERON INDUCTION AND ADJUVANT ACTIVITY. M. R. Hilleman, G. P. Lampson, A. A. Tytell, A. K. Field, M. M. Nemes, I. H. Krakoff, and C. W. Young. Reprint from *Biological Effects of Polynucleotides*, edited by Roland F. Beers, Jr. and Werner Braun. Springer-Verlag, 1971.

THE INDUCTION OF INTERFERON BY NATURAL AND SYNTHETIC POLYNUCLEOTIDES. C. Colby in *Progress in Nucleic Acid Research and Molecular Biology*, edited by J. N. Davidson and W. E. Cohn. Academic Press, 1971.

INTERFERON INDUCTION AND ACTION. C. Colby and M. J. Morgan in *Annual Review of Microbiology*, Vol. 25, page 333; 1971.

BIOCHEMISTRY OF INTERFERON AND ITS INDUCERS. W. J. Kleinschmidt in *Annual Review of Biochemistry*, Vol. 41, page 517; 1972.

INTERFERONS: PHYSICOCHEMICAL PROPERTIES AND CONTROL OF CELLULAR SYNTHESIS. M. H. Ng and J. Vilcek in *Advances in Protein Chemistry*, Vol. 26, page 173; 1972.

INTERFERONS AND HOST RESISTANCE: WITH PARTICU-LAR EMPHASIS ON INDUCTION BY COMPLEXED POLYNUCLEOTIDES. A. A. Tytell and A. K. Field in *Critical Reviews in Biochemistry*, Vol. 1, No. 1; 1972.

THE INTERFERONS AND THEIR INDUCERS: MOLECU-LAR AND THERAPEUTIC CONSIDERATIONS. S. E. Grossberg in *New England Journal of Medicine*, 3 parts, Vol. 287, pages 13, 79, 122; 1972.

15. Pheromones

CHEMICAL SYSTEMS. Edward O. Wilson in *Animal Communication: Techniques of Study and Results of Research*, edited by Thomas A. Sebeok, pages 75–102. University of Indiana Press, 1968.

COMMUNICATION BY CHEMICAL SIGNALS: ADVANCES IN CHEMORECEPTION, Vol. 1. Edited by James W. Johnston, Jr., David G. Moulton, and Amos Turk. Appleton-Century-Crofts, 1970.

THE INSECT SOCIETIES. Edward O. Wilson. Belknap Press of Harvard University Press, 1971.

OLFACTORY COMMUNICATION IN MAMMALS. John F. Eisenberg and Devra G. Kleiman in *Annual Review of Ecology and Systematics*, Vol. 3, pages 1–32; 1972.

II MACROMOLECULAR ARCHITECTURE

16. Giant Molecules

POLYMER SCIENCE AND MATERIALS. A. Tobolsky and H. Mark. Interscience, 1971.

TEXTBOOK FOR POLYMER SCIENCE, 2nd edition. F. W. Billmeyer. Interscience, 1971.

ORGANIC CHEMISTRY OF POLYMERS. J. K. Stille in *Condensation Monomers*, High Polymer Series, Vol. 27. Wiley-Interscience, 1972.

ORGANIC MOLECULES IN ACTION. M. Goodman and F. Morehouse. Gordon and Breach, 1973.

17. Proteins

BIOCHEMISTRY. A. L. Lehninger. Worth, 1970.

BIOLOGICAL CHEMISTRY, 2nd edition. H. R. Mahler and E. H. Cordes. Harper & Row, 1971.

OUTLINES OF BIOCHEMISTRY, 3rd edition. E. E. Conn and P. K. Stumpf. John Wiley & Sons, 1972.

18. The Chemical Structure of Proteins

THE STRUCTURE AND ACTIVITY OF RIBONUCLEASE. W. II. Stein in *Israel Journal of Medical Science,* Vol. 1, pages 1229–1243; 1965.

BOVINE PANCREATIC RIBONUCLEASE. Frederic M. Richards and Harold W. Wyckoff in *The Enzymes,* Vol. 4, 3rd edition, edited by P. D. Boyer, pages 647–806. Academic Press, 1971.

REACTIVATION OF DES(119–124) RIBONUCLEASE A BY MIXTURE WITH SYNTHETIC COOH-TERMINAL PEPTIDES: THE ROLE OF PHENYLALANINE-120. M. C. Lin, B. Gutte, D. G. Caldi, S. Moore, and R. B. Merrifield in *Journal of Biological Chemistry,* Vol. 247, page 4768; 1972.

CHEMICAL STRUCTURES OF RIBONUCLEASE AND DE-OXYRIBONUCLEASE. S. Moore and W. H. Stein in *Science,* Vol. 180, pages 458–464; 1973.

19. The Automatic Synthesis of Proteins

SOLID PHASE PEPTIDE SYNTHESIS, 1: THE SYNTHESIS OF A TETRAPEPTIDE. R. B. Merrifield in *Journal of the American Chemical Society,* Vol. 85, No. 14, pages 2149–2154; July 20, 1963.

THE CHEMISTRY AND BIOCHEMISTRY OF INSULIN. H. Klostermeyer and R. E. Humbel in *Angewandte Chemie: International Edition in English,* Vol. 5, No. 9, pages 807–822; September, 1966.

THE SYNTHESIS OF BOVINE INSULIN BY THE SOLID PHASE METHOD. A. Marglin and R. B. Merrifield in *Journal of the American Chemical Society,* Vol. 88, No. 21, pages 5051–5052; November 5, 1966.

INSTRUMENT FOR AUTOMATED SYNTHESIS OF PEP-TIDES. R. B. Merrifield, John Morrow Stewart, and Nils Jernberg in *Analytical Chemistry,* Vol. 38, No. 13, pages 1905–1914; December, 1966.

20. The Three-dimensional Structure of an Enzyme Molecule

THE MOLECULAR BIOLOGY OF THE GENE. J. D. Watson. Benjamin, 1965.

CATALYSIS IN CHEMISTRY AND ENZYMOLOGY. W. P. Jencks. McGraw-Hill, 1969.

THE STRUCTURE AND ACTION OF PROTEINS. R. E. Dickerson and I. Geis. Harper & Row, 1969.

VERTEBRATE LYSOZYMES. Taiji Imoto, L. N. Johnson, A. C. T. North, D. C. Phillips, and J. A. Rupley in *The Enzymes,* Vol. 7, page 665. Academic Press, 1972.

21. The Insulin Molecule

THE PRINCIPLES OF CHROMATOGRAPHY. A. J. P. Martin in *Endeavour,* Vol. 6, No. 21, pages 21–28; January, 1947.

THE CHEMISTRY OF INSULIN. F. Sanger in *Annual Reports on the Chemical Society,* Vol. 45, pages 283–292; 1949.

THE AMINO-ACID SEQUENCE IN THE GLYCYL CHAIN OF INSULIN. F. Sanger and E. O. P. Thompson in *The Biochemical Journal,* Vol. 53, No. 3, pages 353–374; February, 1953.

22. The Structure and History of an Ancient Protein

THE STRUCTURE AND ACTION OF PROTEINS. Richard E. Dickerson and Irving Geis. Harper & Row, 1969.

THE STRUCTURE OF CYTOCHROME *c* AND THE RATES OF MOLECULAR EVOLUTION. Richard E. Dickerson in *Journal of Molecular Evolution.* Vol. 1, No. 1, pages 26–45; 1971.

FERRICYTOCHROME *c,* 1: GENERAL FEATURES OF THE HORSE AND BONITO PROTEINS AT 2.8 Å RESOLU-TION. Richard E. Dickerson, Tsunehiro Takano, David Eisenberg, Olga B. Kallai, Lalli Samson, Angela Cooper, and E. Margoliash in *The Journal of Biological Chemistry,* Vol. 246, No. 5, pages 1511–1535; March 10, 1971.

CONFORMATIONAL CHANGES UPON REDUCTION OF CYTOCHROME *c.* Tsunehiro Takano, R. Swanson, Olga B. Kallai, and Richard E. Dickerson in *Cold Spring Harbor Symposium on Quantitative Biology,* in press.

23. Enzymes Bound to Artificial Matrixes

MATRIX-BOUND ENZYMES, 1: THE USE OF DIFFERENT ACRYLIC COPOLYMERS AS MATRICES. K. Mosbach in *Acta Chemica Scandinavica,* Vol. 24, No. 6, pages 2084–2092; 1970.

MATRIX-BOUND ENZYMES, 2: STUDIES ON A MATRIX-BOUND TWO-ENZYME SYSTEM. K. Mosbach and B. Mattiasson in *Acta Chemica Scandinavica,* Vol. 24, No. 6, pages 2093–2100; 1970.

AFFINITY CHROMATOGRAPHY. P. Cuatrecasas and B. Anfinsen in *Annual Review of Biochemistry,* Vol. 40, pages 259–278; 1971.

EFFECT OF THE MICROENVIRONMENT ON THE MODE OF ACTION OF IMMOBILIZED ENZYMES. E. Katchal-ski, I. Silman, and R. Goldman in *Advances in Enzymology,* Vol. 34, pages 445–536; 1971.

STUDIES ON A MATRIX-BOUND THREE-ENZYME SYSTEM. B. Mattiasson and K. Mosbach in *Biochimica et Biophysica Acta,* Vol. 235, pages 253–257; 1971.

GENERAL LIGANDS IN AFFINITY CHROMATOGRAPHY. K. Mosbach, H. Guilford, R. Ohlsson, and M. Scott in *Biochemical Journal,* Vol. 127, pages 625–631; 1972.

SYMPOSIUM ON ENZYME ENGINEERING. Edited by L. B. Wingard, Jr. Interscience, 1972.

24. The Structure of the Hereditary Material

THE MOLECULAR BIOLOGY OF THE GENE. J. D. Watson. Benjamin, 1965.

THE GEOMETRY OF NUCLEIC ACIDS. Struther Arnott in *Progress in Biophysics and Molecular Biology,* Vol. 21, edited by J. A. V. Butler and D. Noble, pages 265–319. Pergamon Press, 1970.

25. The Nucleotide Sequence of a Nucleic Acid

LABORATORY EXTRACTION AND COUNTERCURRENT DISTRIBUTION. Lyman C. Craig and David Craig in *Technique of Organic Chemistry,* Vol. 3, Part 1: *Separation and Purification,* edited by Arnold Weissberger, pages 149–332. Interscience, 1956.

SPECIFIC CLEAVAGE OF THE YEAST ALANINE RNA INTO TWO LARGE FRAGMENTS. John Robert Penswick and Robert W. Holley in *Proceedings of the National Academy of Sciences of the United States of America,* Vol. 53, No. 3, pages 543–546; March, 1965.

STRUCTURE OF A RIBONUCLEIC ACID. Robert W. Holley, Jean Apgar, George A. Everett, James T. Madison, Mark Marquisee, Susan H. Merrill, John Robert Penswick, and Ada Zamir in *Science,* Vol. 147, No. 3664, pages 1462–1465; March 19, 1965.

THE STRUCTURE OF TRANSFER RNA. Struther Arnott in *Progress in Biophysics and Molecular Biology,* Vol. 22, pages 181–213. Pergamon Press, 1971.

RECENT RESULTS OF tRNA RESEARCH. D. H. Gauss, F. von der Harr, A. Maelicke, and F. Cramer in *Annual Review of Biochemistry,* Vol. 40, pages 1045–1078; 1971.

26. The Synthesis of DNA

ACTIVE CENTER OF DNA POLYMERASE. A. Kornberg in *Science,* Vol. 163, page 1410; 1969.

ENZYMATIC SYNTHESIS OF DEOXYRIBONUCLEIC ACID, 36: A PROOFREADING FUNCTION FOR THE $3' \rightarrow 5'$ EXONUCLEASE ACTIVITY IN DEOXYRIBONUCLEIC ACID POLYMERASES. D. Brutlag and A. Kornberg in *Journal of Biological Chemistry,* Vol. 247, page 241; 1972.

RNA SYNTHESIS INITIATES *In Vitro* CONVERSION OF M13 DNA TO ITS REPLICATIVE FORM. W. Wickner, D. Brutlag, R. Schekman, and A. Kornberg in *Proceedings of the National Academy of Sciences of the United States of America,* Vol. 69, page 965; 1972.

27. The Recognition of DNA in Bacteria

HOST-INDUCED MODIFICATION OF T-EVEN PHAGES DUE TO DEFECTIVE GLUCOSYLATION OF THEIR DNA. Stanley Hattman and Toshio Fukasawa in *Proceedings of the National Academy of Sciences of the United States of America,* Vol. 50, No. 2, pages 297–300; August 15, 1963.

GENERAL VIROLOGY. S. E. Luria and James E. Darnell, Jr. John Wiley & Sons, 1967.

RESTRICTION OF NONGLUCOSYLATED T-EVEN BACTERIOPHAGE: PROPERTIES OF PERMISSIVE MUTANTS OF *ESCHERICHIA COLI* B AND K12. Helen R. Revel in *Virology,* Vol. 31, No. 4, pages 688–701; April, 1967.

DNA MODIFICATION AND RESTRICTION. Werner Arber and Stuart Linn in *Annual Review of Biochemistry,* Vol. 38, pages 467–500; 1969.

RESTRICTION OF NONGLUCOSYLATED T-EVEN BACTERIOPHAGES BY PROPHAGE P1. Helen R. Revel and C. P. Georgopoulos in *Virology,* Vol. 39, No. 1, pages 1–17; September, 1969.

28. RNA-directed DNA Synthesis

ONCOGENIC RIBOVIRUSES. Frank Fenner in *The Biology of Animal Viruses,* Vol. 2. Academic Press, 1968.

CENTRAL DOGMA OF MOLECULAR BIOLOGY. Francis Crick in *Nature,* Vol. 227, No. 5258, pages 561–563; August 8, 1970.

MECHANISM OF CELL TRANSFORMATION BY RNA TUMOR VIRUSES. Howard M. Temin in *Annual Review of Microbiology,* Vol. 25. Annual Reviews, Inc., 1971.

THE PROTOVIRUS HYPOTHESIS: SPECULATIONS ON THE SIGNIFICANCE OF RNA-DIRECTED DNA SYNTHESIS FOR NORMAL DEVELOPMENT AND FOR CARCINOGENESIS. Howard M. Temin in *Journal of the National Cancer Institute,* Vol. 46, No. 2, pages 3–7; February, 1971.

III CELLULAR ARCHITECTURE

29. The Structure of Cell Membranes

MEMBRANES OF MITOCHONDRIA AND CHLOROPLASTS. Edited by Efraim Racker. Van Nostrand-Reinhold, 1969.

STRUCTURE AND FUNCTION OF BIOLOGICAL MEMBRANES. Edited by Lawrence I. Rothfield. Academic Press, 1971.

MEMBRANE MOLECULAR BIOLOGY. Edited by C. F. Fox and A. Keith. Sinauer Associates, 1973.

30. The Membrane of the Mitochondrion

CHEMIOSMOTIC COUPLING IN OXIDATIVE AND PHOTOSYNTHETIC PHOSPHORYLATION. P. Mitchell in *Biological Reviews of the Cambridge Philosophical Society,* Vol. 41, page 445; 1966.

THE TWO FACES OF THE INNER MITOCHONDRIAL MEMBRANE. E. Racker in *Essays in Biochemistry,*

Vol. 6, edited by P. N. Campbell and F. Dickens, pages 1–22. Academic Press, 1970.

FUNCTION AND STRUCTURE OF THE INNER MEMBRANE OF MITOCHONDRIA AND CHLOROPLASTS. E. Racker in *Membranes of Mitochondria and Chloroplasts*, edited by E. Racker, pages 127–171. Van Nostrand-Reinhold, 1970.

THE COUPLING BETWEEN ENERGY-YIELDING AND ENERGY-UTILIZING REACTIONS IN MITOCHONDRIA. E. C. Slater in *Quantitative Review of Biophysics*, Vol. 4, page 35; 1971.

RECONSTITUTION OF THE THIRD SITE OF OXIDATION PHOSPHORYLATION. E. Racker and A. J. Kandrach in *Journal of Biological Chemistry*, Vol. 246, pages 7069–7071; 1971.

31. The Bacterial Cell Wall

USE OF BACTERIOLYTIC ENZYMES IN DETERMINATION OF WALL STRUCTURE AND THEIR ROLE IN CELL METABOLISM. J. M. Ghuysen in *Bacteriological Reviews*, Vol. 32, pages 425–464; 1968.

PENICILLIN-SENSITIVE ENZYMATIC REACTIONS IN BACTERIAL CELL WALL SYNTHESIS. J. L. Strominger in *Harvey Lectures*, Vol. 64, pages 179–213; 1970.

ROLE OF THE PENICILLIN-SENSITIVE TRANSPEPTIDATION REACTION IN THE ATTACHMENT OF NEWLY SYNTHESIZED PEPTIDOGLYCAN TO CELL WALLS OF *MICROCOCCUS LUTEUS*. D. Mirelman, R. Bracha, and N. Sharon in *Proceedings of the National Academy of Sciences of the United States of America*, Vol. 69, pages 3355–3359; 1972.

IV CHEMICAL BIODYNAMICS

32. Energy Transformation in the Cell

THE MITOCHONDRION: MOLECULAR BASIS OF STRUCTURE AND FUNCTION. A. L. Lehninger. Benjamin, 1964.

BIOCHEMISTRY, Chapters 16–18. A. L. Lehninger. Worth, 1970.

MEMBRANES OF MITOCHONDRIA AND CHLOROPLASTS. Edited by E. Racker. Van Nostrand-Reinhold, 1970.

BIOENERGETICS, 2nd edition. A. L. Lehninger. Benjamin, 1972 (an updated description of energy exchanges in cells—paperback).

33. The Metabolism of Fats

CATABOLISM OF LONG CHAIN FATTY ACIDS IN MAMMALIAN TISSUES. G. D. Greville and P. K. Tubbs in *Essays in Biochemistry*, Vol. 4. Academic Press, 1968.

BIOLOGICAL CHEMISTRY, 2nd edition. Henry R. Mahler and Eugene H. Cordes, pages 583–605. Harper & Row, 1966.

BIOCHEMISTRY. Albert L. Lehninger, pages 417–432. Worth, 1970.

34. How Proteins Start

MECHANISM OF PROTEIN BIOSYNTHESIS. P. Lengyel and D. Söll in *Bacteriological Reviews*, Vol. 33, page 264; 1969.

POLYPEPTIDE CHAIN INITIATION: THE ROLE OF RIBOSOMAL PROTEIN FACTORS AND RIBOSOMAL SUBUNITS. M. Revel in *Bacteriological Reviews*, Vol. 33, pages 87–131; 1969.

PROTEIN BIOSYNTHESIS. J. Lucas-Lenard and F. Lipmann in *Annual Reviews of Biochemistry*, Vol. 40, page 409; 1971.

POLYPEPTIDE CHAIN INITIATION AND THE ROLE OF A METHIONINE tRNA. P. S. Rudland and B. F. C. Clark in *The Mechanism of Protein Synthesis and its Regulation*, edited by L. Bosch. Frontiers of Biology Series, Vol. 27, pages 55–86. North-Holland, 1972.

35. Protein-digesting Enzymes

INDUCTION OF BIOLOGICAL ACTIVITY BY LIMITED PROTEOLYSIS. M. Ottesen in *Annual Review of Biochemistry*, Vol. 30, pages 655–676; 1967.

HOMOLOGY AND PHYLOGENY OF PROTEOLYTIC ENZYMES. H. Neurath, R. A. Bradshaw, and R. Arnon in *Structure-Function Relationship of Proteolytic Enzymes*, edited by P. Desnuelle, H. Neurath, and M. Ottesen, pages 113–133. Munksgaard, 1970.

ACTIVITY AND FLUORESCENT DERIVATIVES OF AMINOTYROSYL TRYPSIN AND TRYPSINOGEN. R. A. Kenner and H. Neurath in *Biochemistry*, Vol. 10, pages 551–557; 1971.

INACTIVATION OF BOVINE TRYPSINOGEN AND CHYMOTRYPSINOGEN BY DIISOPROPYLPHOSPHORYLFLUORIDATE. P. H. Morgan, N. C. Robinson, K. A. Walsh, and H. Neurath in *Proceedings of the National Academy of Sciences of the United States of America*, Vol. 69, pages 3312–3316; 1972.

36. The Genetic Code: II

THE DEPENDENCE OF CELL-FREE PROTEIN SYNTHESIS IN E. COLI UPON NATURALLY OCCURRING OR SYNTHETIC POLYRIBONUCLEOTIDES. Marshall W. Nirenberg and J. Heinrich Matthaei in *Proceedings of the National Academy of Sciences of the United States of America*, Vol. 47, No. 10, pages 1588–1602; October, 1961.

SYNTHETIC POLYNUCLEOTIDES AND THE AMINO ACID CODE, IV. J. F. Speyer, P. Lengyel, C. Basilio, and S. Ochoa in *Proceedings of the National Academy of Sciences of the United States of America*, Vol. 48, No. 3, pages 441–448; March, 1962.

POLYRIBONUCLEOTIDE-DIRECTED PROTEIN SYNTHESIS USING AN E. COLI CELL-FREE SYSTEM. M. S. Bretscher and M. Grunberg-Manago in *Nature*, Vol. 195, No. 4838, pages 283–284; July 21, 1962.

A PHYSICAL BASIS FOR DEGENERACY IN THE AMINO ACID CODE. Bernard Weisblum, Seymour Benzer, and Robert W. Holley in *Proceedings of the National Academy of Sciences of the United States of America*, Vol. 48, No. 8, pages 1449–1453; August, 1962.

QUALITATIVE SURVEY OF RNA CODE-WORDS. Oliver W. Jones, Jr., and Marshall W. Nirenberg in *Proceedings of the National Academy of Sciences of the United States of America*, Vol. 48, No. 12, pages 2115–2123; December, 1962.

THE GENETIC CODE. Marshall W. Nirenberg in *Les Prix Nobel en 1968*. The Nobel Foundation, 1969.

37. The Genetic Code of a Virus

CHEMICAL MODIFICATION OF VIRAL RIBONUCLEIC ACID, 8: THE CHEMICAL AND BIOLOGICAL EFFECTS OF METHYLATING AGENTS AND NITROSOGUANIDINE ON TOBACCO MOSAIC VIRUS. B. Singer and H. Fraenkel-Conrat in *Biochemistry*, Vol. 8, pages 3266–3269; 1969.

SEQUENCE OF THE FIRST 175 NUCLEOTIDES FROM THE 5' TERMINUS OF Qβ RNA SYNTHESIZED *In Vitro*. M. A. Billeter, J. E. Dahlberg, H. M. Goodman, J. Hindley, and C. Weissmann in *Nature*, Vol. 224, page 1083; 1969.

NUCLEOTIDE SEQUENCES FROM BACTERIOPHAGE R17 RNA. P. G. N. Jeppesen, J. I. Nichols, F. Sanger, and C. B. Barrell in *Symposia Quantitative Biology*, Vol. 35, page 13; 1970.

THE REACTIVITY OF FUNCTIONAL GROUPS AS A PROBE FOR INVESTIGATING THE TOPOGRAPHY OF TOBACCO MOSAIC VIRUS. R. N. Perham in *Biochemical Journal*, Vol. 131, page 119; 1973.

38. Chemical Fossils

ORGANIC GEOCHEMICAL STUDIES, 1: MOLECULAR CRITERIA FOR THE HYDROCARBON GENESIS. Eugene D. McCarthy and Melvin Calvin in *Nature*, Vol. 216, pages 642–647; 1967.

BIOCHEMICAL PREDESTINATION. Dean H. Kenyon and Gary Steinman. McGraw-Hill, 1969.

CHEMICAL EVOLUTION. Melvin Calvin. Oxford University Press, 1969.

FATTY ACIDS AND HYDROCARBONS AS EVIDENCE OF LIFE PROCESSES IN ANCIENT SEDIMENTS AND CRUDE OILS. W. Van Hoeven, J. R. Maxwell, and M. Calvin in *Geochimica et Cosmochimica Acta*, Vol. 33, pages 887–891; 1969.

MOLECULAR PALEONTOLOGY. Melvin Calvin in *Perspectives in Biology and Medicine*, Vol. 13, pages 45–62; 1969.

BIOGENIC SUBSTANCES IN SEDIMENTS AND FOSSILS. Pierre Albrecht and Guy Ourisson in *Angewandte Chemie. International Edition in English*, Vol. 10, pages 209–286; 1971.

ORGANIC GEOCHEMISTRY. J. R. Maxwell, C. T. Pillinger, and G. Eglinton in *Quarterly Review*, Vol. 35, No. 4, pages 571–628; 1971.

CHEMICAL FOSSILS: A COMBINED ORGANIC GEOCHEMICAL AND ENVIRONMENTAL APPROACH. G. Eglinton in *Pure and Applied Chemistry*, Vol. 34, No. 3; 1973.

39. The Chemical Effects of Light

LIGHT-SENSITIVE SYSTEMS: CHEMISTRY AND APPLICATION OF NONSILVER HALIDE PHOTOGRAPHIC PROCESSES. Jaromir Kosar. John Wiley & Sons, 1965.

THE MIDDLE ULTRAVIOLET: ITS SCIENCE AND TECHNOLOGY. Edited by A. E. S. Green. John Wiley & Sons, 1966.

ENERGY TRANSFER FROM HIGH-LYING EXCITED STATES. Gisela K. Oster and H. Kallmann in *Journal de Climie Physique et de Physico-Chimie Biologique*, Vol. 64, No. 1, pages 28–32; January, 1967.

PHOTOPOLYMERIZATION OF VINYL MONOMERS. Gerald Oster and Nan-Loh Yang in *Chemical Reviews*, Vol. 68, No. 2, pages 125–151; March 25, 1968.

FLASH PHOTOLYSIS AND SOME OF ITS APPLICATIONS. George Porter in *Science*, Vol. 160, No. 3834, pages 1299–1307; June 21, 1968.

40. How Light Interacts with Living Matter

A REVERSIBLE PHOTOREACTION CONTROLLING SEED GERMINATION. H. A. Borthwick, S. B. Hendricks, M. W. Parker, E. H. Toole and V. K. Toole in *Proceedings of the National Academy of Sciences of the United States of America*, Vol. 38, 662–666; 1952.

FUNCTION OF TWO CYTOCHROME COMPONENTS IN CHLOROPLASTS: A WORKING HYPOTHESIS. R. Hill and F. Bendell in *Nature*, Vol. 186, pages 1936–1937; 1960.

PHOTORECEPTORS in *Cold Spring Harbor Symposia on Quantitative Biology*, Vol. 30; 1965.

LIGHT AND LIVING MATTER. R. K. Clayton. McGraw-Hill, 1970.

LECTURES ON PHOTOMORPHOGENESIS. H. Mohr. Springer-Verlag, 1972.

41. Ultraviolet Radiation and Nucleic Acid

PYRIMIDINE DIMERS IN ULTRAVIOLET-IRRADIATED DNA'S. R. B. Setlow and W. L. Carrier in *Journal of Molecular Biology*, Vol. 17, pages 237–254; 1966.

ADVANCES IN THE PHOTOCHEMISTRY OF NUCLEIC ACID DERIVATIONS. John G. Burr in *Advances in Photochemistry*, Vol. 6, edited by W. A. Noyes, Jr.,

G. S. Hammond, and J. N. Pitts, Jr., pages 193–299. Interscience, 1968.

PHOTOPRODUCTS IN DNA IRRADIATED *In Vivo*. R. B. Setlow in *Photochemistry and Photobiology*, Vol. 7, pages 643–649; 1968.

EFFECTS OF RADIATION ON POLYNUCLEOTIDES. Richard B. Setlow and Jane K. Setlow in *Annual Review of Biophysics and Bioengineering*, Vol. 1, pages 293–346; 1972.

ULTRAVIOLET IRRADIATION OF DNA. Ronald O. Rahn in *Concepts in Radiation Cell Biology*, edited by Gary L. Whitson. Academic Press, 1972.

42. The Path of Carbon in Photosynthesis

THE PATH OF CARBON IN PHOTOSYNTHESIS, 21: THE CYCLIC REGENERATION OF CARBON DIOXIDE ACCEPTOR. J. A. Bassham, A. A. Benson, Lorel D. Kay, Anne Z. Harris, A. T. Wilson, and M. Calvin in *Journal of the American Chemical Society*, Vol. 76, No. 7, pages 1760–1770; April 5, 1954.

THE PATH OF CARBON IN PHOTOSYNTHESIS. Melvin Calvin and James A. Bassham. Prentice-Hall, 1957.

THE PHOTOSYNTHESIS OF CARBON COMPOUNDS. Melvin Calvin and James A. Bassham. Benjamin, 1962.

CONTROL OF PHOTOSYNTHETIC CARBON METABOLISM. J. A. Bassham in *Science*, Vol. 172, pages 526–534; 1971.

PHOTOSYNTHETIC CARBON METABOLISM. J. A. Bassham in *Proceedings of the National Academy of Sciences of the United States of America*, Vol. 68, pages 2877–2882; 1971.

43. The Mechanism of Photosynthesis

PHOTOPHOSPHORYLATION AND THE CHEMI-OSMOTIC HYPOTHESIS. Andre T. Jagendorf and E. Uribe in *Brookhaven Symposia in Biology*, Vol. 19, pages 215–245; 1966.

ELECTRON TRANSPORT PATHWAYS IN PHOTOSYNTHESIS. Geoffrey Hind and John M. Olson in *Annual Review of Plant Physiology*, Vol. 19, edited by Leonard Machlis. Annual Reviews, Inc., 1968.

HAEM-PROTEINS IN PHOTOSYNTHESIS. D. S. Bendall and R. Hill in *Annual Review of Plant Physiology*, Vol. 19, edited by Leonard Machlis. Annual Reviews, Inc., 1968.

44. Molecular Isomers in Vision

MOLECULAR ASPECTS OF VISUAL EXCITATION. R. Hubbard and A. Kropf in *Annals of the New York Academy of Sciences*, Vol. 81, pages 388–398; 1959.

VALENCE, 2nd edition. Charles A. Coulson. Oxford University Press, 1961.

Cis-Trans ISOMERIC CAROTENOIDS, VITAMINS A, AND ARYLPOLYENES. L. Zechmeister. Academic Press, 1962.

THE PHOTOISOMERIZATION OF RETINAE. A. Kropf and R. Hubbard in *Photochemistry and Photobiology*, Vol. 12, pages 249–260; 1970.

CAROTENOIDS. Edited by O. Isler. Birkhäuser Verlag, 1971.

PHOTOCHEMISTRY OF VISION. HANDBOOK OF SENSORY PHYSIOLOGY, Vol. 7-1, edited by H. J. A. Dartnall. Springer-Verlag, 1972.

45. Memory and Protein Synthesis

PROTEIN SYNTHESIS AND MEMORY FORMATION. B. W. Agranoff in *Protein Metabolism of the Nervous System*, edited by A. Lajtha, pages 533–543. Plenum Press, 1970.

EFFECTS OF ANTIBIOTICS ON LONG-TERM MEMORY FORMATION IN THE GOLDFISH. B. W. Agranoff in *Animal Memory*, edited by W. K. Honig and P. H. R. James, pages 243–258. Academic Press, 1971.

THE EFFECT OF PUROMYCIN ON RETENTION OF CONDITIONED CARDIAC DECELERATION IN THE GOLDFISH. W. M. Schoel and B. W. Agranoff in *Behavioral Biology*, Vol. 7, pages 553–565; 1972.

LEARNING AND MEMORY: APPROACHES TO CORRELATING BEHAVIORAL AND BIOCHEMICAL EVENTS. B. W. Agranoff in *Basic Neurochemistry*, edited by W. Albers, G. Siegel, R. Katzman, and B. W. Agranoff, pages 645–665. Little, Brown, 1972.

CAMPTOTHECIN BLOCKS MEMORY OF CONDITIONED AVOIDANCE IN THE GOLDFISH. J. H. Neale, P. D. Klinger, and B. W. Agranoff in *Science*, in press.

46. Human Cells and Aging

THE SERIAL CULTIVATION OF HUMAN DIPLOID CELL STRAINS. L. Hayflick and P. S. Moorhead in *Experimental Cell Research*, Vol. 25, pages 585–621; 1961.

AGEING: THE BIOLOGY OF SENESCENCE. Alex Comfort. Holt, Rinehart and Winston, 1964.

THE LIMITED *In Vitro* LIFETIME OF HUMAN DIPLOID CELL STRAINS. L. Hayflick in *Experimental Cell Research*, Vol. 37, pages 614–636; 1965.

AGING IN CELL AND TISSUE CULTURE. Edited by E. Holečková and V. J. Cristofalo. Plenum Press, 1970.

AGING UNDER GLASS. L. Hayflick in *Experimental Gerontology*, Vol. 5, pages 291–303. Honorary Lecture presented at the 14th Annual Ciba Foundation Meeting on Research on Aging, Leeds, England. Pergamon Press, 1970.

ADVANCES IN GERONTOLOGICAL RESEARCH, Vol. 4. Edited by B. Strehler. Academic Press, 1972.

AN ANALYSIS OF SENESCENCE PHENOMENA EXHIBITED BY CULTURED HUMAN AND ANIMAL CELLS. L. Hayflick. F. K. Schattauer Verlag, in press, 1973.

THE BIOLOGY OF HUMAN AGING. L. Hayflick in *American Journal of the Medical Sciences*, in press; 1973.

47. Free Radicals in Biological Systems

FREE RADICALS IN BIOLOGICAL SYSTEMS: PROCEEDINGS OF A SYMPOSIUM HELD AT STANFORD UNIVERSITY, MARCH, 1960. Edited by M. S. Blois, Jr., H. W. Brown, R. M. Lemmon, R. O. Lindblom, and M. Weissbluth. Academic Press, 1961.

FREE RADICALS IN TISSUE. Irvin Isenberg in *Physiological Reviews*, Vol. 44, No. 3, pages 487–513; July, 1964.

FREE RADICALS. William A. Pryor. McGraw-Hill, 1966.

RADIATION & AGEING: PROCEEDINGS OF A COLLOQUIUM HELD IN SEMMERING, AUSTRIA, JUNE 23–24, 1966. Edited by Patricia J. Lindop and G. A. Sacher. Taylor & Francis, 1966.

FREE RADICAL PATHOLOGY. William A. Pryor in *Chemical and Engineering News*, Vol. 49, pages 34–51; June 7, 1971.

VITAMIN E AND ITS ROLE IN CELLULAR METABOLISM. Edited by P. P. Nair and H. J. Kayden in *Annals of the New York Academy of Sciences*, Vol 203; December, 1972.

INDEX